India and Beyond

Studies from the International Institute for Asian Studies

LEIDEN AND AMSTERDAM

Edited by

Paul van der Velde, General Editor

PUBLISHED

HANI-ENGLISH, ENGLISH-HANI DICTIONARY
Paul W. Lewis and Bai Bibo

INDIA AND BEYOND
Edited by Dick van der Meij

FORTHCOMING

NEW DEVELOPMENTS IN ASIAN STUDIES
Edited by Paul van der Velde and Alex McKay

DYNAMICS IN PACIFIC ASIA
Edited by Kurt W. Radtke, Joop Stam, Takuo Akiyama,
John Groenewegen and Leo van der Mey

INDIA AND BEYOND

Aspects of Literature, Meaning, Ritual and Thought

ESSAYS IN HONOUR OF
FRITS STAAL

EDITED BY
Dick van der Meij

Routledge
Taylor & Francis Group
LONDON AND NEW YORK

IIAS
INTERNATIONAL INSTITUTE FOR ASIAN STUDIES
LEIDEN AND AMSTERDAM

First published in 1997 by
Kegan Paul International

This edition first published in 2009 by
Routledge
2 Park Square, Milton Park, Abingdon, Oxfordshire OX14 4RN

Simultaneously published in the USA and Canada
by Routledge
711 Third Avenue, New York, NY 10017
First issued in paperback 2014

Routledge is an imprint of the Taylor & Francis Group, an informa business

British Library Cataloguing in Publication Data
A catalogue record for this book is available from the British Library

ISBN 978-0-710-30602-9 (hbk)

ISBN 978-1-138-87929-4 (pbk)

Publisher's Note
The publisher has gone to great lengths to ensure the quality of this reprint but points out that some imperfections in the original copies may be apparent. The publisher has made every effort to contact original copyright holders and would welcome correspondence from those they have been unable to trace.

ACKNOWLEDGMENTS

When first being asked to edit the Festschrift for Frits Staal I had no idea what I was getting myself into. I set about editing the numerous contributions delivered in various styles and in three languages. It appeared to be an enormous work, one I could not possibly do alone. It is therefore my pleasant duty to acknowledge the assistance I received from a number of people.

First of all I would like to thank the staff of the Kern Institute at Leiden who checked the Sanskrit transliterations and assisted in correcting and updating the bibliographical data of the contributions: Hanna 't Hart, Marja van Kooij - van der Heijden, Marianne Oort - Lissy, Dory Heilijgers -Seelen, and Jan Houben.

The staff of the Projects Division of the Department of Languages and Cultures of South-East Asia and Oceania, in particular Mary Bakker and Connie Baak; and the staff of the International Institute for Asian Studies, in particular Paul van der Velde, Karin van Belle - Foesenek and Jennifer Trel assisted me in the administrative dealings with the book. Marianne Oort and Rosemary Robson - McKillop assisted with the English of a number of contributions. Thanks are also due to Ilse Lasschuijt for making the Japanese glossary for the contribution of Richard Payne, and to Kitty Yang - de Witte for the Chinese glossary for the contribution of Kristofer Schipper.

Without the patient help of Fediya Andina this book would never have been completed.

I would also like to express my gratitude to the staff of Kegan Paul, and especially to Helen Wilson and Kaori O'Connor, for their patience and support.

Last but not least, I want to thank the Projects Division which gracefully gave me the time to edit the book and especially to its director Wim Stokhof for simply being there.

Dick van der Meij

PREFACE

The International Institute for Asian Studies (IIAS) is pleased to introduce a new series 'Studies from the International Institute for Asian Studies' to the international scholarly community. The IIAS is a post-doctoral institute under the auspices of the Royal Netherlands Academy of Arts and Sciences, founded in 1993. Its main objective is to encourage Asian Studies in the Humanities and Social Sciences. It has built a world-wide network with other institutes in the field and it has developed a number of research programmes. These international research programmes are executed by Asianists from all over the world. Apart from them, on an annual basis the IIAS receives fifty odd visiting fellows. This influx of scholars has made the IIAS into an intellectual beehive to which the many international scholarly meetings it either organizes or participates in also bear witness. Its *IIAS Newsletter* has a global circulation of 20,000 copies, which together with its WWW site, offers an array of information on Asian Studies.

As a consequence of all these international activities, the IIAS has taken the initiative of starting a publication series 'Studies from the International Institute for Asian Studies' in co-operation with Kegan Paul International. This present volume, *India and Beyond; Aspects of Literature Meaning, Ritual and Thought*, contains more than 30 contributions from well-established scholars from different disciplinary backgrounds. These essays are in honour of one of the founding fathers of the IIAS, Frits Staal, Professor Emeritus of Philosophy and South Asian Languages, University of California at Berkeley. This volume is edited by Dick van der Meij, editor of the Indonesian-Netherlands Cooperation in Islamic Studies Programme at Leiden University.

The IIAS will do its utmost to make this series into an ongoing intellectual discovery of the world of Asia.

Paul van der Velde
General Editor

CONTENTS

INTRODUCTION

'Tell me who your friends are and I'll tell you who you are.' A *liber amicorum* of friends and colleagues should tell something about the one to whom the book is dedicated because the contributions are assumed to relate substantially to the activities of the laureate. In this sense, the present volume will tell you something about Frits Staal, but in an unexpectedly intriguing way.

Thirty-four scholars contributed to the present volume. Although they cover a variety of topics - from ancestor worship to transsexual masquerades, from metrical verse in the psalms to temples, from Plotinus to Heidegger, from poetic expression to linguistic diversity - one would expect an easy going categorization along lines drawn by Staal himself, say rituals, grammar, philosophy, religion and science. But whatever we tried to conceptually achieve, we ended up procrustean until we realized that whatever one could do to Frits Staal in terms of trying to understand his unique personality, never should it be possible to categorize him, let alone in terms of compartments.

We realized that a categorization of the contributions of his friends in the present volume would violate one of his most characteristic properties: his fluidity and his ability to melt down all sorts of conceptual barriers characterizing the faculties and departments in academia. This explains the alphabetic order of the authors contributing to this *Festschrift*; only a complex matrix would be able to reveal the connections between them.

Yet, one need not appeal to a matrix in order to understand Frits Staal and his work. In fact, there is a unifying term which really explains what he is doing and what he has done so far: human science. The term 'humanism' is much too western to be useful in the context which he made part of his life, but taken together with the eastern scientific tradition which plays such an important role in his life, the study of humanity is clearly the central force in his existence. Credos for this have been written in Dutch (*The Academic as Nowhere Man*, the Introduction to his 1986 book) and English (e.g. *There is No Religion There*). What we see is someone who does not understand why (and also refuses to accept why) the study of, say, the stratosphere or black holes should be more important than the study of the Harappa writing or the study of rituals or universal grammar. Staal refuses to accept the 'bide your time' model which says that basically human science is a sort of retarded natural science waiting in the stack until more money and intellectual capacity should be available after Nature has failed to continue seduce the brightest, say after the unification of Quantum Physics and Relativity Theory.

For Staal the study of Humanity essentially differs from the study of Nature but it is crucial to add that human science does not require a unique approach such as claimed by phenomenology, hermeneutics and related philosophical movements, which he attacked fiercely in some of his essays. It is easy to see how Staal found his own way between what he rejects as two alternatives. His training as a mathematician and physicist connected him with a western natural science tradition whose philosophy of science has generated very valuable insights in the field of mathematical logic and philosophy, in particular with respect to issues concerning the foundation of scientific knowledge.

At several places in his work Staal pays tribute to one of his teachers, Evert Beth, the famous logician and philosopher whose international network of relations opened up for him a direct access to very important insights that we find back in Staal's work. His work on ritual, for example, cannot be understood without this line. So, logic as a mathematical tool has played a considerable role in at least two fields in which Staal operated: the study of western and eastern grammar and logic, and the study of ritual.

When he decided to go East in the 1950s, Staal's formal training enabled him to make the proper generalizations, such as discovering the importance of the fact that Indian grammar stems from the Vedic science of ritual, as his classical and formal education enabled him to consider Greek mathematics as being fundamentally related to rituals. About ten of the contributions in the present volume focus on rituals, some of them also featuring in the about twenty contributions in which grammar and ritual are related. So in this respect we see some reflection of a deep-rooted interest of Frits Staal in the way Eastern and Western cultures have begun to relate themselves to questions concerning the scientific exploration of the world in which we live. These are questions about how to ask the right meaningful questions and to separate them from the senseless ones.

Basically, Staal finds himself back in the position of an evolutionary biologist studying the present manifestations of species, but always and necessarily focused on their past actualizations. Staal is in essence an evolutionary culturologist, or perhaps better, human-ologist inquiring into the development of humanity by reconstructing it, applying all sorts of methods available among which, as said, formal methods deriving from the scientific tradition mentioned above.

This means that most of his studies concern the past in an attempt to reconstruct the dawn of humanity, not from one specific point of view, not from two or three different points of view, but from the perspective of a (idealized) human trying to come to grips with his position in a senseless universe and giving sense to it by creating a culture asking the proper questions.

It should be observed that Staal's broad education and interest - both in the area in which mathematical tools are employed and in the area of literary and historical approaches - inevitably has led to a position in which the number of people accepting him as an authority diminishes: academia partitioned, heros local. In particular in the Western tradition in which the study of Sanskrit and Vedic verses is determined by philological and anthropological considerations, it is hard for Staal to convince people of the need to employ formal and technical tools. For many topics on which Staal has contributed there will be smaller scale experts claiming that they have a better sight on the topic at issue, but what sets Staal apart and what distinguishes him in the appropriate senses of this word, is his ability to cover a formidable and astonishingly extensive area comprising the study of ritual, grammar, philosophy, religion and science in a very consistent and coherent way without losing the required depth of analysis.

This feature of Staal's work appeals to at least thirty-four scholars, this being only a very tiny part of those relating themselves to his work by reacting to it. What unites the contributors in spite of the alphabetical order in which they appear is that they feel attracted to the unique perspective of the study of human-ness in Staal's work, even though some of them may disagree on specific issues involved or even in the way they operate in the field. This attraction is not restricted to members of his own generation. Among the contributors, there are a number of young scholars. This should be a very encouraging thought for the person at the receiving end of this volume: Frits Staal. Encouraging because it follows that he will not be alone in that receiving position. The contributions constitute a natural part of a lively and ongoing debate in which his ideas are going to be put to the test and we think that it is this part of our tribute that will count: not so much that we contributed in order to pay homage to a friend, but that we contributed to the continuation of a discussion that Staal started forty years ago and that in our opinion deserves to be continued for many more years to come, in fact on our way to the genuine human science that he helped to shape by his work.

The Advisory Board

Prof. J.C. Heesterman
Prof. H. Philipse
Prof. K.M. Schipper
Prof. W.A.L. Stokhof
Prof. H.J. Verkuyl
Prof. T.E. Vetter

I

SOME THOUGHTS ON THE INDIAN CONCEPTIONS OF POETIC EXPRESSION

Kamaleswar Bhattacharya

Among Frits Staal's manifold interests, the problems of language have always occupied a prominent place. Here is a modest contribution in that sphere.[1]

Poetics in India perhaps grew out of Grammar.[2] As we shall see in this paper, some of the finest discoveries in this field, even in later times were made under the influence of Grammar, which indeed provided Indian thought in general with its technical basis as Mathematics did in the West - a subject where again Staal's contributions are among the most valuable.

But, perhaps, Poetics and Grammar go back to a common source: Ritual. Particularly interesting in this respect is the Jñānasūkta (Ṛgveda X, 71) to which Staal has devoted penetrating studies, and about which L. Renou wrote once: 'cet hymne qui fait prévoir pour ainsi dire le développement de la grammaire est en même temps le plus vieux texte de la poétique indienne'.[3] It is interesting to note that this *Sūkta* has been claimed both by grammarians[4] and by poeticians.[5]

Beginning as part of Dramaturgy, around the first century AD, Poetics appears as an independent discipline in India around the seventh-eighth century AD, when it is already constituted (the earlier literature being lost). This discipline is known as *Alaṁkāraśāstra* - a name that will remain attached to it throughout its history. Now what are these *alaṁkāra*s which thus constitute the subject matter of Poetics?[6] The term *alaṁkāra*, literally meaning 'ornament',[7] is used in a wider and in a narrower sense. In its wider sense the term means all that contributes to the beauty of poetry. *Kāvyaśobhākarān dharmān alaṁkārān pracakṣate*, said Daṇḍin in the seventh-eighth century.[8] In the eighth-ninth century, Vāmana clearly defined *alaṁkāra* as 'beauty': *saundaryam alaṁkāraḥ*.[9] In its narrower sense, the word is used to designate the poetic figures: the figures of sound (*śabdālaṁ-*

kāra), alliteration, etc.; and the figures of sense (*arthālaṁkāra*), comparison, and so forth.

Beauty, therefore, is what characterizes poetic language and differentiates it from ordinary language. Vāmana states clearly: 'Poetry is acceptable on the ground of *alaṁkāra*' - *kāvyaṁ grāhyam alaṁkārāt*[10] (*alaṁkāra* here is to be understood in the sense of 'beauty').

Now what is it that imparts to poetic language that beauty which characterizes it, which - let us say - differentiates it from prose?[11] The question is not easy to answer, but the ancient Indians formulated answers to this question which are not different from those found in the Western tradition.

Bhāmaha, the earliest writer on Poetics in India whose work has survived, while emphasizing the role of the poetic figures (*alaṁkāra* in the restricted sense), defined the all-pervading principle of beauty in poetry as *vakrokti*, which literally means 'curved expression', and which we may render by 'deviant expression'. Poetic language therefore, is defined as a deviation from the normal language. Figures like comparison are also found in our everyday speech and in technical literature. But what makes them *poetic* figures is the particular turn that the poet gives to the expressions in which he incarnates them.[12]

Bhāmaha was probably from Kashmir. A little later Daṇḍin, a Southerner proclaimed the metaphor (*samādhi*) - conceived, not as a poetic figure among others but as a poetic excellence (*guṇa*) - as 'the very essence of poetry' (*kāvyasarvasva*) and said that 'the whole host of poets run in pursuit of it'.[13] Sanskrit poetics is usually taxonomic in character. A botany of figures, it resembles very much the classical Western Rhetoric. Starting from the modest four of Bharata, the supposed author of the Nāṭyaśāstra, the number of figures increases gradually. It is significant, in this context, that the earliest writers were not without insights as to the essence of poetic language.

These insights of Bhāmaha and Daṇḍin were however, to some extent obscured by the extremely formal scheme of Vāmana, a Kashmiri of the eighth-ninth century AD. Diction (*rīti*), consisting of a particular arrangement of words (*viśiṣṭā padaracanā*) exhibiting particular literary excellences (*guṇa*), is there declared to be the 'soul of poetry': *rītir ātmā kāvyasya*.[14] It is these 'qualities', equally distributed between the word and the meaning, the two constituents of poetry, which are said to engender the beauty of poetry: *kāvyaśobhāyāḥ kartāro dharmā guṇāḥ*.[15] The figures or ornaments - *alaṁkāra* in the restricted sense - only heighten, according to Vāmana, the beauty thus created: *tadatiśayahetavas tv alaṁkārāḥ*.[16] The qualities or excellences (*guṇa*), he says are permanent (*nitya*) by which he implies of course, that the figures or ornaments are impermanent (*anitya*).[17] As the fifteenth-century commentator Gopendra Tippabhūpāla (Tripurahara) aptly points out, in Vāmana's view the former appertain by the relation of inherence (*samavāya*) to the soul of poetry, while the latter by the relation of contact (*saṁyoga*) belong to the body of poetry.[18] This theory in the same form was also known to Vāmana's contemporary and rival Udbhaṭa, if the tradition that attributes the criticism to him is to be trusted.[19]

For the first time thus, a dualism is introduced between the qualities (*guṇa*) and the ornaments (*alaṁkāra*), between the soul and the body of poetry. It should be admitted that for Daṇḍin too, the qualities are in some way more important than the figures, since it is their presence which characterizes the best poetry: that which follows the Vaidarbha diction (*mārga*, *vartman* in Daṇḍin's terminology, *rīti* in Vāmana's), whereas the figures are common (*sādhāraṇa*) to both the Vaidarbha and Gauḍa dictions, although in the latter as he says, these qualities - sweetness, lucidity, and so on - are generally absent (*eṣāṁ viparyayaḥ prāyo lakṣyate gauḍavartmani*).[20] It should also be admitted that Vāmana is right in holding that the figures or ornaments alone cannot constitute the beauty of poetry. But his radical dualism between the soul and the body of poetry, between the qualities of the soul and the ornaments of the body, leads to an unacceptable consequence. Literally interpreted, his doctrine means this: the figures are simply added to the body of poetry, already ensouled by a superior beauty due to the qualities. And Vāmana himself authorizes this interpretation, by his comparison between the youth of a woman and the poetic qualities on the one hand and between the ornaments that adorn the human body and the figures that adorn the poetic body on the other.[21] The function of the poetic 'ornaments' however is different. There can be indeed good poetry without 'ornaments';[22] but where they exist, they should never appear as something added to a poetic body already produced by a chemical combination of the qualities. Something, in other words, that can be removed or changed. They form an integral part of poetry - at least this is true of all good poetry. The best classical Sanskrit poet, Kālidāsa, perhaps suggested this idea when he wrote: *anyonyaśobhājananād babhūva sādhāraṇo bhūṣaṇabhūṣyabhāvaḥ*, i.e. one cannot tell the 'ornament' from the 'ornamented'.[23] And Kuntaka will say later - expressly taking as his model the idea of Bhartṛhari concerning the grammarian's analysis of sentences into words and of words into the base, the suffix, etc.: it is only for the sake of analysis that the ornament and the ornamented are artificially extracted separately from an indivisible whole; in reality, it is the ornamented which is poetry; there is no addition of ornament to poetry.[24] If therefore any relation is to be conceived between the two, it should also be inherence (*samavāya*) - as Udbhaṭa (?) pointed out.[25] It is also significant that under the name of *vakrokti*, which, as we have seen, was, according to Bhāmaha, the distinguishing characteristic of poetic expression, Vāmana lists a figure among others, a simple 'ornament', which is the same as Daṇḍin's metaphor (*samādhi*) - 'the very essence of poetry'.[26]

Vāmana's mode of thought however, prepared the way for the advent of a theory in the ninth century, which since then has dominated Sanskrit poetics and literary criticism: the theory of suggestion (*dhvani*). Vāmana's dualism between the soul and the body of poetry, between the qualities of the soul and the ornaments of the body, is fully retained.[27] But the soul of poetry is no longer diction but suggestion. This agreement and this difference between the two theories have been clearly put by Vāmana's commentator Tippabhūpāla in his Kāmadhenu.[28]

This theory of suggestion had its limitations. But it seems to me that it proceeded from a genuine, critical appraisal of the poetry available at the time. The earliest Sanskrit poetics, with its classification and definition of the figures, had provided the aspiring poets with a model that they could easily imitate. Of course, such was not the intention of the writers on poetics. All of them had insisted upon the role of creative imagination (*pratibhā*). Bhāmaha, in his characteristic fashion, had said: 'The knowledge of treatises possessed by one who is not a poet is like the pauper's charitableness (*adhanasyeva dātṛtvam*),[29] the eunuch's dexterity in arms (*klībasyevāstra-kauśalam*), and the fool's self-confidence (*ajñasyeva pragalbhatvam*).' 'Even the dull-witted can learn the treatises with the help of a teacher. Poetry however is born to him who has creative imagination, and that only once in a while (*kāvyaṁ tu jāyate jātu kasyacit pratibhāvataḥ*).'[30] However the taxonomy of Sanskrit poetics had a disastrous effect on Sanskrit poetry which, with few exceptions after Kālidāsa, became a verbal acrobatics. This fact was clearly realized by the author of the Dhvanyāloka - again a Kashmiri. So Ānandavardhana says: Among innumerable poets, only two or three, five or six at most (*dvitrāḥ pañcaṣā vā*), beginning with Kālidāsa, could be counted as great poets (*mahākavi*).[31] It seems to me that the Dhvanikāra's purpose, in proclaiming suggestion as the basic principle of poetry to which all the verbal elements are subordinated, was not only to establish the primacy of *rasa* ('sentiment', 'flavour', 'Stimmung'), which is better suggested than expressed, but to save poetry itself.

But this theory of suggestion would perhaps have never been formulated the way it was if the model for this thinking had not been furnished by Grammar in its speculations on *sphoṭa* and *dhvani*, *vyaṅgya* and *vyañjaka*. As Ānanda-vardhana himself says: 'Foremost among the scholars are the grammarians; for all the sciences rest upon Grammar.'[32] In brief, and following Ānanda-vardhana's own words, just as the articulated sound (*dhvani* in the grammarian's sense) with its different variations, reveals an invariable sound-unit (*sphoṭa*), so a poem is said to be *dhvani* when it reveals a meaning over and above the literal meaning.[33] The term *dhvani* therefore, borrowed from the grammatical tradition, is applied to suggestive poetry (it is also used for the suggested sense as well as for the function of suggestion). Similarly the terms *vyaṅgya* and *vyañjaka*, 'manifested' and 'manifester', already used by Bhartṛhari to designate respectively the *sphoṭa* and the *dhvani*, are used respectively, for the meaning 'suggested' and for the words and the meanings which 'suggest' that meaning.

A special function (*vyāpāra*) of word named *vyañjanā* 'suggestion' (literally 'manifestation'), is thus postulated in addition to, but based upon the two generally recognized in the Indian tradition, viz. 'denotation' (*abhidhā*) and 'indication' (*lakṣaṇā*).[34]

Despite the overwhelming importance attributed to suggestion, and above all to the suggestion of *rasa*, the Dhvani theorists could not ignore the question of expression. As the Dhvanikāra states, the great poet should carefully seek the suggested sense and the word apt to suggest it.[35] And all suggestion is not poetic. Dhvani also is beautiful, and it is only a beautiful expression that can bring about a beautiful suggestion.[36] The poetic figures (*alaṁkāra*), it is true, belong in principle to the body of poetry, i.e. to the sound and the literal sense (while the 'qualities' *guṇa*, being particularly suited to the suggestion of different *rasa*-s, belong to the soul);[37] but to the poet gifted with genius and engrossed in *rasa*, they come spontaneously, without any special effort on his part (*apṛthagyatnanirvartya*);[38] and they embellish the *rasa*, the soul of poetry, in so far as they help the suggestion of the *rasa*.[39]

Despite its neat conceptual scheme and great intellectual appeal, the theory of suggestion suffered from a serious limitation. It had to exclude from the realm of best poetry a large amount of poetry simply because it contained no suggestion or because the suggested sense contained in it appeared (to some) less charming than the expressed sense (*guṇībhūtavyaṅgya*). Kuntaka (tenth-eleventh century) - also a Kashmiri - was a more sensitive critic.[40] He did admit suggestion in poetry, but refused to bound poetry within its limits. Reviving Bhāmaha's concept of *vakrokti*, 'deviant expression', and broadening it, he gave a more comprehensive theory of poetic expression.

This 'deviant expression'[41] Kuntaka defines as *vaidagdhya-bhaṅgī-bhaṇiti*, i.e. as explained by himself, a mode of expression depending upon the particular turn given to it by the skill of the poet.[42] He includes suggestion in all its varieties in different types of *vakrokti* - expressly recognizing *vyañjanā* as a particular function of word.[43] Whether the poetic meaning is expressed or suggested, what makes poetry poetry is the manner in which that meaning is expressed or suggested, and not what is expressed or suggested. One is reminded of the famous words of Mallarmé to the painter Degas: 'Ce n'est point avec des idées, mon cher Degas, que l'on fait des vers. C'est avec des *mots*.'[44]

By the term *vakrokti* Kuntaka covers all the poetic figures of sound and sense. All of them represent 'deviations' from ordinary language. The figures of sound - alliteration, etc. - represent what Kuntaka calls *varṇavinyāsavakratā*, 'deviation in the arrangement of the phonemes'. The figures of sense, on the other hand, come under the category of *vākyavakratā*, 'deviation in the sentences'. But Kuntaka's concept of *vakrokti* goes far beyond Bhāmaha's. A single pronoun, a single particle, a single suffix can contain a poetic deviation. Following the lead of the Dhvanikāra who had analysed the diverse suggestive elements in poetry,[45] Kuntaka gives penetrating analyses of the poetic style, where we find - as L. Renou put it - 'les éléments d'une stylistique à base linguistique, très consciente d'elle-même et, somme toute, assez voisine par les intentions de celle que les modernes s'efforcent d'instaurer en Occident'.[46]

But it is not, I think, in the concept of *vakrokti* that lies Kuntaka's greatest contribution. It lies in his theory of *sāhitya*. The idea once again was not new. It was perhaps the poet Kālidāsa who in the first verse of the Raghuvaṁśa, celebrated for the first time 'the divine nuptials of sound and sense in poetry':[47]

vāgarthāv iva saṁpṛktau vāgarthapratipattaye /
jagataḥ pitarau vande pārvatīparameśvarau[48] *//*

Bhāmaha later gave a technical expression to this idea: *śabdārthau sahitau kāvyam.*[49] Probably - as has been thought - it stemmed from the grammatical speculations on the relationship between word and meaning. Other writers expressed similar ideas with their concepts of *pāka*, *śayyā*, *maitrī*. In any case, *sāhitya* came to designate in later times poetry. But no one before Kuntaka had brought out all the implications of this concept - as he rightly claims.[50] In the light of what Kuntaka says on the subject, I would propose to render *sāhitya* by 'complete language' - borrowing the expression from Paul Valéry, one of the most sensitive critics of our time.

It is not enough to define poetic language as 'deviation' from the normal language. A 'deviant expression' can be wild or eccentric.[51] What makes *vakrokti* poetic is *sāhitya*, which transcends it, which, in V. Raghavan's words, is 'the greatest perfection in expression a poet should attain.'[52]
But let us first hear Valéry.

Mère des Souvenirs, Maîtresse des maîtresses ... or
Sois sage, ô ma douleur, et tiens-toi plus tranquille ...

Commenting on these verses of Baudelaire, the poet-critic writes:

Ces paroles agissent sur nous (du moins, sur quelques-uns d'entre nous) sans nous apprendre grand'chose. Elles nous apprennent peut-être qu'elles n'ont rien à nous apprendre; qu'elles exercent, par les mêmes moyens qui, en général, nous apprennent quelque chose, une tout autre fonction. Elles agissent sur nous à la façon d'un accord musical. L'impression produite dépend grandement de la résonance, du rythme, du nombre de ces syllabes; mais elle résulte aussi du simple rapprochement des significations. Dans le second de ces vers, l'accord des idées vagues de Sagesse et de Douleur, et la tendre solennité du ton produisent l'inestimable valeur d'un charme: l'être momentané (author's italics) qui a fait ce vers, n'eût pu le faire s'il eût été dans un état où la forme et le fond se fussent proposés séparément à son esprit. Il était au contraire dans une phase spéciale de son domaine d'existence psychique, phase pendant laquelle le son et le sens de la parole prennent ou gardent une importance égale - ce qui est exclu des habitudes du langage pratique comme des besoins du langage abstrait. L'état dans lequel l'indivisibilité du son et du sens, le désir, l'attente, la possibilité de leur combinaison intime et indissoluble sont requis et demandés ou donnés et parfois anxieusement attendus, est un état relativement rare. Il est rare, d'abord parce qu'il a contre lui toutes les exigences de la vie; ensuite parce qu'il s'oppose à la simplification grossière et à la spécialisation croissante des notations verbales.

Further on in the same text, Valéry calls this indissoluble fusion of the sound and the sense, of the form and the content, 'langage *complet*' (author's italics).[53]

These ideas were expressed in India ten centuries before, by Kuntaka. Elaborating Bhāmaha's definition of poetry: 'Poetry is composed of word and meaning together'[54] Kuntaka finds this definition inadequate. The mere fact that word and meaning exist together - *sāhitya* Kuntaka explains as *sahitayor bhāvaḥ*[55] - cannot be the defining characteristic of poetic expression; for it is what characterizes all linguistic expression; no linguistic expression is possible without it. So, Kuntaka observes, the *sāhitya* which defines poetry is a special kind of *sāhitya* (*viśiṣṭam eveha sāhityam abhipretam*).[56] This speciality consists in the fact that the word and the meaning have an *equal importance*: *anyūnānatiriktatva*.[57] They 'vie with each other' (*parasparas-pardhitva*),[58] are united like two intimate friends (*suhṛdāv iva saṁgatau*), and contribute to the beauty of each other (*parasparasya śobhāyai bhavataḥ*).[59] This accord rests exclusively on the creative imagination (*pratibhā*) of the poet and, realized in a unique poetic instant (*tatkālollikhita*), on the supramundane road of poetry (*alaukike kāvyamārge*), i.e. of poetic activity (*kavikarmavartmani*) it enchants the minds of the sensitive readers or hearers (*cetanacamatkāritā, sahṛdayāhlādakāritā*).[60] Although many other words are available for expressing the same idea, the poetic expression is unique and irreplaceable: *śabdo vivakṣitārthaikavācako 'nyeṣu satsv api*.[61]

Kuntaka beautifully illustrates this accord by analysing a number of stanzas. For instance, in this stanza of Kālidāsa (Kumārasaṁbhava V, 71):

dvayaṁ gataṁ saṁprati śocanīyatāṁ
samāgamaprārthanayā kapālinaḥ /
kalā ca sā kāntimatī kalāvatas
tvam asya lokasya ca netrakaumudī //

he points out how not a single meaning can be expressed by a word other than the one employed (... *ity eteṣāṁ pratyekaṁ kaścid apy arthaḥ śabdāntarābhidheyatāṁ notsahate*).[62] Poetic language, thus, is *sāhitya*, 'accord' ('complete language'). There is accord not only between the words and their meanings, but also between the words themselves and between the meanings themselves - in fact, among all the elements poetry is composed of.[63] By his counter-examples - which he finds even in great poets such as Māgha and Bhavabhūti - Kuntaka shows how a poet, because his concentration has been disturbed, however slightly, can break the accord and thus spoil an otherwise beautiful composition; and how it is sometimes enough to replace a word or a group of words by another to bring the accord to perfection.[64]

It is from this accord that results the musicality of which Valéry speaks, and on which Kuntaka insists, too. Like music, poetry is that which, by virtue of the beauty of its composition, fills with delight the hearts of the connoisseurs, even when its meaning has not been pondered:

aparyālocite 'py arthe bandhasaundaryasaṁpadā /
gītavad dhṛdayāhlādaṁ tadvidāṁ vidadhāti yat[65] *//*

Notes

1. This paper is based on some lectures given in the 1970s. Since then, I have had hardly any opportunity to do further work on the subject, and no significant work on it by others has come to my notice. An earlier treatment is: S.K. De, 'The Problem of Poetic Expression', in: *Sanskrit Poetics as a Study of Aesthetic*, with Notes by Edwin Gerow. Berkeley and Los Angeles: University of California Press, 1963, pp. 18 ff.; K. Kunjunni Raja's *The Language of Poetry*, mentioned in his Preface to the reprint of the second edition of his *Indian Theories of Meaning*. Madras: Adyar Library and Research Centre, 1977, has not been available to me.
2. Cf. De, *op. cit.*, pp. 1 ff.

3. L. Renou, 'Les connexions entre le rituel et la grammaire en sanskrit', *JA*, 1941-42, p. 160 (= Staal, *A Reader on the Sanskrit Grammarians*. Cambridge: MIT Press, 1972, p. 467).
4. Patañjali, Mahābhāṣya, Paspaśā, Kielhorn's edition, vol. I, p. 4.
5. Sāhityamīmāṁsā, Trivandrum Sanskrit Series 114, 1934, p. 161. See also Kunjunni Raja, *op. cit.*, pp. 278-9.
6. On Bharata's elusive concept of *lakṣaṇa*, - about which Abhinavagupta mentions as many as ten different views, - see the exhaustive treatment by V. Raghavan, *Some Concepts of the Alaṁkāra Śāstra*. Madras: Adyar, 1942, pp. 1 ff. Also: P.C. Lahiri, *Concepts of Rīti and Guṇa in Sanskrit Poetics*. Reprint: Delhi, Munshiram Manoharlal, 1977, pp. 8 ff.
7. J. Gonda has devoted a study to this term: *Selected Studies* II. Leiden: Brill, 1975, pp. 257 ff.
8. Kāvyādarśa II,1.I refer to S.K. Belvalkar's edition, Poona, 1924.
9. Kāvyālaṁkāra Sūtra I,1,2. Vani Vilas Press edition, Srirangam, 1909.
10. *Ibid.*, I,1,1.
11. The Indians never made the mistake of confusing verse and poetry. Sanskrit literature contains quite an amount of 'versified prose', as it contains also a number of 'poems in prose'. Versified poetry would perhaps deserve the name 'integral poetry', as prose without beauty would deserve the name 'integral prose'; but the distinction between 'semantic poems' and 'phono-semantic poems', sometimes made in the West does not seem to be applicable here: like the versified poems, the non-versified poems, too, use the poetic figures of sound.
12. Kāvyālaṁkāra I,30.36; II,81 ff.; V,66; VI,23. My references are to P.V. Naganatha Sastry's edition (reprint: Delhi: Motilal Banarsidass, 1970). See also *vṛtti* on Dhvanyāloka III, 36, and Abhinavagupta's *locana* thereon:, pp. 466-7 in Paṭṭābhirāma Śāstrī's edition: Kashi Sanskrit Series 135 (Alaṁkāra Section, No. 5). Benares: Chowkhamba, 1940; Kuntaka, Vakroktijīvita, *vṛtti* on III, 2, pp. 132-3 in K. Krishnamoorthy's edition. Dharwad: Karnatak University, 1977. Also: S.K. De's Introduction to his edition of the Vakroktijīvita, third edition. Calcutta: Mukhopadhyay, 1961, pp. xvii ff.
13. *tad etat kāvyasarvasvaṁ samādhir nāma yo guṇaḥ / kavisārthaḥ samagro 'pi tam ekam upajīvati //* Kāvyādarśa I,100. (The translation follows the reading, *d, tam enam anugacchati.*) See G. Jenner, *Die poetischen Figuren der Inder von Bhāmaha bis Mammaṭa*. Hamburg: Apfel, 1968, pp. 64 ff. - If the introduction into poetry of the secondary function of words 'let open the door to the conception of a third potency of language - the *vyañjanāvṛtti*' (R. Gnoli, *Udbhaṭa's Commentary on the Kāvyālaṁkāra of Bhāmaha*. Roma: IsMEO, 1962, pp. xxiii-xxiv), the credit for it must go, not so much to Udbhaṭa and Vāmana, as one might think on the authority of Abhinavagupta (*locana* [see n. 12] I,1, p. 32), as to this stanza of Daṇḍin, as H. Jacobi perceived (*Schriften zur indischen Poetik und Ästhetik*, mit einer Vorbemerkung von Hans Losch. Darmstadt: Wissenschaftliche Buchgesellschaft, 1969, p. 353).
14. Kāvyālaṁkāra Sūtra I,2,6-8.

15. *Ibid.*, III,1,1. Note the dependence upon, and the difference from, Daṇḍin's formulation: *kāvyaśobhākarān dharmān alaṁkārān pracakṣate* (above), where the term *alaṁkāra* is used to include both the qualities and the figures.
16. *Ibid.*, III,1,2.
17. *Ibid.*, III,1,3.
18. Kāmadhenu (see n. 9), p. 71.
19. This criticism (see below) is said to have been made by Udbhaṭa in his commentary on Bhāmaha. But Udbhaṭa's Bhāmahavivaraṇa is for all practical purposes lost (supposing that the text reconstructed from minute fragments and edited by R. Gnoli [see n. 13 above] really represents it). In any case, the controversy over this point must have acquired some importance in Sanskrit poetics, as is shown by the references made to it in such texts as the Kāvyaprakāśa (VIII, 2, *vṛtti*) and Tippabhūpāla's own commentary on Vāmana (*loc. cit.*). See also, e.g., H. Jacobi, 'Ueber Begriff und Wesen der poetischen Figuren in der indischen Poetik', *GN*, 1908, p. 3 (= *Schriften op. cit.*, p. 295).
20. Kāvyādarśa I,42; II,3. There is some controversy over the interpretation of the term *viparyaya* in I,42 (see Lahiri, *op. cit.*, pp. 59 ff.); but cf. Lahiri, *op. cit.*, p. 22, n. 2.
21. Kāvyālaṁkāra Sūtra, *vṛtti* on III,1,2.
22. As Kuntaka will say, *alaṁkāras* can even 'veil' or 'tarnish' the natural beauty of a subject depicted by the poet. Perhaps the idea was suggested, as Kuntaka thinks, by the poet Kālidāsa:

    tāṁ prāṅmukhīṁ tatra niveśya tanvīṁ
    kṣaṇaṁ vyalambanta puro niṣaṇṇāḥ /
    bhūtārthaśobhāhriyamāṇanetrāḥ
    prasādhane saṁnihite 'pi nāryaḥ // Kumārasaṁbhava VII,13.
    'The matrons placed her facing east
    and stood before her; but they tarried
    with ornaments all ready, for their eyes
    were captured by her natural beauty.'

    Daniel H.H. Ingalls' translation in: *The* Dhvanyāloka *of Ānandavardhana with the locana of Abhinavagupta*, translated by D.H.H. Ingalls, J.M. Masson and M.V. Patwardhan, edited with an introduction by D.H.H. Ingalls. Cambridge: Harvard University Press, 1990 (HOS. 49), p. 705. Kuntaka, Vakroktijīvita, *vṛtti* on III, 1, in connection with the question of whether *svabhāvokti* 'natural expression' ('wenn das wahre, aber nur dem feineren Verständnis zugängliche Wesen eines Dinges naturgetreu geschildert wird': H. Jacobi, *ZDMG* 56, 1902, p. 612 [= *Schriften* . . ., p. 52], n. 1 [Translation of the Dhvanyāloka]) is an *alaṁkāra* or not - a subject on which Kuntaka has most interesting things to say (cf. V. Raghavan, 'The History of Svabhāvokti in Sanskrit Poetics', in: *Some Concepts* ... [n. 6 above], pp. 92 ff.).
23. Kumārasaṁbhava I,42.
24. *alaṁkṛtir alaṁkāryam apoddhṛtya vivecyate / tadupāyatayā tattvaṁ sālaṁkārasya kāvyatā // ... dṛśyate ca samudāyāntaḥpātinām asatyabhūtānām api vyutpattinimittam apoddhṛtya vivecanam. yathā padāntarbhūtayoḥ prakṛtipratyayayor vākyāntarbhūtānāṁ padānāṁ ceti ... tenālaṁkṛtasya kāvyatvam iti sthitam,*

na punaḥ kāvyasyālaṁkārayoga iti. Vakroktijīvita I,6 and *vṛtti* thereon. Cf. *vṛtti* on I,10, I,11 (p. 21) and III,11 (p. 144). Cf. *'Longinus' On the Sublime* with an English translation by W. Hamilton Fyfe, in: *Aristotle, The Poetics*, etc., Cambridge and London, 1946 (The Loeb Classical Library), ch. 17: '... a figure is always most effective when it conceals the very fact of its being a figure', and ch. 38: '... the best hyperbole is the one which conceals the very fact of its being a hyperbole .... Herodotus does not seem to have introduced the incident to justify the hyperbole, but the hyperbole seems the natural outcome of the incident.'

25. See p. 2 and n. 19 above.
26. Kāvyālaṁkāra Sūtra and *vṛtti* IV,3,8.
27. Dhvanyāloka (n. 12 above) II,6.
28. P. 72.
29. Cf. Kuntaka, Vakroktijīvita, *vṛtti* on I,24 (p. 41) and on III,1 (p. 126): *na cāgatikagatinyāyena yathāśakti daridradānavat kāvyaṁ karaṇīyatām arhati; agatikagatinyāyena kāvyakaraṇaṁ na yathākathaṁcid anuṣṭheyatām arhati.*
30. Bhāmaha, Kāvyālaṁkāra I,3 and 5.
31. *Vṛtti* on Dhvanyāloka I,6, p. 93.
32. *prathame hi vidvāṁso vaiyākaraṇāḥ, vyākaraṇamūlatvāt sarvavidyānām. Vṛtti* on Dhvanyāloka I,13, pp. 132-3. Cf. *vṛtti* on III,33, p. 443: *pariniścitanirapabhraṁśaśabdabrahmaṇāṁ vipaścitāṁ matam āśrityaiva pravṛtto 'yaṁ dhvanivyavahāraḥ ...*
33. *Vṛtti* on Dhvanyāloka I,13, pp. 133-5. Cf. J. Brough, 'Some Indian Theories of Meaning', in Staal, *op. cit.*, p. 421.
34. The two primary divisions of *dhvani, avivakṣitavācya* (where the expressed sense is not intended at all) and *vivakṣitānyaparavācya* (where the expressed sense though intended subserves another, i.e., the suggested sense) are based respectively upon 'indication' and 'denotation'. Among other subdivisions of *dhvani* are that which is based on the potency of the word (*śabdaśaktimūla*) and that which is based on the potency of the meaning (*arthaśaktimūla*). However, 'suggestion' is strictly speaking a function of the word. Only there is everywhere an intimate cooperation between the word and the meaning, and the difference is made in order to convey the predominance of the one or the other. See *vṛtti* on Dhvanyāloka III,33, p. 423; Abhinavagupta, *locana* on I,4, p. 53; on I,13, p. 104.
35. Dhvanyāloka I,8.
36. *Vṛtti* on Dhvanyāloka I,5, p. 87, with Abhinavagupta's *locana*, pp. 87-8. Cf. Abhinavagupta: *guṇālaṁkāraucityasundaraśabdārthaśarīrasya sati dhvananākhyātmani kāvyarūpatāvyavahāraḥ* (*locana* on I,4, p. 59); *kāvyagrahaṇād guṇālaṁkāropaskṛtaśabdārthapṛṣṭhapātī dhvanilakṣaṇa ātmety uktam* (on I,13, pp. 104-5).
37. According to Abhinavagupta (*locana* on II,7, pp. 206-7) the qualities really belong to *rasa*, the soul of poetry, but are secondarily transferred to the word and the meaning which suggest the *rasa*. Cf. Kāvyaprakāśa, *vṛtti* on VIII,1. Gopendra Tippabhūpāla evidently follows this idea when he says that although

strictly speaking the qualities belong to the *rīti*, the soul of poetry, by transfer they are said to belong to the sound and the sense. Kāmadhenu, p. 71.

38. Dhvanyāloka II, 16 with *vṛtti* and *locana*.
39. *Ibid*., II,17 with *vṛtti*; *vṛtti* on II,5, pp. 197-8, 204; *locana*, pp. 197-8.
40. Although Kuntaka's influence in ancient times does not seem to have been quite negligible, he fell into unmerited oblivion, apparently because of the overwhelming influence of the Dhvani school, until S.K. De edited for the first time his Vakroktijīvita in 1923.
41. By 'deviant', of course, is meant 'deviant' from what is normal; in Kuntaka's words, *śāstrādiprasiddhaśabdārthopanibandhavyatirekin* (Vakroktijīvita, p. 13), *prasiddhābhidhānavyatirekin* (p. 20), *prasiddhaprasthānavyatirekin* (p. 26), *prasiddhavyavahāravyatirekin* (p. 133).
42. Vakroktijīvita I,10, and *vṛtti* thereon.
43. *Ibid*., *vṛtti* on I,8 and on III,1 (p. 125).
44. Paul Valéry, 'Poésie et pensée abstraite', in: *Variété*: *Œuvres* (Pléiade 127, 148) I. Paris: Gallimard, 1957, p. 1324 (*mots* is in italics in the text). Cf. Rājaśekhara, *Kāvyamīmāṁsā*. Baroda: Oriental Institute, 1934, pp. 45-6.12qw. (GOS. 1)
45. Dhvanyāloka III.
46. L. Renou, 'Le *Dhvani* dans la poétique sanskrite', *Adyar Library Bulletin* 18, 1954, pp. 13-14. Renou does not specifically mention Kuntaka. Speaking of Abhinavagupta, Daniel H.H. Ingalls (*op. cit*., p. 34) observes: 'Such careful aesthetic *explications de texte* (in French in the text) had just come into vogue. We find the fashion also in Abhinava's contemporary Kuntaka. I know of no examples in the older literature. But, once established, it became characteristic of Sanskrit literary criticism and is what gives to that tradition of criticism its great strength. In our Western classical tradition there is nothing to compare with it except pseudo-Longinus.' (It does not seem improbable that Abhinava underwent the influence of Kuntaka: see for example Lahiri, *op. cit*., pp. 16 ff.)
47. I take this from V. Raghavan's quotation, in his *Bhoja's Śṛṅgāraprakāśa*. Madras: Punarvasu, 1963, p. 103, of Wilfred Meynell's comment on Francis Thompson's poem *Sister Songs*.
48. See also Māgha, *Śiśupālavadha* II, 86.
49. Kāvyālaṁkāra I,16.
50. *yad idaṁ sāhityaṁ nāma tad etāvati niḥsīmani samayādhvani sāhityaśabdamātreṇaiva prasiddham. na punar etasya kavikarmakauśalakāṣṭhādhiruḍharamaṇīyasyādyāpi kaścid api vipaścid ayam asya paramārtha iti manāṅmātram api vicārapadavīm avatīrṇaḥ* ... . Vakroktijīvita, *vṛtti* on I,16, pp. 23-4.
51. Cf. *ibid*., *vṛtti* on I,7, p. 13.
52. Bhoja's Śṛṅgāraprakāśa, p. 101.
53. Valéry, *loc. cit*., pp. 1333-4, 1336.; R. Gnoli, *op. cit*., pp. xxvii-ix, quoted a couple of other passages from Valéry in another connection.
54. See above.
55. Vakroktijīvita, *vṛtti* on I,17.
56. *Ibid*., *vṛtti* on I,7, p. 10. Cf. I,16 and *vṛtti* thereon.
57. *Ibid*., I,17.

58. *Ibid.*, *vṛtti* on I,17, I,7 (p. 11) and II,4. Cf. *parasparaspardhādhiroha*: *vṛtti* on I,7 (p. 10).
59. *Ibid.*, *vṛtti* on I,7, st. 18.
60. *Ibid.*, *vṛtti* on I,8, I,9 (p. 16) and I,17 (p. 24). Kuntaka often uses *tatkālollikhita* and allied expressions, in connection with the creative activity of the poet.
61. *Ibid.*, I,9.
62. *Vṛtti* on I,9, pp. 15-16.
63. *Vṛtti* on I,7, pp. 10-11, and on I,17, p. 25.
64. *Vṛtti* on I,7 and 9. - Kuntaka's *Calitāvadhāna* (*etac cātyantaramaṇīyam api manāṅmātracalitāvadhānatvena kaveḥ kadarthitam*, *vṛtti* on I,9, p. 19) recalls what Kālidāsa says in Mālavikāgnimitra II,2: there is an absence of accord between the picture and the model because the painter is *śithilasamādhi*. Note that Vāmana places poetic activity on a par with yogic meditation: Kāvyālaṁkāra Sūtra (and *vṛtti*) I,3, 17-20. Note, on the other hand, Kuntaka's fondness for likening the activity of the poet to that of the painter: Vakroktijīvita III,4 (and *vṛtti*); cf. *vṛtti* on III,1, p. 126. See also Vāmana, Kāvyālaṁkāra Sūtra, *vṛtti* on III,1,25, p. 82 (Raghavan, *Bhoja's Śṛṅgāraprakāśa*, p. 286).
65. Vakroktijīvita, *vṛtti* on I,17, *antaraśloka*, p. 26. For Kuntaka's influence on a twentieth-century (Hindi) theory of poetry, see P. Gaeffke, 'Strömungen der Hindiliteratur im Lichte der einheimischen Literaturkritik', *WZKS(O)* 11, 1967, p. 101.

II

AWAKENING AND ITS IMAGERY IN TAGORE'S EARLY RELIGIOUS POETRY

Victor A. van Bijlert

The past decade has witnessed an increasing revival of scholarly and general interest in Rabindranath Tagore and his writings. Several new translations of some of his Bengali works have recently been published.[1] Among those who have worked on Tagore's poetry, William Radice, Martin Kämpchen and Ketaki Kushari Dyson have been most successful. Their translations possess great literary merit, and have done a lot to revive interest in Tagore.[2] In 1985 a bibliography was published which includes most books and articles written by and on Tagore.[3] In 1990 the Tagore Centre in London has printed an interesting compilation of articles on and by Tagore that have appeared in British newspapers and magazines between 1912 and 1941.[4] In England, as well as in India, a number of authors has recently taken stock of Tagore's significance for contemporary times. Their contributions resulted in two collections of studies. One was published in India in 1988,[5] the other in England in 1989.[6]

These publications show a deep concern for Tagore's popularity in the West at the present time, and highlight the fact that Tagore's fame rests mainly on his poetry. But the significance of certain sources of inspiration behind his poetry (as indeed behind much of his other work as well) and the way these are reflected in his poetic imagery has not been developed nor explored in these publications. Regarding his early religious lyrical poetry, we will try to take a closer look at one of the sources of his inspiration, the manner in which it is symbolized and the role poetry is made to play in conveying the inspiration to an audience.

It is a known fact that as a young man Rabindranath had been heavily influenced by his famous father, Debendranath Tagore (1817-1905) in religious matters. Debendranath had been the second major promoter of the Hindu social and religious reform movement of the Brahmo Samaj founded in 1828 by Raja Rammohun Roy. In the forties of the last century it was Debendranath who, had given the Brahmo Samaj a more or less fixed liturgy[7] and a rapidly expanding adherence. Debendranath's relative modernity lay in his rejection of a literalist authoritarian understanding of scriptures, his emphasis on private conscience and personal experience, and his rejection of many aspects of contemporary 'cultural' Hinduism with its great public rituals and the worshipping of images of various Hindu deities. But in fact much of this Brahmo theological movement does not radically differ from earlier medieval Hindu sects, especially those that cultivated the realization of God without qualities (*nirguna brahma*), for they also often rejected brahmanic social values and rituals, holding personal religious experience in much higher esteem. The major difference with earlier Indian sects are the intellectual debates in the nineteenth century, in which the Brahmos conferred on topics such as social equality, citizenship, freedom of the press, nationalism and the formation of a modern Indian nationhood. Consequently the Brahmos are often regarded as the avant-garde of the cultural constructions of the Indian nationhood. Rabindranath grew up with a deep awareness of these issues, and from his early years onwards, has often taken a position in political matters as well as on religious issues arising within his own community, the Adi Brahmo Samaj.[8]

And yet neither organized religion, nor politics, nor social questions, moved Rabindranath so much as the creation of poetry. This he regarded as his real craft. This was what he had done from early childhood onwards,[9] and between 1878 and 1882 he had already published at least six works.[10] None of them he regarded as good poetry. Something was missing in those works. In his own words in his autobiography, *Jivansmriti*, first published in 1912:

> ... the critics of poetry were circulating a rumour about me that I was a poet of scattered metres and indistinct language. All about me was full of mist and shadow ... In fact, in those poems there was nothing of the firmness of the real world. From childhood I had been a person who had been kept within boundaries, far away from the association with people of the outside [world]; so where could I have got the material to write about? (p. 117)[11]

The backbone of worldly reality, or in Bengali *vastav samsarer dridhatva*, 'the firmness of the real world', he was soon to find and in an unexpected way.

In 1882 Rabindranath had been overcome at least twice by experiences which, being of a mystical nature with strong religious and aesthetic overtones, exerted a great influence on his subsequent poetry. At least we have this on record in his own words, and stated more than once. The first description of these experiences, which is also the most extensive, is found in Rabindranath's autobiography, *Jivansmriti*. There he talks about these experiences at some length to show that they constituted the origins, as it were, of his serious poetry. The first experience came in the evening. In *Jivansmriti* he writes:

> The evening ... that was drawing near and whose light of the setting sun mixed with the gloominess of the end of day, revealed itself to me as particularly charming. Even the walls of the neighbouring house struck me as beautiful. I began to wonder whether it was merely the magic of the twilight illumination that had fully pulled away the covering of banality from before the world as I knew it. This was not so. I could see very well the true cause of this: the evening had entered me and 'I' got covered up ... I was seeing the living world in its own proper form (*svarup*). This form is by no means banal: it consists in joy (*ananda*), it is beautiful (*sundar*) ... (p. 120)

Two important key words occur here which describe the quality of the experience: *ananda*, joy, and *sundar*, beautiful. In it, he loses his small ego, his little self, so to speak. This enables him to see the world in its proper beauty and joy. The 'Me' must have continuously acted as a screen hiding the true world. The experience does not seem to have been consciously brought about by the poet himself. It came unbidden from outside as a force or a real presence, in this instance ostensibly through the glow of the setting sun.

The second and more intense and lasting experience followed some months later. Again the sun is involved, but this time it is the rising sun.

> The sun was just rising from behind the screen of the leaves on the trees there. As I was watching, it suddenly seemed as if a curtain was pulled away before my eyes. I saw the whole world (*visvasamsar*) fully enveloped by an incomparable (*aparup*) majesty, everywhere undulating with joy (*ananda*) and beauty (*saundarya*). The covering of melancholy which lay in layer after layer upon my heart was pierced within a brief moment, and the sparkling light of the world fell in its totality upon my inner being. That day the poem 'The Awakening of the Waterfall' gushed forth and flowed onwards. The writing of it ended, but even then the curtain did not drop before this form of joy (*anandarupa*) of the living world (*jagat*) ... (p. 120-1)

> On that day I wrote all afternoon and evening 'The Awakening of the Waterfall'... I see now that with [it], I had drawn an allegorical map of the path along which my heart was to journey ...[12]

This experience is an intensified repetition of the previous one. Now it is the rising sun which reveals the glory of the world. The moment of its revelation is likened to the pulling away of a curtain. A similar image is also used to describe the first experience. And also now the two major aspects under which the world is seen are joy, *ananda* and beauty, *saundarya*. As this experience is descending into the poet's self, all personal anxieties melt away. The heart is opened, so to say, to admit the exalted guest, the glory of the world. We could say, that these experiences are in the parlance of George Steiner 'real presences', for they are experiences of a deeply religious and aesthetic nature, that have fundamentally transformed the creative 'persona' of Rabindranath.[13] Furthermore, we have a lengthy poetic statement, gushing forth from the immediacy of this 'presence'. This is the poem 'Awakening of the Waterfall', *Nirjharer Svapnabhanga*, the first poem of a volume bearing the fitting title 'Morning Songs', *Prabhat Sangit*. This volume marks the beginning of Rabindranath's serious poetry.

In this particular poem Rabindranath is trying to capture the exuberant quality of the experience, somehow attempting to recreate the experience with the help of words. In the decades to follow, he dismissed this poem as well as the rest of the volume in which it occurred as poetically and artistically immature.[14] This may be so. And yet, the special flavour of this poem depicts the immediacy of unnameable divinity, and of a struggle against all odds to express where words fail and ought to fail.

In the 'Morning Songs' especially, but also in subsequent religiously coloured lyrical poetry, Rabindranath had tried to recover and poetically remodel and transform those experiences. In order to accomplish this, but never definitively, he developed and played with a number of recurring poetic images representing various facets of these experiences or the psychological circumstances before and after their appearance.[15]

The first lines of the 'Awakening of the Waterfall' depict the moment of the redemptive descent of the light.

> What song did the morning bird sing
> this morning!
> From the immensely distant sky
> it came floating.
> I do not know how a tune of its music,
> that lost its way, entered here,
> roaming about in the dark cave,

descending deeper and deeper down the cave,
constantly weeping in anxiety, how
it has touched my heart!
Why did a ray of sunlight that lost its way,
not finding its home,
why did it fall upon my heart
this morning!
After many days a single ray
appeared in the cave,
on the dark waters inside me
a streak of gold has shone.[16] (lines 1-18)

In the poem the light of the rising sun, which in reality caused Rabindranath's spiritual illumination, symbolizes the illuminating experience. It was, after all, the sun that just rose behind the screen of foliage which the poet had been watching, as the curtain of the ordinary world was pulled aside. It was the actual sunlight that gave the poet the illumination. This is the factual background to the image in this poem. The light enters and gradually descends the dark cave of the poet's self, a self hitherto imprisoned in a gloom of its own making. As was said in the autobiography, the illumination takes place suddenly and was not consciously brought about by the poet himself. The illuminating moment is like a phenomenon of nature, it happens outside human control, like the rising of the sun. Therefore the poet says that a ray of sunlight lost its way and asks why it cared at all to enter the dark confinement of his personal self. The illuminating experience has two dimensions: a dimension on a cosmic scale: the light of the sun, and a dimension on a more human scale: a singing bird. The shining of the sun implies no conscious act on a human scale, but the same experience is made more approachable for a human being through the singing of the bird. Singing and making music are acts of a conscious will. Much of Tagore's later poetic imagery consists in a play between things of cosmic and of human dimensions.

The bird is a person and yet not human. It sings the song of dawn, in other words, the rays of the rising sun are the notes, the tunes perhaps, of the bird's shining song. This song as such is always coming and going, for sunrise and sunset are cyclical events bound by time. The event that has immense significance for the poet is the sudden ray falling upon him. The event is so much outside the poet's control that both the tune of the song and the ray of the sun have 'lost their way' when they descended.

The sunlight can be easily associated with divinity itself. The sun as symbol of the supreme deity, as well as the bird as symbol for the supreme self, occur in the *upanishads*, the sacred Hindu texts. Special selections from these works were used in weekly and daily worship by the Brahmo Samaj.[17] The first and most obvious verse that comes to mind is the famous Gayatri mantra in which God is invoked as the solar deity. This verse did form part of regular worship.[18]

If it is true that Rabindranath wished to evoke a notion of divinity by mentioning the sun as well as the bird, he obviously tried to fuse both separate images into one: hence the bird's song is the sunlight, which leaves no other alternative but to believe that the bird actually is the sun. The sun is the cosmic manifestation of divinity, the bird is the divinity in a personal and private manifestation, the only manifestation that can communicate the experience of enlightenment to the poet. For the bird sings, the sun just radiates.

Years later, in the thirties of this century, Rabindranath, still obsessed with this experience as well as the resulting poem, said the following in his book *Manusher Dharma*, 'Human Religion':[19]

> This is the event of that day when from the darkness came the light of the outer world, the light of the unlimited (*asim*). On that day [my] consciousness (*cetana*) abandoned itself and entered the Supreme Being (*bhuma*). (p. 90)
>
> Then I saw clearly, that as the veil of banality dropped away from the world, truth showed itself in incomparable (*aparup*) beauty (*saundarya*). Therein is nothing left of discursive thinking (*tarka*); then I knew this vision as truth. Even now I long to perhaps one day to be able to see again the 'form of joy' (*anandarup*) of the whole world, in that same fullness at some auspicious moment. This is what I had seen one day when I was young, for this reason the words of the Upanishad '[God] possessing the form of joy shines forth as immortality',[20] again and again has issued from my mouth. On that day I had seen that the world is not gross, that in the world there is no thing whose touch will not give joy (*ras*). Why ratiocinate (*tarka*) on what I had seen death belongs to the gross veil, but the truth (*satta*) which is innermost and consists in joy (*anandamay*) knows no death. (p. 93-4)

Almost fifty years later Rabindranath interprets the experience and the poem that flowed from it in unequivocally religious terms, quoting in this connection from the scriptures he and the Brahmo movement to which he belonged held most dear, the Upanishads. More specifically, he is quoting the second line of the Brahmo credo, the verse which summarizes the Brahmo view of God: *satyam jnanam anantam brahma* (Taitt. Up. 2.1), *anandarupam amrtam yad vibhati* (MuU 2.2.7) … 'Brahman is truth, knowledge [or:

consciousness] and endless[ness], which, possessing the form of joy, shines forth as immortality'.

In the poem itself this theophanic rapture is expressed in images of violent movement and drunkenness.

> ... I do not know why, after so many days,
> my heart woke up.
> My heart has awakened,
> the waters have surged up,
> I cannot restrain the longing of my heart,
> the passion of my heart.
> The mountains are violently trembling and shaking,
> rocks are tumbling down in multitudes,
> more and more the waters are swelling up in foam,
> roaring in terrible anger.
> Like madmen they ramble, ...
> (lines 60-71)

> ... bring about today the realization of the heart,
> making wave after wave,
> give blow after blow!
> When the heart rises up in drunkenness,
> what does it care about darkness or stones! ...
> (lines 87-91)

The power of this realization and the joy and restlessness it brings along, rouses the poet to action, to spread and share with others the newly found energy of life. The chosen medium to accomplish this will be music and poetry, in other words, the same medium which the Godhead at the beginning of the poem is said to have used to rouse the poet to spiritual enlightenment, for it was the morning-bird's song of which a fragment entered the cave of the poet's soul. In a way the poet will repeat and imitate the divine creativity and try to accomplish this in the same spontaneous and natural way. In this manner the products of human creativity, especially song and music, though determined and limited by and in human dimensions of time, will be made to contain the unutterableness of the experience of divinity. They perhaps will even aim at re-creating the experience of the divinity, or translating it in a multitude of words and melodies. The poetic experience will become a vehicle for religious experience. The sense of beauty evoked in the audience by a well-executed poem or song will be the shadow of the religious exaltation once felt by the maker of the poem. The performing poet will have temporarily changed into the morning-bird.

As the cosmic creative process encompasses almost infinite spans of time and space measured by human dimensions likewise the human process of artistic creation, once awakened, will seek to imitate this limitlessness, so the poet hopes:

I will pour out a stream of benevolence,
I will break the prison of stone,
I will roam around, singing and flooding
the world as a restless madman;
with loose hair, gathering flowers,
ascending on my wings marked by rainbows,
scattering laughs in the sunlight,
I will pour out my heart ...
(lines 102-109)

Speaking the things of the heart,
I will always sing my songs;
as long as I'll give my heart,
my heart will flow on,
it will not come to an end.
I have so many things to say, so many songs to sing,
I have so much life,
there is so much gladness, so much longing,
this is the dawn of my heart ...
(lines 116-123)

The poet seems to embark on a never ending pilgrimage, back to the source of the experience of grace. The outpouring of creative energy is like a path towards a distant goal, an ever receding destination. Here the endless path is itself the goal. The joy is in the onward movement not in the arrival at a fixed resting place.

Where is so much gladness, so much beauty,
where is so much playfulness to be found!
Who can say towards whom I will drift
in the swiftness of youth!
With unbounded longing, infinite hope,
I wish to see the world!
The desire is aroused, I am flowing on,
flooding all beings and all things.
I can pour out all my heart,
I can endure all times,

I can submerge all lands,
then what else do I desire!
This is the longing of my heart.
(lines 124-136)

These quotations from 'The Awakening of the Waterfall' already show in a nutshell some of the prominent images of awakening, light, music, boundless longing and endless travel that always haunted Rabindranath's later religious lyric poetry. The insistent image of the rising sun, or dawn obviously symbolize the spiritual enlightenment. But there are less predictable images as well. The sun is only one among various possibilities to symbolize the grandeur of the universe. To express his awe before the revealed universe Rabindranath uses images connected with cyclical cosmic or celestial bodies or phenomena: the sun, the sky, the starry night sky, sometimes also huge rain clouds and thunder and lightning. In a way they could all represent the divine in its manifold manifestations and aspects. Time is often a common factor in these cosmic or celestial images for they are also associated with great cyclical movements: the alternation of night and day, or the cyclic change of seasons with its concomitant change of weather. These images seem to represent the cosmic dimensions of the divinity.

The cosmic creative movements of God and its human emulation in artistic creation are expressed in images of song and music. To Rabindranath these represented the highest analogues of divine creativity, for within limited and measured time and rhythm they can temporarily, encapsulate the infinite. Within the limitations of time song and music can truly speak to the listener; they can weave their spell.

The aesthetic joy that music and poetry may give, symbolizes again the joy Rabindranath experienced in the moment of illumination. He believed that aesthetic joy throws forth the shadows of the divine joy. The intensity of this joy is expressed in images of drunkenness, madness, restless movement, and playfulness. To bring about this joy through creation one has to travel along the winding path of perpetual, laborious artistic productivity. This is the image of ceaseless travel, which, at the same time, symbolizes the search for spiritual illumination and the realization of divine presence in this world.

Human gloom, suffering, anxiety, fear, are expressed in the image of the dark cave or the confining stone walls curbing the poet's free movement. This image has direct autobiographical meaning. Because Rabindranath as a young boy was not allowed to go beyond the limits of the house, or if he had to go, only under strict protection. As a child he felt the real world lay beyond the walls. The spiritual illumination seems to have radically ended these feelings of imprisonment. It broke away the walls. This feeling of liberation, as from a prison, may also partly account for Rabindranath later impatience with all kinds of formalism, dogmatism, chauvinism, conservatism, but also violent radicalism.

Rabindranath later expanded and refined the tissue of these images in different lyrical religious poems occurring in among others: *Manasi* (1890), *Sonar Tari* (1894), *Citra* (1896) and *Naivedya* (1901). Ultimately this effort resulted in the unadorned songs of the Bengali *Gitanjali* published in 1910, and the *Gitimalya* and *Gitali*, both from 1914. Some of the most persistent themes in these later poems are the image of light, the restlessness, the endless search for an unnameable God in and behind the manifested world along paths that do not lead to a well-defined last goal, the enduring vision of the beauty of the world which is the beauty of the divine, the joy felt at the constant revelation of natural beauty. But they already belong to the poetical idiom of 'The Awakening of the Waterfall'. It is not surprising therefore that Rabindranath regarded this very poem, immature though it may be, as the preface to his subsequent poetical work, and that with it he 'had drawn an allegorical map of the path along which [his] heart was to journey'.

This does not mean that this poem has already said everything; nor that all subsequent poetry constituted nothing but sheer repetition of the old prototype. 'The Awakening' is the first outburst to be followed by fresh ones. Important developments and innovations in the themes and images followed upon this first attempt. For instance, the divinity who is scarcely directly or explicitly referred to in 'The Awakening of the Waterfall' takes on more tangible characteristics in the later lyrics.[21] There he becomes the awaited friend, or the lover, or the master, or the king, and is often directly addressed with *tumi*, 'you'.

Such personal relationships with the Godhead are a well-known theme in the medieval Vaishnava lyrical poetry of North India and Bengal. The Vaishnava influence on Rabindranath is a common dogma. And yet the poetry of, say, the *Gitanjali*, cannot be read and understood as if it is simply emulating traditional Vaishnava padas. The flavour of the *Gitanjali* is different. In their distinct way, the poems of the *Gitanjali* retain the atmosphere of the 'Awakening of the Waterfall' and share its spirit of revelation of light, theophanic exaltation and delight in natural beauty. The *Gitanjali* songs differ from the 'Awakening of the Waterfall' in their spirit in at least one significant way in that the former are exactly that, namely songs, for Rabindranath composed melodies for almost all of them.

As to their religious ritual use, it should be mentioned that a number of songs from the *Gitanjali* are sung in Brahmo services and form part of Brahmo song repertoire.[22] The following song from the *Gitanjali* might illustrate the similarity and the divergence from, 'The Awakening of the Waterfall'.

6

Flooding the whole sky and the earth
with love, life, song, fragrance, light, joy
your spotless immortal nectar is raining down.
Razing every obstacle in all directions,
joy is awakened, it has assumed a form;
life has been filled with intense sweetness.

My consciousness, rich with the juice of spiritual good,
blossomed like a lotus flower in utmost delight,
laying all its sweetness at your feet.
In a corner of my heart, in soundless light,
the purple glowing loveliness of an exalted dawn awoke,
and the veil before my lazy eyes was pulled away.

This little song[23] seems to summarize in a few words Rabindranath's experience of illumination years back and the 'Awakening of the Waterfall'. The essential elements are still intact. There is the experience of intense joy, delighting in the beauty of the world. Obstacles are being overcome, 'razed' (*tutiya*), by this joy. Towards the end the light of the dawn, of the rising sun that is, appears again along with the image of the veil that is pulled away. Are we reminded of the first illumination? But the poet's consciousness has ripened now, it has blossomed like a lotus in utmost delight, and lays the fruits of its creativity at the feet of God, 'laying all its sweetness at your feet'. This is the major difference with 'The Awakening'; for there God was not spoken to. It is as if over the years the experience has been clothed with the image of a personal God, as if the experience could now be interpreted as having been given by God. This was not so in 'The Awakening'.

In the period of the *Gitanjali*, religious exaltation of vision had been transformed creatively into song and music. The poet returns the old gift to the original giver, but with enhancements. Poetry, music, art, the creative process itself, have become acts of worship, an open-ended dialogue between the human creative personality and the unnameable divinity about Whom the ultimate, definitive word will never be spoken.

Notes

1. Rabindranath Tagore, *Gora*, Roman, Aus dem Bengali von Gisela Leiste. Berlin: Volk und Welt, 1982, reprinted, München: Beck, 1988; Rabindranath Tagore, *Das zerstörte Nest*, Aus dem Bengali übertragen von Gisela Leiste. Berlin: Verlag Volk und Welt, 1985, reprinted: Zürich: Manesse, 1989; *Some*

Songs and Poems from Rabindranath Tagore, translated by Pratima Bowes. London and The Hague: East West, 1985.

2. William Radice, *Selected Poems, Rabindranath Tagore*. Penguin Books, 1987; *Rabindranath Tagore, Selected Short Stories*, Translated with an Introduction by William Radice. Penguin Books, 1991; Ketaki Kushari Dyson (tr.), Rabindranath Tagore, *I Won't Let You Go, Selected Poems*. Bloodaxe, 1991; Martin Kämpchen, *Auf des Funkens Spitzen, Weisheiten für das Leben*, aus dem Bengalischen übersetzt und eingeleitet von Martin Kämpchen. Kösel, 1989; by the same translator: *Wo Freude ihre Feste feiert, Gedichte und Lieder*. Freiburg: Herderbücherei, 1990.
3. Katherine Henn, *Rabindranath Tagore: A Bibliography*. Metuchen, NJ, and London: The American Theological Library Association and the Scarecrow Press, 1985. (ATLA Bibliographical Series. 13)
4. *Rabindranath and the British Press (1912-1941)*, compiled and edited by Kalyan Kundu, Sakti Bhattacharya, Kalyan Sircar, foreword by Mary Lago. The Tagore Centre (U.K.), London N22 4UJ, 1990.
5. *Rabindranath Tagore and the Challenges of Today*, edited by Bhudeb Chaudhuri and K.G. Subramanyan. Shimla: Indian Institute of Advanced Study, 1988.
6. *Rabindranath Tagore, Perspectives in Time*, edited by Mary Lago and Ronald Warwick. London: Macmillan, 1989.
7. This is the well-known *Brahmo Dharma*, a collection of passages taken from the *upanishads* and the later epic texts and from the Dharmashastra of Manu. Debendranath selected these passages in a state of mystical communion with God, cf *The Autobiography of Maharshi Devendranath Tagore*, translated by Satyendranath Tagore and Indira Devi. London: Macmillan, 1916, p. 94.
8. The only recent historian who emphasizes the fact that Rabindranath was, after all, a prominent Brahmo himself, and that his religious, political, and social ideas should also be judged against the background of his Brahmo inheritance, is David Kopf in his book: *The Brahmo Samaj and the Shaping of the Modern Indian Mind*. Princeton: Princeton University Press, 1979, chapter 10.
9. It is well-known that he tried his hand at poetry in 1869 when he was eight, when he was given a notebook, the one he would later refer to as the famous blue notebook. For more details see: Prashanta Kumar Pal, *Ravijivani, prathama khanda*. Calcutta: Bhurjapatra, 1389 (Beng. era), p. 106-8.
10. Many details and background information on these and other works he wrote up to 1889 are gathered together by Pulinvihari Sen in his book: *Ravindra-grantha-panji, prathamkhanda*. Calcutta: Visvabharati Granthan-vibhaga, 1973, p. 1-70, as well as: Svapan Majumdar, *Ravindra-grantha-suci*, pratham khanda, pratham parva. Calcutta: Jatiya Granthagar, 1987, p. 1-29. The latter book is a publication of the National Library in Calcutta.
11. All quotations from *Jivansmriti* are from the following edition: *Jivansmriti*. Calcutta: Visvabharati Granthan-vibhaga, vaishakh, 1387; this edition contains end notes. All translations from Bengali are my own.
12. Pulinvihari Sen, *op. cit.*, p. 82. This passage stems from manuscripts of *Jivansmriti*, but was not included in the final printed version.

13. Cf. George Steiner, *Real Presences, Is there anything in what we say?* London: Faber & Faber, 1989, p. 3-4, 142-3.
14. Rabindranath did not want to completely part with 'The Awakening of the Waterfall', for he preserved it in a much reduced form in the anthology he himself made of his poetry, the widely read *Sancayita*.
15. I am aware that this is hardly a new piece of information, but although it has been formulated before, it was never further explored; cf. Dr. Krishnalal Mukhopadhyay, *Ravindrakavye Rupakalpa*, Calcutta: Abhi Prakashan, 1382, pp. 118, 137 on the poetic image of light connected with this particular experience, and *ibid.*, pp. 176 ff. on the light image in Tagore's poetry in general.
16. The text of Tagore's poetry can be found in every standard edition of his complete Bengali works.
17. This selection is the *Brahmo Dharma* by Debendranath.
18. The Gayatri verse occurs as the meditiation, *dhyanam*, of the Brahmopasana section of the *Brahmo Dharma*. Rabindranath was always very fond of this verse. The passage in which the bird occurs is text 73, p. 59 of the *Brahmo Dharma* corresponding to Rig Veda 1.164.20; Mundaka Upanisad (MuU) 3.1.1; Shvetashvatara Up. 4.6. In this verse, however, there are two birds sitting on the same tree. One bird eats the fruits of the tree, while another bird looks on. The Bengali commentary on this verse by Debendranath interprets the one eating bird as the individual self, *jivatma*, and the other bird as the supreme self, *paramatma*. In the Sanskrit commentary on the verse they are identified as 'knower of the field [of the body]', *ksetrajna*, and the 'Lord', *Ishvara*, respectively.
19. Published by Visvabharati in a separate volume, latest edition Beng. era, 1393.
20. MuU 2.2.7.
21. Here I refrain from discussing Tagore's concept of God as both a cosmic force and a personal deity inspiring the poet's creativity. I have done this in another article entitled *'Awakening the Waterfall': The Wellsprings of Rabindranath Tagore's religious humanism*, which is to appear in the latest volume of the Bengal Studies Conference, held in 1996 in Washington DC.
22. A few random examples might suffice: *Gitanjali* poems 5, 55, 18, 33 correspond to songs 529, 295, 289, 1173 of the Brahmo songbook *Brahma Sangit*, published by the Sadharan Brahmo Samaj, Calcutta, fourteenth edition, Beng. era 1387. This songbook is used in congregational services.
23. It also occurs in the Brahmo songbook as song 266.

III

LIBERATING LANGUAGE: PĀRTHASĀRATHI MIŚRA ON THE SENTENCE AND ITS MEANING

*Purushottama Bilimoria**

It is generally overlooked that despite the intense ritualistic foreground of the traditional Mīmāṁsakas, with Kumārila Bhaṭṭa a shift occurs from the ritualistic to the speculative, epistemological and grammatical, and this is achieved with some degree of dialectical acumen.[1] Pārthasārathi Miśra's discussion of issues in language arises in this context and it underscores an intellectual defence of the Mīmāṁsā position on the authorless character and inviolable authority of *Śruti*. Here is the staunchest possible articulation on *āstikatva* which in principle makes no commitment to the existence of a transcendental (supreme) being, or to any transcendental signified (which is not erasable). Rather curiously, a good part of the erstwhile defence is based on the postulate of the inseparability of word and its meaning, known generally as the *autpattika* thesis. Elsewhere I have discussed *autpattika* in the context of the *apauruṣeyatva* doctrine.[2] Here I shall be concerned mostly with elucidating the Mīmāṁsā view on sentence-meaning in the work of Pārthasārathi Miśra who follows the Bhāṭṭa school even as he brings some new insights to bear on the analysis. Since the *autpattikatva* of the word and its meaning is a necessary presupposition of the theory of *vākyārtha*, I shall

---

* A version of the paper was first presented in the VIIIth World Sanskrit Conference in Vienna (1990); The autpattika portion of the discussion appeared in the following chapter: '*Autpattika*: The originary signifier-signified relation in Mīmāṁsā and Deconstructive Semmology', in: R.R. Diwedhi (ed.), *Mandan Misra Felicitation Volume*. Delhi: L.B.S. Rashtriya Sanskrit Vidyapitha and Motilal Banarsidass, 1994, pp. 187-203. I would also like to take the opportunity to acknowledge gratitude to my teachers of Mīmāṁsā, the late and respected Pandit Pattabhirama Sastri of Varanasi, and Professor K.T. Pandurangi of Bangalore.

be touching on this thesis as well. The discussion overall is apposite in the ambience of this volume as Professor Frits Staal has made several significant observations and even criticisms of the Mīmāṁsā view of language, especially in respect of the belief that language *per se* need not be dependent upon human origins or conventions and the Vedas or *Śruti* being the limiting case of just such a possibility.[3]

What however would be the convincing conditions for the scriptural texts if they are not thought to be dependent on any framework of conventions (*saṁketa*)?

Three moves are made in the attempt to establish *Śruti*'s non-human dependency. I shall flag these in point-form:

1. the relation between word and meaning: *śabdārthasambandha*;
2. relation between sentence and its signification: *vākyavākyārthasaṁyogya*;
3. *dṛśyadarśanabādhitam* or non-perception of an author where such is capable of being perceived (*yogyānupalabdhi*), as, say, with the Mahābhārata.[4]

1 and 2 are generalized in the Bhāṭṭas for both *Vaidika* and *laukika vacanam*,[5] and 3 is substituted with 'reliability' of *laukikavacanam* to fulfill the conditions of *śabdaprāmāṇyam*, while also requiring them to meet the three basic conditions of *prāmāṇyatvam* (*kāraṇa doṣabādhakajñānam*).[6] Since the question of *prāmāṇyatvam* as such of *śabda* is not my concern here, I shall not pursue 3, so I should here like to confine myself to a hermeneutics of 2, bearing in mind also the differences between the *abhihitānyava* (Bhāṭṭa) and *anvitābhidhāna* (Prābhākara) approaches to this very interesting linguistic issue, beginning with only a brief recapitulation of 1. The *adhikaraṇa*s crucial for the discussion are *adhyāya* I: 5,6,7,8, and *adhyāya* II: 14. These are encompassed in about six or so cryptic sūtras in Jaimini, which in turn have generated volumes of pages in Śābarabhāṣya and other commentarial (Vārttika) and sub-commentarial writings. Pārthasārathi Miśra's Śāstra Dīpikā and his commentaries Nyāyaratnamālā and Nyāyaratnākara, with Kumārila Bhaṭṭa's Ślokavārttika, and works such as the Jaiminīyanyāyamālāvistara, the Bhāṭṭadīpikā, and the Mīmāṁsānyāyaprakāśa, as well as a recent independent commentary by Vaidyanātha Śāstrī (on Jaimini's Mīmāṁsā Sūtras) form the important textual references for the ensuing discussion.

In the famous Autpattika Sūtra (I.i.5) Jaimini states that the relation between word and meaning is fixed or permanent (*autpattikaḥ śabdasyā 'rthena sambandhaḥ*).[7] What does this really mean? The standard practice has been to render *autpattikaḥ* as 'eternal', largely because that is the sense Śābara seems to have given it when he commented: *autpattikaḥ - iti nityam brūmaḥ*. But, *autpattikaḥ* is constructed from *utpattiḥ* + *(ḍhak)* = 'origin', which could mean 'originating or arising instantaneously, inseparably, or without interruption'. Hence Jaimini's use of the term *autpattikaḥ* above may be read as 'the relation between word and meaning is originary', in the sense that the presentment of the word and its meaning is simultaneous, so that there is no moment in which they are separable one from the other. In other words, the word and its meaning arise (upon hearing) *as if* psychically or episodically co-present or co-eval. Such is the 'founding binary combination' (to borrow a Derridean expression). I think it is often forgotten that Jaimini is actually referring to the actual instance in linguistic practice in describing this relation and that it is not an abstraction about the temporally primordial source of the word, as if rooted in some prehistorical or cosmological act of origination (*originê*) (not in this verse anyway).

In fact, even in Śābara, despite his use of the term '*nitya*', this reading is quite clear, for consider what Śābara goes on to say in expanding on what he means by *utpatti*: it is *bhāva* (presence) that is spoken metaphorically as *utpattiḥ* ('origin'). The relation between word and meaning is inseparable (*aviyuktaḥ*) by virtue of their (co-eval) presence; the relation is not (constituted) after both are arisen.[8] In other words, their 'origin' rests in their arising together in unison or in mutual presence. This relation is *sui generis*, natural and with a sense of permanency about it.

Pārthasārathi Miśra likewise observes that in ordinary speech there is a relation between word and meaning which is of the nature of *pratyāya* and *pratyāyaka*, what we could also loosely render as the signifier and signified respectively, and that it is by virtue of this intrinsic relation that the word comes to have this significative or denotative character (*pratyāyakatva*); and so the word conveys its meaning (in other uses as well) independently of other relations, such as of sense-organ contact or that involved in inference.[9] Very interestingly, Pārthasārathi considers but rejects the suggestion that the relation might be one of *saṁyoga*, mere conjunction, or even *tādātmya*, identity, for he wants to avoid the possibility of there being, on the one hand, any artificiality about this relation, and, on the other hand, absolute identity; both are qualified by *pratyāyakatva*, in that it holds the signifier and signified together, yet severs them once the signified has surfaced, until its next occurrence. In other words, in the very functioning of the *pratyāyakatva* (signifying) the difference between the signifier and the signified is stamped, otherwise the signifier would be self-reflexive, and it would hang loose as a bit of animated 'sound'; hence there is here a unity but also a differential and deferral function embedded in this relation.

Now the inter-looping or inter-playing of the binary pair of word and meaning, as if in an eternal wedlock is, I believe, a significant indicator at the same time of the differential being impressed upon in the *autpattika* thesis, just as it echoes later in the semiological insights pressed upon in linguistics by De Saussure [see Figure I]. It suggests that the binary is held in tension within the synthetic unity of *śabda* and *artha*, namely, with difference and naturalness.

This disavowal is most important from the point of view of a critique of metaphysics, for by this very recognition the Mīmāṁsakas check the tendency of any self-identification of *vācya-vācaka* (*samjñā-samjñin*) as occurs, say, in Bhartṛhari's Śabdabrahman or in the dissolution of all expressions into the Brahman of Advaita, or a Being that might unite the opposites, overcome the tension, or a *parousia*, as Mahadeva or Īśvara, that governs their presence (*ousia* as presence, essence/existence, substance or subject, which is absent).[10] This would be true even if we admit that Mīmāṁsās hold word as *varṇas*, sound-letters, to be universal and *nitya* (even *kūṭasthanitya*), and its corollorary, that meaning or *artha* as somehow being *nitya* also, but note: it is never *varṇasphoṭa* (meaning saturated-syllables); *śabda* (sound) as *pada* (putative word) and *artha* (sense) remain forever in tension - this position, for all its ambiguity and paradoxicality, is most marked in Kumārila Bhaṭṭa.[11] And if we agree to return to *artha* the magical sense, incantational potency of the *mantras*, then we move it away from 'concept' or the *idea* as the pure signified, the intelligible within spoken sign, and we then have a system of signs - rhythms, silences, gestures, noises - which is not exhausted by logical speech or rational discourse. But I am straying here, for I want to return to the linguistic ramifications of this position, not its anti-metaphysical ruses as such, apart from the agnosticism implied in the paradoxical transcendentalism of the *apūrva*.

Suppose we take the *vidhi*: 'Whoever desires *svarga* should perform sacrifice.' Without concerning ourselves at this stage with questions of syntax, we notice here a string of signs, which breaks up into the signifiers, basically, as the *liṅ+dhatu*, *svargaśabda*, *kāraka* (which is neutral in the instrumental) etc. and the signifieds as the twofold *bhāvanas* or impulses (efficiencies), namely, 'performative action' and 'the *phala* (*svarga*, etc.)'. But the *phala* is not actually observed or is not forthcoming in the sacrifice, for it is deferred or postponed; which means that the two *bhāvanas* in turn become signifiers portending a third signified: namely the *apūrva* (an 'unprecedented' efficiency), this is the 'transcendental signified' (*adṛṣṭārtha*, literally, 'unseen effect or entity').

However, the *apūrva* is 'transcendental' in a modest and then only in a provisional and operational sense, for its significance is not so much in its being *adṛṣṭatva* (as Verpoorten rightly points out)[12] as in being a mysterious result 'which has nothing before it' or being without a preceding instance, i.e. in its novelty and prior unknownness, and so also in its near-empirical givenness for it retains the phenomenality of the *almost certain phala* (result), and further it will erase itself the moment the *phala* has matured later. Hence, once again, *phala* functions as the signifier for the deferred but now-imminent signified, namely, *svarga*, and which too, like the *devatās*, has no ultimate or absolute ontological status in Mīmāṁsā, being possibly a signifier for an inner state of happiness in another birth, or something to be consumed, and so its self-identity is eroded. And, of course, *apūrva* can be annulled by a wrong step (*bādha*) in the sacrifice. What is interesting is that the *apūrva* has no sign (*signé*) for it in the expression itself, and so it cannot achieve reference to itself self-reflexively, as happens when I say 'I'; yet the 'hidden' (*avyakta*), unmanifest and 'not yet visible', does achieve signification, though at the upper split level, i.e. as the pure signified. [See Figure II.] But how can this split occur if *autpattikatva* is to be taken literally, i.e. as prefiguring an inseparable relation? We have to say that this occurs through displacing another signifier, or by metonymy, or (what I will argue later) because the *vācaka* as a syntagm is the signifier in this instance (of performative utterance). And in all this the subject as *auctor* needs hardly be relevant.

Be that as it may, and however philosophically elegant the doctrine might seem, the *autpattika* thesis (on which the doctrine, as we have said earlier, is foregrounded), appears, to common sense at least, to be counterintuitive. One could object that a long process of historical repetition of the pairing of word and meaning in any medium gives the appearance that no personal agency or convention was responsible for fixing the relation. Further, we can only have access to a finite number of words in any given language at any given time and from looking at these we cannot pronounce that *autpattikatva* is true for all words. Also, there are words that have no meaning, and some words have more than one meaning, and many words have the same meaning. We do also observe that, as with names, new words are created with new meanings attached to them. And, as Professor Staal once remarked: *If anything is eternal it would be the lexicon!*[13]

The Indian philosopher, M. Hiriyanna, argued that even if one granted the Mīmāṁsā thesis of the permanence of the word-meaning relation, this does not *eo ipso* establish the eternity of the Vedas. It is the fixing of the particular order of words (*ānupūrvī*) in the text of the Vedas that is meant by eternity of the Text.[14] The permanency of relation thesis is a later development, possibly in response to Patañjali's objection that the order of words in Sanskrit at least, if not in the Vedas themselves, need not be eternally fixed. Patañjali averred more on the side of regarding the sense or meaning of the Vedic texts to be eternal, according to this line of criticism.

Pārthasārathi considers some of these objections (as indeed had Śābara), but the one he chooses to address in rejecting all of them is the questionable assumption that word expresses meaning. His short answer is: it is a fact that immediately upon hearing a word its meaning arises in one who knows the language even if the word is used by an ignoramus.[15] But we also observe that sometimes when words are first heard they convey no meaning. Again, his answer is that this is because certain attendant conditions are not fulfilled, such as absence of impediments in the medium, and so on, or because one has not learnt it from another. This does imply that the meanings of words have to be learnt, from another who learns it from another and so on (the requirement called *vṛddhavyavahāra*). But this does not entail for the Mīmāṁsakas that anyone of the persons in the process of transmission actually fixes signifier-signified relation. More significantly, he says, when a sentence is uttered, one who knows the language apprehends its sense. How is that possible? This and related issues are discussed under *Vedapratyāya katvādhikaraṇa* (#7), to which I now turn.

Pārthasārathi next considers the objection:

Let us grant that *autpattikatva* is true and that the words, meanings and their relations are *apauruṣeya* or *anādi* (of beginningless origin), and awareness of word-meanings is through the usage of elders (*vṛddhavyavahāra*). Even so, what accounts for the generation and understanding of sentences *qua* sentence? For surely sentences have to be constructed by putting together words, and the possibility of such sentences from a finite number of (known and current) words is infinite, and infinitely nuanced;[16] besides, words in lexicons are not sentences, and one might recognize words in an utterance but may fail to comprehend the meaning of the sentence, and yet we understand sentences that we may have never heard before. In short, a sentence appears to be more than a string of discrete words, and *śābdabodha* more than an assortment of *śabdārtha*s, unless they are associated through some convention or other (*pṛthaksaṅghāti/saṁketa*).

Searle has pointed out that linguistic utterances are closely related to the conventions governing the use of words in that language (i.e. the language in which the utterance is made). Linguistic utterances normally rely on the conventional meanings of the words uttered and the hearer's knowledge of these conventions. Accordingly, the analysis of the notion of linguistic meaning must include an account of the connection between the utterance and what the utterance means in the relevant language.[17] Since therefore conventions become inevitable and are added *ab extra* they must be *kṛtrima* (man-made) and therefore cannot be *saṁketānapekṣatva*.

Although the question is raised in respect of *vedavākya*, the ensuing discussion shifts towards *laukikavacanam* (and this extension holds because Śābara had considered word meanings to be the same in both *laukika*- and *vedavacanam*, though he reserved the term *śabdapramāṇa* specifically for *codanā*s which covers *vidhi*s and *arthavāda*s; it is only with the Bhāṭṭas that the term *śabdapramāṇa* gets extended to *laukikavacanam* as well, with *codanā* as the limiting case for the *vijñāna* of *dharma*; it is admitted that certain words in the Veda do not occur in ordinary discourse: so how is their relation fixed?).[18]

All this is the *pūrvapakṣa* elaboration on sūtra 24: *utpattau vā avacanaḥ arthasyāt annimittatvāt, vā kintu: padārthanimikatva abhāvāt*; and rephrasing *Nyāyamālāvistara: pṛthaksaṅghāti apekṣeyam anapekṣatva varjanāt*[19] - since stringing together the separate words is dependent it looses its non-independence (claim with respect to external or human convention); *na kiṁcitmūlaṁ sambhavati*. (I take *prāmāṇya* in this discussion to be simply the fact of valid communication, rather than an epistemological claim.)

Pārthasārathi deals with these objections by considering and rebutting some alternative suggestions as to what gives rise to sentence meaning (*vākyārtham avagamyeyuḥ*), which try to implicate extraneous conventions:[20]

1. *pratekam* : each word separately
2. *padāni* : groups of words
3. *vākya* : through their *(pada)saṅghātaḥ*: association of the words
4. *padārtha* : word-meanings

1. is rejected outright because it is never observed that *vākyārthajñana* arises from isolated individual words; for one, without denying that words have significative capacity (*śakti*), the meanings of words are general (cowhood), while sentences convey more specific references or particulars (*vākyārthaviśeṣaiḥ*); and given that (as we noted earlier), words are finite and sentence-meanings are innumerable/endless, there would be no way of apprehending the relation between words and sentence-meaning in all the variations and permutations, and how can each word distinctly achieve this proliferation of meanings? (... *ca padānāṁ vākyārthaviśeṣaiḥ sambandhagrahanaṁ na asti na ca sambhavati anantatvāt vākyārthasādhāraṇatvāt*).[21]

2. for similar reasons 2 is inadequate as well, for a *pastiche* of words cannot succeed in doing what a well-structured sentence conveys; nevertheless, 1 and 2 are important for teaching language to children and non-speakers (*Prattīvācyārthayāt na ca; bālānāmaprattīteḥ vyutpattivaiyarthāt ca padārtham nānārthika iti ucyate*) and knowledge of conventions may be necessary, viz. that X signifies Y.

3. what about sentence as some kind of association of words? In other words, *padasamūhaḥ* or *padasambandha? na saṁbhavati pratyāyakatva* 'no power to convey sentence meaning'; here again, the question of how the relation to *vākyārtha* is be construed arises. Consider that two sets of the identically associated words can lead to quite different meanings; differently associated

words to the same meaning. And even those who do not know the conventions for such associations is able to comprehend the sentence (*ataḥ tadrahitānām api pratītiḥ na doṣaḥ*).

Under this another possibility might be considered: namely that sentence as a distinct entity gives rise to *vākyārthajñāna*. The argument is that the last syllable (*varṇa*) supported by the impressions (*saṁskāras*) generated by the preceding syllables, pushes forward a new entity, i.e. *vākyasphoṭa pūrvavarṇajanita saṁskāra sahito vā vākyāntyavarṇo vā niravayavo vākyasphoṭaḥ*). Pārthasārathi would accept the analysis of the process involving *varṇas* and *saṁskāras*, but, following Kumārila, rejects that anything like an impartite *vākyasphoṭa* arises,[22] rather he wants to retain some constitutive elements in the sentence, and so he moves swiftly to the last alternative.

Similar issues arise here: what would be the relation between word-meanings and sentence-meanings, and how would this be apprehended in the innumerable instances, as appealed to for 2 and 3? If sentence-meaning is the *vācya* and word-meanings the *vācaka*, there has to be a relation between the two and an apprehension of this relation. You may string together a cluster of word-meanings and nothing might happen; and one often understands sentences even if he is ignorant of meanings of some words. Nonetheless, Pārthasārathi makes comprehension of *padārthas* a prerequisite of *vākyārtha* (+ its *jñāna*) (*vyutpattiḥ ca padārtham* ... cf. 2 above).

So all four alternatives are rejected: agṛhītasambandhatvād eva ca padasaṁghāta vākya padārthānām api na saṁbhavati pratyāyakatvam.[23]

TWO VIEWS OF SENTENCE-MEANING

In this refutation the facticity of relation (*sambandha*) of meanings appears to be a central consideration, which Pārthasārathi does not believe can rest simply on it being established by some external convention. Another rejoinder is considered, viz.: If the sentence is an entity separate from words (i.e. *padas* have nothing to do with formation of sentence and comprehension of its meanings, nor word-meanings bear any relation, nor sentence as an impartite entity to sentence meaning), then why make *padārthajñāna* prerequisite for sentence meaning: *kaḥ tadā padārthavedanasyopayogaḥ*.[24] What logical connection is there? Wouldn't this be superfluous? Or would not convention be required to cement the two?

Pārthasārathi's response to this is that by the same token if conventions are required there would have to be infinite such stipulations corresponding to each sentence. The rebuttal actually turns on rejecting the common view that

1) words remain constant in their meanings. i.e. continue their general meanings as given in lexicons, etymology etc.

2) the relation between word and meaning, and sentence and meaning are of a radically different (grammatical) order: both share the *vācyavācaka* model.

Although, it must be said that Prabhākara who is insistent on 1. and his theory of *anvitābhidhāna* is more consistent in accounting for this, viz. by focusing on the contextual frame of a sentence rather than on its discrete constituents, which have no signification in themselves: contrast the Saussurean interdependency of related signs. While Pārthasārathi holds that words are simply instrumental in yielding word-meanings, which are fixed (though conventional usage will override etymological sense), whereafter words *qua* signifiers erase themselves and the word-meanings *qua* signifieds take over and combine to yield sentence-meaning. Still, the *anvitābhidhāna* is less concerned about mapping the trace of concepts (ideas as signifiers) than fusing the signifieds into a larger whole (which is greater than the sum of the parts). Still, the mark is in the signifier, the sign itself, which is privileged with a force (*bhāvana*) equivalent to that of a perlocutionary speech-act which has the intention of bringing about a particular effect in the hearer - e.g. when I say 'Get out', 'Sit there', 'I launch this ship', 'My signature is on the article', - it is the *act* of saying, the real *bhāvana*, rather than the word-meaning collocation, that produces the effect and ties it to a particular intended perlocutionary effect, or induces the exertion. Similarly, 'You ought to sacrifice' should impel the other immediately towards that act, the exertion, or existential performative, rather than evoke a hermeneutical *vijñāna*; hence also the Prābhākaran insistence on the centrality or *mūkhyaviśeṣatva* of *kriyāśabda*, *ākhyāta*, *proyojana* in all expressions, especially of the *niyoga* (*vidhi*). The communication of the *tātparya* or intentionality occurs through the medium of the signs themselves.

I shall not dwell on their differences much more here, except to say that for both, sentence is the unit of communication; Prābhākarans emphasize a contextual framework, Bhāṭṭas a piecemeal constructed one. In any case, Pārthasārathi claims their views accords with experience.[25] The truth, I believe, as shown also by recent Western linguistic investigations, lies somewhere in between.

We notice that both views in some way or other stress the fact of *relations* - of signifiers, interrelated signs, or of signifieds - but these are not given in the isolated words or word-meanings: where then lie their origin within the sentence? How do they become part of the *vācaka*?

VĀCYAVĀCAKA MODEL FOR SENTENCE-MEANING

First Pārthasārathi agrees that the same kind of *vācyavācaka* relation between sentence-meaning and word-meaning is absent, but he wants to qualify this by suggesting that an analogous model does exist, in the form of *sāmānyaviśeṣa sambandha*, general-to-particular between *padārtha* and *vākyārtha*. But how is this reduction possible without a separate convention?

Here Pārthasārathi argues that this function is simply an extension of the words capacity to disclose its meaning, in that the *abhidhāśakti* will go through to sentence-meaning, i.e. relate the signifieds, by displacing the general (*sāmānya*) with the particular (*viśeṣa*), which, of course, occurs through the *nimitta* of *lakṣaṇā* (secondary signifying capacity, which subdivides into symbolic, figurative, mythic, and metaphoric connotations). *Nimittam* is introduced by Pārthasārathi as an indicator of causal efficiency of words in respect of this capacity. This is not simply a *vyāpāra*, instrumental efficiency, which he grants to the *padas*, but a causal condition, on par with the *kāraṇas*. Under this he brings in *ākāṅkā*, *yogayatā*, *saṁniddhi* etc. as the factors which help the *bhāvana* produce syntactical units. The *bhāvana*, then, is diffused over the related signified rather than centrifugally located in the signifier (as it is for Prābhākara). This shows that there is a shift from the model of *vidhivacanam* as the model of a sentence to *laukika-vacanam* such as '*gām ānaya*'. Our experience seems to confirm the view that as soon as word is heard its meaning is apprehended; as a cluster of words in a sentence is heard, their respective meanings are collected in the mind, and when the last word is heard, a collective meaning is understood. Later when we try to report someone else's utterance, we may use a different set of words to express what was said, and often qualify this by saying, 'I don't remember her actual words ... ', or 'I don't want to put words in his mouth', indicating that we may have forgotten the words but not their meanings in some collocation.

We may note that the definition of sentence is relativized to *padārthas*, in its *abhidhīyate* or significative capacity, and not simply in terms of giving a unitary meaning as with a statement, for not all sentences do so. Connected or related sense is what is given primacy in this view.[26]

What really is being said, in distinguishing this analysis from the earlier alternative of *padārthas*, is that sentence-meaning is the function of *abhidhā*, which is the *nimitta* condition, the indicative force, while *padārthajñāna* is the necessary condition (and *padajñāna* the instrumental, *sādhāraṇakaraṇa* or *vyāpāra*).[27] Sometimes when words are forgotten, this is no great impediment (*doṣaḥ*), for the *padārthas* to be surmised (by *arthāpatti* if need be) and substituted to continue the process. We may ask if there is any distinction then between the kind of connection between *mūkyārtha* and *lakṣyārtha* of particular words and that between *vākyārtha* with *padārtha*? Some Mīmāṁsakas would argue that *vākyaśakti* is distinct from *lakṣaṇā* of *padas* and call this *tātparyaśakti* (or *anvayaśakti*).[28] This *lakṣaṇā* is not related to the *abhidhā* of words; it is, if you like, a second *lakṣaṇā*. Otherwise there is *padapramāṇa* and not *śabdapramāṇa qua vākyavyāpāra*.

Pārthasārathi argues that *abhidhāśakti* is good enough as far as *kāryānvita* is concerned, i.e. where the signified is prescribed activity (more or less suggesting the perlocutionary force), but this is rendered insufficient when extended to ordinary discourse (or illocutionary utterances as Searle also insists, for non-performative intention is essential here), for which *lakṣaṇā* is required. Even Prabhākara will admit that the *tātparya* here arises posterior to *abhidhā* through the *anvaya* of *lakṣaṇā* (almost as though to say that we bypass *abhidhā* of words in hearing a complex utterance).[29] But Pārthasārathi retains the sequentiality of *padārthas* to *vakyārtha* as its means, and accedes to *lakṣyārtha* and the poetical nuance (*vyaṅgyārtha*) in ordinary discourse, but only as posterior to *abhidham*.[30]

Very simply, for Mīmāṁsakas, not only metaphorical words, but sentence-sense as a whole is always *lākṣaṇika* or the indirectly signified.[31] It is this *lākṣaṇika* that completes the *tātparya* of the sentence (the intentional functions as attributed by Grice, Strawson and Searle variously).[32]

EKAVĀKYATVAM, *SĀMĀNYA* AND *LAKṢAṆAM*

Incidentally, I should mention under this that Bhāṭṭas reject *ekārthaikatva*, unity of meaning, as a definition of *vākya*, for it is too wide and was intended in the sūtra for demarcating one *yajus* from another (*mantras*, *brāhmaṇas*, *ṛk*, etc. from *arthavāda*, *yajus*, etc.) hence it discerns a *prayojanam* (the purpose in the sacrifice, which the *prakaraṇa* will yield), rather than a grammatical sentence, for which *ekavākyatva* is deemed sufficient, and which in turn is defined in terms of *abhidhīyate* of words and *sākāṅkṣā* among the *padasamūhaḥ* (a condition *Bhāṭṭadīpikā* takes from Śābara but delimits it to a grammatical unit distinct from 'statement'), or *paryavasāna* of *padārthas* (II. i. 14(46)): meaning that, the absence or severing of one word (word meaning) evokes a sense of incompleteness of the *vākyārtha*, hence there is *ākāṅkṣīta*. Thus, e.g. if one says '*gām*' there is *paryavasānam* for 'what with it?', i.e. for '*ānaya*'. Further, '*gām*' is apprehended as related to the action of bringing (while Prābhākara maintain that action of bringing, the *mūkhya-viśeṣa*, transfigures cow within its unitary sense (*ekasya eva anvitasya*), which is the sentence-meaning) this very fact of relatedness transforms the *anvaya* of *sāmānya* (cow-*ākṛti*) to the *anvaya* of particular (cow *x*). The idea of serving a common purpose is rejected, again, on the same grounds that statements, like '*darśapūrṇamāsābhyāṁ sahadharmāṇām*', make mention of *svargaḥ*, *phalam*, *paśuḥ*, *haviḥ*, *agniḥ*, *devatā*, etc. are taken to serve a common purpose, viz. performance of a sacrifice (*pradhānayāga*), which is inferred by *anumāna* and not perceived as a syntactical unity, whereas each expression might constitute a sentence in its own right, each with its own *viśeṣas* (distinction between *anumitivākya* and *pratyakṣavākya*): *padānāmekasmin nārthe paryavasānam*. *Vākyatā* is the consideration, i.e. where the sense is incomplete if the words are in expectancy of each other. Thus '*go*' must then be connected with the object, *viśeṣa*, of the action of bringing, and *ānaya* by itself cannot encompass this.

It is, argues Pārthasārathi, not the same as simply any two words coming together or being conjoined in a free-standing expression, such as, e.g. in *rājapuruṣaḥ*, and another, say, *atiśobhanaḥ*, for sure there is *anvita* here in both (two words in relation), but there is not the same *anvaya* in either, as there would be if the two expressions are brought together in a sentence whole: '*atiśobhano rājapuruṣaḥ*'. The connection between the two *anvita*-units requires more than what is necessary for each individually, otherwise we could be talking of two different individuals, for the connection of *śobhana* with *rāja* has not been established herewith. The words have to go through to and beyond their (qualified) meanings for the *anvayabodha*, i.e. signifiers are left behind, and signifieds have emerged in the two pairs (*anvita* = complex).

At this point Pārthasārathi points out that *abhidhā* is not sufficient either, for it has already been exhausted in making up each pair; the *signification*, which carries a more composite sense, needs a little more. While the grammarians recognize and restrict the *vṛtti* of *lakṣaṇā* for metaphorical and suggestive connotations (to supplement *tātparyavṛtti*), Bhāṭṭas invoke *lakṣaṇāvṛtti* for eliciting particular from the general denotation.

Tasmād padābhihitaiḥ padārthalakṣaṇayā:[33] therefore sentence-meaning, signified-whole, is conveyed through word-meanings by means of *lakṣaṇā*. This is explained thus: in the expression *'gām ānaya' ānayana kriyā* signified by the word *ānatiḥ*, firstly signifies 'bringing' in a general sense, then by secondary signifying capacity in relation to *gopadam* conveys it as a particularized cow; similarly *gokarmakatvākāreṇa tatsambandhasvarūpena*, through its form (*abhidhāna*) as connected with the action of bringing, again the *pratipādyati* is via *lakṣaṇā*. In other words, the *mūkhyaviśeṣa* is picked out within an *ekavākya* and its *abhidhā* yields the initial signified, which is general, then through this (*viśeṣyatvena*) all other words whose *abhidhāna* have also become active, a connection occurs via *lakṣaṇā*, to yield the composite sense. Where there are many more complex juxtaposed words: adjectives, substantives, etc. there *lakṣitalakṣaṇā* will also come into play, e.g. *atiśobhana/rājapuruṣa* each pair internally is related by *lakṣaṇā*, externally by *lakṣita*. Of course, this opens up the possibility for various interplays, substitutions and supplementations of signifieds (see Figure III).

Ekavākyatva, Pārthasārathi says, helps delimit a general into a particular in terms of the close juxtaposition or mutual presence of the words. This *parampara*, of course, logically can extend indefinitely as long as there is syntactical unity, *vākyatam* (i.e. the mark of which is expectancy if constituents are separated). The problem of infinite regress holds no weight here: *tasmād padārthaiḥ vākyārthabodhana anapekṣitasaṁketa*.

Therefore, sentence-meaning comprehension arises through word-meanings for which there is no dependency on any (external/human) convention.

Finally, if the *bodha* or *grahaṇa* is so possible for *laukikavacanam*, it ought to be so for *vedavacanam* as well; further, the generation of *vākya* by parity is also possible via *anapekṣita*; in fact the former presupposes the latter. The selfsame *śakti* is involved in the *bodha* as in the source; it is only because we are able to apprehend sentence-meanings that we are also able to generate sentences, but neither necessarily entails that *we* are the absolute source of origin of sentence *qua* sentence; perhaps we learnt to do this from the Vedas, or from language itself! *Vākyatvam astu pauruṣeyam māstu* (see Figure IV).

CONCLUDING REMARKS

Mathematical and logical propositions, it would seem, function without any assumption of a subject or intention of law whose source has been buried in historical amnesia; perhaps the Mīmāṁsakas had this inkling, but wanted to make *Vaidikavacanam*, notably *vidhi/niyoga/atideśaka* (mandatory) injunctions, paradigmatic of such impersonal propositions. They would not accept the endless, possibly discourse-dependent if not meaningless link-chain of signifiers, with the possibility of its theoretic closure, as proposed by Derrida and other post-structuralists. From a more analytical position, an account of linguistic meaning is being developed which does not include an account of convention; intention or intentionality is not denied (no more than some cognitivists want to deny intentionality to *AI*), but its intrinsic locus in a subject is.

One might ask why the Mīmāṁsā theory does not go as far as the deconstructionist/semiological theory of language or becomes fragile if pushed to these limits: because in Mīmāṁsā there is a quasi-ontological basis (*arthatvā*) to the relation which accounts for its naturalness, and which is denied or rejected in the semiological models and analyses prevalent in the West. Secondly signs don't bifurcate endlessly or absolutely; it has a non-dual quality about it, more importantly it manifests objects, unconceals something and moves the will, this is its *bhāvana*: in that sense it has power *śakti* or *abhidhāśakti* (*padaśakti* of various kinds), also to create the *apūrva*. And each sign may embed whole clusters of meanings packed into its *mūkhyaviśeṣa*, i.e. entire discourse (just as the term Vedas often conjures in our minds).

And it is through the signs that the texts become the subjectless agency of action; human beings displace these signs for themselves in that empty *topos*, *locus standi*, and act out what is already Work, *kriyā/kartāvyatā*: who is working who? who is acting who? There are really no *kartās*: human beings are merely actors, instruments of sacrifice, for the text is able to iterate itself, inscribe itself in another form. The subject, speaker/doer/actor as well as the object of desire in the act are already in the *text* or in discourse itself, namely the discourse of the Vedas: this was the *epistemé*, in Foucault's sense, of the time. We (post)moderns too have come around to taking word and work in the middle voice, in the form of art, writing, architecture, as the dominant episteme of our age which displaces all previous ones: but here the linguistic/scientific frame has been supplemented with the symbolic and semiotic/semanalyis; like the *padārtho*-centrism (naïve realism) of Nyāya and linguistic analysis (philolocentrism) of grammarians that displaced the *artho*-centricism of Mīmāṁasakas.

However, by making the relation between word and meaning relatively permanent, Mīmāṁsakas ascribed to a certain kind of textuality the power to disseminate just those signs and no other; hence the contingency of Saṁskṛta culture/paradigm were denied; or as Quine has also said, in analysing our language we use tools that already embed those very things we find in the language, such as identity, truth, logical relations and so on. The presencing of the absent *apūrva* gives Vaidika word its power: this is its transgression; the ontic-ontologization of the latter as the Transcendental Signified (which too is *di*-scended) entrenches the connection of the visible with the invisible: that tension is an important one[34] (art exploits that too, as does *via negativa* theology).

Nonetheless, reciting, reiterating the Vedas in the sacrificial discourse, goes on, as if Work all in itself and by itself, as modes of textuality, inscribing and reinscribing itself. Thus writing is referred to in the middle voice, it is used intransitively, like lightening, maybe even sacrificing. The text is never created like a work, but rather it is always there: it is not written, it is speaking, and so it is heard (*śruti*).

Or, as Heidegger has said, 'language is the precinct (*templum*), that is, the house of Being' ... 'Language *speaks*. This means at the same time and before all else: *Language* speaks. Language? And not man? ... '.[35] 'We do not merely speak *the* language we speak by way of it ... In our speaking as a listening to language, we say again the Saying we have heard. We let its soundless voice come to us, and then demand, reach out and call for the sound that is already kept in store for us.' This is how for Heidegger 'language as speaking comes into its own and thus speaks *qua* language'.[36] And so the Mīmāṁsā dictum: 'Thus speaketh the Vedas unaided by Person/human voice or convention' in a minimalist way makes good comparative sense.

ABBREVIATIONS

MS	Mīmāṃsasūtras of Jaimini. Mīmāṃsadarśanam with Śābarabhāṣyam. (edited by Jīvāndanda Vidyāsāgar Bhaṭṭācāryya.) Adhyāya I - VIII. Calcutta: Sārasudhanidhi Press, 1883 (Also, Mīmāṃsādarśanam with Śābarabhāṣyam. Ānandāśrama Sanskrit Series. 97. Poona, 1973-84.)
NMV	Jaiminiyanyāyamālāvistara. (ed. Jīvānanda Vidyāsāgar Bhaṭṭācāryya). Calcutta: Saraswatī Press, 1883
NRM	Nyāyaratnamālā of Pārthasārathi Miśra. Baroda, 1937 (GOS. 75)
SB	Śābarabhāṣa of Śābarasvāmī (in Mīmāṃsadarśanam)
SD	Śāstradīpikā of Pārthasārathi Miśra (edited by Rāmamiśrasāstrī) Kashi, 1881 [So 1949] (Also, Śāstradıpikā with commentary Prabhā by Vaidyanātha Śāstrī [edited by P.N. Paṭtabhirāmaśāstrī] Sri Lalbahadur Kendriya Sanskrit Vidyapitha, New Delhi, 1978)
SV	Ślokavārttika of Kumārila Bhaṭta with commentary Nyāyaratnākara of Pārthasārathi Miśra (edited by Svāmī Dvārikādāsaśāstrī). Varanai: Tārā Publications, 1978

The semiological theory of signification comports a sense (an idea acting as signified) and a 'voice' (acting as the signifier). This indissoluble union of the two primordial ('originary') components is the only essential thing in language, according to this model:

S — Sr / Sd — tree / (signification) ---- / 'tree'

Sr/Sd — signifier ←====→ signified (←====→=binary)

$S_3 \leftarrow S_2 \leftarrow S_1$ ---- Sr / Sd

Figure I

apūrva
(transcendental signified)
(adṛṣṭārtha)

devatān | Being (Presence)

ātman

Sd kriyā | phala (siddhabhāvya) [-ena]

vidhivacana ------------------------------------ dharma

Sr liṅ+ dhatū | 'svarga' [kāraka]

mantra ----------------- thavāda
(supplements)

Sr (transgression)

Figure II

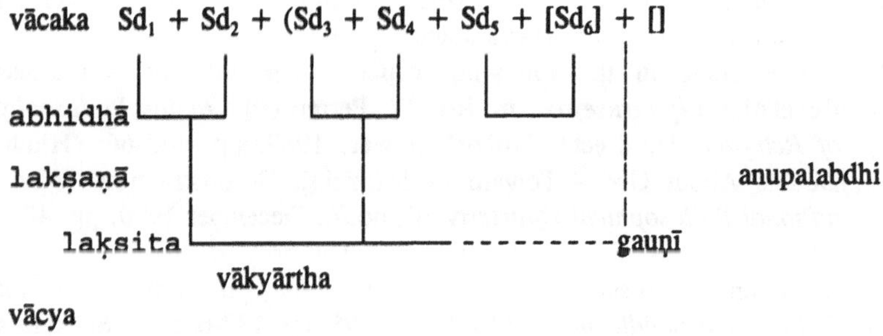

Figure III

qualified viśeṣa (particulars) with relations = vākyārtha

-------lakṣaṇā-------

Sd vācya [padārtha] + [padārtha] + [padārtha] sāmānya =\= arthaikatva

vakya = ―――――――――――――――――――――――――――

Sr vācaka {padasangati . . . ākāṅkṣā-yogyatā-samnidhi. . . } = vākyatā

Abhihitānvaya -- places emphasis on the upper side (signifieds)

Sd_2 vyakti (particular) (related signifieds) = signification

Sr ―――――――――――――――――――――――――――

Sd_1 ākṛti (general) by pratyabhijñā (=\= jātivijñāna)

Anvitābhidhāna -- places emphasis on the lower side (signifiers)

Sr_2 related signifiers = SIGNIFIED/signification

Sr ―――――――――――――――――――――――――――

Sr_1 signifiers

Figure IV

Notes

1. Jean-Marie Verpoorten, *A History of Indian Literature - Mīmāṁsā Literature*. Wiesbaden: Harrassowitz, p. 31.
2. For example in the following papers: 'The Idea of Authorless Revelation (*Apauruseya*)', in: Roy W. Perrett (ed.) *Indian Philosophy of Religion*. Dordrecht: Nijhoff/Kluwer, 1989, pp. 143-66; 'Hindu Doubts About God - Towards a Mīmāṁsā Deconstruction', *International Philosophical Quarterly* 30, no. 4, December 1990, pp. 481-99; 'Authorless Voice, Tradition and Authority in the Mīmāṁsā - reflections of cross-cultural hermeneutics', *Nagoya Studies in Indian Culture and Buddhism-Saṁbhāṣā* 16, 1995, pp. 137-60; see 'Śaṅkara's attempted reconciliation of 'You' and 'I' *Yuṣmadasmatsamanvaya*', in: P. Bilimoria and J.N. Mohanty (eds), *Relativism, Suffering and Beyond, Essays in Memory of Bimal K. Matilal*. New Delhi: Oxford University Press (forthcoming). See also P. Bilimoria, *Śabdapramāṇa: Word and Knowledge*. Dordrecht: Kluwer, 1988, and feature review thereof by Stephen H. Phillips, *Philosophy East and West* 45 no. 2, April 1995, pp. 273-80.
3. A more comprehensive treatment of these more technical issues will be covered in the Mīmāṁsā volume of the *Encyclopedia of Indian Philosophies* which is in progress.
4. Pārthasārathi Miśra's Śāstra Dīpikā (ŚD) (1891 [so.1943] ed., Kashi Rāmamiśraśāstrī) and his commentaries Nyāyaratnamālā (NRM) and Nyāyaratnākara (NRK), with Kumārila Bhaṭṭa's Ślokavārttika (ŚV), and works such Jaiminīya Nyāyamālāvistara (JNMV), Jīvānanda Vidyāsāgar Bhaṭṭācāryya (ed.). Calcutta: Saraswatī Press 1883, Bhāṭṭa Dīpikā, and Mīmāṁsā Nnyāyaprakāśa, as well as a recent independent commentary by Vaidyanātha Śāstrī (on Mīmāṁsā Sūtras (MS), Jīvānanda Vidyāsāgar Bhaṭṭācāryya (ed.), Adhyāya I-VIII. Calcutta: Sārasudhanidhi Press, 1883), are given full bibliographical details under *Abbreviations* below. See ŚD on I.i.5 (1891 ed. p. 28), discussing absence of *purusānupraveśābhāvaḥ*, three kinds of human contact with word; p. 131 *Vedapauruṣeya* #87, p. 67 [*Ven GOS*., p. 123, 230].
5. ŚD I.I.6. #63, p. 51.
6. Namely, *kāraṇa doṣabādhakajñānam*, or *yathārtha* (correctness) (ŚD #12, p. 29), *ahitamagṛhit agrāhijñānaṁ pramāṇam*; or *anādhigatārthakatva* (novelness), *upadeśaḥ = asaṁdigdha* (coherence), and *avyatirekaḥ* (faultless). Śabara I.i.2.2 @ 4, ch. V; ŚD p. 27, #63, p. 96; D'Sa pp. 60, 62. [Ven, p. 50]; cf. *yat śabde vijñāte'rtho vijñāyate* (ŚB, p. 14).

7. *Autpattikastu śabdasyā'rthena saṁbandhas tasya jñānam upadeśo 'vyatirekaścā'rthe 'nupalabdhe, tatpramāṇaṁbādarāyaṇasya anapekṣatvāt. Mīmāṁsadārśanam* MS, p. 161. On various different renderings of *autpattika*, see Biardeau, Gachter, Staal, Clooney, *et al.* (sources referred to in note 12 below).

   Of course, within the Indian tradition, Mīmāṁsakas were not the first to come up with such a thesis that postulates a differential yet inseparable interplay between word and meaning; Patañjali in his Yogasūtra (I.9, 42) admits that one cannot deny that in ordinary human discourse the presence of one (the word) brings about the other (meaning or the conceptual image). Ultimately, of course, Patañjali wants to argue that the relation of words and meanings and their referents are established by convention (*saṁketa*), and that in the practice of meditation the false notion of any identity between them is shattered, revealing what one might say is the inherent *difference* of the *śabda, vikalpa* and *artha*.

   There is also the Grammarian view according to which the relation between the name and the nameable, as between knowledge and the knowable, is *ānādiyogyatā* (a beginningless capacity, although the manifestation of this relation is made co-eval with the origin of the world (which is denied by the Mīmāṁsā). Bhāmaha's definition of *kāvya* also echoes something of this unity, in that for him the 'expression and meaning is combined' (*śabdārthau sahitau*). See Gopināth Kavirāja, *Aspects of Indian Thought*, University of Burdwan, 1984, pp. 14-17; Kāvyālaṁkāra I.16, see A.K. Warder, *The Science of Criticism in India*. Madras: Adyar, 1978, p. 31.
8. *Utpattir hi bhāva ucyate lakṣaṇayā. aviyuktaḥ śabdārthyor bhāvaḥ saṁbandhena, notpannayoḥ paścāt saṁbandhaḥ*. (1883 ed., p. 8) (ŚB, p. 13-14); cf. *yat śabde vijñāte'rtho vijñāyate*.
9. *pratyāyyapratyāyakatvalakṣaṇo'satyeva sambandhāntare svābhāvata*, #89, ŚD, p. 68. cf. p. 28 #3: *padārthasambandhasyanityatvamatrautpattikaśabdenoktam*; Pārthasārathi refers to *autpattikatva*; and *nitya* of *sambandha* again in opening #88, pp. 67-8.
10. Cf. Ferdinand de Saussure, *Cours de linguistic générale*, tr. Wade Buskin as *Course in General Linguistics*. Glasgow: Collins/Fontana, 1974; especially 'The Linguistic Sign', reprinted in Robert E. Innes (ed.), *Semiotics: An Introductory Anthology*. Bloomington: Indiana University Press, 1985, pp. 28-46. On *parousia*, see Derrida, 'Structure, Sign and Play', *Writing and Difference*, tr. Alan Bass, p. 280. The linguistic agnosticism of the Mīmāṁsā has its parallel in the metaphysical agnosticism of Derrida, who insists that what applies to signs applies also to the whole history of metaphysics, which he says,

like the history of the West, is the history of metaphors and metonymies. 'Its matrix ... is the determination of Being as *presence* in all senses of this word. It could be shown that all names related to fundamentals, to principles, or to the centre have always designated an invariable presence *eidos*, *archē*, *telos*, *energeia*, *ousia* (essence, existence, substance, subject), *alētheia*, transcendentiality, consciousness, God, man, and so forth', p. 280. With the concept of the *sign* generalized to all the concepts and all the sentences of metaphysics, Derrida hopes to shake the roots of the metaphysics of presence. The whole Platonist thrust is erased in the process (pp. 281-4); for related discussions, see Derrida, *Of Grammatology*, tr. by Gayatri Spivak-Chakraborti. Baltimore: Johns Hopkins University Press, 1972, pp. 8-9: 1976, pp. 10-11.

11. ŚV 16 (= *sphoṭavāda* #69), stressing the significative capacity of something more than *varṇa*s, which serve collectively as vehicle for the release of the *artha*.
12. Jean-Marie Verpoorten, *op. cit.*, p. 20. Although the *apūrva* may be established via *arthāpatti*, the point is that a deferred signified is being pointed to in the rite, as an incentive or a final cause for Kumārila, and a product or result of exertion (*kārya*) for Prabhākara (p. 34); see also Francis X. Clooney SJ who makes the same point but robs the *apūrva* against Mme Biardeau's better judgement of all essences as such, reducing it to a mere trope within the ritualistic qua liturgical discourse. Clooney argues that in Jaimini at least the *apūrva* as a 'transcendental thing' has no place in his system, while Biardeau considers the *apūrva* to be at the centre of the Vedic revelation for the Mīmāṁsā. I tend to agree with Biardeau. Francis X. Clooney SJ, *Thinking Ritually Rediscovering the Pūrva Mīmāsā of Jaimini*. Vienna: De Nobili Research Library, p. 131. Cf. Mme Biardeau, *Théorie de la connaissance et philosophie de la parole dans le brahmanisme classique*. Paris: Mouton, 1964, pp. 83-93; pp. 156-7. Othmar Gachter, *Hermeneutics and Language in Pūrvamīmāṁsā: A Study in Śābara Bhāṣya*. Delhi: Motilal Banarsidass, 1983, pp. 19-37; p. 44. See also Frits Staal, 'Sanskrit Philosophy of Language', in: Thomas A. Sebeok (ed.) *Current Trends in Linguistics*, vol 5. The Hague: Mouton, and his *Rules without Meaning. Ritual, Mantras and the Human Sciences.* Toronto Studies in Religions. New York-Frankfurt: Lang, 1989.
13. Panel on 'Śabda and Śruti' in the Asian and Comparative Philosophers Research Conference, Honolulu, 1984. (Panel convened and recorded by P. Bilimoria); see also Staal (last item in previous note).

14. K.T. Pandurangi citing M. Hiriyanna in: 'Professor Hiriyanna on Pūrvamīmāṁsā', Oriental Research Institute, Mysore, Vol. V, March-September 1972, p. 3.
15. ŚD see under *autpattika adhikaraṇa*, and p. 125 (under *Vedapratyāyādhikaraṇa*).
16. *Anantavākyā'rthasādhāraṇatvāt* (pp. 125 ff.). A useful discussion is found in John Taber, 'The Theory of the Sentence in Pūrva Mīmāṁsā and Western Philosophy', *Journal of Indian Philosophy* 17, no. 4, December 1989, pp. 407-30, esp. p. 409 and p. 417.
17. John R. Searle, *Speech Acts: An Essay in the Philosophy of Language*. Cambridge: Cambridge University Press, 1969, pp. 43 ff. Cf. P. Bilimoria, *op. cit.*, fn 41, p. 112.
18. See discussion in Francis D'Sa, *Śabdaprāmāṇyam in Śābara and Kumārila. Towards a Study of the Mīmāṁsā Experience of Language*, Vienna: Universität, Institut für Indologie, 1980, p. 46.
19. NMV p. 25, 1883 ed.: *pṛthak saṁketavijñayā manāpekṣatva varjanāt* (#23).
20. ŚD, pp. 125-6.
21. *Ibid.*
22. On *varṇa* vs *sphoṭa* issue: *varṇas* are not substantive, they have constituents like *puḍgala* or *artha* or *sphoṭa*: this is how it works: Auditory sounds (*nāda*) generate *saṁskāras* (impressions), these in turn make manifest (*vyañgyā*) by recalling the *varṇas*, which immediately gives recognition of *padārtha* (not of *sphoṭa* as Grammarians insist). *Śābdabodha* arises from *varṇas* which are the *viṣaya* (object) of indubitable recollection, *pratyābhijña*. Remember *varṇas* disappear as they are heard, as are the *saṁskāras* that manifest them, so there is really no separate word recollection; *varṇas* embed the signifier, *varṇas* are the sound-trace: so a word as a linguistic unit which is spoken does not have to be reconstituted at the hearer's end, unless it is ellipsed and has to be retrieved for the sake of the *varṇas*; of course, the order of *varṇas* is important; *varṇas* then are phonemes but not morpheme, which they help manifest. See also Verpoorten, *op. cit.*, p. 29.
23. ŚD *ibid.*, p. 126.
24. ŚD *ibid.*
25. In NRM (see Taber *op. cit.*, p. 409, n. 6).
26. Although, curiously, Mark Siderits reserves 'related designation' for *anvitābhidhāna*, and 'designated relation' for *abhihitānvaya*, I think he is placing excessive emphasis on 'relation' here, which may be truer for the *abhihitānvaya* of Nyāya, not so much for Mīmāṁsā, who more readily accept the idea that the signifieds, ideas, concepts or meanings (*padārthas* in this sense), get associated and enmesh into each other, or

coalesce, to yield a unified meaning, which they call a sentence-*sense* (*pace* Frege). Because the Nyāya view of sentence approximates better what we now understand as 'proposition' that *relations* between the constitutive terms become rather significant; hence the Nyāya concern with logical relations between the terms of a sentence, by which they mean *padas* which bear direct reference to their *padārthas*, the objective referents; no recourse is made here to the concepts, the *noema*, the intentional. K.T. Pandurangi makes this point, and suggests that Nyāya's *abhihitānvaya* is only in words, not in meanings. (Also, Bhāṭṭas deny that words held recall their meanings, there is no memory function here, rather meanings emerge immediately as the words are heard - see earlier *varṇa* discussion.) See Mark Siderits, *Indian Philosophy of Language*. Synthese Library, Mouton, 1993.

27. *Vākyarthāvagathiḥ* (ŚD, p. 129).
28. K. Krishnamoorty, *Studies in Indian Aesthetics and Criticism*. Mysore: Murthy, 1979, p. 77.
29. NRM, GOS ed., 1937, p. 10<; *ibid*., p. 79.
30. *Ibid*., p. 100.
31. ŚD (Venk. tr. p. 224).
32. Grice's account of meaning and communication as involving three steps is significant in so far as it is tied to intention:
    i. S intends to produce a certain effect in an audience H by uttering U;
    ii. S intends that H shall recognize S's intention;
    iii. S intends that this recognition on the part of H of S shall act as H's reason for the occurrence of such an effect.
    Mīmāṁsā would be non-Gricean, unless S is taken simply as a linguistic utterance regardless of its source. Seems to work better after Searle's criticism and qualification of Gricean intention, although Searle insists that Grice has overlooked what U means in the language in which it is spoken, i.e. connection between U and the convention governing the language is important. (Searle, *Speech Acts*, 1969, p. 43).
33. ŚD p. 129. *Apūrva* is the supplement to divine/Being and so on. It might even be a dangerous supplement for Hinduism, and indeed subversive toward all religions with a Catholic leaning. *Apūrva* is also like the 'spurs' (*Spuren*: traces, in the Heideggerean sense of the gods having fled and left only traces in the infinitely extending abyss). But it keeps the *aśvavāhana* running! So is the Transcendental Signified as a supplementarity, incorporates and displaces, inverts traditional oppositions, and unfolds the indecidables, it intertwines visible and invisible ...
    But the Mīmāṁsā system is a closed one: it interprets rather than produces signs (for a contrasting approach, see Umberto Eco, *Theory*

of Semiotics); its symbolic order has no room to take in what it excludes on the margins: it has, as it were, become unstuck: it has no room for play, desire, the hidden of the unconscious, for it has rationalized all these.

34. Consider *vidhi* or *codanā* to be the trace: it marks the presence of that which is absent, *apūrva* is the supplement: by its mere presencing of the absent, it also displaces the *vidhi*, the text, the sign; more significantly it is a supplement as the erased Transcendental Signified (just as writing is the supplement to speaking).
35. 'What are Poets for', *Poetry, Language, Thought* (tr. by Hofstädter), 1971, p. 32.
36. *On the Way to Language*, tr. by Peter Hertz, 1971, pp. 124 ff.

IV

HINDU REVIVAL IN FIFTEENTH-CENTURY JAVA

J.G. de Casparis

The fifteenth century, which is said to mark the decline of the so-called Indo-Javanese civilization, is more correctly described as a period of transition in which old values of Hindu and Buddhist ideologies undergo profound changes. On the one hand, their influence weakens as an ever-increasing part of the population is converted to Islam; on the other hand, they absorb elements of original Austronesian beliefs and practices. Both factors played an important part in the changes which can be noticed during this period. Yet, there is another factor that is often neglected or ignored: the striking increase in international relations between Java and other parts of southern and eastern Asia. It is true that such relations are often connected with the expansion of Islam, which stimulated trade between Muslim centres. Although this is undoubtedly an important factor, there are other aspects that may be considered.

In some respects it is difficult to distinguish between cause and effect. Thus it can be argued that the increase in commercial relations between various parts of South and Southeast Asia is mainly due to the expansion of Islam, but it can just as well be argued that the expansion is to be attributed to increased commercial relations; in other words, the relations between Islam and trade were reciprocal, one strengthening the other. In this contribution I may, however, call attention to another aspect of this period: new influences in Hinduism in Java. More than once, most clearly in a lecture given in Kuala Lumpur in 1983[1] I emphasized that influence from the Indian subcontinent was not confined to the earliest phase of 'Indo-Javanese' civilization but was a feature which can be observed during ten or more centuries, although in very different forms and intensity. Thus, the impact of Mahāyāna Buddhism in the eighth and ninth centuries (during the reigns of kings of the Pāla dynasty in Bihar and Bengal) or of the Vaiṣṇava revival in South India in the tenth and eleventh centuries have often been noted.[2] These are, however, some of the more striking examples during which these relations appear to be most prominent. I feel, however, that the fifteenth century should be added to this number.

It is well known, but not always sufficiently realized, that Old Javanese 'Hinduism' exhibits a number of features which lend it a special colour. On the one hand, as could have been expected, ancient Austronesian beliefs dating back to before Indian influences made themselves felt, were absorbed into the Indianized beliefs of the Javanese elites;[3] on the other hand, and this may be more significant, these elites were eclectic in that they adopted only those elements of Hindu culture which appealed to them for one reason or another. In many cases the motives of such elites may have been based on considerations of social or political expediency, since the use of foreign customs and habits may have been a means of distinguishing themselves from the agricultural masses. In the opinion of many elites such special features would have commanded certain forms of respect and loyalty. It should however be added that we hardly have any sources to help us understand the true motives behind the adoption of elements of foreign, in particular, Indian culture. As a consequence of this eclecticism we may notice a small number of features which are, except in a few cases, by no means limited to ancient Indonesia but are much more prominent there than elsewhere. One of the most common statues of ancient Java is that of the so-called Śiwa-Guru, also, and perhaps more correctly, described as the *ṛṣi* Agastya.[4] The popularity of this image, regularly placed in the southern niche of Śiwa temples, may reflect the ancient Javanese respect and worship of the teacher, whose statues strike us particularly by their human representation. The worship of the god Haricandana is probably also limited to ancient Java.[5] A number of pre-Hinduistic divine powers occur in the curse pronounced at the conclusion of the ceremonies connected with the inauguration of a *sīma*.[6] The cult of Bhīma, represented by quite a number of statues, is apparently almost limited to Indonesia.[7]

Apart from these few examples we are struck by differences in emphasis on the worship of particular deities. Thus, with reference to India, it has been rightly stated by J.N. Banerjea that 'Brahmā appears to have enjoyed no success as a cult god';[8] yet, statues of this four-faced (*caturmukha*) god are quite common in Indonesia, where even one of the three main temples of the Roro Jonggrang complex at Prambanan is entirely devoted to Brahmā. Images of Durgā, invariably in her Mahiṣamardana form,[9] as well as those of seated Gaṇeśa,[10] are again very common in comparison with Indian iconography in general.[11]

In this short paper, dedicated to a foremost specialist of Hindu ritual, some attention will be given to an interesting, but still somewhat neglected, period of Indonesian history. In fifteenth-century Java, when Islam was expanding in much of eastern Java (the tomb in Gresik of Malik Ibrāhīm, dated AD 1419 is its clearest expression),[12] and the authority of the *kraton* of Majapahit was declining, the court elite would have attempted to revive their authority by renewing their contacts with the civilization which was the main source of their authority: the Indian civilization, in particular Hinduism. As a curious coincidence, similar considerations can be attributed to the Hindu kingdoms in South India. The resistance against the power of the Muslim kingdoms in the Deccan had led to the foundation in 1336 of a strong Hindu kingdom, Vijayanagar, which sought to revive ancient traditions even back to the time of the Ṛgveda. At the height of its power in the latter half of the fifteenth century, the kings of Vijayanagar sought to stabilize their authority by fostering links with other Hindu powers which felt themselves threatened by the expanding Islam. The establishment of relations between Majapahit and Vijayanagar in this period thus becomes understandable.[13]

In this connection it may be of interest to re-examine a small number of Old Javanese inscriptions dated at the end of the fifteenth century on the eve of the rise of Muslim states in the island. The stone inscriptions of Padukuhan Duku and Jiyu, all dated AD 1486, are the last known pre-Muslim documents of this part of the Island. They are four inscriptions in all, about 150 lines of writing, all dealing with perpetual gifts of land to a certain Śrī Brahmarāja Gaṅgadhara (*sic*) carrying important fiscal, judicial, commercial and other privileges. The donor was the last pre-Muslim ruler in Java known from epigraphic sources, viz. Śrī Girīndrawardhana dyah Raṇawijaya.[14] These inscriptions were first edited by Brandes-Krom as *Oud-Javaansche Oorkonden* (*OJO*), Nos. XCI-XCV, studied and presented in a provisional translation by the present author in Muh. Yamin, *Tatanegara Madjapahit (Sapta Parwa)*, 1962, pp. 235-56 and, more recently by J. Noorduyn, 'Madjapahit in the fifteenth century', *BKI* 134, 1978, pp. 207-74, who also bequeathed to us a still unpublished re-edition of these texts. Finally, some aspects of these inscriptions were discussed by Machi Suhadi in his still unpublished Ph.D. thesis on *Tanah Sīma dalam Masyarakat Majapahit*, Universitas Indonesia, Jakarta, 1993.

These texts are of considerable importance for several reasons. The donee, Śrī Brahmarāja Gaṅgadhara, received the above-mentioned privileges on account of his devotion and loyalty, on the occasion of a *śrāddha* carried out on behalf of a deceased ruler twelve years after his death. This *dvādaśa-varṣa-śrāddha* is well known from the Nāgarakṛtāgama, in which no fewer than seven cantos (63 to 69 inclusive) are devoted to this ceremony. Although the performance of *śrāddha*s at regular intervals after the death of a close relative is common in India (Gaya is the most important centre were these rituals are carried out), a grand *śrāddha* after twelve years is unknown to me and seems specific to Java in the fourteenth and fifteenth centuries. Also the ceremony itself contains many elements which do not appear to agree with Indian practice; thus it includes the erection of an elaborate lion-throne (*siṅhāsana*), on which the soul of the deceased queen, known as the Rājapatnī, is invited by the priests who carry out a number of rites leading to the 'deification' of the queen as Prajñāpāramitā.

The fifteenth-century inscriptions show a number of other features which are unknown or unusual. Thus, the inscription of Jiyu (*OJO* XCIII, dated AD 1486, lines 16 to 17) mentions the worship (*kapūjān*) of the 'lord of the *ṛṣis*' Bharadwāja. This saint is known from classical Indian Hinduism, where he is regularly mentioned in the lists of the Saptarṣi, although he is much better known from the Vedic texts, as many Ṛgvedic hymns are attributed to him. On the other hand, he is never mentioned in earlier Old Javanese inscriptions nor, it seems, in literary texts. In another Jiyu inscription of the same year 1486 (*OJO* XCIV, line 11) a sanctuary (*pratiṣṭhā*) of *maharṣi* Bharadwāja is mentioned.[15] Finally, in connection with the precise demarcation of the land to be donated to Brahmarāja Gaṅgādhara the Jiyu inscription No. II (*OJO* XCIV), line b 7/8 mentions a 'war pillar' (*tugu pĕpĕraṅan*) with the emblem of the *maharṣi* Bharadwāja (*maharṣi bharadwāja-lāñcana*).

As to Bharadvāja in classical India, it is interesting to note that the South Indian Pallava kings belonged to the Bhāradvāja-gotra. In the above-mentioned passage the name of the saint is immediately followed by the mentioning of a sanctuary in honour of lord Rāma. This is again a unique reference to the worship of Rāma in ancient Java. Although Rāma, son of Daśaratha, is celebrated in the whole of the Indo-Pakistani subcontinent, yet actual images and temples of Rāma seem to be confined to South India.

At this stage it may be useful to re-examine the titles of Brahmarāja Gaṅgādhara, as given in these inscriptions. His name is preceded by *śrī*, which may not be significant, but also by *pāduka*, which in general appears to be limited to kings and princes, although there are a few examples where the word is applied to very high ecclesiastics in the Nāgarakṛtāgama.[16] This excellent *brāhmaṇa* (*mahādvijaśreṣṭ<h>a*) was not only a supreme royal priest (*paramapurohita*), well versed in the four Vedas (*caturwwedapāraga*),[17] but also a warrior (*samaradhurandhara*) as fierce as (Śiwa), 'Destroyer of the Three Fortresses' (*tripurāntakaprabhāva*). He was clearly a most influential person at the Girīndrawardhana court, but one may wonder whether he may have been an Indian invited to the court of Majapahit at a time when its royal families were facing the victorious march of Islam. The introduction into Hindu religious worship of so many new elements - such as the homage extended to Bharadwāja, the sanctuaries for the worship of Rāma and even of *bhaṭāra* Yama, unprecedented in Java, the performance of the twelve-year *śrāddha* - gives the impression of a last, yet unsuccessful, attempt by the old ruling class to halt the progress of Islam.

Notes

1. 'India and Maritime South East Asia: a lasting relationship', *Third Sri Lanka Endowment Fund Lecture*. Kuala Lumpur, 1983.
2. G. Coedès, *Les états hindouisés d'Indochine et d'Indonésie*. 3rd ed., Paris: Boccard, 1964, pp. 170 and 297.
3. W.F. Stutterheim, 'Iets over prae-hinduïstische bijzettingsgebruiken op Java', *MNKAW*, II, No. 5, 1939.
4. R.Ng. Poerbatjaraka, *Agastya in den Archipel*. Leiden: Brill, 1926.
5. J.G. de Casparis, 'Some notes on ancient Indian ritual in Indonesia', in: A.W. van den Hoek, D.H.A. Kolff and M.S. Oort (eds), *Ritual, State and History in South Asia: Essays in Honour of J.C. Heesterman*. Leiden: Brill, 1992, pp. 480-92.
6. N.J. Krom, *Hindoe-Javaansche Geschiedenis*, 2nd ed., 's-Gravenhage: Nijhoff, 1931, pp. 175-9; F.H. van Naerssen, 'Twee koperen oorkonden van Balitung', *BKI* 95, 1939, pp. 441-61.
7. Hariani Santiko, 'Tokoh Bhīma pada masa Majapahit' in: *Kirana. Persembahan untuk Prof. Dr. Haryati Soebadio*. Fakultas Sastra Universitas Indonesia, Depok, 1992. Cf. also W.F. Stutterheim, 'Een Oud-Javaansche Bhīma-cultus', *Djåwå* XV, 1935, pp. 37-64.
8. J.N. Banerjea, *The Development of Indian Iconography*. Calcutta: University of Calcutta, 1956 p. 513.
9. Hariani Santiko, *Bhatārī Durgā*. Depok: Fakultas Sastra Universitas Indonesia, 1992.
10. Edi Sedyawati, *Gaṇeśa statuary of the Kaḍiri and Siṅhasāri periods*. Leiden: KITLV Press, 1994.
11. Seated Gaṇeśa statues are regularly placed in the Western niches of Śiwa temples in Indonesia, but they are also sometimes found separately, especially at difficult places such as mountain passes and river crossings.
12. Hasan Djafar, *Girīndrawarddhana. Beberapa Masalah Majapahit Achir*. 2nd ed. Jakarta: Yayasan Dana Pendidikan Buddhis Nalanda, 1978, pp. 55-68.
13. See the publication mentioned in note 1 of this article. There are indeed a few indications suggesting such relations, but there is no real evidence.
14. See Hasan Djafar, *op. cit.*, pp. 86-90.
15. The term *pratiṣṭhā*, 'erection, inauguration' etc. may refer to either a building or a statue.
16. The use of *pāduka* for kings and other royal persons is not attested in inscriptions of Java before the Majapahit period. It is occasionally used in the Nāgarakṛtāgama for a few high priests probably related to the royal family such as the Rev. Ratnāṅśa, abbot of Muṅguh in verse 39-1.
17. This epithet, very well known from Indian texts, is rarely mentioned in Old Javanese, except in texts directly based on Sanskrit texts, such as the Brahmāṇḍapurāṇa.

V

A MAN AND A WOMAN: AN ANALYSIS OF A MODERN HINDI SHORT STORY

Theo Damsteegt

The present article proposes an interpretation of a short story entitled Choṭī 'i', by Himāṁśu Jośī. The story appeared in the anthology *Himāṁśu Jośī kī viśiṣṭ kahāniyāṁ* ('Himāṁśu Jośī's best stories'), which was first published in Delhi in 1979,[1] and consists of six pages text (page 82-7 in the anthology).

SUMMARY OF THE STORY

A man called Upendra, who lives alone in Delhi, is visited by a woman in whom he recognizes the girl he used to keep company with when both of them were children. Both she and her sister had the name 'Eelah' (*Ilā*). He used to call her *choṭī 'i'* ('E junior') in order to distinguish her from her elder sister *baṛī 'ī'* ('E senior').[2] She was much older than he was. She would teach him, help him when he had problems, and take him everywhere she would go. When teased because of it she would tell her girlfriends that one day she would marry him, and immediately afterwards she would burst into laughter. Occasionally she would whisper into his ear, or caress him, and he used to be frightened by such behaviour and by the sentimental words she sometimes addressed to him. She lives in Kānpur now. The man is quite surprised when she unexpectedly turns up on his doorstep, and while she talks and steps inside, he cannot find any words to welcome her at first. When he finally manages to speak, telling her that she has not changed at all, she is happy. Even afterwards he does not react much to her questions and remarks, and practically the only thing he tells her is that he usually eats in restaurants instead of cooking dinner for one. While she takes a shower, he inspects the contents of her attaché case. When she returns to the room, he is surprised that even at her present age her face looks so attractive. When she next tells him how a woman they both used to know had embraced her in public at an unexpected meeting, he suddenly asks her why she does not get married. She laughs first, then grows silent. They decide to go out and have food somewhere. While they walk along the street, he wonders what the neighbours will think. They take a scooter rickshaw, which drops them far away, and start walking. She tells him she is on her way to Simla for a teachers' conference, and will have to leave the same night. When she turns down his invitation to drop in again on her way back, he feels offended. They pass by a shop, and she urges him to buy her a present, which he does after some protest. During their meal in a restaurant she is silent, then suggests that they go to see a movie. The man remembers how once, long ago, she had taken him to a movie and had been so close to him that he felt as if he was going to suffocate. Now nothing of the sort happens, and as they leave the cinema hall, she sighs. At his place she packs her luggage, and urges him to visit her, or to write to her, because she is quite lonely since her parents have died. She is crying, and he observes the wrinkles in her face, and notices how more than half of her hair is grey.

BACKGROUND OF THE STORY

Before going into the story itself, some brief observations may be made about its position in the history of Hindi literature. Before 1947 Hindi short stories tended to focus on social problems on the one hand, and psychology on the other. The fifties saw the rise of a generation of authors who themselves acted as literary critics, and strongly criticized earlier short story writing. In their view, several differences could be pointed out between their own stories, which came to be called *naī kahānī* 'new short story', and the earlier ones (cf. Roadarmel, 1974; Jossan, 1981). One of the differences was that whereas earlier authors wrote their stories in order to convey an explicit message to the reader, *naī kahānī* tried to suggest a deeper meaning, leaving it to the reader to interpret the story and focusing on atmosphere rather than plot, action, and climax. Like some stories written before 1947, *naī kahānī* often dealt with inner experiences, but the difference with those stories (for example, those written by Ilācandr Jośī) is that psychology was not emphasized. Another difference was that many 'new short stories' dealt with situations supposedly known to the authors themselves from their own experience and consequently were mostly taken from the reality of daily life. The authors of the new short stories were inspired by their search for new values in an independent India characterized by an increasing Western influence on lifestyles on the one hand and political developments which disappointed many, on the other. From the early sixties onwards *naī kahānī* became increasingly gloomy and pessimistic. Two movements often distinguished after *naī kahānī*, from about the mid sixties onwards, are the 'anti-story' (*akahānī*), which was decidedly pessimistic in its rejection of all traditional values without searching for new ones, and the 'conscious story' (*sacetan kahānī*), which is said to be characterized by a search for meaning in life, however hopeless life may seem (cf. Ansari, 1978). Not everyone agrees, however, that these two movements have in fact existed (e.g., Caturvedī, 1990, p. 12, and Siṁh, 1986, pp. 19, 25). That observation applies even more to subsequent movements like *samāntar kahānī* 'parallel story', which is supposed to focus on the 'common man'. Several authors have been said to belong to more than one movement. Himāṁśu Jośī, who was born in 1935, saw his first short story published in 1954 and his first collection of short stories in 1965, and he continues writing up to this day. He has been associated with both the 'anti-story' (Ansari, 1976, p. 241) and the 'parallel story' (Jośī, 1979a; Vinay, 1977 e.g., pp. 84 f.). If one were to attribute Choṭī 'i' to any movement at all, late *naī kahānī* or possibly its direct successor 'anti-story' would appear to be the most likely categorization. The story does not reflect a search for values and is rather pessimistic. Its contents do remind one of late *naī kahānī* as discussed by Ansari (1975), which shares the *naī kahānī* characteristics mentioned above but tends to emphasize the depiction of frustration, despair, loneliness, and emptiness. And they are also reminiscent of the 'basic ideas' of *akahānī* mentioned by Ansari (1976, p. 242), viz. 'ways leading to no end, helplessness of individuals and losses of values of life'. Even if the above-mentioned movements are indeed assumed to have existed, it is rather doubtful whether descriptions of contents such as those given above allow one to attribute a particular story to a movement with any certainty at all.

INTERPRETING LITERARY TEXTS

Like any act of communication, literature needs to be interpreted. Readers may interpret texts intuitively, but literary research should not be based on such intuitive interpretation because intuition does not lend itself to discussion. In order to be a part of literary research and to make possible any sensible discussion, interpretations need to be based on systematic analysis.

An 'intuitive' interpretation of Jośī's story, partly inspired by its title, would probably focus on the position of unmarried women in Indian society (cf. e.g. Dube, 1988, p. WS-15; Jacobson, 1992, pp. 43, 49). Even though they may be professionally successful, like 'E junior' seems to be, it is not easy for them to make contacts, and meeting intimate friends of old in order to relive a carefree childhood also results in disappointment. Once these women have no close relatives of their own left, they are doomed to be lonely. Men apparently manage rather well, even if being unmarried brings with it some practical problems, like having to eat in restaurants.

Until recently, discussions of modern Hindi literature often tended to be based on impressionistic interpretations such as the one given above of Jośī's story, without much methodological reflection.[3] Only in recent years have several researchers in the field of modern Indian literature started to pay attention to classical Indian literary theories and the methods commonly used in studying European and American literature. Not everyone is convinced that the methods developed in Europe and America can be directly applied to literature written elsewhere, or that classical theories can be used in analysing modern literature, but the proof of the pudding is in the eating. The present article argues an interpretation of Jośī's story on the basis of an analysis of its structure and technique. That analysis will follow the narratological method first developed by Genette (1972), and focus on the 'story' and 'text' levels distinguished by narratology. A systematic and very practical handbook of narratology which partly builds on Genette's method has been written by Bal (1985),[4] who in the fifth Dutch edition of her book (Bal, 1990, pp. 11-15) justly maintains that even when recent developments in literary research theory seem to speak against some aspects of the narratological approach, it still remains a useful instrument for describing the narrative characteristics of a text. It will be shown below that such a systematic description may draw attention to features which often remain unnoticed in an intuitive reading, and may then lead to an interpretation which Kermode (1979, ch. I) calls 'spiritual', in contrast to the 'carnal' (or impressionistic) interpretation which results from reading a text on face value only. As observed by Bal and Kermode, no two 'spiritual' interpretations will ever be exactly the same. Though they are systematic, interpretations based on narratology or any other method will always be subjective and cannot claim to be anything like a 'definitive interpretation'. The important difference with impressionistic interpretations, however, is that they are argued proposals which can be discussed.

ANALYSIS[5]

In discussing Jośī's story,[6] quotations from the Hindi text will be made by means of reference to the page number of the 1979 anthology followed by the paragraph number. The first new paragraph of any page counts as its initial paragraph. Thus, 83:2 indicates the second paragraph of page 83, which starts with *pitājī merā*, and 87:1 is the paragraph starting with *uskā galā*.[7]

Translations from Hindi are not always literal.

1. Some general information about the situation and the characters is as follows. Since Ilā speaks to Upendra about 'your Delhi' (82:6), the story takes place in that city, possibly in some new-built area - 'One cannot find places <in Delhi>. No wretched dead politician or a neighbourhood has been named after him', says Ilā (82:7). No reference is made to historical events and it is not clear in which year the woman's visit takes place, but since Delhi appears as a big city (82:7, see above), it will not be a long time ago. Upendra's father was a lawyer (83:3), and Ilā is the daughter of a doctor (83:2). She appears to be a teacher at a college in Kānpur (84:2, 18; 85:16), and though it is not revealed what Upendra's present position is, Ilā calls him a 'big shot' (*baṛā ādmī*, 84:4). He lives alone (84:11), and so does Ilā (86:17).

In view of the amount of time probably spent on the actions mentioned in the story and because of the fact that Ilā leaves for Simla in the evening (85:22), it may be assumed that she has arrived from Kānpur by night train and stands on Upendra's doorstep in the morning. That would be the reason why she takes a shower at his place.

At one moment Ilā tells Upendra 'You never told me your mother died. You knew, did you not, that I had been living in Kānpur for some years. Have you ever visited me? In those days I was ill - in hospital. You were staying at Nanditā's! Is it not so?' (84:2). Taking into account Upendra's observation of Ilā's emotional state of mind when she speaks these words (84:3), his hesitation to speak about Nanditā (*ibid.*), and the erotic atmosphere of the entire story, this passage suggests that Upendra has had (or possibly still has?) an affair with Nanditā. It may also be referred to in 84:13-15 -. '"May I ask a question?" she said after a short silence, "Will you answer it?" "Yes ... " "Never mind ... !", she looked at me and laughed a bit.'

Characterization will be dealt with in more detail below, in section 5.

2. Gaps are revealed in 85:3, when Upendra asks 'Why don't you get married?', because it is not told how he knows Ilā is unmarried, and in 85:15 in his question 'What are you travelling to Simla for?', because no mention has been made of Simla in the preceding conversations. The gaps are not significant, however.

3. When the story's time structure is considered, an extensive flashback will immediately be noticed in 82:12-83:10. It is an external flashback in which Upendra's memories of his childhood, together with 'E junior' are presented, and it will be seen below that it serves to characterize both the man and the woman. The occurrence of the flashback appears to be motivated by Ilā's 'childish' behaviour mentioned in 82:10 - 'She burst out laughing once again, clapping her hands like a child.' The short flashback contained

in 86:14, which is motivated by space (cf. Bal, 1985, p. 93) viz. the cinema hall, is also external and characterizing. The other flashbacks are all told by Ilā (84:2, 85:1, and 85:5). They serve to characterize, but also to evoke a certain atmosphere which causes a reaction in the man, as will be seen below.

The passages quoted are flashbacks in the sense that events are related in them which have taken place before Ilā's visit to the man. That visit constitutes the primary story-time, which may also be called the 'moment' of experiencing. Besides the time levels of flashback and moment of experiencing, the moment of narrating may be distinguished here is found in the second sentence of 86:5 *ab tak yād hai* 'I still remember', where *ab* refers to the moment the story is narrated.[8]

4. When we next consider the identity of the 'informer' in Jośī's story, we may observe that it is the man himself who informs the reader about almost all the events, while he also plays an important role in them. Thus, he is a character-bound 'informer'.[9] As almost all events are presented from his perspective, he is also the main focalizer.

Occasionally the man comments on the events. 'It was a moment when I could have merrily expressed my pleasure, or somewhat hesitantly could have called her *āp*, or *tum*, or even *tū*'[10] (82:4), for example, is not a thought which he has at the moment of experiencing but a comment which he makes at the time of narrating. And 84:16 is another example. Both in such comments[11] and in the second sentence of 86:5 discussed above, which explicitly shows that he is telling while looking back at the events, the man acts as an external focalizer. However, most of the time the reader is presented with the man's reactions as they were at the moment of experiencing, without any interference of his thoughts or feelings at the moment of narrating. That is clear, for example, when the word *āj* 'today' is used in 82:5, which reads 'As always, she did not wait for my answer today either ... ', or *ab* in 82:11. And another example is seen in 84:18, where the man finds out the contents of Ilā's attaché case item by item, the way he did at the moment of experiencing. It means that the main part of the story is presented with internal focalization (cf. Cohn, 1981, pp. 175 f.).

Even when a character-bound informer focalizes externally, he may still be emotionally involved in the events he has played a role in (cf. Stanzel, 1985, pp. 122, 127 f.). Such a subjective involvement is certain, however, when his focalization is internal, because the reader is then presented with the informer's vision at the moment of experiencing. This observation implies that in any case the man's vision as an internal focalizer is biased (cf. Bal, 1985, p. 104).

5. This bias has its consequences for characterization. The man is in a position to present before the reader, not only the woman's and his own

actions and words but also his own thoughts, interpretations, and feelings, whether they belong to the moment of experiencing or that of narrating. He is thus in a position to influence the reader, as observed by Bal (1985, pp. 109 f.). At first sight his interpretations of the woman's actions and his thoughts about her feelings appear to characterize her, but in reality, because of the bias implicit at least in his internal focalization, those interpretations and thoughts serve to characterize him rather than her.

The characterization of both the man and the woman is largely implicit. It is left to the reader to draw conclusions about the qualities of both characters, on the basis of the man's actions, words, and thoughts, and the woman's actions and words.[12] Such conclusions, presented below, are of course subjective. In the man's case, there is no reason to assume that he misrepresents his own thoughts, feelings or actions, so all information which he furnishes about himself can be assumed to characterize him. In the case of the woman, the only valid sources of information are her words presented in direct discourse, and her actions as far as objectively perceptible. Speech is presented only in direct discourse in the story, and nowhere does the man paraphrase (that is, filter) words spoken by Ilā or by himself. As far as actions are concerned, examples of objectively perceptible actions are found in 82:1 'She stood in front of me with an attaché case in her hand', and 86:12 'While we had food in a restaurant, she would occasionally pick up bits of vegetable which I had left on my plate.' A sentence in 82:2, however, presents a state which is not objectively perceptible, viz. 'There was an expression of a rather strange challenge on her face.' It constitutes a subjective interpretation by the man, and as such it tells the reader more about the man's fear when suddenly confronted with Ilā than about Ilā's face.

The picture of the man and the woman which emerges when the above precautions are kept in mind is best presented in two sections each.

5.1.a In their childhood the *man* was 'weak', in the sense that Ilā took all initiatives and he was even to some extent dominated by 'E junior'. Their relationship was characterized by an atmosphere of erotics which he himself did not understand at the time and which frightened him.

Examples of Ilā's initiatives and of the man's lack of initiative, are found throughout the flashback in 82:12 - 83:10. An indication that Upendra was to some extent dominated by her is found, for example, in 83:8 'Whenever she had to go to the *bāzār* or to some friend's (*sahelī ke ghar*) or somewhere else, she would have me follow her like a shadow', and 83:9, where it is related how she had once fed him a *laḍḍū* which she had brought from a party and had then teased him for days on end that he had eaten her *jūṭhā* 'leftovers'. Eating *jūṭhā* is an action characteristic of married women, who take the leftovers of their husbands (cf. Dube, 1988, WS-17, Jacobson, 1992, p. 56).

An erotic atmosphere is quite visible in 83:2-7 and 10, and in 86:14. Section 83:10, for example, relates how one night when the boy was alone while studying for his exams, Ilā had thrown her shawl over his head like a net (an action symbolic for her power over him), had caressed his head for a long while, and said things like 'Give me these eyes ... Let me keep this nose.' And in 86:14 Ilā comes to sit so close to him in a cinema hall that the boy feels suffocated and punished.[13] And he adds now 'At that time I understood nothing.'

The boy's fear can be seen in 83:7 (*maiṁ ḍar gayā thā* 'I got afraid'), when Ilā laughs loudly after she has told him that his parents are worried that his contacts with her will corrupt him, and at the end of 83:10, and in 86:14 (see above). In 83:10 the boy extricates himself from the shawl and runs away, wondering why she talks like a madwoman.

It is not mentioned what age both characters are at the time of the flashback, but Ilā is of such an age that she visits school friends' homes, and goes to school parties (83:8, 9). She clearly does not behave the way a girl of her age is traditionally expected to, and her uninhibited behaviour provokes a silent comment by Upendra's father (83:2 f.). Girls of Ilā's age who show such a lack of inhibition and play with boys ruin their reputation (cf. Jacobson, 1992, pp. 35, 38 f.). The reason why Ilā is unmarried is not indicated in the text, but one may speculate that her behaviour has played a role in it.

At the time of Ilā's visit, Upendra is initially 'weak', in the sense that it is once again the woman who takes all initiatives. It is her initiative to resume the contact because her visit comes as a total surprise to the man (cf. 82:1 'It did not seem true. She stood facing me, holding an attaché case in her hand. I could not even have imagined her being here in this situation.'). And while Ilā acts and talks, the man takes no initiatives and remains silent (cf. p. 82, and 84:1-7, 14-16).[14] At the moment when Upendra first speaks, and tells her she has not changed (82:10), he is relieved by her reaction (82:11 'It was as though all clouds had been dispersed'). He apparently feels dominated by her. So it also appears from 82:2 ('There was an expression of a strange sort of challenge on her face. She probably wanted to frighten me, or something like that!'), 82:5 ('As always she did not wait for my answer'), 82:6 (where the man is explicitly frightened by her tone of voice), 84:3, 84:6 (*vah bāz kī tarah jhapṭī* 'like a hawk she pounced on <the kettle>'), 84:7 ('touching every object in the room E looked at it with a very strange glance'), 84:16. And in 84:18 he inspects her attaché case when she has left the room, whereas she had inspected the objects in his room in his presence (84:7, see above). At the same time, however, he finds her attractive, cf. 82:8 ('"For some reason she looks quite attractive", at a time when her face shows "deep lines of tension"'), a statement repeated (and thereby emphasized) in 86:16.

Thus, he still experiences both dominance and erotics, but now he is consciously aware of the erotics.

It may be assumed that it is the combination of his childhood memories and his present knowledge of the commonly accepted behavioral norms for girls and women, together with the sexual experience which he has possibly gained in the meantime (84:2, see section 1 above), which makes the man interpret the woman's spontaneous behaviour as erotic and sexually provocative. In 84:19, when she has already been with him for some time, Ilā returns freshly bathed into the room, and Upendra finds her quite attractive. 'Even at her present age her face was amazingly beautiful. White drops of water dripped from her wet hair which touched her hips! And without any reason she started laughing ... !' In the next paragraph (85:1) Ilā braids her wet hair, and tells about a woman whom they both used to know, and who at an unexpected meeting had started to kiss her in public. She adds how the woman used to like Upendra. He reacts by 'looking at her face in a rude way (*śarārat se*)', and smilingly asking her 'Why don't you get married, *choṭī I*?' (85:3), in a question which is at the same time an exhortation. Apparently Ilā's appearance from the bathroom, with her long hair wet and drawing attention to her hips, her laugh which the man may experience as a provocation,[15] and her subsequent talk of kissing and attractiveness make the man experience erotic feelings for her to such an extent that he puts a question to her which refers almost explicitly to erotics. As an exhortation, his question is partly an admonition, which reflects fear on his part[16] - the question may be interpreted as 'You had better get married instead of trying to seduce me.' But whether or not consciously, the question is probably at the same time meant as a challenge to her (something like 'Have you come to seduce me?'), as will be argued below, and it is probably interpreted by her as such. In any case the sudden switch towards the subject of marriage reveals Upendra's sexual desire and his subsequent fear. The question, which is his first real initiative, constitutes a turning point in the story.

5.1.b Initially the *woman* behaves independently and spontaneously during her visit to the man. Not only does she visit him on her own initiative (82:1, 84:1), but she also takes a seat before the man has spoken (82:5), addresses him repeatedly without being withheld by his silence (82:6-7; 84:13), reproaches him for not keeping contact (84:1-2 f.) but is also conciliatory (84:4), does not allow him to fill a kettle (84:6), and puts on his slippers while going to the bathroom (84:17).

Apparently her independent attitude has not changed as she has grown older. The way the woman is characterized as independent and spontaneous does not necessarily imply, that Upendra's interpretation of her visit as erotically provocative reflects her purpose. Because of the narrative situation the reader has to guess at the woman's motives, but the way she behaves in

84:1-4 (see above) and her change of mood after his question in 85:3 (see 5.2.b below) rather suggest that it is her goal simply to re-establish contact with the man. Note, in this respect, Ilā's happiness when Upendra finally speaks after her arrival and tells her she has not changed (82:10). Through his words their childhood comes back to her, and that is probably also the reason why she claps her hands and bursts out laughing, 'like a child'.[17]

5.2.a From the turning point in 85:3 onwards the *man* has grown 'strong'. Now that he has answered Ilā's supposed challenge, he no longer feels dominated. He shows much more initiative than before. When they have left his room,[18] he is still occasionally silent after a remark by the woman (85:6), but that silence no longer seems inspired by an experience of dominance on Ilā's part. And now he repeatedly speaks and acts of his own accord. He 'breaks the silence' (85:7, cf. 86:15), stops a scooter-rickshaw (85:9), reacts to an action by the woman (85:12), questions her (85:15 - 86:2), and protests (86:7-9). At no time, however, he takes any clearly erotic initiative. His attitude towards Ilā grows rather distant and dominant. He thinks that she walks beside him 'silent and as though shrivelled, behaving like a *bahū*' (i.e. a wife, 85:4), is indignant that she does not accept his invitation to visit him on her return from Simla (86:1), takes distance from their past (86:4), thinks that she seems 'very young (*choṭī*) now' (86:5,[19] cf. 86:15) and acts like 'a small kid' (86:6, 87:1).[20] And when on her persistent request he buys her a present, the way one may buy a present for a child (cf. 86:6), it is significantly not identified (*kuch* 'something', 86:11), whereas he does mention the name of the shop where he buys it ('Central Stores').

When they are seated in a cinema hall, he fears that she may grow over-sentimental (*kahīṁ bhāvāveś meṁ vah kuch kar na baiṭhe*), the way she had once done in their childhood. As they leave the hall, he is quite surprised that now she has not acted that way (86:14), and reflects on the difference between her behaviour right now and when she first entered his room (86:16). At the end of the story, when Ilā cries and asks him to keep contact, he finds her unattractive (87:1, cf. the remarks about Ilā's laughs in 86:15 versus 84:19 and 86:16). In contrast to 84:19, when Ilā emerged from the bathroom, he now notices that her face is wrinkled and that her hair, which in 84:19 excited him, is largely grey. The latter observations suggest that the man is disappointed by her lack of erotic initiative in the cinema hall and tries to ignore that feeling by telling himself that Ilā is not really attractive,[21] which, in turn, implies that his question in 85:3 may well have been meant as a provocation to her. At the same time, he also fears her possible response to his challenge.

5.2.b After 85:3 Ilā's behaviour changes. She first laughs at Upendra's question about her getting married, then 'suddenly' grows serious. She does not answer the question. She is less spontaneous, in fact seems increasingly inhibited towards the man as the change in his behaviour discussed above becomes more and more apparent. In 85:5 she still takes the initiative to address Upendra, and mentions a marriage attended by both of them. The initiative also seems to be hers when in 85:11 she gets out of the scooter on her own accord, but both here (cf. 85:13) and in 85:8 it is an initiative inspired by a wish to escape rather than a positive decision.[22] A similar situation is found in 86:14, where Upendra's impression that she does not really watch the movie is confirmed by her own words in 86:15. In 85:20 she declines his invitation to visit him again, while her answers to his questions in 85:15-22 are not exactly forthcoming. Her efforts to re-establish the contact with the man are more artificial now, for example, when she requests him to buy her a present and persists in her request even when he is reluctant (86:6-10), when she eats his leftovers and drinks his water (86:12), and finally when she explicitly requests him to visit her or at least write to her because she often feels lonely (86:17).

The fact that Ilā takes the man's leavings is significant. Her action is the reverse of that in their childhood when she had once made the boy eat her *jūṭhā* (83:9). This reversal signifies that the roles of the man and the woman have shifted round. It is now Ilā who is 'weak'. A similar case concerns the description of Ilā as *pāgal* 'mad'. During their childhood, Upendra experienced her as a madwoman because of her dominating behaviour (83:10, 11), whereas at the end of the story, when she is 'weak', Ilā characterizes herself twice as *pāgal* (86:15, 17).

CONCLUSION

As a young boy Upendra was erotically approached by Ilā, who was his elder by several years and who behaved spontaneously and dominated him. He was too young to understand the erotic nature of her approach, and it inspired him with fear. Now that he is an adult and she visits him at his own place, he first relives his childhood as she behaves independently and dominates him once again. Her present behaviour reminds the man of the childhood erotics which he would now be more conscious of, and when at one moment he suddenly comes to realize his own desire, he interprets her behaviour as intended to provoke him. In the past years the man has apparently gained sexual experience, while Ilā and he are moreover equals now, and therefore he reacts to the supposed provocation. His question both challenges and rebukes her. The effect is that her behaviour towards him loses its spontaneity. He now dominates her and after his question, there is in him both fear and hope that she might respond to his challenge. But his provocative question might have made her realize that he interprets her spontaneous behaviour the same way society at large does, and she behaves with increasing inhibition towards him. She can no longer spontaneously express the purpose which she possibly had with her visit, that is, simply to re-establish the contact with him without any erotic implication. Her efforts to do so grow increasingly strained, and the man gets increasingly disappointed.

On the basis of the above analysis it appears that the man rather than the woman who is the central character of the story. It could be argued that his reaction to the woman's visit reflects men's ideas of feminine sexual abandon which have been discussed among others by Kakar (1989, pp. 12, 17-19, 50f.). It is remarkable that these ideas are here 'justified' by the man's childhood experiences. The element of rebuke in his question why she does not get married refers to marriage as the institution which serves to control female sexuality. At the same time his question challenges her. Her uninhibited attitude attracts the man, because it offers opportunities, but it also frightens him as it threatens male initiative. His lack of erotic initiative even after the turning point could be explained by the rebuke contained in his question, but at the same time it suggests that unconsciously the man may want Ilā to dominate him, a dominance which on a conscious level he fears.

A narratological analysis of Jośī's story allows us to put forward an argued interpretation and leads to a better understanding of the relationship between the two main characters and the motivations behind their actions and reactions which are rooted in Indian culture. As observed above, no two 'spiritual' interpretations will be exactly the same, but the essential point is that they are argued proposals which can be discussed. And as such lead to a better understanding of modern Indian culture and literature.

REFERENCES

Ansari, D.

1975 'Changes in the Hindi New Short Story of the 1960's', *Archív Orientální* 43, pp. 33-52

1976 'Indian social reality of the late 1960's and Hindi Anti-story', *Archív Orientální* 44, pp. 240-52

1978 'The 'Sacetan' Hindi literary trend and contemporary social reality', *Archív Orientální* 46, pp. 320-33

Bal, M.

1978 *De Theorie van Vertellen en Verhalen*. Muiderberg: Coutinho. Fifth, revised edition 1990

1985 *Narratology*. Toronto: University of Toronto Press

Caturvedī, R.
1990 *Hindī kahānī: ek daśak 1976-1985*. Aligarh: Tārāmaṇḍal
Cohn, D.
1978 *Transparent Minds*. Princeton: Princeton University Press
1981 'The encirclement of narrative', *Poetics Today* 2/2, pp. 157-82
Collini, S. (ed.)
1992 *Interpretation and Overinterpretation*. Cambridge: Cambridge University Press. Third reprint
Dube, L.
1988 'On the construction of gender. Hindu girls in patrilineal India', *Economic and Political Weekly*, April 30, pp. WS-11 - WS-19
Gaeffke, P.
1970 *Grundbegriffe moderner indischer Erzählkunst aufgezeigt am Werke Jayaśaṅkara Prasādas (1889-1937)*. Leiden/Köln: Brill. (Handbuch der Orientalistik. II,2)
Genette, G.
1972 *Figures III*. Paris: Seuil
Jacobson, D.
1992 'The women of North and Central India: goddesses and wives', in: D. Jacobson and S.S. Wadley (eds), *Women in India, Two Perspectives*. New Delhi: Manohar. Second, enlarged edition, pp. 15-109
Jośī, Himāṁśu
1979 *Himāṁśu Jośī kī viśiṣṭ kahāniyāṁ*. New Delhi: Śāradā
1993 *Tapasyā tathā anya kahāniyāṁ*. Delhi: Parameśvarī
Jośī, Himāṁśu (ed.)
1979a *Śreṣṭh samāntar kahāniyāṁ*. Delhi: Parāg
Jossan, C.S.
1981 'Hindi short story and the nayi kahani', in: R.J. Crane and B. Spangenberg (eds), *Language and Society in Modern India*. New Delhi: Heritage, pp. 76-118
Kakar, S.
1989 *Intimate Relations. Exploring Indian Sexuality*. New Delhi: Viking. Reprinted 1990
Kapūr, B.
1975 *Lokbhāratī Muhāvrā Koś*. Allahabad: Lokbhāratī
Kermode, F.
1979 *The Genesis of Secrecy. On the Interpretation of Narrative*. Cambridge: Harvard University Press
Mukherjee, M.
1985 *Realism and Reality. The Novel and Society in India*. Delhi etc.: Oxford University Press

Roadarmel, G.
1974 'The modern Hindi short story and modern Hindi criticism', in: E.C. Dimock a.o. (eds), *The Literatures of India, an Introduction*. Chicago: Chicago University Press, pp. 239-48

Śarmā, Ś.
1981 *Vargīkṛt Hindī Muhāvrā Koś*. New Delhi: Takṣaśilā

Schmitt, W.
1986 *Vierzig Ich-Erzählungen Premcands, Beiträge zur Form- und Inhaltsanalyse*. Reinbek: Verlag für orientalische Fachpublikationen

Siṁh, P.
1986 *Samkālīn kahānī: soc aur samajh*. Delhi: Atmaram and Sons

Stanzel, F.K.
1985 *Theorie des Erzählens*. Göttingen: Vandenhoeck & Ruprecht. Third, revised edition. (Uni-Taschenbücher. 904)

Vinay
1977 *Samkālīn kahānī, samāntar kahānī*. New Delhi: Macmillan

Notes

1. 1979 is a *terminus ante quem* for the original publication date of the story, which I have not been able to trace. It is not of direct relevance to the contents of this article, however. The story is also found in Jośī, 1993.
2. *Choṭī i* and *baṛī ī* are also the linguistic terms for the short vowel *i* and the long vowel *ī* respectively.
3. Gaeffke's research, for example his 1970 publication, is an exception. The theoretical views presented in Gaeffke, 1970 have, however, been justly refuted by Schmitt (1986, pp. 5-11, 18, 387, 409).
4. All narratological concepts used in this article are explained in Bal's study. Explicit references to the relevant pages of that book are given in a few cases only.
5. The author is grateful for comments on the analysis which were made by students and colleagues at Leiden University and at the Freie Universität Berlin.
6. The term 'story' is used here in the general sense of 'short story', to be distinguished from 'story' as one of the three layers discussed by Bal (1985, ch. 2).
7. In order to facilitate the use of a different edition, like Jośī, 1993, here follow the initial words of some paragraphs of the 1979 volume: *mujhse choṭī nahīṁ* (83:1), *'thāne meṁ photo* (84:1), *hamāre paṛos meṁ* (85:1), *maiṁ cup rahā* (86:1).
8. Similar expressions are found in 83:7 and the initial sentence of 83:10. Since they are both part of the flashback, however, *ab* in these expressions may also refer to the moment of experiencing. If the man is assumed to remember the events related in the flashback at the moment of experiencing ('Now, during

Ilā's visit, I still remember ... '), the flashback is subjective (cf. Bal, 1985, pp. 56 f.) and *ab* refers to the moment of experiencing. If, on the other hand, the man narrates the flashback events the same way he narrates about the primary story-time ('Now, while I am narrating, I still remember ... '), the flashback is objective and *ab* refers to the moment of narrating.

9. Cf. Bal, 1985, pp. 122-6. In Bal's terminology the 'informer' is called 'narrator'. Because Stanzel (1985) distinguishes between the narrator proper and the reflector, the term 'informer' is here used as a neutral term which comprises both types. Cf. Cohn, 1981, pp. 170-2 on the relationship between Genette's and Stanzel's systems on the point of 'narrator' versus 'reflector'.
10. The words *āp*, *tum*, and *tū* are personal pronouns of the second person in an order of decreasing politeness.
11. Like the expressions discussed in note 8 above, comment included in the flashback, e.g. the second sentence of 83:3 'As a lawyer, father should have argued at least a little!', may belong to the moment of experiencing instead of that of narrating.
    At the end of the story (87:1), when she is tense, the woman has lost her attractiveness for the man. Therefore the second sentence of 82:8 - 'Whenever she is angry, or excited, or deep lines of tension show on her face, she looks very attractive ... ' - would not be a comment made at the time of narrating. It would rather be an interior monologue, that is, a thought which the man has at the moment of experiencing, quoted in direct discourse without a narrative tag indicating who thinks. And sentences like 'Perhaps she wanted to startle me, or had some other intention' (82:2) may well be regarded as Free Indirect Discourse (cf. Cohn, 1978, pp. 166 f.), which is characterized by internal focalization.
12. Cases of explicit characterization of Ilā as *pāgal* 'mad' are found in the final sentence of 83:10 and in 83:11, and as a self-characterization in 86:15 and 86:17. Another example is the initial sentence of 83:1 'Young E was my senior rather than my junior.'
13. According to Kapūr (1975, s.v.) and Śarmā (1981, p. 36) *kān garm karnā* means 'to punish', so *mere kān garam ho āye the* may indicate a feeling of being punished. In view of its literal meaning 'My ears had grown hot', however, one might alternatively assume that it refers to a feeling of excitement.
14. The man's confusion is well expressed in 82:4, where the thought of addressing Ilā with the familiar pronoun *tum* or the even more familiar *tū* contrasts strikingly with his total inability to speak.
15. Cf. Dube, 1988, WS-16 on 'loose unplaited hair' of women as a sign of 'abandon' and a source of attraction, and on 'smiling without purpose' as not becoming a well-bred girl.
16. Cf. section 5.2.a below on the man's fear.
17. In this case the informer's evaluation of her behaviour as childish is justified by the description of her objectively perceptible actions.
18. The change in the characters' relationship coincides with the change in location (cf. Bal, 1985, pp. 43-5).

19. A reference to the title of the story.
20. In 82:10, where the man also regards Ilā's behaviour as 'childish', the expression is a rather neutral *baccoṁ kī tarah*. In the passages quoted here, the word *baccī* or *baccā* is consistently qualified by an adjective, viz. *nanhāṁ* 'small, young' or *nirīh* 'submissive'. Significantly, the latter term is also used for the man himself during his childhood (83:5).
21. It may be no coincidence that the way back home from the cinema is very briefly summarized, *ghūm-phirkar bahut der bād ghar lauṭe to* ... 'When we had roamed about and quite a while later returned home, ...'.
22. Ilā's wish to keep on walking (85:8, 13) may be argued to symbolize her wish to escape her lonely existence and the failed effort to re-establish contact with Upendra, or it may symbolize a wish for real independence, which is as impossible as keeping on walking.

VI

BUILDING BLOCKS OR USEFUL FICTIONS: CHANGING VIEW OF MORPHOLOGY IN ANCIENT INDIAN THOUGHT

*Madhav M. Deshpande**

1. EARLY VEDIC THOUGHT

We have a fair understanding of the history of the formal traditions of Sanskrit grammar and phonetics through the research of modern scholars. However, our understanding of how these formal traditions evolved out of the preceding period is not very clear, in spite of a good beginning made in several small areas. A formal account of the history of morphological conceptions in ancient India may begin with a treatment of formal works like the Vedic Padapāṭhas, Yāska's Nirukta and Pāṇini's Aṣṭādhyāyī. However, it is clear that many of the formal conceptions have an informal and pre-scientific ancestry which goes way back before the emergence of the formal literature on Sanskrit grammar, phonetics, and etymology. To trace this pre-scientific and pre-formal ancestry of morphological conceptions, one needs to carefully search through the Vedic literature. Secondly, one must recognize that conceptions relating to language and linguistic units are simply a special case of some other generic philosophical, magical, and religious conceptions, and thus linguistic conceptions need to be understood in relation to this wider context. In this connection, especially significant are the wider conceptions regarding the relationship between parts and wholes, as well as original and transformed states of entities. We shall begin our quest with such pre-formal and pre-scientific materials.

---

* An earlier version of this paper formed the basis of my Buiskool Lecture, delivered at the Kern Institute, Leiden, in November 1994.

1.1. COMPONENTS OF SPEECH: SYLLABLES AND METRICAL FEET

In the earliest Vedic texts, our search for linguistic units can focus on facts of language use, as well as the explicit terms used to refer to some of these linguistic units. Vedic metrical compositions are marked by two distinctive features. Each verse or verse-form is marked by a fixed number of feet and a fixed number of syllables in each metrical foot. These features of the verse form are found not only in Vedic, but are found in Avestan compositions, and may go back into further antiquity. In Vedic, it is remarkable that these are not just the unconscious features of the metrical use of language, but that these features are explicitly recognized and named already in the earliest known parts of the Ṛgveda (RV). In fact, the Ṛgveda goes one step ahead in naming the different metres, e.g. *gāyatrī*, *triṣṭubh*, and *bṛhatī*. The term *pada*, which literally means 'foot', is used in the extended sense of a metrical foot,[1] and the term *akṣara* is used in the sense of a syllable. Since a prototypical verse-form has four metrical feet, the extension of the term *pada* from foot to a metrical foot is easily understandable. The word *pada* in Vedic occasionally also has its later meaning of 'word'. Somehow, the term *pada*, which in the earlier period refers to a metrical foot, is later used primarily to refer to 'word'. However, by then a new term *pāda* comes to be used to refer to a metrical foot. The cumulative evidence of names of metres and the use of the terms *pada* and *akṣara* shows that a certain stage of conscious analytical understanding of speech and its units was already achieved by the sage-poets of the Ṛgveda.[2]

The use of the names of the metres and of the terms *pada* and *akṣara* indicates the beginning of a metalinguistic tradition. However, besides the structural meanings of these terms, the components of speech units such as metrical feet and syllables are often referred to in mystical terms, indicating further speculative movement:

> *gaurī́r mimāya salilā́ni tákṣaty ékapadī dvipádī sā cátuṣpadī /*
> *aṣṭā́padī návapadī babhūvúṣī sahásrākṣarā paramé vyòman //*
> RV 1.164.41.
> *tásyāḥ samudrā́ ádhi ví kṣaránti téna jīvanti pradíśaś cátasraḥ /*
> *tátaḥ kṣaraty akṣáraṁ tád víśvam úpa jīvati //*
> RV 1.164.42.

These two verses from the Ṛgveda are rendered by Van Buitenen (1988, p. 34) as follows: 'The buffalo cow has lowed, building lakes, having become one-footed, two-footed, four-footed, eight-footed, nine-footed - with a thousand syllables in the highest heaven: on the seas that flow out from her do all four world-quarters live: there from flows the syllable: on it lives everything'. A similar reference to components of speech-units and the activity of 'measuring' larger units in terms of smaller units is seen in the Ṛgveda:

> *gāyatréṇa práti mimīte arkám arkéṇa sā́ma traíṣṭubhena vākám /*
> *vākéna vākáṁ dvipádā cátuṣpadākṣáreṇa mimate saptá vā́ṇīḥ //*
> RV 1.164.24:
>
> With the *gāyatrī* foot, he measures the *arka*, with the *arka* the *sāman*, with the *triṣṭubh* foot the *vāka*, with the two-foot and four-foot *vāka*, the recitation, with the syllable the seven voices (Van Buitenen, 1988, p. 34).

This particular hymn is a significant source for our understanding of the ancient mystery of speech as conceived by the Vedic poets. The reduction of speech-units to their ultimate component, the syllable, is not only an analytical intellectual process, but it is, at the same time, clad in mystery. The Ṛgveda says:

> *ṛcó akṣáre paramé vyòman yásmin devā́ ádhi víśve niṣedúḥ /*
> *yás tán ná véda kím ṛcā́ kariṣyati yá ít tád vidús tá imé sám āsate //*
> RV 1.164.39
>
> What will he do with the hymns who does not know the syllable of the hymn, which is the highest heaven where the gods all live? Only those who know it are sitting together here (Van Buitenen, 1988, p. 34).

The ultimate element of speech is also the ultimate basis of the whole creation: 'The recurring expression *akṣare parame vyoman* is to be explained either as "in the syllable which is the highest heaven" or as "which is in the highest heaven"' (Van Buitenen, 1988, p. 35). Van Buitenen (*ibid.*, p. 160) further clarifies the significance of this verse: 'Since the syllable is the smallest bit of speech that can be spoken and the first that must be spoken, it is conceived at once as the matrix and the embryo of speech and all that can be effected by it', and 'here, as in the Jaim[inīya] Up[aniṣad] Br[āhmaṇa], as in fact already in the Ṛgveda, *akṣara* 'syllable' transcends uttered speech: it is the subtle, germinal principle of the Word, the unborn embryo which when born will be the Word that is creation' (*ibid.*, p. 166). Thus, the creation of a complex linguistic or physical world takes place out of a singular embryonic principle, the basic constituent unit: syllable. The early Vedic notions of creation are closely connected with the primordial rituals performed by the gods, and it is this context of ritual where the creative dimension of speech has a central role.

1.2. HIDDEN VERSUS MANIFEST DIMENSIONS OF SPEECH

The above discussion points to the fact that for the sage-poets of the Ṛgveda the phenomenon of language is not only an object of linguistic/analytical interest, but that it is also of the highest mystical/metaphysical interest, and that these two interests are not consciously separated from each other, but form simply two dimensions of a common quest. The quest leads to a belief that the mystery of language is comprehensible only to a special class of people, the wise Brāhmaṇas, while the commoners have access to and understanding of only a limited portion of this transcendental phenomenon. The mysterious nature of speech and the special category of those who know it are highlighted in the following stanza of the Ṛgveda:

> *catvā́ri vā́k párimitā padā́ni tā́ni vidur brāhmaṇā́ yé manīṣíṇaḥ /*
> *gúhā trī́ṇi níhitā néṅgayanti turī́yaṁ vācó manuṣyā̀ vadanti //*
> RV 1.164.45.

Here we are told that the wise thoughtful Brāhmaṇas know all the four quarters of speech. However, three quarters of speech are hidden from view, and the commoners speak only one quarter of the speech. This may be understood to mean that the form of speech manifest in its normal usage is what the common user recognizes, but that the wise and thoughtful Brāhmaṇas understand the deeper levels. This presumably includes the analytical levels besides the mystical ones. Perhaps, this distinction between 'analytical' and 'mystical' is our modern imposition on the contents of Vedic speculation where such a distinction did not exist.

1.3. POSSIBLE BEGINNING OF SANSKRIT PHONOLOGY IN THE ŚAUNAKĪYA ATHARVAVEDA (1.1.1)

It was stated above that the earliest Vedic compositions already show an awareness of metrical feet and syllables which constitute those metrical feet. However, we are left in the dark as to the exact structure of those syllables. From the point of view of modern phonetics, one may guess that the vowels constitute the peaks of the acoustic waves and can be easily discerned even by an untrained person.[3] However, the boundaries of those syllables are not so easily discernible. Therefore, in practical terms, it was possible for pre-literate poets to compose verses with a certain fixed number of syllables in each line, without being explicitly aware of the exact boundaries of those syllables. What are the constituents of those syllables? This requires an analytical effort to dive beneath the level of a syllable to the level of individual sounds. In the early Vedic literature, we do not as yet have the later term *varṇa* which refers to sounds. This word does occur in the Ṛgveda in the sense of colour and human groups, such as the *Āryavarṇa* and *Dāsavarṇa*, but the term is not yet used in the sense of sound. B.K. Ghosh (1945) says that for the RV poets not the sound (*varṇa*) but the syllable was the irreducible element. This may perhaps be misunderstood to mean that the word *varṇa* in the meaning of 'sound' was known to the poets of the RV, which is indeed not the case. Based on the evidence from the Aitareya Āraṇyaka (3.2.1), I have argued (Deshpande, 1994, p. 3054) that the category of semi-vowels was still not universally accepted until a very late time, and therefore the notions about distinct individual sounds of Sanskrit were in a great flux during this early period.

However, the analytical process of looking at and classifying the constituent sounds of syllables may have begun sometime during the composition of the hymns of the Atharvaveda. While our evidence for such a possibility is indeed very thin at best, such a possibility is argued for by Paul Thieme (1985). He refers to the following enigmatic verse at the beginning of the Śaunakīya Saṁhitā of the Atharvaveda (AVŚ):

yé triṣaptā́ḥ pari yánti víśvā rūpā́ṇi bíbhrataṇ /
vācáspátir bálā téṣāṁ tanvò adyá dadhātu me //
AVŚ 1.1.1.

The thrice seven that go around, wearing all the shapes - let the Lord of speech put their powers into my body's parts today.

What is being referred to by the term 'thrice-seven'? Traditional commentators have proposed various explanations for the referent of the number twenty-one, and it is not immediately evident that the expression 'thrice-seven' refers to sounds. However, after detailed argumentation based especially on the ritual context in which this verse is used, Thieme (1985) concludes that this verse refers to twenty-one sounds of Sanskrit as distinctly conceived by the earliest Vedic thinkers and he lists them as follows:

a i u ṛ e o ai au	8 vowels
y r l v	4 semi-vowels
k c ṭ t p	5 occlusives
ś ṣ s h	4 spirants

Thieme (1985, p. 563) argues that 'the sacred number "thrice seven" could indeed be taken as the number of the abstract forms (*ākṛti-*), of the types, the kinds (*varṇa-*) of sounds of the sacred language'. I have a slightly different view. Such a listing of sounds, if indeed this is what is intended by the verse, would indicate that certain phonetic features were perhaps understood more clearly by this time than others, and that the number twenty-one, in all likelihood, is a reflection of this early pre-scientific phase, rather than an explicit representation of the 'types' (*ākṛti*) of Sanskrit sounds as found in the later developed texts on Sanskrit phonetics. I have argued elsewhere (Deshpande, 1994, p. 3054) that the category of semi-vowels (*antaḥsthā*) was unacceptable as a separate category to the author of the Aitareya Āraṇyaka, and that features of quantity, voicing, and aspiration were not clearly understood during the early Vedic period. In any case, if Thieme's interpretation is correct, then we have here a beginning of an effort to go beneath the level of syllable to the level of constituent sounds. A small number of components combined and recombined in different ways are said to produce all the surface forms of language. Even if this verse did not apply to components of language, it nevertheless points to a notion of a small number of primitive components being used to produce all the surface forms (*viśvā rūpāṇi*).

1.4. BEHAVIOUR AND TREATMENT OF WHOLES EXPLAINED IN TERMS OF PARTS

A general tendency of explaining the behaviour or treatment of wholes in terms of their components is indeed widespread in the middle Vedic literature of the Yajurveda Saṁhitās and Brāhmaṇas. Consider the following passage where the cumulative syllabic value of three words, i.e. *ghṛṇiḥ*, *sūryaḥ*, and *ādityaḥ*, makes them virtually/functionally identical with an eight-syllable foot of the metre Sāvitrī/Gāyatrī.

> *ghṛṇir iti dve akṣare / sūrya iti trīṇi / āditya iti trīṇi / ... etad vai sāvitrasyāṣṭākṣaraṁ padaṁ śriyābhiṣiktam / ya evaṁ veda / śriyā haivābhiṣicyate / tad etad ṛcābhyuktam - ṛcó akṣáre paramé vyòman yásmin devā́ ádhi víśve niṣedúḥ / yás tán ná véda kím ṛcā́ kariṣyati yá ít tád vidús tá imé sám āsate //*
> RV 1.164.39, TB 3.10.9.14.

Ghṛṇiḥ constitutes two syllables. *Sūryaḥ* constitutes three syllables. *Ādityaḥ* constitutes three syllables. This octosyllabic metrical foot of Sāvitrī (= the metre Gāyatrī) is anointed with prosperity. He who knows this is indeed anointed with prosperity. Thus it has been said in a Vedic verse [RV 1.164.39]: What will he do with the hymns who does not know the syllable of the hymn, which is the highest heaven where the gods all live? Only those who know it are sitting together here.

The Taittirīya Brāhmaṇa clearly links the new speculation about the three words *ghṛṇiḥ*, *sūryaḥ* and *ādityaḥ* constituting eight syllables, and therefore becoming a foot of Sāvitrī/Gāyatrī, with the inherited mystical glorification of the knowledge of the syllable in RV 1.164.39. It is important not just to know a foot of a verse, but one must analytically/mystically know its constitution and be able to identify similar patterns elsewhere.

Such counts of syllables contained in the recited words or verses extend to larger numbers as well. Referring to the cumulative number of one hundred syllables in four verses (i.e. *agne tam adya* ..., 24 syllables; *adhā hy agne kratoḥ* ..., 25 syllables; *ābhiṣ ṭe adya* ..., 25 syllables; *ebhir no arkaiḥ* ..., 26 syllables), the Taittirīya Saṁhitā says:

> *śatākṣarā bhavanti śatāyuḥ puruṣaḥ śatendriyaḥ* /
> TS (1.5.2.2)
>
> [These verses together] contain a hundred syllables. A man lives for a hundred years and has a hundred faculties.

A similar treatment of a whole in terms of its different constituents is seen in the following passages of the Śatapatha Brāhmaṇa:

> *athāsmai sāvitrīm anvāha / tāṁ ha smaitāṁ purā saṁvatsare 'nvāhuḥ saṁvatsarasaṁmitā vai garbhāḥ prajāyante / ... atha ṣaṭsu māseṣu / ṣaḍ vā ṛtavaḥ saṁvatsarasya / ... atha caturviṁśatyahe / caturviṁśatir vai saṁvatsarasyārdhamāsāḥ / ... atha dvādaśāhe / dvādaśa vai māsāḥ saṁvatsarasya / ... atha ṣaḍahe / ṣaḍ vā ṛtavaḥ saṁvatsarasya / ... atha tryahe / trayo vā ṛtavaḥ saṁvatsarasya /*
> ŚB XI.v.4.6-11.
>
> Then he taught him [= student] the Sāvitrī [= Gāyatrī]. Previously indeed they used to teach it after a year. Embryos are born after the [gestation] period of a year ... Then [they used to teach it] after six months. There are indeed six seasons in a year. ... Then [they used to teach it] on the twenty-fourth day. There are indeed twenty-four half-months in a year. ... Then [they used to teach it] on the sixth day. There are indeed six seasons in a year. Then [they used to teach it] on the third day. There are indeed three seasons in a year.

tāṁ (sāvitrīm = gāyatrīm) vai paccho 'nvāha / trayo vai prāṇāḥ prāṇa udāno vyānaḥ / tān evāsmiṁs tad dadhāti / athārdharcaśaḥ / dvau vā imau prāṇau prāṇodānāv eva / prāṇodānāv evāsmiṁs tad dadhāti / atha kṛtsnāṁ / eko vā ayaṁ prāṇaḥ kṛtsna eva / prāṇam evāsmiṁs tat kṛtsnaṁ dadhāti/
ŚB XI.v.4.15.

He taught the Gāyatrī [of three metrical feet] one foot at a time. There are indeed three vital breaths: *prāṇa*, *udāna*, and *vyāna*. By that act, he verily installs them in him [= student]. Then he teaches it by half-verse units. There are indeed only two vital breaths, only *prāṇa* and *udāna*. By that act, he verily installs in him *prāṇa* and *udāna*. Then he teaches the entire [verse at a time]. There is indeed one vital breath, i.e. *prāṇa*, complete in itself. By that act, he verily installs the complete *prāṇa* in him.

dvādaśasu vikrāmeṣv agnim ādadhīta / dvādaśa māsāḥ saṁvatsaraḥ /
TB 1.1.4.1.

One should establish the sacrificial fire at the distance of twelve steps. There are twelve months in a year.

This pattern of explaining the treatment or behaviour of wholes in terms of their parts is universally present in the ritualistic literature and its application to linguistic units and their components is simply a special case in point. The above passages also show that a given whole could be segmented differently and these different segmentations may have different justifications and ritual consequences. There is also an implication that one segmentation does not necessarily invalidate other possible segmentations, and that these segmentations are not purely fictional, but that they have real ritual consequences. The above passages indicate that the Vedic tradition looked at the components of wholes as real entities and attempted to explain the treatment or the behaviour of the wholes in terms of their components.

1.5. MATCHING THE STRUCTURE OF LINGUISTIC UNITS TO STRUCTURES IN RITUAL ACTION

There is a notion of *rūpasamṛddhi* 'perfection of ritual form' in the Brāhmaṇas. It refers to an expectation that ideally a verse being recited in a given ritual should express the action which is being carried out in that ritual (*etad vai yajñasya samṛddhaṁ yad rūpasamṛddhaṁ yat karma kriyamāṇaṁ ṛg anuvadati*, AB 1.13). Thus, there is an expectation that the linguistic units and their constituents are somehow to be mapped onto the structures of ritual actions or their results. Consider the following examples:

> *sā [iḍā] abravīt / pratīcy eṣāṁ śrīr agāt / bhadrā bhūtvā parābhaviṣyantīti / ... yasyaivam agnir ādhīyate / pratīcy asya śrīr eti / bhadro bhūtvā parābhavati /*
> TB 1.1.4.4.
>
> [After the demons installed their sacrificial fire incorrectly facing the west], she [= Iḍā] said: 'The prosperity of these [demons] went in the reverse direction [lit. westwards]. [Therefore] having become prosperous [for a short while] they will be defeated.' For him whose fire is thus [incorrectly] installed, his prosperity goes in the reverse direction [lit. westwards], and having become prosperous [for a short while] he is defeated.
>
> *bhṛgūṇāṁ tvāṅgirasāṁ vratapate vratenādhāmīti bhṛgvaṅgirasāṁ ādadhyāt / ādityānāṃ tvā devānāṁ vratapate vratenādadhāmīty anyāsāṁ brāhmaṇīnāṁ prajānām / varuṇasya tvā rājño vratapate vratenādadhāmīti rājñaḥ /*
> TB 1.1.4.8.
>
> Saying *bhṛgūṇāṁ tvāṅgirasāṁ vratapate vratenādadhāmi*, one installs [the sacrificial fire] for the Bhṛgus and Angirases. Saying *ādityānāṁ tvā devānāṁ vratapate vratenādadhāmi*, one installs [the sacrificial fire] for the other brahmanical peoples. Saying *varuṇasya tvā rājño vratapate vratenādadhāmi*, one installs [the sacrificial fire] for the king.

In this passage, one notices that a marginal change in the formula to be recited is matched by a corresponding marginal change in the ritual application, and the marginal changes are clearly correlated in a significant manner. A strong inference can be drawn from this and other similar passages that the authors indeed do not look at the sacrificial formulae as indivisible wholes, but as wholes built up with meaningful components, which can be changed individually without changing the entire structures. Similarly, the ritual contexts are not viewed as indivisible wholes, but as configurations with components which can vary. Often we see that a distinct significance is attached to each distinct component. Consider the following passages:

> *prajāpatir vācaḥ satyam apaśyat / tenāgnim ādhatta / bhūr bhuvaḥ suvar ity āha / etad vai vācaḥ satyam / ... bhūr ity āha / prajā eva tad yajamā-*

> *naḥ sṛjate / bhuva ity āha / asminn eva loke pratitiṣṭhati / suvar ity āha / suvarga eva loke pratitiṣṭhati /*
> TB 1.1.5.1-2.
>
> The Lord of the Creatures saw the real essence of speech and with that he installed the sacrificial fire. [The priest] says: *bhūr bhuvaḥ suvaḥ*. This is indeed the real essence of speech. He [= the priest] says: *bhūḥ*. Thereby, the sacrificer indeed produces progeny. He [= the priest] says: *bhuvaḥ*. Thereby, the sacrificer firmly establishes [himself] in this very world. He [= the priest] says: *suvaḥ*. [Thereby, the sacrificer] firmly establishes [himself] in the heavenly world.
>
> *trīṇi havīṁṣi nirvapati / ... agnaye pavamānāya / agnaye pāvakāya / agnaye śucaye / yad agnaye pavamānāya nirvapati / punāty evainam / yad agnaye pāvakāya / pūta evāsminn annādyaṁ dadhāti / yad agnaye śucaye / brahmavarcasam evāsminn upariṣṭād dadhāti /*
> TB 1.1.5.10.
>
> He [= the priest] offers three oblations, with the formulae: *agnaye pavamānāya*, *agnaye pāvakāya*, and *agnaye śucaye*. The fact that [the priest] offers [an oblation] to fire [described as] *pavamāna*, it indeed purifies him [= sacrificer]. The fact that [the priest offers an oblation] to fire [described as] *pāvaka*, it indeed installs food in him [= sacrificer] who is purified. The fact that [the priest offers an oblation] to fire [described as] *śuci*, it indeed additionally installs in him the brilliance of Brahman.

The substitution of one word with another is matched by a corresponding substitution of one result by another. While in these passages, the abstract logic of substitution is only implicit, it is later made explicit by Patañjali in his Mahābhāṣya (MBh).

1.6. THE USE OF THE ROOTS VI+KṚ AND VI+Ā+KṚ IN VEDIC LITERATURE

Referring to the phrase *tád agnír vy àkarot* in the Taittirīya Āraṇyaka, Jan Houben (1991, p. 98) remarks: 'The activity in relation to the undifferentiated "*sāman*- and *yajus*- milk", here expressed by the verb *vi+kṛ*, is very similar to the activity in relation to undifferentiated speech, expressed by *vy-ā-kṛ*, e.g. in TS 6.4.7.3'. Houben refers to Paul Thieme's (1982-1983, pp. 23-8) study of the use of *vy-ā-kṛ*. Thieme explains the meaning of *vy-ā-kṛ* to be among other things 'to distinguish [by separating similar objects from one another or by separating a seeming whole into its parts]'. While I agree with Thieme's general interpretation, the cautionary phrase 'seeming' is perhaps not necessary. In my view, the Vedic texts seem to suggest that the parts are as real as the wholes. The processes expressed by the verbs *vi+kṛ* and *vi+ā+kṛ* simply refer to ways the parts become distinct or differentiated, with no hint of their unreality.

Van Buitenen (1988, p. 6) referring to the Chāndogya Upaniṣad 6.1.3 (*vācārambhaṇaṁ vikāro nāmadheyam*), points out: 'Name must mean here what later on is termed *nāmarūpe* "name and form", which in 6.3.2. describes the products or creatures that are "separated out of (*vyā+√kṛ* ~ *vikāra*) the materia prima"'. Further (Van Buitenen, 1988, p. 9), the term *vikāra* is explained as 'that which is separated out of the underlying stuff that is the material cause'. It is clear that in the late Vedic texts, the verbs *vi+kṛ* and *vi+ā+kṛ* are used somewhat interchangeably. Both the verbs are very close in meaning with some slight difference. The verb *vi+kṛ* seems to refer more often to a self-induced process of differentiation or transformation. The process of internal self-induced transformation expressed by *vi+kṛ* is further developed in the Sāṁkhya system, and in the doctrine of *Pariṇāmavāda* as seen in Rāmānuja's Viśiṣṭādvaita system. On the other hand, the verb *vi+ā+kṛ* refers to an internally or externally brought about differentiation:

> *sā iyaṁ devataikṣata - hantāham imās tisro devatā anena jīvena ātmanānupraviśya nāmarūpe vyākaravāṇīti / tāsāṁ trivṛtaṁ trivṛtam ekaikāṁ karavāṇīti / seyaṁ devatā imās tisro devatā anena jīvenātmanānupraviśya nāmarūpe vyākarot /*
> ChU 6.3.2-3.
>
> That divinity (= Being) bethought itself: 'Come! Let me enter these three divinities [i.e. heat, water, and food] with this living soul (*ātman*), and separate out name and form. Let me make each one of them threefold.' That divinity entered into these three divinities and separated out name and form (Hume, p. 242).
>
> *tad dhedaṁ tarhy avyākṛtam āsīt / tan nāmarūpābhyām eva vyākriyata asaunāmāyam idaṁrūpa iti / tad idam apy etarhi nāmarūpābhyām eva vyākriyate 'saunāmāyam idaṁrūpa iti /*
> Bṛhadāraṇyaka Upaniṣad 1.4.6.
>
> Verily, at that time this world was undifferentiated. It became differentiated just by name and form, as the saying is: 'He has such a name, such

> a form.' Even today this world is differentiated just by name and form, as the saying is: 'He has such a name, such a form.' (Hume, p. 82)

In any case, both verbs signify the process of differentiation which moves in the direction of parts of a whole becoming more vividly manifest and perceptible.

While the process of internally brought about differentiation of an original unitary principle is seen in the various accounts of the creation of the world, the externally brought about differentiation expressed by *vi+ā+kṛ* is more directly relevant to the development of the systems of grammatical thought (*vyākaraṇa*). This is seen in the story which marks the mythical beginnings of grammatical analysis:

> *vāg vai parācy avyākṛtāvadat / te devā indram abruvann imāṁ no vācaṁ vyākurv iti / ... tām indro madhyato 'vakramya vyākarot / tasmād iyaṁ vyākṛtā vāg udyate /*
> TS 6.4.7.3.

> Speech aforetime spoke without discrimination; the gods said to Indra, 'Do thou discriminate this speech for us'; ... He approaching it in the midst discriminated it; therefore is speech spoken distinctly.[4]

The story suggests an important distinction. The users of language speak unanalysed speech without any comprehension of its parts. Only gods like Indra or wise and thoughtful Brāhmaṇas know how to differentiate the speech into its 'real' components. These components have no hint of unreality. As we shall see, there is a notion that the differentiated form of speech, which is inaccessible to normal speakers, offers a more direct (*pratyakṣa*) view of the true nature of speech, while the undifferentiated form of speech available to the normal users offers only an indirect (*parokṣa*) view of the true nature of speech.

1.7. PRATYAKṢA AND PAROKṢA FORMS OF SPEECH

Several scholars have studied the etymologies offered by the Brāhmaṇa texts, see Gonda (1955-6); also Pandit (1989) for 'Vedic compounds interpreted by Veda'. These etymologies are almost always dubbed as folk etymologies and are often quickly discarded. However, they are extremely important for our understanding of the attitudes of the late Vedic texts and their authors. In the process of offering such folk etymologies, the Brāhmaṇa authors are offering us what they think is the true form of a given expression. Consider the following passage from the Taittirīya Āraṇyaka (Pravargya-Brāhmaṇa) 1.2-4:

> *so 'smayata / ... tasya siṣmiyāṇasya tejo 'pākrāmat / tad devā oṣadhīṣu nyamrjuḥ / te śyāmākā abhavan / smayākā vai nāmaite / tat smayākānāṁ smayākatvam / tasmāt dīkṣitenāpigṛhya smetavyam / tejaso dhṛtyai /*
>
> He smiled. ... Brilliant energy withdrew from him who was smiling. This the gods rubbed upon some herbs. These became the Śyāmākas. These are actually Smayākas by name. That is why Smayākas are called Smayākas. Therefore, an initiate should smile (only) after covering the mouth to preserve his brilliant energy. (Houben, 1991, p. 46)

Here, the 'true' form of the word *śyāmāka* is given as *smayāka*. This true form is either created by the gods and/or is known to the gods, and also to the wise and thoughtful Brāhmaṇas. However, the form known to the common users of the language is a more indirect expression of the true form, and thus hides the truth from its users. Thus, to uncover the truth behind a commonly used word such as *śyāmāka*, one must discover the true word-form like *smayāka* that lies behind it. Several passages in the Brāhmaṇa literature make this distinction even more explicit by using the expression *parokṣa* 'indirect' for the word as it is used in common usage and by saying that the gods love such indirect expressions:

> *tad iṣṭīnām iṣṭitvam / eṣṭayo ha vai nāma / tā iṣṭaya ity ācakṣate parokṣeṇa / parokṣapriyā iva hi devāḥ /*
> TB 1.5.9.2. (Anandashrama ed., vol. I, p. 270)
>
> That is the true nature of the *iṣṭis* 'sacrifices'. The *iṣṭis* are actually *eṣṭis* 'desires'. They are indirectly called *iṣṭis*.

Here, the word *iṣṭi* 'sacrifice' which is actually derived from the root *yaj* 'to sacrifice' is said to be really an indirect representation of the true/original word-form *eṣṭi* derived from the root *iṣ* 'to desire'. The phonetic shape of *iṣṭi* and its meaning seemed closer to the root *iṣ* to the Brāhmaṇa author.

> *daśahūto ha vai nāmaiṣaḥ / taṁ vā etaṃ daśahūtaṁ santam / daśahotety ācakṣate parokṣeṇa / parokṣapriyā iva hi devāḥ /*
> TB 2.3.11.1 (Anandashrama ed., vol. I, p. 481)
>
> He is indeed *daśahūta*. Him really being *daśahūta*, they call him *daśahotṛ* indirectly. Gods indeed love the indirect mode.

Similar other explanations in TB (2.3.11.2-4):

Parokṣa 'surface/indirect' form Non-*Parokṣa* 'true/real/direct' form

saptahotṛ	*saptahūta*
ṣaḍḍhotṛ	*ṣaḍḍhūta*
pañcahotṛ	*pañcahūta*
caturhotṛ	*caturhūta*

Referring to the non-*parokṣa* 'real/true/direct' form, Sāyaṇa says:

vāstavaṁ nāmaviśeṣaṁ ācchādya apareṇa varṇāntaravyavahṛtena daśahoteti nāmnā vyavaharanti
[on TB, Anandashrama edn, vol. I, p. 481]

Having covered up the real/factual specific name [*daśahūta*], they call him by the name *daśahotṛ* uttered with other [=non-original] sounds.

te yan nyañco 'rohaṁs tasmān nyaṅrohati nyagroho nyagroho vai nāma tan nyagrohaṁ santaṁ nyagrodha ity ācakṣate parokṣeṇa / parokṣapriyā iva hi devāḥ /
AB 35.4.30 (Anandashrama ed., vol. II, p. 884).

In that they grew downwards, therefore the Nyagroha grows downwards; its name is Nyagroha; it being Nyagroha the gods call [it] Nyagrodha indirectly [Keith (1920, p. 315), 'mysteriously'], for the gods love the indirect mode [Keith (1920, p. 315), 'mystery'] as it were.

tad vā idaṁ prajāpate retaḥ siktam adhāvat tat saro 'bhavat te devā abruvan medaṁ prajāpate reto duṣad iti yad abruvan medaṁ prajāpate reto duṣad iti tan māduṣam abhavat tan māduṣasya māduṣatvaṁ māduṣaṁ ha vai nāmaitad yan māduṣaṁ [printed text reads: *mānuṣam*] *san mānuṣam ity ācakṣate parokṣeṇa parokṣapriyā iva hi devāḥ /*
AB 13.10.33 (Anandashrama ed., vol. I, p. 377).

The seed of Prajāpati outpoured ran; it became a pond. The gods said, 'Let not this seed of Prajāpati be spoiled' (*mā duṣat*). Since they said, 'Let not this seed of Prajāpati be spoiled (*mā duṣat*)', it became a *māduṣa* [= 'the unspoiled', a human being]. That is how a *māduṣa* has this name. Being *māduṣa* in reality, they call it *mānuṣa* indirectly. Gods as it were love the indirect mode.

sa vā eṣo 'gnir eva yad agniṣṭomas taṁ yad astuvaṁs tasmād agnistomas tam agnistomaṁ santam agniṣṭoma ity ācakṣate parokṣeṇa parokṣapriyā iva hi devāḥ /
AB 14.5.43 (Anandashrama ed., vol. I, p. 407).

The Agniṣṭoma is Agni; in that they praised him, therefore is it the praise of Agni (*agnistoma*); it, being the praise of Agni (*agnistoma*), they call it Agniṣṭoma indirectly [Keith (1920, p. 192), 'mystically'], for the gods love the indirect mode [Keith (1920, p. 315), 'mystery'] as it were.

Similar other explanations in the AB (*ibid.*, p. 407):

Parokṣa 'surface/indirect' form Non-*Parokṣa* 'true/real/direct' form

catuṣṣṭoma catusstoma
jyotiṣṣṭoma jyotisstoma

While most of these folk etymologies may be easily dismissed, they contain important glimpses of the evolution of linguistic thought. For example, the folk etymology of *mānuṣa* as *māduṣa*, suggests that the Brāhmaṇa authors had perhaps not yet made a connection of the word *mānuṣa* with Manu, a connection which is explicitly made by Pāṇini (4.1.161, *manor jātāv añyatau ṣuk ca*). On the other hand, some of these discussions show a certain degree of sophisticated linguistic reasoning. In trying to explain a form like *daśahotṛ* as really being *daśahūta*, the author seems to show an awareness that it is better to regard the entity referred to as 'being invoked' rather than as the invoker. The linking of the surface-form *nyagrodha* to the 'true' form *nyagroha*, also suggests that the word *nyagrodha* as it stood did not make sense in terms of the contemporary meaning of the root *rudh*. It made better sense in terms of the meaning of another root, *ruh*, available in the language. Linking of forms like *agniṣṭoma* to the 'true' forms *agnistoma* etc. seems to suggest that the feature of retroflexion seen in the form *agniṣṭoma* was no longer a productive feature, but a relic of the past, and could not be accounted for in terms of the contemporary norms. While the passages cited above refer to the surface forms of words known to common speakers by the term *parokṣa*, there is a passage in the relatively late Gopatha Brāhmaṇa, which goes a step ahead and uses the term *pratyakṣa* to refer to the 'true' form made accessible through etymology. It not only says that gods love the *parokṣa* mode, but it adds that gods hate the *pratyakṣa* mode:

tasya śrāntasya taptasya santaptasya sarvebhyo 'ṅgebhyo raso 'kṣarat / so 'ṅgaraso 'bhavat / taṁ vā etam aṅgarasaṁ santam aṅgirā ity ācakṣate parokṣeṇa / parokṣapriyā iva hi devā bhavanti pratyakṣadviṣaḥ /
GB (Pūrvabhāga 1.7, p. 6)

From all his exhausted, heated, overly heated limbs (*aṅgebhyaḥ*), a certain essence (*rasa*) came forth. He became the *Aṅgarasa*. Being really *Aṅgarasa*, they indirectly called him *Aṅgiras*. Gods as it were love the indirect mode, and they hate the direct mode.

This discussion sets up the parameters of the central dilemma. The surface-forms of the language which are commonly known to the users/speakers are deemed to be indirect/partial reflections of the true/hidden forms and hence unable to lead to the clearest revelation of the true/hidden meaning. Thus, the surface-forms or the commonly visible forms are the *parokṣa* 'indirect' forms. On the other hand, the forms which are revealed through the offered etymologies are deemed to be the true/real or direct forms (*pratyakṣa*) and are deemed to reveal the true meaning directly. However, these forms are not known and not visible to the users, and hence are mysterious. However, such mysteries of the speech are dear to gods and are known to the wise and thoughtful Brāhmaṇas.

While the Nirukta of Yāska picks up many of these folk-etymologies, it does not pick up the terminology of *pratyakṣa* and *parokṣa* forms. However, the significance of this ancient terminology is not lost on the later tradition altogether. Durgācārya in his commentary on Nirukta (1.1) uses this terminology and refines it further into three categories of words, i.e. *pratyakṣa-vṛtti* 'direct mode', *parokṣa-vṛtti* '[less] indirect mode', and *atiparokṣa-vṛtti* 'most indirect mode'.

tad etad [varṇavyāpattyādilakṣaṇaṁ] parokṣātiparokṣavṛttiṣu śabdeṣu yathāsambhavaṁ draṣṭavyam / trividhā hi śabdavyavasthā / pratyakṣavṛttayaḥ parokṣavṛttayaḥ atiparokṣavṛttayaś ca / tatrotkaṭakriyāḥ pratyakṣavṛttayaḥ / antarlīnakriyāḥ parokṣavṛttayaḥ / atiparokṣavṛttiṣu śabdeṣv eva nirvacanābhyupāyaḥ / tasmāt parokṣavṛttitām āpādya pratyakṣavṛttinā śabdena nirvaktavyāḥ / tad yathā / nighaṇṭava ity atiparokṣavṛttiḥ / nigantava iti parokṣavṛttiḥ / nigamayitāra iti pratyakṣavṛttiḥ / yasmāt nigamayitāra ete nigantava iti nighaṇṭava ity ucyante / ... ta ete nigantavaḥ santo nighaṇṭava ucyante ity evam atiparokṣavṛttayo nirvaktavyāḥ / prāyeṇa ca uṇādiṣu parokṣavṛttayaś śabdāś cintyante / tatra teṣāṁ lakṣaṇam upekṣitavyam / yeṣām api lakṣaṇaṁ nāsti teṣām api tatra kalpyam / aparisamāptā hi uṇādaya iti lakṣaṇavidaḥ pratijānate / sarvathāpi lakṣaṇāsambhave pṛṣodarādipāṭhasiddhir eva draṣṭavyā / tatra hi

yathādhyayanam eva śabdāḥ sādhīyāṁso bhavanti abhivyāhārānabhighātāya iti hi lakṣaṇavido manyante /
Durga on Nirukta, Anandashrama edn., vol. I, pp. 7-8.

All these features of change of sounds etc. should be observed as they occur in words represented in [less] indirect and most indirect modes. Words are classified in three ways. Those in the direct mode, the [less] indirect mode, and the most indirect mode. Those words where the involved actions [= verb roots] are most obvious are in the direct (*parokṣa*) mode. Those words where the involved actions [= verb roots] are inherent [but not directly accessible in the surface] are in the [less] indirect mode. Only for those words which are in the most indirect mode, an etymological analysis is a means [to explication]. Such words, [which are in the most indirect mode] should be [first] paraphrased into the [less] indirect mode and then should be etymologized by means of a word which is in the direct mode. For example, the expression *nighaṇṭavaḥ* is in the most indirect mode. [Its paraphrase by Yāska as] *nigantavaḥ* is in the [less] indirect mode. [Its further paraphrase by Yāska as] *nigamayitāraḥ* is in the direct mode. The [words listed in the Nighaṇṭu and referred to by the word *nighaṇṭu*] are called *nigantu* [first] and [then] *nighaṇṭu*, because they are [really] *nigamayitṛ* 'lead to a proper comprehension [of the Vedas]'.[5] These [words, really] being *nigantu* are [indirectly] called *nighaṇṭu*. The words in the most indirect mode should be explicated in this fashion. In the *Uṇādi* rules, mostly the words in the [less] indirect mode are taken into consideration. One should consult their characterization in that text. For those words which are not taken into consideration [in the *Uṇādi* rules], one should devise their explication in there [= in the mode of the *Uṇādi* rules]. The knowers of rules [of grammar] testify that the *Uṇādi* rules are indeed incomplete. For those words for which no explication is possible, their correctness should be seen only through their inclusion in the lists of [correct, but inexplicable] words like *pṛṣodara* [cf. P.6.3.109: *pṛṣodarādīni yathopadiṣṭam*]. Words included in those [lists] are correct as recited. [They are included in such lists] to make sure that their pronunciation is not altered. This is what the knowers of rules believe.

Durga shows how the ancient categories of the Brāhmaṇas can be brought to bear upon our understanding of the diverse treatment of words in later traditions found in the Nirukta, Aṣṭādhyāyī, Uṇādi Sūtras, etc.; also see: Saroja Bhate (1981).

1.8. LATE VEDIC NOTIONS ABOUT THE SAṀHITĀPĀṬHA AND PADAPĀṬHA

The orally preserved Vedic texts underwent a great deal of consolidation during the Brāhmaṇa period, and scholars like Śākalya and Māṇḍūkeya prepared the Saṁhitās and their Padapāṭhas. The Aitareya Āraṇyaka explicitly refers to the activity of the redactors, but it is not clear how far before the *AĀ* the activity of dividing the Saṁhitā into Padas 'words' actually began. As we have previously noted, the word *pada* has the meaning of a metrical foot in the Vedic Saṁhitās. However, it seems that by the late Ṛgvedic period, the word *pada* was occasionally used to refer to 'word'. Louis Renou (1958, pp. 21-2) points to a few passages in the RV where the word *pada* does not mean a metrical foot, but is more like an equivalent of the term *nāman* 'name', or perhaps 'word' in a non-technical sense [RV 7.87.4: *triḥ sapta yad guhyāni tve it padāvindan nihitā yajñiyāsaḥ*; and RV 10.53.10: *padā guhyāni kartana yena devāso amṛtatvam ānaśuḥ*]. According to Roth and Böhtlingk (PW), the word *pada* in the Śatapatha Brāhmaṇa (10.2.6.16) refers to 'word', where it occurs in the context of reciting a hundred *Ṛks*, *Yajus*, *Padas*, or *Akṣaras*. A clearer instance is perhaps found in the Śatapatha Brāhmaṇa (11.3.8.9, *deva-padam*) where the word *pada* seems to have been used in the sense of a 'word'. The Aitareya Brāhmaṇa (6.33, Anandashrama ed., vol. II, p. 788), referring to a passage called Aitaśa-Pralāpa says that this passage is to be recited *padāvagraham*. Sāyaṇa renders this archaic gerund by *pade pade avagṛhya* 'separating each *pada*' or pausing after each *pada*. If we are to believe Sāyaṇa (*ibid.*, p. 785), the word *pada* here refers to a metrical foot. While the term *avagraha* is used in later grammatical treatises to refer to a pause between the constituents of a word, such as the members of a compound, and is not used to refer to the pause between the separated metrical feet, this passage makes it likely that the procedure of breaking down a continuous text into its constituent units, and reciting these units with separating pauses had begun before the formation of the *Padapāṭhas* or 'word-by-word' versions of the Vedic Saṁhitās. Thus, one can conceive of a stage of development when the textual wholes were broken down into their constituent units such as metrical feet, but that these units were not further analysed into their own constituents. This next stage is found in the available Padapāṭhas for the various Vedic Saṁhitās. It seems likely that the terms *avagraha* and *pada* were inherited by the later tradition, which used them in a somewhat different meaning of 'a short pause between the constituents of a word' (*avagraha*) and a word (*pada*). Sometime during the Brāhmaṇa period, not only did the word *pada* come to mean a 'word', but that another word, i.e. *pāda*, came to be used to refer to the metrical foot. Such a distinction, which is maintained in all the later technical literature, is already evident in the Śāṅkhāyana Brāhmaṇa (26.5). This passage is particularly significant in our understanding of how the notion of various segments may have arisen and how it may have formed a part of the recitational and ritual practice:

> *daivodāsiḥ pratardano naimiṣīyāṇāṁ satram upagamyopāsyadyavicikitsāṁ papraccha yady atikrāntam ulbaṇaṁ sadasyo bodhayetartvijāṁ vānyatamo budhyeta kathaṁ vo'nulbaṇaṁ syād iti ta u ha tūṣṇīm āsus teṣām alīkayur vācaspatyo brahmā sa hovāca nāham etad veda hanta pūrveṣām ācāryaṁ sthaviraṁ jātūkarṇyaṁ pṛcchānīti taṁ ha papraccha yady atikrāntam ulbaṇaṁ kartā vā svayaṁ budhyetānyo vā bodhayeta kathaṃ tad ulbaṇam anulbaṇaṁ bhavet punarvacanena vā mantrasya homena veti punar vācyo mantra iti ha smāha jātūkarṇyas tam alīkayuḥ punaḥ papraccha śastraṁ vā anuvacanaṁ vā nigadaṁ vā yājyāṁ vā yad vānyat sarvaṁ tat punar brūyād iti yāvanmātram ulbaṇaṃ tāvad brūyād ṛcaṁ vārdharcaṁ vā pādaṁ vā padaṁ vā varṇaṁ veti ha smāha jātūkarṇyaḥ /*
>
> Daivodāsi Pratardana having gone to a sacrificial season of the Naimiṣīyas and having glided up asked a question on this point of doubt, 'If the priest in the Sadas should call attention to a flaw passed over or any one of the priests should note it, how would you remove the flaw?' They were silent; Alīkayu Vācaspatya was their Brahman priest; he said, 'I know that not; but will ask Jātūkarṇya, the aged teacher of those formerly'. Him he asked, 'If the performer himself should note a flaw passed over or another should call attention to it, how is that flaw to be made flawless? By repetition of the Mantra or by an oblation?' 'The Mantra should be recited again', Jātūkarṇya said. Him Alīkayu again asked, 'Should one recite in full the Śastra or recitation or Nigada or offering verse or whatever else it be?' 'So much as is erroneous only need be repeated, a verse (*ṛcam*), or half verse (*ardharcam*), or quarter verse (*pādam*), or word (*padam*), or letter (*varṇam*)', Jātūkarṇya replied. (Keith, 1920, p. 498).

This is an extremely important passage which provides a valid pre-grammatical rationale for recognizing the constituents of speech units. Not only do we find here a clear distinction between a *pāda* 'metrical foot' and *pada* 'word', we also find one of the early uses of the term *varṇa* to refer to 'sound', in contrast with the older term *akṣara* 'syllable'. Thus, there is clear conceptual and terminological progress from metrical feet to words, and from syllables to individual sounds. Perhaps, the ritual necessity of making minimum corrections in the recited Mantras may have led to the early efforts in recognizing the constituents in the metrical or non-metrical Vedic compositions. Such efforts on large scale may have eventually led to the formation of what we know as the Padapāṭhas or 'word-by-word' versions of the Vedic texts. Such texts have already come into existence by the time of the Āraṇyakas, where one finds some explicit discussion about their formation.

Another interesting feature of the above passage is the order in which the successively smaller segmentations are listed: a verse (*ṛcam*), half verse (*ardharcam*), quarter verse (*pādam*), word (*padam*), letter (*varṇam*). This would seem to imply the sequence of segmentation into successively smaller units, as well as a notion that first there were the unsegmented texts and then came the successive segmentations. Such an implication, if we can accept it, would lead us to some very ancient ideas concerning the relationship between what came to be later called the Saṁhitās and their Padapāṭhas.

The debate regarding the relationship between the Saṁhitās and their Padapāṭhas as seen in the various texts is indeed confusing at first sight. The oldest complex debate regarding these two sets of texts is found in the Aitareya Āraṇyaka. In this debate, we find the use of two sets of terms. The first set of terms includes *nirbhuja* and *pratṛṇṇa*.[6] The second set is represented primarily by the term *saṁhitā*. Of these two sets of terms, the first set is not only seen here for the first time, it is, strangely enough, a set of terms which appear to be on their way out, and which are explained not intrinsically, but in terms of the emerging term, i.e. *saṁhitā*. It is this term, *saṁhitā*, and its later comrade, *pada*, which appear to be the current paradigm for the author of the Aitareya Āraṇyaka. The term *saṁhitā*, a term related to the more familiar term *sandhi*, suggests a text-form in which constituents are joined or fused together. This contrasts with the text given in the Padapāṭha, where one finds words given in isolation from each other. The text-form which is known commonly as the *Saṁhitā* is referred to by the term *nirbhuja* in the Aitareya Āraṇyaka (3.1.3), while the text-form which is known commonly as the Padapāṭha is called *pratṛṇṇa*. Under the weight and currency of the more common terms *saṁhitā* and *pada*, the older terms *nirbhuja* and *pratṛṇṇa* are misunderstood not only by the traditional commentators, but by the author of the Aitareya Āraṇyaka himself. *Sāyaṇa* takes the word *nirbhuja* to mean that form in which the preceding and the following words are mentioned (*nirdiṣṭau bhujasadṛśau pūrvottaraśabdau yasmin saṁhitārūpa uccāraṇe tad uccāraṇaṁ nirbhujam*, AĀ, Anandashram edn., p. 225). Sāyaṇa says that the Aitareya Āraṇyaka itself explains the meaning of these terms, because they are not well known in the world (*nirbhujādiśabdānāṁ vivakṣitārthe lokaprasiddhyabhāvāc chrutir eva tam arthaṁ darśayati*). The Aitareya Āraṇyaka itself explains the term *nirbhuja* as: *atha yat sandhiṁ vivartayati tan nirbhujasya rūpaṁ*, 'that which carries out the euphonic combination is the form of *nirbhuja*'. This in all probability does not correspond to the original meaning of the term *nirbhuja*, which literally seems to mean 'that which is uncut'. Such a meaning seems all the more likely in view of the other term *pratṛṇṇa*. The terms *pratardana* (MS 3.3.7, and KS 21.10) and *pratṛd* (RV 7.33.14) refer to someone who smashes something to pieces, and the gerund *pratṛdya* is explained by Sāyaṇa as *praviddhaṁ kṛtvā* 'having split up' (on ŚBr 11.7.4.3). Thus, if the term *pratṛṇṇa* refers to something which is 'split up', the term *nirbhuja* most likely refers to something that is not split up. The Aitareya Āraṇyaka says that the form of *pratṛṇṇa* is that in which one utters 'pure' (= euphonically untransformed) sounds (*yac chuddhe akṣare abhivyāharati tat pratṛṇṇasya*). This explanation clearly does not look upon the Padapāṭha as a derived or 'split up' secondary product, but as reflecting a primary or 'original, pure, uncontaminated' (*śuddha*) state of affairs. The Ṛgveda Prātiśākhya (2.1) repeats the same explanation (*śauddhoccāraṇaṁ ca pratṛṇṇam*). It is thus clear that the Aitareya Āraṇyaka and the Ṛgveda Prātiśākhya (RPr) are describing the notions of *saṁhitā* and *pada*, while ostensibly trying to explain the older terms *nirbhuja* and *pratṛṇṇa*. One can thus say with some justification that the terms *nirbhuja* and *pratṛṇṇa* express an older view regarding the relationship between Vedic texts and their Padapāṭhas. In this view, the uncut original Vedic texts were subsequently broken down into their constituent words. This is the historically earlier view.

Already by the time of the Aitareya Āraṇyaka, this old view was no longer held. While the old terms lingered on the horizon, they could be explained only in the context of the new paradigm. In this new paradigm, the Padapāṭha, the word-by-word text, represented the given or the original starting point. It was not to be derived by breaking down anything. On the contrary, it was the starting point for producing the *saṁhitā* through the application of the appropriate sandhi rules. Thus the so-called *saṁhitā*s were viewed as derived products, rather than as the pre-analysis originals. Thus, if the terms *nirbhuja* and *pratṛṇṇa* represent a view where the original uncut texts are subsequently subjected to analysis and segmentation, the terms *saṁhitā* (and *pada*) now represent a new emergent view where the segments are taken for granted, and wholes are built by combining the given segments. It is an atomistic view of language and text. If the earlier set of terms suggests that a science of segmentation has come into being, the term *saṁhitā* suggests a further stage. It suggests that the segments are now so thoroughly understood and so deeply entrenched that they can no longer be viewed as being products of a process of segmentation, but must be viewed as the given starting points or building blocks for a new process, the process of combination (*sandhi*). The Aitareya Āraṇyaka is still at some sort of a transitional stage, where a new paradigm has come into vogue, but the old terms are not yet completely dead.

These two views regarding the relationship between the Vedic texts and the corresponding 'word-by-word' versions persist in the later literature and cause untold problems in our comprehension of specific texts. For example, the well known phrase *padaprakṛtiḥ saṁhitā* which is used by both Yāska (Nirukta, 1.17) and by the Ṛgveda Prātiśākhya (2.1) is interpreted in the later traditions in at least two opposite ways:

The Saṁhitā is the basis (*prakṛti*) of the Padapāṭha.
The Saṁhitā has the Padapāṭha as its basis (*prakṛti*).

The Prātiśākhyas are primarily involved in rebuilding the Saṁhitā by combining the given words of the Padapāṭha, and, to this extent, follow the view (b), although not exclusively. There are sections of certain Prātiśākhyas which provide instruction on how to formulate certain portions of the Padapāṭha. Thus, the general Prātiśākhya tendency is:

PADAs ⇒ [SANDHI RULES] ⇒ SAṀHITĀ

The Prātiśākhyas offer Sandhi rules to combine the given separated words in the Padapāṭha. They also use terminology analogous to that found in the AĀ, namely that the separated words of the Padapāṭha represent the *śuddha* 'pure, uncontaminated, original' shapes of the words, and that the Sandhi rules applied to these words bring about *vikāra* 'transformation'. When the word-final *visarga* does not change into anything else before *ś, ṣ,* and *s*, in the opinion of Śākalya, the Vājasaneyi Prātiśākhya (3.10, *avikāraṁ śākalyaḥ śaṣaseṣu*) says that no transformation or *vikāra* takes place in this instance. For the unchanged original form, the Prātiśākhyas use the term *prakṛti* (cf. VPr 3.11, *prakṛtyā kakhayoḥ paphayoś ca*). Thus, the dominant mode in the Prātiśākhyas is that of building the Saṁhitā by combining the Padas, though the Prātiśākhya tradition is not completely unaware of the derivation of the Padas from the Saṁhitā by separation, and occasionally resorts to rules for the derivation of certain features of the Padapāṭha.

2. THE NIRUKTA OF YĀSKA

The word *nirukta* or *nirvacana* refers to a process of explaining a word by analysing it or breaking it down into its presumed meaningful components. This is the significance of the prefix *nir*. This is the dominant mode of Yāska's explanation. However, Yāska is also familiar with the term *saṁskāra*, which may be best translated by the word 'derivation' (cf. *athāpi ca eṣāṁ nyāyavān kārmanāmikaḥ saṁskāraḥ* ..., NR 1.13, and *padebhyaḥ padetarārdhān saṁcaskāra śākaṭāyanaḥ, ibid.*). Thus, *nirvacana* and *saṁskāra* seem to represent parallel, but opposite, directions. In the *nirvacana* mode, a given word is broken down to its plausible meaningful components. In the *saṁskāra* mode, given components are systematically fused to derive a higher order unit.

However, we may ask an interesting question. Does Yāska think of the original Vedic texts as 'uncut, continuous' units to be subsequently broken down - in line with the older *nirbhuja/pratṛṇṇa* view, or does he think more in terms of 'fused or tied units' versus 'unfused or separated units' - in line with the *saṁhitā* conception? As far as I can see, Yāska is primarily under the influence of the conception of *saṁhitā*. He is no longer in a paradigm where the original Vedic texts could be looked at as 'uncut' sequences.

Yāska's view expressed in his statement *padaprakṛtiḥ saṁhitā* seems to involve a certain degree of apparent contradiction. On the one hand, Yāska is referring to the notion of *padavibhāga* 'segmentation [of a text] into words' (*athāpīdam antareṇa padavibhāgo na vidyate*, NR, 1.17). On the face of it, this may be taken to suggest that before the words are separated, the *saṁhitā* text is uncut or undivided. However, in that case, Yāska's use of the word *saṁhitā* to refer to this uncut or undivided text makes little sense. On the other hand, Yāska defines *saṁhitā* as *paraḥ sannikarṣaḥ* (1.17, also cf. P. 1.4.109, *paraḥ sannikarṣaḥ saṁhitā*) '*saṁhitā* is the maximal proximity of items'. Thus, it is clear that Yāska does not seem to be thinking in terms of an 'uncut' Saṁhitā text, but rather of a text which shows fusion of its parts, while the Padapāṭha is viewed as a text showing their un-fusion or separation. It is too late in the history of linguistic theory to be able to forget about the parts and think of the original Vedic texts as partless wholes.

This relationship of parts and wholes is also clear from the following instances. In discussing the meaningfulness of the *upasargas*, Yāska refers to Śākaṭāyana's view: *na nirbaddhā upasargā arthān nirāhur iti śākaṭāyanaḥ* (NR 1.3). The terminology of this statement is very significant. The *upasargas* are either *baddha* 'tied, fused', or they can be analytically untied or unfused (*nir-baddha*). Śākaṭāyana argued that in their analytically separated state, the *upasargas* do not convey any distinct meanings. However, Yāska disagreed with Śākaṭāyana and argued that they do. We are given here a more definite indication of how Yāska viewed wholes and parts. The wholes are clearly viewed, not as being partless, but as being in a state where their parts are fused together. On the other hand, these parts can be analytically separated by the etymologists and grammarians.

This perception of *saṁhitā* is also seen elsewhere in his work. After pointing out that words (*pada*) are divided into four general classes, i.e.

nouns, verbs, pre-verbs, and particles (*nāmākhyātopasarganipātāḥ*, NR 1.12), Yāska says that, according to Śākaṭāyana and the Nairuktas, all nouns are derived from verb roots (*nāmāny ākhyātajāni*). He also refers to the view of Gārgya and others who argued that not all (*na sarvāṇi*) nouns were derived from verb-roots, suggesting that some were underived. On the one hand, the Nirukta is interested in giving a break-down of words into their components, but on the other hand, by discussing processes such as transformation, augmentation, insertion, deletion, metathesis etc., Yāska is also demonstrating, at least in a rudimentary fashion, the process of deriving or building up words from their components. Thus, the Prātiśākhyas and the Nirukta are aware of both of these dimensions, i.e. breaking down the wholes into their components and rebuilding the wholes by fusing the components. The primary emphasis of the Prātiśākhyas is on rebuilding the wholes, while the primary emphasis of the Nirukta is on giving a break-down of the fused wholes.

In my opinion, Yāska is more closely allied with the tradition of the *Padapāṭhas* and their concerns. While the *Padapāṭhas* break down continuous textual passages into their constituent words, and occasionally the words into their constituent morphological parts, Yāska is in general interested in offering etymological breakdown of words given in the *Nighaṇṭu*, and other incidental words. Essentially, his aim is to account for the meaning of a given word by referring to its possible meaningful constituents, generally a root and an affix. While he shows awareness of processes like substitution, transformation, augmenting, deletion, etc., his main target is not to systematically build up words from their given morphological constituents, but to break down given words into their constituents.

3. PĀṆINI'S VIEW OF LINGUISTIC UNITS

In various ways, in the literature considered so far, both types of processes are manifest, namely a process of analysing wholes into their components and the reverse process of combining the components to build wholes. Pāṇini's grammar seems to have taken a most uniform direction, without any flip-flopping. Pāṇini has chosen the direction of starting with the components and successively combining and processing them until the derivation of whole usable utterances, whether sentential or phrasal, is complete. In this sense, his procedure better fits the word *saṁskāra* as used by Yāska, than the old term *vyākaraṇa*, with its emphasis on analysing wholes into their parts. However, traditionally Pāṇini's grammar is referred to by the analytical term *vyākaraṇa*, rather than by the term *saṁskāra*. The grammar takes meanings of all sorts as given, as well as the morphological segments of different kinds. The grammar makes no effort whatsoever to break down wholes into their components. All the rules given by Pāṇini aim at synthesizing given components into wholes. In this sense, his is a grammar of encoding meaning by means of morphology, syntax, and phonology to produce usable sentential or phrasal utterances, and not a grammar of decoding given utterances either to discover their meaning or to break them down into their morphological or phonological constituents. The three basic morphological categories are verb roots (*dhātu*), nominal stems (*prātipadika*), and affixes (*pratyaya*). Then there are processes such as morphophonemic rules, affixation (*pratyaya*), augmentation (*āgama*),[7] and substitution (*ādeśa*). The processes are usually not context-free, but are context-sensitive. The contexts for different processes are stated in phonological, morphological, syntactic or semantic terms. On the whole, Pāṇini's basic linguistic ontology can be expressed in the following terms:

All components or parts are real.
All processes are real.
All products are real.
The processes are not reversible.

In a given derivation, each step in the process is real. One goes on applying all applicable rules until one cannot find any more rules to apply. Then this grammatical derivation as a transformational process comes to a halt. In a few specific cases, Pāṇini partly or fully suspends the effectedness of certain grammatical processes or their consequences. Such are the cases of *asiddha* 'treating something as non-effected', *asiddhavat* 'treating something as if it were non-effected', *sthānivadbhāva* 'treating a substitute as if it were the original', and *pratyayalakṣaṇa* 'treating a zero replacing an affix as if it is still an affix and carrying out further operations dependent upon an affix'. But these are explicitly stated exceptions to general rules. Thus, unless explicitly stated otherwise, all processes and their products are effected (*siddha*), real and irreversible.

Some of this linguistic ontology can be inferred from Pāṇini's statements. For example, Pāṇini uses the word *prakṛtyā* (as do the Prātiśākhyas) when the phonological shape of a given item is not altered in a given environment. Thus, the shape of a given item, before it is transformed, is regarded as its *prakṛti*, natural or original form. Pāṇini does not use the word *vikṛti* for a transformed shape, but this expression is indeed used by Kātyāyana in his Vārttikas [cf. *ekadeśavikṛtasya ananyatvāt ...*, Vt 4 on Śiva Sūtra 2, MB I, 21; Vts. 9-10 on P.1.1.56]. Indeed, Pāṇini does not seem to be averse to the notion of *vikṛti* or transformation of linguistic units as part of a grammatical derivation. Historically, it seems clear that Pāṇini had no problems with accepting that words had components and that these components may undergo change. To put it in other words, he was not averse to the notion of a *sakhaṇḍa pada* 'a word with components', and a *pada* which could undergo transformations (*vikāra*). It may also be noted that Pāṇini himself made a distinction between primitive nominal stems (cf. P.1.2.45: *arthavad adhātur apratyayaḥ prātipadikam*) and complex nominal stems (cf. P.1.2.46: *kṛt-taddhita-samāsāś ca*), and did not believe that both of these types could be or should be covered by the same definition. This means that he paid special attention to the internal structure of linguistic units. This is also true of his definition of primary and secondary verb roots (cf. P.1.3.1: *bhūvādayo dhātavaḥ* and P.3.1.32: *sanādyantā dhātavaḥ*).

For Pāṇini, the morphological items are without doubt meaningful items [except perhaps a few items like the thematic element *a* in *bhū+a+ti*, which yields *bhavati*], and he raises no question in this regard parallel to Śākaṭāyana's view cited in the Nirukta discussed above or like Kātyāyana's questions which come later (cf. *arthavattā nopapadyate kevalenāvacanāt*, Vt on P.1.4.45). The primary underived nominal stems are meaningful items (cf. P.1.4.45: *arthavad adhātur apratyayaḥ prātipadikam*). While the verb roots are defined by referring to the rootlist (*dhātupāṭha*, cf. P.1.3.1: *bhūvādayo dhātavaḥ*), on numerous occasions Pāṇini refers to meanings of roots (cf. P.1.4.33: *rucyarthānāṁ prīyamāṇaḥ*; P.3.4.72: *gatyarthākarmaka* ...). The affixes, which are always specifically prescribed, are clearly meaningful items. They are not only introduced under specific semantic conditions, Pāṇini often refers to the meaning of specific affixes (cf. *caturthyartha* 'meaning of the dative case endings' in P.2.2.62; *tumartha* 'meaning of the infinitive affix *tum*' in P.2.3.15 and 3.4.9; and *liṅartha* 'meaning of the optative=*liṅ*' in P.3.4.7). The use of the terms *padārtha* 'meaning of the word' (cf. P.1.4.96) and *anyapadārtha* 'meaning of another word, a word other than the ones used in a compound' (cf. P.2.2.24) indicates that for him *pada*s 'inflected words' are also meaningful items. The various operational levels in his grammar require an explicit recognition of successive combinations of items. One level is that of the derivation of *pada*s 'inflected words', which involves a combination or concatenation of nominal stems (*prātipadika*), verbal roots (*dhātu*), bases (*aṅga*, cf. P.1.3.13) and affixes (*pratyaya*) of various kinds including the nominal case-endings (*sup*) and finite verb endings (*tiṅ*). The next level is that of *padavidhi* 'operations relating to inflected words' (cf. P.2.1.1: *samarthaḥ padavidhiḥ*) which rests on these inflected words being semantically and syntactically related to each other (*samartha*). If the notion of *sāmarthya* covers *pada*s 'inflected words' being semantically and syntactically related to each other, Pāṇini uses the term *ākāṅkṣā* to refer to inter-sentential or inter-clausal expectancy (cf. P.8.1.35: *chandasy anekam api sākāṅkṣam*). There is no notion in Pāṇini's grammar of a unitary, indivisible sentence or utterance, as it later appears in Bhartṛhari's works. Pāṇini's linguistic ontology is clearly atomistic.

Pāṇini's treatment of linguistic units being described recognizes their character as stretches within the process of utterance or use (*prayoga*), which is intrinsically sequential. This sequentiality is expressed in many different ways, i.e. through the use of words like *pūrva* 'previous, earlier' and *para/uttara* 'following, later, subsequent' to refer to different phonological and morphological contexts.[8] The rule P.1.4.109 (*paraḥ sannikarṣaḥ saṁhitā*) offers a definition of the notion of *saṁhitā* as the maximal proximity of sounds. It seems to suggest that sounds can be close, closer, or closest to other sounds. This makes sense in view of the temporal succession of utterances. Also note that Pāṇini clearly employs terms like *prayoga* 'use or utterance' (cf. P.2.3.64, *kṛtvo 'rthaprayoge kāle 'dhikaraṇe*) and *anuprayoga* 'use or utterance after another item' (cf. P. 1.3.63, *āmpratyayavat kṛño 'nuprayogasya*). Both of these expressions refer to language as a sequential production and not as a spatial, ideational or conceptual unitary entity. Compare also the expression *aprayoga* 'non-use or non-utterance' in P. 2.1.56 (*upamitaṁ vyāghrādibhiḥ sāmānyā-prayoge*). P.1.4.110 (*virāmo 'vasānam*) defines the term *avasāna* 'end of utterance'. It gives an even clearer view of his notion of speech as a process stretched along time and a process which can be kept going or brought to a halt. The word *virāma* 'cessation, stopping' seems to refer to a notion of bringing the process of utterance to a halt. Kātyāyana in his Vārttika on P. 1.4.110 uses the opposite term *avirāma* 'non-stoppage' or continuity in his expression *hrādāvirāma* 'continuation or non-cessation of voicing'. This seems to suggest a view of speech as a process, rather than as a representation of a timeless non-sequential unit.

4. PERCEPTION OF LANGUAGE IN KĀTYĀYANA'S VĀRTTIKAS AND PATAÑJALI'S MAHĀBHĀṢYA

4.1. ABILITY TO ACCESS PHONOLOGICAL AND MORPHOLOGICAL COMPONENTS ESSENTIAL TO A GRAMMARIAN

Patañjali argues that, as a result of studying grammar, one can become an Ārtvijīna, someone who is expert in analysing language into its constituent words, accents, and syllables (cf. *yo vā imāṁ padaśaḥ svaraśo 'kṣaraśo vācaṁ vidadhāti sa ārtvijīnaḥ / ārtvijīnāḥ syāmety adhyeyaṃ vyākaraṇam /* MBh I, 3).

Wishing to demonstrate the Vedic roots of the science of grammar, Patañjali cites the following verse from the Ṛgveda:

> *catvā́ri śṛ́ṅgā tráyo asya pā́dā dvé śīrṣé saptá hástāso asya /*
> *trídhā baddhó vṛṣabhó roravīti mahó devó mártyā́ ā́ viveśa //*
> RV 4.58.3.
>
> It has four horns, three feet, two heads, and seven hands. The Bull, tied in three places, roars out. The great god has entered the mortals.

This is analysed by Patañjali in terms of constituents of speech: The four horns refer to the four types of words, i.e. nouns, verbs, pre-verbs, and particles. The three feet refer to the three tenses, i.e. past, present, and future. The seven hands refer to the seven cases [cf. *catvāri śṛṅgāṇi catvāri padajātāni nāmākhyātopasarganipātāś ca / trayo asya pādāḥ / trayaḥ kālā bhūtabhaviṣyadvartamānāḥ / ... sapta hastāso asya sapta vibhaktayaḥ* / MBh I, 3]. Historically there is little chance of Patañjali's interpretation being correct. However, besides tracing ideas in grammar to their Vedic roots, it shows his willingness to accept components of speech as significant entities. Citing another Vedic passage [i.e. *catvā́ri vā́k párimitā padā́ni*, RV 1.164.45] which refers to four steps of speech, Patañjali again points out that these four steps refer to the fourfold division of speech into nouns, verbs, pre-verbs, and particles [cf. *catvāri padajātāni nāmākhyātopasarganipātāś ca*, MBh I,3]. This interpretation, though historically unlikely to be valid, again shows his willingness to accept parts of speech as significant components.

4.2. THE NOTION OF VIKṚTI 'TRANSFORMATION' IN KĀTYĀYANA AND PATAÑJALI

While Pāṇini uses the term *prakṛti,* as in *prakṛtyā*, to refer to the derivationally prior state before changes are applied, Kātyāyana and Patañjali use the term *vikṛta* to refer to the changed segment. An argument is offered that a linguistic item which is partially transformed (*ekadeśavikṛta*) is still to be treated like the original item [cf. *ekadeśavikṛtasyā-nanyatvāt plutyādayaḥ*, Vt 4 on Śiva Sūtra 2. MBh I, 21; also *ekadeśavikṛtasyopasaṁkhyānaṁ kartavyam*, Vt 9, on P. 1.1.56 (*sthānivad ādeśo 'nalvidhau*). *kim prayojanam / pacatu pacantu / tiṅgrahaṇena grahaṇaṁ yathā syāt* / MBh *ekadeśavikṛtasyānanyatvāt siddham*, Vt 10 on P. 1.1.56]. The commentators offer an example that a dog with his tail cut off is still a dog, and does not become a horse or an ass. Besides the grammatical relevance of this discussion, it points to issues of change and identity, and their compatibility within a certain frame of reference of commonsense (*loke*).

4.3. THE NOTION OF VIKṚTI LINKED TO IMPERMANENCE OF LANGUAGE, AND RAISES CONCERNS

However, change and identity are not compatible within more rigid metaphysical frameworks, and this becomes apparent in the following discussion. Kātyāyana says that one could have argued that an item partially transformed does not yet lose its identity. But such an acceptance would lead to non-eternality (*anityatvam*) of language, and that is not acceptable. Patañjali asserts that words are eternal (*nitya*), and that means they must be absolutely free from change or transformation and fixed in their nature [cf. *anityavijñānaṁ tu tasmād upasaṁkhyānam*, Vt 11. *anityavijñānaṁ tu bhavati / nityāḥ śabdāḥ / nityeṣu nāma śabdeṣu kūṭasthair avicālibhir varṇair bhavitavyam anapāyopajanavikāribhiḥ / tatra sa evāyaṁ vikṛtaś cetyetan nityeṣu nopapadyate* / MBh I, 136]. If words are truly eternal, one cannot then say that something was transformed and is yet the same. This discussion shows how the common-sense views held by Pāṇini are being challenged by the more rigid ontologies which have developed by the time of Kātyāyana and Patañjali. This points to the emerging ideological shifts in philosophical traditions, which make their headway into the tradition of grammar, and finally lead to the development of newer conceptions within the tradition of grammar.

4.4. PERMANENCE OF LANGUAGE AND PROBLEMS OF PHONOLOGY

In trying to figure out how the emerging doctrine of *nityatva* 'permanence, immutability' of language causes problems with the notion of transformation (*vikāra*) and how these problems are eventually answered by developing new concepts, we should first focus on the phonological issues raised by this doctrine. These issues are minimally of two kinds, i.e. temporal fixity or flexibility of individual sounds, and the compatibility of the notion of sequence of sounds or utterance as a process stretched in time.

From within the new paradigm of *nityatva*, Kātyāyana concludes that the true sounds (*varṇa*) are fixed in their nature in spite of the difference of speed of delivery. The speed of delivery (*vṛtti*) results from the slow or fast utterance of a speaker (*vacana*), though the true sounds themselves are permanently fixed. Here, Kātyāyana broaches a doctrine which is later developed by Patañjali, and more fully by Bhartṛhari. It argues for a dual ontology. There are the fixed true sounds (*varṇa*), and then there are the uttered sounds (*vacana* 'utterance'), [cf. *siddhaṁ tv avasthitā varṇā vaktuś cirāciravacanād vṛttayo viśiṣyante*, Vt 5 on P. 1.1.70, MBh I, 181]. Also we must keep in mind that Kātyāyana is not denying the temporality of these true sounds, but simply saying that they are fixed in their duration, and the differences of speed are caused by secondary features.

Kātyāyana, it must be clearly pointed out for a better understanding of the historical development, is different from Pāṇini in his concerns, and yet does not have the developed terminology of Patañjali to refer to this distinction in linguistic ontology. It is Patañjali who uses, for the first time as far as we know, the term *sphoṭa* to refer to Kātyāyana's 'true sounds which are fixed' (*avasthitā varṇāḥ*) and the term *dhvani* 'uttered sounds'. Kātyāyana's Vārttikas do not use either of these terms.[9] Patañjali adds an important comment to Kātyāyana's discussion. He says that the real sound (*śabda*) is thus the *sphoṭa* 'the sound as it initially breaks out into the open', and the quality [of increase or speed] of the sound is part of *dhvani* 'sound as it continues'. This is my interpretation of these two terms. The term *sphoṭa* refers to something like exploding or coming into being in a bang. Thus it refers to the initial production of sound. On the other hand, the stretching of that sound suggested by the term *śabdaguṇa* seems to refer to the dimension of continuation. This is also suggested by the analogy used by Patañjali. Having struck the kettledrum, a person can walk twenty steps, or thirty steps, or forty steps. The notion of *sphoṭa* is comparable to the initial striking of the drum, which remains identical in all three situations. However, depending upon the energy with which the drum was struck, the reverberations will continue for a shorter or a longer duration. Depending upon the duration of those reverberations, a person may walk twenty, thirty, or forty steps. That does not alter the initial striking of the drum. Similarly, the linguistic sounds are always fixed in their original production, but depending on the habits of the speaker, the sounds may remain audible for longer or shorter duration. Patañjali means to say that it is the same sound, but it may remain audible for different durations [cf. *evaṁ tarhi sphoṭaḥ śabdo dhvaniḥ śabdaguṇaḥ / katham / bheryāghātavat / tadyathā bheryāghātaḥ / bherīm āhatya kaścid viṁśati padāni gacchati kaścit triṁśat kaścic catvāriṁśat / sphoṭaś ca tāvān eva bhavati dhvanikṛtā vṛddhiḥ / dhvaniḥ sphoṭaś ca śabdānāṁ dhvanis tu khalu lakṣyate / alpo mahāṁś ca keṣāṁcid ubhayaṁ tatsvabhāvataḥ //* MBh I, 181].

Thus, Kātyāyana and Patañjali, while accepting the fixed nature of sounds do not deny their intrinsic temporality. This raises the next problem which they must face. Kātyāyana, on P. 1.4.109 (*paraḥ sannikarṣaḥ saṁhitā*), discusses an alternative formulation of the definition of *saṁhitā* as *paurvā-*

paryam akālavyapetaṁ saṁhitā: '*saṁhitā* is the sequence of sounds without the intervention of a duration of time'. Kātyāyana's Vārttika 9 (*paurvāparyam akālavyapetaṁ saṁhitā cet pūrvāparābhāvād asaṁhitaṁ*) says that with such a definition, one can never have *saṁhitā*. Patañjali clarifies that we cannot have an instance of *saṁhitā*, because we cannot have a sequence of sounds to begin with (*na hi varṇānām paurvāparyam asti*). Why can we not have a sequence? Kātyāyana's next Vārttika 10 (*ekaikavarṇavartitvād vāca uccarita-pradhvaṁsitvāc ca varṇānām*) says that one cannot have a sequence of sounds, because the process of speech proceeds sound-by-sound, and that sounds perish as soon as they are uttered. Thus, one cannot have two simultaneous sounds to relate to each other. Patañjali discusses the pronunciation of the word *gauḥ*, which has the three sounds *g*, *au*, and *ḥ*. He points out that one cannot utter two sounds at the same time (*ekaikavarṇavartinī vāk/ na dvau yugapad uccārayati*). While one is uttering *g*, one is not at the same time uttering *au* and *ḥ*. The same is true of the other sounds. Since the sounds perish as soon as they are uttered, a sound cannot have another co-existent companion (*uccaritaḥ pradhvastaḥ / athāparaḥ prayujyate na varṇo varṇasya sahāyaḥ*).

This is a dilemma which Pāṇini did not explicitly face up to and hence did not worry about providing a solution to. Kātyāyana himself points out all these difficulties, which almost remind us of the debates found in the Buddhist texts discussing momentariness of all entities. However, Kātyāyana's solution to these philosophical problems is a return to commonsense: Since the notions of *saṁhitā* and *virāma* are known in the world, one does not need to worry about defining them (*saṁhitāvasānayor lokaviditatvāt siddham*, Vt 8 on P. 1.4.110). Thus, Kātyāyana has not provided a philosophical solution to this dilemma.

Patañjali cites a verse which offers a philosophical solution to this philosophical dilemma. An intelligent person, with his flexible intelligence able to comprehend multiple objects simultaneously, having perceived all the different actions, comprehends their sequentiality in his mind. The sequentiality of sounds is in terms of their comprehension [cf. *buddhau kṛtvā sarvāś ceṣṭāḥ kartā dhīras tanvan nītiḥ / śabdenārthān vācyān dṛṣṭvā buddhau kuryāt paurvāparyam // buddhiviṣayam eva śabdānāṃ paurvāparyam* / MBh I, 356]. Because, the real uttered sounds are only momentary entities, one cannot think of a sequence of any two of them. However, Patañjali seems to suggest that one can pull together impressions of all the uttered sounds and then think of a sequence in this mentally constructed image of a word. Elsewhere, Patañjali says that a word is perceived through the auditory organ, discerned through one's intelligence, and brought into being through its utterance [cf. *śrotropalabdhir buddhinirgrāhyaḥ prayogeṇābhijvalitaḥ ākāśadeśaḥ śabdaḥ*, MBh I, 18]. While Patañjali's solution overcomes the transitoriness of the uttered sounds, and the resulting impossibility of a sequence, there is no denial of sequentiality or perhaps of an imprint of sequentiality in the comprehended word, and there is indeed no claim to its absolutely unitary or partless character. Patañjali in fact means to provide a solution to the perception of sequentiality through his ideas of a mental storage of comprehension. But at the same time, this mental storage and the ability to view this mental image allows one to overcome the difficulty of non-simultaneity and construct a word or a linguistic unit as a collection of perceived sounds or words, as the case may be. Kātyāyana and Patañjali specifically admit the notion of *samudāya* 'collection' of sounds to represent a word and a collection of words to represent compound expressions [cf. on P.2.2.29: Vārttika 7 (*samudāyāt siddham iti cen naikārthatvāt samudāyasya*). MBh says: *samudāyāt siddham iti cet tan na / kiṁ kāraṇam / ekārthatvāt samudāyasya / ekārthā hi samudāyā bhavanti / tad yathā / śataṃ yūthaṃ vanam iti* / MBh I, 432]. The discussion shows that Kātyāyana and Patañjali are not averse either to the notion of sequentiality of sounds in a word and of words in a complex unit, or to viewing them as collections. Contradictory as it may seem, Patañjali's recourse to a mental storage and a mental image allows one to perceive both of these dimensions. Thus, while the ontology of physical sounds does not permit their co-existence, their mental images do allow it, and once they can be perceived as components of a collection, one also recognizes the imprint of the sequence in which they were perceived. Neither Kātyāyana nor Patañjali explicitly claim any higher ontological status to the word-images, and they remain simply an epistemic explanation. However, the very acceptance of such word-images opens up numerous explanatory possibilities.

4.5. NEED TO EXPLAIN SUBSTITUTION WITHIN THE FRAMEWORK OF NITYATVA OF LANGUAGE. PATAÑJALI'S SOLUTION: SUBSTITUTION AS A NOTIONAL CHANGE (*BUDDHIVIPARIṆĀMA*)

On P.1.1.56, Kātyāyana brings up a philosophical objection to the notion of substitution. If substituting *y* for *x* is viewed as a replacement of one item by another, then one may have to say that the substituendum is an item which existed up to a point and then ceased to exist, and that the substitute is an item which did not exist up to a point and came into being at a certain point. Kātyāyana and Patañjali argue that if one accepts the doctrine of permanence (*nityatva*) of linguistic items, these ideas of substituendum and substitute do not hold up. Something that is permanent cannot be said to cease to exist or to come into being [cf. *anupapannaṁ sthānyādeśatvaṁ nityatvāt*, Vt 12, on

P. 1.1.56. *sthānī hi nāma yo bhūtvā na bhavati / ādeśo hi nāma yo 'bhūtvā bhavati / etac ca nityeṣu śabdeṣu nopapadyate yat sato vināśaḥ syād asato vā prādurbhāva iti //* MBh]. One of the explanations provided by Kātyāyana is: *kāryavipariṇāmād vā siddham*, Vt 14 on P.1.1.56. What Kātyāyana means by this Vārttika is not fully clear, though it seems to refer to a notion of transformation of the substituendum into the substitute. Patañjali presumably sees difficulties in the notion of transformation of a word if it is permanent and offers what is probably a new interpretation of the Vārttika in line with the arguments for a level of conceptual construction: 'What is this transformation of the effect? The word 'effect' refers to concepts. It is the concept that is transformed [and not the linguistic unit itself]. ... [Referring to the suppletion of the root *as* by *bhū* taught by P.2.4.52 (*aster bhūḥ*), Patañjali continues:] Here [initially] the root *as* is taught without any specification. Thus, he [= the user of the grammar] came to think that *as* occurs everywhere. Through P.2.4.52, he comes to think of *bhū* instead of *as* [in certain specific contexts]. In its own context, the root *as* is [actually] permanent, and so is the root *bhū* permanent [in its own context]. However, [through Pāṇini's rule of suppletion] his [= student's] notion is transformed', [cf. *kim idaṃ kāryavipariṇāmād iti / kāryā buddhiḥ sā vipariṇamyate / ... evam ihāpy astir asmā aviśeṣeṇopadiṣṭaḥ / tasya sarvatrāstibuddhiḥ prasaktā / so 'ster bhūr bhavatīty astibuddhyā bhavatibuddhim pratipadyate / tataḥ sa paśyati buddhyāstiṁ cāpakṛṣyamāṇaṁ bhavatiṁ cādhīyamānam / nitya eva ca svasmin viṣaye 'stir nityo bhavatir buddhis tv asya vipariṇamyate /* MBh I, 137]. Again what we see is that Patañjali explicitly brings in a two-level ontology: physical and notional.

If words are permanent (*nitya*) can one add an augment to an existing unit? If linguistic units are permanent, then should not the sounds of those units be immutable, unchanging, and not be subjected to loss, additions, or change? The augments are indeed thought to be items which newly come into being. If so, do they not contradict the doctrine of permanence of linguistic units? [cf. *yuktaṁ punar yan nityeṣu nāma śabdeṣv āgamaśāsanaṁ syāt / na nityeṣu nāma śabdeṣu kūṭasthair avicālibhir varṇair bhavitavyam anapāyopajana-vikāribhiḥ / āgamaś ca nāmāpūrvaḥ śabdopajanaḥ /*]. Patañjali then asks whether it is appropriate to have substitutes? The question is answered by saying that substitutes are really independent separate words and there is no real transformation of a word involved, compare 'distribution' in Harris (1954). What is involved is simply a notional change. Instead of thinking about item *x* we start thinking about item *y* [cf. *atha yuktaṁ yan nityeṣu śabdeṣv ādeśāḥ syuḥ / bāḍhaṁ yuktam / śabdāntarair iha bhavitavyam / tatra śabdāntarāc chabdāntarasya pratipattir yuktā /*]. With such a notional explanation of substitution, Patañjali then accounts for augmentation as a form of substitution: substitution of a word which does not have a certain component with a word which contains that component, instead of changing or adding anything to anything [cf. *ādeśās tarhīme bhaviṣyanty anāgamakānām sāgamakāḥ / tat katham / sarve sarvapadādeśā dākṣīputrasya pāṇineḥ / ekadeśavikāre hi nityatvaṁ nopapadyate //* MBh on P. 1.2.20, I,75].

Kātyāyana and Patañjali seem to have believed that the notion of change or transformation of parts of words was contradictory to the doctrine of *nityatva* 'permanence' of language. However, they were not averse to the notion of substitution, if it were understood as a substitution not of a part of a word by another part, but of a whole word by another word, and this especially as a conceptual rather than ontological shift. This leads them to construe Pāṇini's procedures in a very different way. Thus, in going from *bhavati* to *bhavatu*, Pāṇini not only accepts that there are components to the forms *bhavati* and *bhavatu*, but he prescribes the change of the *i* of *ti* to *u* [cf. P.3.4.86: *er uḥ*]. Thus, *i* changes to *u*, leading to the change of *ti* to *tu*, and this consequently leads to the change of *bhavati* to *bhavatu*. Pāṇini does not seem to have been bothered by the kind of philosophical questions which have perplexed Kātyāyana and Patañjali. It is tempting to suggest that the emerging traditions of Buddhism may have forced Kātyāyana and Patañjali to think of new philosophical problems. For Kātyāyana and Patañjali, the above atomistic and transformational understanding of Pāṇini's procedure goes contrary to the doctrine of *nityatva* 'permanence' of words. Therefore, they reconceptualize the procedure to suggest that it is actually the substitution of the whole word *bhavati* by a whole another word *bhavatu*, each of these two words being eternal in its own right. This is simply a notional change and not an ontological change for these two grammarians. However, the discussion seems to imply a sort of unitary value to the words, whether notional or otherwise, and this eventually leads to a movement toward a kind of *akhaṇḍa-pada-vāda* 'the doctrine of partless words'. While one must admit that the seeds for such a conception may be traced in these discussions in the Mahābhāṣya, Patañjali is actually not arguing so much against words having parts, as against the notion of change or transformation. Kātyāyana and Patañjali both in fact accept words as *samudāya*s or collections, and have not offered any arguments against this characterization. In view of the doctrine of permanence of linguistic units, Kātyāyana and Patañjali also reconfigure the relationship between a compound (*samāsa*) and the phrase (*vākya*) to which it is derivationally linked in Pāṇini's system. Kātyāyana argues that there is no reason to explicitly say that a phrase is optionally transformed into a compound, because both the compound and the phrase, as linguistic units, are eternally present, and therefore available to the user [cf. on P. 2.1.1 (*samarthaḥ padavidhiḥ*) *vāvacanānarthakyaṁ ca svabhāvasiddhatvāt*, Vt 2. *svabhāvataś caitad bhavati vākyaṁ samāsaś ca*, MBh I, 364]. Thus, one need not view a compound as a transformation of a phrase. Both the compound and the phrase are *nitya* 'permanent' in their own domains. Pāṇini himself was not averse to the notion of transformation. For him, a compound is derived from the underlying structure of a phrase, and an option simply refers to the option of going forth with the transformation.

4.6. KĀTYĀYANA AND PATAÑJALI: WORDS AND SENTENCES AS COLLECTIONS

To restate the conclusion of the previous section, Kātyāyana and Patañjali view a word as a collection of sounds. Besides using the term *samudāya* for such a collection, they also use the word *varṇasaṁghāta.*[10] [cf. *saṁghātārthavattvāc ca*, Vt 12 on Śiva Sūtra 5, MBh I, 30; and *saṁghātasyaikārthyāt subabhāvo varṇāt*, Vt 13 on Śiva Sūtra 5, MBh I, 31]. In these Vārttikas Kātyāyana uses the term *saṁghāta*, while Patañjali further paraphrases Kātyāyana's arguments by using the terms *avayava* 'part, constituent' and *samudāya* 'collection'. Perhaps there is some terminological shift or preference manifest here. Patañjali also uses the term *samudāya* 'collection' in his discussion of the ontological nature of *dravya* 'substance' as a collection of qualities (*guṇasamudāya*), MBh (on P.4.1.3) II, 200.

This philosophically explicit notion of wholes as *saṁghātas* is not overtly stated by Pāṇini, though it seems to be in agreement with his general procedures. Where does the explicit notion come from? There is a possibility that the grammarian Vyāḍi, who preceded both Kātyāyana and Patañjali, and is said to be the author of the now lost grammatical work *Saṁgraha*, used this notion first, and it was subsequently taken up by Kātyāyana and Patañjali. There are verses quoted in the Vṛtti on the VP (I. 24-26) from a work called Saṁgraha. Whether this Saṁgraha is the same as Vyāḍi's now lost work cannot be verified. The relevant verse is as follows:

> *arthāt padaṁ sābhidheyam padād vākyārthanirṇayaḥ* /
> *padasaṁghātajaṁ vākyaṁ varṇasaṁghātajam padam* //[11]

> On account of its meaning, a word becomes expressive. From [the meaning of] words, one discerns the meaning of a sentence. A sentence comes about from a collection of words, and a word comes about from a collection of sounds.

Kātyāyana and Patañjali argue that words are built by putting together sounds, and that while the words are meaningful, the component sounds are not meaningful. To illustrate the idea that a collection can accomplish a certain purpose which cannot be served by the components of that collection taken severally, Patañjali offers the example of a chariot. While a chariot as a whole is capable of movement, its parts separately are not so capable,[12] [cf. *saṁghātārthavattvāc ceti ced dṛṣṭo hy atadarthena guṇena guṇino 'rthabhāvaḥ*, Vt 11, on P. 1.2.45. *ime punar varṇā atyantāyaivānarthakāḥ / yathā tarhi rathāṅgāni vihṛtāni pratyekaṁ vrajikriyām praty asamarthāni bhavanti tatsamudāyaś ca rathaḥ samarthaḥ evam eṣāṁ varṇānāṁ samudāyā arthavanto 'vayavā anarthakā iti* / MBh I, 220]. The notion of *saṁghāta* is widely used in many grammatical discussions [cf. on P. 6.4.49: Vt 1: *yalope varṇagrahaṇaṁ ced dhātvantasya pratiṣedhaḥ,* and Vt 2: *saṁghātagrahaṇaṁ cet kyasya vibhāṣāyāṁ doṣaḥ*, MBh II, 201. Also see Vārttikas on P. 7.3.50: Vārttika 1: *ṭhādeśe varṇa-grahaṇaṁ ced dhātvantasya pratiṣedhaḥ,* and Vt 2: *saṁghātagrahaṇaṁ ced uṇādimāthitikādīnām pratiṣedhaḥ*, MBh II, 328].

The notion of a word as a collection (*saṁghāta*) applies not only in the sense that it is a collection of sounds, but also in the sense that complex formations are collections of smaller morphological components. Kātyāyana and Patañjali refer to a combination of an *upasarga* and a verb root, and a combination of a stem/root and an affix as a *saṁghāta*, and discuss the question of whether a certain meaning is conveyed by a component or by the collection as a whole. They also discuss the question of whether the morphological components can occur and be meaningful by themselves [cf. *kriyāvacana upasargapratyayapratiṣedhaḥ*, Vt 3. *saṁghātenārthagateḥ*, Vt 4. on P. 1.3.1. *saṁghātena hy artho gamyate saprakṛtikena sapratyayakena sopasargeṇa ca* / MBh I, 254]. At no time does the notion of a partless or an indivisible word come into the picture. Kātyāyana and Patañjali explicitly discuss the meanings of the morphological components of complex formations and the mutual relationships between these meanings. The *upasarga* is said to modify the meaning of a verb-root [cf. *kriyāviśeṣaka upasargaḥ*, Vt 7, on P. 1.3.1, MBh I, 256]. On this Vārttika, Patañjali discusses the distinct meanings of the *upasargas* [cf. *adhir uparibhāve vartate / ... pro 'yaṁ dṛṣṭāpacāra ādikarmaṇi vartate* /].

4.7. DETERMINING THE MEANING OF MORPHOLOGICAL COMPONENTS: AN EXPLICIT PROCEDURE OF ABSTRACTION

We have seen above that Kātyāyana and Patañjali regard words as being collections of sounds as well as collections of smaller morphological components. While the meaning of words as wholes seems more obvious, how does one know that morphological or phonological components are or are not meaningful? An explicit formulation of a method of segmentation of larger wholes into their meaningful components is offered by Kātyāyana and Patañjali for the first time in the history of Indian linguistic thought. Segmentation as such was indeed practised for many centuries before these grammarians, as we have seen earlier. Clearly, Pāṇini's work shows the results of some systematic methods of segmentation. And yet, we need to recognize that Pāṇini does not explicitly describe the method of segmentation which he and his predecessors practised. That honour goes to Kātyāyana and Patañjali, see S. Acharya (1990) for a useful, though historically somewhat uninformed discussion. They inherit the segmentation of language as a *fait accompli*. Since the morphological segmentation is thus inherited, what Kātyāyana and Patañjali offer are not truly methods for arriving at the segmentation *per se*, but a method of determining the meaning to be apportioned to such segments. This is a very important distinction between the kind of methods and discovery procedures which were attempted and developed by the modern structural linguists like Zellig Harris (1951), and the discussions one finds in the works of Kātyāyana and Patañjali. These grammarians rarely, if ever, question the actual segmentation provided by Pāṇini. The primary focus is on the assignment of meaning to these segments, where Pāṇini's rules leave some room for deliberation.

Kātyāyana, for the first time as far as we know, raises some critical questions. If the meaning of words can be recovered from the observation of their usage, how does one know the meaning of the components of words? There is an assumption in Kātyāyana that one can find words as wholes used by themselves at the level of communication, and one can indeed find examples of single-word utterances. However, since the components of words are never used by themselves, how does one know that they have any meaning? The discussion occurs in relation to Pāṇini's definition of a nominal stem (*prātipadika*). A nominal stem is defined as a meaningful item which is not a root or an affix (P.1.2.45: *arthavad adhātur apratyayaḥ prātipadikam*). Here Pāṇini takes for granted the meaningfulness of a nominal stem. However, Kātyāyana wants to know how one arrives at such a conclusion. The prima-facie argument presented by Kātyāyana is that morphological components such as nominal stems have no meanings, because such items are never used by themselves, and hence one can never recover their meanings [cf. *arthavattā nopapadyate kevalenāvacanāt*, Vt 7, on P. 1.2.45]. Patañjali adds a comment that meaning is not understood from a nominal stem like *vṛkṣa*, but only from a stem combined with an affix [cf. *na kevalena vṛkṣaśabdenārtho gamyate / kena tarhi / sapratyayakena* / MBh I, 219]. Since, stems are invariably bound with affixes, neither the stems nor the affixes are ever used by themselves [cf. *na vā pratyayena nityasambandhāt kevalasyāprayogaḥ*, Vt 8, on P. 1.2.45]. Patañjali clarifies this argument by saying that only the wholes built up by combining stems and affixes are used to communicate meanings, and hence the components by themselves are unknown [cf. *samudāyasyārthe prayogād avayavānām aprasiddhir iti* / MBh I, 219]. This is an extremely important discussion. Kātyāyana and Patañjali do not propose that a *pada* is partless (*akhaṇḍa*), but that its components are invariably bound to each other (*nityasambandha*) and therefore do not occur in usage separately. Kātyāyana answers the dilemma involved in discovering the unknown or rather unfamiliar meaning of morphological components by providing a method to isolate the meaning of morphological components from the known meaning of wholes: 'The result is successfully achieved through the method of [observing] concurrent occurrence (*anvaya*) and concurrent non-occurrence (*vyatireka*)', [cf. *siddhaṁ tv anvayavyatirekābhyām*, Vt 9, on P. 1.2.45].[13] Patañjali explains:

> *siddhaṁ etat / katham / anvayād vyatirekāc ca / ko 'sāv anvayo vyatireko vā / iha vṛkṣa ity ukte kaścic chabdaḥ śrūyate vṛkṣaśabdo 'kārāntaḥ sakāraś ca pratyayaḥ / artho 'pi kaścid gamyate mūlaskandhaphalapalāśavān ekatvaṁ ca / vṛkṣāv ity ukte kaścic chabdo hīyate kaścid upajāyate kaścid anvayī / sakāro hīyata aukāra upajāyate vṛkṣaśabo 'kārānto 'nvayī / artho 'pi kaścid dhīyate kaścid upajāyate kaścid anvayī / ekatvaṁ hīyate dvitvam upajāyate mūlaskandhaphalapalāśavān anvayī / te manyāmahe yaḥ śabdo hīyate tasyāsāv artho yo 'rtho hīyate, yaḥ śabda upajāyate tasyāsāv artho yo 'rtha upajāyate yaḥ śabdo 'nvayī tasyāsāv artho yo 'rtho 'nvayī / ... varṇitārthavattānvayavyatirekābhyām eva / ... prakṛtiḥ prakṛty-arthe vartate pratyayaḥ pratyayārthe* / MBh I, 219-220.
>
> [The meaningfulness of uninflected stems] is established. How is this done? [This is done] on the basis of concurrent occurrence (*anvaya*) and concurrent non-occurrence (*vyatireka*). What is this concurrent occurrence and concurrent non-occurrence? When one says *vṛkṣas* 'a tree', one hears an item *vṛkṣa* ending in *a*, and the affix *s*. One also understands some meaning, i.e. something having roots, fruits, leaves etc. and singularity. When someone says *vṛkṣau* 'two trees', some [part of the previous] word is lost, some [new part] is added and some [part] continues to occur. [The affix] *s* is lost, [the affix] *au* is added, and the word *vṛkṣa* ending in *a* continues to occur. [The meaning element of] singularity is lost, [that of] duality is added, and [the meaning element of] something having roots, fruits, leaves etc. continues to occur. Thus, we believe that the meaning which is lost is the meaning of the word which is lost. The meaning which is added is the meaning of the word which is added. The meaning which continues to occur is the meaning of the word which continues to occur.

The above discussion occurs in connection with determining the meaningfulness of a nominal stem, though it has general implications. Are affixes by themselves meaningful? On P.1.2.46 (*kṛt-taddhita-samāsāś ca*) which defines a secondary nominal stem as being either a compound, a *kṛt*, or a *taddhita*, Kātyāyana suggests that Pāṇini should not have just referred to *kṛt* and *taddhita* affixes. He should have said *kṛdanta* and *taddhitānta* 'formations ending in *kṛt* and *taddhita* affixes'. On this discussion, Patañjali says that since the notion of meaningfulness continues from the previous rules, one would naturally think of words ending in such affixes and not of the affixes themselves. The affixes are not meaningful by themselves [cf. *yad apy ucyate kṛt-taddhita-samāsāś cetyantagrahaṇaṁ kartavyam iti / na kartavyam / arthavad iti vartate / kṛttaddhitāntaṁ caivārthavan na kevalāḥ kṛtas taddhitā vā* / MBh I, 319]. Kaiyaṭa's comment on this passage brings out the two levels of meaningfulness involved: conventional level of meaningfulness of items (*laukikī arthavattā*) applies only to wholes and not to components, but the analytical level meaningfulness (*anvayavyatirekagamyā arthavattā*) relates to the morphological components, and that Patañjali's discussion needs to be understood within the context of such distinct levels [cf. *nanu yady arthavattā laukiky āśrīyate sā padasyaiva na tu kṛt taddhitāntasyāpi / padasyaiva loke prayogārhatvāt / anvayavyatirekagamyā tv arthavattā kevalānām api kṛttaddhitānām astīti kimucyate - na kevalā iti / evaṁ tarhy arthavadgrahaṇānuvṛttisāmarthyāl laukikārthapratyāsanno 'bhivyaktataro yo 'rthaḥ pratyayānteṣu lakṣyate sa āśrīyate ity adoṣaḥ /*].

This kind of analysis has been carried out systematically and in broad generic terms by Kātyāyana and Patañjali. Kātyāyana observes that in many forms we have the same stem/root (*prakṛti*), but different affixes (*pratyaya*), while in other forms we have different stem/roots, with the same affix. As examples, Patañjali offers the following cases. In *pacati* and *paṭhati*, we have different roots with the same affix. On the other hand, in the forms *paktā*, *pacana*, and *pāka*, we have the same root with different affixes. Here too, Kātyāyana points out that the meaningfulness of these morphological components is to be established through the method of *anvaya* and *vyatireka* [cf. *pratyayārthasyāvyatirekāt prakṛtyantareṣu*, Vt 6a, on P. 1.3.1. *pratyayārthasyāvyatirekāt prakṛtyantareṣu manyāmahe dhātur eva kriyām āheti / pacati paṭhati / prakṛtyartho 'nyaś cānyaś ca pratyayārthaḥ sa eva*, MBh *dhātoś cārthābhedāt pratyayāntareṣu*, Vt 6b, on P. 1.3.1. *dhātoś cārthābhedāt pratyayānta-reṣu manyāmahe dhātur eva kriyām āheti / paktā pacanam pāka iti / pratyayārtho 'nyaś cānyaś ca bhavati prakṛtyarthaḥ sa eva / katham punar jñāyate 'yam prakṛtyartho 'yam pratyayārtha iti*, MBh *siddhaṁ tv anvayavyatirekābhyām*, Vt 6, on P. 1.3.1. MBh I, 255].

This discussion points to important levels of abstraction. At one stage, only a *pada* is meaningful and the meaningfulness of stems and affixes is secondarily derived through the method of *anvayavyatireka*. The meaningfulness of the components of complex stems is derived by further application of *anvayavyatireka*.

At the beginning of this discussion, I pointed out that Kātyāyana and Patañjali are taking for granted the morphological segmentation as provided by Pāṇini, and discussing only the question of the proper assignment of meaning. They have not questioned the morphological segmentation offered by Pāṇini. For example, in the above discussion, Patañjali takes for granted that in the form *vṛkṣas*, we 'hear' the stem *vṛkṣa* ending in *a*, and the affix *s*. Similarly, in the form *vṛkṣau*, he says that we 'hear' the stem *vṛkṣa* ending in *a*, and the affix *au*. Here, the extent of morphology which is internalized and taken for granted by Patañjali is revealing. He does not, for example, raise the question of alternative possibilities of division for these words, such as *vṛkṣ+au* versus *vṛkṣa+au*, and argue for one and against another. Similarly, the categorization of *s* and *au* as affixes is also taken for granted. We know from other sources that grammarians did differ on such matters from each other. For example, Pāṇini sets up the root *as* 'to be', which is fully preserved in the form *asti* [< *as+ti*]. But, in the form *santi*, the Pāṇinian grammar starts with the sequence *as+anti*, and then deletes the initial vowel of the root [cf. P.6.4.111: *śnasor allopaḥ*]. On the other hand, we are told that the pre-Pāṇinian grammarian Āpiśali set up the same root as a mere single consonant *s*, which is fully preserved in the form *santi* [< *s+anti*]. But, for Āpiśali's system, one needs to add an initial augment *a* to the root *s* in the form *asti* [< *as+ti* < *(a+s)+ti* < *s+ti*]. However, Patañjali does not carry out an open-ended investigation of such alternative possibilities, and accepts Pāṇinian morphology without questioning. Yet there was indeed a residue of words left out by Pāṇini, where either the stem/root or the affix, or both, were not specified by him. In such cases, Patañjali needed to suggest a discovery procedure to find out these morphological components. Here he refers to the older theory found in Yāska's Nirukta, namely that all nouns are derived from verb-roots. Referring to this theory which was accepted by Yāska and Śākaṭāyana, and which apparently formed the basis of the so-called Uṇādi Sūtras, which supplement Pāṇini's grammar, Patañjali says that by looking at the stem, if it is known, one can try to figure out (*ūhya*, *ūhitavya*) the affix, and by looking at the affix, if it is known, one can try to figure out the stem [cf. *nāma ca dhātujam āha nirukte vyākaraṇe śakaṭasya ca tokam /. ... atha yasya viśeṣapadārtho na samutthitaḥ kathaṁ tatra bhavitavyam / yan na viśeṣapadārthasamuttham pratyayataḥ prakṛteś ca tad ūhyam / prakṛtiṁ dṛṣtvā pratyaya ūhitavyaḥ pratyayaṃ ca dṛṣṭvā prakṛtir ūhitavyā / saṁjñāsu dhāturūpāṇi pratyayāś ca tataḥ pare / kāryād vidyād anubandham etac chāstram uṇādiṣu* // MBh II, 139]. Here one sees how the inherited theories shape the perception of Patañjali. Patañjali is not out to provide a brand new theory of segmentation from scratch. He accepts the segmentations provided by Pāṇini, and, for the remainder, he accepts the guidance provided by Yāska's Nirukta. This guidance from Yāska leads him to propose that in trying to find segmentation of those nominal words, which are not derived through Pāṇini's rules, one can look for some verb root in the initial portion of a word and then try to identify the remaining part as some sort of an affix. This general idea is followed by most of the later Sanskrit tradition. Someśvara, the author of the commentary Nyayasudha on Kumārila's Tantravārttika expresses his debt to Yāska in similar terms [cf. *niruktād avagato yo dhātvarthapūrvako nāmārthaḥ / sarvaṁ nāma dhātujam ity arthakalpanādvāraṁ nirukte pradarśitam*, Nyāyasudhā, Fasc. III, pp. 225-226].

Another point to note here are the terms *ūhya* and *ūhitavya* used by Patañjali. The verb *ūh* in this context means something like 'to figure out, to discover', and does not have the meaning of 'to imagine'. However, the later Pāṇinian tradition as seen in Bhartṛhari's works uses the terminology of *prakṛtipratyayakalpanā* 'imagining the stems and affixes'. Thus, we need to trace the movement from the concept of *uha* 'to discover, figure out' to the concept of *kalpanā* 'imagine', which may possibly have been inspired by the Vijñānavāda school of Buddhism.

4.8. FROM WORDS TO WORD-COMBINATIONS IN PĀṆINI

In Pāṇini's grammar, combinations of inflected items (*pada*) leading to higher order linguistic units are brought under the generic rule P.2.1.1 (*samarthaḥ padavidhiḥ*), which says, in brief, that operations prescribed with reference to two inflected words (*padavidhi*) apply only if there is a semantic-syntactic relationship between those words. Many different types of phenomena come under this generic rule, e.g. compounding (*samāsa*), accents of the finite verb in certain position [cf. P.8.1.28: *tiṅṅ atiṅaḥ*], and certain specific sandhi rules [cf. P.8.3.44: *isusoḥ sāmarthye*]. For example, we can derive a compound expression *rājaputraḥ* 'son of a king' from the words *rājñaḥ* 'of the king' and *putraḥ* 'son' only if these words are semantically and syntactically related to each other. If they belong to different phrases or clauses, then a mere phonological sequence is not sufficient to derive a compound. P.8.1.28 (*tiṅṅ atiṅaḥ*) applies to the accent of a verb form in relation to its position in the phrase or sentence. The rule says that a finite verb (*tiṅ*) preceded by a semantically-syntactically related (*samartha*) non-verb word becomes totally unaccented. This notion of semantic-syntactic relationship between various items is a complex topic. I have treated this topic fully elsewhere (Deshpande, 1987), and we shall not go into its details here. Suffice it to say that this is a very elastic concept in Pāṇini's grammar. It covers the direct semantic-syntactic relations between words which are processed into compounds, as well as the sentence/phrase-wide semantic-syntactic network implied in the rule P.8.1.28 discussed above, and is inherently an elastic concept. It is so central to Pāṇini's grammar, that it allows him to design his grammar without formally defining any notion of sentence (*vākya*), see: Deshpande, 1987. My point here is simply that within such complex structures larger than single words, Pāṇini sees complex semantic and syntactic relationships among the components, and there is no notion of partlessness of such wholes. Similarly, within such complex structures, Pāṇini deals with issues of the order of components and other structural matters. For instance, we find terms like *pūrvapada* 'previous word' [e.g. P.6.2.1: *bahuvrīhau prakṛtyā pūrvapadam*], *uttarapada* 'following word' [e.g. P.8.3.45: *nityaṁ samāse 'nuttarapadasthasya*] and *upapada* 'word used in construction' [e.g. P.2.2.19: *upapadam atiṅ*]. Though undefined, Pāṇini uses the term *vākya* 'sentence, utterance' in a few rules which refer to structural dimensions of a sentence such as 'at the beginning of a *vākya*' [cf. P.8.1.8: *vākyāder āmantritasya* ...] and the '*ṬI* (= end portion beginning with the final vowel) of a *vākya*' [cf. P.8.2.82: *vākyasya ṭeḥ pluta udāttaḥ*]. Further he uses the term *sākāṅkṣa* 'with expectancy, dependence' to refer to a semantic relationship between multiple verb-forms [cf. P.8.1.35: *chandasy anekam api sākāṅkṣam*], suggesting his awareness of inter-clausal linkages. Thus, while *sāmarthya* refers to inter-word semantic-syntactic relationships, *ākāṅkṣā* refers to inter-clausal relationships in Pāṇini's grammar.[14] In any case, there is no notion in Pāṇini of partless words, phrases, or sentences, and their singular indivisible meanings.

As we move to Kātyāyana and Patañjali, we find further evolution of different conceptions. First, they discuss the notion of *sāmarthya* 'semantic-syntactic relationship' under two sub-types. According to them [see the discussion in Mahābhāṣya on P.2.1.1: *samarthaḥ padavidhiḥ*], words which are members of compounds etc. undergo a process of unification of their meaning (*ekārthībhāva*, Vt 1 on P.2.1.1), while words in a sentence or a phrase have a semantic-syntactic expectancy for each other (*vyapekṣā*, Vt 4 on P.2.1.1). They try to decide which of these types is intended by Pāṇini in a given rule. I have argued elsewhere (1987) that these sub-types are not intended by Pāṇini himself who had a more elastic notion of semantic-syntactic relationships (*sāmarthya*). However, both of these subtypes recognize relationships between the meanings of items involved, and neither type espouses a notion of wholes having indivisible singular meanings. The semantic-syntactic relationships between words are also not all of equal strength or directness in a given sentence. Some are absolutely direct relationships, while others are more indirect and involve further networks [cf. *yukta-yukta* (Vt 13 on P.2.1.1), *samartha,* and *samarthatara* (Vt 7 on P.2.1.1)]. If *a* and *b* are directly related (*samarthatara* 'more closely related') and *b* and *c are directly related (samarthatara)*, and if no direct relationship holds between *a* and *c*, then the relationship between *a* and *c* is described by the terms *yuktayukta* 'related to the related' and (merely) *samartha* 'semantically-syntactically related'.

4.9. KĀTYĀYANA AND PATAÑJALI: DEFINING A SENTENCE (*VĀKYA*)

As I mentioned above, Pāṇini does not define the notion of sentence (*vākya*), though he uses this undefined term a few times. I have also argued (1987) that Pāṇini manages to do without defining the notion of sentence and uses his elastic notion of *sāmarthya* 'semantic-syntactic relationship'. In contrast with Pāṇini, Kātyāyana defines the notion of sentence explicitly. A sentence is defined either simply as a group of words containing a single finite verb (*ekatiṅ*, Vt 10 on P.2.1.1) or more elaborately as a finite verb along with related adverbs, indeclinables, and words expressing semantic roles in relation to the action of the verb (*kāraka*), *ākhyātaṁ sāvyayakārakaviśeṣaṇaṁ vākyam*, Vt 9 on P.2.1.1. Patañjali explicitly says that Kātyāyana's definitions of sentence represent a new development in the tradition of grammar [cf. *idam adya apūrvaṁ kriyate vākyasaṁjñā* ... MBh I, 367]. These definitions of sentence give us a notion of a sentence as a whole or a collection (*samudāya, samūha*) with a well-defined internal structure.

4.10. SENTENCE AS A GROUPING OF WORDS BY CHOICE

While for Kātyāyana and Patañjali the constituent parts of a *pada* 'inflected word' are *nitya-sambandha* (permanently bound together, Vt 8, on P. 1.2. 45), words in a sentence are freely combined (*yatheṣṭam*). At the level of sentence, Patañjali quite explicitly seems to believe that one freely combines words to form sentences, and that while words are given to the speaker, sentences are formulated by his/her sweet will: 'Moreover, the user has his freedom of will in combining words. For example, consider the sentence *yavāgūr bhavatā bhoktavyā navā*. When the word *yavāgū* 'rice-gruel' is connected with the verb *bhuj* 'to eat', and then the verb is connected with the expression *navā* (to be understood as *na vā*), then the expression *navā* (= *na vā*) is understood as a prohibition. From the sentence *yavāgūr bhavatā bhoktavyā na vā* 'You, Sir, should rather not eat the rice-gruel', one understands the prohibition. When the word *yavāgū* is connected with *navā*, and not with the verb, then the expression *navā* is understood in the sense of 'new, fresh'. From the sentence *yavāgūr navā bhavatā bhoktavyā*, one understands: 'The fresh [rice-gruel should be eaten by you, Sir]', [cf. *api ca kāmacāraḥ prayoktuḥ śabdānām abhisambandhe / tad yathā / yavāgūr bhavatā bhoktavyā navā / yadā yavāgūśabdo bhujinābhisambadhyate bhujir navāśabdena tadā pratiṣedhavācinaḥ sampratyayo bhavati / yavāgūr bhavatā bhoktavyā navā / neti gamyate / yadā yavāgūśabo navāśabdenābhisambadhyate na bhujinā tadā pratyagravācinaḥ sampratyayo bhavati / yavāgūr navā bhavatā bhoktavyā / pratyagreti gamyate* / MBh (on P. 1.1.44), I, 102. This is also expressed in the passage: *saṁskṛtya saṁskṛtya padāny utsṛjyante / teṣāṁ yatheṣṭam abhisambandho bhavati* / MBh I, 39, 'Having been derived, words are released (= made available for use). Their combination (into a sentence or phrase) comes about in accordance with the desire (of the user).' One can see the following terminological and ideological difference in the two processes intended by Patañjali:

P.1. stem/root + affix (through derivation) ⇒ **word** (given, permanent, obligatory)
P.2.word + word (through combination of choice)⇒ **sentence/phrase**

Elsewhere, Kātyāyana and Patañjali clearly refer to a *vākya* 'sentence, phrase' as a *samudāya* 'collection, grouping' of words, and to the constituent words as its components (*avayava*) [cf. *samudāyo 'narthaka iti ced avayavārthavattvāt samudāyārthavattvam yathā loke*, Vt 2 (on P.1.2.45), MBh I, 217]. Patañjali says that not all groupings of meaningful words are meaningful themselves and he offers examples of meaningless groupings [cf. *daśa dāḍimāni ṣaḍ apūpāḥ kuṇḍam ajājinaṁ palaladaṇḍaḥ adharorukam etatkumāryāḥ sphaiyakṛtasya pitā pratiśīna iti / samudāyo 'trānarthakaḥ* / MBh I, 217]. Referring to the same meaningless strings of meaningful words, Patañjali says that there are meaningful sentences and meaningless sentences [cf. *loke hy arthavanti cānarthakāni ca vākyāni dṛśyante*, MBh I, 38].

This discussion brings out another important point, namely that the meaning of the so-called sentence is not solely dependent on its constituents individually being meaningful, but that the sentence-meaning involves something over and above the meaning of the individual words which constitute that sentence. The prima-facie view for Kātyāyana and Patañjali is that there is no meaning in a sentence over and above the meaning of its individual constituents. This is rejected by saying that in a sentence one comprehends the interrelationships between the meanings of the individual words [cf. (*Pūrvapakṣa*): *na vai padārthād anyasyārthasyopalabdhir bhavati vākye* / MBh *padārthādanyasyānupalabdhir iti cet padārthābhisambandhasyopalab-*

dhiḥ, Vt 4, on P. 1.2.45]. Patañjali elaborately explains how the interrelationships between the meanings of the individual words are understood in a sentence:

> *iha devadatta ity ukte kartā nirdiṣṭaḥ karma kriyāguṇau cānirdiṣṭau / gām ity ukte karma nirdiṣṭaṁ kartā kriyāguṇau cānirdiṣṭau / abhyājety ukte kriyā nirdiṣṭā kartṛkarmaṇī guṇaś cānirdiṣṭaḥ / śuklām ity ukte guṇo nirdiṣṭaḥ kartṛkarmaṇī kriyā cānirdiṣṭā / ihedānīṁ devadatta gām abhyāja śuklām ity ukte sarvaṁ nirdiṣṭam bhavati / devadatta eva kartā nānyaḥ / gaur eva karma nānyat / abhyājir eva kriyā nānyā / śuklām eva na kṛṣṇām iti / eteṣām padānām sāmānye vartamānānāṁ yad viśeṣe 'vasthānaṁ sa vākyārthaḥ /*
> MBh I, 218.
>
> [In the context of a sentence such as *devadatta! gām abhyāja śuklām* 'Devadatta, bring the white cow'], when one says *devadatta*, the agent is expressed, but the patient, the action and the specific quality [of the patient] are not expressed. When one says *gām* 'the cow', the patient is expressed, but the agent, the action and the quality are not expressed. When one says *abhyāja* 'bring', the action is expressed, but the agent, the patient and its quality are not expressed. When one says *śuklām* 'the white [cow]', the quality is expressed, but the agent, the patient and the action are not expressed. Now, when one says *devadatta gām abhyāja śuklām*, everything is expressed, namely that Devadatta and none other is the agent, the cow and none other is the patient, bringing and nothing else is the action, only the white [cow] and not a black one. The sentence meaning represents the particularization (= contextualization) of [the meanings of] these words which otherwise have generic [= uncontextualized, lexical] meanings.[15]

This discussion in Kātyāyana and Patañjali clearly demonstrates that they are willing to accept a notion of a sentence as a whole constituted by bringing together words. The component words are meaningful, and yet the whole conveys interrelationships between the word-meanings which are not conveyed by the words taken by themselves. The entire discussion is similar to the notion of *abhihitānvayavāda* of the Bhāṭṭa Mīmāṁsakas and Naiyāyikas (see: K. Raja, 1963 for details), and it is a far cry from the doctrines preferred by Bhartṛhari, i.e. an indivisible sentence has an indivisible sentence-meaning, and that all linguistic units below the level of sentence are merely useful fictions.

5. TRACING THE IDEOLOGICAL DEPARTURE IN BHARTṚHARI

After the Mahābhāṣya of Patañjali (± second century BC), there is a major gap in the tradition of Pāṇinian grammar until we reach Bhartṛhari (± fifth century AD). Passages which occur at the end of the second *Kāṇḍa* of Bhartṛhari's Vākyapadīya, which have been the subject of many lengthy debates, generally tell us that the tradition of the Mahābhāṣya soon fell on hard times because of severe criticisms from authors like Baijin, Saubhava, and Haryakṣa. No work of these authors has survived. The text of Patañjali's Mahābhāṣya lost its importance and was rarely studied for a long time, and was preserved evidently only in manuscript form in the southern regions. Bhartṛhari's grand-teacher Candrācārya, traditionally identified with the Buddhist grammarian Candragomin, recovered the tradition of the Mahābhāṣya from the southern regions, and revived it again. Bhartṛhari's commentary on the Mahābhāṣya is the oldest commentary to have survived for us. He refers to a great many alternative interpretations, but always without naming the persons who held those views. That makes it very difficult to reconstruct an accurate history of interpretation of the Mahābhāṣya before the time of Bhartṛhari. However, Bhartṛhari's own works gained high prestige, both within the tradition of the Pāṇinian commentators, as well as among the philosophical traditions in general. Bhartṛhari's influence on later commentators was so persuasive that the great commentator Kaiyaṭa (eleventh century AD) says in the beginning of his commentary that he has been able to cross the ocean of the Mahābhāṣya by means of the bridge built by Bhartṛhari. Not only Bhartṛhari's commentary on the Mahābhāṣya, but his Vākyapadīya also has had a similar deep influence. While we must join Kaiyaṭa in being grateful to Bhartṛhari for all the assistance he has offered in interpreting the Mahābhāṣya, we must at the same time recognize that the traditional interpretation of the Mahābhāṣya has been so densely overlaid with Bhartṛhari's theories that one rarely gets a discussion in Kaiyaṭa or Nāgeśa where a careful distinction is drawn between the ideas of Kātyāyana and Patañjali on the one hand and Bhartṛhari's ideas on the other. This makes it all the more difficult, though necessary, to draw careful distinctions between Bhartṛhari and his great distant predecessors.

Based on the earlier discussion, we can now draw some broad conclusions. We can say that Pāṇini's understanding of language and its constituents looks like this:

Given	**Real Transformation**		**Real Transformation**	
stems, roots, affixes	⇒	words	⇒	sentence
Real components		**Real Composite**		**Real Composite**

Somewhat in contrast with Pāṇini, Kātyāyana and Patañjali seem to have had the following scheme in their mind:

	derivation	**[Given] free will of the speaker**	
	⇒ *saṁskṛtya* ⇒	⇒ *yatheṣṭam abhisambandhaḥ* ⇒	
stems, roots, affixes	⇔	words	⇒ sentence
	⇐ *anvaya-vyatireka* ⇐		
	discovery		

For Kātyāyana and Patañjali, the level of *padas* 'inflected words' is the basic level for grammar. These *padas* are freely combined by the users to form sentences or phrases. The *padas* are not derived by Kātyāyana and Patañjali by extracting them from sentences by using the method of *anvaya-vyatireka* 'concurrent occurrence and concurrent absence'. On the other hand, a grammarian derives the *prakṛti-pratyaya* 'stem/affix' elements by applying the extraction method of *anvaya-vyatireka* to *padas*, and then in turn puts these *prakṛti-pratyaya* elements through the grammatical process of derivation (*saṁskāra*) to derive the *padas* again. Here, Kātyāyana and Patañjali do make a distinction between the levels of actual usage (*vacana*) and technical analysis. While fully-fledged words (*pada*) occur at the level of usage, the components do not occur by themselves at that level. The morphological components of words are never used by themselves. However, they do not seem to suggest that the stems, roots, and affixes are purely imagined (*kalpita*). It seems that the procedure of *anvaya-vyatireka* for them is a discovery procedure, rather than something leading to imaginary or fictional components.

Bhartṛhari has substantially moved beyond Kātyāyana and Patañjali. For him, the linguistically given entity is a sentence. Everything below the level of sentence is derived through a method of abstraction/extraction referred to by the term *anvaya-vyatireka* or *apoddhāra*. Kātyāyana and Patañjali do not use this latter term. Additionally, for Bhartṛhari, elements abstracted/extracted through this procedure have no reality of any kind. They are *kalpita/parikalpita* 'imagined' [VP III, 14, 75-76]. This looks like Bhartṛhari's paraphrase of Patañjali's words *ūhya/ūhitavya* [MBh II, 139]. Such abstracted/extracted items have instructional value only for those who do not have an intuitive insight into the true nature of speech [V II. 238]. The true speech unit, the sentence, is an undivided singularity and so is its meaning which is comprehended in an instantaneous cognitive flash (*pratibhā*), rather than as a deliberative and/or sequential process. Thus, Bhartṛhari's scheme looks like:

Useful Fiction$_2$		**Useful Fiction$_1$**		**Given Reality**
stems, roots, affixes	⇐	words	⇐	sentence
	⇐ *anvaya-vyariteka* ⇐ **abstraction**		⇐ *anvaya-vyariteka* ⇐ **abstraction**	

The fact that Bhartṛhari has deviated from his predecessors can be established on several different grounds. Consider the following verse of the VP (II.10):

> *yathā pade vibhajyante prakṛtipratyayādayaḥ /*
> *apoddhāras tathā vākye padānām upapadyate //*

> Just as stems, affixes etc. are abstracted from [lit. 'in'] a given word, so the abstraction of words from [lit. 'in'] a sentence is justified.

Here, the clause introduced by *yathā* refers to the older more widely prevalent view seen in the Mahābhāṣya. With the word *tathā*, Bhartṛhari is proposing an analogical extension of the procedure of abstraction (*apoddhāra*) to the level of a sentence. While the *yathā* clause is a statement of the received view, the *tathā* clause is Bhartṛhari's own ideological extension.

Without mentioning Patañjali or Kātyāyana by name, Bhartṛhari seems to treat their views as prima-facie views (*pūrvapakṣa*). Consider the following examples:

a) *sāmānyārthas tirobhūto na viśeṣe 'vatiṣṭhate /*
upāttasya kutas tyāgo nivṛttaḥ kvāvatiṣṭhatām //
VP II.15

> [In moving from the meaning of individual words to the sentence-meaning, if one argues that] the generic meaning [of an individual word taken in isolation] has been set aside, then it cannot be [later] fixed with reference to a particular [=context-specific meaning]. How can [the generic meaning] once admitted be abandoned [later]? Where would [the presumed original generic meaning which is later supposedly] set aside, settle?

This verse appears to criticize the prior discussion in Patañjali's Mahābhāṣya [Vt 4, on P. 1.2.45, MBh I, 218]. For an explanation of Patañjali's arguments, see section 4.10 above.

b) *aśābdo yadi vākyārthaḥ padārtho 'pi tathā bhavet /*
evaṁ sati ca sambandhaḥ śabdasyārthena hīyate //
VP II. 16

> If the meaning of a sentence does not [directly] derive from words [lit. 'linguistic items'], then even the word-meaning could be like that [i.e. unrelated to a linguistic item, *aśābda*]. If such were to be the case, then the relationship of meaning with [the corresponding] linguistic item is lost.

This verse appears to criticize a discussion in the Mahābhāṣya [Vt 4, on P. 1.2.45]. For details of Patañjali's arguments, see section 4.10 above.

c) *kevalena padenārtho yāvān evābhidhīyate /*
vākyasthaṁ tāvato 'rthasya tad āhur abhidhāyakam //
VP II.41
sambandhe sati yat tv anyad ādhikyam upajāyate /
vākyārtham eva taṁ prāhur anekapadasaṁśrayam //
VP II.42

> They say that a word in a sentence expresses exactly the same meaning as is expressed by that word in isolation. The additional [meaning] which emerges, when there is a combination [of a word with other words in a sentence], is said to be the sentence-meaning resting on the group of many words [forming that sentence].

These verses seem to state as a *pūrvapakṣa* 'prima-facie view' the view discussed in the Mahābhāṣya on P. 2.3.46 (*prātipadikārthaliṅgaparimāṇavacanamātre prathamā*), MBh I, 461-2. For the details of Patañjali's view, see section 4.10 above.

I cannot discuss here the very important question of where Bhartṛhari may have derived his inspiration for his doctrine of *akhaṇḍa* 'partless' *vākya* 'sentence' and *vākyārtha* 'sentence-meaning'. However, as demonstrated earlier, it is clear that these ideas do not occur in Kātyāyana and Patañjali, and that the views of these two great grammarians are much closer, though not identical, with the views later maintained by the Mīmāṁsakas. We should also note that when Bhartṛhari states or summarizes different views, he may have unconsciously coloured them by the use of his terminology and his zest for his own views. This, for instance, seems to be the case in his discussion of *Vākyavāda* 'doctrine of sentence as the primary unit' versus *Padavāda* 'doctrine of word as the primary unit':

abhedapūrvakā bhedāḥ kalpitā vākyavādibhiḥ /
bhedapūrvān abhedāṁs tu manyante padavādinaḥ //
VP II.57

> The upholders of the view that sentence is the primary unit of language imagine [subsequent] distinctions [in terms of the constituent words] after [the primary] unity [of a sentence]. On the other hand, those who hold words to be the primary units of language believe that the [comprehension of] unitary [sentences] follows [the experience of the primary] distinctions [i.e. individual words].

Here, the presentation of *Vākyavāda* 'the doctrine that a sentence is the primary unit of language', which is Bhartṛhari's own doctrine, is presented in proper terms. For him, the indivisible singularity of a sentence and its meaning is the real thing, while the segmentation into words and word-components, and their meanings, is simply a subsequently developed useful fiction (*kalpita*). Now, when he presents the view of his opponents, he simply reverses this terminology. He says that the Padavādins give priority to distinctions or segments, and argue that the singularity or indivisibility (*abheda*) of sentence and its meaning is subsequently built. It is not clear that any school which might be included under the broad term *Padavāda* holds such a view.[16] For Patañjali, clearly, the sentence is a *saṁghāta* or *samudāya* 'a group/collection', and the sentence meaning is a network of *vyapekṣā* 'mutual expectancies'. The sentence meaning does not even fit in the category of *ekārthībhāva* 'unification of meaning', let alone any notion of *abheda* 'non-distinction'. The same would be the case with Mīmāṁsakas and Nyāya-Vaiśeṣikas. Thus, while reading Bhartṛhari, one needs to be constantly on guard, and notice the ways in which his presentation of the opposing views is skewed by the use of his own terminology.

6. REJECTION OF BHARTṚHARI'S IDEAS BY THE LATER GRAMMATICAL TRADITION

Finally, I would just like to note that Bhartṛhari's views on the unitary character of a sentence and its meaning were found to be generally unacceptable not only by the opposing schools of Mīmāṁsā and Nyāya-Vaiśeṣika, but by the later grammarian-philosophers like Kauṇḍabhaṭṭa and Nāgeśabhaṭṭa as well. Their discussion of the comprehension of sentence-meaning is not couched in terms of Bhartṛhari's instantaneous flash of intuition (*pratibhā*), but in terms of the conditions of *ākāṅkṣā* 'mutual expectancy', *yogyatā* 'compatibility', and *āsatti* 'contiguity'. In this sense, they are closer to the doctrines of Patañjali himself. On the other hand, what survives from Bhartṛhari into later times is the term *kalpita* 'imagined, fictional' (and the related term *vikalpa*). However, this term is used by these later grammarians almost exclusively to refer to the abstraction of *prakṛti-pratyaya*s from a word, and not to refer to the abstraction of words from sentences, as intended by Bhartṛhari. In this sense, the later grammarian-philosophers are somewhat closer to the spirit of the *Padavāda* as found in Kātyāyana and Patañjali. However, at the level of the abstraction of *prakṛti-pratyayas* from words, and assigning meaning to the abstracted elements, the view of 'useful fictions' comes to prevail among these later grammarians. While for Kātyāyana and Patañjali, the *prakṛti-pratyayas* are more like discovered elements, rather than fictional ones, the later grammarians seem to agree with Bhartṛhari on this count. All Sanskrit grammarians, with the exception of Yāska who occasionally argues for or against a certain breakdown, are generally unwilling to make a principled choice between alternative proposals for morphological or semantic segmentation. Kātyāyana and Patañjali by and large simply accept the segmentation provided by Pāṇini, without discussing alternative possibilities. However, generally they do not seem to accept simultaneous alternative segmentations. On the other hand, later grammarians like Bhartṛhari and Kauṇḍabhaṭṭa are more prone to accept all possible morphological and semantic segmentations as equally fictional, each one acceptable with its limited utility in certain contexts. This view is clearly articulated (by Bhartṛhari?) in the Vṛtti,[17] and later echoed by Bhaṭṭojī Dīkṣita [cf. *Vaiyākaraṇasiddhāntakārikā* (25): *ekaṁ dvikaṁ trikaṁ cāpi catuṣkaṁ pañcakaṁ tathā / nāmārtha iti śāstre 'mī pakṣāḥ śāstre nirūpitāḥ //*]. The acceptance of such multiple alternative possibilities of segmentation may be perfectly in line with a particular philosophy of language, and yet it has prevented the tradition of grammar from developing evaluative techniques to judge the quality of alternative segmentations, Deshpande (1992), pp. 80-1.

7. CONCLUSION

The goal of this paper has been to point out how the doctrines of Pāṇini, Kātyāyana, and Patañjali relate to the previous Vedic epoch, and also how they are substantially different from those of Bhartṛhari. Such a clarification has been made necessary by the general tendency of scholars in the field of Sanskrit grammar to consciously or otherwise take the positions of Bhartṛhari to be representative of the tradition of Sanskrit grammar as a whole. This is not to say that Bhartṛhari's doctrines are without any merit. Their merit needs to be judged on independent philosophical grounds. My focus here has been to place Bhartṛhari in his proper historical position. It is clear that his ideas are not a continuation of the tradition, but represent a radical departure. The reasons for such a radical departure lie beyond the scope of the present study. I hope to return to those reasons in the future.

REFERENCES

Acharya, Sudyumna

1990 'A critical and comparative study of Pāṇini's morphemic principles, in the light of modern linguistics', paper presented at the World Sanskrit Conference in Vienna, September 1990, and published as a booklet. Kolgavan, Satna (M.P.), India: Veda Vani Vitan, Indological Research Institute

Aitareya-Āraṇyaka

1909 With parts of the *Śāṃkhāyana-Āraṇyaka.* Edited and translated by Arthur B. Keith, London: Oxford University Press

Aitareya-Brāhmaṇa

1896 Ānandāśrama Sanskrit Series. 31, Pts. I-II, Pune: Ānandāśrama

Aṣṭādaśa-Upaniṣadaḥ

1958 edited by V.P. Limaye and R.D. Wadekar, Pune: Vaidika Saṃśodhana Maṇḍala

Atharva-Prātiśākhya

1939 edited and translated by Surya Kanta, with critical introduction and notes, Lahore: Mehar Chand Lachhman Das. Reprinted from Delhi in 1968 by Mehar Chand Lachhman Das

Atharvapariśiṣṭāni

1976 edited by Ram Kumar Rai, Banaras: Chowkhamba Orientalia

Atharvaveda (*Śaunakīya*) with the *Padapāṭha* and the commentary by Sāyaṇa

1960-4 edited by Vishva Bandhu. In five parts. Vishveshvarananda Indological Series. 13-17

Bhate, Saroja

1981 'Pāṇini and Yāska: Principles of Derivation', *Annals of the Bhandarkar Oriental Society* LXII, pp. 235-41

Bṛhaddevatā

1904 edited and translated by Arthur Anthony Macdonell. Reprint edition, Delhi: Motilal Banarsidass, 1965. (HOS. 5-6)

Buitenen, van J.A.B.

1988 *Studies in Indian Literature and Philosophy.* Collected articles edited by Ludo Rocher, Delhi: Motilal Banarsidass

Deshpande, Madhav M.

1986 'Some Facets of Pāṇinian Morphology', *Adyar Library Bulletin*, pp. 478-89

1987 'Pāṇinian Syntax and the Changing Notion of Sentence', in: *Annals of the Bhandarkar Oriental Research Institute* LXVIII, pp. 55-98

1992 *The Meaning of Nouns. Semantic Theory in Classical and Medieval India. Nāmārthanirṇaya of Kauṇḍabhaṭṭa translated and annotated.* Dordrecht: Kluwer. (Studies of Classical India. 13)

1994 'Ancient Indian Phonetics', in: Ronald E. Asher (ed.), *The Encyclopedia of Language and Linguistics* 6. Oxford: Pergammon Press, pp. 3053-8

Ghosh, B.K.

1945 'Aspects of Pre-Pāṇinian Sanskrit Grammar', *B.C. Law Commemoration Volume* I. Calcutta, pp. 334-45

Gonda, Jan

1955-6 'The Etymologies in the Ancient Indian Brāhmaṇas', *Lingua* V, pp. 61-85

Gopatha-Brāhmaṇa

1980 ed. by Vijayapal Vidyavaridhi, Calcutta: Savitri Devi Bagadiya Trust

Harris, Zellig

1951 *Structural Linguistics*. Chicago: University of Chicago Press

1954 'Distributional Structure', *Word* 10, No. 2-3, pp. 146-62

Houben, Jan E.M.

1991 *The Pravargya-Brāhmaṇa of the Taittirīya-Āranyaka*. Delhi: Motilal Banarsidass

1993 'Who were the Padavādins?', *Asiatische Studien* 47, Pt. 1

Hume, Robert Earnest

1921 *The Thirteen Principal Upanishads translated from the Sanskrit*. London: Oxford University Press

Kāśikāvṛtti

1965-7 by Vāmana-Jayāditya, with the commentary *Padamañjarī* by Haradatta, and *Nyāsa* by Jinendrabuddhi. In Six Volumes. Banaras: Tara Publications. (Prācya Bhāratī Series. 2)

Kāṭhaka Saṃhitā

1900 edited by Leopold v. Schroeder. Four volumes. Leipzig

Keith, A.B.

1909 *The Aitareya-Āraṇyaka*. [edited and translated into English], Oxford: Clarendon Press. Reprinted in 1969

1914 *The Veda of the Black Yajus School entitled Taittirīya-Saṁhitā*, Pt. I. Cambridge: Harvard University Press. (HOS. 19)

1920 *Ṛgveda-Brāhmaṇas: The Aitareya and Kauṣītaki Brāhmaṇas of the Ṛgveda, translated from the original Sanskrit*. Cambridge: Harvard University Press. Reprinted in 1971, Delhi: Motilal Banarsidass. (HOS. 25)

Laghuśabdenduśekhara

1936 by Nāgeśabhaṭṭa, with six commentaries. Edited by Guruprasad Shastri. Rajasthan Sanskrit College Series. 14. Banaras

Mahābhāṣya
1880-5 by Patañjali. Edited by Franz Kielhorn in 3 volumes, 3rd revisededition by K.V. Abhyankar, 1962-1972, Pune: Bhandarkar Oriental Research Institute

Mahābhāṣya
1967 by Patañjali, with the commentaries *Pradīpa* by Kaiyaṭa and *Uddyota* by Nāgeśabhaṭṭa. In 3 volumes, Delhi: Motilal Banarsidass

Maitrāyaṇī Saṃhitā
1881-6 edited by Leopold v. Schröder. Two volumes in four parts. Leipzig

Nirukta
1921-6 by Yāska. With the commentary of Durga. Pune: Ānandāśrama (Ānandāśrama Sanskrit Series. 1-2, no. 88)

Nyāyasudhā
1901-9 a commentary by Someśvarabhaṭṭa on Kumārila's *Tantravārttika.*Edited by Mukunda Sastri. Chowkhamba Sanskrit Series. 14. Banaras

Pandit, M.D.
1989 *A Concordance of Vedic Compounds Interpreted by Veda (Vol. 1). Publications of the Centre of Advanced Study in Sanskrit,* Class B, No. 10, Pune: Centre of Advanced Study in Sanskrit, University of Poona

Raja, Kunjunni K.
1963 *Indian Theories of Meaning*. The Adyar Library Series 91, Madras: Adyar Library and Research Centre

Renou, Louis
1958 Etudes sur le vocabulaire du Rgveda: 1e serie. Pondichery: Institut français d'Indology. (Publications de l'Institut français d'Indology. 5)

Ṛgveda Prātiśākhya
1959 edited by Mangal Deva Shastri. Vol. I, Critical text of *RPR*, Vaidika Svadhyaya Mandira, Banaras; Vol. II, *RPR* with Uvaṭa's commentary, 1931, The Indian Press, Allahabad; Vol. III, *RPR* in English translation, Punjab Oriental Series. 24, 1937, Lahore

Ṛgveda Saṃhitā
1933-51 with Sāyaṇa's commentary. Edited by N.S. Sonatakke and C.G. Kashikar. Five volumes, Pune: Vaidika Saṃśodhana Maṇḍala.

Ṛktantra
1939 *Prātiśākhya of the Sāmaveda*. Edited by Surya Kanta. Lahore. Reprinted by Meherchand Lachhmandas. Delhi, 1971

Śaisirīya-Śikṣā
1935 ed. by Tarapada Chowdhury. *The Journal of Vedic Studies* 2, No. 2, August 1935, pp. 1-18

Śāṃkhāyana-Brāhmaṇa (= *Kauṣītaki-Brāhmaṇa*)
1977 edited by Gulabrao Vajeshankar. 2nd edition, Anandashrama Sanskrit Series. 65. Pune: Anandashram

Śatapatha-Brāhmaṇa
1849 edited by Albrecht Weber, Berlin. Reprinted in Banaras: Chowkhamba Sanskrit Series. 96, 1964

Scharfe, Hartmut
1961 *Die Logik im Mahābhāṣya*. Berlin: Deutsche Akademie der Wissenschaften zu Berlin, Institut für Orientforschung

Śaunakīyā Caturādhyāyikā
Critical edition with three commentaries. Edited, translated, and annotated by Madhav M. Deshpande. Appearing in the HOS

Siddhānta-Kaumudī
1933 by Bhaṭṭojī-Dīkṣita, with Jñānendra Sarasvatī's *Tattvabodhinī*. Ed. by Wasudev Laxman Sastri Panshikar. 7th ed. Bombay: Nirnaya Sagara

Sphoṭavāda
1946 by Nāgeśabhaṭṭa edited by V. Krishnamacharya, Madras: Adyar, reprinted in 1977.

Taittirīya-Āraṇyaka
1867-9 Ānandāśrama Sanskrit Series. 36, Pts. I-II, Pune: Ānandāśrama

Taittirīya-Brāhmaṇa
1855-70 edited by Rajendralal Mitra. 3 vols. in Bibliotheca Indica, Calcutta

Taittirīya-Prātiśākhya
1868 with the commentary *Tribhāṣyaratna*. Edited and translated by W.D. Whitney. New Haven

Taittirīya Saṃhitā
1871-2 edited by Albrecht Weber. Two volumes, *Indische Studien* XI and XII, Leipzig

Thieme, Paul
1982-3 'Meaning and form of the 'grammar' of Pāṇini', *Studien zur Indologie und Iranistik* 8-9, pp. 23-8
1985 'The first verse of the *Triṣaptīyam* (*AV, Ś 1.1 ≈ AV, P 1.6*) and the beginnings of Sanskrit linguistics', *JAOS* 105, no. 3, pp. 559-65

Vākyapadīya
1977 Critical text - Edited by Wilhelm Rau, *Abhandlungen für die Kunde des Morgenlandes* XLII, 4. Wiesbaden: Franz Steiner

Vājasaneyi Prātiśākhya
1934 with commentaries by Uvaṭa and Anantabhaṭṭa. (University of Madras Sanskrit Series. 5), Madras

Vājasaneyi Saṃhitā
1852 edited by Albrecht Weber, Berlin

Notes

1. This extension of the word for foot to a metrical foot is found in Avestan and Greek and may go back to the Indo-European common period.
2. B.K. Ghosh (1945) has discussed some of these issues and I agree with many of his conclusions, though his statements need to be considerably refined.
3. I discussed this matter with Prof. J.C. Catford, Professor [Emeritus] of Phonetics, University of Michigan, who thought that this was very natural. In support of this, he referred to the historical fact that syllabic scripts preceded the emergence of alphabetic scripts.
4. Keith, 1914, p. 534.
5. This is my guess as to what the word *nigamayitṛ* means.
6. The text also uses a third term *ubhayam antareṇa* which refers to what is conventionally known as the Kramapāṭha. However, we will not go into a discussion of the complexities posed by this term.
7. This term is not used by Pāṇini himself, but is used by his commentators. For details, see: Deshpande, 1986).
8. Compare with this the use of visually oriented terms in modern linguistics like 'right context' and 'left context'. Such usage envisions an entire stretch of items to be processed written on the board, and the reader or the speaker facing this board, because the words 'right' and 'left' refer to the right and the left of the speaker. There is no such spatial perception in Pāṇini's grammar. It is based on the orality of the stretch to be processed, where temporality expressed through words like 'previous' and 'subsequent' plays a dominant role.
9. Later authors like Nāgeśabhaṭṭa claim that the doctrine of *sphoṭa* goes back to the sage Sphoṭāyana mentioned by Pāṇini (6.1.123), cf. final verses of Nāgeśabhaṭṭa's *Sphoṭavàda*, p. 102. This creates an impression of this doctrine being very ancient. However, besides his name, there is no evidence of Sphoṭāyana having composed a work relating to the doctrine of *sphoṭa*.
10. Also compare *Śaiśirīya Śikṣā* (354): *vedo hi varṇasaṁghātaḥ*. It may also be pointed out that while the words *saṁghāta/saṁghāṭa* are rare in old Sanskrit, especially in Vedic, they are very common in canonical Pali texts. For textual references, see the *Pali-English Dictionary* by Rhys Davids and Stede, 1921, London: Pali Text Society. It is conceivable that the philosophical use of this term is borrowed by the grammarians from early Buddhist or similar traditions.
11. It is not entirely clear if the work called Saṁgraha cited by the Vṛtti on VP is the same as Vyāḍi's Saṁgraha presumably belonging to a period before Kātyāyana. There are justifiable doubts in this matter since the Saṁgraha cited by the Vṛtti on VP sometimes states views which are unknown to Kātyāyana and Patañjali. For instance, consider the following verse cited from Saṁgraha in the *Vṛtti* on VP (I. 24-26):
    na hi kiñcit padaṁ nāma rūpeṇa niyataṁ kvacit /
    padānāṁ rūpam artho vā vākyārthād eva jāyate //
    This verse comes much closer to Bhartṛhari's views and seems to make words in a sentence into fictional entities. Also note that the verse *arthāt padaṁ sābhidheyam* is also found in the Bṛhaddevatā (2.117).

12. The example of a chariot and its parts reminds the argument offered by the monk Nāgasena to the Indo-Greek king Milinda in the famous Buddhist text Milindapañha.
13. Similar arguments occur elsewhere in the Mahābhāṣya, cf. *iha pacatīty ukte kaścic chabdaḥ śrūyate pacśabdaś cakārānto 'tiśabdaś ca pratyayaḥ / artho 'pi kaścid gamyate viklittiḥ kartṛtvam ekatvaṁ ca / paṭhatīty ukte kaścic chabdo hīyate kaścid upajāyate kaścid anvayī / pacśabdo hīyate paṭhśabda upajāyate 'tiśabdo 'nvayī / artho 'pi kaścid dhīyate kaścid upajāyate kaścid anvayī / viklittir hīyate paṭhikriyopajāyate kartṛtvaṁ caikatvaṁ cānvayi / te manyāmahe yaḥ śabdo hīyate tasyāsāv artho yo 'rtho hīyate, yaḥ śabda upajāyate tasyāsāv artho yo 'rtha upajāyate, yaḥ śabdonvayī tasyāsāv artho yo 'rtho 'nvayī* / MBh I, 255.
14. The meaning of the term *ākāṅkṣā* substantially changes in later times, see: Deshpande, 1987 p. 63 ff.
15. The same conclusion is reached in other discussions in the Mahābhāṣya, where we are told that the sentence-meaning (*vākyārtha*) is what one understands over and above (*ādhikyam*) the meanings of the individual words. On P. 2.3.46 (*prātipadikārthaliṅgaparimāṇavacana-mātre prathamā*), Kātyāyana raises the question: Vt 1: (*prātipadikārthaliṅgaparimāṇavacanamātre prathamālakṣaṇe padasāmānādhikaraṇya upasaṁkhyānam adhikatvāt*). Patañjali says: *prātipadikārthaliṅgaparimāṇavacanamātre prathamālakṣaṇe padasāmānādhikaraṇya upasaṁkhyānaṁ kartavyam / vīraḥ puruṣaḥ / kiṃ punaḥ kāraṇaṁ na sidhyati / adhikatvāt / vyatiriktaḥ prātipadikārtha iti kṛtvā prathamā na prāpnoti / kathaṁ vyatiriktaḥ / puruṣe vīratvam //* The reply is Vt 2: (*na vā vākyārthatvāt*). Patañjali says: *na vā vaktavyam / kiṁ kāraṇam / vākyārthatvāt / yad atrādhikyaṁ vākyārthaḥ saḥ* / MBh I, 461-2.
16. Houben (1993) extensively discusses the question of who was possibly referred to by the terms *Padavādin* or *Padadarśin*.
17. Compare the *Vṛtti* on VP (I. 24-26): *apoddhāre hi śāstravyavahārārthaṁ samudāyāt saṁsṛṣṭāyāḥ kasyāś cid arthamātrāyāḥ kriyamāṇe taṁ tam avadhiṁ prati nimittatvenārthānāṁ puruṣādhīno vikalpabhedaḥ sambhavati / ... puruṣavikalpeṣv aniyateṣu tathā śāstravyavasthābhedo na bhavati sa vikalpaḥ parigṛhyate / tathaikatvādayo vibhaktyarthāḥ karmatvādayaḥ pañcakaḥ prātipadikārthaś catuṣkas trika iti puruṣādhīna evaṁprakāraḥ pakṣabhedaḥ /*

VII

MYTHS OF TRANSSEXUAL MASQUERADES IN ANCIENT INDIA

Wendy Doniger

1. INTRODUCTION

Frits Staal's interest in India, as in the planet earth in general, is extraordinarily wide. Best known, perhaps, for his studies of such abstract and logical subjects as philosophy, language, and ritual, he has also always maintained a lively interest in earthier matters, such as martial arts, and it is here that his expertise and mine most often overlap. It is to this Frits Staal, the scholar not of the mind but of the body, that this essay is dedicated, in loving friendship.

Many myths in the Epics and *purāṇas* involve gender transformation, either transvestism (masquerading as someone of the other sex) or transsexuality (which I use, in this context, to mean transformation into someone of the other sex). And when someone who has undergone such a transformation is perceived by a sexual partner as being of a sex (and/or gender) other than the one that the subject (and the myth) regards as the true one, a sexual masquerade may take place - one person pretending to be someone else in bed. In contrast with most stories about sexual masquerades, in which we (and the hero/heroine) usually meet the masquerader *in medias res*, in the false form, and only encounter the authentic form at the end, stories of transsexuality usually begin with the authentic, move into the masquerade, and return again to the authentic at the end. We encounter most masquerades at their moment of coding, but we encounter the transsexual masquerade at its moment of decoding. We meet a different gaze, perhaps, and more often hear the transsexual story from the standpoint of the masquerader (as we only occasionally do in non-transsexual masquerades). For the myths usually regard masqueraders as Other, alien, caught in an unauthentic position; while transsexual myths often regard the other as a valid form of the self, authentically constructed from the start. But we may therefore encounter a greater nervousness about what the authentic is, and, sometimes, a greater need to establish it from the start rather than let the reader/protagonist discover it gradually.

Much has been written about transsexual myths in Hinduism,[1] but I wish to concentrate here on transsexual myths that shed light upon the nature of human sexual identity and its revelation in situations which attempt to conceal it. At first glance, Hindu mythology does not seem to regard sex or gender as intrinsic parts of human identity, since these stories were composed within a world in which you probably were recently, and will soon be again, another sex or gender. In some texts, a male is entirely transformed into a female, with a female mentality and memory (aspects of gender rather than of sex), the situation that we might expect from the fluidity of gender. Yet many texts seem to reflect the very opposite view: the male merely assumes the outer form of the female, retaining his male essence, his male memory and mentality, reflecting a view of gender as astonishingly durable. Even the Vedantic theory of illusion, which disparages the body in favour of the soul, implies that you may very well remain a male in some essential way even when you happen to take on a female body. Vedantic philosophy produced many male dream doubles, of whom the most famous is Nārada, who became a woman and lived a full life but eventually returned to his life as a man.[2] These two contrasting views may be correlated with two contrasting attitudes to women and to homosexual love: the texts that view gender as fluid generally depict the transformed male as happy in her female form, while those in which the gendered memory lags stubbornly behind depict him as miserable in her female form.

2. MAGICAL TRANSFORMATIONS: ILĀ

One of the central Hindu charters of the human race, more particularly the lunar dynasty, involves a transsexual masquerade. For Manu, the ancestor of all humans (*mānavas*), the Indian Adam, has a daughter who turns into a son and then back into a daughter; the Mahābhārata tells us, rather cryptically, that Manu had many sons and one daughter named Ilā, who gave birth to Purūravas and became both his mother and his father.[3] (Here, as so often, the myth reifies and embodies a cliché: we often speak of a single parent as being 'both mother and father' to a child; in Indian myths, it actually happens.) Ila's birth is sexually ambiguous, and his/her adult sexual life is so problematic that it sometimes becomes convenient, in discussing this myth, to use at ambiguous moments the otherwise awkward modern non-sexist pronoun s/he to describe him/her and to simplify the vacillating name of Ila/Ilā by failing to designate, as Sanskrit does with a final long vowel, the difference between the male and female forms.

The story of Ila is told in many *purāṇas*, since Ila founded the lunar dynasty and dynastic succession is a central concern of the *purāṇas*. The myth tells of the joining of the descendant of the Sun (Ilā, the grandchild of Vivasvant, the sun) with a descendant of the Moon (Budha, son of Soma, the moon), a very important cosmic and political moment. Indeed, it might be argued that the sexual labyrinths of the text were generated, in part, at least, through a desire to account for the joining of two great dynasties, each claiming descent from a male cosmic body (for both the sun and the moon are usually male in Sanskrit), without demoting either partner to the inferior status of a female. The solution: to imagine two cosmic patriarchs, and to turn one - only temporarily, of course - into a woman. (The parallel desire, to have a child born of both Śiva and Viṣṇu, was resolved by turning Viṣṇu, temporarily, into a woman, Mohiṇī, a story which we will soon investigate.[4])

The *Rāmāyaṇa* contains one of the earliest tellings of this story, in which Ila is said to be the son not of Manu but of a Prajāpati named Kardama:

> When Śiva was making love with Pārvatī, he had taken the form of a woman to please her, and everything in that part of the woods, even trees, had become female. One day, King Ila went hunting and killed thousands of animals, but still his lust for hunting was unsatisfied. As Ila came to that place where Śiva was making love with Pārvatī, he was turned into a woman, and when she approached Śiva to seek relief from her misery, Śiva laughed and said, 'Ask for any boon except manhood.' Ila pleaded with Pārvatī, who said: 'Śiva will grant half of your request, and I the other half. In that way you will be half female, half male.' Rejoicing at this wonderful boon from the Goddess, Ila said, 'If you, whose form is unrivalled by any copy, are truly pleased with me, let me be a woman for a month, and then a man again for a month.' 'So be it,' said Pārvatī, 'but when you are a man, you will not remember that you were a woman; and when you are a woman, you will not remember that you were a man.' 'So be it,' said the king, and for a month she became the most beautiful woman in the world.
>
> During that first month, she was wandering in the forest with her female attendants who had formerly been men, when she came upon King Budha, the son of the moon, immersed in a lake and immersed in meditation. She was struck by his stunning good looks, and started splashing the water; he noticed her and was pierced by the arrows of lust. He thought to himself, 'I have never seen a woman like this, not among goddesses or snake women or demon women or celestial nymphs. If she is not married, let her be mine.' He asked her followers whose she

was, and they replied, 'This woman with superb hips rules over us; she has no husband and wanders with us in the woods.' When he heard this speech, whose meaning was obscure, Budha used his own magic powers and discovered the entire truth of what had happened to the king. Then he said to all the women, 'All of you will become female Quasi-men (*kiṃpuruṣīs*) and live on this mountain.' And as soon as they heard his words, those numerous women became female Quasi-men. And that was the origin of the Quasi-men (creatures half-horse and half-human, literally a 'What'-man - usually horse-headed rather than horse-bodied, a kind of inverse centaur).[5]

When he saw that all the Quasi-men had run away, Budha smiled and said to Ila, 'I am the son of King Soma; look upon me with loving eyes, and make love with me.' In that deserted place, deprived of all her attendant women, she spoke pleasingly to him, saying, 'Son of Soma, I am free to do as I wish, and so I place myself in your power. Do with me as you wish.' Hearing that astonishing speech from her, the king was thrilled, and he caused Ila to enjoy the exquisite pleasures of love-making, for a month which passed like a moment.

But when the month was full, Ila the son of the Prajāpati awoke in the bed and saw Budha immersed in the water, immersed in meditation. He said to Budha, 'Sir, I came to this inaccessible mountain with my attendants. But now I don't see my army; where have all my people gone?' When Budha heard these words from Ila, whose power of recognition had been destroyed, he replied with an effective, conciliating speech: 'Your servants were all destroyed by a hailstorm, and you were exhausted by your terror of the high winds and fell asleep on the grounds of this hermitage. Please be consoled, calm down, and be not afraid. Live here in comfort, eating fruits and roots.'

Though the wise king Ila was encouraged when he heard those words, he was greatly saddened by the death of his servants, and he said, 'I will renounce my own kingdom; I cannot go on for a moment without my servants and wives. Please give me leave to go. My eldest son, named Śaśabindu ('Hare-marked', a name of the moon), will inherit my throne.' But Budha said, 'Please live here. Don't worry. At the end of a year, O son of Kardama, I will do you a great favour.' And so Ila decided to stay there.

Then, for a month she became a woman and enjoyed the pleasure of making love ceaselessly, sleeplessly, and then for a month he increased his understanding of dharma, with the nature of a man. In the ninth month, Ila, who had superb hips, brought forth a son fathered by Budha, named Purūravas. And as soon as he was born, she placed him in the hands of Budha, for he looked just like him and seemed to be of the

> same class. And then when she became a man again, Budha gave him the pleasure of hearing stories about dharma, for a year.
> Then Budha summoned a number of sages, including Kardama, the father of Ila, and asked them to do what was best for him/her. Kardama suggested that they propitiate Śiva with a horse-sacrifice and ask his help. Śiva was pleased by the horse-sacrifice; he came to them, gave Ila his manhood [*puruṣatva*], and vanished. King Ila ruled in the middle country of Pratisṭhāna, and his son Śaśabindu ruled in their country of Bahli.[4]

The form of the curse is no accident; the founding of the lunar dynasty is linked, by natural association, with the monthly vacillation between female and male, a dichotomy doubled, as it were, in the dichotomy between human and equine. Ila's servants, already transformed, like him, into women, through the power of the god, go on to become, through the power of the human king, Quasi-men - self-contradictory as well as liminal creatures, since the word for 'men' [*puruṣa*] defines the male in contrast with the female; to be a female 'quasi-man' is thus to be a kind of Irish bull, or at least an Irish horse, a living oxymoron. Thus time (the months) and space (the halves of the body) provide interchangeable responses to sexual liminality.

There are scattered references here to the problem of recognition, all refractions of the central problem, namely, that when he is transformed into a woman, Ila does not recognize himself. One aspect of recognition is resemblance: does one self resemble another? Thus, in praising the Goddess, Ila says that her form is 'unrivalled by any copy' (*pratimā*, a reflected image), Budha says that he has never seen 'a woman like this', and their child is said to look just like him. When Ila does not recognize himself after he has been restored to his primary form as a man, he is said to be someone whose power of recognition has been destroyed; and as a woman, in his *secondary*, alternative form, she is not himself, but only his (female) shadow or inverted mirror image (*pratimā*).

But recognition also involves memory, and part of the curse (or is it the boon that balances the curse?) is to make Ila forget one gender when s/he is immersed in another. The loss of memory is also hedged about with riddling words: either the servants of the transformed Ila are, like him/her, unable to remember the truth (it is not clear how wide the range of the curse is to extend) or they remember the truth all too well and are embarrassed by it. But in either case, they answer Budha's questions with words 'whose meaning was obscure' (*avyaktapādam*) - that is, they do not tell him Ila's name or history, which is the expected answer to his question. He, in turn, doesn't tell Ila who she is, though he knows this through powers of his own. In this way, he withholds Ila's memory from her/him and keeps him/her in his power. By cutting him/her off from the knowledge of his/her true identity and then seducing him/her, Budha is in effect raping a sleeping woman, engaging in what Manu classifies as the 'marriage of a ghoul' and defines thus: 'The lowest and most evil of marriages, known as that of the ghouls, takes place when a man secretly has sex with a girl who is asleep, drunk, or out of her mind.'[7] In this case, she is quite literally out of her mind, and into someone else's.

But Ila has already been disempowered by the loss of both sex and class. For in one stroke s/he has been deprived of political power, class (servants), and gender. As soon as s/he becomes a woman, even while s/he still has servants (class), s/he loses his/her ownership of his/herself; Budha asks her followers whose (*sic*) she is, the standard way of inquiring about a woman's identity in ancient India (to which the standard answer consists of her father's name, if she is unmarried, or her husband's, if she is married; this is the question that the followers answer 'obscurely'). And of course s/he also loses his/her political power, both because s/he forgets that s/he is a king and because a woman cannot be a king. When Budha pulls out from under him/her the one remaining prop, his/her servants, s/he finds herself alone with him in the middle of the forest, helpless. Naturally, s/he gives in to his sexual demands.

But even when Ila becomes retransformed into a man, he remains helpless, for reasons of class which remain even when gender has been restored. Thus he remarks, 'I cannot live without my servants and wives.' Men, in this world view, are dependent on women, for services, and women are dependent on men, for protection. But they are also mutually dependent for sex. We have already been told that Ila is, as it were, 'asking for it.' We know that she lusted for him before he lusted for her, and, indeed, that even as a man, Ila suffered from the fatal lust to hunt, a lust often connected with sexual excess. Budha keeps Ila captive not only by lying to her but by giving her pleasure, as she gives pleasure to him (the verb for sexual enjoyment, *ram*, is consistently used in the causative both for him and for her). Even when she becomes a man, Budha gives him the pleasure of hearing stories, using the same verb, *ram*, that was used when he gave her sexual pleasure. This is what Barthes has taught us to call the pleasure of the text. And in this mutual dependence, the woman is far more dependent on sex than the man is. Thus, in a parallel story in which a man named Bhaṅgasvana is magically transformed into a female, when Indra asks him/her which sex s/he would like to remain for ever more, s/he says that s/he would prefer to remain a woman, since as a woman s/he had greater pleasure in sex - which also made her love the children she had as a woman more than the children she had as a man.[8] (Elsewhere it is said that a woman has eight times as much pleasure as a man.[9]) Ila, however, despite his/her pleasure in bed, is horrified by the loss of power in his/her transformation, and does not rest until s/he is restored to his/her manhood again.

The myth begins with the paradoxical statement that Śiva himself becomes a woman while making love with Pārvatī. Are we to assume that he goes on making love to her in that form? Or that he stops? That he stops making love is suggested by another text, in which, when the Goddess all by herself creates her peaceable kingdom, no one gets an erection;[10] of such is the Hindu kingdom of heaven. But that he continues is supported by other texts in which the sexual act continues despite the real or apparent change of gender of one of the partners, such as the myth of Mohiṇī that we will soon see. This ambiguity adds yet another nuance to the more general question of the depth of the transformation, for if Śiva is merely superficially transformed into a female, but remains essentially male, we might expect him to continue despite the transformation. This is one Hindu view, the view that, when the body changes, the mind and the memory remain the same and that gender is not fluid or superficial, but embedded in memory. But if it is a complete transformation, Śiva's continued love-making would involve him in a lesbian act, a rare but not unprecedented situation in Hindu mythology;[11] we will encounter several important examples of this exception to the general rule by which the mind remains unchanged, several myths in which, as in the story of Ila, the mind and memory change, too, change their gender when the body changes its sex.

This is the ambiguous situation into which Ila innocently stumbles, and which automatically and magically transforms him. Another text states that it was Śiva, not Pārvatī, who put the spell on the wood, in order to prevent any *other* male (implicitly excepting Śiva) from seeing Pārvatī when she was making love with him.[12] And yet another text explicitly avoids the awkwardness of Śiva himself being transformed into a woman by stipulating a list of exceptions to the condition of femininity:

> One day when Śiva and Pārvatī were making love, Pārvatī said, 'It is the nature of women to wish to hide their sexual pleasure. Therefore, give me a special place, called the Forest of Pārvatī, in which, except for you and Gaṇeśa and Skanda and Nandin [her sons and servants], any man will become a woman.' One day, a female goblin [*yakṣinī*] who wanted to protect her husband from the king took the form of a deer expressly in order to lure him into the magic part of the forest. King Ila entered the wood … .[13]

Hunting, especially hunting people whom you have mistaken for animals, results in sexual impotence for Pāṇḍu elsewhere in the Mahābhārata.[14]

In the Rāmāyaṇa, Ila stumbles into a transsexual situation, a forest already transformed, in this case by Śiva just to please the Goddess; it is no one's fault at all, merely an accidental step onto a sexual land mine that the gods had planted some time earlier. Elsewhere, it is said that someone interrupted Śiva and Pārvatī in their love play, and only then did Pārvatī proclaim that any man entering that place would become a beautiful woman; all the creatures there became women and made love with Śiva like celestial nymphs[15] (a dramatic contrast with the texts which go to such lengths to keep other men from even looking at Pārvatī). Thus Ila becomes innocently infected by someone else's curse. But in yet other texts, Ila himself barges into the divine bedroom and sets in motion the curse that makes everyone (including himself) become female in the magic wood.[16]

In some variants of this myth, Ila begins life not as a male but as a female,[17] which puts a special spin on the story:

> Ila's parents had wanted a boy; but the priest had made a mistake, and so a girl was born instead, named Ila. The priest then rectified his error, and she became a man, named Ila. One day when Śiva and Pārvatī were making love, the sages came to see Śiva. Pārvatī was naked, and when she saw them she became ashamed and arose from Śiva's embrace, tying her waist-cloth around her loins. The sages, seeing that the couple were making love, turned back. Then, to please his beloved, Śiva said, 'Whoever enters this place will become a female.' Some time later, Ila reached this spot and became a woman, and all the men in his/her entourage became women, and all their stallions became mares. Queen Ila, as she had become again, married and gave birth to King Purūravas. Eventually she begged Śiva to change her back to a man, named Sudyumna, and s/he was allowed to be a woman for one month and a man for one month. Finally s/he went to heaven as someone who had the distinguishing signs of both men and women.[18]

It might be argued that, even here, Ila begins as a male, since it was the original desire of his parents (like all Hindu parents) to have a boy. But since his first physical form is that of a female, his final physical transformation (after he has become a man) is in effect a transformation *back* into her original physical nature. The text therefore constantly fights its way upstream against the current of Ila's tendency to revert to female type, and requires constant interventions from male powers (gods or priest) to keep making her male. Even in heaven, s/he still has both sets of distinguishing marks (*lakṣaṇa*), which here cancel one another out and therefore distinguish nothing.

A story in the *Ocean of the Rivers of Story* is related to the story of Ila, and even cites it as a proof text:

> A man named Śaśin, a friend of the great trickster and magician Mūladeva, was in love with a princess who was closely guarded in a harim. Mūladeva gave Śaśin a pill to put into his mouth (not to swallow), which turned him into a woman so that he could gain access to the harim. Mūladeva himself took another pill that transformed him into an old Brahmin. Once inside the harim, Śaśin took the pill out of his mouth, became a man, and made love to his princess. After a while, a prince saw Śaśin when he was in his form as a woman and insisted on taking 'her' as his wife; Śaśin insisted that the marriage not be consummated for six months, during which she lived in the harim with the prince's first wife, the queen. One night she told the queen the story of Ila and the forest of Pārvatī, took the pill out of his mouth, and made love to her, too. Eventually, Mūladeva married the princess secretly, while Śaśin married her officially.[19]

Śaśin, like Śaśabindu, is a name of the moon, appropriate for someone who periodically changes form. Since he remains male inside even when his body becomes female, the text can imagine him making love only to women, never to men.

3. THE ENCHANTRESS AND THE GODDESS: MOHIṆĪ

A related mythological instance of magic transsexuality is the story of Mohiṇī:

> When the demons stole the elixir of immortality from the gods, Viṣṇu took the form of Mohiṇī, a beautiful enchantress, seduced the demons, and returned the ambrosia to the gods. But when Śiva saw Mohiṇī, he was overcome by lust and immediately ran after her, abandoning Pārvatī who stood with her head lowered in shame. Śiva raped Mohiṇī, and his seed fell upon the ground. Mohiṇī disappeared, and Śiva returned to Pārvatī.[20]

Viṣṇu uses sex to destroy a demonic enemy - an old Hindu trick; celestial nymphs like Mohiṇī routinely use their wiles to destroy powerful individual ascetics[21] - just as the male trickster god (Zeus) creates and sends Pandora to punish Prometheus and mankind for the theft of fire (= the elixir).[22] But in the Hindu version of the Indo-European story, the divine trickster, Viṣṇu, instead of creating a woman like Pandora or simply enlisting the aid of one of the always available celestial nymphs, himself masquerades as a woman, Mohiṇī. In many variants of this myth, Viṣṇu really becomes female, and

Śiva's seed - shed on the ground, with no reference to a male or female partner - gives birth to a child, variously identified as Skanda,[23] Hanuman,[24] Aiyanar,[25] or 'Hariharaputra' ('the son of Viṣṇu and Śiva').[26] In this he is like Ila, who forgets his male nature when he becomes a woman. In most texts, however, including the one we have just seen, when Viṣṇu is Mohinī he retains his male memory and his male essence, and so he can be regarded as having male homosexual relations, playing first the active role with the demons (like the demon Āḍi, who takes the form of Pārvatī to seduce Śiva[27]) and then the passive role with Śiva, an inadvertent masquerade in which Mohinī is the victim rather than the aggressor. The extreme case of this occurs in a Telugu variant of the story, in which, when Śiva makes love to Mohinī, in the middle of the act Mohinī turns back into Viṣṇu, and Śiva goes on with it - a very rare instance of a consummated, explicit, male homosexual act in Hindu mythology.[28] This is the male homosexual parallel to the Rāmāyaṇa text of the story of Ila, in which, if Śiva continues to make love to Pārvatī when he has been changed into a woman, he engages in a lesbian act if we regard this as a change of inner essence but not if we see it as a superficial transformation, like that of Ila or Viṣṇu/Mohinī in most texts.

Although Śiva's motives in raping Mohinī seem straightforwardly heterosexual, his own connection with androgyny also facilitates these shifts, and when it comes to transsexuality he can see Viṣṇu and raise him one:

> The gods begged Kālī to rid the earth of demonic kings, and she agreed to become incarnate as Kṛṣṇa. Śiva prayed to Kālī and was given permission to become incarnate as Rādhā, the mistress of Kṛṣṇa, in order to make love in reverse. At Śiva's wish, Rādhā's husband became impotent immediately after marriage.[29]

Śiva becomes a woman - not, this time, to make love to his wife homoerotically, but, on the contrary, to remain heterosexual when she has switched her own sex (like Viṣṇu as Mohinī). The implication is that Śiva wants to experience not only the reversal of being female but the reversal of the reversal, that is, to be a woman in the 'reversed' sexual position (reversed from the missionary position), the man's role, on top.[30] But this would make no sense unless Śiva maintained his own selfconsciousness as Śiva, a male, while experiencing sex in the body of Rādhā, a female; this is another aspect of the 'reversal', the mind as the reverse of the body.

Viṣṇu - more precisely his avatar, Kṛṣṇa - engages in transsexual behaviour on other occasions, too, one of which is used as a charter myth for the Hijras, or transvestite eunuchs, of India:

> Krishna takes on the form of a female to destroy a demon called Araka. Araka's strength came from his chasteness. He had never set eyes on a woman, so Krishna took on the form of a beautiful woman and married him. After three days of the marriage, there was a battle and Krishna killed the demon. He then revealed himself to the other gods in his true form. Hijras, when they tell this story, say that when Krishna revealed himself he told the other gods that 'there will be more like me, neither man nor woman, and whatever words come from the mouths of these people, whether good [blessing] or bad [curses], will come true.'[31]

The vulnerable man 'who had never set eyes on a woman' is a theme taken from other sources in Indian mythology, such as the famous 'transformation' of the courtesan that Ṛṣyaśṛṅga mistakes for a man who has funny things on her chest.[32]

Another version of the story of the Goddess and the transsexuals makes the demon Araka into the virtuous warrior Aravan, a son of Kṛṣṇa:

> Aravan agreed to sacrifice himself for the sake of the rest of his family, but only after he had been married. The only one willing to be widowed in this way was Kṛṣṇa, who became a woman, married Aravan, made love to him all night on the wedding night, saw him beheaded at dawn, and, after a brief period of mourning, became a man again.[33]

Aravan thus one-ups Oedipus, by sleeping with (a transformation of) not his mother, but his father.

The philosophical underpinnings of the stories about Śiva and Viṣṇu are shared by Hinduism and Buddhism. Thus, just as certain Hindu men became women in order to have the god Kṛṣṇa as their lover, so the Buddhist monk Soreyya was so taken by the beauty of the elder Mahakaccayana, whom he glimpsed at the bath, that he wished to marry him, and his genitals were instantaneously transformed from male to female; like Bhaṅgāśvana, Soreyya has children both as male and as female, and prefers the children of his female persona to those of the male. (No one seems to have dared to ask him which way sex was better.)[34] Other Buddhist variants often express a darker underlying world view:

> A bodhisattva should regard all living beings as a wise man regards the reflection of the moon in water or as magicians regard men created by magic … like the track of a bird in the sky; like the erection of a eunuch; like the pregnancy of a barren woman … .
> The goddess employed her magic power to cause the elder Śāriputra to appear in her form and to cause herself to appear in his form. Then the

goddess, transformed into Śāriputra, said to Śāriputra, transformed into a goddess, 'Reverend Śāriputra, what prevents you from transforming yourself out of your female state?' And Śāriputra, transformed into the goddess, replied, 'I no longer appear in the form of a male! My body has changed into the body of a woman! I do not know what to transform!' The goddess continued, 'If the elder could again change out of the female state, then all women could also change out of their female state. All women appear in the form of women in just the same way as the elder appears in the form of a woman. While they are not women in reality, they appear in the form of women.' With this in mind, the Buddha said, 'In all things, there is neither male nor female.' Then, the goddess released her magical power and each returned to his or her ordinary form. She then said to him, 'Reverend Śāriputra, what have you done with your female form?' Śāriputra: 'I neither made it nor did I change it.' Goddess: 'Just so, all things are neither made nor changed, and that they are not made and not changed, that is the teaching of the Buddha.'[35]

This is the ultimate philosophical support for the fluidity of gender - though that is, we have seen, just one half of the argument.

Serial androgynes sometimes become simultaneous androgynes, like Śiva as Rādhā when Kālī becomes Kṛṣṇa, or the Quasi-men, or the prince (in a story retold by A.K. Ramanujan) who married his own left side.[36] The world title in narcissism surely belongs to the Indian saint Caitanya, in whom Kṛṣṇa and Rādhā became simultaneously incarnate; this happened because Kṛṣṇa looked into a mirror and said, 'How handsome I am! I wish I were a woman and could fall in love with me', and so he became Rādhā, and was able to make love with himself.[37] The seeds of the transformation lie back in the transformation of Kṛṣṇa himself, and his worshippers, the cow-herd women, or, in some cases, cow-herd men: the cow-herd women, abandoned by the teasing Kṛṣṇa, fantasize that half of them have been sexually transformed into men, and make love with one another,[38] while the men wish to become women (and the male worshippers of Kṛṣṇa sometimes dress as women) in order to make love with Kṛṣṇa. Similarly, it is said that the men who saw Rāma wanted to become women to make love with him, and the women who saw Draupadī wanted to become men to make love with her;[39] and, to close the circle, it is said that the men who saw Rāma and wanted to become women to make love with him became reincarnate as the cow-herd women and made love with him when he had taken the form of another incarnation of Viṣṇu, Kṛṣṇa.[40] The Tantric yogi, too, copulates with the female principle inside his own body (the Kundaliṇī), thus becoming his own sexual partner - a neat but, as we have seen, not a unique trick.

4. LETHAL TRANSSEXUALITY

The story of Ila is an instance of the transformation of a man into a woman, the predominant form of the transformation in India (and, indeed, in most of the world). The corresponding transformation of a woman into a man is both rarer and, usually, more dangerous. The most elaborate Epic episode of lethal transsexuality is the story of Ambā, who became a transsexual in order to kill the man (Bhīṣma) who had done her wrong when she was a woman:

> Ambā was abducted and then rejected by Bhīṣma; when she tried to return to her chosen husband, he, too, rejected her. She propitiated Śiva and obtained the promise that she would be reborn as a man who would kill Bhīṣma, and then she entered the fire and died.
>
> Now, King Drupada, whose wife had no sons, asked Śiva for a son, but Śiva said, 'You will have a male child who is a female.' In time the queen gave birth to a daughter, but they pretended that it was a son and raised the child as a son, whom they called Śikhaṇḍin. Only Bhīṣma knew the truth, from his spies and from the sage Nārada, who had told him about Śiva's promise to Ambā in response to her asceticism. When the child reached maturity, 'he' married a princess; but when the princess found out that her husband was a woman, she was humiliated, and her father waged war on King Drupada. When Drupada's daughter, Śikhaṇḍinī, saw the grief and danger of her parents, she resolved to kill herself, and she went into the deserted forest. There she met a goblin (*yakṣa*) named Sthūṇa who agreed to lend her his male sex for a while. They made this agreement and exchanged sexual organs. When Drupada learned from Śikhaṇḍin what had happened, he rejoiced and sent word to the attacking king that the bridegroom was in fact a man. The king sent some fine young women to learn whether Śikhaṇḍin was female or male, and they happily reported that he was absolutely male. The father of Śikhaṇḍin's bride rebuked his daughter and went home.
>
> Bhīṣma, who knew that Śikhaṇḍin had been born a female, insisted that he had made a vow, as a warrior, never to take arms against 'a woman, or anyone that was a woman before, or who has a woman's name or form.' Bhīṣma told the Pāṇḍavas about this vow, and Arjuna and the rest of the Pāṇḍavas used Śikhaṇḍin as a shield in their vanguard. Bhīṣma fell under the rain of their arrows; later, Śikhaṇḍin was killed by Karṇa.[41]

Ambā was caught in limbo between two men, her beloved and the man who abducted her (Bhīṣma); she was socially, if not physically, raped by Bhīṣma (for his abduction of her made her second-hand goods from the standpoint of the man she loved) and then rejected by Bhīṣma as well as by her betrothed lover. Ambā thus epitomizes the no-win situation of a woman, tossed like a shuttlecock between two men, each of whom ricochets between inflicting upon her sexual excess or sexual rejection. And when she becomes a man, that is precisely the sort of doubly hurtful man she becomes: the liminal Śikhaṇḍin/Śikhaṇḍinī rejects her bride, who is humiliated as Ambā (and Ila's bride) had been, and unsexes (and humiliates) a helpful goblin (who has to remain a female far longer than he had originally intended). His/her sexual ambivalence is itself ambivalent, or at least doubled: s/he is androgynous not only in rebirth (a woman reborn as a man) but also within the second birth (a female masquerading as a male - and then transformed, as well).

Śikhaṇḍin does not seem to remember that she was Ambā. There is therefore something anticlimactic about the killing of Bhīṣma by Śikhaṇḍin, an event which is further blurred by its diffusion: Śikhaṇḍin does not kill Bhīṣma outright, but merely functions as a human bulwark for Arjuna (or, in Robert P. Goldman's nice phrasing, Bhīṣma is slain by 'Arjuna hiding, as it were, behind the skirts of his "mother" Ambā in her sexually ambiguous form of Śikhaṇḍin').[42] And Bhīṣma does not die immediately of his wounds but withdraws and dies long, long afterwards (after declaiming thousands of verses of renunciant philosophy, lying on a paradigmatic bed not of nails but of the arrows shot into his body).

But if Śikhaṇḍin does not remember, Bhīṣma certainly does; he has the whip hand over her in this, too, though it seems to be more important to him that Śikhaṇḍin was born a woman (Śikhaṇḍinī) in this life than that she was a woman in a former life (Ambā), let alone a woman who died cursing his name and vowing to kill him. Indeed, it is imperative for him that Śikhaṇḍin is in essence a woman, despite her outer male form, for Bhīṣma has vowed never to fight with a woman. This vow leaves a loophole: only a woman can kill him, precisely because he regards a woman as so lowly that he would not stoop to defend himself against her. (Similarly, the Buffalo Demon can only be killed by a woman, inspiring the gods to create the Goddess to do the job by releasing the female parts of theirselves, their *śaktis*, rather like the right half of the prince.) This mythological oversight is the parallel to the political observation that those in power, by ignoring the differences between those whom they dominate, can be tricked and overcome by them.[43] It is also related to the liminal riddle which appears elsewhere in the Hindu Epics (and in Epics from other cultures): the villain demands, and gets, the assurance that he can be killed neither by day nor by night, neither on land nor on sea, neither by a god nor by an animal, and so forth, and the gods must create a liminal situation (twilight, on the shore, a creature half man, half lion) to kill him.[44] In this case, the perfect solution, a creature with the technical status of a woman but the power of a man, is a murderous transsexual.

Ambā lives on in contemporary Indian politics, as Lawrence Cohen has noted in a cartoon that was plastered onto walls near a big political rally in 1993:

> A male figure representing the common man and labelled the *Sikhandin janata* (*janata* means the people and Sikhandin is the gender-bending warrior from the *Mahābhārata* epic, who for most Banarsis is thought to be like a *hijra* or eunuch) is shown bent over and raped at both ends by two other male figures, orally by a *gandu neta* or politician-bugger and anally by a *jhandu pulis* or useless policeman.[45]

This image is classical in two senses. First, it draws upon a political insight already documented in an ancient Brāhmaṇa text about the horse-sacrifice, which speaks of a male who 'thrusts the penis into the slit, and the vulva swallows it up', and glosses this statement: 'The slit is the people, and the penis is the royal power, which presses against the people, and so the one who has royal power is hurtful to the people.'[46] Second, Śikhaṇḍin himself was, in his previous life as Ambā, un-raped, as it were - sexually rejected - at both ends: by Bhīṣma and by her betrothed lover. A very apt image indeed, but transformed, like Śikhaṇḍin himself, from the image of a woman to that of a man. To make the metaphor powerful and meaningful the authors of the Banarsi cartoon had to transform the double-raped Ambā into Śikhaṇḍin - a man who, in the Epic, is not raped at all, but, rather, takes revenge, and hence is more raping than raped.

5. TRANSSEXUALITY AND HOMOSEXUAL DESIRE IN ANCIENT INDIAN MYTH

Freudians see latent homosexual impulses lurking under the covers even of ostensibly heterosexual acts like cuckolding, which mask a sexual attraction between the man who seduces another man's wife and the man whom he cuckolds. According to this interpretation, a cuckolder is a man who wants to get at another man through his sexual partner; the Hindus speak of weakening a man by destroying the shield constituted by his wife's chastity, rather like the shield that Śikhaṇḍin became for the Pāṇḍava brothers. Unwilling to engage in an overt homosexual act, the cuckolder takes the indirect route via the woman, who may be regarded as the facilitator in a transaction between two men. This is hardly a homosexual act in the strict sense of the word, but it does depict a world in which the sexual tension, if not the desire, is between members of the same sex - or, indeed, within a single person, at war with his or her changing self. Robert P. Goldman has seen this scenario at work in certain myths of the transsexual transformation of a man into a woman, which 'takes place as the consequence of a desire to avoid or defuse a potential sexual liaison with a prohibited female seen as the property of a powerful and revered male and/or the desire to be passively enjoyed sexually by such a male'.[47] The repression of a homosexual impulse may account for the violence in so many of these myths: some, such as the stories of cuckolding, may be motivated not only by lust, but by hatred and the desire for revenge.

Part of the psychoanalytic hypothesis is substantiated by several different sorts of myths: realistic stories in which men dress as women to seduce other men (like Bhīma - with Kīcaka - in the Mahābhārata,[48] or Kavikumar, who dresses as a woman to seduce and kill his brother[49]); fantastic stories in which men become magically doubled and the homosexual fantasy is enacted in a conveniently simplified form, by eliminating the woman (indeed, any separate partner) altogether (such as the prince who married his own left half); and magical stories in which the fantasy is actually acted out by a man who transforms himself into a woman and consummates the heterosexual act with the man (the demon Āḍi, Viṣṇu as Mohiṇī). The most direct variant is also by far the most rare: stories in which men or women, untransformed and undisguised, actually do consummate a homosexual act. In this way, myths may express homosexual fantasies that until now only psychoanalysts have read in (or into) more realistic stories.

On a repressed level, available to a hermeneutics of suspicion, many of these myths do indeed depict masked homosexual encounters. The homosexual themes in traditional myths are seldom overt, because such myths almost always have, as a latent agenda, the biological and spiritual survival of a particular race, in both senses of the word: race as contest and as species ('us against them'). Such myths regard homosexual acts as potentially subversive of this agenda (or, at the very least, irrelevant to it, perhaps not part of the problem, but certainly no part of the solution). The ascetic aspects of Hinduism create a violent dichotomy between heterosexual marriage, in which sexuality is tolerated for the sake of children, and the renunciant priesthood, in which asceticism is idealized and sexuality entirely rejected, or at least recycled. In this taxonomy, homosexual love represents what Mary Douglas has taught us to recognize as a major category error, something that doesn't fit into any existing conceptual cubbyhole, 'matter out of place' - in a word, dirt.[50] Traditional Hindu mythology regards homosexual union not, like heterosexual marriage, as a compromise between two goals in tension (procreation and asceticism), but as a mutually polluting combination of the worst of both worlds (sterility and lust). The myths therefore seldom explicitly depict homosexual unions at all, let alone sympathetically.

Moreover, all sexual acts, homosexual or heterosexual, are regarded with a jaundiced eye by mainstream Hindu mythology. The 'sweet death' or 'little death' of the orgasm or the romantic *Liebestod* becomes a bitter, full-sized, and most real death in many of these stories. The Hindu boundaries of identity are fluid; acts of eating and sex further blur those boundaries by transgressing the limits of the human body. This is surely one of the factors contributing to the great danger that is felt, in India, to accompany the sexual act (and, indeed, eating): if you are not sure where your body ends, you will be very uneasy about exposing it to intimate contact with someone else's body. This anxiety hedges the openings of the body, the things that fall off the body (nails, hair, mucous, and, of course, semen), and, ultimately, sexual intercourse. It is revealed not just in myths that depict sadistic sexual acts or lethal love; it is a part of natural, everyday sex.

The sexist assumption is that every woman is lethal (a poison damsel) and a fake. But why is this fear of women not averted when the partner is imagined, on some level, as male? Instead, the homophobic paradigm of the murderous transvestite goes on to argue that a fake woman (that is, a man pretending to be a woman - like Bhīma with Kīcaka - , or, less often, a woman a man - Śikhaṇḍin) is doubly lethal. Any woman corrupts; a transvestite woman corrupts absolutely. This is one reason why transvestites are more dangerous than transsexuals. For while men who are magically transformed into women may be ashamed, or disenfranchised, they are at least said to experience greater sexual pleasure and, perhaps because of this, do not usually kill anyone. Ambā, who is elaborately and explicitly denied any sexual pleasure at all, becomes predictably murderous. But when men merely *pretend* to become women, they usually experience no sexual pleasure at all but become vicious killers.

6. BISEXUALITY AND ANDROGYNY

But a homophobic Freudian analysis is of only limited relevance to these myths. Not all the homosexual desire in these myths of transvestism and narcissism is depicted as perverse or destructive. Nor should we be too quick to see homosexual desire as an evitable component of the myths of sex change. Often the change is effected in the service of heterosexuality, and, even more often, in the service of a kind of androgyny or bisexuality. And many of these myths are about bisexual desire rather than homosexual desire *tout court*; they are stories about men and women who enjoy sex with both men and women on different occasions, offering, in subversion of the dominant homophobic paradigm, closeted images of a happily expressed and satisfied homosexual desire. They are not all happy stories, nor charters for the affirmation of a polymorphous, Jungian androgyny. Some of the myths about androgynes attempt to eliminate the woman, to eliminate the other, to produce the only truly safe sex - when you are alone (as in the Woody Allen joke about masturbation: you meet such nice people that way). And it would certainly be simplistic to overlook the misogynist implications of the argument that women enjoy sex more than men do. But these texts do tell us that sexual pleasure is a serious goal for both sexes (it influences the preference for one set of children over another, which is certainly significant) and that the desire for pleasure with members of both sexes is real, though ultimately out of reach for all but the magically gifted - or cursed.

Margaret Trawick has pointed out some of the complexities of transvestism in South India:

> Transvestism among men is very common, and even 'normal' boys can get away with dressing up like girls, just for fun, and learn to move their bodies in convincingly female ways. Śiva is a hermaphrodite; pictures of Krishna make him look like a girl. Does this mean that masculine and feminine are not valid categories to South Indian people? Are the women who plow or the men who dress up as women questioning the essential opposition between male and female? No. ... We might consider the proliferation of androgyny there to be one aspect of a pleasure in sexuality in its original polymorphous nature that we ourselves miss, together with an intellectual enjoyment of paradox, which, also, we fail to share. This is not to say that sexual oppression, and repression, do not occur in India. But too often, this is all that ethnographers see.[51]

These myths may express a desire that transcends both sex and gender, a desire that desires the mind no matter what bits of flesh may be appended to various parts of the body. Moreover, they remind us of two truths in tension, a paradox: one Hindu view of gender makes it as easy to slough off as a pair of pants (or a dress), but this view is often challenged by myths in which skin is more than skin deep, in which the mind and the memory, too, are gendered, an intrinsic part of the mortal coil that is not quite so easily shuffled off.

Notes

1. See, most recently, the excellent survey and analysis by Robert P. Goldman, 'Transsexualism, Gender, and Anxiety in Traditional India', *JAOS* 113.3, 1993, pp. 374-401.
2. See Wendy Doniger O'Flaherty, *Dreams, Illusion, and Other Realities*. Chicago: University of Chicago Press, 1984, pp. 81-9.
3. Mahābhārata. Poona: Bhandarkar Oriental Research Institute, 1933-69, 1.70.16.
4. Brahmāṇḍa Purāṇa. Delhi, 1973, 4.10.41-77.

5. Liṅga Purāṇa 1.65.19-23; Matsya Purāṇa. Poona, Anandasrama Sanskrit Series 54, 1907, 11.44-8; Viṣṇu Purāṇa, with the commentary of Śrīdhara. Calcutta: Sanatana Sastra, 1972, 4.1.10-15; Devībhāgavata Purāṇa 1.12.35; Padma Purāṇa. Anandasrama Sanskrit Series 131, Poona, 1893 1.8.82-116.
6. Rāmāyaṇa. Baroda: Oriental Institute, 1960-75, 7.87-90.
7. *The Laws of Manu*. Translated by Wendy Doniger, with Brian K. Smith. Harmondsworth: Penguin, 1991, 3.34.
8. Mahābhārata 13.12.1-49. See also Wendy Doniger O'Flaherty, *Women, Androgynes, and Other Mythical Beasts*. Chicago: University of Chicago Press, 1981, pp. 305-6.
9. Garuḍa Purāṇa. Bombay, 1892, 109.33.
10. Skanda Purāṇa 1.3.1.10.1-69. See Wendy Doniger O'Flaherty, *Hindu Myths*. Harmondsworth: Penguin, 1975, p. 245.
11. Padma Purāṇa, *Svarga Khaṇḍa* 16.6-24; *The Svarga Khaṇḍa of the Skanda Purāṇa*, edited by A.C. Shastri. Varanasi, 1972. Translated in Wendy Doniger O'Flaherty, *Textual Sources for the Study of Hinduism*. Chicago: University of Chicago Press, 1990, pp. 98-100.
12. Sāyaṇa on Ṛgveda 10.95; in: Ṛgveda, with the commentary of Sāyaṇa. 6 vols., London, 1890-92, vol. 4, pp. 639-40.
13. Brahma Purāṇa. Calcutta, 1954, 108.26-30.
14. Mahābhārata 1.109-111.
15. Brahmāṇḍa Purāṇa 3.60.23-7.
16. Bhaviṣya Purāṇa. Bombay, 1959, 3.4.17.23-7.
17. Liṅga Purāṇa. Calcutta, 1812, 1.65.19-20.
18. Bhāgavata Purāṇa, with the commentary of Śrīdhara. Benares: Pandita Pustakalaya, 1972, 9.1.18-42. See also Devībhāgavata Purāṇa. Benares, 1960, 1.12.1-35. Wendy Doniger O'Flaherty, *Śiva: The Erotic Ascetic*. London: Oxford University Press, 1973, pp. 304-5.
19. Kathāsaritsāgara. Bombay: Nirnaya Sagara Press, 1930, chapter 89 [12.15].
20. Brahmāṇḍa Purāṇa 4.10.41-77. See O'Flaherty, *Śiva*, pp. 228-9.
21. O'Flaherty, *Śiva*, pp. 87-90.
22. Hesiod, *Theogony* and *Works and Days*.
23. R. Dessigane and Jean Filliozat. *Les légendes çivaïtes de Kāñcipuram*. Pondichéry: IFI, 1964 (Publication de l'Institut français d'indologie No. 27) No. 59, pp. 76-7.
24. Śiva Purāṇa. Benares: Pandita Pustakalaya, 1964, 3.20.3-7.
25. Bhāgavata Purāṇa 10.88.14-36.
26. R. Dessigane and P.Z. Pattabiramin, *La légende de Skanda selon le Kanda-puranam tamoul et l'iconographie*. Pondichéry: IFI, 1967 (Publications de l'Institut français d'indologie No. 31), pp. 84-5 (2.32.6-47).
27. Skanda Purāṇa 1.2.27-29; O'Flaherty, *Hindu Myths*, pp. 251-61.
28. This story was told to me by V. Narayana Rao.
29. Mahābhāgavata-Purāṇa 49-58, cited by Rajendra Chandra Hazra, *Studies in the Upapurāṇas*. Calcutta: University of Calcutta, vol. 2, 1963, pp. 272-3.
30. O'Flaherty, *Women*, p. 116.

31. Serena Nanda, *Neither Man Nor Woman: The Hijras of India*. Belmont: Wadsworth, 1990, pp. 20-1.
32. O'Flaherty, *Śiva*, pp. 42-51.
33. Kavita Shetty, 'Eunuchs: A Bawdy Festival', *India Today*, 15-6-1990, pp. 50-5.
34. Dhammatthakathā 3.9, on *Dhammapada* 43, cited by Goldman, 'Transsexualism'.
35. *The Holy Teachings of Vimalakīrti*, tr. by Robert A.F. Thurman. University Park & London: Pennsylvania State University Press, 1976, pp. 56, 61-2.
36. A.K. Ramanujan, 'The Prince who Married his Own Left Side', in: Margaret Case and N. Gerald Barrier (eds), *Aspects of India: Essays in Honor of Edward Cameron Dimock, Jr*. New Delhi: American Institute of Indian Studies and Manohar, 1986, pp. 1-16.
37. Caitanyacaritāmṛta of Kṛṣṇadāsa, translated by Edward C. Dimock, ms., my paraphrase.
38. Bhāgavata Purāṇa 10.30.
39. Govindarāja on Rāmāyaṇa 2.3.39, cited by Goldman, 'Transsexualism', p. 383.
40. Padma Purāṇa 6.272.165-67.
41. Mahābhārata 5.170-87, 189-93; 6.103-114; 8.59-60; O'Flaherty, *Women*, p. 307.
42. Goldman, 'Transsexualism', p. 392.
43. James C. Scott, *Weapons of the Weak: Everyday Forms of Peasant Resistance*. New Haven and London: Yale University Press, 1985.
44. Thus Indra kills the demons Namuci and Vṛtra (Mahābhārata 5.9-13; O'Flaherty, *Hindu Myths*, p. 83), and Viṣṇu as the Man-Lion kills Hiraṇyakaśipu.
45. Lawrence Cohen, 'Semen Gain, Holi Modernity, and the Logic of Street Hustlers', paper presented at the annual meeting of the Association of Asian Studies, Boston, 25-3-1994, p. 3.
46. Śatapatha Brāhmaṇa. Chowkhamba Sanskrit Series, 96, Benares, 1964, 13.2.9.6-9. See O'Flaherty, *Textual Sources*, p. 17.
47. Goldman, 'Transsexualism', p. 391.
48. Mahābhārata 4.21.1-67.
49. Avadānakalpalatā, Bodhisattva Kalpalata, Akanda, Ralana, Kshemendra, Anubada (tr.): Saruccandra Basu; Sampadana: Bishnu Basu, 1981.
50. Mary Douglas, *Purity and Danger*. London: Routledge and Kegan Paul, 1966.
51. Margaret Trawick, *Notes on Love in a Tamil Family*. Berkeley: University of California Press, 1990, p. 253.

VIII

NOTES ON THE TSHECHU FESTIVAL IN PARO AND THIMPHU, BHUTAN

Jan Fontein

Throughout the region in which Tibetan Buddhism once flourished we come across the ancient tradition of monastic festivals in which sacred dances and edifying pantomimes are performed. The first traveller to report in writing on these 'devil dances' appears to have been H.H. Godwin-Austen in 1865.[1] For the next hundred and thirty years these masked mystery plays have fascinated many Western and other travellers to these regions. The Chinese did their best to stamp out this form of ritual and religious performing art in Tibet during their infamous Cultural Revolution and its aftermath. Although there are scattered bits of information on performances of this type coming out of China, it is unclear to what extent similar rituals have survived as a genuine tradition in any of the monasteries in China proper, its Northeastern Provinces and in Mongolia. It may, therefore, be only in the southern borderlands of Tibetan Buddhism - in Ladakh, Nepal, Sikkim and Bhutan - that we can still observe these monastic dances being performed as part of an authentic, living religious tradition. Numerous early descriptions and even occasional colour photographs, such as, for example, those taken before the Second World War in a Mongolian monastery in Gansu Province by the Japanese Buddhologist Hashimoto Kōhō[2] reveal all kinds of differences as well as striking similarities with what can still be seen today in Bhutan. One is inclined to conclude, therefore, that the monastic dances performed in these regions were all part of the same Tantric ritual tradition, but that they must once have displayed all kinds of interesting regional and perhaps even local variations.

It was not long after the Himalayan kingdom of Bhutan first opened its doors to small numbers of tourists in 1974 that the American public was given its first glimpse of the Bhutanese tradition of monastic dances. This happened in 1980, when a group called 'The Royal Dancers and Musicians from the Kingdom of Bhutan' toured the United States under the auspices of the Asia Society's Performing Arts Program.[3] The spirited performances in

New York, Boston and Cambridge drew great crowds and were warmly received. Unavoidably, however, dances had to be performed on a modern stage that could not convey any impression of the monumental architectural backdrop against which these dances are usually seen in Bhutan. For not just the dances themselves, but also their impressive monastic setting are essential for a true appreciation of the grandeur of this art form.

Two recent visits to Bhutan have provided me with a rare opportunity to attend two Bhutanese monastic festivals of the type locally known as *Tshechu* in the setting for which they were created, the Buddhist monastic fortresses or *dzong* of Bhutan. After having first attended the *Tshechu* at the Tashichoedzong in the Bhutanese capital Thimphu two years ago, I recently had an opportunity to compare the monastic dances that I had seen there with those performed at the *Tshechu* of the Rinpungdzong at Paro. The following notes, which merely represent my first impressions and preliminary observations, were written immediately after my return, while the solemn sound of the telescopic trumpets still echoed in my ears. They are offered here as a token of my appreciation to my compatriot and fellow traveller across the Atlantic, Frits Staal, whose profound interest in and vast knowledge of all kinds of Asian rituals is well known to scholars all over the world.

In the traditional training of monks in Bhutan the first elementary classes concentrate on the novice's mastery of three basic skills: the drawing of *mandalas*, the chanting of the liturgy, and the choreography of the sacred dances. At festivals one can see the young monks leaning out of the windows of the monastery, keenly observing every step and whirling movement made by their seemingly indefatigable seniors in the courtyard below. But even though careful observation, followed by close imitation, is an essential part of this learning process, it is not deemed sufficient by the Bhutanese to ensure the ritual perfection that is required for such monastic ceremonies. Each *dzong* or monastery, therefore, has a Master of Dance or *Chhampon*, a senior monk and experienced dancer, who hands down the skills of his traditional techniques to a younger generation. A recent visit to the new monastery at Ura (Central Bhutan) gave me an unexpected opportunity to observe such a dance class. There the monks rehearsed, under the strict supervision of their teacher, a complex sequence of dance movements, all of them counting their steps aloud in unison. All this was done barefoot on a wooden floor, under the watchful eye of a huge, recently modelled statue of Padmasambhava, who is credited with bringing Buddhism to Bhutan.

Depending on their individual ability, other monks may receive instruction from the Master of Trumpets or *Dongpi Lhopen* in playing the wind and percussion instruments that are used during the ceremonial dances. These are the long, telescopic horns and the smaller trumpets, drums and cymbals. A few weeks before each festival there is one official public rehearsal or 'dance test' (*chhamgyu*) in which monks and laymen perform without either the colourful costumes or the sacred masks. This rehearsal, often almost as well attended as the *Tshechu* itself, serves not only the purpose of ironing out the last logistical and choreographical wrinkles of the entire ritual, but also gives the audience an opportunity to actually see the *dramatis personae*, whose identity remains hidden behind the masks during the actual performances.

Unlike many other traditional or ritual dances in other parts of the world, the dances (*chham*) performed at the *Tshechu* festivals are not considered anonymous creations of Buddhist folklore. Although the historical truth about the origins of these dances is often difficult to establish, many of them have been traditionally attributed to specific choreographers, often famous saints like the *Terton* or 'Finder of Treasure' Pema Lingpa (1450-1521) or the Shabdrung Ngawang Namgyal (1594-1651). Even the founder of Bhutanese Buddhism himself, the Indian saint Padmasambhava (ca. AD 800), known in Bhutan as Guru Rimpoche or 'Precious Teacher', in whose honour the festival in Paro is held, is among those who are credited with the choreography of some of these dances. Their divine origin is indicated by the fact that the Tantric deity Heruka, the 'Drinker of Blood', is widely believed to have created the first of these dances.

The timing of the festivals varies in different parts of the southern border regions. For reasons that are obvious in a strictly agricultural community, almost all monastic festivals in Ladakh take place during the winter months, when all farming activity ceases. For example, the Great Festival of Leh, the capital of Ladakh, coincides with the end of the twelfth moon according to the Tibetan lunar calendar. In Ladakh the festival at Hemis, called Mela, constitutes the only exception; it takes place around the month of June. It was probably this timing, during the season when all the passes leading into Ladakh are open, and the greater accessibility to foreign travellers resulting from it, which account for the early discovery of this festival by the Western world.

It was at the Hemis festival that as early as 1856 Emil and Hermann von Schlagintweit made the first plastercasts of the masks used in these dances. These casts have been preserved in the National Museum of Ethnography, Leiden.[4] At Wutaishan in China, where such a festival is thought to have been held ever since the reign of the Fifth Dalai Lama, the ceremonies are reported to have lasted ten days, ending on the fifteenth day of the sixth lunar month. At the large monastery Yonghegong in Beijing similar dances were performed on the first day of the second month of the lunar calendar.[5]

In Bhutan fifteen major monasteries (*dzong*) hold monastic festivals of the *Tshechu* type, of which six were open to foreign visitors in 1995. Although these Bhutanese festivals vary greatly in date, many seem to begin on the tenth day of a lunar month. The name *Tshechu*, i.e. 'ten', suggests that this may originally have been the rule, but at least in recent years festivals of the *Tshechu*-type have been held on other dates as well. In the minds of the Bhutanese the tenth day of the lunar month is associated with some of the great moments in the life of Guru Rimpoche. The tenth and the twenty-fifth day of the month according to the Bhutanese calendar became designated as days on which Padmasambhava is worshipped with special ceremonies. *Tshechu* festivals that last five days, such as the Paro *Tshechu*, reach their climax in the early hours of the fifteenth day, which coincides with the full moon. That day is also an important anniversary, for the fifteenth day of the second lunar month is the day on which the Buddha is believed to have entered *nirvāṇa*.

One of the features of the *Tshechu* festival that is thought to be uniquely Bhutanese is the participation of laymen in the monastic performances. This sharing of duties with laymen does not occur at all *dzong*: in the festivals at the *dzong* of Punakha and Lhuntse only monks are permitted to dance. Whenever allowed, however, the role of laymen is strictly limited to that of dancers and actors in the edifying plays that are part of the standard repertoire. All laymen are required to dance barefoot, except when performing the role of a nobleman. In Thimphu they enter and leave the courtyard in which the dances and pantomimes are staged through an entrance clearly separated from that used by the monks. In an obvious concession to modern times the performances at Paro, in which all female roles have traditionally been performed by men, are now interspersed with performances of folk dances and folk singing by a choir of women. This innovation was not yet in evidence at Thimphu two years ago.

The repertoire of dances and pantomimes is largely the same in Thimphu and Paro and consists of approximately twenty different pieces, three or four of which are performed on a given day and several of which are repeated once or twice on different days of the festival. The repertoire can be roughly subdivided into plays of three distinct types. If we can give credence to the ancient legends, the first and original type of dance serves to magically cleanse of all evil influences the place, where the dances are to be performed, as well as the minds of participants and spectators. Such dances, which are called *mechham*, have an unmistakable exorcist function and are exclusively performed by monks, who often wear huge demonic masks and who sometimes inwardly repeat *mantras* while they are dancing. By mounting these masks high up on their face, the dancing monks cannot look through the eyes of the mask, but only through its open mouth. The result is that these masks make the participants look taller than they are, an impression which enhances the terrifying aura of their masks and costumes, especially when they face an audience seated on the floor of the paved courtyard.

The second type is of a more popular, unmistakably edifying character. It is called *bodchham* and is largely performed by the laymen or *gomchen*, who play an important role in the monastic life of Bhutan. These dancers often wear animal masks. Most of the diverse subject matter of their performances seems to have been borrowed, albeit far from literally, from Buddhist literature such as the tales of previous rebirths of the *jātaka* and *avadāna* type. Some of these stories have been modified to such an extent that only the barest minimum of a connection with the original tale has been preserved.

The third type of dance, in which all of the performers consist of masked drummers, glorifies the Guru Rimpoche and symbolizes the triumph of his doctrine over his demonized shamanistic adversaries.

One of the most spectacular dances of the first type, that seems to have been performed in Mongolia in much the same manner, is the *Durdag* or Dance of the Lords of the Cremation Grounds.[6] Dressed in white skirts and boots and wearing white skull masks, the four dancers carry between them in scarves a small box containing a figurine made of dough symbolizing the essence of evil. This dance seems to be the one that is most characteristic for Tibetan Buddhism. These dancing skeletons, who are believed to have been converted from enemies to defenders of the faith, haunt the cremation ground, the favourite location for tantric meditation. At the same time they symbolize the graveyard of all desire that lies at the root of the chain of rebirths. From an illustration in a recently published booklet on Wutaishan, the great Buddhist centre of pilgrimage in China, written mainly for the benefit of overseas Chinese and Japanese tourists, it seems that a version of this type of dance is still being performed there.[7]

Another dance of the exorcist type is the Dance of the Terrifying Deities or *Tungam*, in which the demonic attendants of the terrifying manifestation of Guru Rimpoche as Dorje Dragpo or 'Fierce Thunderbolt' make their appearance. They encircle the evil spirits, forcing them all into a small box, after which Dorje Dragpo kills them with his *phurbu*, his ritual dagger. The splendid, colourful costumes, the large demonic masks and the long sequence of solo dances by each of the sword-carrying demons make this one of the most impressive dances in the monastic repertoire.

Not all dances of the exorcist type require the wearing of masks. The Black Hat Dance (*Shanag*), attributed to Lhalung Palgi Dorji (tenth century), named after the large hats worn by all participants, is danced without masks. The costume (*chhamgo*) is believed to be reminiscent of the costumes worn by priests of the pre-Buddhist Bonpo religion.

Among the most popular and spectacular plays, lasting at least three to four hours, is the Judgement of the Dead or *Raksha Marcham* (*Raksa mar'chams* or litt.: 'the Dance of the Demons from Below'). It depicts in colourful detail the popular belief, ultimately derived from the descriptions in the well known *Tibetan Book of the Dead*, how the spirits of the deceased have to appear in the court of the Supreme Judge of Hell, Shinje Chhogyel (Gshin rje Chos rgyal), to account for the good and bad deeds that they committed during their lifetime. After the animal-headed monsters, who constitute the helpers of the court, have danced their long introductory dance, a puppet representing Shinje Chhogyel is carried into the courtyard. In Paro this huge puppet, made of woven basketry covered with cloth, can easily be carried by six or seven monks and is placed, after having made a solemn round of the courtyard, with its back to the wall facing the audience in the courtyard.[8] In Thimphu this figure is of such enormous size that it can only be pushed from the hall inside the monastery onto the platform at the top of the double flight of stairs in front of the building.

The white god and the black demon, who accompany all beings through life, taking note of their positive and negative karma, also make their appearance at the court of Shinje Chhogyel, one pleading for a rebirth in heaven, invoking all the good deeds performed by the deceased, while the black demon points to the sins that can only be expiated by a rebirth in hell. The supreme judge holds a mirror in which all the deeds of the person appearing before him are faithfully reflected. A sinner is the first to appear in court, and after a great deal of argument in pantomime a long strip of black cloth is laid out. It symbolizes the road to hell that the condemned sinner now has to travel, dragged and surrounded by triumphant demons. After this episode a pious person makes his appearance, armed with a prayer flag. In spite of repeated and forceful interventions by the black demon a shorter, white strip of cloth is eventually laid out and the Judge of Hell and all of his animal acolytes join the saved soul in a final dance around the courtyard.

There appears to be no solid evidence that this dance was ever performed in Tibet. A recent photograph taken of a procession at Wutaishan includes several participants who belong to the cast of characters that can be associated with the Judgement of the Dead. This may perhaps be interpreted as an indication that a play of a similar type was once performed there.[9] However, there is no trace of the huge puppet of Shinje Chhogyel and an earlier description by the Japanese scholars Ono and Hibino of the dances at Wutaishan makes no mention of such a figure at all.[10]

One of the edifying pantomimes that seems to have enjoyed great popularity throughout the region is the Dance of the Stags and the Hounds or *Shawa Shachhi*. Enlivened by comic interludes in which clowns participate, and often divided into two separate parts, the lengthy story recalls a theme often found in *jātakas*: the conversion of a hunter dedicated to the killing of animals, to a non-violent lifestyle. In Bhutan the hero of this edifying tale is not the Buddha himself in one of his previous rebirths, as he always is in the *jātaka* stories, but the saint Milarepa (1040-1123). Also, it is not only the hunter who finally accepts the precepts of *ahiṁsā*, as happens in other versions of the story: here even his two hounds are converted to a strictly vegetarian way of life!

How even a well known story can be changed almost beyond recognition is vividly illustrated in the dance called *Pholey Moley*, the Dance of the Princes and Princesses, highly popular among the Bhutanese because of its rather crude and earthy slapstick humour. The story, in Bhutan usually associated with a legendary Indian king named Norzang, is actually the *avadāna* of Sudhana and the Nāga-princess Manoharā, well known from Sanskrit literature. The story has been illustrated *in extenso* on the reliefs of Borobudur and was until recently still performed in Burmese theatre.[11] In Thailand it can still be seen in the traditional theatre today.[12] Strangely enough, however, the only motif of the original story that has been retained in the *Pholey Moley* dance is that the princes go off to the war. Instead of being innocently accused, as in the original story, here the princesses blatantly commit adultery with the clowns during their husbands' absence. When the heroes return from the war and discover the infidelity of their spouses, they make them suffer for their indiscretions by cutting off their noses. Later a doctor is called in to reattach the noses, a delicate surgical operation that is crowned with success only after a string of comic failures.[13]

After the performance of the Judgement of the Dead, the most edifying of all the pantomimes, on the fourth day, the *Tshechu* of Paro reaches a dramatic climax on the fifth and last day of the festival. At three o'clock in the morning, just as the full moon appears above the mountain ranges, a procession of monks, preceded by trumpeters, emerges from the *dzong*, carrying on their shoulders a huge *thangka*, wrapped in red cloth. They place, unwrap and fold out the *thangka* in front of the four-storied building overlooking the courtyard in which the *Tshechu* is celebrated during the daytime. The huge, square *thangka*, each side of which measures more than sixty feet, is attached by means of loops all along its top to a long steel tube. Two monks, standing on the roof of the adjacent building, sound their large trumpets to mark the beginning of the ceremony. Directed by a senior monk, who nowadays gives his orders through a public address system, other monks slowly hoist the *thangka* up to the level of the roof of the building. During the entire operation great care is taken to keep the steel pipe, from which the *thangka* is suspended, in the required horizontal position. After the *thangka* has been hoisted to its proper height, adjustments are made to eliminate folds and creases. It is undoubtedly no mere coincidence that the *thangka* fits exactly in the available space, its upper edge close to the eaves, its lower edge touching the slightly projecting stone base of the four-storied building.

As soon as the ropes of the *thangka* have been tied down to keep it in its unfurled position, and the image of Padmasambhava and his retinue manifest themselves in all their splendour, large numbers of Bhutanese begin to stream towards it, all prostrating themselves in front of it and receiving the blessings from the assembled clergy in a ceremony called *Shugdrel*. Huge scrolls of this particular type are called *thongdrol*, which literally means: 'whose very sight produces enlightenment', and the reaction of the crowds to the unfurling of this *thongdrol* clearly illustrates that in Bhutan this term still retains its profound, literal meaning for the faithful. According to the hagiographic *Life of Padmasambhava*, the Buddha Śākyamuni predicted the birth of Padmasambhava prior to his entrance into *nirvāṇa*. The Bhutanese believe that Padmasambhava appeared in this world in order to continue enlightening all human beings after the Buddha had entered *nirvāṇa* and could no longer perform this task himself. The magic appearance of Padmasambhava on the *thongdrol*, exactly on the anniversary of the Buddha's entrance into *nirvāṇa*, seems to symbolize this tenet in a most spectacular and dramatic manner.

After receiving the abbot's blessing and after having made an appropriate cash contribution to a pair of monks holding up a large plastic moneybag, all participants are invited to file through the ground floor storage room of the building from which the *thongdrol* has been suspended. In this storage room the masks and various other paraphernalia that are used for the *Tshechu* are on display. The masks, made of carved wood or papier mâché, are treated as sacred heirlooms and do not leave the storage except to be worn in the performances during the *Tshechu*. Included among these objects on display is the huge basketry frame in the rough shape of a human head and torso that is used to dress up the giant figure of the Judge of Hell.

During my visit a single floodlight, installed by the monastery, shone on the surface of the painting. To this source of light was added intermittently the much more intense glare of the lights of a Japanese television crew that seemed intent on demonstrating its utter disdain for religious ceremonies other than their own. When they were occasionally distracted or otherwise engaged one could still visualize the mysterious effect that the unfurling of this giant *thongdrol* must have made at a time when butter lamps and the full moon provided the only sources of illumination. Although one would be inclined to regard the unfurling of the *thongdrol* on the day of the Buddha's entrance into *nirvāṇa* and on the night of the full moon as deliberately timed in order to provide lighting conditions that are appropriate for the solemn occasion, such a supposition is not supported by other evidence. By far the largest of all *thongdrol* in Bhutan is preserved at the Kurje Lakhang monastery in Central Bhutan, where an impression of Guru Rimpoche's body is worshipped. There, however, the *Tshechu* ceremony and the unfurling of the *thongdrol* take place on the tenth day of the lunar month, i.e. five days before full moon, on the day designated for ceremonies honouring Padmasambhava. What is claimed to be the largest of all these *thangkas*, although no precise measurements seem to be available, is one that is kept in the Hemis monastery in Ladakh. This *thangka* is said to be shown only once every eleven years.

The tradition to display huge hanging icons on the occasion of temple festivals may have its origins in Lhasa, where a huge thangka, sixty metres wide, was hung from the Potala at the conclusion of the ceremonies connected with the New Year celebrations.[14] At one time this tradition may have been observed throughout the Buddhist world, but at the present time this custom survives only in Ladakh, Bhutan, Korea and, perhaps, in some areas of Tibet. Judging from the photographs I have seen, it seems that not all monasteries display icons representing the same figures from the Buddhist pantheon. It is possible that these differences reflect the varying sectarian affiliations of these monasteries. For example, the painting displayed in the Likir monastery of Ladakh, belonging to the Sect of the Yellow Hats (Gelugs-pa), shows Tsongkhapa, the founder of this sect.[15]

The Korean tradition differs from the Bhutanese in several respects. In Korea huge representations of the Historical Buddha, preaching the *Lotus Sūtra* at Vulture Peak, are unfurled in the temple precincts on the occasion of his birthday, the eighth day of the fourth lunar month. Of these paintings, called *kwaebul*, litt.: 'Hanging Buddha', at least thirty-six datable examples exist, ranging in date from AD 1622 to 1892. As hardly any multi-storied buildings have survived in Buddhist temples in Korea, the large paintings are usually hoisted on a pair of tall masts. The largest painting of this type in Korea is perhaps that of the temple Ssanggyesa (Hadong-gun, Kyōngsan Namdo Province). It measures 45 by 23 feet and bears an inscription datable to AD 1799.[16] Although embroidered examples, called *subul*, do exist, most of these Korean paintings are painted on cloth. Nothing is known about the origin of this Korean tradition. The most plausible hypothesis seems to be that the custom to display these large paintings came to Korea with the Mongols during the Koryō period.

The Bhutanese *thongdrol* all seem to have been made using an appliqué technique, creating an ingenious patchwork of many different types of textiles. Instead of the Historical Buddha, the Bhutanese *thongdrol* invariably represent the 'Second Buddha', Guru Rimpoche, i.e. Padmasambhava, surrounded by all of his Eight Manifestations. A monastic dance in which these same Eight Manifestations make their appearance is scheduled for later on the last day of the Paro festival, thus complementing the static two-dimensional rendition of the *thongdrol* with a live, dynamic, three-dimensional representation of the same deities. The composition of the huge *thongdrol* follows the standard iconography known from *thangkas* of much smaller size. Several examples of such *thangkas* of modest size can be seen in the old watchtower above the Paro monastery which has in recent years been converted into the National Museum of Bhutan. The poetic but sparse lighting conditions under which I was able to study the *thongdrol* made it difficult to assess its age. While some of the colours of the patchwork looked fresh and new, the careful handling and, above all, the minimal exposure to daylight may well have contributed to its almost perfect preservation and account for the fresh look of a picture that may be quite a bit older than it looks. For the rays of sun are not permitted ever to touch this *thongdrol*, and shortly after daybreak the painting is ceremonially cleansed, lowered, rolled up and folded up again to be escorted back to storage, where it will be kept wrapped until the next year.

Illustration 1.

The giant temple hanging (*thongdrol*) during its nocturnal display at Paro Monastry (author's photograph).

Although some of the mystery plays are spiced with earthy humour, most of the plays are solemn and moving at a slow pace. An age-old, universal device to break the monotony of solemn, protracted performances and to keep alive the attention of the public is the introduction of comic interludes, performed by clowns. Like everywhere else in the world dressed up in tattered, patched-up clothes and wearing long-nosed or other types of caricatural masks, the clowns or *atsara*, as the Bhutanese call them, who work at the festivals four or five at a time, originally seem to have performed a dual function. After a particularly long and solemn dance performance they take over centre stage and perform their own satirical and slapstick version of the same dance to lighten the atmosphere. They also take an active part in some of the edifying plays such as the Hunter and the Stag (*Shawo shachhi*) or the Princes and Princesses (*Pholey Moley*). They originally seem to have had the additional duty of keeping the public at the proper distance from the performers, chasing away unruly youths with a baton in the shape of a huge phallus. Nowadays this last function seems to have been largely taken over by uniformed police carrying batons of a more conventional design. Even though huge crowds attend these festivals, the public behaves in a happy and relaxed, but orderly manner. Only rarely does the police actually have to intervene. Foreign photographers are the most frequent offenders.

Between carrying out these duties, the clowns mingle with the public, making young girls giggle and blush with their dirty jokes and unmistakably obscene gestures, in which their phallic badge of office invariably plays a prominent role. Older women tend to give them tit for tat and treat their sexual innuendos with undisguised scorn. On the side the clowns do a little bit of fund raising, not for themselves, but strictly for the monastery. As the roles of the *atsara* are less precisely prescribed than those of the dancers, it is in the behaviour of the clowns that the differences between the Thimphu and Paro festivals became most apparent. The comic talents of the Thimphu clowns were unquestionably superior. With their outrageous antics they diverted the attention of the public time and again. Much more so than the clowns in Paro, who made a more restrained and rather subdued impression. It is evident that the masks of the clowns, just like those of the actors in the principal animal, human and demonic roles, represent types rather than individuals. For example, a group of *atsara* masks from Bhutan, now in the Museum of Ethnography in Heidelberg, represent exactly the same types as those still used by the clowns in Thimphu and Paro.

Another important difference between the *Tshechu* festivals in Thimphu and Paro seems to be the result of the architectural setting. At Thimphu the entire festival is conducted inside the monastery, in the spacious main courtyard, surrounded on all sides by tall buildings. At Paro, on the other hand, the dances are performed in the much more confined space inside the monastery only on the first day of the festival. The next day the performances shift to an open space outside the monastery itself, to a square in front of a tall, four-storied building. The road leading from the main entrance of the *dzong* to the square where the dances are performed is lined with stalls, creating the atmosphere of an annual fair. In Thimphu no commercial activity occurred at all. The absence of commerce and the presence there of so many high government and ecclesiastic authorities added to the solemnity of the occasion.

While the remoteness of some of the *dzong* and the government policy to restrict tourism still afford a measure of protection to this vibrant type of religious performances, it is obvious that the *Tshechu* of Bhutan should be placed on the list of endangered cultural manifestations. Surrounded by countries that have woefully mismanaged their ecological and cultural resources the Bhutanese are determined not to make these same mistakes, and in many areas their carefully considered measures to bring Bhutan into the twentieth century without losing their national cultural identity seem to meet with success. It remains to be seen, however, whether a policy can be successful that aims on the one hand at doubling the annual number of tourists to five thousand and at the same time keeps foreign visitors away from the greatest masterpieces of indigenous architecture, the *dzong*. As grateful as any scholar of ancient Buddhism should be for an opportunity to experience this fascinating Buddhist culture that is still sparkling with life and largely untouched by Western influences, we should also exercise the utmost restraint in order to make our presence there as unobtrusive, as non-disruptive as possible. That is the least we can do to help the Bhutanese protect and preserve their ancient, rich and fragile cultural and religious heritage.

Except for the modest first efforts by Ono and Hibino, to date no scholarly effort seems to have been undertaken to compare systematically the differences in programming, choreography, in costume and in the styles of carving of the masks in the various regions where these monastic rituals have been observed through the ages. If such a project should turn out to be still feasible now, it should be carried out in the near future, before it is too late.

Illustration 2.

The puppet of the Judge of Hell Shinje Chhogyel (Gshin rje Chos rgyal) during the performance of the Raksha Marcham (Raksa mar'chams) or The Dance of the Judgement of the Dead (photograph by Bradford M. Endicott).

Notes

1. See H.H. Godwin-Austen in the *Journal of the Asiatic Society, Bengal Branch*, 1865, p. 71 ff.
2. Hashimoto Kōhō, *Mōko no Ramakyō* (The Lamaism of Mongolia). Tokyo: Bukkyō Kōronsha, 1942, colour plates 2-10;, pp. 266-74.
3. For the occasion the Asia Society produced a most informative brochure entitled *The Sacred Dance-Drama of Bhutan*, written by the Director of the National Museum of Bhutan, Mynak R. Tulku. In these pages grateful use has been made of this brochure.
4. P.H. Pott, *Introduction to the Tibetan Collection of the National Museum of Ethnography, Leiden*. Leiden: Brill, 1951, p. 118.
5. Ono Shōnen and Hibino Jōbu, *Godaisan* (Wutaishan). Tokyo: Zauhō Press, 1942, pp. 192-212.
6. Hashimoto Kōhō. *op. cit.*, plate 6.
7. *Fojiao Shengdi Wutaishan* (Wutaishan, Holy Land of Buddhism). Beijing: Huayi, 2nd. ed., 1994, p. 85 under.
8. In her important thesis *Les revenants de l'au-delà dans le monde tibetain*. Paris: Editions du CNRS, 1989, Françoise Pommaret provides the best description of this play as well as two illustrations (figs. 10 & 11).
9. *Fojiao Shengdi Wutaishan*, pp. 86-7.
10. Ono and Hibino, *op. cit.*, pp. 196-8.
11. *Borobudur, History and Significance of a Buddhist Monument*, ed. by Luis Gomez and Hiram W. Woodward, Jr. Berkeley: Asian Humanities Press, 1981, pp. 89-94.
12. Jean Drans, *Histoire de Nang Manora et Histoire de Sang Thong*. Tokyo: Presses Salesiennes, 1947, pp. 17-39.
13. A number of Bhutanese pamphlets, published in English for the benefit of foreign visitors by different travel agents, contain more or less identical excerpts from all the plays. All are based on a publication of the Bhutan Tourism Corporation, Ltd. One of these excerpts was published April 8, 1994 as a *Tshechu Supplement* of Bhutan's only weekly newspaper, the *Kuensel*. I want to thank Mr. Kinley Dorje, Editor in Chief of the *Kuensel*, for the candid remarks on the cultural future of Bhutan that he provided during my stay in Thimphu.
14. A brief description is provided in Thubten Dschigme Norbu, *Tibet verlorene Heimat*, erzählt von Heinrich Harrer. Wien etc.: Ullstein, 1960, p. 199. For an illustration of the Potala *thangka* see Heinrich Harrer, *Seven Years in Tibet*. London: Pan Books, 1953.
15. Siddiq Wahid, *Ladakh, Between Earth and Sky*. New York/London: Norton, 1981 p. 83; *ibid.*, p. 69 shows the huge *thangka* from Thikse monastery representing the Buddha Śākyamuni with two of his disciples.
16. I am grateful to Mr. Kang Woo Bang of the National Museum of Korea, Seoul, for drawing my attention to a recent study on the topic: Yun Yolsu, *Kwaebul*. Seoul: Taewonsa, 1990.

IX

VEDISCHE WEISUNG
WAS VERSTAND KUMĀRILA BHAṬṬA UNTER EINER VEDISCHEN WEISUNG (*CODANĀ*)

Lars Göhler

Die wichtigsten modernen Erklärungsmuster für das Verständnis des vedischen Rituals, die auch in heutiger Zeit immer wieder diskutiert werden, verdanken wir wohl den Arbeiten von Professor Staal.

Für das Verständnis des Veda mag es aber auch von Interesse sein, welche sprachwissenschaftlichen, sprachphilosophischen und textexegetischen Theorien die Inder selbst für diesen entwickelt haben. Die Mīmāṁsā war wohl dasjenige System, daß vor allem zu solchen Theorien beigetragen hat, da sie darin ihre eigentliche Domäne sah. Sie entwickelte ein kompliziertes System sprachphilosophischer und hermeneutischer Ideen, das ein adäquates Verständnis der vedischen Texte garantieren sollte.

Für sie wie auch für die anderen indischen philosophischen und religiösen Systeme war die *selektive Rezeption* vedischer Texte charakteristisch. Viele der indischen orthodoxen philosophischen Systeme meinen, wenn sie 'Veda' sagen, nur die philosophischen Passagen der Upaniṣads. Es waren im wesentlichen die Grammatiker und die Mīmāṁsā-Anhänger, die sich in ihren Erklärungen auf den größeren, rituellen, Teil des Veda stützten. Wie schon Jacobi bemerkt,[1] fühlen sich beide Richtungen auch eng verwandt und entlehnen Ideen voneinander. Und wenn es eine 'Arbeitsteilung' zwischen beiden gegeben hat, dann ist es wohl die, daß die Grammatiker sich mehr für das korrekt gesprochene Wort (*sādhuśabda*) und die Mīmāṁsakas für das korrekt durchgeführte rituelle Werk (dies ist vor allem gemeint, wenn von '*dharma*' die Rede ist) verantwortlich fühlten. Diese Unterteilung beschreibt nur die Tendenz, tatsächlich gibt es viele Überschneidungen im Untersuchungsgebiet der beiden Richtungen, die in der späteren Entwicklung dieser Theorien auch zu Kontroversen führten.

Im folgenden soll vor allem erörtert werden, wie sich die Mīmāṁsā in einem wichtigen Punkt zum Veda verhält und welche Kategorien sie zu dessen Exegese entwickelt. - Das Programm, dem sich die gesamte Mīmāṁsā verpflichtet fühlt, findet sich bereits in den ersten beiden Sūtras des Jaimini formuliert. Das erste Sūtra erklärt die Untersuchung des Dharma zum Hauptgegenstand.[2] Das zweite Sūtra sagt, wodurch der Dharma charaktierisiert ist: Er trägt Weisungscharakter.[3] Damit ist auch ein weiterer Unterschied zu dem Untersuchungsgebiet der Grammatiker gegeben: Während die Grammatiker sehr wohl auch Aussagen, also indikative Sätze untersuchen, findet die Mīmāṁsā ihr Hauptuntersuchungsgebiet in Weisungen, also vor allem Sätzen mit einer optativischen oder imperativischen (*liṅādi*) Verbform. Sie sah ihre Aufgabe im weiteren Sinne darin, aus Texten, die als maßgeblich angesehen wurden, den *normativen Gehalt* zu extrahieren. Dadurch wurden ihre Methoden auch u.a. zur Untersuchung von Rechtstexten interessant.[4]

Insgesamt fanden sich die indischen Traditionen, die sich als Śāstra (das man nur mit großer Vorsicht als 'Wissenschaft' zu übersetzen sollte) verstanden, mehr der Proklamation eines Sollens als der Repräsentation eines Seins verpflichtet. Kumārila Bhaṭṭa (7. Jh.), der wohl einflußreichste Vertreter der Mīmāṁsā meint, daß Śāstra, worunter sich auch die Mīmāṁsā zählt, den Menschen mit ewigen (vedischen - *nitya*) oder menschlichen (hervorgebrachten - *kṛtaka*) Worten vorschreibe, was zu tun (*pravṛtti*) und was zu unterlassen (*nivṛtti*) sei. Beschreibungen von Dingen (*svarūpakathana*), die man in ihr finde, haben gegenüber den Vorschriften sekundäre (*aṅga*) Bedeutung.[5]

Damit mußte sich die Aufmerksamkeit der Mīmāṁsā in sprachwissenschaftlicher Richtung, wie bereits bemerkt, ganz auf injunktive Formen konzentrieren. Diese werden einerseits häufig ihrer grammatikalischen Form nach als Sätze mit einem injunktiven Verbalsuffix charakterisiert. Andererseits wird auch die pragmatische Rolle von Weisungen beschrieben: Nach Śabara ist eine Weisung ein Ausspruch, der eine Tätigkeit bewirkt.[6]

Es ist wohl zuerst Kumārila, der die Auffassung vom Satz und die Auffassung von der Weisung zusammenführt: - Als eine Weisung wird in der 'Wissenschaft' ein (eine Tätigkeit) bewirkender *Satz* verstanden. Bei einer solchen Weisung ist das Suffix (des Verbes) erst vollständig fähig (diese auszudrücken), wenn es durch die erforderlichen (Modi des) 'Was' usw. [also auch des 'Wie' und 'Wodurch'] ergänzt ist.[7] Diese Bestimmung der Weisung durch diese drei Momente, die Kumārila an anderer Stelle (ŚV, Vākyādhikaraṇa 289 ff.) als 'Mittel' (*sādhana*), Zweck (*sādhya*) und Art und Weise (*itikartavayatā*) bezeichnet, dürfte für die Satzanalyse, zumindest in der Mīmāṁsā ebenso grundlegend gewesen sein, wie die 'westliche' Unterteilung in Subjekt, Prädikat und Objekt.[8]

Was wird von einer Weisung nun aber ausgedrückt? Welches ist der Gegenstand einer Weisung? - Kumārila folgt hier Śabara wenn er als

Gegenstand eines (Weisungs-) Satzes eine 'bewirkende Kraft', ein 'Bewirken' (*bhāvanā*)[9] bezeichnet. Er unterteilt dieses Bewirken jedoch weiter in ein 'gegenständliches Bewirken' (*arthātmikā bhavanā*) und ein 'verbales Bewirken' (*śabdātmikā bhāvanā*). Das 'gegenständliche Bewmrken' ist die Tätigkeit, das Bewirken eines Subjekts (etwa: *Devadattaḥ pacati*), das vom Satz *beschrieben* wird. Das verbale Bewirken entsteht durch die sprachliche Form (Weisung) selbst, diese veranlaßt eine Tätigkeit. Freilich ist, wenn eine Tätigkeit veranlaßt werden soll (*Devadattaḥ paceta*), die Beschreibung der Tätigkeit mit enthalten. Sätze, die also *beide* Formen des Bewirkens zum Gegenstand haben, sind für Kumārila vor allem interessant. Ein reiner Aussagesatz hat nur dann eine Bedeutung im eigentlichen Sinne, wenn er einem Weisungssatz zugeordnet werden kann, d.h. wenn er dessen Weisungskraft unterstützt, indem er eine der oben erwähnten Komponenten (*kim*, *kena*, *katham*) genauer spezifiziert.[10]

Kumārila exemplifiziert, was er unter den 'drei Faktoren' des Bewirkens versteht, an einer sogenannten 'ursprünglichen Weisung' (*utpattividhi*), die in verschiedenen Varianten vor allem in den Brāhmaṇas zu finden ist, etwa: 'Wer den Himmel erreichen möchte, opfere mit dem Agniṣṭoma' (*svargakāmo 'gniṣṭomena yajeta*). Bei genauerer Betrachtung, so Kumārila, sei dies eigentlich falsch im Sinne der drei Faktoren der Bhāvanā formuliert. Um alle drei Faktoren, die miteinander verwoben (*anusyūta* - ŚV, Vākyādhikaraṇa 266) sind und in einer Relation gegenseitiger Hilfe[11] stehen, korrekt auszudrücken, muß der Satz aus den Brāhmaṇas umgestellt werden: Er muß in den richtigen Kasusverhältnissen und Wortarten das Mittel (Instrumental: *yāgena* - nicht mehr *yajate* als Verb), das Ziel (Akkusativ als Zielkasus: *svargam*) und den Prozeß (natürlich in Verb-form: *bhāvayet*) ausdrücken (ŚV, Vākya-, 255). Da sich jeder Weisungssatz auf eine solche Form zurückführen läßt, sei, so Kumārila, die invariante Bedeutung eines Weisungssatzes jene schon erwähnte, auf das '*bhavet*' zurückgehende Bhāvanā.

Nun ist in groben Zügen deutlich geworden, was eine Weisung ausdrückt. Wie aber wird eine solche Weisung aus den Worten des Veda erkannt?

Nach Auffassung Kumārilas konstituiert sich die Bedeutung eines Satzes aus der Bedeutung seiner Worte plus deren syntaktischer Verbindung. Die Erkenntnis der Worte geschieht durch eine Potenz, die ihnen innewohnt (*abhidhāna-śakti*, *vācakasāmarthya*) und die in der Lage ist, ihre Bedeutung zu offenbaren. Dabei ist die Verbindung des Wortes mit seinem Gegenstand ursprünglich (*autpattika* - MS I.1.5) oder in der Terminologie von Śabara Kumārila: beständig, bzw. ewig (*nitya*).[12] Diese Feststellung schließt ein, daß sowohl das Wort, seine Bedeutung, als auch die Verbindung beider ewig ist. Demzufolge kann die Bedeutung des Wortes nicht ein Einzelding sein, das vergänglich ist, sondern seine Bedeutung ist das Universale (*ākṛti*, *jāti*, *sāmānya*). Dieses Universale ist jedoch nicht völlig getrennt von den Besonderheiten (*viśeṣa*) oder von den Einzeldingen (*vyakti*, *piṇḍa*). Es befindet sich zu diesen in einer Relation der 'Identität in der Unterscheidung' (*bhedābheda*). Universale, Besonderheiten und Einzelnes sind also sowohl identisch als auch nicht identisch. Sie gehen einher miteinander. Damit verweist ein Wort vermöge seiner Bezeichnungskraft auf das Universale, in sekundären Sinne (*lakṣaṇā*) aber auch auf das Einzelne und Besondere.

Dabei sind es die Verben, die eine Tätigkeit in allgemeiner Weise ausdrücken, wogegen Substantiva und Adjektiva eher zum Ausdruck von Substanzen (*dravya*) und Qualitäten (*guṇa*) dienen.

Diese jeweils von den Worten bezeichneten Universale werden durch ihren syntaktischen Zusammenhang derart spezifiziert, daß eben jenes 'Bewirken' entsteht, das ja den Gegenstand (oder das Ziel - *artha*) des Satzes darstellt. Kumārilas Präferenz der Weisungen hat nicht nur Bedeutung für die Analyse eines sprachlichen Ausdrucks, sondern auch für die Analyse resp. Klassifikation vedischer Texte. Der 'Veda' im engeren Sinne besteht für Kumārila nur aus den dort aufgefundenen Weisungen. Wenn er von der absoluten Autorität des Veda spricht, meint er nur diese. Alle anderen Textteile bestimmen sich durch ihre Nähe bzw. Ferne zu diesen Weisungen.

Sprachphilosophisch aus heutiger Perspektive vielleicht interessanter als die Weisungspassagen, die Kumārila zumeist kurzerhand mit Brāhmaṇa bezeichnet, sind die Mantras des Veda. An diesen ihrem semantischen Status nach eher zweifelhaften Teilen des Veda entfaltete sich der sprachwissenschaftliche Scharfsinn fast der gesamten Grammatiker- und Mīmāṁsā-Traditionen. Man stellte sich die Frage, ob die im Ritual zu rezitierenden Mantras 'wörtlich zu nehmen' seien, oder ob sie ihre Bedeutung nur aus ihrem Platz im Ritual erhalten. Ein schon bei Yāska (Nirukta 1.15) erwähnter Kautsa muß bezweifelt haben, daß die Mantras das meinen, was sie sagen. Yāska selbst wie auch die gesamten uns heute textlich verfügbaren Grammatiker- und Mīmāṁsā-Traditionen sprechen sich dagegen *für* eine direkte Bedeutung der Mantras aus.

Kumārila schließt sich bei der Diskussion dieses Themas zunächst seinem Vorgänger Śabara an und schreibt den Mantras erinnernde (*smāraka*) und beschreibende (*abhidhāyaka*) Funktion, jedoch *keine* Weisungsfunktion zu.[13] Mantras sollen also an Weisungspassagen erinnern, und sie sollen bestimmte Dinge beschreiben. An einer anderen Stelle im Tantravārttika im Kommentar zu MS II.1.30 ff. greift er jedoch das Thema noch einmal auf und meint, daß Mantras durchaus Weisungsfunktion haben können, wenn sich in ihnen eine imperativisches oder optativisches Suffix eines Verbes findet. Dann mache es keinen Unterschied, ob sich ein anweisender Satz in einem Brāhmaṇa oder einem Mantra befinde.[14] Bei einem Widerspruch zwischen Textpassagen aus den Mantras und solchen aus den Brāhmaṇas seien jedoch letztere zu bevorzugen, da sie der vornehmliche Platz für Weisungen seien. Damit sind die Mantras, indem sie also beschreiben, erinnern und anweisen können, in den Augen der Mīmāṁsā die komplexesten sprachlichen Gebilde des Veda. Es scheint sogar, als hätte man ihren Inhalt nicht nur auf diese drei Funktionen beschränkt.

Der Vṛttikāra, der von Śabara bei verschiedenen Gelegenheiten (wie mir scheint, vor allem in Momenten der Ratlosigkeit) hervorgeholt und zitiert bzw. paraphrasiert wird und dem Kumārila in aller Regel folgt, erwähnt folgende Mantratypen: Solche, die 'asi' enthalten (Identifikationen?), solche, die '*tvā*' (oder *tva*) enthalten, Segnungen (*āśis*), Lobpreisungen (*stuti*), Aufzählungen (*saṁkhyā*), Geschwätz (*pralapta*), Wehklage (*paridevita*), Aufforderungen (*praiṣa*), Suche (*anveṣaṇa*), Frage (*pṛṣṭa*), Antwort (*ākhyāna*), Ergänzung (*anuṣaṅga*), Anwendungen (*prayogita*) und Fähigkeiten (*sāmarthya*).[15] Es scheint, als sollten hiermit solche textlichen Intentionen aufgelistet werden, die keine direkte Weisungskraft haben.

Arthavāda-Passagen, Legenden und Erklärungen des Rituals verfügen jedoch über keinerlei selbständige Bedeutung.[16] Sie entlehnen ihren Bedeutungsgehalt vollständig den Weisungspassagen, auf welche sie sich beziehen. Sie bringen auch nur insoweit Resultate hervor, wie sie die Befolgung von Weisungen erstrebenswert machen, zu den im Veda angewiesenen Tätigkeiten veranlassen.

In diesem Zusammenhang ist es nun interessant, daß der 'Weisungszentrismus' Kumārilas nicht nur zur Textexegese und - klassifikation herangezogen wird, sondern auch, um grammatikalische Kategorien, wie die des Satzes, zu bestimmen. Die Satztheorie der Mīmāṁsā ist in der Fassung von Jaimini auch unter den anderen indischen Philosophien recht bekannt geworden. Sie besagt, daß eine Wortgruppe, die *einen* Sinn ausdrückt, als ein Satz anzusehen sei. Wenn diese Wortgruppe wiederum geteilt würde, entstünde eine Erwartung nach Vervollständigung der Satzglieder (MS II.1.46 - *arthaikatvād ekaṁ vākyaṁ sākāṅkṣaṅ ced vibhage syāt*). Schon bei Jaimini ist unklar, ob sich diese Definition auf *alle* Sätze oder nur auf bestimmte Teile des Veda beziehe. Śabara und auch Kumārila meinen, daß sie sich *nur* auf die Yajussprüche beziehen, deren Satzteilung im Veda nicht deutlich gekennzeichnet sei.

Im strengen Sinne jedoch trifft das Prinzip der Satzteilung für Kumārila nur auf Weisungen zu, bei anderen Passagen (Arthavādas) können in beliebiger Weise mehrere Gegenstände gleichzeitig vorkommen, ohne daß dies den Satz spalte.[17] In Anlehnung an Kātyāyana ist man damit versucht, Kumārila eine Satzauffassung zu unterstellen, die den Satz als eine Wortgruppe mit *einem* injunktiven Verb ansieht (etwa: *ekaliṅ vākyam*).

Die Frage, *weshalb* für Kumārila die Weisungen eine derart zentrale Rolle spielen, ist nicht leicht zu beantworten. Große Teile des Veda, vor allem die rituellen Teile, können nur dann als sinnvoll angesehen werden können, wenn man die dort enthaltenen Ideen als *Magie* interpretiert, die in sich die Keime für das spätere spekulative Denken tragen.[18] In der vedischen Magie finden wir eben die gleiche Absicht, vermittels einer sprachlichen Äußerung etwas zu *bewirken*, wie in Kumārilas Interpretation des vedischen Rituals. Kumārila fühlt sich also hier, wenn auch sein philosophisch-exegetisches System mit dem Opfersystem des Veda kaum noch vergleichbar ist, der 'Logik' der vedischen Opfermagie verpflichtet.

Die Rolle von Weisungen in der Auffassung von Kumārila sei hier noch einmal kurz resümiert. Die Tatsache, daß für Kumārila ein 'normaler' Satz eine Weisung ist, hat sowohl Konsequenzen für die Exegese vedischer Texte wie für die Bestimmung sprachwissenschaftlicher Kategorien. Die vedischen Texte werden hierarchisiert. An der Spitze dieser Hierarchie stehen die Weisungen, und die Bedeutung aller anderen vedischen Passagen bestimmt sich nach dem Verhältnis zu diesen. Aussagen, die wohl für die westliche Sprachphilosophie grundlegend waren (und es vielleicht noch sind) und die in irgendeiner Weise ein 'Sein' beschreiben, sind bei Kumārila nur sinnvoll, indem sie sich auf ein 'Sollen' beziehen. Folglich ist der Gegenstand eines Satzes bei Kumārila primär ein 'Bewirken' und daraus abgeleitet erst die 'Repräsentation eines Sachverhaltes'.

Daraus leiten sich auch die Kriterien für die Bewertung von sprachlichen Äußerungen ab. Weisungen können nicht wahr oder falsch sein, da sie keinen 'äußeren' Gegenstand repräsentieren, sie können nur gültig oder nicht gültig (*prāmāṇyāprāmāṇya*) sein.[19]

Mitunter wurden derartige Überlegungen der Mīmāṁsā mit wohlwollend-entschuldigender Nachsicht als apologetische betrachtet. Sicherlich sind sie auch größtenteils apologetisch motiviert. Aber gerade nach indischer Auffassung ist ein Gedanke nur dadurch, daß er auf einen bestimmten Zweck hin entworfen ist, noch lange nicht entwertet. Und soweit ich sehe, finden sich vergleichbare Ideen in der 'westlichen' Sprachphilosophie wohl erst in diesem Jahrhundert.[20]

LITERATUR

Apte, Vinaya Ganesh (ed.)

1929-34 *Mīmāṁsādarśanam*. 7 Vols. *Ānandaśrama Sanskrit Series* 97, Poona.

Jaimini

1929-34 '*Mīmāṁsāsūtra*', in: V.G. Apte (ed.), *Mīmāṁsādarśanam* (zitiert als MS [Adhikaraṇa]. [Pāda]. [Sūtra])

Kumārila Bhaṭṭa
1978 *Ślokavārttika of Sri Kumārila Bhaṭṭa. With the Commentary Nyāyaratnakāra of Pārthasārathi Miśra.* Hrsg. Svāmī Dvārikādāsa Śāstrī. Benares (Zitiert als ŚV, [Adhikaraṇa] [Nr. der Kārikā]).

Kumārila Bhaṭṭa
1929-34 '*Tantravārttika*', in: V.G. Apte (ed.), *Mīmāṁsādarśanam*. (zitiert als TV, [Seite], [Zeile]

Śabara
1929-34 '*Śabarabhāṣya*', in: V.G. Apte (ed.), *Mīmāṁsādarśanam*. (zitiert als ŚBh, [Seite], [Zeile]; Zu MS I.1.1.-5: Erich Frauwallner. Materialien zur ältesten Erkenntnislehre der Karmamīmāṁsā. Wien, 1968 (zitiert als ŚBh (F) [Seite]: [Zeile]

Notes

1. Herman Jacobi: 'Mīmāṁsā und Vaiśeṣika', in: *Indian Studies in Honor of C.R. Lanman*. Cambridge: Harvard University Press, 1929, besonders S. 147 ff.
2. MS I.1.1.: *athāto dharmajijñāsā* - im Gegensatz zu den Brahma Sūtras, die die Untersuchung des Brahman (*brahmajijñāsā*) zum Hauptgegenstand haben.
3. Um die genaue Übersetzung des Sūtra I.1.2.: *codanālakṣaṇo 'rtho dharmaḥ //* kann man sich streiten. Frauwallner ŚBh (F) S.17 übersetzt: "Der Dharmaḥ ist etwas Nützliches, dessen Kennzeichen die (vedischen) Weisungen sind." - Gleichermaßen gerechtfertigt wäre etwa: "Der Dharma ist ein Gegenstand, der Weisungscharakter besitzt." Frauwallners Übersetzung trägt der eher hedonistischen Interpretation der Sūtras durch Śabara Rechnung, die letzte wäre eine eher rationalistische Interpretation.
4. Vgl. P.V. Kane, *History of Dharmaśāstra*. Vol. V.2, Poona: Bhandarkar, 1962 S. 1283-338.
5. ŚV, Śabdapariccheda 4, 5: *pravṛttir vā nivṛttir vā nmtyena kṛtakena vā / puṁsāṁ yenopadiśyeta tac chāstram abhidhīyate // svarūpakathanaṁ yat tu kasyacit tatra dṛśyate / tadaṅgatvena tasyāpi śāstratvam avagamyate //* Vgl. dazu auch S. Pollock: 'The Theory of Practice and the Practice of Theory', *JAOS* 105.3/1985 S. 501.
6. ŚBh (F) 19.9: *codanā iti kriyāyāḥ pravartakaṁ vacanam āhuḥ*.
7. ŚV, Codanā Sūtra 3: *kimādyapekṣitaiḥ pūrṇaḥ samarthaḥ pratyayo vidhau / tena pravartakaṁ vākyaṁ śāstre 'smin codanocyate //*
8. Nach ŚV, Autpattika Sūtra 11cd sind *codanā*, *upadeśa* und *vidhi* synonym, 'drücken den gleichen Gegenstand aus': *codanā copadeśaś ca vidhiś caikārthavācinaḥ //* Auf diese Stelle bezieht sich Kumārila wahrscheinlich weiter unten (ŚV, Śabdapariccheda 12cd), wenn er meint, daß '*codanā*' und '*upadeśa*' eigentich '*śāstra*' meinen: *codanā copadeśaś ca śāstram evety udāhṛtam //* Dabei ist hier mit '*śāstra*' möglicherweise, wie man es gelegentlich findet, der Veda und weniger die Mīmāṁsā selbst gemeint.
9. ŚV, Vākyādhikaraṇa 330: *bhāvanaiva ca vākyārthaḥ* ..., wiederholt in TV 445. 19 f.

10. Die bisher wohl genaueste Erörterung der Kategorie Bhāvanā bringt Erich Frauwallner: 'Bhāvanā und Vidhiḥ bei Maṇḍana Miśra', etwa in (ders.:) *Kleine Schriften*. Wiesbaden 1982, besonders Seiten 164-79. Weiteres findet sich bei F. Edgerton: 'Some Linguistic Notes on the Mīmāṁsā System', *Language* 4/1928.
11. ŚV, vākyādhikaraṇa 265a,b: *ekabhāvanayopāttās trayo 'py aṁśāḥ parasparam/*
12. Im Śabdanityatādhikaraṇa von ŚV widmet Kumārila diesem Thema außerordentlich viel Aufmerksamkeit. Mit 444 Kārikās ist dies das größte Kapitel von ŚV. Siehe dazu auch E. Abegg: 'Die Lehre von der Ewigkeit des Wortes bei Kumārila', in: *Antidoron. Festschrift für Jakob Wackernagel*. Göttingen: Vandenhoeck & Ruprecht, 1923 S. 255-64.
13. TV 144. 6: *tatra vidhistutitve tāvat svayam eva nirākariṣyati, avidhisarūpatvāt* / - Daß sie Weisungen oder Lobpreisungen sind, wird zurückgewiesen, weil sie nicht die Form von Weisungen haben.
14. TV 433. 9 f.: *vidhāyakatvāvidhāyakatve tu tayor yathoktam eva karaṇam iti na mantrabrāhmaṇatvayor vyāparaḥ* / Mit dem anweisenden oder nicht anweisenden Charakter dieser Beiden (Mantra oder Brāhmaṇa) verhält es sich wie aus bereits erwähnter Ursache ersichtlich: sie ergibt sich nicht aus ihrer Funktion als Mantra oder Brāhmaṇa.
15. hier der Kürze halber in der Wiedergabe von Kumārila TV (zu MS II.1.32) 434. 15 ff.: *vṛttau lakṣaṇam eteṣām asyantatvāntarūpatā / āśiṣaḥ stutisaṁkhye ca pralaptaṁ paridevitam // praiṣānveṣaṇapṛṣṭākhyānānuṣaṅgaprayogitāḥ / sāmarthyaṁ ceti mantrāṇāṁ vistaraḥ prāyiko mataḥ* // Diese Passage scheint sich auf Yāskas Klassifikation von ṚV-Passagen (NR VII.1-3) zu stützen.
16. TV 123. 3 f., *asmākaṁ tu punar ya eṣāṁ śabdānāṁ śrauto 'rthaḥ sa naiva vivakṣitaḥ*. - Nach unserer Auffassung jedoch ist die Bedeutung, die diesen Worten entnommen wird, nicht die eigentlich gemeinte (Bedeutung).
17. TV 551.20: *bahavo 'pi hy arthā yugapad ekena saṁbadhyante / na ca tāvatā vākyaṁ bhidyate/ anekavidhito hi vākyabheda uktaḥ* /
18. Prof. Staal wird hier mit Sicherheit einwenden, daß man den Weisungspassagen des Veda nicht unbedingt einen Sinn und schon gar keine Bedeutung zuschreiben müsse. Kumārila wendet aber nahezu sein gesamtes Werk darauf, die Sinnhaftigkeit dieser Passagen zu erweisen.
19. Im Gegensatz zum Begriff der Wahrheit (*satya*), der in seinen Überlegungen kaum eine Rolle spielt, widmet Kumārila dem Begriff der Gültigkeit (*prāmāṇya*) eine umfangreiche Untersuchung in seinem Ślokavārttika zum Codanā Sūtra (MS I.1.2).
20. Etwa in John Langshaw Austins epochemachenden *How to do Things with Words*. Oxford, 1962.

X

HEAVEN ON EARTH: TEMPLES AND TEMPLE CITIES OF MEDIEVAL INDIA

Phyllis Granoff

1. INTRODUCTION: IN SEARCH OF A NEW MODEL

Discussions on the religious meaning of the Hindu temple have tended to draw almost exclusively upon two different types of texts. The first group consists largely of technical handbooks on architecture and ritual texts, which themselves may be divided into those associated with the building and consecration of the temple and other forms of architecture, and those that deal with the building and consecration of the Vedic fire altar. The second major group of texts to which scholars have had recourse are of a very different nature: they are the very early philosophical passages and some of the cosmogonic speculations in the Brāhmaṇas and Upaniṣads. This emphasis on Vedic and ritual material has led to a number of interesting and widely accepted hypotheses: the temple is a model of the cosmos and depicts in visual form the creation of the universe; within, in the 'womb chamber' lies the god in his or her most subtle essence from which the teeming images on the temple walls are manifested. The goal of the devotee is to retrace the cosmic evolution in a reverse process and merge with the subtle godhead; the ascending spire of the temple symbolizes this process, for it leads to the crowning *āmalaka*, itself a symbol of the formless heaven, and in effect a symbol of release.[1]

Without rejecting entirely earlier hypotheses about the religious meaning of the Hindu temple, I would like to turn to a very different body of evidence that in fact leads us to almost the opposite conclusions. I would like to turn to purāṇic stories and descriptions of the abodes of the deities, which I shall argue, point us towards a much more concrete and a much less abstract understanding of the nature of the god within the temple and indeed the various gods and goddesses, mortals and immortals, displayed on the temple walls. These texts do not talk of meditation and denial of sensual pleasures in quest of union with a formless absolute; they speak instead of paradises that abound in all manner of delights, and of gods that worship and are worshipped, all of radiant unparalleled beauty and all of visible and concrete, sensible form. Before I begin to examine in some detail how the Purāṇas conceptualize the worlds of the gods, I should like to review briefly some of the reasons why I feel that a new model might be useful for understanding temple Hinduism.

My desire for a different model to understand the temple and its sculpture stems from what I have come to perceive as a major gap between the purāṇic accounts of temples and the merits of worship in temples, on the one hand, and the descriptions in the Āgamas of the rituals for worship in a temple and consecration of temples, on the other hand.[2] There is in fact a growing tendency on the part of specialists of ritual texts themselves to acknowledge the presence of an unbridgeable gap between the temple in its visual complexity and the various rituals of worship with which the temple is associated in the Āgamas and Tantras. The building of the temple begins in ritual, and throughout its lifetime the temple continues on a daily basis to be the focus of rituals, particularly in the form of the cult of the main deity enshrined in its innermost recess. In a recent article, one of the leading scholars on Śaiva ritual texts, Hélène Brunner, emphasized the major disparity that is apparent between both the rituals for consecrating the temple and worshipping the main image in the sanctum of a temple, and the actual sculptural programme of the temple.[3] In her article Brunner noted repeatedly that none of the deities lodged at various parts of the temple in the consecration ritual is actually figured in the sculptural programme; she also stressed that the ritual of the main deity in the sanctum does not in any way help to explain the profusion of sculptures along the temple walls. The images figured on the walls are not representations of the various deities that the officiant meditates upon in worshipping the main image of the sanctum. There is also no ritual continuity between worship of the main image in the sanctum and the worship of deities on the walls; the worship of the deities on the walls begins as it were totally anew. The various rituals are totally independent of one another and do not suggest any symbolic underlying unity connecting inner deity and outer wall figures.

One of the assertions most commonly made about the proliferation of sculpture on temple walls is that the outside images should be seen as visible manifestations of the inner god, the subtle god, who becomes the manifest world. In this understanding, the temple becomes a visual metaphor for a particular understanding of reality, in which multiform life is seen as a coalescence of a subtle ultimate reality, which literally becomes the world. In my mind, Brunner's important observation that there is no ritual continuity between the worship of the main deity and the worship of the subsidiary deities and thus that there is nothing that connects them in any way in the ritual texts, strongly casts doubt on this standard interpretation. In addition, a careful reading of the ritual texts themselves advises caution. The groups who wrote these ritual texts for temple worship did not all subscribe to a non-dualist cosmology that many modern scholars have insisted lies behind the visual form of the Hindu temple. The Śaiva Siddhāntins and the major Vaiṣṇava groups in medieval times denied vigorously that material creation proceeded directly and uniquely from god; they did not regard the physical world as simply the gross manifestation of a non-dual single, subtle deity. In their cosmologies material creation remains material and proceeds from some type of primordial matter.[4]

Much of the modern philosophical discussion on the inner meaning of the temple in fact mixes statements from ritual texts like the later Śaiva Āgamas with concepts derived from much earlier texts like the Upaniṣads; some discussions even refer back to ritual texts that deal with the building of the Vedic fire altar. The homologies between temple and fire altar, where they exist at all, are limited to certain rituals in the consecration and building of both structures. These similarities in fact may do more to mislead us than enlighten us if we assume that they have something fundamental to tell us about the symbolism of the temple itself; in interpreting the similarities between rituals devoted to fire altar and temple, we need to keep in mind the fundamental disparity noted above between rituals of temple consecration and worship in the temple and the actual temple and its sculpture. I believe with Brunner and other scholars of ritual that the gap between the rituals and the actual temple suggests that the priestly regulation of temple ritual was in fact later than and foreign to the original temple worship.[5] The priests who sought to regulate temple rituals, particularly those associated with consecration and building, were schooled in *śrauta* rituals; it would have been completely natural for them to couch the rituals they sought to impose on temple construction and consecration in the language of the earlier Vedic sacrifice that they knew and supervised. This Vedic language, with its allusions to the building of the sacrificial fire altar and to speculation on the mystical meaning of the fire altar that we find in the Brāhmaṇas and Upaniṣads, tells us everything about the world of the priests who wrote those texts, but we are left still to grapple with the gap between that world and the visual reality of the temple itself.[6]

That gap is only highlighted by the evidence of inscriptions and stories, which became my starting point in my own search for a different model to understand the Hindu temple. The evidence of the stories actually told about temple worship and its merits, in which the god is a real concrete physical presence and not some abstract subtle essence, must make us suspicious of accepting ritual handbooks and monistic Upaniṣads as the primary clues to deciphering the meaning of the temple. The stories never mention the temple or the deity in it as an aid to meditation, which is one of the prime emphases of the ritual handbooks. They also never refer to the images inside the temple or on the walls as manifestations of the god in any cosmogonic process. The only process of origin in these stories concerns the descent of the God, who is the main deity of the temple. The stories tell us how the God came to reside in this particular locale. These stories take place after creation; the world and its inhabitants are a given, and the climax of the story is how a God comes to act in that created world. Furthermore, the benefits to be gained from temple worship are not at all consistent with the strictly non-dualist philosophical interpretations of the temple: the benefits are primarily this-worldly, either rebirth as a wealthy individual, or rebirth in heaven, a land of infinite pleasures of the senses. Liberation or *mokṣa*, while possible as a result of piety and worship, is understood as an eternal sojourn in paradise.[7] One might cautiously add here the evidence of modern ethnography, which reveals increasingly that individuals come to the temple as concrete individuals with specific wants and do not view themselves as abstractions or as involved in a process in which the phenomenological self is gradually or abruptly denied.[8]

The evidence from the Purāṇas and the inscriptions, then, suggests that the conceptual world of the temple, images and temple worship, is different from that of the ritual texts. While Brunner does not make the criticism explicit, her article in pointing out the disparities between ritual and visual forms of the temple, also calls into question many of the currently held theories about the meaning of the visual form of the Hindu temple. As we have seen above, her observations strongly cast doubt on the commonly accepted hypothesis which views the many outer deities as grosser forms of an inner subtle deity, and the profusion of forms on the temple walls as a metaphor or an indicator of some larger cosmogonic process. Brunner's article also calls into question the understanding of temple worship as a meditative process in which the devotee seeks to be united with a formless absolute. Brunner states clearly that the temple is not a three dimensional representation of any of the various internal visualizations or meditations that are enjoined in the rituals associated with the āgamic worship of or in the temple. Implied by this is the warning that we are not to use these rituals as the sole guide to understanding the motivations, goals or processes of worship in the temple. If the priest in the ritual becomes Śiva and uses the image as a basis or support for meditation, we are not thereby to conclude that this was the normative or normal way of understanding the image and its function in the temple. The meditations in the priestly worship do not seem in any way to be organically tied with the temple and its iconography; they do not seem to have guided the sculptural programme and are unlikely to have served as its motivating force.

In another article, Brunner also called into question the burden of meaning that has been placed on the *vāstumaṇḍala*, the diagram that precedes the construction of the temple and indeed is an essential starting point for all types of building. This article also makes another point that is relevant to the present discussion. Brunner argues that a *maṇḍala* in general has no unique cosmic symbolism but functions in a way analogous to a *liṅga* or any seat of a god. In fact the *maṇḍala* can be seen as a temple in miniature; some *maṇḍalas* are called *bhavana* or 'palace', while others are called cities or *pura*.[9] In this paper I will be arguing something similar; namely that the temple itself has no unique cosmic symbolism, but that it is the city and palace of the god, and that its special visual features follow from that observation.

Like Brunner, but using a different body of evidence, I shall conclude that the temple seems to belong to a totally different conceptual world from that of the ritual texts. Brunner, whose interest is after all in the ritual texts themselves and in the development of Śaiva ritual, finally uses this conclusion to reflect back on the ritual texts. She suggests that the disparity between temple and ritual points to the fact that the rituals of the Śaivas are probably not homogeneous in nature. She further hypothesizes that they must derive from two fundamentally different sources. One, which provides the ritual rules as we know them in the Āgamas, was esoteric and pertained to the quest of the individual for divine powers. This individual, the *sādhaka*, was a special seeker, who conducted his worship largely in private. A second source of the rituals she associates with a public, popular cult of the temple, which she sees as persisting in village festivals today. Brunner stresses that the attitude of the devotee in this popular cult is totally at variance with that of the *sādhaka*; the *sādhaka* seeks an inner vision and uses images as a basis for meditation; the popular devotee, according to Brunner, sees the image as the god himself or the goddess herself. Brunner adds that one might assume that the priests of the original temple cults, prior to the overlay of āgamic ritual, must have had an attitude towards images that is closer to what we see today among the simple village worshippers. Brunner implies in her analysis that the primary meaning of the temple and its natural interpretation must be found outside the abstract speculations of the Āgamas.[10] In this paper I would like to see if we can recover what that primary understanding of the temple might have been, and if the purāṇic texts, despite their priestly overlay, can tell us something of the context, actual and figurative, in which temples and temple worship flourished.

If it was my reading of some of the purāṇic stories on the merits of worship and of poems describing temples and images that first led me to want to find another possible metaphor or image that lay behind the medieval

Hindu temple, it was one of the texts on a specific temple site that also suggested to me what that metaphor might be. The Ekāmra Purāṇa, which celebrates the famous site Bhuvaneśvara, first suggested to me a different way in which we might conceptualize the temple.[11] In part II, chapter 11 we hear this: '*martyaloke svargarūpaṁ viddhi kṣetram anuttamam*' (verse 40b), 'Know that this holy place is heaven on earth.' The text continues, describing the topography of the holy site, in much the same terms as heavens are described in the purāṇic descriptions of them. There is first a forest, and then a mountain. On the mountain is a great tree. Śiva is directly present there '*liṅgavyaktiḥ*', *in the manifest form of the liṅga* (verse 49). It is worth noting that the language of this text makes clear that the *liṅga* is not at all a subtle form of the god but a manifest form of the god. To the East, West, North and South of this *liṅga* are millions of other *liṅgas*. In the next chapter, verse 34 we are told:

> *kṛttivāsāḥ svayaṁ sākṣād ekāmravanottame /*
> *koṭiliṅgodbhavaṁ kṣetraṁ devabhūmiḥ sanātanīḥ //*

> Śiva himself resides in this most excellent Ekāmra forest, in physical form. This holy site, the source of a million *liṅgas*, is really the eternal abode of the gods.

And furthermore, in the very next verse we are warned:

> *bhūmīti nāvamantavyaṁ svarṇakūṭaṁ harāśrayam* 35a

> Do not disparage the mountain, 'Gold Peak', thinking it to be just some mountain here on earth. It is the abode of Śiva.

Here, although the text is speaking not of a temple but of a topographical feature of the holy site, the mountain 'Gold Peak', the statement is also relevant to the man-made temples this paper considers. It is well known that temples were often considered as mountains, as the names given to temples in the architectural texts indicate. While previous scholars have tended to focus on the cosmological significance of the temple as world-mountain, this paper begins from a different starting point: the importance of the fact that the temple is Meru, the Golden Peak, lies in the very simple fact that Meru is the beginning of heaven. In trying to read the temple as heaven on earth, then, I am really returning to a very old understanding of the temple with a new emphasis. I focus on the Gold Mountain, Meru, archetype of the temple, not as a cosmic symbol, but as the locus of the heavens, the abode of the gods.[12]

The text makes other efforts to ensure that we understand that the holy site is heaven. The river that waters it is the Gaṅgā, the river of heaven, and again and again we are told that the holy site is heaven on earth:

> *mama sthānaṁ paraṁ viddhi martyaloke hyanuttamaṁ*
> *bhaumasvargam idaṁ samyak pravadanti manīṣiṇaḥ* // 14.8
>
> Know this to be my most important and sacred abode here on earth. The wise call this place 'Heaven on Earth'.

Like heaven, the holy site of Ekāmravana is untouched at the time of the destruction of the world and is peopled with all manner of living beings, gods, siddhas, yakṣas, vidyādharas, apsarases, and those who seek release. It is also filled with palaces of the gods.

Although the Ekāmra Purāṇa is considered to be a late text, written several centuries after the first round of temple construction at Bhuvaneśvara, I decided to pursue the clues that it gave me.[13] If the temple site is heaven and the temple is the abode of the god, I wondered, could that explain any of the actual visual features of the medieval temple? I began to look at descriptions of heavens in a variety of purāṇic sources. I focused on a few major texts: the Devībhāgavata Purāṇa, 12:10-12 which contains an elaborate description of the world of Durgā, Maṇidvīpa; the Mahābhāgavata Purāṇa, which in chapter 59 describes the abode of the goddess Kāmākhyā; the Padma Purāṇa Uttarakhaṇḍa, chapters 255-6, which is considered by most Vaiṣṇavas to be the *locus classicus* for the description of Vaikuṇṭha, Viṣṇu's heaven; the Bhāgavata Purāṇa, *skandha* 2, *adhyāya* 9, in which the heaven of Viṣṇu is described; the Brahmavaivairta Purāṇa, 4.4, in which the heavens of Viṣṇu and Kṛṣṇa are described, and the Vāyu Purāṇa Uttarārdha, adhyāya 39 ff., which describes all the various heavens of the different gods. In the sections that follow I look first at some descriptions of heaven in the Purāṇas (section 2) and then ask what relevance these descriptions might have for a reading of actual temples (section 3).

2. THE CITY OF GOD: HEAVEN IN THE PURĀṆAS

The first thing that is apparent, even from a cursory reading of the texts, is that the descriptions of a god's heavenly abode are extremely varied and fluid. The descriptions of heaven in the Purāṇas lack the more rigid standardizations of the descriptions of hell, for example. In fact in their variety they mirror the temples themselves, which are never exact reproductions of each other, but differed greatly from region to region, time period to time period, according to religious affiliation, and even within a single site, single religious affiliation and for a single time period. Nonetheless, there are some recurrent features in these descriptions: for example, heaven is always a vast metropolis, with numerous concentric areas all crammed with buildings and peopled by gods and other creatures who have come to serve the main deity. The city is watered by a river, more often by two rivers. The descriptions all proceed from the outermost precincts of this city inward. The city is surrounded by a series of walls with gateways that are carefully guarded. Sex is a major preoccupation of the inhabitants, who are often accompanied by their wives. The main god whose heaven is being described lives in the innermost region of the city, called the *antaḥpura*. This *antaḥpura* itself is usually described as if it were a series of detached buildings, and the god resides in the innermost building of this innermost quarter of the city, surrounded by his devoted servants, who usually look exactly like the god himself/herself, but who may also be of the most varied forms. Indeed every conceivable creature appears in heaven and it is this I think that is the most striking correspondence between our textual descriptions and actual temples: just as temple walls teem with all sorts of living beings, heavenly damsels, sages and their wives, sages practising austerities, copulating couples, gods, incarnations of gods, animals and plants, so is heaven full to the brim with every conceivable life form. Let us look now at some of these textual descriptions.

One of the most elaborate descriptions of the abode of any god is the abode of the goddess in the Devībhāgavata Purāna. Maṇidvīpa is an island, on which the goddess resides. But it is not at all a wild and uncultivated place; it is rather a royal city, with carefully planned areas and buildings. The island of Maṇidvīpa is surrounded by an ocean, whose waves gently lap its shores (10.8). The island is beautifully forested and carefully guarded by armed guards (10.9-10). All of the gods come to see the goddess, and so the island is filled with them, with their attendants and their animal mounts (10.12). There are ponds and tanks and trees of jewels. (10.16). The island is covered with buildings that are arranged in concentric areas, each surrounded by its own walls. As we move inward from the outermost precincts, the building materials are more and more costly. To reach the goddess in her innermost palace, we must pass through courtyard after courtyard, each encircling the other. There are courtyards governed by the different seasons, all magnificently in flower, for example. The seasons are all there as men, who enjoy sensual delights with their two wives. And there are rivers, followed by further courtyards. These courtyards are all packed with gods and siddhas, who are there as servants of the goddess (10.56). In some courtyards the major activity seems to be enjoying sensual pleasures, for as we move inward towards the residence of the goddess we come to areas that are peopled by the siddhas, who had made gifts to the goddess and who now enjoy all kinds of erotic delights (10.69). They are accompanied by women and experience great bliss; they are surrounded by numerous attendants and servants who wait upon them (10.70). At each direction we find the heaven of the various gods, who are traditionally regarded as the guardians of the quarters, Indra, Agni, and so on. There too the gods are enjoying sexual delights with their wives. Again, as we continue to move inward, we come to an area surrounded by gates and filled with pillared *maṇḍapas* (11.2). Here we begin to meet the female attendants of the goddess (11.5 ff.). Again, there are animals and chariots and attendants of the gods who have come to see the goddess (11.21).

As we approach the actual residence of the goddess, the servants begin to look exactly like the goddess herself. It is worth noting that these goddess look-alikes are not manifestations of the goddess or forms of the goddess; one of the rewards of the pious life is to be reborn in heaven as an eternal servant of the god, and when you are reborn in this way you assume the appearance of the god you have served (11.75).[14]

The goddess has eight counsellors, just as a king might have counsellors, and they bring her news of the outside world. They also know everything that is going on in the world (11.78). Each courtyard of the goddess' palace swarms with the animal mounts and chariots of the gods and their attendants. Finally after passing through courtyard after courtyard, past Brahmā and Viṣṇu in all his incarnations, we reach the innermost recesses of the palace, the *Cintāmaṇi-gṛha*, or 'Wishing stone chamber' where the goddess actually lives. This chamber is made up of wishing stones, as the name implies, and its pillars are of sunstone and moonstone, so dazzling that they blind you (11.110). One thinks of an actual temple in which after you transverse frieze after frieze along the walls and see god after god, all servants of the god within, you are left unable to see that god hidden from sight by the superstructure and its walls, which seem to grow in impenetrability.

We are not, however, finished with our journey. The actual residence of the goddess is still within this *Cintāmaṇi-gṛha*, which itself seems to consist of four *maṇḍapas* or pavilions (12.8). The *Cintāmaṇi-gṛha* floats in the sky; it changes in shape at the creation and dissolution of the world, and is the actual final resting place of those who have worshipped the goddess (12.41-44). There is no old age and sickness here; everyone here is a young man who is accompanied by his wife. Perhaps this is why we see amorous couples sometimes high up on the temple structure. Every god in the universe is present there in the innermost palace of the goddess.

To summarize, the abode of the goddess in this text has several characteristic features. It is arranged in concentric regions; it is teeming with life, particularly with gods come to serve the goddess; it is an abode of sensual delights, where men enjoy their two wives (the seasons and their consorts) and devotees are reborn as men who are never alone, but are always accompanied by a female companion.

While few descriptions of heaven in the Purāṇas are as elaborate as this one, most seem to have the same general outline. In the Mahābhāgavata Purāṇa, for example, an Upapurāṇa devoted to the goddess Kāmākhyā of Assam, in chapter 59 we are admitted into the palace of the goddess.[15] Her abode is difficult of access, even for the gods (59.4-5). It is surrounded by a vast ocean (59.6), but is in fact a part of a large city. The abode of the goddess is made of shining jewels (59.6), and is surrounded by a jewelled wall. There are four jewelled gates to this city, festooned with garlands of pearls (59.7) and in the middle of the city is the jewelled palace in which the goddess lives (59.10). The palace is a pillared structure, covered in gold. In the middle of the palace is a jewelled throne, on which sits the goddess, atop a corpse (59.11). The goddess in her throne room enjoys sensual pleasures with her consort, Mahākāla (59.15). The palace in which the goddess and her lover engage in these delights is called the *antaḥpura* (59.16). The description then proceeds with further details about the surroundings of this innermost palace. The *antaḥpura*, or intimate living quarters of the goddess, is ringed by an outer courtyard which is protected by jewelled walls (59.19). The walls have four gateways, one at each of the cardinal directions, and each gateway is adorned with a decorated *toraṇa* (59.19). The area is protected by the most prestigious of Śiva's hosts (*gaṇanāyakas*). In this outer courtyard all the gods, thousands of Viṣṇus, all the minor Yoginīs, wait to be admitted into the presence of the goddess. All the creatures of the universe are there, meditating, waiting to see the goddess (59.20-21). In the next enclosure, also walled, are thousands of Indras, and hosts of other gods, also waiting to see the goddess (59.23). To the north of the palace and its surrounding enclosures is a marvellous garden, in which it is always spring (59.26-27). Viṣṇu, Brahmā, and the other gods, having assumed the form of birds, sing the glories of Mahākālī in the garden (59.27-28). There is also a lotus lake, wooded and serene (59.28). The text tells us that each of the nine Mahāvidyās, goddesses who wait on the main goddess of this text, has her own city, just like this one (59.31). And Śiva is by the side of each of these nine goddess, who make love to him (59.32).

This description of the abode of Mahākālī shares several obvious features with the longer description of Maṇidvīpa. The goddess lives in a vast city, surrounded by an ocean. The outer precincts of the goddess' palace are teeming with thousands of gods and every manner of living creature, all of whom wait to see the goddess. The goddess herself is to be found in the innermost recesses of the palace, the *antaḥpura*, which corresponds to that part of a royal palace where the king slept with his consorts. The goddess, like the royal ruler, spends her time in the *antaḥpura* making love with her lover, Mahāśiva. In addition, the palace of the goddess is surrounded by other palaces of some of the minor goddesses, who also are engaged in the pursuit of sensual pleasures. We shall see that these are features that recur in descriptions of the heavens of other gods.

Turning now to Vaikuṇṭha, the heaven of Viṣṇu, for the Bengali Vaiṣṇavas the authoritative description of Vaikuṇṭha is contained in the Uttarakhaṇḍa of the Padma Purāṇa (chapters 255-6), although other descriptions are scattered throughout the Purāṇa. It is these passages that Jīva Gosvāmī cites in his Bhagavatsandarbha.[16] Vaikuṇṭha, like Maṇidvīpa, is surrounded by an ocean, this time by the milk ocean (282.42). Viṣṇu actually lives in the centre of this city, the *antaḥpura*, in a radiant palace (282.44).

In chapter 255 we are told that Viṣṇu plays in the world, but heaven is where he enjoys sensual pleasures:

> *bhogārthaṁ paramaṁ vyoma līlārtham akhilaṃ jagat* (255.9)
>
> The highest heaven is for his pleasure, while he has the entire world to play in.

This highest heaven is actually described in the next chapter, 256. It has countless worlds in it, all made of pure consciousness, as opposed to material substance (256.1). All these worlds are eternal and golden, and shine with the radiance of billions of suns (256.2). All the worlds are made of the Vedas, divine, without any trace of lust or anger, and all are watered by rivers (256.5). In them reside divine men and women, themselves made up of the same substance as the Vedas (256.5). Again in verse 10 we are told that the highest abode of Viṣṇu is filled with many different countries, *nānājanapadākīrṇa*, in verse 10. It is all surrounded by walls, and by pavilions and palaces made of jewels. In the midst of all of this luxurious splendour is the city Ayodhyā (256.11). The city of Viṣṇu, like the medieval temple itself, is an intricately self-replicating structure. This inner city of Ayodhyā is itself surrounded by walls of gold and jewels and gateways. It has four entrances and has high gopuras (256.12). In the midst of this inner city we now find the *antaḥpura* of the god, his personal residence. It too is surrounded by jewelled walls and beautiful gateways. It is surrounded again by magnificent homes and pavilions and many palaces. Everywhere there are divine women, and apsarases. (256.18). In the midst of all this splendour we come to the actual palace of Viṣṇu, which is described as a divine palace (*maṇḍapa*) and the abode of the king, *rājasthāna*. It has thousands of jewelled pillars and is made of radiant jewels (256.19). It is filled with released souls and gods. In its centre is a lion throne (256.20). What the god does here in his palace is enjoy himself with his wife (256.47). He is constantly surrounded by beautiful women (256.51-53) and his main pastime seems to be enjoying sexual delights (256.55).

Around this highest heaven are disposed other worlds. To the north is Aniruddha and his world; the northeast is Śāntiloka and the east is for Vāsudeva. Lakṣmī is at the southeast, while Saṃkarṣaṇa is in the south. Sarasvatī is in the southwest and Pradyumna in the west, while Rati is in the northwest. This seems to form the first outer circle. In the second outer circle are Prahlāda and other great devotees, while the third circle seems to be reserved for the *avatāras* of Viṣṇu, the turtle, fish and so on. The fourth contains the treasures, *śaṅkhā* and *padma*. There are seven such circles in all.

The Bhāgavata Purāṇa, 2.9.9 ff. also contains a description of Viṣṇu's heaven and its residents. All of those reborn in this heaven look exactly like Viṣṇu; they have four arms and wear yellow garments. The palace of Viṣṇu is surrounded by other buildings. Viṣṇu is attended by the goddess Śrī and numerous other attendants.

Vaikuṇṭha is described in the Brahmavaivarta Purāṇa, *khaṇḍa* 4, chapter 4. The earth is distressed at her great burden and seeks aid from the gods. We are first introduced to Brahmā in his court, and then to Śiva, who is sitting on a tiger skin on the banks of the heavenly river, Ganges. It is worth noting that neither god is alone; both are surrounded by a huge retinue and entertained by dancing apsarases and singing celestial musicians (4.5; 4.43). All the gods then go to Viṣṇu in Vaikuṇṭha. Viṣṇu's abode, like the *Cintāmaṇi-gṛha* of the goddess of Maṇidvīpa, hangs in the air, supported by the wind (4.53). The whole heaven is made of jewels; even the streets, called '*rājamārga*', the term normally given to the broad cross streets in the king's capital city, are made of jewels. The god is to be found in his *antaḥpura,* which is consistent with the descriptions of heavens we have seen thus far (4.56). He is accompanied by an ample retinue and by his wife Lakṣmī (4.59-60).

Viṣṇu tells the gods to go to Goloka, the world of Kṛṣṇa. This world lies above Vaikuṇṭha and is also hanging in the air (4.77). First the gods must cross a river, which is the source of numerous jewels (4.79-83). They next come to a mountain, on top of which is the actual abode of Kṛṣṇa, the *rāsamaṇḍala* (4.86). This *rāsamaṇḍala* is enclosed by walls, and has gardens and numerous pavilions (4.87-88). It is approached by jewelled staircases and has numerous jewelled pillars (4.89-90). There are four gateways in the jewelled walls (4.91). Numerous beautiful women, all beautifully adorned, people the many pavilions around this central palace (4.90-113). We are later told that this enclosed world also contains numerous hermitages and residences of the gopas or cowherds and the attendants of Kṛṣṇa; it also contains residences of the devotees of Kṛṣṇa who had worshipped him faithfully during their lives, and homes of all the gopīs, Kṛṣṇa's bucolic lovers (4.135-140). Kṛṣṇa's attendants look exactly like Kṛṣṇa himself (4.136), and the abodes of his devotees seem to look very much like temples: they are crowned with jewelled pots (*kalaśa*), 4.144). The various hermitages are enclosed in jewelled walls. The hermitage of Rādhikā, for example, is round and made of jewels. It consists of hundreds of jewelled buildings and is surrounded by moats and gardens and jewelled walls (4.160-167). Even into this bucolic splendour intrudes the more urban architecture of the standard heaven.

When we stop to analyse some of these descriptions from the Vaiṣṇava texts, it is clear that there are a number of features that are common to these descriptions of Viṣṇu's and Kṛṣṇa's world and the worlds of the goddesses of Maṇidvīpa and Kāmākhyā. Goloka, Kṛṣṇa's heaven is somewhat aberrant in that it follows the dominant theme of the country idyll of Kṛṣṇa in Vraj; it is more rustic and bucolic than other heavens. Nonetheless it shares many features with our other texts. In all of these descriptions, the god's world is a vast city, surrounded by walled area after walled area, all peopled with living beings and packed with palaces. There is an emphasis in all of these descriptions on sensual pleasures: devotees are there with their wives, and the god and goddess both spend their time in love-making. It is worth noting here that Viṣṇu's heaven is bordered not only by the ocean, but also by rivers. Rivers figure prominently on the gateway of temples, and much labour has been spent by modern scholars trying to understand their role and their position there.[17] In fact they are a common feature of purāṇic heavens, where they often serve as boundary markers. In the Vāyu Purāṇa, chapter 42, for example, the river of every god's heaven is considered to be a form of the Gaṅgā, and the point is stressed that the sacred topography is surrounded by a river.

Another point worth noting is the curious feature of the central palace in heaven: it floats in the sky. This was also the case with the palace of the goddess in Maṇidvīpa. That the temple on earth was the god or goddess' palace, floating in the sky is suggested by a moving verse from a Jain text that was written to celebrate the building of a famous Jain temple. This is the Kumārapālaviharaśataka of the Jain monk Rāmacandragaṇi. Here is how the poet describes the temple to the Jina:[18]

> And the crowds of people who were constantly coming from afar to see the temple thought in truth that they had reached heaven. For the temple indeed seemed to float in space, as the rays of light coming from its radiant walls made of moonstone spread out in every direction, concealing from view the temple's solid foundation on earth.

Without reference to the purāṇic texts one might have been tempted to take this verse as poetic fancy. In the context of the Purāṇas, it seems to be a clear reference to the heavenly palace which floats in the ether.

Buddhists as well as Jains could see temples as embodiments of heaven. The paradigmatic temple for Buddhists was the Jetavana vihāra, and in a late version of the story of Anāthapiṇḍada's gift we find this verse,[19]

bhaktyutsāhād athārambhaoṛtasāhāyyakaiḥ suraiḥ /
vihāraṁ tridivikākāraṁ cakārānāthapiṇḍadaḥ //

> And stirred on by his great devotion for the Buddha, aided in his undertaking by the gods, Anāthapiṇḍada built a vihāra that was a likeness of heaven.

The parallel between the earthly temple and heaven is stressed when Śāriputra tells Anāthapiṇḍada that as soon as construction was begun on the Jetavana vihāra on earth, a gold vihāra suddenly materialized in heaven (verse 72). The vihāra, like the temple, is really the god's heavenly palace, and what we see on earth is but the likeness of that heavenly structure. Anāthapiṇḍada in our story is spurred on by this account of the appearance of the heavenly Jetavana, and inspired to make his vihāra of gold and the most excellent jewels (verse 73). No doubt he is attempting to replicate as closely as possible the heavenly vihāra of gold about which he has just heard.

In another Buddhist *avadāna* told in the Mahākarmavibhaṅga, a young man visits heavens and then hells.[20] The heavens are beautiful cities, each surrounded by a gateway. Each one is more beautiful than the next. The first is of gold, the second of silver, the third of cats-eye gems, and finally the highest heaven is of crystal. They are each equipped with pleasure gardens, groves, and ponds, and fragrant incense wafts from them. Flowers abound and there are garlands everywhere. In each case the young man is ushered into the city by heavenly damsels who come out to greet him; the number of the damsels increases as we move from heaven to heaven. What the young man does in the cities with his divine hosts is also explicitly stated, for at one point when he leaves the crystal heaven, to which thirty-two heavenly damsels have invited him, we are told that he left '*ratikhinna*', 'totally exhausted from sex'. Heaven here is very much like some of the more exuberant temples of Khajuraho with all their erotic sculpture; what the literary description had abbreviated, the visual descriptions of the temple walls tell us in luxurious detail.

That the temple was heaven on earth thus seems to find ample support from Jain and Buddhist texts. In fact I find much in all of these descriptions, from the Purāṇas, the Buddhist and the Jain sources, that is reminiscent of the medieval temple, with its walls jammed with figures, often in erotic poses and explicitly sexual acts. Even the abundant use made of architectural ornament suggests to me a shorthand for the crowded urban network of the purāṇic paradise. I should like in what follows to consider even briefly how we might make a beginning of relating these poetic descriptions to concrete art monuments.

3. THE TEMPLE AS MAP OF HEAVEN

Given the fact that as we have seen above even the purāṇic descriptions of heaven themselves exhibit a wide range, we cannot expect to find an exact correspondence between temple and text. But I would like ultimately to leave you with this suggestion: the temple city, a not infrequent feature of the medieval Indian landscape, was indeed heaven on earth. Like heaven, it was populous and built up, containing numerous buildings which were the abodes of the different gods who lived there. But even more suggestive, I would like to ask us to use our imagination and read the individual temples too as heavens on earth. For if we turn vertical elevation into horizontal extension, the temple like heaven was surrounded by outer solid walls (the base); it was watered by two rivers, whom scholars have thus far seen as guardian deities, but who are a consistent feature of the natural topography in descriptions of heavens. If the temple is heaven, we are also in a better position to understand the presence of the *navagrahas* or nine planets on the door lintels of so many medieval temples. Like the river goddesses, the *navagrahas* tended to be seen as guardian figures. In fact in purānic descriptions the nine planets circle the heavens, and their presence on the temple would be a simple topographical marker, telling us that we have entered the lands of the gods.

If we turn from these initial markers and then go from the base to the panels of sculpture on a temple like the fully developed temples at Khajuraho, could not these panels also be those concentric regions in which dwell the various living beings: yakṣas, apsarases, gods, sages, and so on, all in their own palatial homes? Architectural ornament on the temple walls then becomes not mere ornament but shorthand for the palaces in which the citizens of heaven live in the purāṇic descriptions. As the concentric panels end and we move up the *śikhara* or spire of the temple, imagine instead moving inward through the various outlying areas to the centremost area where resides the main god.

The god lives in the temple in the *garbhagṛha*, which is in fact an interesting word. While earlier interpretations have focused on the term *garbha*, one meaning of which is womb, and have stressed the procreative aspects of god and developed the theory of the temple as the unfolding universe, I would like to keep to the model of heaven: god lives in a great palace structure or *antaḥpura*, and the *garbhagṛha* must be part of that structure.[21] In fact the term does appear in secular architecture as well as in temple architecture, and it can refer to the innermost room in the *antahpura* where the king has his private quarters and normally his sleeping chambers. In an early medieval Jain text, the Jñātādharmakathā, chapter 8, which deals with the life of the tīrthāṅkara Mallīnātha, the princess Mallī has a palace which has in its innermost recesses a number of pavilions or buildings. At one point, wishing to teach her importunate suitors a lesson, she embarks on a new building project and has a number of buildings made. There is first the *mohanaghara*, and then inside that are six *garbhagharas*; these are disposed around a central *jālaghara* in which the princess orders a golden statue of herself to be placed. As its name implies, this chamber is surrounded by latticed walls so that the statue can be seen from outside.[22] From a thorough study of relevant texts on palace architecture Roth concluded that the *garbhagṛha* was part of the inner apartments of a ruler in which he slept. The *garbhagṛha*, moreover, was often described as surrounded by a labyrinthine structure, the *mohanagṛha* of the Jain text, designed to confuse any late night would-be assassins.

Another early text, the Caraka Saṁhitā, suggests that *garbha* or *garbhagṛha* simply means a room inside a larger structure. The person who would practice alchemy or *rasāyana* is instructed to build a hut that is described as *trigarbhā*, 'having three *garbhas*'. The commentary explains that this is a structure in which there is a series of rooms, one inside the other. There is first a large room, and then the second room is inside this first room. The third room in turn is inside this second room. *Garbha* is then glossed in the commentary simply as 'interior space'.[23] What seems to me particularly important about this information from the Caraka Saṁhitā is the conclusion that *garbha* refers simply to any interior chamber; interpretations of the symbolic meaning of the temple have constantly taken the *garbhagṛha* as the innermost area, 'the womb', a unique physical structure having its symbolic counterpart in the human uterus, and implying that the god within this room is somehow in the process of creating the multiple universe that is his temple. This correspondence is more difficult to support, when we consider texts that describe multiple '*garbhas*'. In addition the *garbha* of the Caraka Saṁhitā and its commentary corresponds perfectly to the *garbhaghara* of the Jain text and both find a close parallel in the *antaḥpura*, the inner living quarters of the god in his palace in heaven in the Purāṇas. Other uses of the word *garbha* in medieval texts indicate that its basic meaning is simply 'room' of a house or building, without any further symbolic import.[24] Indeed the diversity of texts in which *garbha* is used for 'room' indicates that this was a common usage of the word and does not reflect a specialized or technical usage of the term.

Other parts of the temple, often infused with symbolic meaning by modern scholars, similarly appear as symbolically neutral components of secular buildings. In a fourteenth century description of a palace given by Vidyāpati in his Kīrtilatā, the *śikhara* of the palace is crowned by a golden pot or *kāñcanakalaśa*.[25] As late as the eighteenth century Pṛthivīcandracarita we hear of palaces with the spires crowned with *āmalaka*, the crowning member of the temple superstructure.[26] From the innermost chamber, to the uppermost ornament, architectural elements that scholars have seen as carrying the most important religious symbolism, we have to do with structures that throughout the history of secular architecture were components of ordinary palace buildings.

If we return to the purāṇic descriptions of heaven, where God lives in a vast palace, I would like to try roughly to follow a generalized purāṇic description of heaven as we look at the temple. The spire now goes not only upward but indicates an inward motion; even its lattice tracing recalls the labyrinthine approach to the king in the innermost chamber of the palace. As we make this imaginative journey through heaven to reach the god in his sleeping quarters where he is served only by his most faithful devotees and where he sports with his wife, we must remember that this is only one of the many metaphors for the temple. To accept its plausibility, and you may indeed not wish to do that, by no means excludes other even conflicting interpretations. Like heaven, the temple was not one single thing, but a fluctuating reflection of the worshipper's religious needs. As one text tells us, people were born in heaven in a form that reflected their dying wishes and that reflected the way in which they worshipped god. Some are there as sages, some as gods, some as humans, some as various incarnations of Viṣṇu, the fish, turtle, boar, and so on; some are four-faced like Brahmā, some are in the form of the sun god Sūrya, or Agni, the fire god.[27] Perhaps the temple was equally fluid in the religious imagination, sometimes the cosmic man, sometimes the cosmos, and sometimes heaven on earth.

4. CONCLUSIONS: SOME FURTHER SUGGESTIONS AND A NOTE OF CAUTION

Potentially, I think, looking at the temple as heaven can help us put sacred architecture in medieval India back into context. A text like the eleventh century AD Samarāṅgana Sūtradhāra of king Bhoja makes it clear to us that rituals for building temples did not differ from rituals for any other type of building.[28] Chapter 8, for example, on the examination of the ground of a potential building site, lists as attributes of the ground appropriate to building a city the very same qualities that scholars have discussed in terms of the building of temples: the area chosen should have rivers nearby and gardens with flowers and birds; the sites should also bring pleasure to those who wish to engage in love-making (8.27-35; 42-43). Again, you are to avoid the very same types of sites for the building of a city or a temple. You must avoid ground that has ashes or bones on it, for example (53). A medieval Jain story describes the trials and tribulations of one would-be temple builder who built his temple over some human bones.[29] Rituals in the Samarāṅga Sūtradhāra for the laying of the first stone are also given as exactly the same for all types of building. In chapter 35 we are told that one should follow the rituals enjoined in the chapter equally for building private homes or temples. A medieval text on the settlement of Brāhmin villages also gives rituals for the founding of the village that parallel those for temple building.[30] If we understand the temple as heaven and realize that heaven itself is a vast and populated city in the Purāṇas, I think we are in a better position to begin to understand the context of temple building: temple building is city building, and many of the rituals for building a city and its individual structures apply to the building of a temple. I think that considering the building of temples, abstracted from this context, has led scholars to overemphasize the potential symbolic meaning of the ritual acts that accompany the building process. A detailed study of the similarities and differences in the rituals for building cities, secular buildings and temples would contribute greatly to our understanding of the context and the uniqueness of sacred architecture in India.

Considering the temple as heaven can also lead to further important questions. We can look afresh not only at individual temples, but also at sites as a whole. The map of medieval India was dotted with 'temple cities', from Bhuvaneśvara in the east to Pattadakal and Aihole further south, Osian in the northwest and Khajuraho in central India. Like these temple cites, the heaven of the Purāṇas is also a giant lush and densely populated temple city. The gods never live alone in heaven; they reside at the centre of a complex metropolis that is crowded with temple buildings that house the lesser gods and the devoted worshippers who have come to serve the main deity. In addition, if temple cities are indeed heaven on earth as the Ekāmra Purāṇa with which I started this essay calls Bhuvaneśvara, then one might also expect the different temples at a single site to be dedicated to different gods. This is in fact what we observe at the known temple cities from medieval India. The existence of temples dedicated to various gods at a single site has often been taken as a sign of medieval Indian 'religious tolerance'; I would like to suggest that the temple site as heaven implies something quite different. It implies that we are wrong to look at the presence of different temples to different gods as the result of an absence of sectarian affiliation at a given site or as the end-product of a spirit of broad-mindedness in which there was an equal emphasis on all the different deities honoured in the various temples. Heaven, after all, was where all the gods came to do homage to a single god. While it is true that these other gods lived in their own temples, their presence in heaven was a submissive gesture. They are there as servants of a central deity.

To verify these suggestions we need to see if it is possible to find a pattern in the arrangement of the temples at a given site and to see if the disposition of temples there indicates that one particular deity was at the centre of a hierarchy. The temple city as heaven thus asks us to pose new questions about temple sites as organized wholes. A consideration of temples as groups at a given site might well help test the limits of the validity of what I have proposed here. If we cannot find any site that reflects some organizational system, then I should myself consider the translation of textual reference, holy site as heaven on earth, to actual site a more complicated process.

A further body of evidence that might be considered is pilgrimage maps of temple sites that were made during the late medieval period. While research on maps in India is at an early stage, several maps have been studied and published.[31] While such maps indicate a clear differentiation between surrounding environs and the *antaḥpura*, the walled city of the main deity to whom the site is dedicated, it is the rare map that depicts the outlying area in neatly arranged concentric pattern, something that might correspond more closely to the purāṇic descriptions of heaven.[32]

At the same time, the depiction on the maps of numerous temples and worshippers reflects the general spirit of the purāṇic paradise. Indeed, purāṇic paradise, temple walls and map are alike in eschewing a rigid iconographic program and limiting representational possibilities to a strictly prescribed schema.

This last remark might occasion some hesitation. Scholars have always considered that the arrangement of sculptures on a temple was indeed fixed in a pattern that was governed by some text. The problem was to discover the correct text. I think I first began to be suspicious of this idea when I read in one of the iconographic manuals lists of alternatives for the placement of deities on the temple walls: one could put god x in place y or indeed one could put god a, b, or c there. There was much greater flexibility than I thought should be possible if there was a fixed relationship between images and a symbolic pattern to the whole that was determined by the position of the individual images. The descriptions of heaven in the Purāṇas can provide at least a pseudo-explanation for this phenomenon: in heaven gods appear nearly everywhere, and in no clear relationship to each other. Besides, what looks like gods are often not gods, but are worshippers reborn in heaven in the form of the deity whom they worshipped. For these divine look-alikes, there is no question of any relationship of one to the other. In heaven it would seem there are few deities whose place is fixed; the only fixed deities would seem to be the guardians of the quarters, the *dikpāla*, who must appear in the right direction of the compass, and the rivers, which are to be found towards the outskirts of heaven, flowing along its borders, a place I have sought to consider homologous with the doorways to the temple. The other fixed deity, of course, is the main god to whom the temple is dedicated; he or she should appear in the innermost sanctum and if we are reading up as inwards, might also appear on the top of a gateway or the top of a spire. The rest of heaven is a delightful hodge-podge of every living creature imaginable. Perhaps the most important change in the way we might view temples once we see them as miniatures of heaven is in our appreciation of the flexibility and creativity of the overall iconographic schema.

There are other perhaps minor details that the model of temple as heaven might help us to understand. Numerous prescriptive texts praise the temple made of jewels as the most worthy gift to the god; stories tell us that in the past temples were made of jewels but that given the depredations of this Age of Evil, the Kali Yuga, men have found it more prudent to construct temples of less valuable materials, in an effort not to invite plunder and despoiling.[33] In addition late medieval inscriptions from temples in Bengal typically refer to the temple with the modifier *navaratna*, perhaps meaning made of nine jewels; in one inscription the temple is described in these words: *saudhaṁ sundararatnamandiram* 'a magnificent palace with beautiful jewelled pavilions'.[34] In this case the description in the inscription might even seem closer to the purāṇic texts we have been examining than to the actual temple. The temples of heaven are all made of jewels in the purāṇic descriptions; the jewelled temple on earth would most closely reproduce the heavenly paradise of the gods for all men, and even where a temple was obviously not constructed of precious materials to call it a jewelled temple might be sufficient to evoke its heavenly prototype.

In the final words of this paper, I should like to acknowledge the imprecision that accompanies the model I am proposing. Once we agree that heaven itself is a place of infinite possibilities, we have shrugged off the burden of finding a precise, one-to-one correspondence between textual model and actual temple. Whether the proposed model will prove liberating or stultifying, I think will depend to a great deal on its usefulness as a research tool. If it provokes discussion and fruitful inquiry, then I think that this heaven, at any rate, will have shown its own rewards.

Notes

1. The classic statement of these theories is Stella Kramrisch, *The Hindu Temple*. Calcutta: Calcutta University Press, 1946. More recent renditions include Michael Meister, 'Fragments from a Divine Cosmology: Unfolding Forms on India's Temple Walls', in: Vishakha N. Desai and Darielle Mason (eds), *Gods, Guardians and Lovers: Temple Sculptures from North India, A.D. 700-1200*. New York: The Asia Society Galleries, 1993, pp. 94-116, and Heinrich von Stietencron, *Gaṅgā und Yamunā: zur symbolischen Bedeutung der Flussgöttinnen an indischen Tempeln*. Wiesbaden: Harrassowitz, 1972, particularly chapter VIII, 'Exkurs über den Tempel als "Leib der Gottheit"', pp. 80-101.
2. I surveyed and translated some of the purāṇic stories on temple worship in my article, 'Halāyudha's Prism: The Experience of Religion in Medieval Hymns and Stories', in: Vishakha Desai and Darielle Mason (eds), *op. cit.*, pp. 66-94.
3. Hélène Brunner, 'L'image divine dans le culte āgamique de Śiva. Rapport entre l'image mentale et le support concret du culte', in: André Padoux (ed.), *L'image divine: culte et méditation dans l'hindouisme*. Paris: Éditions du CNRS, 1990, pp. 9-31.
4. This very important point has in fact already been made by Richard Davis, in: *Ritual in an Oscillating Universe: Worshipping Śiva in Medieval India*. Princeton: Princeton University Press, 1991, p. 64.
5. On this point see the article by Colas, 'Le dévot', cited below in note 10.
6. In a recent article, 'Changes in the Vedic Priesthood', in: A.W. van den Hoek, D.H.A. Kolff and M.S. Oort (eds), *Ritual, State and History in South Asia: Essays in Honour of J.C. Heesterman*. Leiden: Brill, 1992, pp. 556-78, R.B. Inden suggests that a new group, the astronomers, became particularly concerned with the rituals centring around images. The old *śrauta* priests played only minor roles in these rituals, according to texts like the Viṣṇudharmottara Purāṇa. In this scenario we might well view the overlay of *śrauta* symbolism as an effort to legitimate the new rituals by cloaking them in the familiar language of the building of the fire altar. In any case we are still faced with the same gap between ritual and temple that scholars working on ritual texts have highlighted. See also J.C. Heesterman, *The Broken Sacrifice, An Essay in Ancient Indian Ritual*. Chicago: University of Chicago Press, 1993, p. 220, where the fundamental difference between the world of the *śrauta* ritual and the temple is stressed.
7. Perhaps the clearest statement of the concrete nature of the god in the temple has been made for Buddhist images by Gregory Schopen in his article, 'The Buddha as Owner of Property and Permanent Resident in Medieval Indian Monasteries', *Journal of Indian Philosophy* 18, 1990. I have discussed the language of some of the stories about the *jyotirliṅgas* in the Śiva Purāṇa in the article, 'Halāyudha's Prism: The Experience of Religion in Medieval Hymns and Stories', in: Desai and Mason (eds), *op. cit.*, pp. 66-94, particularly p. 71 and note 6.

8. See the interesting comments of Jackie Assayag, *La colère de la déesse décapitée: traditions, cultes et pouvoir dans le sud de l'Inde*. Paris: Éditions du CNRS, 1992, pp. 55 ff.
9. Hélène Brunner, 'Maṇḍala et yantra dans le śivaïsme āgamique: définition, description et usage rituel', in her: *Mantras et diagrammes rituels dans l'hindouisme*, Paris 21-22 Juin 1984. Paris: Éditions du CNRS, pp. 11-37, particularly, pp. 17, 30-1.
10. In his article in the volume *L'image divine* cited above, 'Le dévot, le prêtre et l'image vishnouite en Inde du Sud', pp. 99-115, Gérard Colas suggests that there exist convergences between the attitudes of the devotee and priest, despite their many differences. Colas, like Brunner, suggests that the rituals of installation may be late accretions to temple worship.
11. The text is edited by Upendra Nath Dhal, Delhi: Nag, 1986.
12. There is an informative article on Mount Meru as the underlying form of Southeast Asian architecture by Robert von Heine-Geldern, 'Weltbild und Bauform in Südostasien', *Wiener Beiträge zur Kunst und Kultur Asiens* 4, 1930, pp. 28-78. Heine-Geldern emphasizes particularly in his discussion of Burmese city planning and palace and temple architecture that the temple, palace or city was heaven on earth. In this paper I develop this widely accepted hypothesis by considering in some detail actual purāṇic descriptions of heavens.
13. On the date of the Ekāmra Purāṇa, see Chandra Panigrahi, *Archeological Remains at Bhubaneswar*. Cuttack: Kitab Mahal, first edition 1961, second edition, 1981, p. 21.
14. This is in fact a widespread belief in purāṇic religion. The Saura Purāṇa, which is a Śaiva text, tells the history of a king of the gods named Śibi. The divine king worships Śiva assiduously and as a reward is promised that at the end of his life he will become a *gaṇa*, one of Śiva's close associates. The account ends with a description of Śibi, now the *gaṇa* named Caṇḍa. He looks exactly like Śiva: he has three eyes, carries a trident, wears a tiger skin and has four arms. See chapter 32, verses 53-54, p. 107 in the Ānandāśrama Sanskrit Series, vol. 18, Poona, 1924.
15. This text is virtually unique in explicitly describing the form of the goddess who resides in her palace as the subtle form. Despite that characterization, however, the goddess appears there in a physical form (atop a corpse) and engaged in sensual pleasures.
16. Edited by Chinmayi Chatterjee. Calcutta: Nadavpur University. 1972, p. 24, 26.
17. See for example Heinrich von Stietencron, *Gaṅgā und Yamunā: zur symbolischen Bedeutung der Flussgöttinnen an indischen Tempeln*. Wiesbaden: Harrassowitz, 1972.
18. The text is published in the Ātmānanda Sabhā in Bhavnagar, 1909. I have translated some of the verses in my essay, 'Halāyudha's Prism' cited above. The present verse concludes a chapter on pilgrimage in Jainism that I wrote for the catalogue of the Jain exhibition held at the Los Angeles County Museum of Art in 1993, 'Jain Pilgrimage: In memory and Celebration of the Jinas', in: Pratapadiya Pal (ed.), *The Peaceful Liberators*. Los Angeles County Museum

of Art, 1994, pp. 63-77.

19. Avadānakalpalatā 21, verse 56, edited P.L. Vaidya, *Buddhist Sanskrit Texts*. 22, Darbhanga: Mithila Institute of Post-Graduate Studies and Research in Sanskrit Learning, 1959 p. 156.
20. The text is edited by P.L. Vaidya in the *Buddhist Sanskrit Texts*. 17, Darbhanga: Mithila Institute of Post-Graduate Research and Studies in Sanskrit Learning, 1961, pp. 189-91. A variant version of the story figures as the Maitrakanyakāvadāna, number 92 in Kṣemendra's collection.
21. The most recent reworking of this older theory can be found in Jackle Assayag, *op. cit.*, pp. 382-3. Here Assayag refurbishes the older interpretation of Kramrisch in more current language, suggesting that the worshipper views the *liṅga* in the *garbhagṛha* from the radically new perspective of being in the womb that the male organ penetrates. 'Synonyme d'un "voir dedans", le *darśana* découvre l'envers de l'enveloppe corporelle dans son étrangeté radicale. Altérité d'une intériorité analogue à la sienne, mis que redouble celle de l'autre sexe et qui n'est pas limitée par un point de vue strictement mondain.' The interpretation depends to a large extent on the interior chamber being the womb, something I argue is not substantiated in the texts. Assayag also notes that the temple she is studying has two such 'womb' rooms and a 'womb' temple (p. 117); she does not discuss the symbolism or meaning of the three wombs, however.
22. Gustav Roth, *Mallī-Jñāta, das achte Kapitel des Nāyādhammakahāo im sechsten Aṅga des Śvetāmbara Jainakanons, hrsg., übers. und erl.* Wiesbaden: Steiner, 1983, pp. 202-20 on *mohanaghara*.
23. Caraka Saṁhitā, Cikitsā Adhyāya 1.19, edited with commentary of Cakrapāṇidatta, by Vaidya Jādavji Trikamji Āchārya. Bombay: Nirṇaya Sāgara Press, 1941, p. 377.
24. See for example the Pāli Manorathapūraṇī, vol. 1, edited Dr. Nand Kishore Prasad. Nālandā: Nava Nālandā Mahāvihāra, 1976, p. 192, where Bhadrā, the future wife of the monk Mahākāśyapa, is said to be so beautiful that she is capable of dispelling the darkness with the radiance of her body; even in a room as large as twelve *hastas*, she has no need of a lamp: *dvādasahatthe gabbhe nisinnāya padīpakaccaṁ nathi, sarīrobhāseneva tamaṁ vidhamatīti*. The word used for room here is *gabbha* or *garbha*.
25. The text is cited by V.S. Agrawala, 'Palace Architecture in Bāṇa's *Harṣacarita: Skandhāvāra, Rājakula, Dhavalagṛha*', in: *Mélanges d'indianisme à la mémoire de Louis Renou*. Paris: Boccard, 1968, pp. 7-23. (Publications de l'Institut de Civilisation Indienne, Fascicule 28)
26. *Ibid.*, p. 19.
27. The Bṛhadbhāgavatāmṛta of Śriam Sanātana Goswāmī, edited by Śrī Śyāma Dās. Vṛndāvan: Śrīharināma Saṁkīrtana Maṇḍal, 1975, *Khaṇḍa* 2, *Adhyāya* 4, verses 139-144.
28. Edited by T. Gaṇapati Śāstrī, rev. by Vasudeva Saran Agrawala, Baroda: Oriental Institute, 1966. (GOS. 25)

29. This is the story of Lalla, which is told in the biography of the Jain monk Jīvadevasūri. I have translated one version of the biography of Jīvadevasūri in the book that I edited, *The Clever Adulteress: A Treasury of Jain Stories*. Oakville: Mosaic Press, 1990, pp. 149-53.
30. Dr. Gaya Charan Tripathi, *The Ritual Founding of a Brahmin Village*. Delhi: GDK Publications, 1981.
31. The major work on the subject is Joe Schwartzberg, *History of Cartography, South Asia*. Chicago: University of Chicago Press, 1992. See also Susan Gole, *Indian Maps and Plans. From Earliest Times to the Advent of European Surveys*. Delhi: Manohar, 1989.
32. See Gole, plate 28, p. 67.
33. I have touched on the subject of the jewelled temple briefly in an article 'Tales of Broken Limbs and Bleeding Wounds: A Study of Some Hindu and Jain Responses to Muslim Iconoclasm', *East and West* 41, 1991, pp. 189-205.
34. A.K. Bhattacharyya, *A Corpus of Dedicatory Inscriptions From Temples of West Bengal (From c. 1500 to c. 1800)*. Calcutta: Navana, 1982, inscription 55. For a note on the use of the term *navaratna* see inscription 17, note 8.

XI

HAVE: LINGUISTIC DIVERSITY IN THE EXPRESSION OF A SIMPLE RELATION

Ken Hale and Jay Keyser

Putting aside the use of *have* as an auxiliary verb, itself a matter of enormous typological interest, the use of this element and its equivalents in expressing possession represents an area of impressive cross-linguistic morpho-syntactic diversity. One popular type of possessive predication (*x* has *y*) is expressed by a suffixal noun in Warlpiri (of Central Australia) and by null (zero, nothing) in 'O'odham (Pima-Papago of Arizona and Sonora), by a verb in English, by a preposition-like verb in Yoruba, by a true preposition in Bantu, and by a verbal suffix in Hopi. This diversity is underlain by an extremely simple system of structural relations. In an important sense, there is but one *have*. The observed cross-linguistic variation in the expression of the predicational relation involved here can be attributed to the flexibility which is inherent in the language-specific morpholexical realization of universal lexical and functional categories (the parts of speech). The underlying structural categories themselves are invariant, but their overt expression is highly variable. We will be concerned with *diversity* as much as, perhaps more than, with *universal* aspects of linguistic expressions in the domain at issue. In this, we wish to recognize our friend Frits Staal, who has contributed greatly to our appreciation of linguistic diversity.

Linguistic diversity is necessary. First, it is necessary to us, as linguists and as people, for the very business of doing our professional work, which many of us would argue is impossible without it, as well as for the fulfilment of the greater purpose of providing, together with cultural diversity, the enabling condition for the maximal exercise of human intellectual capacities in the creation of the most precious products of human labour. Second, linguistic diversity is necessary in the sense that it is inevitable, a natural consequence of the very nature of linguistic structure. We will be concerned here with this latter, fundamentally linguistic sense in which diversity is 'necessary'.

Our own interest in linguistic diversity stems, in fact, from our interest in universal grammar, an aspect of language which is not, and cannot be, diverse. We want to know how it can be that there is universal grammar beside almost unimaginable diversity in the actual languages of the world. The answer is perhaps complex, but we think there is an idealized answer that is very simple. The answer is this: linguistic diversity stems from the fact that there are no stipulations in grammar, no inviolable principles; all invariant aspects of grammar are due to the fundamental nature of the *elements which define, or project, linguistic form*. While this is an idealization, we think it is closer to the truth than anything else, and we will pretend that it is absolutely true for present purposes.

Consider the projection of syntactic structure from lexical items, or 'heads', taking these to be the *basic nuclear elements of grammar*. And suppose that it is in the *nature* of these entities to enter into, or not to enter into, certain linguistically fundamental structural relations, of which there are just two: the 'head-complement' relation, and the 'subject-predicate' relation. A head *x* will therefore belong to one of just four classes, according to its participation in these relations, as depicted in (1a-d):

(1) Head (*x*), complement (*y* of *x*), predicate (*x* of *z*):

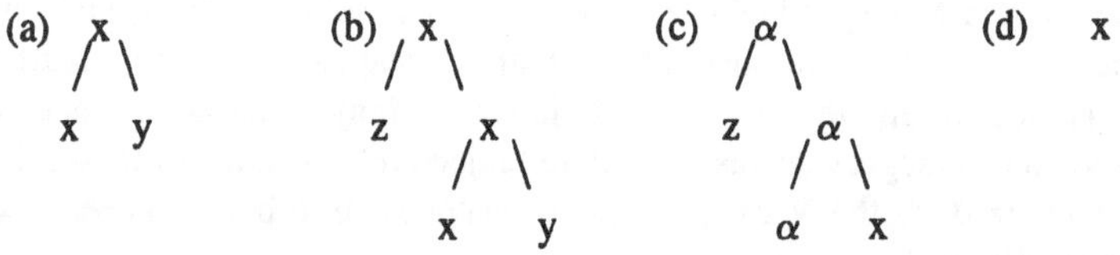

In (1a), *x* belongs to the class whose members take a complement (*y*), thereby defining the simple syntactic structure given - the category of the head projects, by definition of the relation expressed. This class does not project a 'specifier' position, since it does not participate in the subject-predicate relation. By contrast, in (1b), *x* takes a complement (*y*), forming a predicate therewith; and it therefore necessarily further projects a subject (*z*) - in short, this class enters into both fundamental relations. The third class of heads, represented by *x* of (1c), participates in the subject-predicate relation, but not the head complement relation. The logic of the system requires that the subject *z* be projected external to *x* itself - if *z* were presented internal to the *x*-projection, it would be a complement, not a subject. Thus, to satisfy the predication required in (1c), there must be some host category, represented here by α. Finally, *x* of (1d) represents the class of nuclear elements which participates in neither one of the two basic structural relations - it selects no complement and it is not a predicate, hence it has no subject.

We are assuming that this is an invariant aspect of grammar - there is no sense in which diversity is possible here. However, when the nuclear elements are associated with specific morphosyntactic categories, diversity ensues, since there are no stipulations in grammar. Thus, while the four-class scheme inherent in the basic system of relations in (1) is neatly replicated in the English part-of-speech repertoire, it is not perfectly replicated there, as is evident from (2), where the b-class is realized as P (as *on* in *(put) the book on the shelf*) or V (as *learn* in *she learned the language* and in dialectal *we learned her the language*); similarly, the c-class is prototypically A in English, but V is also found in that structure (as *grow* in *the corn grows*). This is not surprising, since, by hypothesis, nothing forces a one-to-one correspondence between the structural projection of a nuclear element and the morphosyntactic category which realizes it. There are tendencies, and certainly ones which are linguistically significant, but there are no hard and fast laws of association.

(2) English:

(a) x = V (b) x = P/V (c) x = A/V (d) x =N

Although the a- and d-configurations are predominantly the morphosyntactic categories V and N respectively, even this is not absolutely fixed, and our assignments here represent just the 'favourite' realizations. The greatest variety is found in the b- and d-classes. Many languages lack a morphosyntactic category corresponding to 'adjective' - in some of these, the c-class is realized by the V-category (as in Navajo), in others it is realized by N (as in Warlpiri).

(3) Navajo:

(a) x = V (b) x = -P/N (c) x = V (d) x = N

(4) Warlpiri:

(a) x = V (b) x = -P/(-)N (c) x = N (d) x = N

While the b-class is commonly represented by an 'adposition', category P (preposition or postposition), sometimes bound (as in Navajo and Warlpiri), it is commonly represented by the morphosyntactic category N as well (as in Warlpiri, where some are bound and some are free). In Lardil, the favourite realization of the b-class is by suffixal verb (cf. also (8)):

(5) Lardil:

(a) x = V (b) x = -P/-V (c) x = N (d) x = N

In some languages, exemplified here by Salish, a single open class is associated with the four structural classes of (1a-d). The nuclear elements themselves have no fixed categorial association (cf. Kinkade, 1983; Jelinek and Demers, 1994). The distinct categories of phrasal syntax - e.g. clause, noun phrase - are defined by categorially distinct extended projections headed by the Open Class Roots, INFL for the clause, DET for nominal arguments, as in (6):

(6) Salish:

(a) x = INFL (b) x = INFL/OBL (c) x = INFL (d) x = DET

Cross-linguistic diversity, and intra-linguistic diversity as well, is present not only in the *morphosyntactic* realization of the basic lexical-syntactic configurations but in their *morphophonological* realization as well. English is extraordinarily replete with denominal and deadjectival verbs. Thus, while the class of structures corresponding to (1a) is productively represented by verbs taking a free-standing phrasal complement, like *make trouble, do a jig*, it is also abundantly represented by denominal verbs, like *sneeze, laugh, dance*, which involve the 'merger' of a phonologically empty verbal head with a nominal complement - the head of *y* 'merges with' or 'incorporates into' the head *x* which selects it; on this view a verb like *sneeze* has the same basic structure as hypothetical *do a sneeze*. In many languages, the same class of structures - i.e. (1a) - involves conventional, or standard, incorporation, in which a noun is incorporated, that is to say 'adjoined', to the phonologically overt head of the governing verb. This is exemplified here by the Tanoan language Jemez. In other languages, like Basque, for example, the a-class structure is overtly a transitive light-verb construction (cf. Grimshaw and Mester, 1988; Laka, 1993):

(7) English (a): *laugh, sneeze, sing, work; make trouble, do a jig, have puppies*.

Jemez (a): *sae-'a* (work-do) 'work', *záae-'a* (song-do) 'sing', *see-'a* (word-do) 'speak'; *shíl-'a* (cry-do) 'cry', *tún-'a* (whistle-do) 'whistle'.

Basque (a): *lan egin* (work do), *hitz egin* (word do), *negar egin* (cry do), *ziztu egin* (whistle do), *barre egin* (laugh do), *usin egin* (sneeze do).

These expressions, we assume, share the abstract argument structure configuration (1a). And the three languages agree, furthermore, in associating the head x with the category V and the complement y with the category N, as depicted in (8), order of terminals immaterial:

(8)

The languages differ, however, in the morphosyntactic form which these items take. In English, V is empty and hence must fuse with, or 'incorporate', its complement, thereby acquiring phonological constituency, resulting in a denominal verb. In Jemez, the noun incorporates into the verbal head in the standard manner (cf. Baker, 1988). In Basque, the verb-complement syntactic structure remains essentially unmodified (cf. Laka, 1993).

English-style 'fusion' or 'incorporation' is also found in deriving location and locatum verbs from P-based structures corresponding to (1b). English *put apples in the box* involves a b-class structure embedded as the complement of an a-class structure, as depicted in (9):

(9)

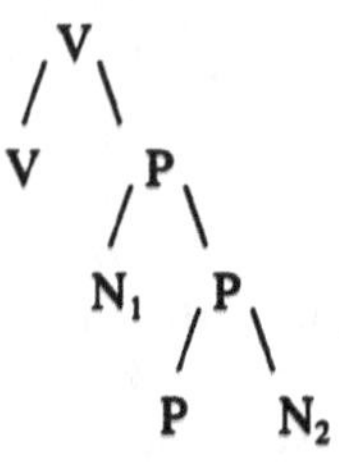

In the form just cited, all heads are overt, with V = *put* and P = *in*. But English permits non-overt P and non-overt V as well. Suppose N_2 is *box*. Successive cyclic incorporation of that nominal into the empty governing heads (P, then V) yields the denominal verb *box*, as in the verb phrase *box apples*. This derivation is responsible in general for English denominal location verbs, like *shelve*, *corral*, *bottle*, *bag*, etc. and for locatum verbs, like *saddle*, *blindfold*, *shoe*, *harness*, etc. The fundamental basis of these English verbs, we maintain, is the category P (adposition), the supreme b-class item, inasmuch as it expresses a relation between two entities, hence two arguments, a complement and a subject. In English, these acquire their verbal category through incorporation, as suggested. They are generally transitive in English, because the verbal component is the upper head, not the lower. A few English verbs have V, rather than P, as the lower head - as expected, these have intransitive as well as transitive uses, e.g. *land* (as in *the plane landed*), *centre* (e.g. *the cursor centred*), *front* (e.g. *the vowel /a/ has fronted in Danish and Greenlandic*).

We will not repeat here the full range of arguments in favour of the idea that the classification embodied in (1) is the fundamental basis of predicate argument structures - in general, it is supported empirically by observed limitations on the variety and complexity of argument structures (as outlined, for example, in Hale and Keyser, 1992, 1993). Let us look now at some examples of the 'possessive relation' in some of the languages mentioned at the outset:

(10) (a) *The coyote has a tail.* (Cf. Coyotes have tails.)

(b)	*Ban 'o*	*(ge)*	*bahï.*	[O'odham]
	coyote	AUX3	(Aff)	tail
(c)	*Warnapari*	*Ø*	*ngirnti-parnta.*	[Warlpiri]
	dingo	AUX3	tail-WITH	
(d)	*ajá*	*n'írù* (< ní ìrù)		[Yoruba]
	dog	HAVE.tail		

We will begin the discussion with the Yoruba construction. The nuclear element *ní* 'have, central coincidence' belongs to the Yoruba morphosyntactic category V (but see Manfredi, 1995 for much relevant discussion of the origins of this element). The structure it projects can appear in the 'causative' serial construction, as in (11), indicating that it forms a predicate and therefore projects a subject, the s-structure object of the causative verb *fún* 'give, benefactive':

(11) *Ó* *fún* *mi* *n'íwé* (< ní ìwé).
he give me HAVE.book
'He gave me a book.'

Since *ni* takes a complement and forms a predicate, requiring a subject, it belongs to the structural class (1b). This is prototypically the class associated with the morphosyntactic category P - as in English and many other languages, but not in Yoruba, which uses the category V for (1a) and (1b) alike. Assuming this to bc essentially correct, the structure of (10d) is that depicted in (12), in which N_1 and N_2 represent the nominal expressions corresponding to the subject (*ajá* 'dog') and object (*ìrù* 'tail'). The notation V represents the verbal nuclear element *ní* 'have', whose category projects to the phrasal level in the expected manner, so that the whole belongs likewise to the verbal category:

```
(12)        V
           / \
         N1   V
             / \
            V   N2
```

In the Warlpiri expression (10c), the nuclear element is realized by the so-called 'proprietive' suffix *-parnta* 'with, having' (cf. Dixon, 1976). The Warlpiri construction differs from (11) in two ways. First, its head (*-parnta*) belongs to the major non-verbal category of the language, namely N, participating in that morphosyntactic category clearly and fully, in relation to inflection and to derivation. Second, the head is phonologically dependent. It is a suffix - which is to say, it is preceded by an empty phonological matrix, as in (13):

```
(13)       N
          / \
        N1    N
             / \
          []N   N2
```

This forces 'incorporation', permitting the phonological matrix of N_2 to fuse with the empty matrix of the head of the construction, giving the observed surface form.

As in Yoruba, so also in Warlpiri, this possessive construction may appear embedded as the complement of a higher transitivizing verb whose s-structure object is the 'internal subject' N_1, as in the following:

(14) *Ngarrka-patu* *ka-rna-lu-jana* *yimi-parnta-ma-ni.*
man-PL PRES-1-PL-3PL speech-WITH-CAUSE-NPST
'We give the men speech (e.g. remove speech ban).'

The causative verb *-ma-*, is itself a suffix, forcing incorporation of the possessive expression *yimi-parnta* 'having speech', resulting in the complex verb stem *yimi- parnta-ma-* 'cause *x* to have speech, enable *x* to speak'.

The O'odham possessive predication, exemplified by (10b), shares with the Warlpiri construction the requirement that the complement, N_2, incorporate to license an underlying empty phonological matrix. Here, however, the head of the construction is entirely empty in the initial representation - the head is not a suffix as in the Warlpiri case:

(15)

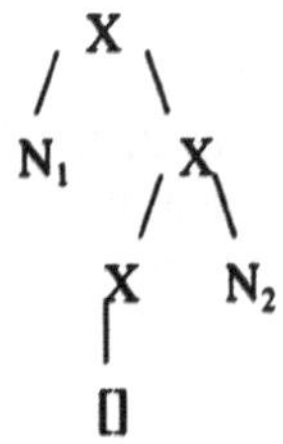

Incorporation will permit the empty head to acquire phonological substance by replacing the empty matrix with that of the noun N_2, *bahï* 'tail' in the example given. The empty matrix of (15) is used regularly where N_2 denotes an entity which is 'dependent' (e.g. a part, a kinsman, a product of human culture). Where N_2 denotes an entity which is 'independent' (e.g. an animal, plant, phenomena in the landscape or the heavens), the head is a suffix (*-ga*, glossed -GA), as in (16):

(16) *Heg* *'o* *(ge)* *gogs-ga.*
he AUX3 (Aff) dog-GA
'He has a dog.'

The distinction resembles closely, but not entirely, the traditional 'alienable/inalienable' distinction (cf. Bahr, 1986). We will use that familiar terminology here, though it is somewhat inexact in my view. The structure projected in the alienable construction is as shown in (17):

(17)

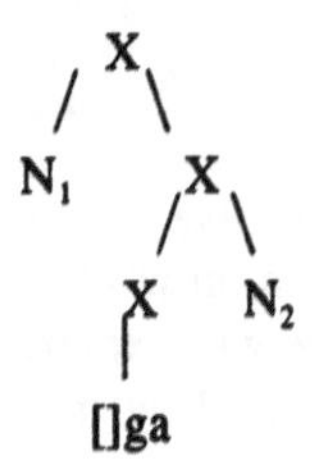

Inalienables (but not alienables) can embed as the complement of a transitivizing verb, as in (18) where, as expected the inner subject N_1 appears as the s-structure object of the derived transitive verb. The matrix verb is, again, a suffix:

(18)	*Heg*	*'at*	*o*	*xu:xk-rad*	*g*	*kawyu.*
	he	AUX3T	FUT	shoe-PUT	ART	horse

'He is going to shoe the horse.'
(Cf. *Kawyu 'o (ge) xu:xk.* 'The horse has shoes (on).')

The morphosyntactic category of the head in (15) and (17) is not obvious. Since there is nothing to contradict the view that it is P, the 'favourite' realization of (1b), we will assume that it is indeed a member of that category. In this respect, O'odham agrees with the Bantu pattern, exemplified by the Xhosa construction (19):

(19) *Ndi-néncwadí* (< na-íN-ncwadí).
1sg-WITH.CL5sg.book
'I have a book.'

The Bantu element *na* 'with' is a preposition, not a prefix, though it is proclitic to its complement in the same way that Yoruba *ní* 'have' is proclitic. The Bantu construction perfectly represents the favourite realization of (1b):

(20)
```
       P
      / \
    N1   P
        / \
       P   N2
```

The precise nature of the English possessive predication, exemplified by (10a), is not altogether clear. If English *have* projects the structure (12), then we must explain its failure to transitivize:

(21) **The Creator had the coyote a tail.*
(Cf. The Creator gave the coyote a tail.)

Of course, we might assume that *give* is the transitive counterpart of *have*, in which case (19) occupies the complement position in (1a). While this is entirely possible, and may in fact be correct, we will explore an additional possibility as well. Whatever the structure projected by *have*, it is almost certainly the same as that of *cost*, *weigh*, and *resemble*, as in (22):

(22) (a) *The book costs five dollars.*

(b) *The book weighs five pounds.*

(c) *The book resembles a brick.*

Like *have*, the verbs of (22) resist passivization. And, like *have*, they resist straightforward transitivization: **We cost the book five dollars* (i.e. priced it at five dollars); **we weighed the book five pounds* (i.e. got it to weigh five pounds); **We resembled the book a brick* (i.e. made it resemble a brick). In this case, however, there is no suppletive counterpart like *give*. There is simply no transitive counterpart of these verbs. It is possible, therefore, that *give* is merely a 'spurious' transitive of *have*. If so, we are left without an analysis of English *have*.

Recent work by David Pesetsky suggests a possible solution. Nominals derived from the verbs of (22) exhibit a behaviour - exemplified by the ungrammaticality of the hypothetical nominalizations of (23) - which begs to be understood in terms of 'Myers' Generalization' (Pesetsky, 1995; Myers, 1984):

(23) (a) **the book's cost of five dollars*

(b) **the book's weight of five pounds*

(c) **the book's resemblance of a brick*

If *have* belongs to the same class of verbs, then its failure to furnish a Myers' Generalization violation is merely accidental, resulting from the fact that there is simply no noun derived from *have*. Let us assume that this is the case, and assign to all of these verbs the structure depicted in (24), in which V corresponds to a verb of the type represented by *have, cost, weigh,* or *resemble*;

(24)

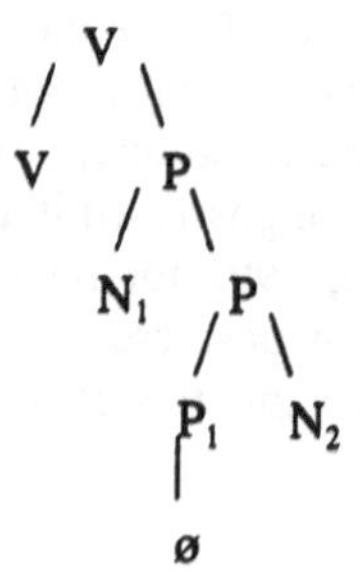

The head of the inner projection is a 'zero morpheme', not merely an empty phonological matrix. It is a true element, with properties. The relevant property here, in addition to its membership in the structural class (1b), is the fact that it must raise and adjoin to V. This prevents nominalization, by virtue of whatever principle underlies Myers' Generalization. Nominalization would involve further adjunction of [ø[V]] to a nominalizing head, NOM, resulting in the following configuration, ruled out by Myers' Generalization which allows ø only at the outermost layer of derivation:

(25) [[ø[V]]NOM]

Here, NOM is the outermost derivational head, asymmetrically c-commanding ø. The projection (24) fails to transitivize for the same reason, although an alternative account also exists. Since V of (24) belongs to class (1a), it projects no subject. It therefore fails to provide a source for the s-structure objects in (21) and (23) - N_1 is not a candidate in this case, as it is too deeply embedded in the hypothetical transitive, in which (24) is the complement in yet another V-projection.

In sentential syntax, V of (24) is intransitive, forcing the DP corresponding to N_1 to raise to an appropriate external subject position, hence the observed word order in (10a) and (22). In this respect, English *have* and the verbs of (22) differ from dialectal *learn*, which has both transitive and intransitive alternants (suggesting that it is fundamentally verbal, not prepositional).

If this is correct, then the languages we have examined here all make use of structure (1b), with the possessor (N_1) in the *z*-position and the possessed (N_2) in the *y*-position. The languages have this in common, though as we have seen, we find much diversity even here, as permitted by the anti-stipulation thesis. This arrangement is quite natural in expressing the possessive relation, as it embodies the asymmetry according to which the possessor, freely specific and definite, is structurally higher than the possessed, typically non-specific and indefinite. In some languages, this is a formal property of the favourite possessive construction - in O'odham, for example, the nominal corresponding to the possessed cannot be specific or definite; the structure cannot be used to say *I have the book you gave me*.

But here again we encounter diversity. Many languages express the possessive relation using this same basic structure, (1b), but with the arguments reversed, so to speak - with the possessed in a structurally higher position than the possessor (see Freeze, 1992 for related observations and discussion). In these constructions, the possessor is in the complement position (*y*), while the possessed occupies the subject position (*z*), as in (26):

(26) (a) *Yesh* *l-i* *sefer.* [Hebrew]
is to-1SG book
'I have a book.'

(b) *Kyún* *un-k'á.* [Tanoan (Towa)]
child 1SG.DAT-be
'I have a child.'

(c) *Tá* *leabhar* *ag-am.* [Irish]
is book at-1SG
'I have a book.'

In many such languages, there exist grammatical devices which have the sentential syntactic effect of reversing the asymmetry, resulting in the possessor being in a structural position which is higher than that of the possessed. In Tanoan, this is accomplished by incorporation of the inner subject (possible under government in sentential syntax):

(27) *Un-kún-k'a.* [Tanoan (Towa)]
1SG.DAT-be
'I have a child.'

In Irish, the N_2, the possessor, forms a chain with a higher 'subject' (McCloskey and Sells, 1988). Since the head of this chain is structurally higher than N_1, the possessed, the favoured asymmetry is achieved, i.e. with possessor higher than possessed. Evidence for this arrangement comes from the control relation, among other things - the possessor is the controlled argument:

(28) (*Caithfidh mé*)
(I must)

[*PRO*$_i$ *airigead* *a* *bheith* *ag-am*$_i$].
[PRO$_i$ money PRT be.INF at-1SG$_i$]
'... have (some) money.'

REFERENCES

Bahr, Donald
1986 'Pima-Papago-ga, 'Alienability', *IJAL* 52, pp. 154-71

Baker, Mark C.
1988 *Incorporation. A Theory of Grammatical Function Changing*. Chicago: University of Chicago Press

Dixon, R.M.W.
1976 *Grammatical categories in Australian languages*. Canberra: Australian Institute of Aboriginal Studies; New Jersey: Humanities Press

Freeze, Ray A.
1992 'Existentials and other locatives', *Language* 68, pp. 553-95

Grimshaw, Jane, and Armin Mester
1988 'Light Verbs and q-Marking', *LI* 19, pp. 205-32

Hale, Ken, and Jay Keyser
1992 'The Syntactic Character of Thematic Structure', in: Roca, I.M. (ed.) *Thematic Structure: Its Role in Grammar*. Berlin [etc.]: Foris, pp. 107-43
1993 'On argument structure and the lexical expression of syntactic relations', in: Hale and Keyser (eds) *The View from Building 20: Essays in Honor of Sylvain Bromberger*. Cambridge: MIT Press

Jelinek, Eloise, and Richard A. Demers
1994 'Predicates and pronominal arguments in Straits Salish', *Language* 70, pp. 697-736

Kinkade, Dale
1983 'Salish evidence against the universality of 'Noun' and 'Verb'', *Lingua* 60, pp. 25-40

Laka, Itziar
1993 'Unergatives that assign ergative', in: J.D. Bobaljik and C. Phillips (eds), *Papers on Case & Agreement* 1, Cambridge: MIT Press, pp. 149-144. (MIT Working Papers in Linguistics 18)

McCloskey, James, and Peter Sells
1988 'Control and A-Chains in Modern Irish', *NLLT* 6, pp. 43-189

Manfredi, Victor
1995 'Syntactic (de)composition of Yoruba 'be' and 'have'', in: L. Nash and G. Tsoulas (eds) *Langues et Grammaire* 1. Paris VIII

Myers, Scott
1984 'Zero-derivation and inflection', in: M. Speas and R. Sproat (eds) *Papers from the January 1984 MIT Workshop in Morphology*. Cambridge: Department of Linguistics and Philosophy. (MIT Working Papers in Linguistics 7)

Pesetsky, David
1995 *Zero Syntax*. Cambridge: MIT Press

XII

METRICAL VERSE IN THE PSALMS

*Morris Halle**

About fifteen years ago, John McCarthy and I noticed that psalm 137 was composed in accordance with a syllable-counting metrical scheme that is quite similar to that used in many of the modern Romance languages, and we published this observation in Halle and McCarthy, 1981. In the years that have elapsed since, I have continued to study metrical properties of biblical poetry, and I have found a number of additional texts in the Old Testament that are composed in accordance with the same metrical scheme as psalm 137. In addition to psalms 54 and 23, which I discuss below, these metrical texts are psalms 2, 24 and 114, the curse of Lamek (Gen. 4, 23-24), and one of the prophecies of Amos (3, 3-8). I have little doubt that there are additional texts in the Old Testament that are composed with the same syllable-counting metrical scheme, but I have not succeeded in finding them. My metrical analyses of the texts just mentioned except those of psalms 2 and 24 were published in Halle, 1989. Since the latter paper was written well over a decade ago, I have had occasion to revise somewhat my analyses of these texts.[1] I believe, moreover, that I have discovered several further properties of these texts that, if correct, shed provocative light on a number of peripheral topics, such as the influence of Greek models on Hebrew poetry, the use of numerological devices in the Old Testament and, last and most speculative, the date of composition of psalm 137.

---

* I am grateful to Elan Dresher and Robert Hoberman for general advice, to Alan Prince for a suggestion about a numerological aspect of psalm 137, and to Israel Shahak for critical remarks on realistic vs. ideal representation of objects in classical art. The papers of David N. Freedman (see Freedman, 1980), and especially his study of the structure of psalm 137 provided the original impetus for my study of biblical versification. I am indebted to Professor Freedman also for his kindness in discussing with me by letter most of the matters addressed below. None of the above necessarily shares any of my views, and I alone bear responsibility for errors or other inadequacies in the paper.

1. ON THE NATURE OF METRICAL VERSE

Metrical verse is distinguished from prose in that it obeys special conditions on word sequences above and beyond those that sequences in prose texts must obey.[2] Some of the additional conditions are illustrated in example (1).

```
      1     2     3   4   5    6  7    x
(1)  Lives of great men do  remind (us),
      1 2     3   4    5    6   7
     We can make our lives sublime,
      1    2 3 4    5      6 7    x
     And departing leave behind (us,)
      1     2    3    4 5    6   7
     Footprints on the sands of time.
```

Thus, if we stop the count with the last stressed syllable in the line, then as shown above all lines in (1) consist of seven syllables.[3] Moreover, the stresses by and large fall on the odd-numbered syllables. And each of lines ends with a syllable (or syllable sequence) that rimes; i.e. *(sub)lime - time*; *(re)mind us- (be)hind us*.

1 Poems with a different metre are composed by placing different numerical restrictions on the lines. Thus, in (2) each line consists of ten syllables, the stresses fall on even-numbered syllables, and the lines do not rime.

```
     1 2 3  4 5    6 7 8 9 10
(2)  It little matters that an idle king,
      1  2   3  4     5 6    7   8 9    10
     By this still hearth, among these barren crags,
     Matched with an aged wife, I mete and dole
     Unequal laws unto a savage race,
     That hoard, and sleep, and feed, and know not me.
```

The distribution of stresses in the line is clearly an important factor in English metrical verse and in the poetic traditions of many other languages, but there are numerous poetic traditions that disregard the placement of stresses and are based exclusively on syllable counting. Such purely syllable-counting metres are employed in the Japanese haiku, in the major verse forms of Polish, Italian, Spanish, French, and - as noted above - in some of the poetry of the Old Testament.

To get some feeling for what poets do when they write syllable counting verse we examine the well-known lines by Verlaine in (3).

(3)

De la musique avant toute chose	Music above everything,
Et pour celà préfère l'Impair	And for this prefer the uneven
Plus vague et plus soluble dans l'air,	Vaguer and more soluble in air
Sans rien en lui qui pèse ou qui pose.	With nothing in it that weighs and that counts.

In standard French pronunciation the reduced *schwa* vowel, which is represented by the letter *e* in the French orthography, is not pronounced in a great many contexts, most notably not at the end of a polysyllabic word. In French grammars this vowel is often referred to as *e-muet* 'mute *e*'. Thus, in (3) the words *musique, toute, chose, prefère, vague, soluble, pèse, pose* all end with a 'mute *e*' that is not pronounced. If we now count the syllables that are actually pronounced we find there are eight such syllables in the first three lines and nine in the fourth.

1 2 3 4 5 6 7 8
De la musique avant toute chose,
1 2 3 4 5 6 7 8
Et pour celà préfère l'Impair
1 2 3 4 5 6 7 8
Plus vague et plus soluble dans l'air,
1 2 3 4 5 6 7 8 9
Sans rien en lui qui pèse ou qui pose.

This is a somewhat irregular and implausible distribution of line lengths in this kind of lyric poem. It suggests that there is something wrong in our way of counting syllables. In fact, we can readily regularize the line lengths by assuming that for purposes of French metrics not all *e-muets* count equally. Specifically, as books on French versification standardly tell us, for metrical purposes a word final *e-muet* counts when followed by a consonant, but not otherwise. Hence in the first stanza of (3) the *e-muet* in *toute*, *préfère*, *soluble* counts, but not in *musique*, *chose*, *vague*, *pèse*, *pose*. As a result, the number of metrical syllables in the first three lines of (3) increases from eight to nine, and the stanza is now made up of lines of equal length. To obtain this result, however, we must admit that the rules of French metrics are based not on the facts of pronunciation directly, but on the essentially non-phonetic principle stated just above. To support further the proposition that the rule for counting metrical syllables is not directly based on French pronunciation I cite the lines in (4), which are taken from scene iv, act 1 of Molière's play *Tartuffe*. At this point in the play there is a series of exchanges between the character Orgon, and his wife's maid, Dorine.

(4)

1 2 3 4 5 6 7 8
O. Et Tartuffe? D. Tartuffe? Il se porte à merveille, 233
1 2 3 4 5 6 7 8
O. Le pauvre homme! D. Le soir, elle eût un grand dégoût, 235
1 2 3 4 5 6 7 8 9
O. Et Tartuffe? D. Il soupa, lui tout seul, devant elle, 239
1 2 3 4 5 6 7 8
O. Le pauvre homme! D. La nuit se passa toute entière 241
1 2 3 4 5 6 78
O. Et Tartuffe? D. Pressé d'un sommeil agréable, 245
1 2 3 4 5 6 7 8 9
O. Le pauvre homme! D. A la fin, par nos raisons gagnée, 249
1 2 3 4 5 6 7 8 9
O. Et Tartuffe? D. Il reprit courage comme il faut, 251
1 2 3 4 5 6 7 8
O. Le pauvre homme! D. Tout deux se portent bien enfin, 256

Like most French classical plays of the seventeenth century, *Tartuffe* is composed in the *Alexandrin* metre. For present purposes we can say simply that a line in this metre consists of twelve syllables.[4] Each of the eight exchanges in (4) begins with Orgon saying either *Et Tartuffe?* 'And Tartuffe?' or *Le pauvre homme!* 'The poor fellow!' This is followed by Dorine's response, which makes up the rest of the line. Notice now that both of Orgon's phrases end with an *e-muet*, which, as we know, is metrically ambiguous: it counts if followed by a consonant, but not otherwise. In the present instance, whether or not Orgon's mute *e* counts metrically will depend on whether Dorine's response begins with a consonant or with a vowel. Specifically, Orgon's *Et Tartuffe?* and *Le pauvre homme!* will count as quadri-syllabic if Dorine's response begins with a consonant, but as tri-syllabic if her response begins with a vowel. By the same token, since Dorine's response in lines 239, 249, 251 begins with a vowel, the response should be 9 syllables long, and since in the other five lines her response begins with a consonant, the response must be 8 syllables long. As shown in (4) both of these predictions are correct.

In speaking of Orgon's phrases above I said that they must '*count* as tri-syllabic' rather than '*be pronounced* as a tri-syllabic sequence'. I used this formulation to bring out clearly that what we are dealing with here cannot be a fact of pronunciation, for at the point of uttering his words, Orgon cannot know whether Dorine will begin her response with a vowel or a consonant and cannot therefore know whether to pronounce the *e-muet* that ends his response.

In sum, we have in (4) an inside joke that is likely to have been noticed only by a few of Molière's fellow actors, and perhaps by one or two literary critics. We turn next to the metrical verse in the Bible, where, as I shall try to show, 'inside jokes' similar to those just examined can be found.

2. ON SYLLABLE COUNTING VERSE IN THE BIBLE

As remarked above, in Halle and McCarthy, 1981 and other papers, it has been argued that certain poetic texts of the Old Testament are composed in a syllable counting metre that is similar to that of French. Like French, Hebrew has reduced vowels - i.e. *schwas* - which I have represented in the transcriptions below with the capital letter E. Like in French, the *schwas* count for purposes of metre only in certain contexts but not elsewhere. In the transcription of the biblical texts below the uncounted schwas have been omitted. The conditions under which vowels are counted for metrical purposes are detailed in (5).

(5)

i. Vowels following the last stress are not counted.

ii. Secondary *Hătēpîm* are omitted; thus, instead of Massoretic *?a9ălê* 'I shall ascend' we read *?a9lê*.

iii. The schwa is omitted in 'doubly open' syllables of the form $VC_1\_\_\_C_2V$ where C_1 and C_2 are not identical. We therefore read *?omrîm* 'say' (pl.) and *bin?ôt* 'in pastures' rather than *?omErîm* and *binE?ôt*. If the consonants flanking the schwa are identical, the schwa counts metrically; i.e. we read *lEsōrErāy* 'to my enemies' rather than *lEsōrrāy*.

iv. The *pataH* associated with word final gutturals is omitted. We therefore read *koH* 'strength' rather than *koaH*.

v. Massoretic *yErûšālāim* and *yEhwāh* is read systematically as *yErûšālem* and *yahweh*.

Like in English and in French, the syllable count in biblical verse stops with the last stressed syllable: the syllables that follow the last stress are not counted and have therefore been enclosed in parentheses.

2.1. METRICAL ANALYSIS OF PSALM 54

As an example consider vv. 3-9 of psalm 54 given in (6).[5]

(6)

?ĕlōhîm bEšimkā hôšî9ē(nî)	9	God, save me by thy name
ûbigbûrātkā tEdînē(nî)	8	and judge me by thy strength.
?ĕlōhîm šEma9 tEpillāt	9	God, hear my prayer,
ha?zînâ lE?imrê-pî	7	give ear to the sayings of my mouth.
kî zārîm qāmû 9ālay	7	For strangers have risen against me,
wE9ārîcîm biqšû napšî	8	and oppressors seek my soul:
lō? śāmû ?ĕlōhîm lEnegdām	9	they have not set God before them.
selâ		Selah.
hinnê ?ĕlōhîm 9ōzēr lî	8	Behold, God is my helper,
?ădōnāy bEsōmkê napšî	8	the Lord is with them that uphold my soul:
yāšîb hāra9 lEsōrErāy	8	he shall return evil to my enemies,
ba?mittEkā hacmîtēm	7	cut them off in thy truth.
bindābâ ?ezbEHâ-llāk	7	Freely I will sacrifice unto thee
?ôde ššimkā yahweh kî-Tôb	8	I shall praise thy name, o Lord, for it is good
kî mikkol cārâ hiccîlā(nî)	8	for he has delivered me out of all trouble.
(ûb?oybay rā?tâ 9ênî)	7	(and to my enemies my eye has seen)

The psalm is composed of two stanzas that are separated by the word *Selah*, which is a Hebrew word whose exact translation is a matter of dispute, but which appears to be used as an exclamation like *Amen*. The first stanza consists of seven lines whereas the second seems to have eight lines. However, the eighth line is somewhat problematical. The King James version translates it 'and mine eye hath seen *his desire* upon mine enemies', emending the text by inserting the phrase *his desire* that is not in the Hebrew text. A better translation is 'My eye will gloat over my enemies', as suggested to me Robert Hoberman. But neither of the two modifications eliminates the abrupt and unmotivated shift in subject matter. I propose therefore that this line was added by a later editor and is not part of the original poem, and I shall assume that the psalm consists of two stanzas each seven lines long. Below we shall see a modicum of additional motivation for this proposal.

That psalm 54 was subject to later editing is almost certain. It is generally supposed that the psalter is composed of five separate collections. The divisions between the collections are marked by the appearance of the formula

> bārûk yahweh ?ĕlōhê yiśrā?ēl mēhā9ôlām wE9ad hā9ôlām ?āmēn wE?āmēn
> Blessed be Yahweh God of Israel for ever and ever Amen and Amen'

which appears in this form at the end of ps. 41, and in somewhat modified form, after pss. 72, 89 and 106. That pss. 41-72 represent a separate collection from the rest is made even more likely by two further facts. First, among the texts in pss. 41-72 there are three passages that appear with slightly varied readings elsewhere in the other books. I have listed these duplicated texts in (7).

(7) ps. 53 = ps. 14 ps. 57:8-12 + ps. 60:7-14 = ps. 108
ps. 70 = ps. 40:14-18

The hypothesis that our text of the Psalms consists of five separate collections provides a plausible explanation for these duplications: the same poems were chosen for inclusion in more than one collection.

Moreover, as the article on the Psalms in the fourteenth (1928) edition of the *Encyclopedia Britannica* states:

> We have clear evidence of other editorial work in the overwhelming predominance of the name 'Elohim' for God in Psalms xlii-lxxxii as compared with the personal name 'Yahweh' (the occurrences are 200:43; in psalms 1-xli 15:272); this is confirmed by the fact that Ps. xiv ... appears as Ps. liii, with its fourfold 'Yahweh' changed into 'Elohim' (so also Ps. xl, 13-17 and lxx).[6]

Since the psalms included in Book Two thus underwent important editing it is not implausible that the last line in (6) is a later emendation that was not in the original text, and that the original text consisted of two stanzas each seven lines long.

We now turn to the syllable count of the lines. In the text given in (6) the lines composing the two stanzas have the syllable counts in (8), where that of the first syllable is noticeably different from that of the second.

(8) 9 8 9 7 7 8 9 8 8 8 7 7 8 8

At this point the fact mentioned in the quoted passage that part of the editing that psalm 54 has undergone consisted of replacing *yahweh* by *?ĕlōhîm* becomes important. It tells us that not all occurrences of *?ĕlōhîm* need to have been in the original text; at least some of them could be replacements of the original *yahweh*. I propose that this is the case in lines 1, 3 and 7 of (6). As a result the three lines in the first that were nine syllables long are now shortened to eight syllables. The emended text will then have instead of (8) the line length distribution below

8 8 8 7 7 8 8 8 8 8 7 7 8 8

With these emendations psalm 54 consists of two identical stanzas composed in syllable counting metre. I also draw attention to the fact that the total number of syllables in each stanza adds up to 54 = 3 x 18, a number which, in the light of what follows, may not be altogether accidental.

2.2. GEMATRIA IN PSALM 23

Another syllable-counting poem is the famous psalm 23 'The Lord is my shepherd'. The Hebrew text and a somewhat modernized translation of the King James text is given in (9).

(9)

yahweh rō9î lō? ?eHsār	Yahweh is my shepherd, I shall not want.
bin?ôt deše? yarbîcē(nî)	He makes me lie down on grassy pastures,
9al-mê mEnūHôt yEnahlē(nî)	He leads me beside still waters,
napšî yEšôbēb	He restores my soul,
yanHēnî bEma9gElê-ce(deq)	He leads me into paths of righteousness
lEma9an šEmô	For the sake of his name.
gam kî-?ēlēk bEgê-calmā(wet)	Though I walk through the valley of the shadow of death
lō?-?îrā? rā9â	I shall fear no evil,
kî-?attâ 9immādî	For you are with me.
šibTEkā umis9antē(kā)	Your rod and your staff
hēmmâ yEnaHmū(nî)	They comfort me,

ta9rōk lEpānay šulHān	You prepare a table before me
neged cōrErāy	In front of my enemies,
diššantā baššemen rō?šî	You have anointed my head with oil
kôsî rEwāyâ	My cup is full.
?ak Tôb wāHesed yirdEpū(nî)	Only goodness and mercy shall pursue me
kol-yEmê Hayyāy	All the days of my life,
wEyāšabtî bEbêt-yahweh	And I shall dwell in Yahweh's house
lE?ōrek yāmîm	To the end of the days.

I have emended the official (Massoretic) text in two places. In the last line of the first stanza I have replaced the *rā9* 'evil', whose grammatical gender is masculine, by its feminine counterpart *rā?â*. In the penultimate line of the poem, I have replaced the MT *wEšabtî* 'I returned' with *wEyāšabtî* 'I dwelled, sat', which fits better not only metrically, but also semantically, and which has also been adopted by many other writers.

The line lengths of the poem are then as given in (10).

(10) 7 7 8 5 8 5 8 5 6 7 5 7 5 8 5 8 5 8 5

As shown in (10) the poem is readily sub-divided into two eight-line stanzas separated by a three-line stanza. The two eight-line stanzas consist of three couplets of an 8-syllable line paired with a line of 5 syllables preceded by a couplet that is composed of two 7-syllable lines in the opening stanza, and of a 7- and a 5-syllable line in the closing stanza. We shall return to these facts below.

Bazak, 1987 has suggested that psalm 23 has a covert structure somewhat paralleling that of Molière's *Tartuffe* passage (4). In order to appreciate Professor Bazak's proposal it is necessary to recall that the Hebrew letters standardly serve as numerals. For example, in the Jewish calendar the present year 5755 <since the creation of the world (= 1994/95 CE)> is designated by the letter sequence *taw šin nun heh*, which respectively stand for 400 + 300 + 50 + 5 and add up to 755. (The millennia are usually omitted.)

Letter sequences representing certain numbers make up recognizable words. For example, 14 is represented by the letter sequence *yod dalet*, which stands for the Hebrew word *yād* 'hand'. Following this procedure we should expect the following integer 15 to be represented by the letter sequence *yod heh*. This letter sequence, however, stands for *yah*, one of the appelations of Yahweh. Since there is a tabu on using the name of Yahweh, the number 15 is represented not with the letter sequence *yod heh*, but with the letters *Tet wāw* i.e. 9 + 6.

The fact that letters also serve to represent numbers has led to certain numbers being replaced by words whose letter sequence has the same value. Thus, the word for 'dog' *keleb* has the numerical value of 20 + 30 + 2 = 52,[7] and I recall my father telling me on his fifty-second birthday that he had now reached a 'dog's age'. My father was thus referring to a number by means of a word whose letters add up to the number.

The reverse of this procedure is even more widely used: a number is used as a kind of code to refer to a word whose letters add up to the number in question. The latter type of numerology has a very ancient history. According to the *Encyclopaedia Judaica* ('Gematria') its first use 'occurs in an inscription of the Assyrian king Sargon II (727-707 BCE), which states that the king built the wall of Khorsabad 16,283 cubits long to correspond to the numerical value of <the letters in> his name'.

This type of numerology was explicitly recognized also as a valid principle of exegesis in the Mishnah (R. ?eli9ezer b. Yosi HaggElili, second century CE). 'According to this (principle) words with identical numerical value can be substituted for one another' (Dornseiff, 1925, p. 95). An example commonly quoted is the Talmud's interpretation of Genesis 14:14, which reports that Abram armed 318 of his servants to fight against the kings who had captured his kinsman Lot. The Talmud explains that in fact Abram only sent his retainer named *?ĕlî9ezer* into battle and justifies this assertion with reference to the fact that the numerical value of the letters that make up the name *?ĕlî9ezer* is 318; i.e. that 318 is a code for *?ĕlî9ezer*.

Bazak points out that the text of psalm 23 is 55 words long. He notes that if we divide these 55 words into 26 + 3 + 26, the poem will then consist of two large stanzas separated by a line of three words, which in translation reads 'For you are with me' (cf. (9).). This division, as Bazak notes, has the effect of highlighting the 3-word line 'For you are with me'. This is highly plausible since the line epitomizes the ideological basis of the poem: the author's faith in God. More important for what follows, Bazak remarks that the word beginning the poem is *yahweh*, whose numerical equivalent is 26, and argues that this explains the division of the poem into two 26-word sequences separated by a 3-word sequence.

I believe that Bazak's lead can be pursued even somewhat further with results that are at least suggestive. Examining the stanzaic organization in (9), we see that the total number of syllables in the first stanza is 53 = 2 x 26 + 1 whereas the last stanza is 51 = 2 x 26 - 1. The middle stanza contains 18 syllables.

We know that the number 26 is a code for *yahweh*. The number 18 is an equally well known code, it stands for *Hay*, which is a form of the stem meaning 'life' or 'be alive'.[8] Thus, ps. 23 can be viewed as encoding the message

Yahweh Yahweh Hay Yahweh Yahweh

a cheer of the sort that was heard at a visit of a pope to France, where the crowd is reported to have spontaneously shouted 'vive Dieu!' (H. Cartier-Bresson *The Decisive Moment*.)

It will be recalled that in the discussion of ps. 54 it was noted that each stanza was composed of 54 or 3 x 18 syllables. If 18 is the code for *vivat!* 'cheers!' then 3 x 18 is the equivalent of '3 cheers'.

2.3. GEOMETRY AND GEMATRIA IN PSALM 137

I conclude with a discussion of the formal properties of psalm 137, which, if I am correct, include not only numerological devices quite similar to those of psalms 23 and 54, but also additional covert features of a rather surprising kind. The text of the psalm is given in (11). The text reproduced below is essentially identical with that published in Halle and McCarthy, 1981.

(11)

9al nEhārôt bEbābel	7	By the rivers of Babylon
šām yāšabnû gam bākî(nû)	7	There we sat and wept
bEzokrēnû ?et ciyyôn	7	As we remembered Zion.
9al-9ărābîm bEtôkāh	7	By laurels in its midst
tālînû kinnorôtē(nû)	7	We hung up our harps.
kî šām šE?ēlû(nû)	5	For there they asked of us,
šôbênû dibrē-šîr	6	Our captors, words of song,
wEtôlālênû śimHâ	7	And those who mocked us, rejoicing:
šîrû lānû miššîr ciyyôn	8	'Sing for us of Zion's song.'
?ēk nāšîr ?et-šîrê-yahweh	8	How can we sing Yahweh's songs
9al ?admat nēkār	5	On alien soil?
?im-?eškāHēk yErûšālēm	8	If I forget thee, Jerusalem,
tiškaH yEmînî	5	May my right arm wither.
tidbaq-lEšônî lEHikkî	8	May my tongue stick to my palate
?im-lō? ?ezkErē(kî)	5	If I remember thee not,
?im-lō? ?a9le yErûšālēm	8	If I fail to ascend to Jerusalem
9al rō?s śimHātî	5	As my chief joy.

zEkōr yahweh libnê-?ĕdôm	8	Remember, Yahweh, to Edom's sons
?et yEmê yErûšālēm	7	The days of Jerusalem,
hā?ōmrîm 9ārû 9ā(rû)	6	Who say: 'Strip bare, strip bare,
'9ad hayyEsôd bāh	5	To its very foundation!'
bat-bābel haššEdûdâ	7	Daughter of Babylon, the doomed,
?ašrê šeyyEšallem-lāk	7	Happy he who renders you
gEmûlēk šeggāmalt lā(nû)	7	The payment you paid us.
?ašrê šeyyō?Hēz wEnippēc	8	Happy he who grasps and shatters
9ōlālayik ?el-hassā(la9)	7	Your babes upon the cliff.

The psalm is thus made up of five stanzas in a chiastic arrangement: ABCBA. The first and fifth stanzas consist each of five lines that with one exception are all 7 syllables long. The second and fourth stanzas are made up of four lines that increase in length in the second stanza and decrease in length in the fourth: 5-6-7-8 vs. 8-7-6-5. Finally the middle, third stanza is composed of four couplets of two lines each, of which one is 8 and the other 5 syllables long. I have summarized these findings in (12).

(12) 77777 5678 85858585 8765 77787

I conjecture, following a suggestion made to me by John Hollander, that the extra syllable in the last stanza is an irregularity introduced by the poet on purpose, reflecting his belief that only God can create a perfect thing and that all things made by men must therefore be imperfect.

In addition to the symmetries already noted, to which we return below, the poem has a number of numerical properties that rather resemble those noted above in psalm 23. First, the total number of lines in the poem is 26, which, it will be recalled, is the code number for *yahweh*. The syllable count in the three innermost stanzas is 26 + (52 = 2 x 26) + 26 = 104. Disregarding the extra syllable in the last stanza the total number of syllables in the poem is 174, which equals 18 + 156. We already know that 18 is the code for the blessing *Hay* - 'vivat!' 156 is the sum of the letters of the word *cywn* 'Zion'. These numbers therefore encode the message 'long live Zion!'[9]

The number 174 may also be analysed as 34 + 140. 34 is the sum of the letters in the word *bbl* 'Babylon'. If this is correct then 140 must encode some curse, but unfortunately I have been unable to find a plausible biblical curse word whose letters add up to 140.

The preceding does not exhaust all formal properties of psalm 137. In Halle, 1987, I posed the question as to the motivation of the verse lengths in the poem. Why would a poet choose to assemble lines with such odd numerical properties? One part of the answer was given above: the numbers were chosen for numerological reasons. A second part of the answer might be found, I suggested, in the graphic interpretation (13) of the metrical structure of the poem.

(13)

	X	X	X	X	X	X	
XXXXX	XX	X	X	X	X	XX	XXXXX
XXXXX	XXX	X	X	X	X	XXX	XXXXX
XXXXX	XXXX	XX	XX	XX	XX	XXXX	XXXXX
XXXXX	XXXX	XX	XX	XX	XX	XXXX	XXXXX
XXXXX	XXXX	XX	XX	XX	XX	XXXX	XXXXX
XXXXX	XXXX	XX	XX	XX	XX	XXXX	XXXXX

In (13) each column of X's represents the metrical syllables in a line of the poem. The poem has been rotated through an angle of ninety degrees so that the seven X's of the first column in (13) represent the seven syllables that compose the first line of the poem; the seven X's of the second column stand for the seven syllables of the second line, etc. In Halle, 1987 I conjectured that the graphic shape (13) represents a building with two wings, a sloping roof and a façade with columns, and I speculated that it might be the Temple in Jerusalem. We return to this speculation below.

'Pattern poems'; i.e. poems whose graphic shape represents a physical object, were known already in classical antiquity. The direct antecedents of 'pattern poems' were formulas of Orphic magic of classical Greek antiquity 'whose graphic shape imitated specific objects. Most frequent among them are instances of wings ...' (Wojaczek, 1969, p. 60) Subsequently the idea was adopted by the poets of the so-called Bucolic group, of whom the best known is Theocritus, (c. 310-250 BC). The technical term for such poems in Greek literature was *technopaignia*. This word means 'playful creations in which the art of the poet is shown off ... . Often these poems were called simply *paignia*; i.e. literary jokes (*literarische Scherze*)' (Wojaczek, 1969, p. 56).

In view of the existence of the formulas of Orphic magic whose graphic shape imitated the shape of particular objects there is no reason to suppose that the Bucolic poets borrowed the idea of composing pattern poems from a Jewish source, especially since except for psalm 137 no Hebrew pattern poems from that time has been found. That, on the other hand, the Greek poems should have provided a model for a Jewish poet is fully to be expected in view of the tremendous Greek cultural influence over all peoples living in the area that was (to be) occupied by the Roman empire, and this included the Jews, both in Palestine and elsewhere, both before and after the Maccabean uprising.[10]

For matters under discussion here perhaps the most striking of all is the statement in the Palestinian Talmud quoted by Rosén, 1979, p. 47.

(14) Four languages are appropriate for the world's use: the language of the foreigners (*l9z*) for poetry, the language of the Roman for combat, the *swrsy* for lamentation, and the Hebrew (*9bry*) language for conversation. (SoTa 21c14).

Rosen comments: 'That "the language of the foreigners" to be used for poetry is Greek, has never been in doubt, if only because Greek is not otherwise mentioned and it is clear that it should have been.' We thus have it on the authority of the Talmud that Greek was at one time the language of poetry for Jews. In light of this it should hardly be surprising to find a Hebrew poem - i.e. ps. 137 - imitating a Greek model.

As I noted in Halle, 1989 additional corroborative evidence for the suggestion that ps. 137 represents the Temple in Jerusalem is provided by the existence of coins with representations of the façade of the Temple. One such coin has been reproduced in (15) and it shows a façade made up of four columns. It is not far-fetched to suppose that each of the four columns is represented by the 8-syllable line in each of the four couplets that constitute the central portion of the psalm.

(15)

The coin in (15), reproduced from Williamson 1894, is a silver tetradrachm struck by the organizers of the Bar Kokhba uprising between 132-135 CE. Several thousand specimens of this tetradrachm have been found. Many writers have assumed that the façade of the building shown on the coin is that of the Temple destroyed by the Romans; e.g. Muehsam, 1966 and Avi-Yonah, 1968. Mildenberg, 1984, the author of the definitive study of the coinage of the Bar Kokhba revolt, is somewhat less categorical: 'In AD 132 ... the Temple was only a memory, but a memory to be cherished and kept alive until the Temple stood again. The rebels of the Bar Kokhba revolt, therefore, chose to propagate with their new tetradrachm design ... the plan of the Temple to be re-erected in the future, *a plan that conformed either to their memory of the old Temple or to their dream of a new Temple.*' (p. 68 - emphasis supplied).

(16)

In (16) I have reproduced from Rossi 1882 a fourth century (CE) gold glass fragment of a Jewish drinking cup that was found in the catacomb of SS. Pietro and Marcellino in Rome and is now in the Vatican.[11] What is especially interesting to us about this representation of the Temple is that it suggests that each of the six 8-syllable lines in the psalm - cf. (13) - stands for an actual pillar in the Temple. In the words of Narkiss, 1974 the fragment 'gives a bird's eye view of the Temple surrounded by the Tabernacle curtains (Exodus 26). The sanctuary in the centre symbolizes that of Solomon with the two pillars, *Jachin* and *Boas*, on either side (I Kings 7:21).' Rossi, the scholar who first described the glass fragment, concluded that 'the glass reproduces in a summary and very imperfect fashion a picture or model of the Jerusalem temple preserved by the Jews after the destruction of the holy city and represented in accordance with the memories and traditions of their ancestors. That it was not altogether imaginary (*fantastique*) is shown by the two isolated columns and their position relative to the building, for we never find these shown in this way in the representations of the temple that is furnished to us by ancient Christian art ...' . (p. 153).

The pattern in (13) therefore reproduces salient features of the Temple found also on the Bar Kokhba coins and on the Vatican glass. In all of them we find a façade with four columns. Moreover, if we assume that each of the six eight-syllable lines in the psalm represents a column, then both the psalm and the Vatican glass fragment show the two additional columns that were originally in Solomon's Temple.

Since the Bar Kokhba coins and the Vatican glass are posterior to the destruction of the second Temple it is most probable that they represent features of the second Temple. We recall that the first Temple was destroyed by the Babylonians in 586 BC. Upon the return from the Babylonian exile seventy years later, the second Temple was constructed, but because of the difficult conditions that prevailed at the time the reconstructed Temple was a rather 'unpretentious building' (*Encyclopaedia Judaica*, 'Temple') and remained such for many centuries. It is unlikely to have had the magnificent four-column portico present in the representations under discussion here.

Concerning the structure represented in the glass fragment (16) Rossi 1882 wrote that 'even admitting that these (two) famous columns were erected by Solomon ... our glass represents them *as they were in the temple of Herod*'. (emphasis supplied, p. 148). Thus, if the structure represented in psalm 137 is the same as that depicted on the glass fragment, then psalm 137 represents Herod's Temple. This would date the composition of ps. 137 after 19 BC, i.e. after Herod's reconstruction of the Temple got under way. Such a late date of composition for the psalm, however, goes counter to generally held views. The fact that psalm 137 is included in the Psalm Scroll of Qumram Cave 11 (Sanders, 1965) would not of itself cut the ground out from under the proposed late dating of the psalm, since many of the Qumran documents are of the first century CE and even later. More problematic, perhaps, is the fact that psalm 137 is included in the Septuagint, which is generally held to have been produced before Herod's time. In view of the great complexity of the issue the date of composition of psalm 137 must be left open here.

REFERENCES

Avi-Yonah, M.

1968 'The Façade of Herod's Temple, an Attempted Reconstruction', in: J. Neusner (ed.), *Religions in Antiquity, Essays in Memory of Erwin Ramsdell Goodenough*. Leiden: Brill, pp. 327-35

Bazak, J.

1987 'Numerical Devices in Biblical Poetry', *Vetus Testamentum* 38, pp. 333-7

Bickerman, E.

1989 *Jews in the Greek Age*. Cambridge: Harvard University Press

Cartier-Bresson, H.

1952 *The Decisive Moment\Photography by Henri Cartier-Bresson*. New York: Simon and Schuster

Dornseiff, F.

1925 *Das Alphabet in Mystik und Magie*. Leipzig: Teubner

Freedman, D.N.

1980 *Pottery, Poetry, and Prophecy. Studies in early Hebrew Poetry*. Winona Lake: Eisenbrauns

Halle, M.

1987 'A Biblical Pattern Poem', in: N. Fabb *et al.* (eds), *The Linguistics of Writing*. Manchester: Manchester University Press, pp. 67-75

1989 'Syllable-Counting Meters and Pattern Poetry in the Old Testament', in: P. Wexler, A. Borg and S. Somekh (eds), *Studia Linguistica et Orientalia Memoriae Haim Blanc Dedicata*. Wiesbaden: Harrassowitz, pp. 110-20

Halle, M. and S.J. Keyser

1980 'Metrica', in: *Enciclopedia* IX. Torino: Einaudi, pp. 254-84

Halle, M. and J. McCarthy

1981 'The Metrical Structure of Psalm 137', *Journal of Biblical Literature* 100/2, pp. 161-7

Mildenberg, L.

1984 *The Coinage of the Bar Kokhba War. Typos: Monographien zur alten Numismatik* VI. Aarau: Verlag Sauerländer (Typos. IV)

Muehsam, A.

1966 *Coin and Temple, a Study of Architectural Representation on Ancient Jewish Coins*. Leeds-Leiden: Leeds University Oriental Society

Narkiss, B.

1974 'A Scheme of the Sanctuary from the Time of Herod the Great', *Journal of Jewish Art* 1, pp. 6-27

Rosén, H.B.
1979 *L'hébreu et ses rapports avec le monde classique*. Paris: Geuthner
Rossi, Jean Baptiste de
1882 'Sur un verre colorié représentant le temple de Jerusalem', separate reprint from *Bulletin d'Archéologie chrétienne* 4-ième Série, 1-ère Année, Livraison 4, pp. 141-63
Sanders, J.A.
1965 *The Psalms Scroll of Qumrân Cave 11 (Discoveries in the Judean Desert* 4). Oxford: Clarendon
Williamson, George C.
1894 *The Money of the Bible*. London: The Religious Tract Society
Wojaczek, G.
1969 *Daphnis: Untersuchungen zur griechischen Bukolik*. Meisenheim am Glan: Hain

Notes

1. Halle, 1989 unfortunately contains a number of typographical errors of which the most important are corrected herewith:
   p. 112: replace the two lines above ex. (4) reading 'Since there ... to read' with 'I have emended the second line of couplet IV to read'.
   p. 112: in the second line below ex. (4), replace 'seven' with 'eight'.
   p. 112: in the third and fourth lines below ex. (4), delete 'and in the fourth couplet the second line has 10 syllables.'
   p. 118: in the last line of the quoted text insert 'roar' after 'the forest'.
   p. 119: in the quoted text move the three lines '*wEterep ?ên lô* .. . if he have taken nothing?' to the top of the page.
2. For a general treatment of poetic metre, see Halle and Keyser, 1980.
3. Parentheses enclose the syllables after the last stress in the verse. These syllables are extra-metrical, do not count for metrical purposes.
4. See also Halle and Keyser, 1980, where the passage in (4) was first analysed.
5. I have omitted vv. 1-2, because these are clearly introductory remarks in prose. They read in the King James Version: 'To the chief musician on neginoth maschil, a psalm of David when the Ziphim came and said to Saul: 'Doth not David hide himself with us?'' I have enclosed the last line of the psalm in parentheses for reasons to be detailed below.
6. The *Britannica* follows the verse numbering of the King James Version, whereas in (7) I use the verse numbering of the *Biblia Hebraica Stuttgartensia*. This accounts for the difference in the reference to the verses of ps. 40 in (7) and in the quoted passage.
7. Since the vowels of this word are not represented in the Hebrew orthography, only the consonants are taken into account.

8. The significance of the number 18 is further highlighted by the fact that the prayer that is 'the core and main element of each of the prescribed daily services' of the Jewish liturgy is 'known popularly among Ashkenazim as *Shemoneh-Esreh* ('Eighteen') because of the 18 benedictions which it originally comprised.' (Quotations from the article 'Amidah' in the *Encyclopaedia Judaica*).
9. Mark Aronoff has pointed out to me that 156 is also 6 x 26 (=yhwh), but I think my guess above is to be preferred.
10. Indications of the extent of this influence are provided not only by the fact that the Gospels were written in Greek, but also by such additional facts as that in Palestine, of 168 grave inscriptions of this period 114 are exclusively in Greek, that some of the letters from leaders of the Bar Kokhba rebellion found in the Qumran caves are in Greek, that Jews widely adopted Greek names, so that even rabbis had names such as Alexander, Antigonus, Pappus, Symmachus, Tarphon, etc. (For additional information see *Encyclopaedia Judaica*, articles 'Hellenism'; 'Greek and Latin Languages, Rabbinical Knowledge of', 'Symbolism, Jewish (in the Greco-Roman Period)' and also Bickerman, 1989.)
11. I am indebted to Mrs. Robert Hoberman for drawing my attention to this material.

XIII

THE LOSING OF TAPAS

Minoru Hara[*]

I. (1-1) In contrast to several important contributions to ancient Indian asceticism, which is the positive way of accumulating (*ci-*, *arj-*, *vṛdh-*, etc.) *tapas*,[1] little has been done for its negative aspect of loss and expenditure (*vyaya*), which is no less interesting and important to studies of Hindu religion and literature.

As is indicated by the Sanskrit compound *tapo-dhana* or *tapo-nidhi*, *tapas* has an aspect comparable to wealth (*dhana*).[2] Its pecuniary aspect as indicated by the word *-dhana* and *-nidhi*[3] (wealth, property)[4] may be further illustrated by such compounds as *tapo-mūlya*[5] and *tapaḥ-krīta*.[6] This aspect of *tapas* was observed long ago by M. Monier-Williams who says as follows,

> According to Hindu theory, the performance of *tapas*, or austerities of various kinds was like making deposits in the bank of heaven. By degrees an enormous credit was accumulated, which enabled the depositor to draw to the amount of this saving without fear of his drafts being refused payment. The merit and power thus gained by weak mortals was so enormous, that gods as well as men were equally at the mercy of these all but omni-potent ascetics.[7]

---

* The gist of this article was originally prepared for the Radhakrishnan Memorial Lectures delivered at the University of Oxford in 1978-1979. The lectures were entitled 'Aspects of Ancient Indian Asceticism', and consisted of four parts, of which this paper formed part four. I am grateful to the late Professor B.K. Matilal who invited me to Oxford and also to Professor R.F. Gombrich for his kind arrangements and for presiding over the lectures during the period of 15-22 June, 1979.
Thanks are due to Dr. R.F. Young who was kind enough to read this paper and suggest corrections to my English. It should be noted here that this study was made possible with the Mitsubishi Foundation Subsidy for the Human Sciences 1994.

R. Gombrich also remarks in passing that merit is 'a kind of spiritual money' and that good action is the 'building up a kind of spiritual bank account from which one can make payments at will in case of need'.[8] All these statements point to the fact that, as money has an economic value, so also *tapas* is furnished with spiritual value.

The possessor of *tapas* (*tapasvin*) is further compared to a battery. The spiritual battery is charged by *tapas* from time to time, and thus its electric power is increased. The amount of this electricity may differ from one person to another, depending upon how much *tapas* each person has stored up within himself.[9] The spiritual electricity, so to speak, is intensified (*vṛdh-*) by the practice of austerity and virtuous conduct, but weakened (*vṝ-*, *kṣar-*, *kṣi-*) by the discharging factors.

No matter whether it is compared to money or electricity, the ancient Indian religious concept *tapas* is a power-substance (*Daseinsmacht*),[10] which can be earned (*arj-*), gained (*āp-*), accumulated (*ci-*), and increased (*vṛdh-*) by mortals, particularly brahmin ascetics, in the course of the practice of severe asceticism.[11] Occasionally, one could add to his storage of *tapas*[12] or to his spiritual bank-account, an amount of *tapas* transferred from other people.[13]

All that we usually say about ancient Indian asceticism is, however, the credit side of the *tapas*-bank account, and we must not overlook the presence of its debit-side, for *tapas* is liable to decay and loss (*vyaya, kṣaya*). It is even subject to theft (*haraṇa*), if one is careless.[14] The spiritual bank sometimes even suffers from a sort of bankruptcy. Here, in this paper we shall investigate the debit side of the *tapas*-bank account, and ascertain what are the moments that are responsible for the discharge of *tapas*-battery.[15]

(1-2) The most dangerous factor that is responsible for the discharge of *tapas* is for the *tapas*-possessor to submit himself to two sorts of violent temper, anger (*krodha*) and desire (*kāma*). The former results in his act of cursing (*śāpa*) and the latter in his seminal effusion.[16] Both of them are contrary to his life-mottos: the former against his abstention from idle speech (*saṁyata-vāc*) or vow of silence (*mauna-vrata*), and the latter against his life-long vow of celibacy (*brahmacarya*). Though contrary to his life-mottos, and much against his will, but simply motivated by violent temper or lust, both curse and seminal effusion that happen are nothing but outward manifestations of *tapas*, which is supposed to be kept inside of the ascetic body. Instigated by anger, it bursts out of his mouth in the form of curse, and urged by desire, it leaks out against his vow from his sexual organ in the form of semen. Hence, a warning in a didactic verse:

jihvopastha-nimittaṁ hi patanaṁ sarva-dehinām
tasmād amitravat paśyej jihvopasthaṁ hi mānavaḥ

> The fall of all men is caused by the tongue and the sexual organ. Therefore, the wise man indeed should regard the tongue and the sexual organ as an enemy.[17]

Despite an apparent difference between the two, that is, the curse being destructive and seminal effusion productive, both are equally efficacious as an incarnation of inner spiritual energy, for the former never fails to be true and the latter always promises the birth of extraordinary human beings. Centring upon these major discharging factors, in the following pages we shall discuss these negative aspects (*vyaya*) of *tapas* and its relevant problems.

II. (2) Students of Sanskrit literature are familiar with the literary motif of an irritable ascetic and his curse. The most famous example is undoubtedly that of Durvāsas' curse put on Śakuntalā. The heroine of this famous drama was in deep thought of her separated husband, and being absent-minded, she failed to show proper respect to the ascetic when he came to her hermitage. He, then, felt offended and being a captive of anger he cursed her to the effect that her husband would not remember her even when he was reminded.[18]

> *vicintayantī yam ananya-mānasā*
> *tapo-nidhiṁ vetsi na māṁ upasthitam*
> *smariṣyati tvāṁ na sa bodhito 'pi san*
> *kathāṁ pramattaḥ prathamaṁ kṛtām iva* (Śakuntalā 4.1)
>
> He of whom you were thinking with mind on nothing else, so that you did not notice that I, a treasury of austerity, had approached, - he, even though he is reminded, will not remember you, as a man who was drunk does not remember the conversation that took place before. (Emeneau)

Since Sanskrit literature is never short of stories about an irritable ascetic's curse, here we shall dwell upon some characteristic features of Hindu curses.
(2-1) Throughout Hindu curse-stories, we notice a sort of revenge, an eye for an eye, in function.[19] In the example quoted above, Durvāsas cursed Śakuntalā tit for tat (*vetsi na māṁ upasthitam: smariṣyati na bodhito 'pi*). Here, let us see the requital-construction of the Hindu curse by quoting several examples.
(2-1-1) An erotic story is given in Gautama's curse upon Indra. The mischievous god once seduced Ahalyā, wife of the sage Gautama, assuming his form. The sage in rage then cursed Indra as follows.

akartavyam idaṁ yasmād viphalas tvaṁ bhaviṣyasi (R 1.47.26cd)

Because of this improper behaviour, be impotent!

Cursed by Gautama, Indra was instantly deprived of his manhood. The sage cursed Ahalyā also and departed for a peak in the Himālaya.
(2-1-2) King Parikṣit met with a mortal bite of the snake Takṣaka under the curse of a young ascetic, Śṛṅgin. The king placed a dead snake upon the old ascetic's shoulder, when the latter in his vow of silence (*mauna-vrata*) did not respond to the king who was in quest of a deer while hunting. Deeming his father slighted, Śṛṅgin cursed the king as follows,

yo 'sau vṛddhasya tātasya tathā kṛcchra-gatasya ca
skandhe mṛtam avāsrākṣīt pannagaṁ rāja-kilbiṣī (12)
taṁ pāpam atisaṁkruddhas takṣakaḥ pannagottamaḥ
āśīviṣas tigma-tejā mad-vākya-bala-coditaḥ (13)
sapta-rātrād ito netā yamasya sadanaṁ prati
dvijānām avamantāraṁ kurūṇām ayaśaskaram (MBh 1.37.14)

That foul and evil king who has thrown a dead snake on my father's shoulder, aged and feeble though he is, him the great snake Takṣaka, enraged and virulent with all the fury of his venom, shall hurl into the kingdom of Yama within seven nights from now, at the prompting of my word - that despiser of the brahmins and disgrace of his line of Kurus.[20]

(2-1-3) King Pāṇḍu's romantic death in his embrace with Mādrī was also the result of a curse by an ascetic Kiṁdama. The ascetic, assuming the form of a buck, was shot by the king while he was mating with its doe. In his curse upon the king, the ascetic said,

mṛga-rūpa-dharaṁ hatvā mām evaṁ kāma-mohitam
asya tu tvaṁ phalaṁ mūḍha prāpsyasīdṛśam eva hi (27)
priyayā saha saṁvāsaṁ prāpya kāma-mohitaḥ
tvam apy asyām avasthāyāṁ preta-lokaṁ gamiṣyasi (MBh 1.109.28)

Since you have injured me in this situation, in the form of a deer and overcome by love, you shall find the same fate: when you are lying with a woman you love, blinded by your passion, you too in that very same state will depart for the world of the dead.

(2-1-4) More tragic was the death of king Daśaratha, who breathed his last, longing for the sight of his beloved son Rāma. He was under the curse of an old, blind ascetic whose only son had been killed by an arrow of the king. The old ascetic in grief ascended the funeral pyre, laying a curse on the king:

> *putra-vyasana-jaṁ duḥkhaṁ yad etan mama sāṁpratam*
> *evaṁ tvaṁ putra-śokena rājan kālaṁ kariṣyasi* (R 2.58.46)
>
> May you suffer the same grief that you have caused me, through separation from your son at the last moment of your life.[21]

(2-2) However, this process of the ascetic's act of cursing occasionally takes the form of the theft of *tapas* by the hand of Indra. As is well known, Indra is afraid of an ascetic's *tapas*[22] and ponders over the means of depriving him of *tapas*.[23] One of his devices is to provoke the ascetic to anger so that he may be induced to utter a curse. As a matter of fact, designed by Indra, Gautama uttered such a curse, as we shall see presently.

(2-2-1) When the emasculated Indra came back to the hosts of gods and seers, he said to them as follows,

> *kurvatā tapaso vighnaṁ gautamasya mahātmanaḥ*
> *krodham utpādya hi mayā sura-kāryam idaṁ kṛtam* (2)
> *aphalo 'smi kṛtas tena krodhāt sā ca nirākṛtā*
> *śāpa-mokṣeṇa mahatā tapo 'syāpahṛtaṁ mayā* (3)
> *tan māṁ sura-varā sarve sarṣi-saṁghāḥ sacāraṇāḥ*
> *surasāhyakaraṁ sarve saphalaṁ kartum arhatha* (R 1.48.4)
>
> In arousing the anger of the great Gautama and thereby creating an obstacle to his austerities, I have accomplished the work of the gods. For, in his wrath he has emasculated me and repudiated her, and so, in provoking this great outpouring of curses, I have robbed him of his ascetic power. Therefore, all of you - great gods, celestial bards, and hosts of seers - should restore my testicles to me, for I have aided the gods.

It is the business of the gods (*sura-kārya*) to create an obstacle (*vighna*) and to deprive him of *tapas* (*tapo...apahṛta*) by leading him to hurl a curse (*śāpa-mokṣa*) in anger (*krodham utpādya*).

(2-2-2) Another example is met in Viśvāmitra's curse upon Rambhā. When he underwent severe penances, Indra in fear of his *tapas* asked an apsaras to seduce the ascetic to sensuality. By Indra's order, she went forth to allure him. Unfortunately, however, Viśvāmitra was attentive enough to detect Indra's design, and, being irritated, he petrified her by his curse. But, by this very act of cursing he lost *tapas*. On realizing his folly, he resolved to speak no word to anybody and never give way to anger. The relevant passages read as follows:

kopena sa mahā-tejās tapo'-paharaṇe kṛte
indriyair ajitai rāma na lebhe śāntim ātmanaḥ (R 1.63.15)
pūrṇe varṣa-sahasre tu kāṣṭha-bhūtam mahāmunim
vighnair bahubhir ādhūtaṁ krodho nāntaram āviśat (R 1.64.3)

Now that the mighty sage, being deprived of *tapas* through anger, did not find peace of mind with his senses uncontrolled. But, when the thousand years had passed, anger no longer seized this great sage who had become just like a piece of wood, passing through various obstacles.

(2-2-3) From these examples it becomes clear that the ascetic's anger leads him to utter a curse upon the man who wrongs him. His curse is infallible and never fails to be true, but this is made at the expense of the *tapas* he has so far stored up. It was also Indra's trick to steal (*hṛ-*, *apahṛ-*) the *tapas* of mighty ascetics. Hence, the warning to ascetics not to be enraged (*akrodha*) and the recommendation of patience (*kṣamā)*, as we shall see later.

(2-3) At this point, mention must be made of another peculiarity of the Hindu curse. In Epic and the classical Sanskrit literature, the curse is often conditioned by a specific duration,[24] and we seldom find an eternal curse. That is to say, the curse hurled by the ascetics is mitigable, when they are pacified by the people concerned.[25] In the story of Śakuntalā as we have seen above, her friends begged the pardon of Durvāsas, who in turn set a term to his curse,[26] although it was impossible to alter what he had proclaimed before,[27] because each and every word he has once uttered is loaded with truth (*satya-vacana*).

(2-3-1) When Gautama cursed Ahalyā, he promised that she will be free of the curse when Rāma comes to the place. The text reads as follows,

yadā caitad vanaṁ ghoraṁ rāmo daśarathātmajaḥ
āgamiṣyati durdharṣas tadā pūtā bhaviṣyasi (R 1.47.30)

When Rāma, the invincible son of Daśaratha, will come to this dreadful forest, you will be purified (of the curse).

(2-3-2) Rambhā's curse is also destined to terminate, when a great ascetic rescues her.

brāhmaṇaḥ sumahā-tejās tapo-bala-samanvitaḥ
uddhariṣyati rambhe tvāṁ mat-krodha-kaluṣīkṛtām (R 1.63.12)

A mighty brahmin furnished with ascetic power will save you, who have been defiled by my wrath, O Rambhā!

(2-3-3) Another example of absolution of curse through death (*vadha-nirdhūta-śāpa* Raghuvaṁśa 12.57) is the case of Kabandha, who was cursed by the sage Sthūlaśiras (R 3.66-69, MBh 3.263.25-43). The demon Kabandha pleaded for relief and the sage promised that he would regain his original form (*prakṛti*) when Rāma cut his hands off and left him in the desolate forest.

sa mayā yācitaḥ kruddhaḥ śāpasyānto bhaved iti
abhiśāpakṛtasyeti tenedaṁ bhāṣitaṁ vacaḥ (5)
yadā chittvā bhujau rāmas tvāṁ dahed vijane vane
tadā tvaṁ prāpsyase rūpaṁ svam eva vipulaṁ śubham (R 3.67.6)

But I begged the enraged seer to set a limit to the curse provoked by my insulting him. And these were the words he spoke: 'When Rāma cuts off your arms and cremates you in the desolate forest, then you shall regain the grand and lovely form that is properly yours.' (Pollock)

(2-3-4) More technically it is called *śāpa-mokṣa* (MBh 1.57.52, cf. MBh 3.176.21, 3.178.40, 14.96.14). Besides we have a number of passages where one propitiates ascetics (*prasādaya-* MBh 5.193.44) and entreats (*yāc-* MBh 3.178.39) the enraged seers to limit the duration of the curse (*śāpasyānta*: MBh 3.176.12, 18, 12.3.22, 14. 96.12) as we have seen in the above Rāmāyaṇa example (R 3.67.5).
(2-4) No matter whether the ascetic himself is enraged or instigated by Indra's trick, the curse is an outburst of *tapas* which he has so far laboriously accumulated. Since it is hurled at the expense of *tapas*, it is desirable for him not to submit himself to anger which necessarily results in cursing others. Hence the teaching of non-anger (*akrodha*, or *kṣamā*) especially for celibate ascetics. This expenditure, however, is technically called *vyaya*, and thus below we shall list the passages where the idea of *tapaso vyaya* appears.
(2-4-1) Dharma was born as Vidura by virtue of Māṇḍavya's curse, which is styled as an expenditure of *tapas*.

tapo-bala-vyayaṁ kṛtvā sumahac cira-saṁbhṛtam
māṇḍavyenarṣiṇā dharmo hy abhibhūtaḥ sanātanaḥ (MBh 15.35.14)

For (even) the eternal Dharma was prevailed over by the sage Māṇḍavya who spent a great amount of *tapas* that he had obtained after a long time.

(2-4-2) The pious Vedavatī who was Sītā herself in a previous birth threw herself into the fire, when she was disgraced by Rāvaṇa. She refrained from laying a curse upon him with the view of treasuring her *tapas* up even into the next world. She said to Rāvaṇa as follows,

śāpe tvayi mayotsṛṣṭe tapasaś ca vyayo bhavet (R 7.17.26cd)

If I curse you, (my) *tapas* will be spent.

(2-4-3) In Kālidāsa's Raghuvaṁśa king Raghu makes an inquiry to a young ascetic, Kautsa, about his teacher's tidings. There we read,

kāyena vācā manasāpi śaśvad
yat saṁbhṛtaṁ vāsava-dhairya-lopi
āpadyate na vyayam antarāyaiḥ
kaccin maharṣes trividham tapas tat (Raghuvaṁśa 5.5)

I hope, that threefold *tapas* of the great sage, which has been accumulated constantly by the body, speech and mind, and which (as its result) is apt to deprive Indra of steadfastness, does not get into trouble of spending because of the obstructions.

Mallinātha paraphrases the relevant portion as *indra-preritāpsaraḥ-śāpair vyayaṁ nāśam nāpadyate*.

(2-4-4) Another verse of Raghuvaṁśa describes how sages resorted to Rāma in harness by the demon Lavaṇa.

avekṣya rāmaṁ te tasmin na prajahuḥ svatejasā
trāṇābhāve hi śāpāstrāḥ kurvanti tapaso vyayam (Raghuvaṁśa 15.3)

Seeing Rāma (as their protector), these ascetics who have curse as their weapon did not hurl (their curse) upon him by means of their own splendour, for they spend their *tapas* only in the absence of a protector.[28]

(2-4-5) Refraining from abuse (*vyaya*) of *tapas* is tantamount to protection (*anupālana*) of *tapas*. Sītā declined to curse Rāvaṇa, seeing the danger of losing her *tapas*. Thus, we read,

asaṁdeśāt tu rāmasya tapasaś cānupālanāt
na tvāṁ kurmi daśagrīva bhasma bhasmārha tejasā (R 5.20.20)

I do not reduce you to ashes by means of (my own) splendour, because I am not instructed by Rāma to do so, and for the sake of protecting my *tapas* (from unnecessary waste).

(2-4-6) The same waste is expressed by the verb *khaṇḍ-*. The sages in the Daṇḍaka forest did not curse the demons who killed them. Though they were in a position to kill them by means of curse-power, they esteemed *tapas* more than their lives. Thus, we read,

kāmaṁ tapaḥ-prabhāveṇa śaktā hantuṁ niśācarān
cirārjitaṁ tu necchāmas tapaḥ khaṇḍayituṁ vayam (13)
bahu-vighnaṁ tapo nityaṁ duścaraṁ caiva rāghava
tena śāpaṁ na muñcāmo bhakṣyamāṇāś ca rākṣasaiḥ (R 3.9.14)

True, we could kill the night stalkers with our ascetic powers, but we are reluctant to squander what took so long to amass. Ascetic practice is ever beset by obstacles, Rāghava, and so difficult to perform. That is why, although devoured by Rākṣasas, we do not unleash our curse. (Pollock)

(2-4-7) As cursing means the waste of *tapas* on the part of the ascetic, a kind and thoughtful person appeases the anger of the ascetic who is about to curse him. Uttaṅka, when he had heard from the mouth of Kṛṣṇa of the destruction of the whole Kuru family, became angry and blamed him for negligence in his duty and prepared to curse him. Kṛṣṇa, seeing this, appeased Uttaṅka and dissuaded him from the violent act, lest he should lose his *tapas*.

śrutvā tvam etad adhyātmaṁ muñcethāḥ śāpam adya vai
na ca māṁ tapasālpena śakto 'bhibhavituṁ pumān (24)
na ca te tapaso nāśaṃ icchāmi japatāṁ vara
tapas te sumahad dīptaṁ guravaś cāpi toṣitāḥ (25)

kaumāraṁ brahmacaryaṁ te jānāmi dvija-sattama
duḥkhārjitasya tapasas tasmān necchāmi te vyayam (MBh 14.52.26)

After having heard my words relating to the soul, you may utter your curse. No man is able by a little ascetic power to put me down. O, foremost of ascetics, I do not wish to see the destruction of all your penance. You have a large measure of blazing penance. You have gratified your preceptors and seniors. O foremost of brahmins, I know

> that you have observed the rules of *brahmacarya* from the day of your infancy. I do not, therefore, desire the diminution of your penance achieved with so much pain.[29]

Note here the words *cira-saṁbhṛta* or *cirārjita*, *duḥkhārjita* which modify the word *tapas*. It is a pity for the ascetic to lose it by the single act of cursing others.

(2-5) Throughout all these curse-stories, there is presupposed the belief in the 'power of speech'. This is to be cultivated in the course of the ascetic's abstention from idle speech and strengthened by his practice of the vow of silence. Each and every word he utters is loaded with *satya* (*na me moghaṁ vaco bhavet* MBh 3.13.117, 5.80.48) in his full conviction that he has never uttered an untruth before (*na me vāg anṛtaṁ prāha* MBh 1.43.28, 1.49.19, 1.189.29, 1.188.13, *na me vāg anṛtaṁ bhadre* MBh 5.188.11) even in his unreserved conversation (*svaireṣv api*).[30] The presence of true speech is tantamount to the absence of idle speech, not to mention untruth. Supported by this firm conviction, his speech never fails to be true and thus the cursed persons are instantly transformed into anything the ascetic specifies, that is, into a stone, tree, or sometimes even into a goblin. However, prior to entering into details of this curse-power, a glance should be made at the general problem of cursing and blessing.

III. Before moving from curse to another discharging factor of *tapas*, mention must be made of its counterpart, that is, blessing.

(3-1) By the side of, yet in sharp contrast to, the curse (*śāpa*) hurled upon lay-people by the enraged ascetic, we have a recurring literary motif of the boon (*vara*) in Epic and classical Sanskrit literature. Brahmin ascetics[31] are often depicted as pleased (*prīta, tuṣṭa, prasanna*) by lay-people for their devoted service (*satkāra, pūjā*), and inclined to grant them a boon (*vara*). We notice here that *vara* and *śāpa* are both sides of one and the same coin. That is to say, if the lay-people are successful in pleasing the ascetic, the latter grants them a blessing by his true speech in the form of *vara*, whereas, if they fail in pleasing him or even offend him, the latter condemns them by the same true speech in the form of *śāpa*. In both cases[32] it is presupposed that the solemn statement made by ascetic brahmins always proves true (*satya-vacana*). The only difference lies in that blessing is motivated by satisfaction, while cursing is preceded by anger. No matter whether urged positively or negatively, his speech is ever loaded with truth. We shall investigate the functional equivalence of *vara* and *śāpa* by quoting textual evidence.

(3-2) When Vyāsa proposed to perform spiritism in order to please Dhṛtarāṣṭra and others, he says to them as follows,

> *ime ca deva-gandharvāḥ sarve caiva maharṣayaḥ*
> *paśyantu tapaso vīryam adya me cira-saṁbhṛtam* (20)
> *tad ucyatāṁ mahābāho kaṁ kāmaṁ pradiśāmi te*
> *pravaṇo 'smi varaṁ dātuṁ paśya me tapaso balam* (MBh 15.36.21)
>
> Let the deities and Gandharvas, and all these great seers, behold today the efficacy of *tapas* which I have acquired for a long time. Therefore, o, king, tell me what desire (of yours) shall I fulfil, for I am inclined to grant you a boon. Behold the power of my *tapas*.

The phrase *paśya me tapaso balam* thrice appears otherwise in MBh 1.37.4, 1.46.10 and 1.101.23. Of these, the first two are found in Śṛṅgin's curse against Parikṣit and the last in Āṇi Māṇḍavya's curse against Dharma. Here the same phrase appears in the context of *vara* and this testifies to the equivalence of *vara* and *śāpa*.[33]

(3-3) The equivalence of *vara* and *śāpa* is further ascertained by the juxtaposition of both words in an Epic passage:

> *kāmam eṣa varo me 'stu śāpo vāpi maheśvarāt*
> *na cānyāṁ devatāṁ kāṅkṣe sarva-kāma-phalāny api* (MBh 13.14.104)
>
> A boon or a curse from Maheśvara; this is my desire, for I would not wish the fruition of my wishes from any other deity.[34]

Both ideas are further expressed in the form of such *dvandva*-compounds as *śāpa* and *anugraha*, *nigraha* and *anugraha*. The word *vara* is replaced not only by the word *anugraha* (MBh 3.32.12), but also by *prasāda* (Avimāraka 6.16). Occasionally, both *śāpa* and *vara* are paraphrased by *krodha* and *prasāda*. For clarity's sake, we shall list below the examples so far collected by the present writer.

śāpa	: *anugrahaṇa*	(MBh 3.32.12) (Tantravārttika 135.8-11)[35]
śāpa	: *prasāda*	(Avimāraka 6.16)
nigraha	: *anugraha*	Brahmin's *prabhāva*, Ganapati Sastri on Pratijñāyaugandharāyaṇa TSS 16, p. 83, line 29)[36]
krodha	: *prasāda*	(MBh 3.197.27)

(3-4) Despite the functional similarity between *śāpa* and *vara* as we have seen above, there exists an aspectual difference between them. In the case of granting *vara*, the ascetic has enough time to consider carefully the balance, *tapas* as 'deposit' against *vara* as 'expenditure', whereas in the case of a curse he has no time to reflect upon the balance under the pressure of a violent temper, that is, anger. Once he has cast a curse, the bank-account of spiritual 'currency' (*tapas*) is often reduced to zero and he has to start accumulating *tapas* again from the very beginning. As an instant sparking discharge of the *tapas*-battery, his curse is certain, but it means a serious expenditure on his part.

(3-5) Here, mention must be made of what ensures the truth of the Hindu ascetic's speech. As we have seen above, the solemn statement of a brahmin, no matter whether it functions positively (*vara*) or negatively (*śāpa*), always proves to be true. But, we must ask why his speech is always so true and certain. In a dialogue between the young Śṛṅgin and his father Śamīka we read,

> *yady etat sāhasaṁ tāta yadi vā duṣkṛtaṁ kṛtam*
> *priyaṁ vāpy apriyaṁ vā te vāg uktā na mṛṣā mayā* (1)
> *naivānyathedaṁ bhavitā pitar eṣa bravīmi te*
> *nāhaṁ mṛṣā prabravīmi svaireṣv api kutaḥ śapan* (2)
> *jānāmy ugra-prabhāvaṁ tvāṁ putra satya-giraṁ tathā*
> *nānṛtaṁ hy ukta-pūrvaṁ te naitan mithyā bhaviṣyati* (MBh 1.38.3)

> [Śṛṅgin] 'If I have acted rashly, father, or if I have done wrong, or whether it pleases or displeases you, the word I have spoken will be not belied! It shall never be altered, here, I stand and tell you, father, I do not speak idly even when joking, let alone when cursing.' [Śamīka] 'I know you have awesome power, my boy, and that you are true to your word. Never before have you spoken a lie, nor let this curse be belied.'[37]

As is stated here, a brahmin ascetic never speaks a lie even in his free and unreserved speech (*svaireṣv api*). His true speech is confirmed by his firm belief that he has never spoken falsely in his life. His speech-power is further strengthened by the fact that he always refrains from idle speech (*saṁyata-vāk*), which culminates eventually in his vow of silence (*mauna-vrata*). We have a number of passages where an ascetic speaks of himself as immune of falsehood (*anṛtaṁ nokta-pūrvaṁ ... svaireṣv api*[38] *na smarāmy anṛtaṁ kiṁcit*,[39] and the like). Here, *anṛta* is occasionally replaced by *mṛṣā*[40] or *mithyā*[41] or even *vitatha*,[42] and *svaireṣv api* is by *hasatāpi*.[43] This Hindu belief may be summarized in the following half verse.

na śakyam anyathā kartuṁ yad uktaṁ brahma-vādinā (MBh 1.91.13cd)

What the utterer of brahman pronounces cannot be undone.

This is the reason why he has to pronounce another true speech, when he was requested by his lay-devotees to set a limit on the curse, which he had uttered upon them.

(3-6) The ascetic is portrayed as a man who possesses his speech (*vāg-astra*),[44] or occasionally even curse (*śāpāstra*)[45] as a weapon, which excels in strength that of the arms which the Kṣatriyas possess (*bhujāstra, bāhv-astra*).[46] Herein lies the absolute superiority of the brahmin caste to the warrior caste in ancient Indian society.[47] But his speech-weapon (*vāg-astra*) is supposed to be sharpened constantly by the restraint of speech.

No matter whether he uses it positively (*vara*) or negatively (*śāpa*), the true speech of the ascetic is uttered at the expense (*vyaya*) of his *tapas*.[48] He is in a position to resort to his powerful weapon (*vāg-astra* = *śāpa*) in case of need, but it is desirable not to use it unnecessarily and inattentively.

IV. Another violent form of temper beside anger (*krodha*) which deprives the ascetic of self-control and thus is responsible for a serious discharge of *tapas* is sexual desire (*kāma*), the indulgence in which results in the breaking of his vow of celibacy (*brahmacarya*). His indulgence is preceded by the woman's act of tempting him, a beautiful and charming woman, divine as well as human, who catches the ascetic's heart, and deprives him of self-control. As is the case with anger, carnal desire (*kāma*) prevails over his self-control and brings about a loss of his spiritual treasure. Below we shall quote some examples.

(4-1) The most interesting story of this sort is found in Daṇḍin's Daśakumāracarita, where we read how the great ascetic Marīci lost his clairvoyance (*divya-cakṣus*) because of his indulgence in carnal pleasure with a courtesan. Apahāravarman, one of the ten princes, heard a rumour among the people of the Aṅga country as follow:

kaś cid asti tapaḥ-prabhāvotpanna-divyacakṣur marīcir nāma maharṣiḥ (p.78, line 1)

There lives a certain great sage named Marīci, who possesses a divine sight produced by the power of *tapas*.

He wanted to meet this sage because he was at that time in search of the lost hero. Now, he happened to meet the sage who, however, was destitute of his reputed clairvoyance as the result of his indulgence in carnal pleasure with a courtesan named Kāmamañjarī. The sage confessed the whole sad story, but promised the prince that once he could accumulate anew a sufficient amount of *tapas*, he would certainly find the lost hero by his clairvoyance. In due course the sage recovered his clairvoyance, thanks to that fact Apahāravarman could locate the lost hero. Meeting again the lost hero, he told him whole the story as follows,

> *atha bhagavantaṁ marīciṁ veśa-kṛcchrād utthāya punaḥ pratitapta-tapaḥ-prabhāva-pratyāpanna-divyacakṣuṣam upasaṁgamya tenāsmy evam-bhūtaṁ tvad-darśanam avagamitaḥ* (p. 135, lines 5 ff.)
>
> Then, I approached once again the holy Marīci, who, recuperating from the trouble with the courtesan, recovered divine sight by the power of renewed *tapas*, and through him I was thus made to see you.

Here *veśa-kṛcchra* (trouble with the courtesan) means the ascetic's indulgence in sexual union with her which deprived him of ascetic merit (*tapas*). The ascetic's submission to carnal desire, which is nothing but the breaking of the vow of celibacy, nullifies his asceticism in the form of a lapse of supernatural power (*divya-cakṣus*).

Here we must notice that ascetic merit (*tapas*), which enables him to obtain supernatural power, is subject to both increase and decrease. One is endowed with ascetic merit by the practice of asceticism, but it easily melts away in carnal enjoyment. Hence, the repeated warning for ascetics to avoid a woman who may become a serious danger to his *tapas*-property.

(4-2) In the Epic literature this motif of the ascetic's seduction is associated with a heavenly nymph (*apsaras*) who is sent by Indra. As was the case with anger (*krodha-śāpa*), here also it is Indra's trick, because he fears *tapas*. This is testified to by the fact that the concept of *sura-kārya* 'god's business' that we have seen in Gautama's curse (R 1.48.2) appears here in seduction story of Viśvāmitra (R 1.62.10) in the form of *surāṇāṁ karma*. Seminal effusion is fatal to the ascetic, for it reduces his spiritual energy to naught. However, since the present writer had another occasion to discuss this point,[49] here one example suffices to be quoted.

(4-2-1) The most famous story of this sort is that of the seduction of Viśvāmitra by the apsaras Menakā. In the well known story of Śakuntalā Indra requests Menakā as follows,

> *sa māṁ na cyāvayet sthānāt taṁ vai gatvā pralobhaya*
> *cara tasya tapo-vighnaṁ kuru me priyam uttamam* (25)
> *rūpa-yauvana-mādhrya-ceṣṭita-smita-bhāṣitaiḥ*
> *lobhayitvā varārohe tapasaḥ saṁnivartaya* (MBh 1.65.26)

> Lest he topple me from my status, go to him and seduce him. Obstruct his asceticism, do me the ultimate favour. Seduce him with your beauty, youth, sweetness, fondling, smile and flatteries. Turn him away from his austerities.

Menakā's mission was successful and Viśvāmitra lost his *tapas* completely in the course of indulging himself in sensual pleasure with her.

V. (5-1) Whether it takes the form of a curse or seminal effusion, the losing of *tapas* on the part of the ascetic is nothing but an outward manifestation of *tapas* which is supposed to be kept inside of the ascetic. The one, being instigated by anger (*krodha*) bursts out of the mouth (*jihvā*), which should be restrained (*saṁyata*), and the other, urged by carnal desire (*kāma*) leaks out of the sexual organ (*upastha*), which is expected to be kept in strict control (*brahmacarya*). Yet, both are powerful and efficacious as an embodiment of inner spiritual energy (*tapas*).

(5-2) Generally speaking, however, all the ascetic's performance of supernatural power, including spiritism and the granting of boons (*vara*), is an outward manifestation of his *tapas*. As a matter of fact, clairvoyance and clairaudience are a proof of the possession of ascetic merit (*tapas*), and the ascetic's performance of magic acts, such as spiritism, is also possible at the price of ascetic merit. Those who have not accumulated a sufficient amount of *tapas* are not entitled to perform miracles. This ability of performing supernatural power differs from one person to another, depending upon how much they have stored up within themselves the power-substance, that is, *tapas*. That is to say, the more deposit he has made in his spiritual bank, the greater privilege he can enjoy.

Yet all the exercises of supernatural power, major or minor, are the same in their being an expenditure out of his spiritual property (*dhana, nidhi*).

(5-3-1) The difference between curse and seminal effusion on the one hand and the exercise of supernatural power for the sake of lay-devotees on the other lies in degree of *tapas*-expense. Wrath and lust demand a major expense, whereas only small portion of his *tapas*-deposit is needed in the case of the granting of *vara*.

(5-3-2) Another difference between the two consists in whether or not the ascetic is conscious of himself at the time of its expenditure. The performance of miracles such as spiritism may need a great amount of *tapas* and it costs a major expense of *tapas* on the part of the ascetic, but even at that time he is conscious of the expenditure. Whereas in the case of curse and seminal effusion, he is unaware of the *tapas*-discharge, being a captive of violent temper, wrath and lust. The former is an intentional withdrawal of spiritual money from the *tapas*-bank, while the latter, so to speak, is an unintended abuse.

VI. (6-1) Since *krodha* and *kāma* result in major discharges of *tapas*, the ascetic is warned not to yield himself to these two sorts of violent emotions and recommended to be steadfast even when exposed to maltreatment and temptation. Hence the various didactic verses which teach him the virtues of non-anger (*akrodha*) and endurance (*kṣamā*).

> *yajño naśyaty asatyena tapaḥ krodhena naśyati*
> (Cāṇakya Rājanīti-Śāstra 8.60ab)
>
> Sacrifice perishes by falsehood, *tapas* perishes by anger.
>
> *tad eṣa tapasāṁ śatruḥ śreyasaś ca nipātanaḥ*
> *nigṛhīto mayā roṣaḥ śrutvaiva vacanaṁ tava* (MBh 12.348.18)
>
> Upon hearing your word, I restrained my anger, which is the enemy of *tapas* and destroys fortune.
>
> *krodho hi dharmaṁ harati yatīnāṁ duḥkha-saṁcitam* (MBh 1.38.8)
>
> Anger takes away the merit of ascetics, which they laboriously worked for.

One must protect *tapas* from anger.

> *nityaṁ krodhāt tapo rakṣet chriyaṁ rakṣeta matsarāt*
> *vidyāṁ mānāvamānābhyām ātmānaṁ tu pramādataḥ* (MBh 3.203. 40 = 12.182.10)
>
> Constantly one should guard *tapas* from anger, fortune from greediness, learning from respect and disrespect and the self from carelessness.

Patience is, then, called the ornament of ascetics.

> *kṣamā śatrau ca mitre ca yatīnām eva bhūṣaṇam* (IS 2012ab)
>
> Patience both to friend and foe is the ornament of ascetics.

The following verse regards wrath and lust as the major obstacles.

> *atha vā daiva-saṁsiddhāv āśṛṣṭer viduṣām api*
> *kāma-krodhau hi viprāṇāṁ mokṣa-dvārārgalav ubhau* (Kathāsaritsāgara 20.130)

> But the fact is, lust and wrath are appointed in the dispensation of fate, from the very birth even of wise brahmins, to be the two bolts on the door of their salvation. (Tawney)

(6-2) Whether the ascetic is enraged (*kruddha*) or pleased (*prīta*), in consequence of which he is inclined to cursing (*śāpa*) or blessing (*vara*), both are equal in their being the outgo of *tapas*-storage. Thus, the harbouring of *tapas* within, without making a display, is highly valued by Hindu saints. Brahmin ascetics are looked upon with a kind of fearsome wonder as more holy and awesome than other people, as long as they remain in full possession of *tapas* without displaying it. An Indian scholar compared *tapas* to a bird's act of incubation,[50] and L. Renou used the term 'couvaison'.[51]

Under such a circumstance, it is considered as a meagre act of showmanship to wilfully exhibit one's supernatural power. Then, it becomes desirable to conceal inside one's property of spiritual energy as much as possible. To display one's spiritual possession is condemned for serious aspirants as a disgraceful act. Hindu saints are respected as invested with mysteries, as long as they are looked upon as possessing their verbal weapon (*vāg-astra*) in its potential form.

(6-3) This condemnation as a frivolous act of intentional exhibition of spiritual property reminds us of *Yoga-Sūtra* 3.37 (*te samādhav upasargā vyutthāne siddhayaḥ*), where the supernatural activities are said to be obstacles to Yogic concentration. Since the present writer wrote an article on this sūtra once,[52] here we only refer to the symbolism of a pitcher which is full to the brim (*pūrṇa kumbha*) as an ideal of Hindu saints (*kumbhaka*).[53] Serious Hindu ascetics considered *siddhi* as an ancillary by-product in the course of their endeavour to the final goal (*mokṣa*), but never aimed at it as their ultimate object.[54] Though Hindu fiction is full of stories of magical performance, they are simply the acts of so-called false ascetics and nuns.[55]

(6-4) Here at this stage, *tapas* is elevated from a simple magical concept (*ein Weg zur niederen Seligkeit, Kasteiung des Körpers und Geistes um magischer Wirkung willen*) to a highly refined ethical concept (*Selbstbeherrschung, Enthaltsamkeit*).[56] It is on this elevated level that *tapas* is equated with patience (*kṣamā*) as is indicated in the following passage.

> *kṣamā-tulyaṁ tapo nāsti na saṁtoṣāt paraṁ sukham*
> *na ca tṛṣṇā-paro vyādhir na mokṣāt paramaṁ padam* (IS 2011)

> There is no *tapas* equal to patience, no happiness higher than contentment, no disease more serious than longing and no place higher than the final emancipation.

This semantic adaptation of *tapas* from the old concept (*magisch*) to new one (*ethisch*) has been observed through the penetrating insight of P. Hacker, who attributed the innovation to the Buddhist influence of *mettā* and *karuṇā* (*Freundlichkeit und Mitleid*).[57] It was, however, A. Wezler who critically reviewed Hacker's proposition and suggested this semantic shift as *autochthon-hinduistische Entwicklung*.[58] In view of *akrodha* and *kṣamā* (here including also *śama*), which are highly recommended to Hindu ascetics as the means to guard against unconscious expenditure of *tapas* caused by violent temper of *krodha* and *kāma*, as we have seen above, the modification of the semantic content of *tapas* from magical to ethical seems to be explained as a natural development.[59] At the same time, here we notice that the Sanskrit word *tapas* acquires finally the meaning original to the English word 'ascesis'.[60]

REFERENCES

Al-George, S. et A. Roşu

1957 'Pūrṇa Ghaṭa, et le symbolisme du vase dans l'Inde', *Arts Asiatiques* IV, pp. 243-54

Bedekar, V.M.

1968-9 'Yoga in the Mokṣadharmaparvan of the Mahābhārata', *WZKS(O)*12-13, pp. 43-52

Bhagat, M.G.

1976 *Ancient Indian Asceticism.* New Delhi: Munshiram Manoharlal

Bloomfield, M.

1924 'On False Ascetics and Nuns in Hindu Fiction', *JAOS* 44, pp. 202-42

Bronkhorst, J.

1993 *The Two Sources of Indian Asceticism.* Bern: Lang

Caillat, C.

1981 'The Rules concerning speech (*bhāsā*) in the Āyāranga- and Dasaveyāliya-suttas', *Aspects of Jainology* III, *Pt. Dalsukh Bhai Malvania Felicitation volume* I. Varanasi

1984 'Prohibited Speech and subhāsita in the Theravāda Tradition', *IT* 12, pp. 61-73

Čičak-Chand, R.

1974 *Das Sāmajātaka, Kritische Ausgabe, Übersetzung und vergleichende Studie*. Bonn

Gössel, H.

1914 'Indische Strafrechtstheorien', *Festschrift für Ernst Windisch.*

Leipzig: Harrassowitz, pp. 78-84

Gombrich, R.F.

1972 'Merit Transference' in Sinhalese Buddhism: A Case Study of the Interaction between Doctrine and Practice', *History of Religions* 11, pp. 203-19

Gonda, J.

1952 *Ancient-Indian "ojas", Latin "*augos" and the Indo-European Nouns in -es/-os*. Utrecht: Oosthoek

1991 'Nidhi-pati', *Selected Studies* VI/2. Leiden: Brill

Goudriaan, T.

1978 *Māyā Divine and Human*. Delhi: Motilal Banarsidass

Hacker, P.

1978 *Kleine Schriften*, hrsg. von L. Schmithausen. Wiesbaden: Steiner (Glasenapp-Stiftung. 15)

1983 'Inklusivismus', in: G. Oberhammer (ed.) *Inklusivismus. Eine indische Denkform*. Wien: Gerold, pp. 11-28

Hara, M.

1970 'Tapo-dhana', *Acta Asiatica* 19, Tokyo, pp. 58-76

1975 'Indra and tapas', *Adyar Library Bulletin* 39, pp. 129-60

1983 'Rāma Stories in China and Japan: A Comparison', in: K.R.S. Iyengar (ed.), *Asian Variations in Rāmāyaṇa*. New Delhi: Sahitya Akademi

1985 'Yoga Sūtra III-37', *Festschrift S. Kumoi*. Kyoto, pp. 41-56

1994 'Transfer of Merit in Hindu Literature and Religion', *Memoirs of the Toyo Bunko* 52, pp. 103-35

Hopkins, E.W.

1932 'The Oath in Hindu Epic Literature', *JAOS* 52, pp. 316-37

Kirste, J.O.F.

1993 *Kleine Schriften*, hrsg. von W. Slaje. Stuttgart: Steiner (Glasenapp-Stiftung. 33)

Kloppenborg, R.

1974 *The Paccekabuddha. A Buddhist Ascetic*. Leiden: Brill (Orientalia Rheno-Traiectina. 20)

Köhler, H.-W.

1973 *Śrad-dhā in der vedischen und alt-buddhistischen Literatur*, hrsg. von K.L. Janert. Wiesbaden: Steiner (Glasenapp-Stiftung. 9)

Laine, J.W.

1989 *Visions of God. Narratives of Theophany in the Mahābhārata*. Vienna (Publications of the De Nobili Research Library. 16)

Larivière, R.

1989 *The Nāradasmṛti*. Philadelphia: Department of South Asia Regional Studies, University of Pennsylvania (University of Pennsylvania

Studies of South Asia. 4-5)
Norman, K.R.
1993 *Collected Studies* IV. Oxford
Pischel, R.
1922 *Kālidāsa's Śakuntalā*. Cambridge: Harvard University Press (HOS. 16)
Renou, L.
1958 *Études sur le vocabulaire du Ṛgveda*. Pondichéry: Institut français d'indologie (PIFI. 5)
Sharma, R.K.
1982 'Siddhis in the Yogasūtras and Saundarya Laharī', *IT* 10, pp. 193-8
Singh, S.P.
1967 'Conscious Introversion and Divine Intimation in the *Ṛgveda'*, *Kaviraj Abhinandana Grantha*. Lucknow, pp. 45-54
Smith, W.L.
1986 'Uses of the curse in Rāma literature', in: E. Kahrs (ed.), *Kalyāṇa-mitrārāgaṇam. Essays in Honour of Nils Simonsson*. Oslo: Norwegian UP, pp. 261-76
Thieme, P.
1984 *Kleine Schriften*. Wiesbaden: Steiner (Glasenapp-Stiftung)
Wezler, A.
1979 'Śamīka und Śṛṅgin. Zum Verständnis einer askese-kritischen Erzählung aus dem Mahābhārata', *WZKS* 23, pp. 29-60

Notes

1. As is the case with other Sanskrit abstract nouns, the word *tapas* means both process (ascetic performance, *tapas car-, kṛ-, tap-, āsthā-*) and result. Here in this paper, *tapas* is mainly used in the sense of ascetic merit, as the result of asceticism. For a bibliography of ancient Indian asceticism, see the most recent contribution on the subject by Bronkhorst (1993).
2. Cf. Hara, 1970.
3. Cf. Gonda, 1991, pp. 441-2.
4. Other compounds comparable to them, containing *-dhana* or *nidhi* as the last member of the compound are as follows:

ahaṁkāra-dhana	: Harṣacarita 221, 21
abhimānaika-dhana	: Śiśupālavadha 1.67 (used in a bad sense)
atharva-nidhi	: Raghuvaṁśa 1.59
audārya-dhairya-vīrya-nidhi	: VC p. 229 (Story 32.22)
caritra-dhana	: KSS 29.196
dayā-nidhi	: VC p. 230 (Story 32.70)
dayā-dhairya-nidhi	: KSS 90.106
guṇa-nidhi	: Mahāvīracarita 4.17, 5.47

jñāna-nidhi	: Mahāvīracarita 1.5, Śiśupālavadha 1.11
kalā-nidhi	: VC p. 229 (Story 32.9)
kāruṇya-nidhi	: KSS 22.219
maṅgala-nidhi	: Mahāvīracarita 4.14
mano-dhana	: Veṇīsaṁhāra 4.1
māna-dhana	: Raghuvaṁśa 5.3, Veṇivāsavadatta 3.12
nṛśaṁśa-dhana	: Mārkaṇḍeya Purāṇa 8.46
parākrama-nidhi	: VC p. 220 (Story 31.9)
prajñā-dhana	: VC p. 8 (Frame Story 2.14)
pratāpa-nidhi	: Raghuvaṁśa 5.71 (sun)
sāhasa-dhana	: KSS 27.208, Pañcatantra 3.3
saṁyama-dhana	: Śakuntalā 4.17
satkāra-dhana	: Mṛcchakaṭika 2.15
sauhārda-nidhi	: Raghuvaṁśa 14.15
śarīra-dhana	: JM 6.16, 6.30 (used in a bad sense)
śīla-dhana	: JM 10.17, 20.5
śīla-jñāna-nidhi	: MBh 1.125.12
śrauta-dhana	: MBh 3.198.84
tapaḥ-śruta-jñāna-dhana	: JM 12.20
tapaḥ-parākrama-nidhi	: Mahāvīracarita 2.22
tejo-nidhi	: Harivaṁśa 90.4
(sarva-)vidyā-nidhi	: VC p. 221 (Story 31.12)
yaśo-dhana	: Raghuvaṁśa 2.1, 3.48, 14.35
yaśo-nidhi	: Mahāvīracarita 2.30
nidhir vratānām	: Mahāvīracarita 2.24.

5. Saundarānanda 10.33 (*tapo-mūlya-parigraheṇa svarga-krayārtham*). Cf. Saundarānanda 10.59 (*vidhatsva śulkārtham ihottamaṁ tapaḥ*).
6. Kumārasambhava 5.86 (*tavāsmi dāsaḥ krītas tapobhir* ...) and Nāradasmrti 18.23 (ed. Larivière) (*tapaḥ-krītāḥ prajā rājñā* ...).
7. Monier Williams, *Indian Wisdom* (Reprint, Varanasi, 1963), p. 344, note 2.
8. R.F. Gombrich, 1972.
9. Hara, 1970, p. 75.
10. Gonda, 1952.
11. Hara, 1970, p. 62.
12. Cf. Kādambarī 89.8 (*kośas tapasaḥ*).
13. Hara (1994).
14. Thus, one must be careful in protecting (*rakṣ-*) it from abuse or unnecessary expenditure. *tapo ... rakṣan* (Raghuvaṁśa 17.65).
15. These aspects of *tapas* to be earned (*arjana*), increased (*vardhana*), protected (*rakṣaṇa*) are also shared by property (*artha*) in general.
arthas tāvad arjana-vardhana-rakṣaṇātmakaḥ, kṛṣi-pāśupālya-vāṇijya-saṁdhi-vigrahādiparivāraḥ tīrtha-pratipādana-phalaś ca (Daśakumāracarita, p. 88, lines 6-7, NSP 1951).
arthānām arjane duḥkham arjitānāṁ ca rakṣaṇe nāśe duḥkhaṁ vyaye duḥkhaṁ dhig arthāḥ kaṣṭa-saṁśrayāḥ (PPT 1.123)
tathā arjana-rakṣaṇa-kṣaya-saṅga-hiṁsādayo doṣāḥ (Pañcārthabhāṣya ad

Pāśupata Sūtra 5.35, TSS 143 p. 133, lines 8 ff.)

16. Cf. Bhagat, 1976, pp. 267-70.
17. Ordinarily, however, *upastha* (*śiṣṇa*) is construed with *udara*, indicating two principal instincts, appetite and sex.
śiṣṇodaraṁ rakṣet (MBh 5.40.2.2, 12.232.6, 12.317.28, 13.110.28)
jita-śiṣṇodara (MBh 13.110.131, 14.45.20)
upasthodarayor-vega (MBh 12.152.8)
udaropastha-vega (MBh 12.269.15, 12.288.14)
śiṣṇodara-parāyaṇa (MBh 12.287.25)
śiṣṇodare (MBh 12.288.36),
upastham udaram (MBh 12.288.28).
18. This motif does not appear in the corresponding portion of MBh where the king apparently kept memory (*smarann api* 1.68.18).
19. For the idea of *lex talionis* in general, cf. Gössel, 1914, pp. 79-80.
20. Cf. Wezler, 1979.
21. Cf. Čičak-Chand, 1974, and Hara, 1983, p. 356 notes 17-21.
22. Hara, 1975, pp. 130 ff.
23. Indra's fear is not only caused by *tapas*-practice, but also *samādhi*. Cf. the compounds *samādhi-bhīta* (Raghuvaṁśa 13.39) and *samādhi-bhīrutva* (Śakuntalā 1.24.51).
24. Smith, 1986, pp. 261-4 (background curse).
25. For example, the Śakuntalā passages read as follows: *tuha aviṇṇāda-pahāva-paramatthassa duhidājaṇassa bhaavadā aaṁ avarāho marisiddavvo* (= *tava/ avijñāta-prabhāva-paramārthasya/duhitṛjanasya/bhagavatā/ayam/aparā-dhaḥ/marṣitavyaḥ*) (4.1.16-17) (Pischel, 1922).
26. *kiṁ tu ahiṇṇāṇāharaṇa-saṁsaḍādo se savo ṇiattissadi* (= *kim/tu/abhijñānā-bharaṇa-darśanāt/ asyāḥ/śāpaḥ/nivartiṣyate*) (4.1.19-20).
27. *tado so ṇa me vaaṇaṁ aṇṇadhā bhaviduṁ arihadi* (= *saḥ/na/me/vacanam/anya-thā/bhavitum/ arhati*) (4.1.19). Cf. MBh 1.91.13 (*na śakyam anyathā kartuṁ yad uktaṁ brahmavādinā*).
28. However, in Raghuvaṁśa 15.37 the same compound *tapo-vyaya* is used in a different meaning (obstruction of disturbance of asceticism).
29. Laine, 1989, p. 219.
30. Hopkins, 1932, p. 328.
31. Buddhist equivalent of Hindu ascetic is the *paccekabuddha* in a position to curse and bless. For this, see Kloppenborg, 1974, pp. 67-9.
32. Smith, 1986, p. 265. However his *ERE* quotation (*cursing and blessing*) should be 4.367 instead of 376.
33. The phrase *paśya me tapaso balam* never appears in the Rāmāyaṇa, whereas *paśya me tapaso vīryam* is met once in R 1.59.12.
34. Laine, 1989, p. 203.
35. *śāpānugraha-samarthā maharṣayaḥ śrūyante*. I owe this reference to Kirste (1993) p. 202.
36. Cf. A. Padoux, *Le Coeur de la Yoginī, Yoginīhṛdaya*. Paris: Boccard, 1994, p. 106, note 55.
37. Cf. Thieme, 1984, p. 623.

38. The phrase not only occurs in a curse-context, but also in oath and *satya-kriyā* contexts. Cf. MBh 1.38.2, 1.49.19, 1.209.10, 3.194.23, 5.120.5, 12.49.24, 14.56.10. Cf. also Hopkins, 1932 p. 328.
39. Cf. MBh 1.44.11, 3.71.13, 7.12.13, 17.2.20. Cf. also MBh 3.281.97 (*satya-kriyā*).
40. Cf. MBh 1.38.2 quoted above.
41. *nokta-pūrvaṁ mayā mithyā svaireṣv api kadā cana* (*satya-kriyā*) (MBh 14.68.19).
na ca mithyā-pralāpo 'tra svaireṣv api (MBh 1.57.10).
nāhaṁ mithyā-vaco brūyām (MBh 13.51.17).
42. *vitathaṁ nokta-pūrvaṁ me svaireṣv api kuto 'nyathā* (MBh 1.107.17).
svaireṣv api na tenāhaṁ smarāmi vitathaṁ kvacit (MBh 1.44.11).
43. *anṛtaṁ nokta-pūrvaṁ me hasatāpi kadā cana* (MBh 1.209.10).
44. *vāg-vajrā brāhmaṇāḥ proktāḥ kṣatriyā bāhu-jīvinaḥ* (MBh 12.192.45ab).
bāhubhiḥ kṣatriyāḥ śūrā vāgbhiḥ śūrā dvijātayaḥ (MBh 7.133.23).
brāhmaṇasya ca vāg-balam (MBh 5.142.21b). Cf. also Caillat, 1984, p.64, note 11.
45. Cf. Raghuvaṁśa 15.3 quoted above.
46. *na balaṁ kṣatriyasyāhur brāhmaṇo balavattaraḥ brahman brahma-balaṁ divyaṁ kṣatrāt tu balavattaram* (R 1.53.14).
kva ca te kṣatriya-balaṁ kva ca brahma-balaṁ mahat (R 1.55.4ab).
dhig balaṁ kṣatriya-balaṁ brahma-tejo-balaṁ balam ekena brahma-daṇḍena sarvāstrāṇi hatāni me (R 1.55.23).
47. Hara, 1970, p.68. *nṛpa-śrīr brahma-śāpāntā* (Cāṇakya-rāja-nīti-śāstra 8.58c).
ekaḥ kruddho brāhmaṇo hanti rāṣṭram (MBh 5.40.7d).
hanti vipraḥ sarāṣṭrāṇi purāṇy api kopitaḥ (MBh 1.76.24cd).
48. Cf. Hopkins, 1932, p. 328. 'An oath (*śapatha*) by *tapas* is similar, in that it invites the loss, in case of perjury, of hard-earned asceticism here and in the hereafter'.
49. Hara, 1975. Cf. the birth story of Māndhātṛ (MBh 3.126.18 ff.).
50. Singh, 1967, pp. 47-8.
51. Renou, 1958, pp. 55-6.
52. Hara, 1985.
53. Al-George et A. Roşu, 1957, pp. 243-54.
54. Bedekar, 1968-9, p. 45, Sharma, 1982, p. 196, Goudriaan, 1978, pp. 230 ff.
55. Bloomfield (1924).
56. Köhler, 1973, p. 44, note 80.
57. Hacker, 1978, p. 344, 1983, p. 27, Norman, 1993, p. 279. Köhler shared the same view with Hacker (Köhler, *loc. cit.*).
58. Wezler, 1979, pp. 54-7.
59. However, one is recommended to take into consideration the rules concerning speech (*bhāṣā*) in Jaina and *subhāsita* in the Theravāda tradition (*ariya-vohāra*) as carefully investigated by Caillat (1981 and 1984).
60. Greek *áskēsis* (= exercise), 'training, discipline'.

XIV

RITUAL AND RITUALISM: THE CASE OF ANCIENT INDIAN ANCESTOR WORSHIP

J.C. Heesterman

1. In the world of sacrifice it is the gods who hold the superior position. Themselves fervent sacrificers they fight their perennial adversaries, the asuras, at the place of sacrifice. They also fight amongst themselves for leadership and precedence while making deals and repartition for the goods of life. As it is with the gods, so it is with the men who invite the gods to sit together with their human hosts at the sacrifice and take part in their contest, sharing in the ensuing feast.

Yet for all their high profile and conspicuous deeds, the gods cannot obscure the power of another category of beings. These beings are the departed ancestors, the *pitaraḥ* or 'fathers'. They clearly lack the spectacular persona of the gods and are mostly viewed as an indistinct collectivity. Even when the three immediate ascendants - father, grandfather and great grandfather - are invoked by their names, they are still no more than impersonal shadows. Nonetheless, their presence and power to bring good as well as evil are felt to be everywhere. In their inconspicuous way they are ubiquitous, as the gods are not. After slaying Vṛtra and defeating the asuras, when the warrior god Indra came back at the time of the new moon to the place of sacrifice, the fathers were already there, having arrived the day before the new moon. They had stolen a march on Indra and the gods with the serious consequence that the sacrifice was now with them. When the gods demanded the fathers to return the sacrifice, they refused to comply. So there was no alternative but to strike a deal. The fathers could have their sacrificial feast on the eve of the new moon before the gods have theirs on the next day.[1] This episode from the Taittirīya Brāhmaṇa is meant to justify the worship of the *manes* on the eve of the new moon sacrifice. But beyond that it succinctly depicts the power of the departed ancestors who jealously hold the key to the sacrifice. Only by redeeming it from them can the human sacrificer perform his sacrifice.[2]

Although the ancient Indian ritualists took every care to keep the gods and the ancestors separate, the whole sacrificial ritual remains suffused with the presence of the manes to the point that the carefully elaborated divide becomes rather blurred. Consequently the fathers can be designated Viśve Devāḥ, 'All-Gods', or *devāḥ pitaraḥ*, 'Gods-Fathers'.[3] Fittingly they drink the soma beverage together with the gods: 'This is their symposion (*sampā*); of old they drank together visibly, but now they do so unseen'.[4]

The way in which the unseen ancestors share in the soma is by means of the so-called *nārāśaṁsa* cups used after the libation in the fire and subsequent drinking of the soma, when the soma liquid left in the cups is made to 'swell' again (*āpyāyana*) - either by refilling them or simply by a mantra invoking soma's blessings - and the cups are put down again.[5] Not only are the *nārāśaṁsa* cups the father's part of the soma beverage, the fathers themselves are now the divine 'King Soma' and as such are designated *nārāśaṁsāḥ pitaraḥ*.[6] In other words, the fathers, identified with the divine 'King Soma', are again seen to hold the key of sacrifice. In fact, since 'King Soma' is viewed as the divine epitome of sacrifice,[7] they are the representatives of the sacrificial institution. In passing it may be noted that the designation *nārāśaṁsa*, 'related to the praise of heroes', for both the soma cups and the fathers puts the latter's importance further into relief. This designation originally involved more than the ritualistic manipulation of the cups' *āpyāyana*. It suggests an exuberant original scenario involving bardic songs in praise of the ancestors and their heroic deeds.[8] That this was the case finds a late confirmation in the Mānava Dharmaśāstra, where such performances are part of the sacrificial feast in honour of the fathers, the *śrāddha*.[9]

If, then except for the few separate sacrifices specifically designed for them (generically known as *pitṛyajña*), the ancestors have only a rather unassuming, almost surreptitious, part to play in the standardized ritual, the ritual as it stands still contains telling indications of a once more lively and elaborate celebration of the manes in accordance with their overall importance to the sacrifice.[10] One is led to wonder whether they do not rank before the gods who, as sacrificers themselves, utterly depend on the sacrifice to which the fathers hold the key. Indeed Manu ranks the rites in honour of the fathers above those for the gods.[11] One even comes to suspect that the manes have preceded the gods who appear to be late-comers in sacrificial thought.[12] Manu still seems to hark back to such precedence of the manes when he calls them 'the previous gods' (*pūrvadevatāḥ*), apparently belonging to a primordial world antedating the gods.[13] That the manes are qualified as 'All Gods' does not mean that they were amalgamated with the more clearly profiled and distinct gods for the sake of convenience but rather the other way round. The gods seem to have arisen from an indistinct but dynamic mass of vital powers, epitomized by the manes, the 'All-Souls', that was later to find its doctrinal apogee in the *brahman* power and its identification with the 'soul', the *ātman*.

At any rate, it is clear that the manes held a power that far exceeded their shadowy existence - the power of the sacrifice.

2. The outstanding potency of the otherwise indistinct ancestors can hardly cause surprise. Having passed the barrier of death they hold the secret of the life-and-death nexus which is the pivotal issue of sacrifice. In sacrifice the insoluble riddle of life and death is enacted;[14] it is there that the mysterious power of the ancestors resides.

Against this background we can readily understand that with the rise of ritualism the cult of the fathers must have been of particular interest to the ancient Indian ritualists when they set about to reform sacrifice, and construe their rigid system of ritual. Although the fathers belong to another sphere, whether aerial or subterranean, they are in their ghostly way involved all the time in the fortunes and misfortunes of the living. They are astride the divide between life and death, at the synapse of sacrifice. Consequently it is at the behest of the fathers, not the gods, that the epic Pāṇḍavas had to perform the royal sacrifice, *rājasūya*, which was to bring about the annihilation of the kṣatriya race.

How was the power of the sacrifice - the enigmatic nexus of life and death where the fathers had their being - to be mastered? This was the task the ritualists had set for themselves. It meant that they had to aim for a static, ultramundane order beyond the dynamic alternation of life and death. But sacrifice was based on the enigmatic alternation which is perennially enacted in the renewed contest for the goods of life and which, once won, had to be staked again and possibly lost in the next round. This was the 'everlasting path of fortune' (*śāsvataḥ ... bhūtipathaḥ*), as the despondent *Yudhiṣṭhira* is reminded by Arjuna, when he wants to quit the catastrophic cycle.[15]

It was on this point of cyclical repetition that ritual came in. Even though moored in mundane society, its conventions set sacrifice apart on an autonomous stage, in its own space and time. This autonomy opened the way for the ritualists to definitively lift sacrifice out of its worldly context, so as to create a sovereign realm of ritual transcending the mundane order ruled by life and death. Consequently the dynamic flexibility and ambivalence of sacrificial rites were turned into the static rigidity of ritualism which propounds the ultramundane law of ritual.[16]

Contrary to the uncertainty of the sacrificial contest the ritualists posited the dead certainty of a rigid ritualism that ruled out the contest and straight away promised the goods of life to the now unopposed solitary sacrificer who, for the duration of the ritual submitted to its absolute rule. In this way the ritualists broke the unremitting cycle of sacrificial contests that each time threatened to turn into uncontrollable devastation and collapse. The segments of the cyclical pattern were remoulded into separate lineal sequences, rigidly regulated according to a standardized paradigm. The dark mystery of sacrifice was turned into the ratiocinative clarity of doctrinal ritualism, demoting the sacrality of the sacrificial contest to the mundane order, in favour of the transcendent rule of ritual.

Such was the high achievement of the ancient Indian ritualists as well as their failure. For the transcendent order of ritualism meant an unbridgeable chasm that separated it from the mundane world.[17] Ritualism had desocialized sacrifice and thereby created, as a matter of principle, the insoluble problem of again relating sacrifice to worldly reality.

One would perhaps think that the gods and their cult raised a formidable barrier for the ritualistic enterprise. In fact this was not the case. The classical view which wants the destruction of the oblation in the fire to be an invigorating gift to the gods - *do ut des* - offers at best a tenuous explanation for the paradox of sacrifice: destroying food, the substance of life, in order to produce more of it. The institution of sacrifice did not depend on the gods. It can do without them.[18] Indeed the mīmāṁsaka theoreticians of ritual reduced the role of the gods to mere names. Not the gods but the act of offering for destruction (*tyāga*) is the single operative factor that brings about the transcendent 'fruit' of sacrifice.

The case of the fathers was entirely different. They are directly and pivotally concerned with sacrifice. Although separated from the living, they are still involved in the fate of their descendants. 'The fathers share in the world of men because of their offspring.'[19] They can even be said to have power over the household (*gṛhāṇām īśate*).[20] How much they were tied in with the life of their descendants can easily be seen in the stipulations regarding preferential marriage, the *sapiṇḍa* rules on forbidden degrees of kinship, succession and rights of inheritance. The latter are bound up with the duties of ancestor worship and, conversely, the due performance of their duties documented the successor's rights.[21]

It was precisely the involvement in the world of the living that made the fathers essential to the sacrifice, far more so than the gods. It resisted their being cut off from their worldly moorings through the desocialization that resulted from transcendent ritualism. Consequently the fathers were a serious problem for the ritualists.

3. In what way was the social significance of the fathers now reflected in sacrifice? Their part in sacrifice found its clearest expression in a communal meal in which the sacrificial viands are consumed. This meal - known as

śrāddha, from *śraddhā*, trust, bond - is offered to brahmins who are invited to represent the fathers and are treated as such by their host. Separated from the meal, lumps of the same sacrificial food (*piṇḍa*) are placed in furrows (or holes) near the cooking fire, as an exposure sacrifice to the fathers, more specifically to the father, grandfather and great grandfather (corresponding with the standard number of three *piṇḍas*).[22] Similarly, as we already saw, the fathers share in the drinking of soma, while they are also offered *piṇḍas* of sacrificial food.[23]

What is striking here is the emphasis on food and feeding - a social value if ever there was one. Generally speaking, the sacrifice - and Vedic sacrifice is no exception - now revolves around food, the substance of life, resulting from the immolatory killing. In its classical form Vedic sacrifice puts almost exclusive stress on the destruction of food in the oblational fire at the neglect of its consumption. We are not explicitly told so, but the bulk of the sacrificial food is removed from the place of sacrifice, presumably to be consumed afterwards, outside the space and time of the ritual in an unspecified way by equally unspecified brahmins. There is separately cooked food, known as *anvāhārya* ('served up afterwards'), that is offered to the brahmin officiants. But again, after having been served up and divided, the *anvāhārya* is removed to be eaten after the ritual is completed. The only appreciation of the value of food and the meal is the prescribed consumption of the so-called *iḍā* portions by the sacrificer and the brahmin officiants during the ritual (after the burnt oblation). But this is not a meal in any real sense, as little as the Eucharist is, nor is it (in contradistinction to the Eucharist) communal.[24]

By contrast, in the *śrāddha*, it is the meal that is the focus of the proceedings. The Baudhāyana Śrauta Sūtra acquaints us with an elaborate version of such a feast, where a cow is sacrificed in honour of the fathers (*gopitṛyajña* or *upavasathagavī*) on the eve of establishing the fires for the 'solemn' (*śrauta*) ritual. Here the social values of food and the communal meal are given full rein. Aside from the curious insertion of a game of dice for the parts of the victim - usually a sequence in the subsequent setting up of the *śrauta* fires, the interesting point is that the sacrificial meal concluding the *gopitṛyajña* sacralizes a bond of companionship. 'All those who partake of the [cow's] meat become his [the sacrificer's] *gobhājaśaḥ*, eaters - that is participants - of the cow sharing.' In other words they become the sacrificer's co-sharing partners.[25]

4. The *gopitṛyajña* illustrates in exemplary fashion the problem the ritualists encountered in dealing with the worship of the ancestors. Given the ancestors' involvement in human society it stands to reason that their worship is specially related to the sacrificial meal taken in common. But for the same

reason one also readily understands that the ritualists' view of transcendence prompted them to discard the meal from the *śrauta* ritual.

But then, what of the *śrāddha*? The first point to be noted is that the *śrāddha* is *not* a part of 'solemn' *śrauta* ritual. Similarly the *gopitṛyajña* takes place *before* the setting up of the fires for the *śrauta* ritual and so falls outside the transcendental *śrauta* realm. The ritualists handled the problem by taking the meal in honour of the fathers out of the *śrauta* ritual and relegating it to the domestic or *gṛhya* ritual which is by nature more open to the social aspects of sacrifice than the desocialized *śrauta* ritual. In fact, whether specifically in honour of the fathers or not, the festive meal, mostly in its classical form of the meritorious feeding of brahmins, became the hallmark of *gṛhya* ritual, where it is an integral part of all domestic ceremonies.

Because the *śrāddha* was essentially a sacrificial meal, it could not very well stand by itself. It logically required a sacrificial context. The ritualists, therefore, construed the *śrāddha* as a fully-fledged *gṛhya* sacrifice in accordance with the standard paradigm prominently featuring the burnt oblation in the domestic fire (in this case addressed to the 'Gods-Fathers', the *devāḥ pitaraḥ*) and concluded by a full-scale meal that was all but completely discarded from the *śrauta* ritual. The distinctive feature of the *śrāddha* meal was the offering of the small lumps of food (*piṇḍa*) to the three immediate ancestors.

Moulded as a standard domestic sacrifice, the *śrāddha* is clearly a ritualistic construction. In this sense Caland will have been right when he argued in his basic work on ancient Indian ancestor worship that the first part of the *śrāddha* known as *daiva* or *vaiśvadeva*, which focused on the burnt oblation, was a later addition.[26] Typical for the cult of the ancestors is the exposure (or burying) of the oblation instead of burning it. The burnt oblation rather seems to double the exposure of the *piṇḍa*s. That the sacrificial cult of the ancestors by itself did not require the burnt oblation is shown by the alternative offering the oblation 'in the hand' of the brahmin guests who represent the fathers, instead of in the fire.[27] 'The mouth of the gods is the fire, the mouth of the fathers is the brahmin.'[28] By replacing the fire with the hand of the brahmin the value of food and a meal is restored with a vengeance!

This does not necessarily mean that from the beginning there was a sharp divide between the sacrificial cult of the gods stressing the fire and that of the ancestors centred on food and meal, the one featuring the burnt oblations, the other exposure of the offering. The all but exclusive prominence of fire and burnt oblation at the expense of the role of other elements (the waters, the earth, the air) appears to have been an innovative development. It seems likely that we have to start from an original, loosely integrated pattern of sacrifice combining the burnt oblation with exposure (or other ways of offering) and giving special prominence to the manes in a communal sacrificial meal. Such an integral pattern, which gives due attention to the unavoidable presence of the fathers can still be discerned in the already mentioned *nārāśaṁsa* rite of the soma ritual.

When the archaic pattern of socially embedded sacrifice was broken up in favour of transcendent ritualism, the sacrificial meal in which the manes had their *locus standi* was made into a fully-fledged separate domestic sacrifice, be it that they were represented by the brahmin guests or were fed with exposed lumps of food, or in both ways. Put differently, the meal and manes were enclosed in a special niche, the *śrāddha* sacrifice, which was generalized as the monthly celebration of the fathers and even came to function as a paradigm for occasional feasts. The *śrāddha* became the last stronghold of the societal significance of sacrifice and the sacrificial meal. Even though safely pigeon-poled, its societal potential kept irking the ritualists and prompted them to unrealistically proscribe the *śrāddha* for worldly purposes, that is, more specifically for building and maintaining networks of alliance and patronage.[29]

5. If the sacrificial meal could with relative ease be confined to the less arcane and more open sphere of the domestic ritual, the manes were not so easily restrained. Their power which resided in the secret of life was essential to the sacrifice and could not be contained in the carefully construed niche of the domestic *śrāddha*. Therefore, the ritualists provided them with separate niches in the *śrauta* ritual as well. Those were the two *pitṛyajñas*, namely 'the father sacrifice of food lumps' (*piṇḍa-pitṛyajña*) and 'the great father sacrifice' (*mahāpitṛyajña*). The first is the one which is associated with the monthly new moon sacrifice (*darśa*) and usually takes place on the eve of the latter. The more elaborate *mahāpitṛyajña* is part of the *Sākamedha* sacrificial complex that marks the end of the year.[30] Both have the same pattern, combining the burnt oblation with the *piṇḍa* offerings. But significantly the sacrificial meal is conspicuously absent, as is the rule for a *śrauta* sacrifice.

Yet it is equally significant that in the *śrauta* form of the *pitṛyajña* the meal could not simply be cut out. As a telling Brāhmaṇa passage has it: 'The experts (*brahmavādinaḥ*) argue, whether [the food] should be eaten or not. If he - the sacrificer - would eat it, he would eat 'alien' (*janya*) food - that is, food related to the alien realm of the fathers - [and so] he would die. If he would not eat it, he would be deprived of the sacrificial food [and consequently] would be cut off from the fathers. He should only smell at it. In that way it is eaten and yet not eaten.'[31] In their peculiar scholastic way the ritualists knew how to circumvent the either-or dilemma by arguing that inhaling the smell is both eating and not eating. However, it does clearly show that the meal had somehow to be accounted for. The more elaborate *mahāpitṛyajña* even has the rice mess passed round the participants (before the *piṇḍas* are taken out), each of whom in turn smells it.[32] Here the original common meal taken still shines through.

The ritualists managed to set apart and isolate the fathers. In the first place they relegated the sacrificial meal to the worldly oriented *gṛhya* ritual as a separate sacrifice, and then also in the *śrauta* ritual, where the fathers were equally enclosed in a similar sacrifice, the *pitṛyajña*, but without the common meal.[33] In other words, they were subjected to the formal rule of ritualism and thereby deprived of their diffuse and unpredictable sacral power.

6. Though neutralized, the ancestors did not suffer any loss in their social and legal importance. They kept defining lineage, rights and duties. As we saw, preferential marriage, succession and inheritance are underwritten by the offering of lumps of food (*piṇḍadāna*). In this respect ritualism even enhanced the formal importance of the fathers. On the other hand, depriving the fathers of their sacral power it formalized and specified the rules of *piṇḍadāna* to the three immediate patrilineal ascendants - to which later the matrilineal ones were added - thus judicially defining the sacrificer.

Ritualistic formalism and its concern with the legal status of the sacrificer comes out clearly in the ample consideration given to the case of the sacrificer whose father is still alive. Is he qualified to offer the *pitṛyajña* or its domestic counterpart, the *śrāddha*? The questions obviously involves the social and legal relationship of father and son. To what degree is the son an independent agent *vis-à-vis* his father? So much is clear that to be a sacrificer he must maintain his own domestic fire and so be qualified to establish his own sacrificial fires. Put differently, like his father he must be a *gṛhapati* in his own right, at the head of a separate household and managing his own property acquired, if not by inheritance, by his own efforts or through partition of the patrimony during the father's life time. As a propertied householder he is bound to fulfil his duties to the gods and fathers. But then what of the still living father?

The marked formalism of the *śrauta* system of ritual gives the question a special edge, and it is the *śrauta* rather than the *gṛhya* manuals that give it ample attention in connection with the *piṇḍapitṛyajña*, which is associated with the monthly new moon sacrifice. In general the *śrauta* manuals agree on the various possible answers. After reproducing the standard discussion, only the Kātyāyana Śrauta Sūtra adds a different line of reasoning which will concern us further on.

The standard discussion offers four various possibilities, none of them entirely satisfactory. In the first place there is what we may call the 'fundamentalist' position that the three *piṇḍas* should be offered in any case, whether the father is alive or not. The opposite position is to rule out the *piṇḍapitṛyajña* altogether if the father is alive. Though more realistic, this opinion impairs the son's qualification as an independent sacrificer in his own right. In the third place a compromise is brought forward. The sacrificer passes beyond the living father and offers the *piṇḍa*s to the same ancestors as the father. This oddly makes the son step in the place of his still living father. But this compromise is rejected by the rule stated in the sūtras that 'one should not offer [*piṇḍa*s] beyond the living'.[34] In accordance with this rule one only offers the so-called *daiva* or *vaiśvadeva*, that is, the burnt oblations to the gods associated with the ancestors, so that the three immediate fathers are not invoked by name. This seems to be the generally preferred answer. In so far as it maintains the *iṣṭi* paradigm of the *pitṛyajña* is ritualistically an acceptable solution but it does away with the otherwise essential *piṇḍa*s. In fact the problem cannot be definitively solved.

What this tells us is that there is an unbridgeable gap between an original ritual, organically embedded in living reality, and the mechanistic formalism of ritualism, irrespective of actual circumstances. In accordance with the latter viewpoint - strictly individualistic to the point of being desocialized - the discussion is exclusively concerned with the single independent sacrificer. Even though it revolves around the societal question of the relationship of father and son, the discussion has to stay within the bounds of ritualistic discourse and must consequently fail to bridge the gap, other than by unsatisfactory compromise.

Strictly speaking, the question of the father being still alive should not impinge on the transcendent realm of ritualism. In order to be ritualistically meaningful the performance or non-performance of the *pitṛyajña* had to be deprived of its social content and turned into a purely ritualistic question. Only on condition of offering an orientation point transcending worldly society could ritualism objectively define the position of the sacrificer as an independent agent regardless of his social ties.

7. How this could be achieved is shown by a passage Kātyāyana appended to the usual discussion of the sacrificer whose father is still alive.[35] The passage is clearly an interpolation, and a rather awkward one at that. The intention, however, is perfectly clear. The socially relevant question is replaced by one that is internal to the Mīmāṁsā system of ritualistic analysis. How is this question framed?

The starting point is that the *piṇḍapitṛyajña* coincides with the monthly new moon celebration, the *darśa* or *amāvāsya*. The two clearly belong together and, although the *pitṛyajña* provided the fathers with a separate sacrifice, it remained associated with the *darśa* sacrifice with which it shares the juncture of the waning and waxing moon phases.

But then the question arises in what way the two sacrificers are precisely related to each other. In the analytic language of the Mīmāṁsā the question is whether the *piṇḍapitṛyajña* is an *aṅga*, a fixed and unchangeable attribute of the obligatory (*nitya*) new moon sacrifice or an independent sacrifice (*anaṅga*). In this way the critical problem of the still living father could be replaced by a purely ritualistic one to be discussed in equally ritualistic terms.

The decisive point is that if the ancestor sacrifice is an *aṅga*, it has to be performed unchanged, including the three *piṇḍas*, irrespective of any or all of the three immediate ancestors being alive. On the other hand, if it is decided that it is *anaṅga*, an independent sacrifice, it can be handled separately and subjected to certain rule-bound changes such as ending the ancestor sacrifice, after the burnt oblations, with the *iḍā* rite, and leaving out the *piṇḍas*. It can even be completely cancelled, since it is technically not counted among the *nitya* sacrifices.

In both cases the performance of the *piṇḍapitṛyajña* is effectively lifted out of its social context and transferred to the transcendent realm of ritualism, the basic questions now being whether it is *aṅga* or *anaṅga*. There are then two schools of thought; the one holding the *pitṛyajña* - more specifically its distinctive feature, the offering of the three *piṇḍa*s - to be an *aṅga* of the new moon sacrifice, the other deciding that it is an independent (*anaṅga*) sacrifice. The latter is the dominant view which has the authority of Jaimini's Mīmāṁsā Sūtra and its commentator Śabara,[36] characteristically without going into a consideration of the case of the father (or another of the three ascendants) being still alive. Strictly speaking, from the purely ritualistic point of view such consideration would be irrelevant.

This brings us to the passage inserted in Kātyāyana's Śrauta Sūtra. Rejecting Jaimini's view it boldly, albeit somewhat shakily, holds that the *piṇḍapitṛyajña* is an *aṅga* of the new moon sacrifice. Interestingly, while Jaimini does not even refer to the matter, Kātyāyana brings his opposite view to bear directly on the case of the still living father in order to defend the 'fundamentalist' position that the *piṇḍapitṛyajña* must at all events be performed to the letter.

Brought forward to combat Jaimini's view, Kātyāyana's argumentation is not a novel one. Āśvalāyana Śrauta Sūtra already records it as the opinion of one Taulvali who also wanted the three *piṇḍas* to be offered on the grounds that they are an unchangeable attribute of the ancestor sacrifice (*kriyāguṇatvāt*).[37] Āśvalāyana rejects Taulvali's view in favour of offering *piṇḍas* only to the deceased fathers and, in line with the other *śrautasūtras*, prefers to limit the *pitṛyajña* to the *daiva* part in case they (or anyone of the three) are living.[38] It shows, though, that the purely ritualistic line of reasoning was already present at a fairly early stage.

This line of reasoning reached its most radical form in Kātyāyana's polemic against Jaimini's more moderate and more reasonable view. The latter refrains from entering into the discussion of the case of the living father and leaves the matter, which is no longer relevant to him, entirely open. Implicitly Jaimini's teaching rather seems favourable to the practice preferred by the *śrautasūtras* of limiting the *pitṛyajña* to the *daiva* part in case the father is still living.[39] Kātyāyana, by contrast, explicitly reopens the old and by now superseded discussion when he links it directly, and, as we shall see, rather ineptly, to the *aṅgatva* argument.

In the end, however, whether one decides in favour of *aṅga* or *anaṅga*, the discussion was definitively transferred from its mundane context to the abstract realm of ritualism. And so were the fathers themselves.

8. We might end our inquiry here with the triumph of ritualism over reality if it were not for the distinct awkwardness of our chief witness, the Kātyāyana passage. Following hard on the usual discussion of the sacrificer whose father is still alive (*jīvapitṛka*) our passage strikingly differs from the preceding sūtras in its wording, which mirrors Jaimini's usage. Moreover, although it is formally linked to the preceding discussion by the particle *vā*, in Jaimini's sense of rejecting the immediately preceding statement, in this case the statement that no *piṇḍa*s should be offered, 'beyond the living',[40] no further reference is made to it.

The inserted passage (KŚS 4.1.28-30) runs as follows: *pūrvo vāṅgatvāt piṇḍapitṛyajñaḥ*, the *piṇḍapitṛyajña* precedes [the new moon sacrifice] because it is an *aṅga* [of it]' (S. 28). Then (S. 29) in Mīmāṁsā style the arguments for the opposite *anaṅga* view follows, somewhat changing and adding to those adduced by Jaimini: context (*prakaraṇa*), (separate) time (*kāla*), indicative wording (*liṅga*), (separate) mentioned elsewhere (*anugrahavacana*), authoritative statement of performance by one not in possession of the *śrauta* fires (and so barred from performing the *śrauta* new moon sacrifice *anāhitāgniśruti*).[41] The next sūtra (30) rejects these arguments and decides for the *aṅga* thesis on the single ground of its being prescribed in direct connection with the new moon sacrifice, *aṅgaṁ vā samabhivyāhārāt*.[42]

The crux is the first sūtra (28), 'the *piṇḍapitṛyajña* precedes because it is *aṅga*'. Although directly connected with the previous sūtra by the particle *vā*, it is hard to see how it can relate to the latter which states that no *piṇḍa*s are to be offered 'beyond the living', let alone reject this statement. The crucial sūtra merely states that the *piṇḍapitṛyajña* precedes the new moon sacrifice because it is an *aṅga* of it. Not its being *aṅga* or *anaṅga* is the issue but its being *pūrva*, preceding the new moon sacrifice.

The easy way out would be to take *pūrva* to mean the previous opinion, given right at the start of the discussion and then rejected, that the *piṇḍapitṛ*-

yajña including the three *piṇḍa*s should in any case be performed to the letter, also by the *jīvapitṛka* (5.24 f.). This would require a shift from the last word of the crucial sūtra, *piṇḍapitṛyajnaḥ*, to the next sūtra. In itself this would be perfectly feasible and would result in a better reading. Sūtra 28 would than mean: 'On the other hand (*vā*) the previous [rule, viz. unchanged offering of the *piṇḍapitṛyajña*,] is valid on the grounds of being an *aṅga*'. Though arguable, taking *pūrva* to refer to a previously stated but rejected *pūrvapakṣa* opinion in order to bring it back again as the final *siddhānta* opinion would be rather unusual. Moreover, the commentary does not even mention this possibility but, as we shall see, takes a different line of explanation.

The crucial sūtra (28) does not address the matter of the *piṇḍa*s. The question that is being answered is that of the time of the ancestor sacrifice. Should it be offered before or after the new moon sacrifice? The answer is that it should precede the new moon sacrifice because it is an *aṅga* of the latter. This is the line taken by the commentator. Astutely but unconvincingly he then interweaves the question of timing with that regarding the fathers.

His starting point is that there are two possible days for the new moon sacrifice; either the proceedings start on the eve of the new moon day (the 14th of the dark half of the month), or all is done on the single new moon day itself (the fifteenth).[43] The time generally prescribed for the *piṇḍapitṛyajña*, however, is the afternoon of the new moon day.[44] So, if the proceedings of the new moon sacrifice start on the eve, as they usually do, the *piṇḍapitṛyajña* must needs come after the new moon sacrifice, in the afternoon the fifteenth day. However, according to our commentator, this option is only available to the *jīvapitṛka*. On the other hand, the *mṛtapitṛka*, whose father is deceased, must stick to the rule that the *piṇḍapitṛyajña* precedes the new moon sacrifice as stated by the sūtra. In this case it is enclosed between the preparation of the new moon sacrifice and its main part which follows on the same afternoon, so that the *piṇḍapitṛyajña* is then practically an *aṅga* of the new moon sacrifice.

One may admire the commentator's serendipity in tying together the timing of the new moon sacrifice, the question of the living or deceased father and the *aṅgatva* of the *piṇḍapitṛyajña*. But for all its inventiveness his construction does not carry any conviction. At the very least it does not settle the question of the *piṇḍa*s to be offered or not offered by the sacrificer whose father is alive, which was the purpose for introducing the sūtra.

Our commentator is not even interested in this question. The only point he is interested in is the time of the *piṇḍapitṛyajña*, before or after the new moon sacrifice. He must have been aware that being an *aṅga* is by itself far from a watertight argument for the priority of the ancestor sacrifice. Strictly speaking, *aṅgatva* has no direct bearing on it. So he feels free to draw the, unwarranted, conclusion that the difference between the *mṛtapitṛka* and *jīvapitṛka* sacrificer is in the time of the ancestor sacrifice; the former must perform it before the new moon sacrifice, the latter may also do so after it.

On the other hand, he takes the *aṅgatva* as a foregone conclusion and consequently holds on to the integral performance of the *piṇḍapitṛyajña*. In his view both the *mṛtapitṛka* and the *jīvapitṛka* offer it in full. The whole matter is neatly summarized by the paddhati to Kātyāyana.[45] Contrary to Kātyāyana and the commentary, it first states that the *jīvapitṛka* does not perform the *piṇḍapitṛyajña* and then states that nevertheless all the commentators of Kātyāyana, from Karka onwards, hold the view of the ancestor sacrifice being an *aṅga*. However, we are told that this is not the authoritative tradition as represented by Jaimini who teaches that it is *anaṅga*, independent. Finally, to clinch the matter the *paddhati* refers to the independent offering of the *piṇḍapitṛyajña* by the *anāhitāgni* who has not solemnly installed the *śrauta* fires and so cannot perform the *śrauta* new moon sacrifice.

What we retain from all this is Kātyāyana's and his commentator's concern with the proper time of the *piṇḍapitṛyajña*. This, after all, is the actual content of the crucial sūtra, 'it precedes [the new moon sacrifice] because it is an *aṅga* [of it]', irrespective of the sūtra's use or abuse. But then the conclusion must be that the sūtra is simply out of place. So where did it originally belong? Its proper place would seem to have been where Kātyāyana prescribes the time for the *piṇḍapitṛyajña*, that is at the very beginning of its exposé of this sacrifice. 'The *piṇḍapitṛyajña* [is to be performed] in the afternoon, when the moon is invisible, on the new moon day.' However, this creates a problem regarding its connection with the new moon or *darśa* sacrifice which according to Kātyāyana is to take place either on the eve of the new moon day or on that day itself.[46] If the first alternative, the eve of the new moon day, is adhered to, the *piṇḍapitṛyajña* must also fall on the eve of the new moon day. Then, in all likelihood, our sūtra was originally a gloss meant to offer the same alternative for the *piṇḍapitṛyajña* too. It would then mean: 'Or the *piṇḍapitṛyajña* precedes [sc. the new moon day; namely when performed on its eve together with the *darśa* sacrifice], because it is an *aṅga* [of the *darśa*].'

The reason for shifting the sūtra, or rather the gloss, to its present place is clear enough. Even though it originally did not refer to the matter of the still living father, it illustrates the *aṅga* thesis (which then is argued in the next two sūtras, 29-30) and brings it to bear on the case of the *jīvapitṛka*. At the same time it enabled the ritualists to move away from the worldly order to the ultramundane realm of ritualism.

9. This still leaves us with a double question. How did being an *aṅga* come to be adduced as a conclusive argument for settling the time of the *piṇḍa-pitṛyajña*? And then why the insistence on the option of having it precede the day of the new moon? In the context of Kātyāyana taken in isolation the answer is fairly straightforward. If the *darśa* sacrifice is performed on the eve of the new moon day, the *piṇḍapitṛyajña* is also due on that day. Here the *aṅga* thesis comes in handy. It legitimizes the shift of the ancestor sacrifice to the day preceding the new moon day, in opposition to Kātyāyana's rule that it should fall on the afternoon of the new moon day.

However, being an *aṅga* does not by itself determine the time of the *piṇḍapitṛyajña*. It only posits its dependence on and hence contiguity with the new moon sacrifice. But this does not necessarily mean that they should both fall on one single day as Kātyāyana's followers would have it. It would also be possible to have the ancestor sacrifice on the eve of the new moon day and the *darśa* sacrifice on the new moon day itself. This is indeed the practice prescribed by the Taittirīya Brāhmaṇa.[47] Although the Taittirīyakas do not adhere to the *aṅga* thesis, their *piṇḍapitṛyajña* is embedded in the new moon sacrifice and on that score might be considered an *aṅga* of the latter.

But, irrespective of the *aṅga-anaṅga* discussion the interesting point is the close relationship of the two sacrifices. As we already surmised, it suggests an original pattern combining the fire sacrifice with the sacrificial meal and the exposure of the lumps of food for the ancestors. The ritualistic *aṅga* thesis then maximizes and rigidifies the original integrated pattern.

On this point we come up against a difficulty. Given the special connection of the ancestor cult with food and a meal one would expect them to be worshipped at the end of the new moon sacrifice when it is concluded by a sacrificial meal. As we already noticed, the ancestor sacrifice when construed as a separate *iṣṭi*, reproduced this pattern by first offering burnt oblations to the gods associated with the fathers. Similarly for good measure the *śrauta* celebration of the new moon (including the *piṇḍapitṛyajña*) is again followed by a domestic *śrāddha*.[48] But, as we have seen, the *piṇḍapitṛyajña* generally comes first, albeit on the same day. The older Taittirīya tradition even wants to have it on the previous day (*pūrvedyuḥ*). How is this apparent deviation to be explained?

10. So far we have only considered the relationship of the *piṇḍapitṛyajña* with the new moon sacrifice. But the latter cannot be isolated from its counterpart, the full moon sacrifice. Ritualism even homogenized them, albeit as separate sacrifices, and made them the basic paradigm of the *iṣṭi*-type of sacrificial ritual. So ancestor worship has to be viewed in the context of the pair formed by the full moon and the new moon sacrifices. In that case, the question is, which place do the fathers occupy in the sacrificial scenario of

the waning and waxing moon.

Here we come upon a peculiar feature of the new and full moon sacrifices, a feature which the standardized ritual hardly makes us expect. This is the connection with strife, rivalry and bloody conflict.[49] 'He who has a rival should only perform the full moon sacrifice, not the new moon sacrifice; [in that way] having slain his rival he does not make him wax [i.e. revive] again [through the new moon sacrifice].'[50] Only the *piṇḍapitṛyajña* should be offered at the new moon. For greater efficacy it is further recommended to have the full moon sacrifice for Agni and Soma immediately followed by another *iṣṭi* comprising an offering to Agni and Viṣṇu, one for Sarasvatī and one for Sarasvat, her male counterpart. With the full moon offering he hurls the *vajra* weapon at his rival, with the Agni-Viṣṇu offering he deprives him of gods and the sacrifice, with the two offerings for Sarasvatī and Sarasvat of his paired [i.e. male and female] cattle. Thus he deprives him of all his possessions.[51] But then, how can one have the new moon sacrifice at all? Given the surfeit of rivals cropping up at every turn in the brāhmaṇa expositions as well as in the *śrautasūtra* it would take an implausibly irenic sacrificer who has no mind to contend with anybody for the goods of life.

At this point we should recall the role of the fathers. We already learnt that according to the Taittirīya Brāhmaṇa the gods had to redeem the sacrifice from the fathers before they could have their new moon feast. Therefore, the fathers should be honoured with sacrifice on the eve of the new moon. Another version of the Taittirīyaka tradition gives some further background to the priority of the fathers. After his awesome deed of slaying the Vṛtra monster on the full moon day Indra fled far away and went into hiding, fearing that he had missed his aim.[52] The fathers then found him and so obtained the 'first share' of sacrifice, that is the *piṇḍapitṛyajña* on the eve of the new moon day. It was not fortuitous that the fathers found Indra, or, according to the Brāhmaṇa version, that it was from them that the sacrifice had to be redeemed.

Indra's fear and anxiety, which caused him to flee and go into hiding, underline the awesome ambiguity surrounding the sacrificial killing. Vṛtra who holds in himself Agni and Soma, the igneic and the aquatic element, is the sacrifice embodied. Slaying him is sacrilege. The Taittirīya Saṁhitā tells us that when Indra perpetrated his sacrilegious deed his detractors (*mṛdhaḥ*) raged against him.[53] Similarly, when he slew the three-headed Viśvarūpa - another embodiment of the sacrifice - Indra is loudly accused of brahman-killing.[54] Defamated and burdened by the guilt of killing, Indra is under the sway of the realm of the dead. That is why it is the fathers who found him, or rather held him ransom as they held the sacrifice, and would not relent without having their 'first share'.

Actually it was not so much the sacrifice but Indra, the archetypal sacrificer, who had to be redeemed. Having slain Vṛtra and thereby taken hold of Agni and Soma, Indra was now the embodiment of sacrifice. Thus, when Vṛtra was struck down and Agni and Soma left him for Indra, he stopped Indra from delivering the *coup de grâce* by pleading: 'Do not hurl your weapon; you are now what I (was before).'[55] Having taken Agni and Soma into himself Indra became his own enemy, Vṛtra.[56] Exhausted, all passion spent, he must be redeemed and made to 'swell' again in phase with the waxing moon.

This ambiguous tension is at the core of all sacrifice. The core is the enigmatic nexus of life and death. Sacrifice is the riddle of death and renewal put into action. Striving for the goods of life the sacrificer must pass through the realm of death. This is what myth tells us the archetypal sacrificer Indra had to undergo. No wonder he was terrified at slaying Vṛtra. It is the double face of sacrifice for which the phases of the waning and waxing moon offer a compelling paradigm. Following the slaying of Vṛtra on the full moon day Indra, defamated and hiding, enters a period of interdict under the rule of the fathers during the waning phase of the moon. Only by redeeming Indra by honouring the fathers on the eve of the new moon, can the feast of renewal in honour of Indra and Agni be celebrated.

Understandably the transition from the realm of the fathers to the renewal of life was far from unambiguous. This, it would seem, is tellingly illustrated by a Ṛgveda passage eulogizing Indra and Agni, the dual divinity of the new moon sacrifice. 'I will now proclaim the heroic deeds you two performed. Your fathers, whose enemies are the gods (*deváśatravaḥ*), have been slain; you, Indra and Agni, are both alive.'[57] Although smoothing over the role of the fathers, the Śatapatha Brahmaṇa equally refers to the violent strife the Ṛgveda passage evokes. Discussing the 'great father sacrifice' (*mahāpitṛyajña*) on the second (the main) day of the Sākamedha celebration at the end of the year the Śatapatha tells us that the gods first slew Vṛtra on that day with the main sacrifice (the *mahāhavis*). 'Then they revived through the *pitṛyajña* those of them that they [i.e. the enemies] had slain in that battle; they [i.e. the slain] were indeed the fathers; hence the name *pitṛyajña*.'[58] Although the Śatapatha takes the (divine) fathers for the gods slain in the Vṛtra battle - the enemies of the gods being apparently the asuras, as usually in the Brāhmaṇas - the Ṛgveda passage suggests otherwise. There the fathers are the enemies from whom the goods of life have to be wrested. At any rate, the transition is not a smooth and peaceful one.

11. Against this background one easily understands that the ritualists took the *piṇḍapitṛyajña* to be *pūrva*, preceding the new moon sacrifice. It was not meant to be the conclusion of the new moon feast of renewal but of the Vṛtra killing at the preceding full moon. It marks the end of the period of violent

strife and disorder under the sign of the Vṛtra battle. In that sense the *piṇḍapitṛyajña* falls into the place we expected, that is, as the sacrificial meal that brings the contending parties together and initiates renewal, till the next round. In the same way the *mahāpitṛyajña* is due at the end of the outgoing year, where it is equally preceded by a sacrifice, the *mahāhavis*, celebrating the slaying of Vṛtra.

In this connection it is significant that the gods, to whom the burnt oblations of the first (or *daiva*) part of the ancestor sacrifice are offered, are Soma and Agni. Between them they epitomize sacrifice and by leaving Vṛtra they signal his defeat at the hands of Indra. The ancestor sacrifice, comprises both ends of the period of violent disorder, from its onset with the Vṛtra battle on the full moon day till its resolution on the new moon day.

Although construed as a separate sacrifice, the *piṇḍapitṛyajña* still bears witness to the central role of the fathers in sacrifice and renewal. Being essential to sacrifice ancestor worship was an integral part of the original unitary pattern of sacrifice. As we already concluded earlier, the ritualistic *aṅga* thesis of Kātyāyana's followers harks back to such a unitary pattern. This pattern did not, however, mean that the *piṇḍapitṛyajña* depended on the new moon sacrifice as an adjunct to it. Rather the relationship was the other way round. It was the new moon sacrifice that depended on the cult of the ancestors.[59] From this point of view the question whether the *piṇḍapitṛyajña* precedes or follows the new moon sacrifice is irrelevant, as in fact it is according to the *anaṅga* thesis of Jaimini. We have seen that both are possible. The question could only arise when the worship of the fathers and the new moon sacrifice were split into two separate sacrifices.

More important, however, is that the original unitary pattern did not concern the single new moon sacrifice. The pattern included the full moon sacrifice as well. It was the concern with the ancestors and their worship that permeated and held together the full and new moon sacrifices, or in other words the two phases of sacrifice, violence and destruction as opposed to restoration and renewal. In the final analysis it was the ancestors who presided over the crisis that was the sacrifice.

12. Ritualism fundamentally changed all that. Although it remained in all its labyrinthine detail indebted to the archaic world of sacrifice, the concept of sacrifice was turned into an abstract system of ritual. What the ancient ritualists still knew as the *yajñāraṇya*, 'the wilderness of sacrifice',[60] was broken under the plough of rational analytic thought and reconstructed as an artificial maze, where only the learned expert can find his way. But there was more to it then analytic system building *per se*. It aimed at the exact opposite of the archaic sacrifice. Where the uncertain confrontation with death was heightened and acted out in the sacrificial arena in periodic, never definitive

contests for the goods of life ritualism propounded the certainty of its static system of rules, the very image of transcendence. Out of the violence and ambivalence of sacrifice - as pictured in the mythic Vṛtra fight - the ritualists had won a vision that held out to man the promise of a transcendent world beyond life and death.

As the price of sacrifice had been crisis and uncertainty, so the absolute certainty of ritualism had to be paid for by the divorce from mundane reality. Geared to transcendence, ritualism was as a matter of principle out of joint with the mundane world. Therefore, it could not and did not take the place of the sacrificial festivals in the life of society. Instead it posited a new understanding of sacrifice that did not address society but the individual Self. It did not seek to control the unstable relationship with the unseen but always present powers of gods and fathers. On the contrary, it was meant to break up this uneasy and threatening intimacy, so as to make room for transcendence. The ritualists understood sacrifice to be the structuring of the Self as the fulcrum of transcendence. In the language of the Mīmāṁsā theoreticians sacrifice is *puruṣārtha*; it has man himself as its purpose and, conversely, man's purpose is sacrifice.[61] In the last resort ritualism called upon man to realize through sacrifice the integration of the world and transcendence in the ultimate arena of the inner Self.

Notes

1. Taittirīya Brāhmaṇa (TB) 1.3.10.1. Cf. Taittirīya Saṁhitā (TS) 2.5.3.6; Aitareya Brāhmaṇa (AB) 3.15.1.
2. TB *loc. cit.*, *pitṛbhya eva yajñaṁ niṣkrīya yajamānaḥ pratanute*.
3. Cf. Manu 3.284, where the fathers are equated with three groups of gods, the Vasus, the Rudras, and the Ādityas.
4. Śatapatha Brāhmaṇa (ŚB) 3.6.2.26.
5. W. Caland - V. Henry, *L'Agniṣṭoma*. Paris: Leroux, 1906-7, pp. 219 f. (§ 147e).
6. ŚB 12.6.1.33.
7. *Ibid.*, 12.6.1.1. Apart from their intimate connection with Soma the fathers are equally connected with Agni, the (sacrificial) fire, one of whose epithets is precisely *nārāśaṁsa* (A. Hillebrandt, *Vedische Mythologie*, vol. I, pp. 108-18; also P. Horsch, next note).
8. On *nārāśaṁsa*, see P. Horsch, *Die vedische Gāthā- und Śloka-Literatur*. Bern, 1966, pp. 11, 411-16.
9. Manu 3.231.
10. Thus the basic paradigm of sacrifice (*iṣṭi*) contains the invocation of the ancestors (*pravara*), immediately preceding the main part of the sacrifice (the burnt oblations).
11. Manu 3.203; P.V. Kane, *History of Dharmaśāstra*. vol. IV, p. 483.

12. Cf. J.C. Heesterman, *The Broken World of Sacrifice*. Chicago: University of Chicago Press, 1993, pp. 11-14, 17.
13. Manu 3.213.
14. On sacrifice as enacting the existential riddle, cf. Heesterman, *op. cit.*, pp. 2, 17, 74 f.
15. Mahābhārata 12.8.34-37.
16. In this way the Mīmāṁsā also understands *dharma*; that is, as 'the purpose (*artha*) defined by the transcendent (Vedic) prescripts' (Pūrva Mīmāṁsā Sūtra 1.1.2, *codanālakṣaṇo 'rtho dharmaḥ*).
17. Cf. Heesterman, 'The Ritualists' Problem', in: S.D. Joshi (ed.), *Amṛtadhārā* (R.N. Dandekar Fel. Vol.). Delhi: Ajanta, 1984, pp. 167-79; Same, *op. cit.*, ch. 2.
18. Heesterman, *Sacrifice*, p. 17.
19. ŚB 13.8.1.6.
20. *Ibid.*, 2.4.1.24, 6.1.42.
21. Cf. Kane, *op. cit.*, vol. III, 735; vol. IV, 510; L. Rocher 'Inheritance and śrāddha', in: A.W. van den Hoek, *et al.* (eds), *Ritual, State and History in South Asia*. Leiden: Brill, 1992, pp. 637-49.
22. Caland cogently argued for the three *piṇḍas* and the named immediate fathers to be a later development, the original addressees being the manes in general (W. Caland, *Altindischer Ahnenkult*. Leiden: Brill, 1893, pp. 153, 175.
23. Caland - Henry, *Agniṣṭoma* § 231, p. 350.
24. Cf. Heesterman, *Sacrifice*, ch. 7, esp. p. 208 f.
25. Baudhayana Śrauta Sūtra (BŚS) 2.11: 52.4. Cf. H. Krick, *Das Ritual der Feuergründung*, Vienna: Verlag der Österreichischen Akademie der Wissenschaften, 1979, p. 81 n. 210; Heesterman, *Sacrifice*, p. 203. Though explicitly called the *upavasatha* (BŚS 2.8:45.6;2.11:52.7), the *gopitṛyajña* is not the actual eve of the *agnyādheya* (which on its eve has yet another cow sacrifice not specifically destined for the fathers). Possibly it is an archaic form of setting up the fire as may expected, at a new settlement or, otherwise, a periodic feast of renewal, that is, a social occasion in which, as may be expected, the fathers had to be prominently present. The standardized *agnyādheya* by contrast does not feature a specific worship of the ancestors. It does involve, though, a (reduced) meal offered to the officiating brahmins, at its beginning as well as at its end (resp. the rice mess or *brahmaudana* and the *caru* for Aditi).
26. Caland, *Ahnenkult*, pp. 60-5, 181-5. Caland is concerned with the *daiva* or *vaiśvadeva* as such, not with the offering in the fire (as opposed to the exposure offering of lumps of food). The important point is that the *Devāḥ* or *Viśve Devāḥ* to whom the burnt oblations are offered are no other than the divine fathers, the *devāḥ pitaraḥ*. How far their being 'divine' required offerings in the fire instead of by exposure on the ground, is not discussed by Caland. Rather they seem to be the manes in general who may also have been the original addressees of the exposed lumps of food. The reason for introducing the burnt oblations would seem to be the (re)moulding of the ancestor cult as an *iṣṭi* according to the standardized paradigm centred on the fire and the burnt oblation. No *iṣṭi* without burnt oblation. Interestingly Caland also argued for the

three *piṇḍas* offered to the three 'human' fathers (father, grandfather, great grandfather) as being an addition. Both arguments point at a thorough going remoulding rather than at fortuitous additions.

27. Āśvalāyana Gṛhya Sūtra 4.7.21-24. Cf. Manu 3.212.
28. It would seem that here we touch upon the origin of the extraordinary prominence of the brahmin; their connection with the world of the dead. Thus, while Manu calls the manes 'the previous gods' (*pūrvadevatāḥ*, 3.192), the brahmin guests at the *śrāddha* are similarly said to be the ancient divinities of the *śrāddha* (*purātanāḥ ... śrāddhadevatāḥ*, 3.213). As the representatives of the realm of the dead, impersonating the ancestors, they hold the enigma of life and death - that is, the *bráhman*.
29. Hiraṇyakeśi Gṛhya Sūtra 2.10.3; Manu 3.139.
30. The *Sākamedha* concludes the last *caturmāsa* or four-month period of the year. This period is generally connected with the cult of the manes who are honoured in the domestic *aṣṭakā* celebrations, on the eighth of each waning half of these months. The generalized *śrāddha* appears to be derived from the *aṣṭakās* (on which see Caland, *Ahnenkult*, pp. 166-72).
31. TB 1.3.10.6; cf. Āpastamba Śrauta Sūtra (ĀpŚS) 1.9.17.
32. ĀpŚS 8.15.23-16.3. Cf. ŚB 2.6.1.33.
33. For a point by point comparison of the domestic *śrāddha* and its *śrauta* equivalent, the *piṇḍapitṛyajña*, see Caland, *Ahnenkult*, pp. 154-6. To Caland (over more than a hundred years ago) their similarity suggested that the latter is the origin of the *śrāddha* which would have been a simplified version of the *piṇḍapitṛyajña* adapted to the *gṛhya* ceremonial (p. 156). It seems odd, though, that the truncated *śrauta* ritual, which lacks the meal and does not quite make clear what is to be done with the substantial amount of food remaining after the offerings, should be prior to and the source of the fuller *śrāddha*, which includes the meal in which the sacrificial food is consumed by the participants. The relationship of the two versions rather seems to be the other way round. The full pattern, preserved in the *gṛhya* version, killing (grinding grain or pressing the soma plant are also viewed as 'killing'), oblation, and a meal, was, for the reasons discussed, deprived of its logical conclusion, the sacrificial meal, in the *śrauta* ritual. Caland was, of course alert to the fact that the *śrāddha* prominently featured a meal offered to brahmin guests as representatives of the ancestors, but viewed this as no more than a minor variation of the meritorious feeding of brahmins which is usually part of domestic ceremonies (p. 157). This, however, is the critical point; the meal is indeed an integral part of the *gṛhya* ritual, but is purposely cancelled in the *śrauta* ritual. The reason for assigning priority to the *śrauta* version seems to be the prestige of the highly detailed and systematical elaboration of the *śrauta* ritual in general.
34. *na jīvantam atidadāti* (*-dadyāt*); ĀpŚ 1.9.8, HŚS 2.7:251 f., Bharadvāja Śrauta Sūtra 1.8.11, BŚS 24.32:217.9-12, Mānava Śrauta Sūtra 1.1.2.21, Vārāha Śrauta Sūtra 1.2.3.21, Kātyāyana Śrauta Sūtra (KŚS) 4.1.27, Āśvalāyana Śrauta Sūtra (ĀsvŚS) 2.7, Śāṅkhāyana Śrauta Sūtra 4.4.7.
35. KŚS 4.1.29-31.

36. Pūrva Mīmāṁsā Sūtra 4.4.19-21.
37. ĀsvŚS 2.6.17. Āśvalāyana's discussion refers only to the *piṇḍadāna*, not to the ritual as a whole. It does show, though, that the 'fundamentalist' view was a long-standing one, as was its rejection.
38. *Ibid.*, 2.6.20-23.
39. Āpastamba's paribhāṣā all but literally agrees with Jaimini's view (ĀpŚS 24.2.36-38).
40. KŚS 4.1.27.
41. Jaimini only mentions its own separate time (*svakālatva*), equal status in the enumeration of sacrifices (*tulyavacca prasaṁkhyānāt*) and occurrence in cases where the new moon sacrifice) is prohibited (*pratiṣiddhe ca darśanāt*).
42. Confusingly a final sūtra is added stating that the *piṇḍapitṛyajña* is also performed, that is, in the domestic *śrāddha* form by the domestic sacrificer who has not established the fires for the *śrauta* ritual (*anāhitāgnī*) and consequently cannot perform the *śrauta* new moon sacrifice. This clearly argues against the *aṅga* view. Most probably this sūtra is spurious, taken from the commentary or the paddhati (see A. Weber's edition, pp. 296, l.1 and 299, l.13; and Weber's note).
43. KŚS 4.2.1 *śvo noditetyadṛṣṭe vā.*
44. *Ibid.*, 4.1.1 *aparāhṇe piṇḍapitṛyajñaścandrādarśane 'māvāsyāyām.*
45. *jīvapitṛkasya pitṛyajnābhāvaḥ; piṇḍapitṛyajño darśāṅgam iti karkādayaḥ sarve bhāṣyakārāḥ, neti saṁpradāyaḥ; svakālatvād anaṅgaṁ syād iti Jaiminiḥ; anāhitāgner apyayaṁ piṇḍapitṛyajño bhavati.* (Weber's edition of KŚS, p. 299, ll. 11-13).
46. Both statements - option of newmoon's eve or the day itself for the *darśa* sacrifice, no options for the *piṇḍapitṛyajña* (only on new moon day) - is common to the *śrautasūtras*.
47. Notwithstanding their adherence to the general rule - *piṇḍapitṛyajña* on the new moon day - the sūtras of the Taittirīyakas clearly have the ancestor sacrifice on the eve. Hence, it would seem, Āpastamba's paribhāṣa ruling that the new moon celebration can fall on either the new moon day itself or on its eve (ĀpŚS 24.2.24f.). Baudhāyana's Karmānta specifies that the new moon day starts at noon on the previous day (in contradistinction to the full moon day which is said to start at midnight); any way, both celebrations are said to comprise two days (BŚS 24.20: 204.6-13). Interestingly, Kātyāyana also says that the new moon celebration comprises two days - is *asadya*, not on one single day - and so would, in principle have room for the *piṇḍapitṛyajña* on the previous day (KŚS 4.2.44, in contradistinction to the full moon celebration which either comprises both days or only the full moon day itself, 2.1.16). His followers prefer, however, to reserve the eve for the preparatory activities of the new moon celebration (readying the fires, undertaking the fast and the like) and have the main part of the new moon sacrifice together with the *piṇḍapitṛyajña* on the new moon day (originally, though, both could also be performed on the previous day, this being the original purpose of the sūtra we discussed: *pūrvo vāṅgatvāt*).

48. Thus the commentary and the *paddhati* to Kātyāyana, see Weber's edition, pp. 297, l.2 and 299, l.11.
49. On the theme of conflict in sacrifice, cf. Heesterman, *Sacrifice*, pp. 2 f., 39-44.
50. TS 2.5.4.3; ĀpŚS 3.16.6-10; Śabara ad Pūrva Mīmāṁsā Sūtra 4.4.21 (explaining the occurrence of the *pitṛyajña* separate from the *darśa* sacrifice).
51. TS 2.5.4.2; ĀpŚS 3.16.5.
52. TS 2.5.3.6. Cf. AB 3.5.1; ŚB 1.6.4.1. Also TS 2.5.5.6, saying that the newmoon sacrifice is 'razor-edged' (*kṣurápavī*), the sacrificer either is successful or perishes; in other words, it is all or nothing.
53. TS 2.5.3.1 *Indram, Vṛtraṁ jaghvāṁsā, mṛdho 'bhi prāvepanta.*
54. *Ibid.*, 2.5.1.2 *taṁ bhūtānyabhyakrośan brahmahann iti.*
55. ŚB 1.6.3.17; cf. TS 2.4.12.5 f.
56. In this connection it is significant that when the *dīkṣita* in anticipation of the Soma sacrifice comes to the fire hut, he is in danger of being overwhelmed by Agni and Soma. He must then redeem himself by offering them a he-goat (which later, on the eve of the actual soma feast, will be immolated), see Maitrāyaṇī Saṁhitā 3.7.8:87.9; cf. Kāṭhaka Saṁhitā 21.7:97.12-14, TS 6.1.11.6, ŚB 3.3.4.21. (further J.C. Heesterman, 'La réception du roi Soma', in: A.M. Blondeau and K. Schipper, *Essais sur le rituel*, vol. III (Bibliothèque de l'École des Hautes Études, Sciences Sociales), forthcoming.
57. Ṛgveda Saṁhitā 6.59.1b *hatā́so vāṁ pitáro devā́śatrava, índrāgnī, jī́vatho yuvā́ṁ*. The *vāṁ* - 'of you both' - in *hatāso vāṁ pitaraḥ*, might also be understood as 'slain *by* you'.
58. ŚB 2.6.1.1.
59. The fact that the *pitṛyajña* can occur even then when the new moon sacrifice does not take place, as also the rules which fix a definite day and time for it (afternoon of the new moon day), while the *darśa* sacrifice may be either on that day or on its eve (so that the *darśa* sacrifice can be performed either before or after the *pitṛyajña*, as is indeed apparent from Kātyāyana's commentator), suggest that the sacrificial worship of the ancestors may well have been the primary celebration.
60. For this expression see J.C. Heesterman, 'Opferwildnis und Ritual-Ordnung', in: G. Oberhammer (Hrsg.), *Epiphanie des Heils*. Wien: Institut für Indologie der Universität Wien, 1982, pp. 13-25, esp. p. 19.
61. Cf. J.C. Heesterman, 'Puruṣārtha', in: F.X. D'Sa - R. Mesquita (eds), *Hermeneutics of Encounter*. Vienna, 1994, pp. 137-51.

XV

SŪTRA AND *BHĀṢYASŪTRA* IN BHARTṚHARI'S MAHĀBHĀṢYA DĪPIKĀ: ON THE THEORY AND PRACTICE OF A SCIENTIFIC AND PHILOSOPHICAL GENRE

Jan E.M. Houben[*]

1. INTRODUCTION

1.1 The earliest traceable roots of sciences in South Asia - if we take 'science' in a broad sense (cf. Staal, 1993, pp. 6-7) - lie in the works of the Vedic ritualists and the Sanskrit grammarians. Especially the grammarians attained widely acclaimed success in the development of a science, and it was their system that started to play a paradigmatic role in practically all areas of the South Asian scientific and philosophic literary production. All this is well known and widely accepted since Staal compared Pāṇini's method and his influence in the Sanskit tradition, with the method of Euclid and his influence in the Western tradition (Staal, 1965). Much of Staal's remarkable research career has been devoted to the investigation of the precise conditions and factors leading to the development of a science in one area of scholarly activity and in one part of the world, but not in other areas of scholarly activity or in other parts of the world (e.g. Staal, 1982, 1993, 1995).

[*] While working on this article I profited from discussions with Dr. Herman Tieken (Leiden). I am also indebted to Dr. Harunaga Isaacson (Oxford), Dr. Konrad Klaus (Bochum), Professor T. Vetter (Leiden) and Professor A. Parpola (Helsinki) for comments on an early version of this article, and to Mrs. Drs. Hanna 't Hart and Drs. Dick van der Meij for further suggestions for improvement. Thanks are due to the International Institute for Asian Studies (IIAS) and the Netherlands Organization for Scientific Research (NWO) for financial support for the research on which the article is based.

1.2 One of the preconditions for the development of science seems to be the development of systematic and reliable ways of representing knowledge. In order to be able to represent knowledge and in order to work with represented knowledge, a scholarly community may or may not develop an artificial language - if it does this seems to augur well for the emerging science (cf. Staal, 1995) - but it always has to deal with texts which occupy different positions (varying from central to peripheral) in a given science. While the fields of reality which are the subject of the sciences differ (even when sometimes they overlap considerably), they all have to deal with texts in one way or the other. It is not surprising that the ancient scholars who were working in the various fields of the emerging South Asian sciences and had to deal with texts in a systematic and responsible way developed some terminology and notions of their own in order to distinguish different styles of composition as well as different types and functions of texts.

1.3 Of major importance in this context is the notion of a text consisting of a series of very brief 'aphoristic' expressions which together give a complete treatment of the field. The Sanskrit name nowadays generally accepted for such a compact scientific text is *sūtra* and the stock example of a successful *sūtra*-text is Pāṇini's grammar. Like most other *sūtra*-texts, Pāṇini's grammar needs a commentary because it would otherwise remain incomprehensible. In the course of time, all self-respecting sciences, disciplines and philosophical-religious systems in the South Asian traditions (especially as far as their Sanskrit literature is concerned), created basic *sūtra*-texts and accompanying commentaries. It is usually this *sūtra*-text which, at least in name, occupies a central place, while the commentaries and sub-commentaries, being more peripheral, derive their authority to a great extent from their claim to be faithful to the statements and intentions of the *sūtra*-author. Western works which in their brevity and organization are somewhat similar to the Sanskrit *sūtra*-genre are Wittgenstein's *Tractatus Logico-Philosophicus* and Spinoza's *Ethica more geometrico demonstrata*.

The term *sūtra* (the literal meaning of which is 'thread') is not only used to refer to a scientific text (e.g. Pāṇini's grammar or Patañjali's Yoga Sūtra- as a whole, it also refers to each of the individual expressions which constitute this text.[1] When our points of departure are Pāṇini's grammar and modern (generative) grammar, it is therefore natural that the sūtras which constitute Pāṇini's grammar are associated and equated with grammatical 'rules'.[2]

1.4 Important contributions to our understanding of the literary genre of the Sūtra have been provided by Louis Renou (Renou, 1947; Renou & Filliozat, 1947, pp. 59-60, 301; Renou, 1956, pp. 53-61; 1957; 1963). There are some indications which point to Vedic (so-called Śrauta) ritual, and grammar as the areas where the Sūtra-style first developed (Renou, 1963, pp. 168, 175-81). Renou distinguishes two basic types of Sūtras: an older one, which he calls type A, and a younger one, which he calls type B (Renou, 1963, pp. 181-98). Type A consists first of all of the Śrauta Sūtras, which describe in detail the 'solemn' Vedic ritual of which the earlier Brāhmaṇa-texts provide speculative interpretations. Next, also the Gṛhya Sūtras (dealing with 'domestic' rituals) and Dharma Sūtras (dealing with rules of conduct and legal matters) would come in this category. This type can be considered to be descriptive-normative (Renou, 1963, pp. 182-3), in the sense that these Sūtras generally deal with their subject of norms for (ritual or social or linguistic) behaviour in a descriptive way.

The most important representatives of the newer type B are the Darśana Sūtras, i.e. the Sūtras of the philosophical schools, which are, according to Renou's examples, primarily the Sūtras of the six 'orthodox' systems: Vedānta, Mīmāṁsā, Sāṁkhya, Yoga, Nyāya, and Vaiśeṣika. At one place, Renou says that the type B Sūtras originate 'perhaps' from Kātyāyana's Vārttikas - usually not considered to be sūtras in the strict sense of the word - and becomes fixed in the Darśana Sūtras (Renou, 1963, pp. 181, also 169). At another place, Renou speaks of Kātyāyana's work as a full representative of type B Sūtras (1963, p. 191). Unlike those of type A, these Sūtras usually have an argumentative, dialectical structure. They also distinguish themselves from type A in some striking stylistic features (avoidance of the use of finite verbs, frequent use of abstracta and verbal nouns, and of compounds with oblique case relations between their members). A class apart, according to Renou (1963, pp. 198-9), are the grammatical Sūtras and especially Pāṇini's grammar, in which formalization is at its maximum. For the purpose of the present exposition, I would like to associate Pāṇini's work with the type A Sūtras as a special extension, taking into account its age and its descriptive-normative character as a grammar.[3]

Although type-A Sūtras take on type-B features and vice versa, there seem to be two distinguishable families here. Renou's categorization, proceeding as it does partly by enumeration and partly by characterization, is not without problems, but in the absence of a better elaborated proposal I take it here as my starting point. However, we will see that, even if the available material does support a non-absolute distinction between the type-A and type-B Sūtras, the latter are conceptually even less well-established in the tradition than one would expect. At any case, I will make the following distinctions:

Type-A Sūtras:	descriptive-normative, brief.
Extended type-A:	type-A plus Pāṇini's grammar, extremely brief.
Type-B Sūtras:	Darśana Sūtras; allow for argumentation.

1.5 Renou gives a few, sometimes quite diverging, suggestions concerning the original meaning of the term *sūtra*. In the 1941-1942 issue of the *Journal Asiatique* (published 1947), he suggested, without giving a detailed explanation, that the term stems from the terminology of weavers (1947, p. 113 note 1). In *l'Inde Classique* (1947) this proposal is ignored and Renou briefly explains the meaning of sūtra as "(guiding) thread', hence 'rule'."[4] In 1956 we find for the first time a more elaborated proposal when Renou explains that the earliest Sūtra-texts, the Śrauta Sūtras, originated as brief comments to the Brāhmaṇa-texts. These comments described, step by step, the ritual. Next, these brief comments were placed in a continuous order, like the pearls on a string, hence the word *sūtra* (1956, pp. 53-4). In 1963, he first poses the question whether the employment of the term for the individual aphorism precedes that for the text as a whole, or whether it is the other way round. Referring to an older discussion on this problem,[5] he initially says to prefer the former possibility. Next, he comes with a more subtle proposal: *sūtra* was originally the term for the implicit 'thread' (which he calls a 'leitmotiv' at one place) formed by the formal presuppositions on which the aphorism of a chapter are constructed; next it means these aphorisms themselves; next the coherent treatise, the text (Renou, 1963, pp. 166, 174).

1.6 Renou's survey and his suggestions concerning the origin and development of the Sūtra-literature not only contribute to a better understanding, they also lead to important further questions. The 'aphoristic' character of the Sūtras (i.e. their utter brevity and systematic organization) is least obvious for the oldest texts classified as such. That their brevity is not a fundamental, but rather a resultant feature not applicable to the oldest Sūtras is, in fact, admitted by Renou (1963, p. 166). But can it not be argued that the same applies to the systematicity which he claims to be an essential characteristic of the Sūtra? After all, what seem to be the oldest or at least very old specimens of this genre, viz. the Baudhāyana and the Vādhūla Śrauta Sūtra, contain a number of narrative passages which interfere strongly with the systematic, descriptive approach one would expect in a Śrauta Sūtra.[6] Apart from this, it will be difficult to decide just when a text has more systematicity than other texts; usually, some degree of systematicity can be perceived in almost any text, including even anthologies and collections of narratives. What makes the earliest Sūtras Sūtras?

Renou tried to assimilate also the Jaina and Buddhist Suttas/Sūtras, which often contain extensive narrative material, to the Brahmanical Sūtras, but recognized that they are rather 'diffuse' in character - which would amount to admitting that they lack the supposed typical Sūtra-character. For a certain group of ancient Jaina Suttas, nevertheless, Colette Caillat (1994) showed that they do have several properties which make them similar to Brahmanical

Śrauta and Dharma Sūtras (in other words, to Renou's type A Sūtras).

The term Sūtra in the classical Brahmanical sense of the word suits the genre of Buddhist texts which are called Suttas so badly that an alternative explanation has been proposed for this Pāli term: it would correspond to Sanskrit *sūkta* rather than to *sūtra*. The designation Sūtra for later Mahāyāna texts would be a wrong translation of the Pāli *sutta*. Nevertheless, the relatively ancient Pātimokka Sutta would contain early layers which are very close to the Brahmanical Sūtra (again, Renou's type A), which, according to von Hinüber (1994, pp. 131-2 and n. 27), is one of the reasons to be hesitant to accept the *sūkta-sutta-sūtra* hypothesis for the Buddhist Suttas and Sūtras.[7]

Another point where Renou's views of the nature and development of the Sūtra as a literary genre in ancient South Asian scientific literature are problematic, is his suggestion that a Sūtra text would normally be free from versified passages (Renou, 1963, pp. 167). It has been pointed out that one of the early Sūtra texts, viz. the Mīmāṁsā Sūtra, contains a considerable number of metrical or almost metrical passages.[8] Another example which speaks against Renou's suggestion is the Yuktidīpikā (YD), commenting on the Sāṃkhyakārikā. Kārikās in the Āryā-metre, and parts of varying length of these kārikās are discussed and treated as sūtras. In referring to the kārikās or their parts the word *sūtra* is occasionally used (YD, pp. 9, 10). There is even an explicit justification of the status of the kārikā-text as a Sūtra (YD p. 2). In a study of Jaina ideas concerning the 'Perfect Sūtra' Balbir has collected statements about the 'virtues and flaws' of a Sūtra, and one of the flaws mentioned is the employment of the wrong metre (Balbir, 1987, p. 9 under B3) - which goes to show that metrical *sūtras* were by no means considered uncommon.

1.7 All this raises serious questions with regard to the understanding that authors in the first millennium AD and earlier had of the nature and characteristics of the Sūtra as a genre. While Renou has given a good survey of the available Sūtra-literature, his attempts to understand the concepts which guided their production remained only very preliminary. For instance, his suggestion to interpret the term *sūtra* as a 'guiding thread' or a 'leitmotiv' may be enlightening to a Western public which is familiar with the Greek legend of Theseus and Ariadne in the labyrinth and with modern theories of literature,[9] but it is very doubtful whether it evoked similar ideas to an ancient student or composer of Sanskrit texts. What is very much needed, as von Hinüber rightly observed in a recent paper,[10] is a study of the use of the word *sūtra* in the Sanskrit as well as the Prākrit-Pāli traditions.

The present article intends to make a small contribution to such a study which eventually should comprise important representative basic texts and commentaries of the different traditions and schools of thought. It is only on such a basis that we can hope to improve our understanding of the origins and development of the theory and practice[11] of one of the most significant literary genres in the South Asian intellectual traditions.[12]

2. THE USE OF SŪTRA AND RELATED TERMS IN THE MAHĀBHĀṢYA DĪPIKĀ: GENERAL OBSERVATIONS

2.1 A valuable testimony on these matters is available to us - even though only in one incomplete manuscript - in the form of Bhartṛhari's Mahābhāṣya Dīpikā (ca. fifth century AD.), a sub-commentary on Patañjali's commentary (Mahābhāṣya) on Pāṇini's grammar. Also Bhartṛhari's predecessors in the Pāṇinian tradition, especially Patañjali, Kātyāyana and Pāṇini himself, provide important material concerning the use of the term *sūtra* (cf. below, section 2.2). But the author Bhartṛhari is of special interest because, as a fifth century author, he was contemporaneous with or immediately followed by authors who gave to the philosophical and scientific systems the classical shape with which modern students of early South Asian thought are most familiar. Apart from the Mahābhāṣya Dīpikā (MBhD) Bhartṛhari wrote at least one other work, viz. the Vākyapadīya, a work which has been much better conserved than the MBhD.

In two recent studies, Bronkhorst (1990, 1991) has shown that pre-classical authors like Bhartṛhari had conceptions of the basic categories of scientific texts which were not quite as we would have expected. Bronkhorst concentrated on two types of commentaries, the Bhāṣyas and the Vārttikas, and distinguished between what he called the Bhāṣya-style (1991) and the Vārttika-style (1990 and 1991). The Bhāṣya-style, according to Bronkhorst, consists of a 'tendency ... to swallow up the *sūtras*, or verses, on which they comment, so that together they come to look like one single work' (Bronkhorst, 1991, p. 218). The Vārttika-style is the 'style in which ordinary prose and short nominal phrases alternate' (idem). Both the Bhāṣya and the Vārttika comment on or deal with a segmented 'basic text' (*mūla*); the segments commented upon are usually called *sūtras*, propositions which may or may not be in verse form (idem).

Moreover, in the Mahābhāṣya, in which the main distinction nowadays accepted is the one between Kātyāyana's Vārttikas and Patañjali's comments (*bhāṣyas*) either on the Vārttikas or directly on Pāṇini's sūtra, Bhartṛhari distinguished at least four major components (Bronkhorst, 1990, p. 138): (a) Pāṇini's sūtras; (b) the *vākyas*, or nominal phrases which often correspond with what we tend to regard as Kātyāyana's Vārttikas; (c) *vārttikas*, corresponding to portions of the Bhāṣya, often those which directly discuss the *vākyas*; (d) remaining Bhāṣya portions, attributed to a different author.

In the remainder of this paper I will briefly discuss some general features of the use of the term *sūtra* in the Mahābhāṣya Dīpikā (MBhD) (section 2.2-), as well as one peculiar compound ending in *sūtra*, viz. *bhāṣya-sūtra* (section 3). An exhaustive treatment of the topic is here not attempted.[13] In a very preliminary way, section 4 will place the results found for the MBhD in a larger historical context of literary practice and conceptual reflection. In section 5, finally, I will formulate some conclusions on the basis of my findings and hint at further implications and questions.

2.2 In the MBhD the term *sūtra* very frequently refers to the individual statements which make up the Sūtra text of Pāṇini, as in 'Now he (Patañjali) introduces the view of another commentator, according to which, if it is adopted, there is no flaw in the sūtra (i.e. Pāṇini's statement under discussion).'[14]

Bhartṛhari shows to be clearly aware of the complementary usage of the term for Pāṇini's grammar as a whole when he says: '*sūtra* (in MBh 1:11.15): its meaning is "collection of *sūtras*" with regard to the Aṣṭādhyāyī' (MBhD 1:32.8, *sūtram / sūtrasamudāyo 'ṣṭādhyāyyām asyārthaḥ*).[15] The term is also used to refer to other texts, as in MBhD 1:11.15, *āśvalāyana-sūtre* and 1:11.18 *āpastambasūtre*, which are apparently references to the well known Śrauta Sūtras of Āśvalāyana and Āpastamba. The term *bahvṛca-sūtrabhāṣya* in MBhD 1:11.15 must refer to a commentary on a Ṛgvedic Śrauta Sūtra, probably the Āśvalāyana Śrauta Sūtra, as pointed out by Bronkhorst MBhD 1:122 note 36. In MBhD 7:12.16 we read *dharmasūtra-kāra*, a reference to the authors of the Dharma Sūtras. In addition, we come across expressions like *pāṇinīyasūtra* (MBhD 1:34.25), *pāṇinisūtra* (? MBhD 6a:9.11-12[16]), *smṛtisūtra* (MBhD 1:34.24), all in the context of grammar. All these references concern Renou's type A (basically the Śrauta, Gṛhya and Dharma Sūtras) extended with grammar.

In this respect, Bhartṛhari is not different from his predecessor Patañjali, whose references also seem to be confined to texts in the field of Vedic ritual[17] and grammar. For Patañjali, the main grammatical Sūtra is of course Pāṇini's Sūtra. Hence, also the referents of Patañjali's *sūtra* coincide with the extended type A Sūtras, except for a few cases of which one - Kātyāyana's Vārttikas (a type B Sūtra-text according to Renou) regarded as sūtras - will be discussed more elaborately below.[18] While Pāṇini provides the stock-example of a successful Sūtra-text, he also provides us with what seem to be the earliest roughly datable occurrences of the term *sūtra* as a literary term, viz. those in P 4.2.110-111, where mention is made of two (classes of) texts, the Bhikṣu Sūtra and the Naṭa Sūtra. Although these texts are not extant, their titles suggest that they deal with a normative description of areas of behaviour (religious begging and dance), and belong to the descriptive-normative type, hence to Renou's type A.

2.3 Are there no references in Bhartṛhari's work to Sūtras of Renou's type B? Bhartṛhari is among the earlier authors referring to various philosophical systems such as Mīmāṁsā, Vaiśeṣika and Sāṁkhya, and he is one of the first to employ the word *darśana* 'philosophical view', 'philosophical system' in this context (Halbfass, 1988, pp. 268-9). At one occasion he even quotes a piece of text which we know from the first sūtra of the Vaiśeṣika-system.[19] Nevertheless, it is not quoted as a sūtra or part of it, but as a statement on a par with the opening sentence of Patañjali's Mahābhāṣya, and this sentence is neither by Bhartṛhari, nor by later authors in the grammarian's tradition, regarded as a sūtra in the strict sense of the word.[20]

Bhartṛhari also refers to Mīmāṁsā and quotes from their sources, but there is no reference to or quotation from the Mīmāṁsā Sūtra. That Bhartṛhari was familiar with Mīmāṁsā sources no longer available to us, and that the Śābara Bhāṣya - nowadays the earliest extant complete commentary on the Mīmāṁsā Sūtra - was not necessarily known or considered exclusively authoritative by Bhartṛhari was already pointed out by Bronkhorst (1989, p. 114). However, the Mīmāṁsā Sūtra should be expected to be even more authoritative than Śabara's Bhāṣya. One would therefore first of all expect references to this work. Instead, Bhartṛhari quotes, from one or more unknown sources, verses and metrical lines which correspond to ideas expressed in different sūtras of Mīmāṁsā.[21] From a historical perspective we may expect Bhartṛhari to have been familiar with the text which is now known as the (i.e. Jaimini's) (Pūrva) Mīmāṁsā Sūtra,[22] though we find no references to this text even if we do find references to other Mīmāṁsā-works. This leaves one wondering what authority was attributed to the Mīmāṁsā Sūtra.[23]

Another work can be mentioned which, at least in later times, is regarded as a Sūtra-text, and which from a chronological perspective probably formed part of the literary field in which Bhartṛhari was working, namely the Jaina Tattvārthādhigama Sūtra. Elsewhere (Houben, 1994), I argued that Bhartṛhari was familiar with Jaina doctrines of which we find the earliest exposition in Sanskrit in this text. However, we do not find direct quotations from or explicit references to this text.

2.4 The way Pāṇini's statements, referred to as sūtras, are critically discussed by Bhartṛhari, implicitly shows what criteria should be met by a sūtra to be a good sūtra. Sometimes, the expectations are more explicit, e.g. in MBhD 4:9.13 *sopaskāratvāt sūtrasya vākyaśeṣo 'dhyāhriyate* 'because a *sūtra* needs to be supplemented we supply a remaining part of the sentence' (in a similar phrase also in MBhD 7:1.8[24]). This statement points to some of the aspects of the Sūtra-genre as described by Renou, namely that the *sūtra* should be extremely concise, and needs additions (usually to be supplied by a commentator) to become a satisfactory sentence. From the discussions it becomes moreover clear that apart from economy of expression (brevity) also descriptive exactness, both for each single *sūtra* and for the entire text, are of the utmost importance.[25] These characteristics particularly suit the Sūtras of the extended type A (including grammar).

2.5 The term *sūtra,* however, is not the only term used to refer to statements of Pāṇini. In addition we find *lakṣaṇa* ('characterization', 'rule', viz. with regard to *lakṣya* 'object', 'target', i.e. the correct word-form to be described),[26] *śāstra* ('command', 'precept', 'rule')[27] and *yoga* ('application', 'rule').[28] Moreover, we find terms for subsets of grammatical rules, for instance *saṁjñā* referring to a statement which defines a term, *paribhāṣā* or 'metarule', *adhikāra* or 'heading', *vidhi* 'injunction', *pratiṣedha* 'prohibitive rule' (I noticed two occurrences of *pratiṣedhasūtra,* MBhD 4:35.5 and 24), *utsarga* 'general rule', *apavāda* 'exception'. In the use of these terms, Bhartṛhari mainly accords with the Mahābhāṣya and with later grammarians, as well as with the modern understanding of Pāṇinian grammar.[29]

A relatively elaborate discussion of the special properties of a grammatical rule we find in MBhD 3:13.10-14.1. The Sanskrit word used in this passage for 'grammatical rule' is *lakṣaṇa,* in accordance with the term used in the MBh. Palsule translates here *lakṣaṇa* with another Sanskrit term, *sūtra* (MBhD 3:36).[30] Commenting on *lakṣaṇaṁ nāṃa dhvanati bhramati muhūrtam api nāvatiṣṭhate,* Bhartṛhari explains that a grammatical rule (*lakṣaṇa*) such as *neṭi* (P 7.2.4) expresses itself indistinctly (*dhvanati*), in the sense that it is not specifically clear whether the prohibition to substitute Vṛddhi applies to the one or the other verb form. The rule is general and applies to several particular cases.[31]

While the requirement of general applicability favours a general, in some respects even 'indistinct' formulation, the requirement of descriptive exactness works in a different direction. When it is proposed to replace three elaborate sūtras (P 1.1.34-36), in which a number of nominal bases are explicitly enumerated together with specific semantic conditions, by a single compact sūtra in which only the first element is explicitly mentioned, the response is that this would lead to confusion (*saṅkara*).[32]

Another basic aspect of a sūtra, in addition to and closely connected with its brevity, general applicability and descriptive exactness, is its authoritativeness. As was rightly emphasized by Ojihara (1978, pp. 226-7) and by Staal (1993, p. 28), the authoritativeness of the sūtra does not mean that one is expected never to challenge its formulation. Rather on the contrary: the commentarial tradition is full of attempts to show defects, followed by attempts to do away with the objections. In some instances, especially in the case of authors before Patañjali, this led to generally accepted amendments to the grammar. To illustrate this aspect in Bhartṛhari's MBhD we may refer here to the discussion occasioned by Patañjali's statement that 'if one would say something that goes beyond the rule, this cannot be accepted' (*yo hy utsūtraṁ kathayen nādo gṛhyeta*, MBh 1.12.27).[33] What is at stake in this discussion is the acceptance or otherwise of a word as a correct word (*śabda*) rather than an incorrect word (*apaśabda*) on the authority of Pāṇini's sūtras or those of other grammarians: 'When someone says "this is a correct word" he will certainly be asked "How is that to be known by us?"' If [then that person] utters [the appropriate] sūtra from the traditional science, people agree and understand that his is no idle talk' (MBhD 1:34.23-25, tr. Bronkhorst).[34] The discussion ends with the statement that examples etc. (probably: examples and other commentarial matter), as well as the employment of correct words, are only derived from the sūtra the meaning of which is understood.[35]

3. THE TERM BHĀṢYASŪTRA

3.1 An unexpected and more problematic expression is the term *bhāṣyasūtra*. The most elaborate discussion of this term was the one by Ojihara (1978), but he was aware of only one of the two passages in which it is used.[36] Both passages are further briefly referred to by Bronkhorst (1990, p. 137). Joshi & Roodbergen, 1986, p. 169 refer to the first passage when translating the corresponding discussion in the MBh.

Bhartṛhari uses the term *bhāṣyasūtra* first when he comments on the MBh sentence *na cedānīm ācāryāḥ sūtrāṇi kṛtvā nivartayanti* 'Now, it is not the case that teachers, after having phrased *sūtras*, take them back' (MBh 1:12.9-10).[37] The sentence is part of a discussion in the context of a questioning of two Vārttika-statements, the first of which would have become superfluous in the light of the second, more comprehensive one (MBh 1:12.1 and 4; cf. Joshi & Roodbergen, 1986, pp. 165-9; Ojihara, 1978, pp. 219-22). The expectation of the objector is apparently that in view of the brevity which is required for sūtras, the first, less comprehensive statement should be suppressed. The answer is that once a teacher (here especially the author of the Vārttikas) has phrased a sūtra, he does not suppress it at a later time. What is remarkable, of course, is that the term sūtra and the expectations belonging to it are applied not to Pāṇini's sūtras (at least not primarily so), but to statements of the Vārttika-author.[38]

Bhartṛhari comments on this passage as follows[39]:

> *bhāṣyasūtreṣu gurulāghavasyānāśritatvāt*
> *lakṣaṇaprapañcayoś ca mūlasūtreṣv āśrayaṇād ihāpi lakṣaṇaprapañcābhyāṁ pravṛttiḥ* /
>
> Since [questions of] prolixity and brevity are not taken into consideration with regard to *bhāṣyasūtras*, and since both general rules and specific amplifications are resorted to in the basic *sūtras*, here too [the author of the *bhāṣyasūtras*] proceeds by general rule and specific amplification. (MBhD 1:32.27-33.1)[40]

There can be little doubt that Bhartṛhari's expression *bhāṣyasūtresu* explains *sūtrāṇi* in the MBh-passage, and refers to Kātyāyana's Vārttikas.[41] We see that Bhartṛhari considers 'general rules and specific amplifications' to be present in both the basic sūtras and in the *bhāṣyasūtras*. What is implicitly accepted in his remarks (and amply demonstrated in his discussions elsewhere) is that sūtras are expected to be very brief. However, with regard to the *bhāṣyasūtras* this does not apply to the same extent as it does with regard to the *mūlasūtras*.

By accepting that Bhartṛhari's *bhāṣyasūtra* refers to Kātyāyana's Vārttikas, however, the term is not yet sufficiently understood. As pointed out above (section 2.1), the term Vārttika in Bhartṛhari's MBhD does not correspond precisely with what we would presently regard as Vārttikas. Whereas Kātyāyana's Vārttikas are nowadays usually considered to be the brief nominal expressions which are further commented upon in the Mahābhāṣya,[42] Bhartṛhari's *vārttika* rather refers to a combination of both the brief statements in the nominal style and some of the accompanying prose elaborations. To these statements, as pointed out above (section 2.1), Bhartṛhari usually refers as *vākyas*.[43] The question therefore arises whether Bhartṛhari's *bhāṣyasūtra* is here equivalent to Bhartṛhari's *vārttika* or rather to his *vākya*. Because the brief nominal expressions (*vākyas*) are in style closest to the sūtras, it would seem natural to assume that Bhartṛhari's *bhāṣyasūtra* refers to only these, and not to the accompanying prose elaborations which Bhartṛhari also attributed to Kātyāyana.

It is to be noted, that by extending the term *sūtra* to include the *vākyas*, Bhartṛhari for the first time attributes the Sūtra-status to (at least part of) a work of Renou's type B, namely Kātyāyana's Vārttikas. However, Renou's understanding of the extent of this work, corresponding with Kielhorn's understanding of it, differs considerably from the more elaborate passages which Bhartṛhari apparently attributed to a 'Vārttika-author'.

3.2 The second time the term *bhāṣyasūtra* is used (twice in this passage), the discussion concerns P 1.1.34, which lists a series of words, *pūrva, para, avara* and others, which are pronominal under a certain semantic condition (they should not be proper names).[44] Next, in a Vārttika an objection is made to this rule: since *avara* etc. are already listed in the *gaṇapāṭha* - i.e. in the lists of lexical items (mostly nominal bases) - it is useless to mention these again in this rule.[45]

After having explained the intention of the Vārttika-author in formulating his statement (*vākya*),[46] Patañjali first asks a question about something which is apparently presupposed in the Vārttika: how can we know that the *gaṇapāṭha* has been established before Pāṇini formulated his sūtra? After a short reference to the two possibilities - either the sūtra is before the *gaṇapāṭha*, or the *gaṇapāṭha* is before the sūtra - Patañjali accepts the presupposition of the author of the Vārttika, namely that the *gaṇapāṭha* indeed precedes Pāṇini's sūtra. However, unlike the author of the Vārttika, Patañjali does find a reason to mention the words one by one (together with certain semantic conditions) both in the *gaṇapāṭha* and in Pāṇini's sūtra.

Now, at a certain point in his detailed discussion of Patañjali's Mahābhāṣya, Bhartṛhari emphasizes that in Pāṇini's sūtra there are not only cases where the sūtra is apparently formulated before the *gaṇapāṭha* enumerates several examples to which the sūtra applies, but also cases where the *gaṇapāṭha* apparently contains a pre-established list, and the sūtra, formulated later than this list, mentions the operation which applies to the nominal bases of this list. Next, Bhartṛhari further develops the point that in some cases a sūtra precedes the statement of the *gaṇa* or list of nominal bases to which the sūtra applies. Here, Bhartṛhari argues that, if a sūtra mentions some forms just by way of example, the list in the *gaṇapāṭha* is to be seen merely as a (later) amplification of this example (MBhD 6a:26.1-3). The list in the *gaṇapāṭha* thus has a status similar to that of the illustrations to Pāṇini's sūtras given in certain basic commentaries (*vṛtty-udāharaṇa-vat*).[47] Next, Bhartṛhari says:

> *na ca teṣu bhāṣyasūtreṣu gurulaghu prati yatnaḥ kriyate / tathā ca 'na cedānīṁ ācāryāḥ kṛtvā sūtrāṇi nivartayanti' iti / bhāṣyasūtrāṇi hi lakṣaṇaprapañcābhyāṁ nidarśanasamarthanaparāṇi /*
>
> And with regard to these *bhāṣyasūtras*, no effort concerning prolixity and brevity is being made. And thus [it has been said]: 'Now, it is not the case that teachers, after having phrased *sūtras*, take them back.' (MBh 1:12.9-10).[48] Indeed, the *bhāṣyasūtras* intend to illustrate and corroborate by means of general rule and specific amplification. (MBhD 6a:26.4-5)

Contextually, *bhāṣyasūtra* here evidently refers to expressions in the *gaṇapāṭha*, perhaps implicitly also to those in the *vṛttis*. Since the phrase *na cedānīm* ..., as we have seen (section 3.1), refers to the *vākyas* of the Vārttika-author and was also understood as such by Bhartṛhari, the fact that it is cited here associates the expressions in the *gaṇapāṭha* and *vṛttis* with those of the Vārttika-author.

A highly corrupt passage follows, which in the reconstruction of the editors of the recent Poona-edition (for this part: V.B. Bhagavat and S. Bhate) contains references to *gaṇasūtras* (the manuscript contains only the sequence ... *gasūtra* ... in a few lines which are erroneously repeated), and which would thus focus the attention again on the expressions in the *gaṇapāṭha* which were also the starting point of the discussion.

Further down, Bhartṛhari reverts to the same problem: if the sūtras are formulated in a later period after the *gaṇapāṭha* (MBhD 6a:26.12: *sūtrapāṭhas tv avarakālaḥ*), one may expect that the sūtra is as brief as possible (*laghu sūtraṁ kartavyam*), and one might find fault with the full enumeration of all nominal bases which were already enumerated in the *gaṇapāṭha* (MBhD 6a:26.12-13). However, if the *gaṇapāṭha* is formulated later than the sūtra, it may contain enumerations which overlap with those in the sūtras. Then, an author may just have collected in the *gaṇas* or 'lists' some elements from the *sūtrapāṭha* (i.e. the sūtras), or else skilled composers may have drawn up these lists for the sake of memorization: hence with regard to these elements enumerated in the list we need not consider the prolixity vs. brevity of the expressions.[49]

3.3 This last-mentioned passage (MBhD 6a:26.12 ff.) is important because it contrasts, in these very terms, *sūtras* with *gaṇas*, and the *sūtrapāṭha* with the *gaṇapāṭha*. In this respect, it is more in accordance with the usage of these terms elsewhere in the MBhD. Hence, we can infer that the application of the term *sūtra*, and specifically *bhāṣyasūtra* (perhaps also *gaṇasūtra* if we accept the reconstruction in the recent Poona-edition), to the lists in the *gaṇapāṭha* in the immediately preceding passage (MBhD 6a:26.4-8) is to be considered a matter of secondary usage.

Also the other occurrence of the term *bhāṣyasūtra* in Āhnika 1 of the MBhD (above, section 3.1) is to be regarded as an exceptional usage occasioned by the MBh-passage commented upon: Bhartṛhari's usual term for the Vārttika-statements to which *bhāṣyasūtra* is applied here, is *vākya*.

3.4 To sum up, it is found that in Bhartṛhari's MBhD only Sūtras of the extended type A (Śrauta and Dharma Sūtras, grammatical Sūtras) are referred to as Sūtras (and isolated statements in them as sūtras). In conceptual reflections on the sūtra, we find that mention is made of properties which best suit the descriptive-normative character of the extended type A. In other words, both conceptually and in terms of the examples cited, it is only the extended type A of Sūtra-texts, the type of descriptive-normative Sūtras, which is well-established.

We also found that, even if, on the basis of historical considerations, we have to assume that Bhartṛhari must have been familiar with at least some of the Darśana Sūtras which form the main example of type B, he never refers to these texts as Sūtras/sūtras. Thus, in spite of the conspicuous absence of any quotation or reference, one would expect Bhartṛhari to be familiar with the Sūtra-text of the (Pūrva) Mīmāṁsā-system. Bhartṛhari does refer to part of what we know as the first sūtra of the Vaiśeṣika-system, but he does not refer to it explicitly as a sūtra: he discusses it as being on a par with a statement of Patañjali - which is usually not considered as a sūtra. The Vaiśeṣika-quotation may or may not have been known to him as a sūtra, but he does not explicitly acknowledge its status as such.

A partly exception to this observation is formed by two passages in which the term *bhāṣyasūtra* is employed. In the first passage, this term is apparently used with reference to Kātyāyana's Vārttikas, a text which Renou considered to be either a precursor or one of the earliest examples of the type B Sūtras. In the second passage, the term *bhāṣyasūtra* refers to all kinds of commentarial statements, especially those of the Vārttikas, and those in the *gaṇapāṭha*, which is primarily a list of nominal bases to which Pāṇini's sūtras apply. (Another *possible* case, the uncertain occurrence of *gaṇasūtra* in a highly corrupt passage immediately after the second *bhāṣyasūtra*-passage, may be left out of consideration here, as it merely continues the *sūtra*-usage of the expression *bhāṣyasūtra* in this passage.)

However, the application of the term *bhāṣyasūtra* to commentarial statements concerning Pāṇini's sūtras is clearly secondary in Bhartṛhari's prose. The first passage is occasioned by a statement of Patañjali in which he uses the term *sūtra* with regard to a Vārttika-statement. Patañjali may have been familiar with a quite broad concept of the Sūtra, as he not only speaks of *vārttikasūtrika*, but also of *sāṁgrahasūtrika* (ref. to Vyāḍi's Saṁgraha?). Not only does Bhartṛhari usually employ other, specific terms to refer to the different types of commentarial statements (*vākya* for the Vārttika-statements, *gaṇa* or *gaṇapāṭha* for the statements in the *gaṇapāṭha*), even in the passages where he does accept, apparently on a secondary level, the designation *sūtra* for commentarial statements he contrasts these with the statements of Pāṇini which are sūtras in the full sense of the term. The character of Pāṇini's sūtra is thus clearly brought to the fore by Bhartṛhari in a way which fully matches Renou's descriptive-normative type A Sūtras: they are extremely concise (*laghu*), and consist of general rules (*lakṣaṇa*) and specific amplifications (*prapañca*).

In immediate connection with this, we are allowed to make another observation. Even if Bhartṛhari is willing to apply the term *sūtra* to the commentarial statements only in a secondary sense, we find that he already reflects - be it only in a vague and preliminary way - one of the main distinguishing features of the type B Sūtras. As we have seen, in the first passage the only difference between sūtras in the full sense and the *bhāṣyasūtras* was that brevity need not be striven after in the latter to the same extent as in the former. But in the second passage he gives a more positive characterization of *bhāṣyasūtras*: they 'intend to illustrate and corroborate' the basic sūtras. Even if the argumentative element is absent from the specific sūtras and statements in the *gaṇapāṭha* under immediate discussion in this passage (MBhD 6a:26.4-5), it is clear that it is the aim of 'corroboration' (*samarthana*) which opens the door for the discussions and argumentations (with statements, denied statements, and causal ablative compounds in *-tvāt*) which are comparatively frequent in Kātyāyana's Vārttikas and in the Darśana Sūtras, but which are not usually found in the older type A Sūtras.

4. THE SŪTRA IN THE FIELD OF SOUTH ASIAN LITERARY PRODUCTION BEFORE AND AFTER BHARTṚHARI: SOME OBSERVATIONS

4.1 Although, as pointed out in the introduction (section 1.7), much more detailed study is needed before we can aspire to attain a full understanding of the origins and development of the theory and practice of the Sūtra as literary genre, we will here attempt to place our findings with regard to Bhartṛhari's MBhD in the larger historical context of the Sanskrit tradition. Even if it is too early for definitive and comprehensive conclusions, this may help to focus future research.

4.2 If we first limit ourselves to the grammatical tradition, we see that, at least as far as Pāṇini's grammar is concerned, Bhartṛhari's concept of the 'perfect sūtra' (to use here Balbir's expression) is not substantially different from the concept which later grammatical authors had. The criteria of descriptive exactness, generalization and brevity, and, last but not least, authoritativeness, remain valid in the discussions of later authors. It seems, however, that Patañjali could speak of a sūtra in a slightly looser sense, and Bhartṛhari followed Patañjali in this when he coined the term *bhāṣyasūtra* to explain a Mahābhāṣya-passage which posed considerable problems to later commentators and interpreters (above, section 3.1 and notes).

If we take Bhartṛhari's characterization of statements which are not sūtras in the full sense of the word seriously, we arrive at an interesting result with regard to a passage which has since long been suspected to be an interpolation in Pāṇini's grammar, viz. P 1.2.53-57.[50] The sūtras in this passage are clearly argumentative in character (note, for instance, the frequent use of compounds in *-tvāt*), in contradistinction to all other sūtras of Pāṇini. Hence, they suit an aim which Bhartṛhari mentions only for the *bhāṣyasūtras*, namely that of corroboration (*samarthana*). In P 1.2.53-57 we find this corroboration together with its negative counterpart, refutation. If sūtras in the full sense of the term should only give general rules and specific amplifications, and otherwise be as brief as possible, this passage should certainly be relegated to the commentarial statements.

4.3 According to the same criterion, the Kātyāyana Śrauta Sūtra would contain much that can hardly be considered to be strictly sūtra-like in character. The Kātyāyana Śrauta Sūtra is a late text among the Śrauta Sūtras, but should nevertheless be expected to be considerably earlier than Bhartṛhari (who, however, nowhere refers to it). Parpola (1994, pp. 298-305) has recently argued for identical authorship of the Kātyāyana Śrauta Sūtra, the Vārttikas on Pāṇini and the Vājasaneyi Prātiśākhya of the Śukla Yajurveda. A reason for caution to accept this conclusion lies in the fact that the author of the Vārttikas, just as the author of the extant Pūrva Mīmāṁsā Sūtra, seems to have a preference for metrical octosyllabic sequences[51] of which there is, as far as I could see, no trace in either the Kātyāyana Śrauta Sūtra or the Vājasaneyi Prātiśākhya. Be that as it may, it can be said that the Kātyāyana Śrauta Sūtra, even if it existed in Bhartṛhari's time, did not influence his idea about what a *sūtra* was and should be. (Nor did occasional argumentative phrases in the older Śrauta Sūtras [such as the Āśvalāyana Śrauta Sūtra which he definitely knew] do so.)

4.4 The term *sūtra* is employed quite differently in another text which was important in the time of Bhartṛhari, whether or not he was directly acquainted with it. In the Abhidharma Kośa Bhāṣya (AKBh), the term *sūtra* refers to the discourses of the Buddha[52] usually called Sutta in Pāli,[53] and which Renou, from his point of view, considered 'diffuse' in character. We may accept with von Hinüber (1994) that at an earlier stage in the Buddhist tradition the use of the name Sutta is not too far removed from the use of the name Sūtra for the Brahmanical Śrauta and Dharma Sūtras, but the use of the term *sūtra* in the AKBh clearly reflects a stage in which the Buddhist concept of the *sūtra* had evolved in a direction of its own. It is interesting to see that the element of authoritativeness of the sūtra is also present here, and that in this context also the term *utsūtram* 'going beyond the sūtra', which we know from the MBh, is employed (AKBh, p. 60).

4.5 Above, I pointed out that on the basis of historical considerations one would expect Bhartṛhari to be familiar with what we presently know as the Sūtra of the (Pūrva) Mīmāṁsā-system. Especially in the light of the unitary nature of Mīmāṁsā as a hermeneutic undertaking (as emphasized by Parpola, 1981 and 1994), we would expect that the same applies to the system of Vedānta or Uttara Mīmāṁsā. Elsewhere (Houben, 1995, p. 59), I observed that, even if Bhartṛhari does not show clear evidence of familiarity with a well-established school of Vedānta, it seems likely, in view of his Brahmanical orientation and his chronological position, that he was familiar with the *prasthānatrayī* of all Vedānta-schools, i.e. the Upaniṣads, the Brahma Sūtra and the Bhagavad Gītā. As for this latter text, since it forms part of the Mahābhārata, and since Bhartṛhari seems to refer directly to characters of the Mahābhārata (VP 3.7.4-5, cf. MBh 2:34.17), one would assume the Bhagavad Gītā to antedate Bhartṛhari unless there are strong reasons to consider it a later insertion.[54] In the present context this is interesting for two reasons. First, the Brahma Sūtra - in the form we know this text now - has a very strong argumentative character, and is as such a good example of Renou's type B. Second, the Bhagavad Gītā evinces not only a familiarity with (some version of) the Brahma Sūtra, it also explicitly recognizes its argumentative character in the following verse:

ṛṣibhir bahudhā gītaṁ chandobhir vividhaiḥ pṛthak /
brahmasūtrapadaiś caiva hetumadbhir viniścitaiḥ //

The seers have chanted about [the *kṣetra* 'field'] severally in various metres as well as in definite statements corroborated by arguments in the Brahma Sūtra. (BhG 13.4, Van Buitenen's translation, slightly adapted)

According to Van Buitenen, the reference to the Brahma Sūtra must be to 'some collection antedating Bādarāyaṇa, perhaps an appendix to the *Karmamīmāṁsāsūtras*' (Van Buitenen, 1981, p. 11). In any case, Bādarāyaṇa's work as we have it contains two sūtras (2.3.45 and 4.1.10) which would refer to the Bhagavad Gītā according to the earliest available commentaries which otherwise contradict each other so much (Van Buitenen, 1981, p. 11).

4.6 Early references to the Sūtra as a text apparently characterized by brevity are already found in the Nāṭya Śāstra (not easily datable, but generally held to be pre-fourth century AD - so pre-Bhartṛhari - for its main parts), in the introductory section of the Rasa-chapter (NāṭŚ 6.8, 9, 11, 31).[55] Unfortunately, the term *sūtra* is here used to define other terms without having been explicitly defined itself. It is interesting to note that the term *sūtra* is already used in juxtaposition to *bhāṣya* (NāṭŚ 6.9). Also the compound *sūtragrantha* is attested here (NāṭŚ 6.8, 31). Abhinavagupta (ca.

AD 1000) explains *sūtra* as *lakṣaṇa* (e.g. NāṭŚ p. 98; cf. MBh 1:12.17), and *grantha* as *bhāṣya* (NāṭŚ p. 106)

4.7 Characterizations of the Sūtra which give a definite place to argumentation we find in some of the Jaina verses which were studied by Balbir (1987). These verses, in Prākrit,[56] occur in commentaries (the Āvaśyaka Niryukti, Viśeṣāvaśyaka Bhāṣya and others) which are heterogeneous in the sense that they contain material from different periods. Balbir mentions the first century as the traditional dating of the Āvaśyaka Niryukti, but estimates that it may have incorporated new material up to the eighth century (Balbir, 1987, p. 5).

4.8 Even if we find early recognition of the argumentative character of the Brahma Sūtra in the Bhagavad-Gītā and an acceptance of this character in Jaina definitions of the Sūtra, Madhvācārya, the thirteenth-century founder of the school of Dvaita Vedānta and commentator of the Brahma Sūtra, defended the status of this text as a Sūtra-text by referring only to the characteristics of the type A Sūtras. This he did by means of a Purāṇic quotation containing the following verse[57]:

> *alpākṣaram asaṃdigdhaṁ sāravad viśvatomukham /*
> *astobham anavadyaṁ ca sūtraṁ sūtravido viduḥ //*
>
> A *sūtra* consists of a small number of syllables, (of) intelligible (sentences), contains the essence, 'faces all sides' (does not focus attention on one subject), is free from (exclamatory and laudatory) insertions, (and) irreproachable (mainly following Gonda's translation, 1977, p. 466 n. 4).

What is emphasized here is nothing more than the brevity and descriptive-normative character of the type A Sūtras. The *hetu*s and, more generally, the argumentative character which the Brahma Sūtra clearly possesses is entirely neglected.[58] In spite of the difficulties in deciding on the historical details, it seems that with regard to Vedānta, as in the case of the Kātyāyana Śrauta Sūtra and the Buddhist Sūtras, there was a development in the practice of Sūtra-writing antedating Bhartṛhari but not having any influence on his concept of the sūtra, at least that of the grammatical sūtra. Judging after Madhvācārya's much later commentary, it seems that also the Vedāntins remained oriented to the type A Sūtras instead of developing a clear conception corresponding to type B Sūtras of which their own Sūtra-text is such a good and comparatively early example.[59]

4.9 The verse quoted by Madhvācārya occurs already in the Yuktidīpikā, a Sāṁkhya text which is quite early but definitely later than Bhartṛhari as it contains quotations from his Vākyapadīya (YD 7, 34; on the problem of the date: Larson & Bhattacharya, 1987, p. 228). In the Yuktidīpikā, too, the verse is quoted in a passage (YD 2) which defends the Sūtra-status of the Sāṁkhyakārikā, the text on which the Yuktidīpikā comments. Here it is followed by another definition of the sūtra:

laghūni sūcitārthāni svalpākṣarapadāni ca /
sarvataḥ sārabhūtāni sūtrāṇy āhur manīṣiṇaḥ //

The wise say that sūtras are brief, with meanings which are (merely) indicated (*sūcita*), with words that have few syllables, essential in all respects.

What is emphasized in this verse, is again nothing more than the brevity and descriptive-normative character of type A Sūtras. In fact, if we take a closer look at the Sāṁkhyakārikā, it is to be admitted that the argumentative element is practically absent in this text (cf. also Renou, 1963, p. 171; Larson and Bhattacharya, 1987, p. 149). Passing mention can here also be made of the Yoga Sūtra, a text which has always been closely connected with the Sāṁkhya-system, and which also lacks an argumentative character.[60] Quite different in this respect is the Sūtra-text which was written in the Sāṃkhya-school at a much later date (Larson & Bhattacharya, 1987, p. 327: 'it first appears in the Sāṁkhyasūtravṛtti of Aniruddha some time in the fifteenth century'). It is interesting to see that the need for such a Sūtra-text somehow arose in this school, as it were in spite of the defence of the Sūtra-status of the Sāṁkhyakārikā by the author of the Yuktidīpikā. Some major stylistic features which distinguish the Sāṁkhyakārikā from the Sūtra-texts of other philosophical schools are its consistent metrical character (in the Āryā-metre), the absence of polemics, and its *amaṅgala*-beginning with *duḥkha-*. In all these respects, the Sāṁkhya Sūtra seems to have followed the practice of the other well-established Darśana Sūtras, and this makes it, especially on account of its polemical sections, a later example of the type B Sūtras.[61]

4.10 Here, I will only briefly mention that the same two verses quoted in the YD appear also in Vācaspati Miśra's Nyāya Vārttika Tātparyaṭīkā (on NS 1.1.2). Vācaspati Miśra (ca. tenth century, Potter, 1977, p. 454) quotes them to explain why the author of the Nyāya Sūtra has merely indicated a certain point instead of being explicit: '[what is indicated] need not be explicitly stated by a Sūtra-author, for Sūtra-authors do not explicitly state what is indicated through meaning'[62] (follow the two above-cited verses). What is emphasized, both in Vācaspati Miśra's argument and in the verses, is brevity and description.

4.11 Later than the YD but before Vācaspati Miśra is the Mīmāṁsā-author Kumārila (seventh century, Verpoorten, 1987, p. 22), who quotes a verse which emphasizes that 'everything that is found in the commentaries (*vṛtti* and *vārtika*) is in the sūtras':

> *sūtreṣv eva hi tat*[63] *sarvaṁ yad vṛttau yac ca*
> *vārttike/*[64] *sūtraṁ yonir ihārthānāṁ sarvaṁ sūtre pratiṣṭhitam //*

> Indeed, everything that is in the *vṛtti* and in the *vārttika* is present in the sūtras. Here it is the sūtra which is the source of the meanings. Everything finds its basis in the sūtra. (Kumārila's TV on Mīmāṁsā Sūtra 2.3.16)

This verse is put in the mouth of someone questioning the authority of a certain interpretation of a sūtra of Jaimini, the author of the Mīmāṁsā Sūtra. According to this interpretation, the sūtra refers to a specific instance, but this instance is not mentioned or hinted at in the sūtra. Kumārila defends the acceptance of this instance as the one meant in the sūtra on the authority of the commentators against those who would like to apply the idea expressed in the cited verse too strictly. According to Kumārila, the supplying of material from outside the sūtra is prohibited only if a sūtra would be quite intelligible without it. Thus, while accepting that the sūtra has considerable authority, Kumārila emphasizes - here as elsewhere - the responsibility and scope for the interpretation and elaboration of later commentators like himself.

4.12 Finally, a brief reference may here be made to a work on poetics, Ruyyaka's Alaṅkārasarvasva (twelfth century, De, 1960, p. 181). This work is written in a sūtra-style, and can be considered to be mainly descriptive-normative in character. At the end, after elaborate treatment of the *alaṁkāra*s ('poetic embellishments') based on the form of the words, those based on meaning, and those based on both, the last sūtra plus accompanying commentary (the *vṛtti*, also by Ruyyaka) clearly shows what the author's concept of a sūtra was:

> *evam ete śabdārthobhayālaṁkārāḥ saṁkṣepataḥ sūtritāḥ //*
> *... sūtritā alaṅkārasūtraiḥ sūcitāḥ saṁkṣepeṇa pratipāditāḥ /*

(last sūtra:) Thus have these *alaṅkāras* of the word, meaning and of both (word and meaning) been briefly described in sūtra-form.
(commentary:) ... *sūtritā* 'have been described in sūtra-form' (means that) they have been indicated (*sūcita*), (i.e.) briefly stated, by the Alaṁkāra Sūtras.'

5. CONCLUSION

On the basis of our survey we may conclude that Bhartṛhari must already have been familiar with quite divergent practices of the *sūtra*, in which the only constant element seems to be the authoritative status attributed to the *sūtra* in relation to commentaries and interpretations. Bhartṛhari, one of the earliest authors to offer explicit reflections on the character and purpose of the *sūtra*, remained focused on what we have called, with reference to Renou's discussion of the genre of the Sūtra, the extended type A Sūtras. It would seem that Bhartṛhari's orientation towards the type A Sūtras remains typical for much of the later (especially the Brahmanical) Sanskrit tradition.

While there were apparently periods, before and after Bhartṛhari (cf. the early Brahma Sūtra and the late Sāṁkhyasūtra), when the practice of writing Sūtra-texts allowed these to be strongly argumentative in character, and while occasionally we do find verses which recognize the argumentative character of Sūtras, there seems to be no well-established conceptual reflection of this character at the place where we would expect it most: the traditions of the philosophical systems which possess some of the best and earliest examples of the type B Sūtras. The *Nyāyakośa* of the traditional nineteenth-century Sanskrit scholar Jhalakīkar does make a sharp distinction between Sūtras of the philosophical systems and other Sūtras, but his distinction is merely taxonomic, while his characterization of the *sūtra* is clearly oriented to type A and disregards the argumentative element.[65]

If we look at the Sanskrit authors' conceptual understanding of the genre of the Sūtra (rather than at the practice of Sūtra-writing), type B appears to be much less well-established as a separate category than one would expect on the basis of Renou's overview. The type-B Sūtras appear more as a special development within type A. In these type-B Sūtras descriptiveness and brevity remain of central importance, whereas the normative character recedes into the background, and more room is allowed for justification and argumentation. On the basis of the available evidence (and lack of it) it may be surmised that the writer of the late Sāṁkhya Sūtra was more influenced by the canonical status of (especially) the Nyāya and Brahma Sūtra, than by any explicit and well-defined conception of a philosophical or type-B kind of Sūtra.

In a 'universal history' of scientific and scholarly thought - still to be written, but to which Staal has already contributed much valuable material - the development of the practice and theory of writing Sūtra-texts in South Asia will occupy an important place. The attempts to store, in memorizable and hence reproducible texts, a large amount of descriptive-normative knowledge, sometimes together with abbreviated arguments and counter-arguments to defend the knowledge polemically, have no doubt greatly contributed to a tradition of knowledge which was to a considerable extent objectified (or 'exosomatic', to speak with Popper),[66] which, at least in some periods, allowed for dynamic progress through continuing dialectics and refinements, and which may be part of an explanation for 'the Wonder that was India' (or, rather, South Asia) which need not have recourse to notions like 'the Indian mentality' or 'the Indian genius', notions which are both scientifically worthless and politically dangerous.

ABBREVIATIONS OF SANSKRIT SOURCES

A — Aṣṭādhyāyī of Pāṇini. Edition and translation: Böhtlingk 1887. (At places the German translation needs revision in the light of modern research.) Translation: Katre, 1989 (to be used in tandem with the above edition because of the considerable number of printing errors in the latter)

AKBh — Abhidharmakośa-Bhāṣya of Vasubandhu. Edition of books 1 and 2: Dwarikadas Shastri, *Abhidharmakośa & Bhāṣya of Acharya Vasubandhu with Sphuṭārthā commentary of ācārya Yaśomitra*, Part I (I and II Kośasthāna), Varanasi: Bauddha Bharati, 1970

MBh — Patañjali's Mahābhāṣya. References to (number of volume):(page).(line) in F. Kielhorn's edition (Poona: Bhandarkar Oriental Research Institute, 1880-85; 3rd rev. ed. by A.V. Abhyankar, 1962-1972)

MBhD — Mahābhāṣya-Dīpikā. Ref. to the recent Poona edition by a team of scholars, Bhandarkar Oriental Research Institute, 1985-1991. The only available manuscript was reproduced in Mahābhāṣyadīpikā of Bhartṛhari, Poona: Bhandarkar Oriental Research Institute, 1980. Earlier edition: Abhyankar & Limaye, 1970; partial edition: Swaminathan, 1965

NS — Nyāyasūtra. Edited (together with Pakṣilasvāmi Vātsyāyana's Nyāya-Bhāṣya) by Digambara Sharma, Poona, 1985 (Reprint, Ānandāśrama Sanskrit Series. 91)

NāṭŚ — Nāṭya Śāstra. Edited (together with the commentary Abhinavabhāratī on Adhyāya 6 only) by M. Ramakrishna Kavi; Revised and critically

	edited by K.S. Ramaswami Sastri, Baroda: Oriental Institute, 1980
NVT	Nyāya-Vārttika-Tātparyaṭīkā by Vācaspati Miśra, edited by Rajeswara Sastri Dravid, Varanasi: Chaukhamba Sanskrit Sansthan, 2nd ed. 1989
SD	Sāṃkhya-Dīpikā. Edition in Venkatanathacharya, 1982
SS	Sāṃkhyasūtra. Edition in Venkatanathacharya, 1982
SSV	Sāṃkhyasūtra-Vṛtti. R. Garbe, *Sāṃkhyasūtravṛtti, Aniruddha's commentary and the original parts of Vedāntin Mahādeva's commentary to the Sāṃkhyasūtras*, edited with indices. Calcutta: Asiatic Society of Bengal, 1888
TV	Kumārila's Tantravārttika. In *Mīmāṃsāsūtra, edited together with Śābara-Bhāṣya, Prabhā-commentary, Kumārila's Tantravārttikam and Ṭupṭīkā* by K.V. Abhyankar - G.A. Joshi and M.C. Apte, Ānandāśrama Sanskrit Series. 97 [First edition vol. 1-7: 1930-1934, vol. 2, third edition 1981
VP	Bhartṛhari's Vākyapadīya. References (with two or three arabic numerals separated by periods) follow Rau's critical edition of the kārikās (W. Rau, *Bhartṛhari's Vākyapadīya*, Wiesbaden: Steiner, 1977)
YD	Yuktidīpikā, an ancient Commentary on the Sāṃkhya-kārikā of Īśvarakṛṣṇa. Edited by Ram Chandra Pandeya. Delhi: Motilal Banarsidass, 1967. (Ref. to page number)
YS	Yogasūtra. Edited (together with the Vyāsa-bhāṣya) by Ram Shankar Bhattacharya, Varanasi: Bhāratīya Vidyā Prakāśan, 1963

REFERENCES

Abhyankar, K.V. and V.P. Limaye

1970 *Mahābhāṣyadīpikā of Bhartṛhari*. Critically edited, Poona: Bhandarkar Oriental Research Institute

Abhyankar, K.V. and J.M. Shukla

1977 *A Dictionary of Sanskrit Grammar*. Second Edition, Vadodara: University of Baroda. Reprint 1986. (Gaekwad's Oriental Series. 134)

Balbir, Nalini

1987 'The Perfect Sūtra as Defined by the Jainas', *Berliner Indologische Studien* 3, pp. 3-21

Bannanje Govindacharya (ed.)

1969 *Sarvamūlagranthāḥ, Prasthānatrayī of Sri Ānandathirtha Bhagavatpāda* (With the versions of the oldest manuscript of Sri Hrishikesha

Thirtha, one of the direct disciples of Sri Anandathirtha). Bangalore: Akhila Bhārata Mādhwa Mahā Mandala, Poorna Prajna Vidyapeetha

Böhtlingk, Otto

1887 *Pāṇini's Grammatik. Herausgegeben, übersetzt, erläutert und mit verschiedenen indices versehen*. Leipzig. (Reprint 1977: Hildesheim, New York: Olms)

Bronkhorst, Johannes

1987 *Three Problems Pertaining to the Mahābhāṣya*. (*Post-Graduate and Research Department Series* 30, 'Pandit Shripad Shastri Deodhar Memorial Lectures' [Third Series]) Poona: Bhandarkar Oriental Research Institute

1989 'Studies on Bhartṛhari, 2: Bhartṛhari and Mīmāṃsā', *Studien zur Indologie und Iranistik* 15, pp. 101-117

1990 'Vārttika', *WZKS* 34, pp. 123-146

1991 'Two literary conventions and their consequences', *Asiatische Studien/Etudes asiatiques* 45.2, pp. 210-27

Buitenen, J.A.B. van

1981 *The Bhagavadgītā in the Mahābhārata*. A Bilingual Edition. Translated and edited, Chicago and London: University of Chicago Press

Caillat, Colette

1994 'Le genre du sūtra chez les jaina', in: *Genres littéraires en Inde*. Volume collectif sous la responsabilité de Nalini Balbir, URA 1058, Université de Paris 8I / CNRS. Paris: Presses de la Sorbonne, pp. 73-101

Cardona, George

1976 *Pāṇini. A survey of research*. The Hague, Mouton & Co. (Indian reprint: Delhi, 1980)

1988 *Pāṇini: His work and its traditions. Volume 1: Background and Introduction*, Delhi: Motilal Banarsidass

Chomsky, Noam

1965 *Aspects of the theory of syntax*. Cambridge: MIT Press

1980 *Rules and Representations*. New York: Columbia University Press

De, Sushil Kumar

1960 *History of Sanskrit Poetics* (In two volumes.) Second revised edition, Calcutta: Mukhopadhyay

Faddegon, Barend

1936 *Studies on Pāṇini's Grammar*, VKNAW, deel 38.1. Amsterdam: Noord-hollandsche Uitgeversmaatschappij

Filliozat, Pierre

1975 Le Mahābhāsya de Patañjali avec le Pradīpa de Kaiyaṭa et l'Uddyota de Nāgeśa, Adhyāya 1 Pāda 1 Āhnika 1-4. Traduction par Pierre

Filliozat. Pondichéry Institut Français d'Indology. (Publications de l'Institut Français d'Indology. 54,1)

Garbe, Richard (ed.)

1895 *The Sāṃkhya-pravacana-bhāṣya or commentary on the exposition of the Sānkhya Philosophy by Vijñānabhikṣu* 2 Cambridge, Mass.: Harvard University. (HOS. 2)

Goldstücker, Theodor

1861 *Pāṇini: His place in Sanskrit Literature*. London: Trübner. (Reprint: Varanasi: Chowkhamba 1965)

Gonda, Jan

1977 *The Ritual Sūtras*. A History of Indian Literature, Vol. 1, Fasc. 2. Wiesbaden: Harrassowitz

Halbfass, Wilhelm

1986 Review of Frauwallner, *Nachgelassene Werke I Aufsätze, Beiträge, Skizzen*. Herausgegeben von Ernst Steinkellner, Wien: Verlag der Österreichischen Akademie der Wissenschaften, 1984, in: *JAOS* 106, pp. 857-58

1988 *India and Europe. An Essay in Philosophical Understanding*. Albany: State University of New York Press

Hinüber, Oskar von

1994 'Die Neun Aṅgas', *WZKS* 38, pp. 121-35

Houben, Jan E.M.

1994 'Bhartṛhari's familiarity with Jainism', *Annals of the Bhandarkar Oriental Research Institute* 75 (parts 1-4), pp. 1-24

1995 *The Saṃbandha-Samuddeśa (Chapter on Relation) and Bhartṛhari's Philosophy of Language*. Groningen: Egbert Forsten. (Gonda Indological Studies. 2)

Jhalakīkar, Mm. Bhīmācārya

1928 *Nyāyakośa or Dictionary of technical terms of Indian Philosophy*, third edition, revised by Mm. V.S. Abhyankar. (The fourth edition of 1978 is a reprint of the third edition) Poona: Bhandarkar Oriental Research Institute

Joshi, S.D. and J.A.F. Roodbergen

1983 'The Structure of the Aṣṭādhyāyī in Historical Perspective', in: S.D. Joshi and S.D. Laddu (eds), *Proceedings of the International Seminar on Pāṇini*. Poona: Centre of Advanced Study in Sanskrit, pp. 59-95

1986 *Patañjali's Vyākaraṇa-Mahābhāṣya, Paspaśāhnika*. (Publications of the Centre of Advanced Study in Sanskrit. Class C, No. 15) Introduction, Text, Translation and Notes, Poona: University of Poona

1992 *The Aṣṭādhyāyī of Pāṇini*. Volume 1 (1.1.1-1.1.75), New Delhi: Sahitya Akademi

1993 *The Aṣṭādhyāyī of Pāṇini.* Volume 2 (1.2.1-1.2.73), New Delhi: Sahitya Akademi

Katre, Sumitra M.

1989 *Aṣṭādhyāyī of Pāṇini*. Delhi: Motilal Banarsidass

Kevalānandasaraswati

1952-66 *Mīmāṃsākoṣaḥ I-VII*. Wai: Prājña Pāṭhashālā Maṇḍala

Larson, G.J. and R.S. Bhattacharya (ed.)

1987 *Encyclopedia of Indian Philosophies*, vol. IV. *Sāṃkhya*. Delhi: Motilal Banarsidass

Manné, Joy

1990 'Categories of Sutta in the Pāli Nikāyas and their Implications for our Appreciation of the Buddhist Teaching and Literature', *Journal of the Pali Text Society* 15, pp. 29-87

Ojihara, Yutaka

1978 'Sur une formule patañjalienne; « *na cedānīm ācāryāḥ sūtrāṇi kṛtvā nivartayanti* »', *Indologica Taurinensia* 6 (Proceedings of the Third World Sanskrit Conference, Paris, 20-25 June 1977), pp. 219-34

Panchamukhi, R.S. (ed.)

1980 *Brahma Sūtra Bhāṣhya of Sri Madhvāchārya*, with the commentary Tatva-Prakāśikā of Sri Jayatirtha and a gloss thereon Bhavadipa of Sri Raghavendratirtha. Dharwad: Raghavendra Tirtha Pratishthana, Karnataka Historical Research Society

Parpola, Asko

1981 'On the formation of the Mīmāṃsā and the problems concerning Jaimini. Part I', *WZKS* 25, pp. 145-77

1994 'On the formation of the Mīmāṃsā and the problems concerning Jaimini. Part II', *WZKS* 38, pp. 293-309

Pollock, Sheldon

1985 'The theory of practice and the practice of theory in Indian intellectual history', *JAOS* 105.3, pp. 499-519

1989a 'The idea of Śāstra in traditional India', in: A.L. Dallapiccola (ed.), *The Shastric Tradition in the Indian Arts*. Stuttgart: Steiner, pp. 17-26

1989b 'Playing by the rules: Śāstra and Sanskrit literature', in: A.L.Dallapiccola (ed.), *The Shastric Tradition in the Indian Arts*. Stuttgart: Steiner, pp. 301-12

Popper, Karl R.

1994 *The myth of the Framework. In defence of science and rationality*. Edited by M.A. Notturno. London: Routledge

Potter, Karl H.

1977 *Encyclopedia of Indian Philosophies*, vol. II. *The tradition of Nyāya-Vaiśeṣika up to Gaṅgeśa*. Delhi: Motilal Banarsidass

Renou, Louis
1947 'Les Connexions entre le rituel et la grammaire en sanskrit', *JA* 233 (1941-1942), pp. 105-65
1956 *Histoire de la langue sanskrite*. (Collection 'les Langues de Monde'), Lyon: Editions IAC
1957 'Les divisions dans les textes sanskrits', *IIJ* 1, pp. 1-32
1963 'Sur le genre du sūtra dans la littérature sanskrite', *JA* 251, pp. 165-216

Renou, Louis et Jean Filliozat
1947 *L'Inde classique. Manuel des études indiennes*. Tôme Premier, Paris: Payot

Robins, R.H.
1979 *A Short History of Linguistics*. Second edition. London and New York: Longman. (Longman Linguistics Library. 6)

Scharfe, Hartmut
1977 *Grammatical Literature*. A History of Indian Literature, Vol. 5, Fasc. 2. Wiesbaden: Harrassowitz

Shastri, Mangal Deva
1928 'Metrical Basis of the Mīmāṃsā-Sūtras of Jaimini', *Proceedings of the Fifth Indian Oriental Conference* (Lahore, 1928) (vol. 2), pp. 842-54

Smith, Helmer
1951 *Retractationes Rhythmicae*. Studia Orientalia 16.5, Helsinki: Societas Orientalis Fennica
1953 *Inventaire Rythmique des Pūrva-Mīmāṃsā-Sūtra*. Uppsala/Wiesbaden: Harrassowitz

Speijer, J.S.
1886 *Sanskrit Syntax*. Leyden: Brill. Reprint: Delhi: Motilal Banarsidass 1980.

Staal, J.F.
1965 'Euclid and Pāṇini', *Philosophy East and West* 15, pp. 99-116. (Also in Staal 1988, pp. 143-160.)
1982 *The Science of Ritual*. Pune: Bhandarkar Oriental Research Institute
1988 *Universals: Studies in Indian Logic and Linguistics*. Chicago/ London: The University of Chicago Press.
1993 *Concepts of Science in Europe and Asia*. Leiden: International Institute for Asian Studies
1995 'The Sanskrit of Science', *Journal of Indian Philosophy* 23, pp. 73-127

Swaminathan, V.
1965 *Mahābhāṣya Ṭīkā by Bhartṛhari*. Varanasi: Hindu Vishvavidyālaya

Thieme, Paul
1931 'Grammatik und Sprache, ein Problem der altindischen Sprachwissenschaft', *Zeitschrift für Indologie und Iranistik* 8, pp. 23-32. (*Kleine Schriften*, ed. G. Budruss, Wiesbaden 1971, 2, pp. 414-523.)
Varma, K.M.
1958 *Seven words in Bharata: What do they signify?* Calcutta: Orient Longmans
Venkatanathacharya, N.S. (ed.)
1982 *Sankhyadarshanam, with Sridhar's Sankhyadeepika vritty and Ishwarakrishna's Sankhyakarika with Bhavaprakasha*. Mysore: Oriental Research Institute. (Oriental Research Institute Serie. 134)
Verpoorten, J.M.
1987 *Mīmāṃsā Literature*. Wiesbaden: Harrassowitz. (A History of Indian Literature. 6, fasc. 5.)

Notes

1. For the text as a whole we will write 'Sūtra', for the individual expression 'sūtra' (*sūtra* - in italics - will be used for the Sanskrit term and in direct citations).
2. A brief remark to place the term 'grammatical rule' as it is often used in modern linguistics in a historical perspective will not be out of place. While the term was consciously dropped from 'Bloomfieldian' linguistics 'in order to avoid any suggestion of normative or prescriptive bias getting in the way of objective description' (Robins, 1979, p. 227) it was reintroduced by Chomsky in his work from, 1957 onwards when he elaborated his ideas about a 'grammar of rules' instead of a 'grammar of lists', the latter being the ideal of his immediate predecessors. In the older European tradition, 'rules of grammar' form the core of the normative grammars written for didactic purposes. For Chomsky and other transformational-generative grammarians, however, 'grammatical rules' are claimed to describe the native speaker's competence, i.e. his creative capacity to produce and understand an infinite number of sentences (e.g. Chomsky, 1965; 1980, esp. p. 90). As we will explain below, Pāṇini's grammar, just like the ritual Sūtras, should be regarded as descriptive-normative in character. While Pāṇini's system of description surpasses by far that of the older normative grammars of the European tradition in sophistication, it would be incorrect to interpret him directly in the light of a Chomskian framework of theories and presuppositions. Even before Chomsky, Faddegon saw in Pāṇini's grammar the work of someone who 'searches the rules that are unconsciously and instinctively active in the minds of the people as a linguistic unity' (Faddegon, 1936, p. 59), a statement for which there is no scratch of evidence in the ancient Pāṇinian sources, but which foreshadows the transformational-generative research program with so much precision that we can say that a Chomskian approach was apparently already 'in the air' in the time of Faddegon, who refers to the Indo-europeanists of the 19th century, Delbrück's

Syntax of Sanskrit ('Old-Indian'), Herbart's mechanistic psychology, and especially de Saussure.

3. In an earlier remark in which his division into type A and type B sūtras is foreshadowed, also Renou placed the grammatical and ritual Sūtras into one category: 'La différence principale entre l'un et l'autre type de *sūtra* est que ceux de la grammaire et en général ceux du rituel se bornent à décrire (ou à prescrire), ceux de la philosophie justifient, et le raisonnement causal y tient une place considérable' (Renou in Renou & Filliozat, 1947, p. 60, §80).
4. Renou in Renou & Filliozat, 1947, p. 301 §593: 'le sens propre est « fil (conducteur) », d'ou « règle » — peu vraisemblablement « fil » servant à attacher les feuillets manuscrits' in this last phrase rejecting a proposal first made by Goldstücker 1861, p. 27.
5. Renou (1963, p. 200 note 5) refers to A. Weber, Max Müller, M. Winternitz and T. Goldstücker, without adding new arguments to this discussion.
6. As observed by Caland and also mentioned by Renou (1963, p. 180 and notes).
7. In a so far unpublished lecture, the text of which was kindly made available to me by the author, Dr. Konrad Klaus shows in more details what the problems are in the *sūkta-sutta-sūtra* line of argument.
8. Reference may here be made to Smith (1951, 1953). Smith seems to assume a tendency on the part of the author to use certain octosyllabic sequences. Twenty-three years earlier, however, M.D. Shastri, also studied the metrical aspects of the Mīmāṁsā, and concluded that the presently available Mīmāṁsā Sūtra must be a recast of an older, metrical work (Shastri, 1928, p. 854). Even if Smith's assumption is probably to be preferred unless some better evidence is found for the one of Shastri, the latter's paper has been undeservedly ignored by later scholars on the subject.
9. The dependence of the idea of a 'guiding thread' on the legend of Theseus and Ariadne was probably rightly pointed out and criticized by Scharfe, 1977, p. 87 note 49. Scharfe, as well as Gonda, 1977, pp. 465-6 who refers to Scharfe's discussion, gave preference to an interpretation of the term sūtra as a weaver's term. Even if Scharfe's explanation (1977, p. 87), for which he finds support in the interpretation offered by the Tamil grammar Naṉṉūl, seems to be suitable for the earliest application of the term, this does not exclude the possibility that later authors associated the term with other images, as was indeed the case in the verse in which Pāṇini's sūtra is compared with a thread in a necklace (Staal, 1965, p. 115 [1988, p. 159]). Added note: In a recent discussion in the Indology List (March-April 1997 on indology@liverpool.ac.uk, subscription through listserv@liverpool.ac.uk) S. Palaniappan argued that the term *sūtra* is best understood through its equivalents in Dravidian languages (e.g. *nūl*, *panuval* in Tamil) which evoke the image of 'producing speech from one's mouth like a spider produces a thread from his mouth'.
10. Von Hinüber, 1994, p. 131 n. 27: 'Die Wortgeschichte von Sanskrit *sūtra-* im Sinne von Text bedarf noch einer genauen Untersuchung vor allem hinsichtlich der Chronologie der Verwendung des Wortes'. Attention should be paid, of course, not only to *sūtra* in the sense of an entire text or a genre of texts, but also to *sūtra* in the sense of a segment of a text, and to the relation between

these two senses.

11. For the importance to consider both theory and practice in tandem cf. Pollock, 1985, 1989a and b. It is to be noted, however, that especially Pollock, 1985 probably overstates the contrast between 'Sanskritic culture' and 'the West' in their attitudes towards 'rules and practice' by focusing, as far as Sanskritic culture is concerned, on Mīmāṁsā and especially Kumārila as providing the main examples of Sanskritic attitudes, and ignoring the dynamics between and within the Brahmanical, Jaina and Buddhist Sanskritic traditions.
12. Cf. Renou in Renou & Filliozat, 1947, p. 59: 'le style des *sūtra* est l'une des grandes innovations littéraires de l'Inde ancienne; il n'a sa pareille nulle part.' And in 1956 Renou observed that it has been 'la grande tentation du génie indien de formuler en expressions quasi-mathématiques le contenu de sciences humaines, comme la grammaire ou la métrique' (Renou, 1956, p. 71).
13. For instance, the interpretation of the terms *vṛtti-Sūtra* and *vṛttigrantha-Sūtra* (in Āhnikas 6 and 7 of the MBhD) would certainly deserve a special study. I hope to discuss the problems and possible solutions regarding these terms at another occasion.
14. *idānīṁ vṛttikārāntaramatam upanyasyati yasmin pakṣe kriyamāṇe sūtre tu na kaś cid doṣo bhavati*, MBhD 6b:9.4-5; other examples are 7:2.21; 20.1; 25.4, 13. This use of the term sūtra is dominant in the Vākyapadīya (e.g. VP 1.70, 3.7.138, 3.9.93).
15. See also the more elaborate discussion concerning the two meanings of *sūtra* in a later paragraph (MBhD 1:34.9-13). Here too we find the idea that *sūtra* may refer to the whole and to the part. In addition, Bhartṛhari proposes as another possible explanation of these two terms that they express the general (*sāmānya*) and particular (*viśeṣa*) respectively (cf. discussion Joshi & Roodbergen, 1986, pp. 177-83). Apart from the idea of *sūtra* as expressive of a general or universal aspect, it is difficult to find anything in Bhartṛhari's work that could bring us closer to Renou's 'guiding thread' or 'leitmotiv'.
16. The editors follow here Abhyankar & Limaye's suggestion for emendatmon, Abhyankar & Limaye, 1970:197.18 and note 10.
17. In MBh 2:284.3-4 Patañjali mentions *kālpasūtra* (i.e. the Sūtra belonging to the field of Kalpa, or ritual) to illustrate a certain grammatical formation.
18. Patañjali mentioned the terms *vārttikasūtrikaḥ* and *sāṁgrahasūtrikaḥ* (MBh 2:284.3-4), which may be taken to refer respectively to Kātyāyana's Vārttikas (referred to as sūtras in another expression as well, see below section 3.1) and to Vyāḍi's Saṁgraha (not extant). Both are works in the field of grammar, but they are not usually regarded as Sūtra-texts. Apart from the great number of references to Pāṇini's sūtras, there are passages which suggest that the works of other grammarians are also considered as Sūtras (e.g. MBh 1:12.5-6, cf. Abhyankar & Shukla, 1977, s.v. Āpiśali and Kāśakṛtsna). Although these latter grammatical Sūtra-texts are not extant it may be assumed that they are to be categorized, just as Pāṇini's grammar, with the type A Sūtras. An exceptional reference like the one in MBhD 1:34.25 (*nyaṅkoḥ pratiṣedhaḥ*, probably referring to a sūtra of Āpiśali, cf. Bronkhorst in note 28 MBhD 1:139 and Palsule in MBhD 3:127) would confirm this categorization.

19. *yathā dharmaṁ vyākhyāsyāma iti*, MBh 1:2.6-7; cf. Halbfass, 1986; Houben, 1995, p. 48.
20. On the basis of the quotations and references to the Bhāṣyakāra preceding and following the quotation from the first line of the MBh (*śabdānuśāsana*, MBhD 1:1.22), it would seem that Bhartṛhari treats the words discussed as those of Patañjali.
21. For instance, in MBhD 3:3.19-20 he quotes a metrical line (Anuṣṭubh) which is similar to part of Mīmāṁsā Sūtra 4.3.11 (as pointed out by Palsule, MBhD 3:66), and even more similar to a verse quoted in a later grammatical commentary (viz. Vaidyanatha's *Chāyā* to the Mahābhāṣya: Bronkhorst, 1989, p. 111). See further discussion in Bronkhorst, 1989.
22. On the basis of several considerations such as its similarity with a work like Kātyāyana's Vārttikas, a relatively early date has been proposed for the Mīmāṁsā Sūtra. Parpola has argued that the Mīmāṁsā Sūtra must have been the earlier work (Parpola, 1981 and 1994).
23. In the case of the Śrauta Sūtras Bhartṛhari first of all refers to these texts themselves; he refers to a commentary only as a commentary on a Sūtra: MBhD 1:11.15. Much earlier also Patañjali referred to Mīmāṁsā, but the only authority mentioned is Kāśakṛtsni (MBh 2:206.8, 249.17, 325.14). This does not imply that Jaimini's Mīmāṁsā was not known to Patañjali, but only that one of the sources of alternative statements to which Jaimini refers (Parpola, 1994, p. 294) was probably still available. This, in turn, implies that Jaimini's text did not occupy the unique authoritative position it acquired in later times.
24. *sopaskāratvāt sūtrasya 'bhavati' vākyaśeṣatvena samarthayiṣyāmahe* 'because a *sūtra* needs to be supplemented we corroborate it by [the word] "is" as a supplement to the sentence.'
25. Thus, to mention a single example, in a long discussion on the preliminary *pratyāhārasūtras* 3 and 4 of Pāṇini's grammar (together with the comments on these in the MBh) it is investigated whether the required phonemes (*e, o, ai, au*) are in all cases correctly referred to in their proper duration (i.e. two mora apart from some exceptional cases), or whether they should be affixed with a *t* in order to achieve this (MBhD 2:16-21, notes by Palsule 2, pp. 178-212).
26. E.g. MBhD 3:13.10-14.1, discussed below.
27. E.g. MBhD 5:2.9ff.
28. E.g. MBhD 6a:29.16-20, discussed below.
29. Cf. Cardona, 1988, pp. 3-4, 655-71. Cf. also the verse *saṁjñā ca paribhāṣā ca vidhir niyama eva ca / pratiṣedho 'dhikāraś ca ṣaḍvidhaṁ sūtralakṣaṇam*, often quoted with regard to grammatical sūtras (Abhyankar & Shukla, 1977, p. 432 s.v. *sūtra*, Jhalakīkar, 1928, p. 1030 s.v. *sūtra*). In a few details, however, Bhartṛhari's understanding of these terms seems to have been slightly different from that of his predecessors, cf. on the expression *vyākaraṇasya sūtram* Joshi & Roodbergen, 1986, pp. xx, 177-83.
30. Although translating one Sanskrit term with another in an English rendering of a Sanskrit text is not a practice which can be generally recommended, in the present case some justification for this may be found, first, in the fact that the term sūtra has almost become a special technical term in English, and, second,

in the explicit explanation of *lakṣaṇa* as *sūtra* in the MBh (MBh 1:12.17 on the Vārttika-statement that *vyākaraṇa* is *lakṣyalakṣaṇe*).

31. In another interpretation of the verb *dhvanati* this word indicates a 'repeated activity'. This implies again the applicability of the rule to several individual cases.

32. MBhD 6a:29.17-18: *vibhaktāś caite yogeṣv arthā na śakyā ekasmin yoge saṃkaram antareṇopādātum* 'And these meanings separated in the rules (viz. P 1.1. 34-36) cannot be included in a single rule without confusion.'

33. None of the available translations (Filliozat, 1975, p. 128; Ojihara, 1978, p. 230; Joshi & Roodbergen, 1986; Bronkhorst in MBhD 1:100) gives a satisfactory rendering of the construction *yo ... na-adaḥ gṛhyeta* in which the relative pronoun in the clause is to be taken as equivalent to *yadi kaś cit* (Speijer, 1886, p. 356 §459) as it obviously does not refer to the pronoun *adas* (nor to *nādaḥ* as noun if we assume a pun with Ojihara, 1978, p. 230, whose suggestion on this point is accepted by Joshi & Roodbergen, 1986, p. 184 note 782).

34. ... *yadā kaś cid evam eva brūyāt ayaṁ śabda iti so 'vaśyaṁ pṛcchyeta katham asmābhiḥ pratyetavyam iti / yadi smṛtisūtram āha saṁdhīyate 'thāpralāpas tasya gṛhyate /*

35. Similar statements are found in a passage which immediately precedes (MBhD 1:34.18-21), but which contains some highly corrupt parts. A sentence which is not very doubtful is: *sūtrād eva tv abhivyaktārthāc chabdapravṛttir iti* 'the employment of correct words is only based on the sūtra of which the meaning is made manifest'. If the reconstruction of Bronkhorst is accepted the next sentence is: *yad etad udāharaṇādīnām upādānam etan nānutantrāṇāṁ bhāṣyasya vā brūyāt śabdāntaraṁ pratipadyeteti* 'This use of examples etc., one should not say that it belongs to the additional texts (*anutantra*) or the commentary (*bhāṣya*) in order that one should understand another correct word (which is not referred to by the sūtra).' The discussion continues: *kiṁ tarhi / anabhivyaktaṁ sūtre abhivyañjayed iti / etam eva cārthaṁ samarthayate /* 'What then? In order that what is unmanifest in the sūtra is made manifest. And it is that very meaning (which is in the sūtra) that [the commentary or example] corroborates.'

36. From note 13 on, p. 222 of his article we may infer that Abhyankar & Limaye's complete edition of 1970 was not available to him, as he only refers to an earlier one by the same authors which covered not more than Āhnikas 1-5, as well as to the incomplete edition of Swaminathan 1965 covering Āhnikas 1-4.

37. I prefer a comparatively simple and straightforward interpretation of this term (similar to the one P. Filliozat has given in his French translation, 1975, p. 124), to the more sophisticated but not entirely convincing interpretations of Ojihara (1978) and Joshi & Roodbergen (1986, p. 168). *idānīm* need not be taken as 'nowadays' like Joshi & Roodbergen did. As in other occurrences of *idānīm* in the MBh, it emphasizes here more the sequence of argumentation than that a strictly temporal 'now' would be contrasted with the past (hence Filliozat's 'Or' ['Well'] is sufficient to render *ca-idānīm*). *nivartayanti* is rather not to be taken as a term from the Vedic ritual as proposed by Ojihara, but in its basic sense 'to remove' or 'withdraw', as argued by Joshi and Roodbergen,

1986, p. 168 note 692.

38. Ojihara found it difficult to believe that Patañjali assimilated Kātyāyana's Vārttikas to Pāṇini's sūtras (e.g. Ojihara, 1978, p. 226). However, since Patañjali mentions the term *vārttikasūtrikaḥ* (MBh 2:284.3-4, see also note 19 above), probably referring to the work of one of the Vārttika-authors who preceded him (on the problem of Vārttikas in Patañjali's MBh most recently Bronkhorst, 1990, p. 128), there is no basis for Ojihara's hesitation.
39. MBhD-quotations are from the recent Poona-edition (MBhD 1-7) without change of peculiarities like the absence of sandhi etc., unless indicated otherwise.
40. There is no reason to divide this passage into two contrasting segments, as done by Ojihara. This will become more clear when the second passage is taken into account.
41. This has been accepted by all scholars from Kaiyaṭa and Nāgeśa onward. That Bhartṛhari would have misunderstood Patañjali on this point, as Ojihara believed, is not likely. Cf. above, note 39.
42. Cf. Bronkhorst, 1987, pp. 1-13 and 1990, p. 128 and the references mentioned in his notes.
43. To add one clear instance to those quoted by Bronkhorst (1990, p. 137): in the first Vārttika on P 1.1.34, it is asked why the words *avara* etc. are again enumerated in the sūtra, as they have already been enumerated in the *gaṇapāṭha*, the list of nominal bases which accompany the grammar. Here Bhartṛhari comments: *tad iha vākyakāro na sūtrārambhaṁ paryanuyuṅkte / sarveṣāṁ tu svarūpeṇa pratipadam uccāraṇaṃ tribhiḥ sūtraiḥ kimarthaṁ kriyate ity etad vākyenopanyasyati* / 'So here the author of the *vākyas* (i.e. our 'Vārttikas') does not question that sūtras are being formulated. But [the problem] that is referred to through this *vākya* (Vārttika on P 1.1.34) is: Why do these three sūtras (P 1.1.34-36) make mention of each nominal base in its own form, word by word?' (MBhD 6a:25.13-15)
44. P 1.1.34: *pūrvaparāvaradakṣiṇottarāparādharāṇi vyavasthāyām asaṁjñāyām.*
45. Vārttika 1 on P 1.1.34, MBh 1:92.18: *avarādīnāṁ ca punaḥ sūtrapāṭhe grahaṇānarthakyaṁ gaṇe paṭhitatvāt.*
46. See passage quoted and translated in note 44.
47. Although Bhartṛhari frequently refers to such commentaries (*vṛttis*), no early pre-Bhartṛhari *vṛtti* has survived, the Kāśikā-vṛtti being considerably later than Bhartṛhari (Cardona, 1976, pp. 278-82).
48. In the MBh the word *kṛtvā* is placed after *sūtrāṇi*. On the interpretation of this expression see discussion above, section 3.1.
49. ... *ācāryeṇa* ... *gaṇeṣu pāṭhāt kecit samuccitāḥ, kuśalair vā praṇetṛbhiḥ smṛtyartham uparacitās, tatas teṣu gurulaghubhāvaṁ praty anādaraḥ* (MBhD 6a:26.13-15).
50. Cf. Böhtlingk 1887 *ad loc.*; Thieme, 1931, p. 24; Faddegon, 1936, pp. 57-9; Renou, 1947, p. 115 note 3; more recently Joshi and Roodbergen, 1983, 1992, pp. 101-2 and 1993, pp. 92-100.

51. This applies especially in the introductory section where there are no interfering quotations from Pāṇini's sūtras; cf. above, note 9 and especially Smith, 1951, pp. 31-3, 1953; so far I could not get hold of a copy of V.G. Paranjpe's, 1922 thesis (Paris-Heidelberg) on the style of the Vārttikas.
52. Cf. AKBh p. 56, 60, 115, 150, 161 etc.
53. On different types to be distinguished among the Pāli Suttas see Manné, 1990.
54. Van Buitenen (1981, p. 5), however, argued that the Bhagavad Gītā 'was not an independent text that somehow wandered into the epic. On the contrary, it was conceived and developed to bring to a climax and solution the dharmic dilemma of a war which was both just and pernicious.' Of course, the BhG itself may still contain different layers and later insertions, but I am not aware of independent philological reasons to consider the verse under discussion particularly late.
55. I am grateful to Dr. Herman Tieken for having kindly drawn my attention to this passage, as well as to K.M. Varma's *Seven words in Bharata: What do they signify?*, which discusses in detail the relevant passages in the NāṭŚ and in Abhinavagupta's commentary on *sūtra*, *bhāṣya*, and *kārikā*. According to Varma, his predecessors (mainly P.V. Kane and S.K. De who follow Abhinavagupta to a great extent) interpreted *sūtra* as a part of the prose passages of the NāṭŚ, while it refers in fact to some separate, pre-NāṭŚ work (Varma, 1958, pp. 80-3). The author's views on the strong separation of *sūtra* and *bhāṣya*, and their chronological distance have now become unconvincing (in the light of e.g. Bronkhorst, 1990 and 1991).
56. The expression in the definition of the sūtra which interests us here is *heu-juttam* 'provided with reasons' (an alternative verse has *heū-kāraṇa-coiyaṁ*), which is explained in a sub-commentary as *sāhammêyara-heū sa-kāraṇaṁ vā* 'with a homogeneous or heterogeneous reason, or also with an efficient cause' (text and translation following Balbir, 1987, pp. 7-9).
57. Madhvācārya, introduction to his Brahma Sūtra Bhāṣya, in *Sarvamūlagranthāḥ*, ed. Bannanje Govindacharya, 1969, Sūtrasthāna-section, p. 2. Cf. also the discussion of this passage on the sūtra-status (*sūtratvam*) of the Brahma Sūtra in the commentary Tatvaprakāśikā (sic) and the gloss Bhāvadīpa, in the edition of Panchamukhi, 1980, pp. 14-17.
58. Interestingly, Balbir (1987, p. 8) refers to a Sanskrit verse, quoted in some Jaina Digāmbara sources, which looks like a variant of the *alpākṣaram asaṁdigdham* verse and does give a place to argumentation: *alpākṣaram asandig[d]haṁ sāravad gūḍhanirṇayam / nirdoṣaṁ hetumat tathyaṁ sūtraṁ sūtravido viduḥ*.
59. In the *Nyāyakośa* (Jhalakīkar, 1928 s.v. sūtra), the verse quoted by Madhvācārya is referred to as the Vedāntin's definition of a sūtra.
60. Renou, 1963, p. 196: 'En revanche, les *YoSū* et (pour autant qu'on puisse présumer) les *Sāṁkhyasū* (authentiques) sont dénués de controverse'. Renou's speculations on an ancient but lost Sūtra-work in the Sāṁkhya-school (also, 1963, p. 171) show that he was too much thinking by analogy and was perhaps too much influenced by later systematizations of the 'Six Darśanas' in the Sanskrit tradition. The evidence of the early sources rather suggest that

'Sāṃkhya probably did not have an ancient *sūtra* collection' (Larson & Bhattacharya, 1987, p. 85).

61. The Sūtra-text contains references to the systems of Vedānta, Nyāya and Yoga, made explicit in a commentary by a certain Śrīdhara (to be placed in the 16th century with Venkatanathacharya, 1982, p. 36 ?) with quotations of relevant sūtras (cf. SD on SS 1.153, 5.51, 5.128). The earlier commentary by Aniruddha refers to sūtras of Nyāya, Vaiśeṣika and Yoga (SSV on SS 5.85, 86; 6.19), and it refers to statements in the Sāṁkhya Sūtra explicitly as sūtras (introduction to SS 1.1, comm. on SS 6.40). Note also that SS 2.33 is identical with YS 1.5, and that Aniruddha's commentary on this sūtra starts with YS 1. 6-8. Vijñānabhikṣu's Sāṁkhyapravacana Bhāṣya refers to sūtras of Yoga, Nyāya and Vedānta, as well as to sūtras of the Sāṁkhya Sūtra on which it comments, as sūtras (Garbe 1895, e.g. p. 3.36-38, 41.15, 127.5-6).
At around the same time as the Sāṁkhya Sūtra we find the first references to another Sūtra-text within the Sāṁkhya-school, namely the brief Tattvasamāsa Sūtra (Larson and Bhattacharya, 1987, pp. 315-20). The text consists mainly of a simple enumeration of topics. It is neither argumentative, nor can we detect any stylistic sophistication in the description (e.g. with 'definition and specific amplification' as Bhartṛhari's characterization of the sūtra would require) which would associate it with the main examples of the type A Sūtras.

62. ... *kāryakāraṇabhāva ākṣipta iti nāsau sūtrakāreṇa darśanīyo, na hy arthākṣiptaṁ sūtrakārā darśayanti* (NVT 82). The term *ākṣipta* reminds one of the device of *arthāpatti*, but in the former compound *artha* would rather be the directly expressed meaning on account of which one infers a more distant one, in the latter it would be the meaning or matter inferred.

63. Read this with Kevalānandasarasvatī, Mīmāṁsākoṣa vol. 7, p. 4331, instead of *sat* of the Ānandāśrama edition (see abbreviation, under TV).

64. Read this instead of *vārtike* of Ānandāśrama edition and Mīmāṁsākoṣa vol. 7, p. 4332.

65. Jhalakīkar, 1928, s.v. *sūtra* (sandhi as printed): '*śāstrīyabahvarthapratipādakasaṁkṣiptavākyaviśeṣaḥ / śāstrīyasūtrāṇi ṣaḍvidhāni brahmamīmāṁsāsūtraṁ dharmamīmāṁsāsūtram nyāyasūtraṁ vaiśeṣikasūtraṁ sāṁkhyasūtram yogasūtraṁ ceti / ... / pāṇinyādipraṇītāni vyākaraṇasūtrāṇi tu vedāṅgāny eva / na tu śāstrīyasūtraṣaṭkāntarbhūtāni / evam āśvalāyanāpastambādimaharṣipraṇītāni dharmagṛhyaśrautasūtrāṇi bahūni santi /*'

66. Popper *A pluralist approach to the philosophy of history* revised version with some additional remarks in Popper, 1994, pp. 130-53, esp. 134, 149-50 and note 35.

XVI

BECOMING A VEDA IN THE GODAVARI DELTA

David M. Knipe

In the second volume of Frits Staal's monumental *Agni* appears an article by C.G. Kashikar and Asko Parpola, 'Śrauta traditions in recent times'.[1] It includes a state-by-state catalogue of *āhitāgni*s and *śrauta* sacrifices with maps locating their villages and towns. Described by the compilers as a pilot list, this indispensable resource is a measure of the extent of surviving Vedic ritual and textual traditions toward the close of the twentieth century.

Of the total of 626 *āhitāgni*s listed for India and Nepal nearly one-third (200) were residing in Andhra. Concerning records of those *somayāji*s who performed two or more soma sacrifices, Andhra again leads this set of statistics with 62, or slightly more than half of the total of 120 for India and Nepal.[2] This essay is a sketch of life expressions from a half-dozen Andhra Vaidika Brahmans located inside the delta of the Godavari River, a wavy triangle known locally as Konasima, approximately eighty kilometres to a side, where the river ends its 1,450-kilometre journey to the Bay of Bengal. The focus is on perceptions of the Vedas (*vēdam* in the singular in Telugu[3]) and *śrauta* sacrifices as cumulative interior composition of the Vaidika Brahman, that rare individual who has abandoned his life to Vedic text and ritual. Fieldwork[4] for this essay was conducted periodically between September 1980 and March 1995, spread out over various months within nine of those years, and including at present writing approximately sixty recorded interviews of Vaidika Brahmans in agrahāras in the Mukkamala - Nedunuru area of Amalapuram Taluk, in the west-central part of Konasima, directly on the Kauśika River.

Little has been written about the lives of these elite Brahmans, their opinions of Vaidika life and - particularly relevant today - their reflections on the current generation's radical shift away from traditional ways in general and Vaidika life in particular. This essay is part of a larger work surveying and assessing the careers, beliefs, and attitudes toward the changing times of the individuals cited in this essay as well as others not mentioned. When photographed for the standard *Commemorative Volume for Pandita XYZ* the veda panditas and their wives sit bolt upright side by side, not touching, not smiling, not being anything but austere, dutiful, attentive pinnacles of brahmana character. But my camera has caught them fervently teaching, laughing and joking, gesticulating in argument, lost in reading, gazing pensively, walking with grandsons. Similarly, while it is important to know about them bolt upright in their expected accomplishments, with all due honour to their learning, dignity, magnanimity and courage, we should also see them as complicated, free-spirited individuals, sometimes quixotic and elusive, often self-contained and rivalrous, infrequently venal and duplicitous. And it is equally important to attempt to situate them in a complex, cumulative and changing traditional space, a world loaded with concern for the effects of malevolent planets, an evil eye, or sorcery (as well as the rising price of a wooden ploughshare or petrol for a son's motor scooter), a world of factions, adversaries and compromises (as well as shared seats on a public bus), a world of karmic accountability and divine grace (as well as government pension plans).

In his portrayal of the Nambudiri Brahmans of Kerala Frits Staal found them to be 'sincere, straightforward, and disinclined to take themselves too seriously. After initial reluctance, they are eager to explain the intricacies of their recitations, chants, and ceremonies; they never claim knowledge they don't really possess; they will not preach or become pompous, and express no interest in coming to America. Though no longer averse to modernization, they remain attached to their simple habits.'[5] The description is entirely apt for the Konasima Vaidika Brahmans. In 1980 few knew about their Kerala counterparts and none knew of the Nambudiri *śrauta* performances. Although separated by fewer than 900 kilometres, their Vaidika ways reveal as many differences as similarities.

As in Kerala and other pockets of South India where Vedic traditions survive, there is an aching sense of vulnerability hovering over particular texts, recitation patterns, the *śrauta* ritual schedule as a whole, and a certain Vaidika ethos that already now rapidly resists articulation. Therefore we turn to the older men of the agrahāras to learn from their life histories and their current reflections on the passing of generations. The observations of these extraordinary individuals is precious from an ethnographic standpoint. They lived through nearly the whole of this century in self-sufficient enclaves unknown to urban India, places where little regard was given to events distinguished only by their remoteness - two world wars, the disappearance of the British, subsequent changes of the guard in Delhi, wars with Pakistan. Their views on Vedic spirituality can be instructive, particularly at this crucial juncture in Andhra history when the coming turn of a millennium accompanies their re-examination of a regional textual and sacrificial tradition, one that by our outside reckoning may have been in Andhra for more than two thousand years, and by their inside reckoning is *anādi*, without beginning, something that has always been there.

TWO KONASIMA AGRAHĀRAS

The two agrahāras described in this essay are less than five kilometres apart. They bracket the Kauśika River, demythologized in modern engineering terms as the Amalapuram Canal. The delta is known locally to comprise seven rivers, each connected to one of the seven *ṛṣis*, the Kauśika being the important stream for the Mukkamala region, or Viśvāmitra-*kṣetra* (Kuśika being another name for the great *ṛṣi* Viśvāmitra). The larger of the two agrahāras, a *dāna* for Brahman panditas in the 1940s, contains sixteen houses, and the smaller one, donated for veda panditas in 1960, has five. These two lanes, each isolated from non-Brahman village life and commerce, have been the homes of four *āhitāgnis* for the most productive parts of their careers, with three in the larger agrahāra, a fourth in the other. A son-in-law of the fourth one is an *āhitāgni* born and raised in Konasima and interactive with the other four throughout his life. Although currently residing in another taluk on the fringe of Konasima, his views factor into the perspectives of this localized community of Vaidika specialists. Numerous other Konasima Vaidika Brahmans, including three *āhitāgnis*, are within walking, cycling, and more recently, bussing distance, and this larger clan has made it possible over the decades for these five *āhitāgnis* to collaborate in sacrificial traditions and share textual expertise.

Of the current veda panditas, the earliest pair of families to settle in the larger agrahāra arrived in 1954 and the others quickly followed. Each family received from the donor a house site, one thousand rupees with which to build a house, and one acre of agricultural land with an assured yield of fifteen bags of paddy. In 1960 another veda pandita, soon to be *āhitāgni*, relocated only a few hundred metres from his family home to the other isolated agrahāra, built in the crop fields beyond the borders of his village, where he was given a house and half an acre of land elsewhere for income from a tenant cultivator.

As in other parts of traditional India these Vaidika Brahmans carry long names that document their ancestries, including long past accomplishments or aspirations of textual and ritual expertise. But names may also record their own personal construction by text, wider knowledge, and sacrifice. Some wear accumulated names that become entire litanies. Here in this essay, with no respect withheld, each will be mentioned without the conversationally obligatory Telugu honorary suffix *-garu*, and by a single name frequently applied informally. For example, one *āhitāgni* in the larger agrahāra, now 80 years old, performed the *agnicayana* and is known for that event as Cayanulu, although his eminent surname (family name) is Bulusu. Another is known as Yajulu (surname the well known Duvvuri, also aged 80), named for his performance of the basic *agniṣṭoma*, although he went on to perform *agnicayana* and the *pauṇḍarīka* and bears another name for the latter as well. A third in the same agrahāra died at the age of 89 on 6 August 1993. His family name is Bhamidipati but he was known by an affectionate nickname, Baballa, rather than by the *āruṇaketuka* form of the *cayana* that he performed 35 years ago. In the smaller agrahāra lives a celebrated *āhitāgni* well known throughout Andhra, Lanka Venkatarama Shastri Somayajulu, now 82. He will be mentioned here by his family name, Lanka. His son-in-law, age 65, has the same famous surname as Baballa, Bhamidipati, but is distinguished here by his second name, Mitranarayana. The surname Samavedam will be our reference to a 63-year-old *ghanapāṭhi* in the larger agrahāra, one who does not aspire to the *agniṣṭoma*, but rather currently devotes his time to teaching the Taittirīya *śākhā* to a grandson, having already long since passed it on to his sons. The personal names of wives are never mentioned, although an *āhitāgni*'s wife is known as Sōmidāvamma or Sōmidēvi for her ritual connections to one who offers soma, i.e. the *somayāji*.

Our cast for this essay will thus include five aging *āhitāgnis*, by seniority, Baballa, Lanka, Yajulu, Cayanulu, and Mitranarayana,[6] and one *ghanapāṭhi*, Samavedam. The ethnographic present will be March 1992 when Baballa was alive at the age of 88. At that point all five *āhitāgnis* had ceased to maintain the three fires. One was about to rekindle, one was considering such, and the other three deemed themselves retired in old age and failing health. One was deaf, bedridden and considerably weakened, one suffered from diabetes and arthritis and was unable to walk, one was virtually blind with glaucoma, and the youngest of the five had a seriously ill wife. Such 'retirement' from maintaining the three fires is legitimate in their Āpastamba tradition. By continuing to do the mantra *pāṭha* for the morning and evening *agnihotra*, unaccompanied by *kriyās* due to physical incapacity (Telugu *aśaktata*), one still retains the status of *āhitāgni*. Baballa was cremated in *brahmamedha* in 1993 as an *āhitāgni* who had completed 35 years of *agnihotra*. Three of the four remaining *āhitāgnis* are still in the rarefied company of those considered proficient in all seventeen *ṛtvij* roles for *śrautins* within the *ādhvaryava*, the *hautra*, and the *audgātra*.

THE SELF AS TEXT

Baballa (passing by during an interview with Yajulu): 'What's happening?'
Yajulu: 'Oh, nothing much, just discussing the texts'.
Baballa (in mock horror): 'What do you mean '*just* the texts'? *You* are the text!'

The influence of the veda and its ethos in the twentieth century can be gauged in part by the number of professionals actively known as veda panditas. Volume 2 of a directory of *Veda Pandits in India*, compiled and published by Tridandi Srimannarayana Ramanuja Jeeyar in 1976, was devoted to Andhra with a total number of 409.[7] Of the 14 listed under the most distinguished category, Veda-*bhāṣya* scholars, four are from Konasima. Of the 12 *salakṣaṇa ghanapāṭhis* ranked for Andhra, three are from Konasima and another two are across the Vasistha branch of the river on the Godavari's west bank. As for Andhra's total of 51 *ghanapāṭhis*, or specialists in the *ghana* recitation pattern, Konasima is particularly strong in this specialty, providing 17, with another five located in West Godavari. The largest single category in this directory is that of *kramapāṭhi*, the paṇḍita who has been successfully examined in the *krama* recitation pattern of veda. There are 344 in this list, with 71 in Konasima and another 65 on the west side of the river, a strong representation in the latter from Tanuku Taluk. Thus to sum up these numbers for the Godavari Delta, including both East and West Godavari Districts, by the reckoning of this list of twenty years ago a relatively small portion of the state of Andhra contributed almost 40% of all veda panditas. And some can name eight generations of distinguished Vaidika ancestors in both paternal and maternal lines.

The Konasima-Vaidikas are Taittirīyas, with Āpastamba as basic sūtra.[8] The Taittirīya Śākhā of the Kṛṣṇa Yajurveda is the hereditary basis for all but a few. *Veda Pandits in India* listed in 1976 only three veda panditas resident in Konasima who were not Taittirīyas, all of them *kramapāṭhis* in Kothapeta Taluk belonging to the Śukla Yajurveda, Kāṇva recension. Another two Vājasaneyis were located in West Godavari, where there was one Ṛgveda pandita also listed as a resident in Tanuku Taluk.

For Konasima the syllabus for the Vedic student, anticipating the examination system, consists of a standard 82 *pannas* (classical Telugu *pannam*, 'question', for Sanskrit *praśna*, indicating a chapter or section): Taittirīya Saṁhitā, Brāhmaṇa, Āraṇyaka, Upaniṣad. The seven *kāṇḍas* of the Saṁhitā are taken in five portions, making up 44 *pannas*, or slightly more than half of the syllabus. The 'dependent' (Telugu *parāyatta*) portion of 38 pannas is the second half of the *brahmacārin*'s work, i.e. the Brāhmaṇa of

28 pannas followed by the ten pannas of the Āraṇyaka and Upaniṣad. Only the Saṁhitā is employed in *padapāṭha* and the *vikṛtis* (modifications) known as *krama*, *jaṭā* and *ghana* for those capable of proceeding into the most complicated recitations and willing to devote extra years to the task. Few advance beyond the *kramapāṭha* and the majority who do are devoted to the local specialty, *ghana*. The *rathapāṭha* known among the Kerala Nambudiris[9] and in North India[10] is not a part of the Konasima teaching tradition.

Some *brahmacārins* turn to the Ṛgveda as second veda without further attention to *vikṛtis* of the Taittirīya Saṁhitā, and this may frequently be according to family tradition. Lanka chose the Ṛgveda as second text and completed 64 *adhyāyas* or about three-fourths of the hymns. The Atharvaveda, interestingly enough, was resuscitated into oral transmission during the 1970s for active use in personal, not public, rituals by several individuals in the agrahāras. The beneficial reasons given by Lanka for turning to the Atharvaveda are *śānti*, *puṣṭi*, and *abhicāra*.[11] Mostly this trend is small-scale, quiet, almost secretive acquisition of the fourth veda, but a few have gone on with an unquiet aspiration to 'know' the Atharvaveda in order to claim the status that attends the *caturvedin*. The Sāmaveda tradition for *śrauta* is a *prayoga* compilation said to be from the Rāṇāyanīya. Traditional panditas outside of Konasima frequently contribute to a legendary view of this Konasima compilation by referring to four *śākhās* (Jaiminīya, Kauthuma, Rāṇāyanīya and 'the one from Konasima', the last sometimes labelled 'Gautama'), whereas cosmopolitan scholars have sometimes treated Kauthuma-Rāṇāyanīya as a lone *śākhā* alternative to the Jaiminīya.[12] In Konasima itself the Rāṇāyanīya is praised and negative views are expressed regarding 'attempts to introduce the Kauthuma tradition from "the South"' (*dakṣiṇa* being a geographical reference vaguely indicating Tamilnadu and southern regions of Andhra more or less under Tamil influence).

As far as ritual manuals are concerned, shortcomings of the requisite Āpastamba are made up by recourse to Baudhāyana and Hiraṇyakeśin. Even if it is a family tradition to enter *śrauta*, the study of *smārta* and *apara* is recommended as a part of the Vaidika curriculum. Although there may be no intention of practising domestic ritual or funerary traditions, respectively, a rudimentary knowledge of such is held significant. Some students may be genuinely intrigued by these ritual studies and enter employment as *karma-kāṇḍis* specializing in *saṁskāras* or as domestic *purohitas*. One local authority estimates that circa 120 panditas now living have completed the Taittirīya texts including Āpastamba. Many of them, professionally active performing *saṁskāras* for clients, are nevertheless capable of stepping out of *smārta* and into *śrauta* to perform as *ṛtvij*-priests in a *yajña* or recite Taittirīya Saṁhitā in a *sabha*. Certainly their long training in veda assures a high degree of textual competence in routine Hindu rituals, and such pilgrimage centres as

Rajahmundry and Draksharama point with pride to a Vaidika background for a select few panditas who have become *karmakāṇḍis*.

To return to the agrahāra, after *upanayana* at the age of seven or eight the Taittirīya Saṁhitā is begun, usually with the father or grandfather, beginning before dawn and continuing daily with the exception of specified days of *anadhyāya* such as new and full-moon. ('Learning on new-moon day', says Lanka, 'is like fetching water in a sieve'.) Instruction is usually one on one, although occasionally two students may share a guru. Unlike local veda *pāṭhaśālās*, more than two is not permitted. A diligent student is said to be able to complete the 82 segments of the Taittirīya in eight years, an average student ten to twelve years, and thus age 15 to 20 is the normal range for the succession of *samāvartana*, *snāna*, *vivāha*, and the assumption of life as a householder. Since many *brahmacārins* learn from their fathers or older brothers and remain in their own homes, eating their mothers' cooking, their performance of the two traditional options for food, *mādhukara* or *vārālu*, becomes merely symbolic. To go collecting food in a group of *brahmacārins* from house to house in the village is *mādhukara*, each student carrying the donations in a cloth sling held out from the shoulder. Like bees collecting pollen in a sack they are 'honey-makers'. Alternatively, taking cooked food in a different house each of the 'days of the week' (Telugu *vārālu*) is considered less demanding. As a small boy, Lanka was up at 3 a.m. for the three-kilometre walk from his village to Mukkamala, where the remainder of the morning was devoted to *adhyāya*, veda instruction. His lunch was provided by the wife of his guru. The afternoon went into review of the morning lessons followed by the walk home, again in the dark, for supper from his mother and bed immediately at 8 p.m. To this day Lanka has kept to the same schedule of rising and retiring.

In general, those who learn from their fathers (or grandfathers) in their own homes appear to complete their initial veda somewhat sooner than those who walk from home to guru each day or live in another village or agrahāra. Having a biological father as guru is no invitation to laxity or informality. On the contrary, the transferring of veda from father to son may be even more disciplined. A middle-aged veda pandita who learned everything from his father daily for eight years as a boy expressed their current relationship: 'Only now that I am the father of sons learning veda can I approach my father to say something. I still cannot sit down with him for anything other than learning veda.'

A feature of traditional agrahāras that cannot fail to impress the outsider is an ebullient dedication to life-long learning. In a traditional Vaidika family, *snāna* after passing of the basic Taittirīya exams in *mūla* or even *kramapāṭha* may be considered merely the equivalent of a high-school diploma, the expectation from the student being adherence to a continuous advance through further veda and *śrauta*, and on to 'graduate school' with electives from the *dharmaśāstras*, *vedāṅgas*, philosophical *darśanas* (particularly Mīmāṁsā and Vedānta), *kāvya*, and other genres. Indeed the agrahāra may have the guise of a small college with specialists in texts and rituals available to the eager student for short or long-term study, their libraries of moldering books and pamphlets ready for 'book-learning' to supplement the many years of oral teachings. Book learning includes instruction in reading and writing the mother tongue, Telugu. For traditional families, until the current generation of school attendance, this meant approximately the age of 16 for a first acquaintance with an alphabet.

Famous teachers will draw students from other agrahāras, other districts. There is constant praise for those elders known for studying late into the night with oil lamps, who dedicated an ideal 17 or 18 hours a day to learning. After *upanayana*, age is no barrier. Lanka began learning the Atharvaveda from another Konasima pandita during his sixties, and frequently he himself has taken on older students in refresher courses or in specializations. In 1989 he was teaching Taittirīya Saṁhitā in *jatapāṭha* to a student who had long since completed *kramapāṭha* with his home village guru. Mitranarayana, age 63, is now working through the fourth *kāṇḍa* of the Taittirīya Saṁhitā in *kramapāṭha* with a 60-year-old 'student' at the rate of eight hours every day. Even when the performance of a great soma sacrifice occurred early in life, the *āhitāgni* maintains constant pursuit of new texts and refined expertise in those already learned or studied.

All Vaidika Brahmans over the age of forty speak with nostalgia of the old days when the agrahāra projected a constant 'tumult of mantras', the only period of possible silence being that between one and three a.m. The *āhitāgnis* shake their heads and remember those bygone years of obsession for, even addiction (*vyasana*) to, the perpetual learning of new texts, new recitation patterns, new commentaries until each of them, in a distinctive and entirely personal way, felt himself becoming a veda, becoming the veda.

THE SELF AS SACRIFICE

Baballa: 'Putting soma into the fire makes one a *somayāji*.
Then one travels in a well-grooved path, like a cart moving easily along a track, not like one on an untrodden field.'

Yajulu: '*Yajña* means rules, physical strain, ordeal. It is *tapas*, putting up with physical strain, all part of an intense desire to see it through … to be appreciative of the correct procedures and the aesthetics of the rite.'

The *brahmacārin* who completes his four-day *upanayana* and initial course through the Taittirīya *śākhā*, then his *samāvartana*, *snāna*, and five-day *vivāha*, establishes with his bride their household fire and begins the life of the householder-sacrificer. He may also be in a family tradition that encourages the next giant step of the sacrificer, taken either early in marriage or after a shorter or longer period of child-raising, the decision to become a soma sacrificer, one who establishes the *śrauta* fires and then within three months (within one year according to some) performs the *agniṣṭoma*. The former procedure (known in Konasima as *ādhāna* rather than *agnyādheya*) is a two-day ritual, beginning on either *amāvāsya* or *pūrṇimā* of a month astrologically appropriate for the *yajamāna*, and involves four *ṛtvij*-priests serving the candidate: *adhvaryu*, *āgnīdhra*, *hotṛ*, and *brahman*. Although there are textual provisions for *paśubandha* on the *upasad* day before the churning of new fire, the Konasima *ādhāna* omits it. Then within months the *agniṣṭoma* itself, being the foundational soma rite, requires 16 *ṛtvij*-priests: *adhvaryu*, *hotṛ*, *udgatṛ*, and *brahman*, each with three acolytes. These are chosen by the *somapravaka* who acts as broker and settles the *dakṣiṇā* honoraria with the invitation. The long strands of soma, a green leafless creeper with sectioned stalks known in classical Sanskrit as *lata*, Telugu *sōmalata* or *sōmatīga*, are purchased and the seller driven out of the field by a formulaic throwing of stones, the *mahāvedi* is laid out, fire transferred, a goat is sacrificed, and on the fifth and final day the soma is pressed under a small hanging 'soma cart' and then offered and consumed in conjunction with the offering of a second goat.

A distinctive feature of the Andhra Vaidika tradition, by comparison with other parts of India, is the retention of the *paśubandha* along with soma as undebated and indispensable core of the ritual pattern. Quite apart from the *agniṣṭoma* and *kāmya* soma schedules, an annual *nirūḍha-paśubandha*, known in Konasima as *śrāvaṇapaśu*, is considered as normative as the annual *āgrāyaṇa*. When shown a photograph of the 1975 *sāgnicitya atirātra* in Panjal, Kerala (a ritual Nambudiris label '*agni*' and Konasima Vaidikas '*cayana*'), the *āhitāgnis* looked at the row of eleven banana-leaf-wrapped packets of rice flour substituting for goats and asked in amazement how anyone could perform a sacrifice without a killing and still call it *yajña*. Whether it is the *śrāvaṇapaśu* or the *agnīṣomīya paśus* within the soma *yajñas*, the work of the priests on the sacrificial victim is intense. Cayanulu compares the *ṛtvij*-priests to surgeons at work on the body. After its smothering by the Potter (Kummari) *śamitṛ*, the goat is dissected with eleven parts considered essential offerings,[13] beginning with the *vapā* that is withdrawn by the *adhvaryu*. He raises it, stretches it out in front of his chest, and turns to display it to all before it is offered into the fire. So sensitive have *yajamānas* become to being interrupted and 'attacked' by animal-rights' activists, increasingly the *paśubandha* is put off until the final day, sometimes in a different, secluded locale from that of the preceding days of sacrifice.

Aspiring to the *agniṣṭoma* is not solely a matter of will, for one must be declared 'eligible' by *saṁskāras* and collegial recognition, and also be financially capable of bearing expenses that include *dakṣiṇā* to sixteen *ṛtvij*-priests. The achievement of *agniṣṭoma* - or simply *yāgam* or *yajñam*, these synonyms in Telugu stating the paradigmatic role of the *agniṣṭoma* - labels the performer a *somayāji*. He then has unrestricted access to the other *saṁsthās* or *yajñas*, with no particular order required, although 'family knowledge/ practice' (Telugu *kutumba vidya*) often serves as template. In Konasima the post-*agniṣṭoma* choices are generally one or another form of *agnicayana*, including the 40-day *pauṇḍarīka*, the rite of choice for many, or the *vājapeya*. The astonishing *sarvatomukha* is the rite of ultimate extravagance and a demonstration that one has been able to command the presence of 72 *ṛtvij*-priests to construct this fire altar 'facing in all directions'.

It is understandable for Baballa to express the cart-in-the-track metaphor quoted above. Born in 1904 into a five-generation family of soma sacrificers, he was one of six brothers who aspired to the *agniṣṭoma*. One, unfortunately, died before completing it but the other five became distinguished veda panditas and then *somayājis*, Baballa in 1959 at the age of 55. Yajulu, the son of an *āhitāgni* and also descended from a strong *vaṃśa* of *somayājis*, believes the proper time for the *agniṣṭoma* is the middle years from the age of 33 to 61, although his own was close to the time that he turned 30, and he recounts with pleasure his supervision of one for a relative who was 80. The youngest of the five was Mitranarayana, who proceeded at age 22 to the *agniṣṭoma* soon after his marriage to Lanka's daughter. Again, as with veda, a certain contagious passion (*vyasana*) may flow in the family and agrahāra, the ardour now directed toward *śrauta*. Of course, to aspire to perform a great sacrifice is also to accomplish new textual involvement, at least a year or two of intense study of ritual manuals and reviewing with new purpose those portions of the *śākhā*, already committed to memory, that detail the targeted *yajña*. In traditional language, this means to go beyond the mantras and ritual precepts (*vidhi*) to the meaning and contextualization (*arthavāda*) of mantra and *kriyā*.

Baballa went on to perform the *āruṇaketuka* form of *agnicayana* in 1960, only a year after his *agniṣṭoma*. Yajulu achieved the *sarvapṛṣṭha agnicayana* (*aptoryāma* with 'all [six] *pṛṣṭha samans*') in 1949, four years after *agniṣṭoma*, and then 20 years later the 40-day *vyūdha* form of *pauṇḍarīka*. In that same year, 1969, Cayanulu accomplished the *sarvapṛṣṭha agnicayana* that his neighbour across the lane had done two decades previously. Since all three were mutually exchanging *ṛtvij* roles it is easy to see how the enthusiasm of one great *yajña* would carry over into the next. Mitranarayana, another son of an *āhitāgni*, had an early start with *agniṣṭoma* only six years after *snāna*, and managed the *sāgnicitya aptoryāma*, *pauṇḍarīka*, and *sarvatomukha* with 72 officiating priests. This was the high-point of his career and a ritual so demanding that it appears nearly to have exhausted his intrigue with *śrauta*. There are so many forms of *cayana* that it is possible to perform several in combination with other named ritual sequences. Āpastamba cautions, however, that after a third the *yajamāna* may not have sexual intercourse with his wife and co-sacrificer, an indication that such an accomplished *śrautin* has approached *brahmacarya* of the *saṁnyāsin*.

Being an *āhitāgni* brings responsibility for a heavy routine of ritual activity aside from duties of a householder (supervising the cultivation of crops), preceptor and examiner of veda, and participant in veda *sabhas* or group recitations such as the *svasti* on festival occasions. The ritual bath three times daily, the evening and morning *agnihotra* with milk from the family buffalo, one or another *pākayajña*, the fortnightly *iṣṭi* of *darśapūrṇamāsa*, the annual *āgrāyaṇa* and *śrāvaṇapaśu*, and other scheduled rituals. Having sons accomplished in veda and remaining *alaukika* lightens the burden of providing *ṛtvij*-priests. For example, Yajulu's two sons, one cycling from nearby Mukkammala, the other bussing from Rajahmundry where he is studying Mīmāṁsā with a guru, provide two for *iṣṭi* every two weeks for new and full moon. With Yajulu himself serving as one *ṛtvij* as well as *yajamāna*, he need seek only a fourth from outside his family. The *āgrāyaṇa* in the autumn links each *āhitāgni* directly to his crop fields as well as to the larger community that labours in yellowing paddy fields lining the banks of every river and canal. 'The newly harvested rice,' says Yajulu, 'cannot be eaten until it is offered in the *āgrāyaṇa* to the appropriate deity. I do this either on Dipavali *amāvāsya* or on Karttika *pūrṇimā*. 80% of the results of this ritual go to benefit the community.' There is some relaxation from the rigor of the traditional calendar in that the three 'four-monthly' *cāturmāsya* are regarded locally as so complex and demanding, with a patchwork (Telugu *atukulu* is Cayanulu's word) of mantras taken from disparate sources, that this five-day sequence of the *vaiśvadeva*, *varuṇa-praghāsas*, and *sākamedha* seems rarely to be performed now, once by some, only a few times by others.

Beyond such *nityakarmas* there are certain *kāmyeṣṭi* rituals that can be quite spectacular and may rival classical soma rites in duration and expense. An example is the *nakṣatreṣṭi* or *nakṣatra homa* that cannot be less than 27 days (an *iṣṭi* outside the house for each *nakṣatra*) and usually lasts 40 days. A recent one in Vijayawada was 36 days from the first day of the dark half of Karttika (22 November to 26 December 1991), the *agnihotrin yajamāna* being assisted by 13 priests. Baballa once performed this propitiation (some say 'subjugation') of the constellations / lunar mansions in a ritual that may be seen in part as a Vaidika precursor to the *navagrahā* (planetary) *pūjās*, *abhiṣekas*, and *dānas* of popular Hinduism (Knipe, 1995: Part 1).

THE SELF AS PROVIDER

> Samavedam: 'Teaching veda is very demanding. I must always think of gaining money for survival. How can I go out and look for money and also stay home and teach every day? If I don't earn money I cannot maintain this *saṃsāra*. If I do go out and earn money, teaching suffers. The two are incompatible.'

The *dāna* to each of the Vaidika Brahmans included not only a house and house-site but also a half or a full acre of agricultural land as subsistence for a life devoted to veda and *śrauta*. Each of the *āhitāgnis* has had a different economic history during the last four decades. The label veda pandita, it must be remembered, hangs over one who is also husband, father, grandfather, lifelong householder, and responsible provider.

Lanka, for one, has been the true hands-on farmer among them, taking close personal interest in every crop, every season. His half-acre *dāna* he has never seen, for it is land worked by a tenant who sends paddy or cash in payment. But the six and a half acres he inherited surround his house at the agrahāra lane's end. For years he kept his 'office' in the crop field where he could supervise the labour crews and at the same time receive his students for veda, his colleagues for consultations, or unannounced villagers requesting astrological advice from a *pañcāṅga* or general counsel on conceiving a male child. His attitudes toward the earth he lives on are traditional and environmentally reasoned. He refuses to use new hybrid seed on the grounds that it is not native, and he quotes *śāstras* to support his contention that chemical fertilizers extract the essence of the soil and that food grown by chemical and not bovine fertilizer is unhealthy. He has always kept a buffalo sufficient for *agnihotra* and family milk only. Selling milk one time, he says, is the karmic equivalent of torturing fish for a year.

Today, unable to walk to the fields, Lanka sits surrounded by stacks of books and papers on his verandah where he can survey the changing greens of his rice fields and coconut palms flourishing only an arm's length away. And yet, despite the considerable income from these crops, and minor earnings in honoraria for serving as *ṛtvij*, or as veda and *śrauta* examiner, or a recent (May 1994) presidential award from Delhi honouring his accomplishments, including published books and pamphlets, he reports without complaint that he has been in debt his entire life, the two burdens being the cost of sacrifices and dowries for three daughters.

Others have had mixed results with agricultural profits. By the end of the 1980s Yajulu was the only one in the larger agrahāra to retain his original acre. Cayanulu was forced to sell off his land a piece at a time until it was gone. Samavedam sold his plot and bought a small one in hopes of a better yield. After successfully working it for decades, Baballa sold his land because of encroachment by squatters on his roadside holdings.

The dowry system and traditional marriage ceremonies take a financial toll. Samavedam speaks of paying out Rs. 23,000 in 1982 for the marriage of his first daughter, but figures it well spent since he found a veda pandita for her. But after four years of fruitless searching for another veda pandita before she reached puberty, he finally spent Rs. 20,000 on the marriage of his second daughter to a *laukika* Brahman. Every two-hour conversation with Samavedam slides easily into monetary matters without a word on the experience of mantras or a reflection on the significance of a Vaidika life. He carries in his head a wonder of details for the annual schedule of veda *sabhas* and the amounts to be paid to him and his sons for appearing as *ghanapāṭhis* in Vijayawada, Guntur, Eluru, Tenali, Gudiwada, Machilipatnam, Kaikaluru, and Bhimavaram, towns in other districts, or closer to home in Amalapuram, Razole, Ambajipeta, or Rajahmundry. His quote above is a significant one: Of his four sons, he taught veda to all, but stopped his lessons to the first and the fourth for economic reasons. Thus two became *ghanapāṭhis* in their father's footsteps, while two entered *laukika* life as a Sanskrit teacher and an engineer, respectively.

One of the most significant changes for Andhra Vaidika Brahmans in modern times occurred with the active entry of the Tirupati-Tirumalai Devasthanam (T.T.D.) into panditas lives and fortunes. The officers of this temple famous for its wealth have channelled funds for decades into promoting 'Vedic culture' in the form of salaries, pensions, and honoraria to Veda panditas for special or daily recitations (*pārāyaṇa*) or, more recently, for their publications and other accomplishments.[14] Opinions are split between those on one hand who see what is known as the 'T.T.D. Scheme' as salvific of a withering Vedic tradition, and those on the other hand who blame it for fragmenting and dividing a cultural legacy that has survived only by remaining reclusive, out of the public eye and institutional control. Certainly the T.T.D. has been successful in achieving its aim of injecting the veda into popular culture. In every important temple in Andhra Vaidika Brahmans may be seen and heard reciting their veda.[15] In Srisailam, for example, there are two each for the Ṛgveda, Yajurveda (Taittirīya), Sāmaveda, and Atharvaveda, reciting under signboards identifying their veda for all passing temple visitors from 7:30 to 11:30 a.m. Over the years several of them have been drawn from the agrahāras of Konasima, usually at a young age, and at times their elders point to this transfer of residence and exposure to a wider non-Brahman world as simultaneously the selling of veda (veda *vikraya*) and the destruction of *śrauta*. 'Only in the isolation of an agrahāra,' they say, 'away from worldly distractions, can an *āhitāgni* survive'. The same suspicion greets an increasing number of lucrative invitations that come to agrahāras from Hindu temples in Pittsburgh, Chicago, Atlanta, Houston, and other American cities.

Baballa, always an independent rustic, earthy, good-humoured, modest, but stubborn, scoffed at repeated invitations and bellowed his reply to the scheme: 'Not for *āhitāgnis*!' The T.T.D. eventually sent him pension checks anyway. Lanka, in his own quiet, unassuming way, also declined stipends. But the other three *āhitāgnis* and Samavedam and his sons found in the *pārāyaṇa* offer an appealing way to climb out of debt. Samavedam was early to sign on in 1963 at the age of 30 and today, wiry, driven with nervous energy, a healthy but nearly toothless 60-year-old, he is still involved and will go anywhere, anytime to a recitation or examination centre if a fee is forthcoming. Yajulu served the T.T.D. for two full decades until he retired at age 70 in 1985, Cayanulu participated until pensioned in 1993, and Mitranarayana, who enrolled at age 49 in 1979, is still active with daily *pārāyaṇa* in a Veṅkateśvara temple he reaches by bus from his house. Employment by the T.T.D. for them as established *āhitāgnis* did not remove them from their residences but rather allowed them to recite their veda in small local temples. Their rank at the top of the Brahman hierarchy also meant a higher salary. For Yajulu this *pārāyaṇa* was no more than a one-minute walk to the end of the agrahāra lane and a tiny private temple of Rādhākṛṣṇa where he spent three hours in recitation of the same texts he was teaching in *adhyāya*, while Cayanulu could cycle each morning after *agnihotra* to a public temple in the nearby village of Vyaghreshvaram and do the same.

None of the *āhitāgnis* feels comfortable discussing the economic side of Vaidika life and each places financial insecurity high on the list of reasons for the precipitous decline in *śrauta* in the late twentieth century. And this despite the good-will offerings of honoraria, pension plans, and other subsidies from the T.T.D., gratuities unimaginable to the *āhitāgnis* when they and many of their brothers and cousins were small boys enchanted by a tumult of mantras.

WIVES AS CO-SACRIFICERS

Baballa: 'Without a wife neither *nityakarma* nor *śrautakarma* can be performed. One obtains the authority of a ritual life (*karmādhikāra*) only if she is there.'

Yajulu: 'My wife delivered 14 times, always in our house .... She is always with me. She is absolutely essential to me. If I am away overnight she accompanies me. We take *agnihotra* with us.'

Among the many rules guarding the purity of Vaidika sacrificial traditions (and in the present day further shrinking the already tiny pool of those eligible to continue them) is the injunction that a potential *āhitāgni* must marry in a full five day wedding a pre-pubertal girl, and further, this girl must be the daughter of a woman who married before puberty. Since Vaidika males usually marry upon completion of veda study in their teens, their wives are usually five to ten years younger. Mostly they are from nearby villages and frequently they are cousins. Baballa was being raised as a child in the house of his maternal uncle when his future wife was born. She was three years old when he left that house, but they lived only half a kilometre apart until their marriage in 1922 when she was twelve. When he died after 71 years of married life and 35 years of doing *agnihotra* together she survived for only ten months alone.

For half of those married years she was the wife of an *āhitāgni* and therefore co-sacrificer, responsible alongside him for tending the fires, accompanying the morning and evening milk offerings, contributing certain mantras to rituals, and participating in the fortnightly and seasonal rites. She had married a man in a family famous for hosting and feeding large crowds of guests at sacrifices, festivals, and assemblies (*sabhas*), so the couple was given the house at the roadside end of the agrahāra. Although much of her life as a Sōmidēvamma was the steady pressurized labour of the hostess, her only complaint was directed not at her exhausting schedule but at the softness and negligence of the current generation of women. 'They are responsible,' she insisted, 'for the lack of continuity in *śrauta* today.'

Several indicators define the married couple as a ritual pair, as co-sacrificers. The wife is involved with engendering the single household fire (*aupasana*) subsequent to marriage, and should there be *ādhāna*, the construction of the *śrauta* fires, she is there as well. In order to accomplish the *agniṣṭoma* both husband and wife must undergo the *dīkṣā* consecration ritual. Yajulu was married at the age of '13 or 14' when his wife was age seven. They underwent *dīkṣā* several times together, the first time when he was 30 and she was 23, each of them subsisting on warm milk or curds with hands tied together. And, 'the husband being guru,' he taught her the special mantras for each *yajña*. Upon completion of the bath that closes each *yajña*, as he would be Somayāji, she would be Sōmidēvamma, which meant, Yajulu was careful to point out, not that she actually drank soma, but was ritually linked to one who did. Both wife and husband must remain healthy enough to function in the normal *agnihotra* and *iṣṭi* routines. Therefore, if either one is seriously incapacitated by illness or accident, the fires are vulnerable. If *agnihotra* is not done in three successive days the fires are *vicchina* (interrupted).

Perhaps the most graphic mark of ceremonial co-dependency is the fact that if the wife dies first, those ritual implements retained by the couple for life (and not, like the soma implements, thrown into the river after *yajña*) are burned with her body. The *brahmamedha* funeral, followed by *loṣṭacayana* (=*loṣṭaciti*) according to Āpastamba,[16] is for her, exactly as it would have been had he died first. Thus quite dramatically the widower, who is also the *karta* for her cremation, experiences while contemplating the fire a closure to the sacrificial life that began with their marriage. Even though he remarries, the widower is no longer considered eligible to perform additional soma or *agnīṣomīya* animal sacrifices. Cayanulu's first wife died, he married again, rekindled the fires in *punarādheya*, and continued the *nitya* (obligatory) *vidhi* of the *agnihotrin*, but a *kāmya* (optional) *yajña* such as the *pauṇḍarīka* or *vajapeya* was technically no longer a choice for him.

A constant concern for maintaining ritual purity (Telugu *maḍi*) is present for all members of a household, but among mature females it is the sacrificer's wife in particular who must be alert. During her period she must withdraw until her seclusion-ending bath, away from fire, the ritual complex, and kitchen to guard them from pollution (*aśauca*, Telugu *maila*). There being no precise way to predict the onset of menses, the timing of rituals other than daily ones entails uncertainty. In *yajña*, however, it is the *dīkṣā* that assures defense against *aśauca* for both the sacrificer and his wife. Until the final bath (*avabhṛtha*) both are free from the threat of bodily impurities or the pollution of someone's death. If the wife begins her period during the course of the *yajña* she may remain in the *yajñaśālā*, to one side for three days where no one may touch her, but continuing with her normal ritual performances. Aside from the *dīkśā*-protected *yajña*, a death or a birth in the household (delivery by a daughter-in-law for example) would be possible causes for interruption of ritual purity. One of the reasons frequently given for a lapse in maintaining the fires is a lack of home ownership, and the security from ritual pollution that only a free-standing, single-family, controlled dwelling can provide.[17]

SONS, GRANDSONS, AND A WORLD TRANSFORMED

Samavedam: 'Parents today don't want their sons in an agrahāra. They want them to go to an office and sit under a fan.'

> Mitranarayana: 'Vaidika life has disappeared, the spirit has gone, it is a period of *hūṇavidya* … . The generation of accomplished elders is over … . There are no *ṛtvij*-priests, there is no eligibility (*adhikāra*) remaining for someone to become a sacrificer (*yajamāna*).'

> Yajulu's third son, who passed examinations in veda and *śrauta* in 1969 at age 16: 'Since I cannot live like my father, the best path open to me is to try to minimize my needs and interaction with others. To the extent that it is possible I will try to emulate the ideals of Vedic life. But I am conscious of the society in which I live.'

> Lanka: 'I cherished a wonderful dream: four of my sons would know the four Vedas and the other three would learn *śāstras* … . But all seven are in *laukika* (worldly) pursuits.'

Samavedam and the five *āhitāgnis* discussed here have a total of 41 children,[18] 23 sons and 18 daughters. The number of sons per family ranges from Lanka's seven to Baballa's one.[19] Not one of them has established fires (*ādhāna*) or announced preparations for the *agniṣṭoma*. Only two of them are still eligible to do so. Whereas Baballa's generation saw all six brothers in his family alone aspire to the *agniṣṭoma*, the succeeding generation appears to have another agenda. The reasons for this precipitous generational shift are many and varied, and are of course the subject of consuming discussions. Over a decade and a half many different replies were offered on this question. The following is only a sample.

First, it is necessary to document for both veda and *śrauta* the situation of our two agrahāras, one that mirrors the overall trend in the Konasima Vaidika tradition. As far as the sharp decline in the teaching of veda is concerned, Samavedam remembers the village of Mukkamala when he was a boy. There were 20 to 30 *brahmacārins* then, he says, whereas at the present time there are only two. Yet he himself managed to duplicate exactly the feat of his father, seeing that two of his four sons completed veda, and marrying a daughter to a veda pandita as well. Samavedam's veda pandita brother has no sons in veda, but both of Samavedam's sons have advanced to *ghanapāṭhi*. Yajulu also managed to teach two, the other three sons having gone into business careers. Baballa's only son learned some veda but his career was undistinguished, his life was plagued with ill health, and he died of cancer only five months after his famous father.

Of the sons of Cayanulu, Lanka and Mitranarayana, all thirteen prepared themselves for worldly careers. One of Lanka's seven sons says 'To be able to be a veda pandita is a matter of fate (Telugu *yōgam*). It must be written [on the forehead]. Such was not my destiny.' This was one of the two sons

Lanka tried to teach. 'But both complained,' recalls Lanka. 'So I said to them why not be cultivators *and* veda panditas? But there was pressure from relatives who said to me 'Why do you keep them captive? Release them! Let them study what they want, they are good students!' It was not an easy decision.' Lanka's consolation for the failure of his dream is the fact that he was successful as spiritual guide for all seven sons and then his grandsons, each staying with him for a time 'to interiorize *sandhyāvandana*, to prepare for the *upanayana*', and as Lanka adds with his own quiet grace, 'to learn to do things independently.'

Samavedam and Yajulu have both been persistent not only with teaching two sons each but also in selectively pursuing *adhyāya* with grandsons in their own agrahāra. For Samavedam it is presently the son of his daughter and veda pandita son-in-law, who live in another district, and for Yajulu it is the eldest son of his own third son. Both of them are on track to complete the 82 *panna*s at some point in 1995. Presumably all four of their veda-accomplished sons have the capacity to teach when Yajulu and Samavedam are no longer able to do so, and therein exists a delicate hope of continuity in the local Taittirīya tradition.

For his part, Cayanulu has a disturbing Bulusu family legend to fall back on. One of his ancestors, a life-long devotee of Vighneśvara (Gaṇeśa), was approached in his old age by the cherished god and given a reward for his many years of faith and service, a boon in the form of distinguished pandita descendants in three successive generations. Cayanulu is one of the third and final generation.

To turn now to *śrauta*, there is an even more precipitous decline in the great ritual tradition than in the teaching of veda. Samavedam's two veda-certified sons aim to be professional *ghanapāṭhi* like their father and are not taken by the idea of performing *yāgas*. Of the five *āhitāgni*s it is only Yajulu who has a son eligible for *ādhāna* / *agniṣṭoma*, and in fact he has two. Therein lies the problem, for the first son, the proper successor to his *āhitāgni* father, is not interested in *śrauta*, although he completed veda. The third son has the will to follow his father into *agniṣṭoma* but cannot establish fires if his older brother has failed to do so.[20] Aside from this sibling blockade, he is also doubtful if the 'atmosphere' in which he grew up, the agrahāra support system for daily *agnihotra* and fortnightly *iṣṭi*s, could ever be recreated when the elders are gone. Therefore this remarkable concentration of *āhitāgni*s that enlivened the heart of Konasima for the whole of the twentieth century may disappear. At the turn of the century the youngest *āhitāgni* in the area will be 70 years of age.

The situation of veda teaching is not as drastic at present, but since the *śrauta* tradition is naturally dependent upon a large pool of veda panditas, many with expertise in *śrauta* and a few intrigued into energetic pursuit of *yajñas*, any decline at all in veda undermines the entire edifice. In assessing the dwindling Vaidika presence in Konasima, multiple explanations are given. Some blame the fathers for sending their sons into English medium schools so that they will go on to college degrees and earn better salaries. In this new economy that constantly demands cash outflow everyone can relate to financial stress as an explanation of the shift into *laukika* existence, and it is always the first rationalization supplied. But Samavedam, like several older women in the agrahāras, blames the younger more status-conscious mothers: 'Even if fathers want to send their sons into veda the women do not cooperate. They want their sons to be modern, to wear shirts and slacks, to have normal haircuts, not a *cūḍa* (Brahman hair-tuft, Telugu *pilaka*), and to use shaving lotion. What mother wants her son to look like this [pointing to his hair-tuft and traditional clothes] rather than a college lecturer in a town?' Lanka refers rather vaguely to social or political interference in Vaidika life, including animal-rights activists who bus themselves to scheduled *yajña*s and interrupt them with protest demonstrations.

Mitranarayana, who discounts the over-all significance of his father-in-law's remark, sees wider implications, beyond mere economics, in the trend toward *hūṇavidya*. This Telugu term, literally barbarian practice (from *hūṇuḍu*, the Hun or barbarian), is used in the sense of *laukika* culture, especially the English-speaking world that is, to the traditional panditas, foreign, threatening, and divisive. There is a clear line between the older two generations of agrahāra dwellers who speak no English whatever and their grandsons presently in English-medium schools who see a wider world open to them because of their second language. The elders are correct in their fear of the links between language and culture.

Another, less obvious rupture between the generations has to do with spiritual life. The *āhitāgni*s appear in many ways to be almost exclusively oriented to Vaidika existence. Yajulu's third son has by now lived half of his life in a town far from the agrahāra where he was raised. Although he aspires some day to *ādhāna* and *agniṣṭoma*, speaks no English and deliberately limits his *hūṇavidya*, he observes the major Hindu festivals, confesses Añjaneya (Hanuman) as *iṣṭadevata*, and peppers his discourse with reference to mokṣa. Concerning the *āhitāgni*s, he remarks, noting their distance from his perspectives, 'they do not believe in the existence of a deity in a temple somewhere', adding with regard to his father, 'his only god is *agnihotra'*. It is a provocative observation, for it is true that the *agnihotrin*'s ritual life is so full that there is little time left for observation of festivals such as Divāli or Kṛṣṇā's birthday, or *pūjā*s to the deities who live in the temples where they perform daily *pārāyaṇa*. The reservations expressed by this middle-aged veda pandita are genuine; in his own experience he is tugged between the old gods and the new, and his father cannot advise him on the outcome, or even appreciate his quandary.

A final comment here comes from a local authority on *śrauta* who is not a sacrificer. He believes that part of the current decline of *yajña* in Konasima is to be attributed to a fear of imprecision in ritual, a reluctance in a tradition-bound territory to make mistakes. 'In Krishna District,' he says, 'they are not afraid to make mistakes, so they go right ahead with complicated *yajñas*.' There is something to be said in support of this view. An example might be the *cāturmāsya* schedule, even the condensed one that nevertheless is rarely performed because it is considered so difficult that errors might erupt and damage it.

FRUITS AND THE FUTURE

> Lanka: 'In the old days people skipped meals in order to do obligatory rituals. Now it's the reverse.'

> Yajulu: 'My son-in-law asked me to arrange the marriage of my granddaughter. She is not a BA and cannot go out and earn money, which is what parents want for their sons now. Values have shifted. Family tradition, a Vaidika heritage, generations of veda teaching and soma sacrificing, etc, etc. - all that falls on deaf ears. My presence will not make any difference in the prospect of her marriage.'

The protagonists in this sketch of two agrahāras are all householders. The *āhitāgnis* are *yajamānas* in the daily sense of the *agnihotra*, with or (at the end) without fire, and all have been *yajamānas* for the great world-maintaining *yajñas* that construct Agni anew and regenerate the cosmos. When asked now, in their declining years, about the personal results of a lifetime of veda and *śrauta*, about 'the cumulative spiritual result or merit due to a man's performance of sacrificial and charitable acts' (Kane's apt definition of *iṣṭāpūrta*[21]), they sometimes invoke the doctrine of *tyāga*, abandonment to the gods of the results (*phala*, fruit) of their ritual labour (*karman*). But the notion of striving toward immortality, transcendent rewards, a place in heaven (*svarga loka*), remains a key element of the discussion. In answer to the question of why he elected the *āruṇaketuka* after his *agniṣṭoma*, Baballa replied: 'One who performs *āruṇaketuka* is able to sit in *brahmasthāna*, communicating with Brahma without hindrance. He will have no obstruction when he enters *brahmaloka*.' On the other hand, when asked after Baballa's *brahmamedha* if Baballa had ceased to be reborn because of cumulative fruits of all of his *yajñas*, Yajulu had this to say: 'Merely for doing *yajñas*, *janmarahita* (absence of births) will not occur.

Performing a *karma* does not bring freedom from rebirth. Whatever Baballa has done, that is for *cittasuddhi* and for the pleasure of Parameśvara. This *janmarahita* is inconceivable for those of us in *saṁsāra* with wives, children and grandchildren. Those *saṁnyāsins* who renounce everything and obtain *siddhi*, they alone are eligible for *janmarahita*.' The *āhitāgni*'s status as householder, short of *samnyasa*, is simultaneously a joy and a burden. He is mobile between two poles, the ritually created *svarga loka* and the household world of family, crop fields and cattle. He is the sacrificer who becomes immortal, yet remains in *saṁsāra*, who confesses an *alaukika* self, yet reluctantly must trade in *laukika* terms, and his conflicted language often illustrates the kinds of tensions articulated by Heesterman in *The Broken World of Sacrifice*.

Brian Smith,[22] stimulated by penetrating observations of Gonda on Vedic concepts of heaven, such as recognition that 'the condition which for convenience may be called "immortality" belongs to the person concerned already in his earthly existence, before his removal to *svarga loka*',[23] goes on to write convincingly about the sacrificial journey, including correspondences of the *yajña* to a bird, cart, ship or chariot carrying the sacrificer to heaven. This *svarga loka* obtained by the sacrificer's ritual efforts is 'a realm which man cannot, in his present mortal state, survive ... The sacrificial journey places the sacrificer in the other world only temporarily, just long enough to mark out and reserve a space for the next life. The ritual journey to heaven taken by human sacrificers, in other words, must be a round trip.'[24] One of the Konasima veda panditas has put a slightly different spin on the metaphor, noting that '*svarga loka* is like a summer resort in which a tourist can have a good time as long as his purse is full. When the money is gone he has to return to his home town, for *svarga loka* provides only temporary pleasures.'

In sum, the contemporary Vaidika Brahman of Konasima, and the *āhitāgni* in particular, experiences enhanced contradictions. Like his father and grandfather, he has always known that 'one should encompass the world and at the same time keep out of it.'[25] But unlike his forebears he has suddenly discovered that his cart (on Baballa's well-grooved track) is on a collision course with a seemingly unstoppable *laukika* juggernaut headed in the opposite direction. In response he may continue, defensively, to advocate the *agrahāra* alaukika life that has been his heritage. At the same time his language solicits refuge in the renunciant life that has always been the relevant impossible ideal of Indian civilization. It is the renouncer's vow that solves all conflict between *alaukika* and *laukika* householding, simply by eliminating the house.

It has taken time for some of the harsher realities of change to catch up with agrahāra life. The *āhitāgnis* have by and large made accommodations to the modern world. Their houses have television sets, though without colour or cable, a doctor comes by routinely to give injections against body weakness, they are driven by car to the hospital for surgery or emergency treatment, and they receive, like it or not, pension checks from the T.T.D. But it has been difficult to adjust to a steadily diminishing status. Yajulu laments the fact that his presence, the sum total of *āhitāgni* life and tradition, the legacy of generations of veda teaching and soma sacrificing families, with all the intricate rules obeyed and all the lofty values sustained and transmitted on to the next two generations, all is meaningless outside the tiny world of veda panditas and an arbitrary public accolade from the T.T.D. Suddenly it is dawning on these soft-spoken, reclusive, now decrepit, but still elegant and always cultivated scholars, that the two agrahāras have already lost one *āhitāgni*, no new one has appeared in a whole generation, and the possibility looms that *śrauta* in action rather than recitation may cease to exist in Konasima. And yet Yajulu is also the 80-year-old blind man teaching his grandson the final *panna*s of the Taittirīya *śākhā* that this English-speaking, college-bound young man has taken more than the normal decade to become. Yajulu became that veda a phrase at a time seventy years ago, as did his father and his father's father. This grandson, married years ago to his mother's brother's nine-year-old daughter, is a last link of continuity.

However diminished, the obsession (*vyasana*) of the older generation for texts and sacrifices is not dead. Several times one young potential *āhitāgni*, currently employed in a temple town by the T.T.D., but projecting his *agniṣṭoma* for post-retirement years, has detailed his deep affection for the *aśvamedha*. When its royal character and extreme rarity in historic documentation is pointed out to him, his ardour for the two-year *yajña* is undiminished. 'I don't necessarily have to do it,' he explains, 'I just have to know everything about it.' The Taittirīya Saṃhitā ends, as he knows, with a dazzling portrayal of the bodily parts and actions of the horse as the full cosmic panoply of being, an account that may have inspired the poet in the famous opening lines of the Bṛhadāraṇyaka Upaniṣad. If this young hopeful does achieve the *agniṣṭoma* subsequent to thirty years of service as a temple veda pandita, the pattern of his life will be entirely different from that of the agrahāra-bound *āhitāgni*s of the past, those who managed the rigors of *agnihotra* life in youth or at the peak of veda-teaching careers. For that event in the distant future he will need the services of at least 16 veda panditas capable of serving as *ṛtvij*-priests, individuals who are already now as rarely sighted in Konasima as horses.

REFERENCES

Duvvuri, Vasumathi K.

1991 *Play, Symbolism and Ritual. A Study of Tamil Women's Rites of Passage*. New York: Lang

Gonda, Jan

1966 *Loka: The World of Heaven in the Veda*. Amsterdam: N.V. Noord-Hollandsche 1966

1977 *The Ritual Sūtras*. Wiesbaden: Harrassowitz

Heesterman, Jan C.

1993 *The Broken World of Sacrifice. An Essay in Ancient Indian Ritual*. Chicago: University of Chicago Press

1987 'Vedism and Brahmanism', in: Mircea Eliade (ed.), *Encyclopedia of Religion*. New York: Macmillan, vol. 15, pp. 217-42

1983 'Other Folk's Fire', in: Frits Staal (ed.), *Agni. The Vedic Ritual of the Fire Altar*. Berkeley: Asian Humanities 1983, vol. 2, pp. 76-94

Howard, Wayne

1977 *Sāmavedic Chant*. New Haven: Yale University Press

1986 *Veda Recitation in Varanasi*. Delhi: Motilal Banarsidass

Jeeyar, Tridandi Srimannarayana Ramanuja (compiler)

1976 *Veda Pandits in India. Volume 2 (Andhra Veda Pandits)*. Srirangam-Hyderabad: Published by compiler

Kane, P.V.

1968-75 *History of Dharmaśāstra*. Rev. ed., 5 vols. Poona: Bhandarkar Oriental Research Institute

Knipe, David M.

1995 'Softening the cruelty of god: Folklore, ritual and the planet Sani (Saturn) in southeast India', in: David Shulman (ed.), *Syllables of Sky: Studies in South Indian Civilization (in honour of Velcheru Narayana Rao)*. Delhi: Oxford University Press, ch. 9

Leslie, I. Julia

1989 *The Perfect Wife. The Orthodox Hindu Woman according to the Stridharmapaddhati of Tryambakayajvan*. Delhi: Oxford University Press

Reddy, R. Soma

1984 *Hindu and Muslim Religious Institutions. Andhra Desa, 1300-1600*. Madras: New Era

Smith, Brian K.

1989 *Reflections on Resemblance, Ritual, and Religion*. New York: Oxford University Press

Smith, Frederick M.
1987 *The Vedic Sacrifice in Transition. A Translation and Study of the Trikandamandana of Bhaskara Misra.* Poona: Bhandarkar Oriental Research Institute

Staal, Frits
1961 *Nambudiri Veda Recitation.* The Hague: Mouton
1983 *Agni. The Vedic Ritual of the Fire Altar.* 2 vols. Berkeley: Asian Humanities

Notes

1. Berkeley, 1983, pp. 193-251.
2. The principal Andhra informant for the list compiled by Kashikar and Parpola resides in the Krishna District. The Krishna and East Godavari Districts together account for well over half (114) of the circa 200 *āhitāgnis* listed for Andhra. There being no available informant from either of the two Godavari districts, the Godavari region of Andhra appears to have been underreported in the number of *āhitāgnis* as well as the various types of soma sacrifices performed in the twentieth century.
3. For the most part this essay will employ the traditional singular veda for 'the Vedas'. For ease in recognition, terms will be in standard Sanskrit, unless noted as Telugu, with omission of a final nasal from Telugu *vēdam, śrautam, agrahāram, agnihotram*, etc.
4. I am indebted not only to the Vaidika Brahmans and their families for their gracious hospitality and long days of conversations, but also to my long-time friend and indispensable colleague, Professor M.V. Krishnayya of Andhra University in Waltair, my research assistant Dr. K.V.L. Narasamamba of Rajahmundry, and many others. This work was supported in part by a Senior Research Fellowship from the American Institute of Indian Studies and in part by the Graduate School of the University of Wisconsin.
5. *Agni*, vol. 1 p. xxii.
6. They may be identified in this order in Kashikar and Parpola's 1983 list as numbers 43, 22 and 55, 45, 44, 31. The name of the larger agrahāra is Śrīrāmapuram and includes the houses of Baballa, Yajulu, Cayanulu and Samavedam, while Lanka's house is outside of Nedunuru in Kāmeśvari agrahāra, named after the donor's mother. Mitranarayana now lives in Kakinada.
7. I am grateful to H. Daniel Smith for sending me this publication in the year it appeared. The list includes only those who responded to a request for capsule biographies and photographs. It should be noted that there are some duplications of names and a substantial number of errors in headings, titles, and locations. Residence at the time of compiling this list is not an indication of the site of learning. Many panditas received veda, particularly in more difficult *praśnas* and *pāṭhas*, in more than one village or agrahāra, and in more than one district. The recent division of the old Godavari District of the Madras Presidency into

East and West is an example of an artificial boundary where Vedic traditions are concerned. Vaidika Brahmans within the delta intermarry and teach across the river's outer branches.

8. The Āpastamba's school may have developed in this part of Andhra. See Kane, *History of Dharmaśāstra*, vol. 1, pt. 1, Poona: Bhandarkar Oriental Research Institute, 1968, p. 67.
9. Frits Staal, *Nambudiri Veda Recitation*. The Hague: Mouton, 1961, pp. 47-9.
10. W. Howard, *Veda Recitation in Varanasi*. Delhi: Motilal Banarsidass, 1986, pp. 121 f., 148 ff.
11. Respectively, tranquillity, prosperity, and mantric power in the form of charms and spells. The last term calls for explanation. Atharvaveda mantras are commonly used for healing (one's self or another) and for sorcery or counter-sorcery. Ritual texts at times provide alternative or tangential procedures to be adopted if the performer has a personal need, for example, a score to settle with one's enemy, the fact of having an adversary being taken for granted. Any discussion of *abhicāra* in the agrahāra reveals the information that the discussant employs mantras solely for defensive or counter-sorcery, while his enemy is prone to use them in sorcery. Both are *abhicāra*. Inevitably, common Telugu concepts slip into the conversation - *cētabaḍi, kṣūdra vidya,* and other terms for sorcery, the evil eye, and malevolent arts - but Vaidika Brahmans make a special reference unmentioned by non-Brahman ritualists. In addition to mantra *vidya* they speak of *śrauta kakṣa*, '*śrauta* rivalry', the Telugu word *kakṣa*, allowing stronger tones than the Sanskrit, pointing toward enmity, spite, and malice between ritualists. It should be noted that rivalries certainly occur within an agrahāra, despite the necessarily close cooperation ritually and socially between families who are all near or distant relatives, and hostilities can appear between Vaidika and *laukika* sides of agrahāras and families, but the true enmities of *śrauta kakṣa*, even to the point of accusations of actual murder and attempted murder by abhicāra, seem to have regional bases.

    One reflects here on Heesterman's well delineated theories on the agonistic structure of Vedic religion and his interpretation of sacrifice as a competition, a rivalry involving hostility and conflict in the continuing ritual redistribution of the material goods necessary for life. 'In the ritual texts the *bhrāṭrvya* (rival kinsman), or the *dviṣat* (the foe), is all but ubiquitous. There clearly is the idea of a stable unalterable order ... but this order is destabilized from within by the dualism of conflict for the goods of life' ('Vedism and Brahmanism', *Encyclopedia of Religion*. New York 1987, vol. 15, p. 226). See also *The Broken World of Sacrifice*. Chicago: University of Chicago Press, 1993, pp. 48 ff., and his essay 'Other Folk's Fire', in: *Agni,* ed. by Frits Staal, Berkeley: Asian Humanities, 1983, vol. 2, pp. 76-94.
12. There are different opinions. Howard notes distinctions between the Kauthuma and the Rāṇāyanīya *śākhās*, and also points out reasons to believe that the 'Kauthumas' of Tamilnadu are actually Rāṇāyanīya. See his *Sāmavedic Chant* (New Haven: Yale University 1977). On the other hand, Heesterman notes that 'the Sāmaveda boasts two *śākhās* that in fact differ only minimally, the Kauthuma-Rāṇāyanīya and the Jaiminīya' ('Vedism and Brahmanism',

Encyclopedia of Religion, vol. 15. New York: Macmillan, 1987, p. 220).

13. Contra the statement of Kashikar & Parpola that 'in Karnataka, Andhra, and Tamil Nadu the body of the immolated animal (goat) is not dissected, but the organs are extracted by making an aperture in the dead body of the animal' (*Agni*, vol. 2 p. 248), in Konasima the hole in the abdomen is only for obtaining the omentum (*vapā*). The goat is then completely dismantled for the requisite offering portions. The labour of dissecting is one that entirely exhausts the staff in a *yajña* such as the Vājapeya that requires 23 goats on the final day.
14. Due to the Andhra Pradesh Charitable and Hindu Religious Institutions and Endowments Act of 1966 subsidies have also come from the state government.
15. Temple support of veda panditas for recitations, rather than *pūjā* staff, is of course not an innovation. For example, an inscription in the temple of Simhachalam dated 1383 notes that payment from the temple treasury goes to an individual for *vedādhyāya* to local Brahmans. See Reddy, *Hindu and Muslim Religious Institutions. Andhra Dēsa, 1300-1600* (Madras 1984) p.154.
16. See ch. 4, the Pitṛmedhasūtras, in: Gonda, J., *The Ritual Sūtras*. Wiesbaden: Harrassowitz, 1977, p. 619.
17. For further details on women and ritual see F.M. Smith, *The Vedic Sacrifice in Transition*. Poona, 1987, pp. 86-90, including references to the Trikāṇḍamaṇḍana of Bhāskara Miśra. Leslie, *The Perfect Wife*. Delhi: Bhandarkar Oriental Research Centre, 1989, pp. 102 ff., provides an eighteenth-century Thanjavur view of a wife's roles in serving the sacred fire, rules for the menstruating woman (pp. 283-7), and other matters. See also Duvvuri, *Play, Symbolism and Ritual*. New York, 1991, particularly ch. 6 on women's rites of passage in an Aiyar (Tamil Smārta Brahman) agrahāra in the Bangalore area.
18. Lanka: 'There was no concept of family planning in our time. In fact the ideal of marriage is to have children ... . Anyway, those who observe family planning methods today are physically weaker. Nature takes care of excess population with train and motor accidents and other kinds of unnatural deaths.'
19. Counting only those children who survived into adulthood, Baballa has one son and one daughter, Yajulu five each, Cayanulu two sons and four daughters, Samavedam four sons and two daughters, Lanka seven sons and three daughters, Mitranarayana four sons and three daughters.
20. This stressful situation can devastate a family tradition. In his own prime, Yajulu was insistent enough to 'inspire' his older brother, as he puts it, to perform *agniṣṭoma* so that he could follow. It is said that if an older brother gives consent the younger may proceed. However, if the older brother changes his mind and does *agniṣṭoma* after all, the younger then forfeits all fruits of his *yajña*. (The matter of *tyāga*, the ritual endowment of fruits to the deity, is conveniently unmentioned here.)

    There are two prominent examples of this dilemma of sibling succession. Kapilavayi Yajnesvara Agnihotra Sastri was the fourth of five sons of an *āhitāgni* in Kakinada who performed *agnicayana*. From the age of six he learned the Taittirīya *śākhā* from his older brother in Kakinada and upon completion went on to other Vedas, the Atharvaveda in Kapileśvarapuram,

Ṛgveda in Vijayawada, Sāmaveda in Kanci, Śukla Yajurveda in Kasi, Mīmāṁsā in Nagpur, a lifetime of travelling and learning until he settled in Krishna Lanka in Vijayawada in 1975 and died there in 1983. His aim was *yajña*, and he remained a great *śrauta* expert and promoter of *yajña*s all his life, but could never establish fires because his older brother, his own guru, remained a bachelor and was therefore ineligible. A second example is the family of Renduchintala Venkatachala Yajulu who died in Kasi in 1987. Perhaps the most prolific *yajamāna* in twentieth-century India, his full name is a *yajña*-litany of more than thirty terms that many in Konasima recite with awe and affection. Like Kapilavayi, he also grew up in Konasima but moved on, maintaining his life as an *agnihotrin* for almost half a century, performing his first *yajña* in 1936, his last in Kasi, where he spent most of his years, in 1980. For the last, semiparalysed years of his life he accomplished *agnihotra* by holding onto the hands or back of his wife and reciting as she performed the *kriyas*. His entire life was given to *śrauta*, performing as a *yajamana* or as *ṛtvij*, the range of rituals being, according to his widow, 'like a sport to him. Fire rituals were as challenging to him as the taming of a tiger.' Again like Kapilavayi, he had five sons and all became veda panditas. But by 1988, despite the constant urging of their father while he was alive, only the first son had established fires. (This eldest son went on after *agniṣṭoma* to perform *sarvatomukha sāgnicit* in April 1980.) The fifth son desired to do *agniṣṭoma* but could not because the second, third and fourth sons had not. I am grateful to Professor M.V. Krishnayya of Waltair and Kapilavayi Venkatesvara Sastri of Simhachalam for interviewing on my behalf in Vijayawada the widow of Renduchintala Yajulu, June 1988, and to Venkatesvara Sastri for biographical details on his father.

For more on *pariṣṭadoṣa*, the *doṣa* of supersession of an older by a younger brother's *ādhāna* or *yajña*, see F.M. Smith, *The Vedic Sacrifice in Transition* (Poona 1987) p. 98 as well as I.39, 67-9 in the text of the Trikandamandana.

21. *History of Dharmaśāstra*, vol. 2, pt. 2, pp. 843 ff.
22. Smith, Brian K. *Reflections on Resemblance, Ritual, and Religion*. New York: Oxford University Press, 1989.
23. Jan Gonda, *Loka: The World and Heaven in the Veda*. Amsterdam: N.V. Noord-Hollandsche 1966, p. 97, cited in *Reflections on Resemblance, Ritual, and Religion*. New York: Oxford University Press, 1989, p. 103.
24. *Reflections*, p. 109.
25. Heesterman, *The Broken World of Sacrifice*, p. 71.

XVII

THE FOCUS ON THE HUMAN BODY: TWO ICONOGRAPHIC SOURCES ON THE ORIGINS OF INDIAN ART

Karel R. van Kooij[*]

INTRODUCTION

Even a cursory look at an exhibition of Indian art would be sufficient for the visitor to realize that what he or she actually saw first of all was a collection of human figures, sometimes rather fantastic and imaginative, but nevertheless human.

Indian sculptors and painters set a high value on the expressive qualities of the human body, at the cost of landscape painting which is one of the great achievements of Chinese art, or of the mathematically exact decorations which are so characteristic of Islamic art. It may be an overstatement to say that Hindu and Buddhist art are the most 'iconic' of all Asian art - yet it is certainly true that India has developed a sophisticated iconographic language which has proved capable of communicating nearly everything: myths, abstract concepts, emotions, and actions. Iconism is a characteristic of all Indian art.

Indian artists used the potential of the human body so exclusively that one is tempted to ask what this fascination is all about. In my contribution to this volume in honour of Dr. Frits Staal, two iconographic texts are consulted to find answers to this question. One account is a moving tale of the untimely death of a boy, which resulted in the first portrait ever made, and the other

[*] I am thankful to the Rijksmuseum for giving me permission to publish the photograph. I am grateful to Mrs. drs. P. Lunsingh Scheurleer for reading the manuscript of this paper and to Mrs. R. Robson-McKillop (B.A. Hons) for correcting my English.

discusses how to interpret icons as religious metaphors. Both sources deal with the 'origin' of pictorial art and moreover reveal the capacity to take an astonishing intellectual distance towards image-making, which often passes unnoticed in art historical or religious studies with regard to Indian art.

Contrary to what is generally stated by art historians, the authors of the iconographic works presented here do not address themselves to artists. Their context leaves no doubt that these texts were aimed at the patrons, mostly royal patrons, those who usually 'ordered' works of art. The kind of knowledge which is imparted to the reader does not relate to the technical skills which artists should have acquired. It is more the kind of artistic knowledge which any well-educated member of ancient Indian society should acquire:

> When you have acquired, through me, the knowledge of measures, characteristics, proportions, ornaments and the ideals of beauty, you will be familiar with all arts and skills, and you will become an expert of painting.[1]

This context should be noted carefully, because it has a bearing upon the evaluation of the data we are confronted with.

CITRALAKṢAṆA

The first account is the Citralakṣaṇa, from which the above words are quoted. The Sanskrit original of this work is lost but it is usually dated to about the 5th century AD on various grounds.[2] Although no indications can be found that this work has anything to do with Buddhism, the text was incorporated into, and only survived in, the Tibetan Buddhist canon. The first and up till now only translator of this work, Berthold Laufer, suggested a Jain origin.[3]

The work opens with a tale about the first drawing ever made. The main protagonist in this story is a king, whose name is Nagnajit. Brahmā and Viśvakarmā are the main interlocutors. The latter explains the rules of painting to the king in the chapters to follow. It is worthwhile summarizing the tale in some detail:[4]

> In an age long gone by when everyone lived as long as one hundred thousand years, and when untimely death simply did not occur, it so happened that, for the first time, the son of a brahmin died young and unexpectedly. The father could not accept this loss and he personally carried the body to the royal palace. Standing before the king in the

audience hall his complaint was eloquent and full of anger. He blamed, quite frankly, the king himself for this tragic death. He said he wanted his son back, otherwise he did not wish to live himself. The king used his extraordinary powers to summon the god Yama to his presence and demanded that he would restore the life of the brahmin's son. But Yama refused and declared that such a course of action was beyond his jurisdiction because untimely death was dictated not by him but by the law of *karma*.

Stubbornly the king did not accept this answer and a dispute ensued, which ended in a bitter fight.[5] The battle raged until the god Brahmā intervened. He advised the king to make a painting that would be an exact likeness of the deceased son.[6] The king obeyed. First he wanted to view the body and then he made the painting.

After it was finished, Brahmā let the painted body arise and gave it back to his father 'as if it were alive'.[7] The father's eyes opened wide with joy, colour came back into his face, he bowed before Brahmā and the king, and he left the palace taking his son home.

In this tale the very first drawing is that of a boy. The message seems to be that this first work of art was created in order to serve as a 'stand-in' for someone who had died. For this reason, the image had to be a faithful likeness of the person represented. The father considered it a satisfactory substitute of his son who had passed away and could not be brought back to life again.

In our own times this conception of the image is comparable to a portrait or a photograph of someone who died and was dear to us. We cherish the portrait because it is the only visible token left of a relative. The drawing of the Citralakṣaṇa may not have been a photographic portrait in our sense of the word, but it was meant to be a faithful 'likeness' and it did serve the same purpose.[8]

This faithful imitation involved the representation of an ideal human body, not of a dead body of course, which is set out in detail in the next chapter of the Citralakṣaṇa in which all the measures of this ideal body are meticulously summed up, from the crown of the head to the soles of the feet. When we compare these ideal measures with those of the healthy body as they are given in the Suśruta Saṁhitā,[9] which is a medical text dating from about the same period, we see a complete correspondence. There, exactly the same measures are given for every smallest part of the body. The conclusion must be that the concept of the ideal body in the aesthetic sense was derived from that of the healthy body in a medical sense. The drawing of the brahmin's son had to comply with these standards of health and beauty. The same standard measures have been incorporated into other iconographic works, Hindu as well as Buddhist, and were followed whenever the Buddha image or any other religious image was made.[10]

The motif that an ideal image could represent a person who is not present is reminiscent of two well known groups of legends which can be found in Chinese sources with regard to the first Buddha images. The sources where they are found are relatively late and may date from about the same time as the Citralakṣaṇa or somewhat earlier.[11] The legends are often quoted in art historical literature in order to prove that images of the Buddha already existed in his lifetime. This is precisely the sort of question that modern scholarship so much wants to have answered. No matter how desirable this may be, it is hazardous to try to elicit any historical evidence from these legends.[12] But this observation should not prevent us from examining the issue which is actually raised in these accounts, namely the question of whether or not the Buddha could be represented in his absence through an image.

The stories are too well known to be repeated here in full. The first group contains legends about the first sandalwood images of the Buddha and runs briefly like this: When the Buddha was preaching in the Heaven of the Thirty-Three Gods King Udyāna 'was stricken with grief at the Master's absence'[13] and ordered a statue of him to be made. The legend tells us that after the Buddha had come down from heaven, this statue began to move, walked towards the Buddha, and greeted him respectfully.

The second group consists of rules which can be found in the Chinese Vinaya. These case histories are based on the standard model of Buddhist virtue, the rich merchant Anāthapiṇḍada who 'asks permission to have a sandalwood image made, so that the monks may be "disciplined" by its majestic appearance when the Buddha himself is not present'.[14]

The legend and the Vinaya rule demonstrate that these 'first' Buddha images served as a replacement for the physical body of the Buddha for the time that he was absent. Surprisingly, our sources make clear that the physical *absence* of the Buddha was the actual reason of creating these images, as was the case in the first drawing of the brahmin's son.

Another common motif in these stories is that the image almost seems to come to life. The painting of the brahmin's son acts as if it were a living being. The same can be said of the Buddha images in the legends just quoted. Literary sources affirm that 'true to life' (*sādṛśya*) was considered a quality of good painting.[15] It does not need to have anything to do with miracles.

In view of what these passages reveal to us with regard to the questions raised at the beginning, we first take note of the accent they lay on the human body. They provide us with one of the motivations for the Indian predilection for the human body. After all, the first attempts to create a drawing or a sculpture regarded the faithful likenesses of a boy who died or of a master who went away. Secondly, we observe an intellectual distance towards visual representations which should not be underestimated. This attitude allowed people to abstract from the image and to consider it as a token, a 'sign' replacing someone who is not, or no more, present.[16]

VIṢṆUDHARMOTTARA PURĀṆA

Our second source is the Viṣṇudharmottara Purāṇa, a work which is usually dated in the 8th century AD. The instruction about the rules of painting and the making of images forms part of the third Khaṇḍa. Again, this instruction is given to a royal patron not to a painter or a sculptor. It is not the painter who is expected to steep himself in literature and allied arts like music and dance,[17] but the well-educated gentleman, someone who takes part in the life of the wealthy elite. The text itself makes this abundantly clear: the person asking the questions is a mythical king, Vajra, who is the representative of the *kṣatriya* caste and of the class of the patrons at the same time. The patron, of course, is the one who is to be persuaded to build a Viṣṇu temple, and to order all the appropriate sculptures and paintings. Therefore, he should know about the meaning of the temple and its decorations, and he should be aware of the theological viewpoints of the religious sect he is favouring with an important donation.

It is this royal patron to whom the brahmanical 'master of theology' (*ācārya*) is speaking, through the mythical sage Mārkaṇḍeya. This 'master' was usually the man who supervised the construction of the temple as far as ritual and iconography were concerned.

Besides, a particular grammatical formula used in this text should warn us that the information is not meant for the actual maker of the images or paintings, but for the one who commissioned them. The causative *kārayet*, or *kārya*, which is regularly used in these chapters, only makes sense if the instruction is directed to someone who orders an image be made.[18] The theological character of the explanations of the meaning of the images (see below) is another indication that the author wanted to write a religious work first and foremost, not a sculptor's or painter's manual.

Like the first account discussed in this paper, the text deals with the origin of painting but in a way that is entirely different from the approach of the Citralakṣaṇa. At the very beginning of his treatise the author is adamant that a painting cannot be understood properly without a thorough knowledge of dance. A training in the rules of dance is an absolute prerequisite, he says, for everyone who wants to understand pictorial art.

> 'Tell me, blameless one, about the making of the forms (*rūpa*) of the deities, in order to lay down their outward form in the handbooks forever,' Vajra said. Mārkaṇḍeya answered, 'Who does not fully know the rules of painting, O king, can never understand the characteristics of images' (III 2,1-2).[19]

When Vajra wants to know about painting, Mārkaṇḍeya says:

> 'Without a book on dance, the rules of painting are very difficult to understand, because the imitation (*anukriyā*) of the world should be accomplished by means of both, O king' (III 2,3).

Before this can be done, the text continues, it is necessary to know about instrumental music (*ātodya*) and then about singing (*gīta*), which includes the use of Sanskrit and Prakrit, prose and poetry, metres, and such matters. The king has a long way to go before his guru is ready to answer his first question about the rules of painting and sculpture.

The close relationship between painting and dance is stressed again just before the chapters on painting begin, 'In painting reality (literally: 'the three worlds', *trailokya*) is imitated (*anukṛti*) in the same way as in dance' (III 35,5).[20]

It is remarkable that in this hierarchical order, as we would prefer to call it,[21] the place of honour is given to the performing arts, in particular to dance. The hierarchy of arts which is offered here is in striking contrast to the Chinese and Islamic scales of arts in both of which calligraphy is valued as the ultimate artistic form.[22] In India, the human body is placed in the foreground.

In another passage we are told that the art of dancing was invented by Viṣṇu:

> In times of yore when the world was one ocean of water and nothing inanimate or animate existed, and when Madhusūdana was asleep on a bed of Śeṣa, while his feet were supported by Lakṣmī, a lotus sprung from his navel. On this lotus, the god Brahmā was born spontaneously, pure and having four faces, together with the gods who were provided with bodies (III 34,2-4).

In order to kill two demons, Madhu and Kaiṭabha, who had snatched Brahmā's Vedas away:

> Viṣṇu rose from the water and started moving on his bed of water. Watching him moving with lovely bodily movements and dance steps, the longeyed Lakṣmī felt an extraordinary passion for him (III 34,8-10).

Heedless of her desire, the god immediately assumed a terrifying appearance, assumed the head of a horse and descended into the netherworld to kill the demons Madhu and Kaiṭabha.[23] His dancing on this occasion was the very origin of this noble art, which was then handed over to Brahmā, then to Śiva, and afterwards to mankind (III 34,19-22).

In spite of the mythological colours which the author has chosen to brighten his story, his remarks on the superiority of the performing arts compared to the plastic arts are illuminating. 'The imitation of reality is based upon dance', Viṣṇu explains (III 34,17). This statement means that the plastic arts are completely dependent on choreography and gesture-language as far as forms of expression are concerned.

The author of the Viṣṇudharmottara Purāṇa apparently had his own 'agenda'. Being a Vaiṣṇava theologian, it was his aim to explain the ideas and concepts of *Pāñcarātra* Viṣṇuism. The author's own devotion towards the god Vāsudeva as the supreme deity is clear throughout the whole work. In his eyes, the arts serve merely as instruments of worship. Song and dance are classified as forms of religious worship, more costly perhaps than a bunch of flowers or a basket of food offerings, but all the same elements that form part of a ritual (III 34,25-7).

As we study this exposition on the meaning of the images of the gods of the Vaiṣṇava pantheon, we are systematically confronted with religious metaphors. One of the attractive sides of this text is precisely the way in which every characteristic of a religious image is connected with a specific theological concept. Everything seems to signify something else, something beyond its primary meaning. The number of faces or heads, the attributes, hand gestures and body postures, and the like - every iconographic feature is interpreted in terms of Vaiṣṇava theology. The text uses the term *hetu*, 'motivation' or 'reason', to indicate this metaphorical stratum.

To illustrate this point Brahmā's image and its symbolic meanings will be briefly discussed. In Chapter 44,3, right at the beginning of his treatise on the 'characteristics of images' (*pratimālakṣaṇa*), the author mentions that Viṣṇu has three bodily appearances, which are identified with the three cosmic qualities (*guṇa*) of Sāṁkhya philosophy. The Brahmā-form represents 'cosmic energy' (*rajas*), which sets the universe in motion; his Viṣṇu-form is identical with 'cosmic light' (*sattva*), by which the created world is preserved; and the third Śiva-form is equal to 'cosmic darkness' (*tamas*), which brings about destruction. In this way the three cosmic qualities (*guṇa*) are introduced to explain Viṣṇu's three manifestations, namely Brahmā, Viṣṇu, and Śiva.

After having explained Brahmā's iconography in Chapter 44, the text gives us instructions how to interpret Brahmā's features symbolically in Chapter 46. The passage opens with a lesson in Viṣṇuitic theology,

> Vajra said: 'If Puruṣa (=the transcendent Being) is without form, smell, taste, sound or touch, as you said, how can he have an outward form?' //1//
> Mārkaṇḍeya said, 'The Highest Being has two forms (*rūpa*), a basic mode of being (*prakṛti*) and a modified mode of being (*vikṛti*). The basic mode of being is known as his form without characteristics, //2//
> his modified mode of being should be understood as being provided with an outward appearance; it is known as the universe. Worship, meditation, etc. can only be performed of the form with an outward appearance. //3//
> The outward appearance of the deity himself can be worshipped according to the rules. But his unmanifested mode of existence can hardly be reached by those who have a body. //4//
> Therefore, the outward appearance which is shown in its manifestations by the Lord at will, that form is worshipped by the gods. //5//
> Because of this, worship is performed of the Lord in his outward appearance.'

After setting the tone with this theological statement, the author proceeds by giving a metaphorical meaning to the iconographic features of Brahmā,

> 'Listen to me when I speak about his outward form and its reasons (*hetu*). //6// Red is the colour of energy (*rajas*), therefore Brahmā, the excellent god, honoured by all beings, should be known as being similar to the points of the red lotus flower. //7//[24] His eastern face is the Ṛgveda, his southern face the Yajurveda, his western face the Sāmaveda and his northern face the Atharvaveda; //8// his faces are to be understood as the Vedas,[25] his four arms as the directions of space. The whole universe, inanimate and animate, is identical with the waters; //9// these waters Brahmā is carrying, therefore a watervessel is in his hand. The rosary in Brahmā's hand is indicating time. //10// It is called "time" (*kāla*) because of its "urging" (*kalana*) all living beings. Each sacrifice is arranged in accordance with the light and the dark offering, //11// therefore Brahmā's garment which is the hide of the spotted antelope should be understood as representing the white and the dark (rites). The seven worlds are known as *bhū-*, *bhuvas-*, *svar-*, *mahat-*, //12// *janas-*,

tapas- and *satyaloka*. These worlds are Brahmā's, the sacrificer's, geese in front of his chariot. //13// The lotus that springs from Viṣṇu's navel, that is the earth. The pericarp of this lotus should be understood as Mt. Meru, O best of kings. //14// This mountain means stability of the earth; because of this, the Lord assumes a stable meditation posture by means of the lotus posture which is the earth. //15//[26] He is meditating on the highest 'heaven' (*dhama*) of himself as being without a form; in order to see the worlds (at the same time) he is sitting with his eyes half closed in meditation. //16// The long strings of ascetic hair of Brahmā, the holy one, who is in everything, should be understood as the vegetation (*oṣadhayas*) which causes the world to be preserved, O king. //17// The strings of ornaments of the supreme Lord should be understood as the branches of knowledge which illuminate the world. //18// This, then, is the outward form which I have explained to you of the incomparable (*apratima*) one, the one in whom the whole world is existing. In this way the world supporter supports the complete world with his body.' //19//

In the next chapters the author proceeds to enumerate the *hetu*s of Viṣṇu and then of Śiva. We shall follow him no longer. Our example shows that he is teaching us an ingenious, though somewhat arbitrary, piece of theology, using the bodily form of Brahmā as a device for creating his metaphors.

When we try to see through his scholarly discourse on the 'motivations' of Hindu iconography, we discover that he is applying various categories of brahmanical theology to the various parts of Brahmā's image: cosmological elements such as water, earth, Mount Meru, the four points of the compass, the seven 'worlds' (*loka*), time; philosophical concepts like the three cosmic qualities, transcendent and immanent, manifested and unmanifested; and terms referring to sacrifice, the four Vedas, the branches of knowledge.[27] The image becomes a complicated metaphor of a theological discourse, a text in visual form. In this metaphorical treatment of the icon, the deity as a personal god is very distant. Devotion belongs to other occasions or other levels.

It would be dangerous to generalize the metaphorical meanings presented in this text. The present author would not subscribe to Shah's view which regards them as 'keys to unlock the meaning of the various forms and symbols found in Hindu Sculpture'.[28] The statement made by some historians of Western art that 'symbolic motivations are absolutely not generalizable'[29] seems to be equally valid for this type of interpretation. They do represent the views of an influential medieval brahmanic stream which prefers to look at the image of the supreme deity as a religious metaphor.[30]

Recent research has made some progress in tracing down other levels of meanings of the same images which were relevant to other groups of people in the same period. For the meanings the royal patrons themselves would prefer to be attributed, we have to look at the phenomenon of 'royal iconography', not of religious iconography.[31] In the case of the iconography of Viṣṇu, it has become clear that the attribute of the conchshell is simply a martial attribute, a sort of trumpet which was blown to signal the commencement of a battle. The discus, too, was a real weapon which was used in epic as well as in real battles, as was the club.[32]

An eighth century image of Viṣṇu from Kashmir, now in the Rijksmuseum in Amsterdam, makes us aware of the royal features of the figure sculpted in stone (see photograph). The posture, the crown, the ornaments, the garments, the dagger in the belt, even the mythic-heroic connotations of the heads of a lion and a boar represented on its right and left side - everything about this sculpture is reminiscent of a king who is posing before his people during a great ceremony, like royal officials still do to this day when they present themselves at ceremonial occasions in full-dress military uniform. Kings need to show physical and political strength, and provide their gods with the same traits.

CONCLUSIONS

This interpretation of two iconographic sources from different periods has provided us with various 'motivations' for the Indian tendency to put the human form at the centre of artistic expression.

The author of the Citralakṣaṇa answers our initial question with a tale about a boy who died unexpectedly. The first painting ever made was a faithful likeness of his idealized body. It was meant to represent a dear relative who is not alive anymore. The Buddhist legends about the first Buddha images in human form are telling us that these images were created as a token of the master during his absence. These answers suggest that the emphasis on the human body in Indian art has something to do with a need of representing persons who died or are elsewhere by making pictures of them.

The opinion of the author of the Viṣṇudharmottara Purāṇa is that the language of the performing arts, dance in particular, should be understood first, before the art of painting can be grasped. In saying this, he is implying that the expressive qualities of the human body form the core of all pictorial art. It is significant that in his theological discussion of the religious image, he forgets all about dance and gesture language but still regards the human body as an appropriate means to convey abstract ideas.

These diverse answers only partially explain why the human body is so important in Indian art. On the other hand, they all reveal a high degree of intellectual freedom to interpret the image in their own abstract terms.

Illustration

Photograph. Three-headed Viṣṇu. Stone sculpture, 8th century, from Kashmir. H. 64 cm. Rijksmuseum Amsterdam RAK 1990-3.

REFERENCES

Freedberg, D.
1989*The Power of Images. Studies in the History and Theory of Response*. Chicago and London: University of Chicago Press

Gail, A.J.
1989 'Iconography or icononomy? Sanskrit texts on Indian art', in: A.L. Dahmen-Dallapiccola (ed.) *Shāstric Traditions in Indian Arts*. Stuttgart: Steiner, Vol. I, pp. 109-14

Goswamy, B.N., and A.L. Dahmen-Dallapiccola
1976 *An Early Document of Indian Art. The 'Citralaksana of Nagnajit'*. New Delhi: Manohar

Laufer, B.
1913 *Das Citralakshaṇa, nach dem tibetischen Tanjur herausgegeben und übersetzt*. Leipzig: Harrassowitz (Dokumente der indischen Kunst, erstes Heft, Malerei)

Lüders, H.
1911 *Bruchstücke buddhistischer Dramen*. Berlin: Reimer (Königlich Preussische Turfan-Expeditionen, Kleinere Sanskrit-Texte, Heft I)

Pollock, S.
1989 'The idea of Śāstra in Traditional India', in: A.L. Dahmen-Dallapiccola (ed.) *Shāstric Traditions in Indian Arts*. Stuttgart: Steiner, Vol. 1, pp. 17-26

Roth, G.
1990 'Notes on the Citralakṣaṇa and Other Ancient Indian Works on Iconometry', in: M. Taddei (ed.) *South Asian Archaeology 1987*. Rome: Istituto Italiano per il Medio ed Estremo Oriente, Part 2, pp. 979-1028

Ruelius, H.
1974 *Śāriputra und Ālekhyalakṣaṇa. Zwei Texte zur Proportionslehre in der indischen und ceylonesischen Kunst*. Göttingen: Dissertation

Shah, P.
1958,61 *Viṣṇudharmottara-Purāṇa, Third Khaṇḍa*. Vol. I. Textcritical edition. Vol. II. Introduction, Appendices, Indexes. Baroda: Oriental Institute (GOS. 130 and 137)

Sivaramamurti, C.
1978 *Chitrasūtra of the Vishṇudharmottara*. New Delhi: Kanak

Suśruta
1835 *The Suśruta or System of Medicine*. Vol. I. Text edited by Madhusūdana Gupta. Calcutta: The Education Press
1883 *The Suśruta Saṁhitā. The Hindu System of Medicine according to*

Suśruta. Translated by U.C. Dutt. Calcutta: Asiatic Society (Bibliotheca Indica N.S. No. 500. Fasc. 2)

Taddei, M.

1993 'Reflections on *śaṅkhā* in Vaiṣṇava Iconography', in: A. Gail and G.J.R. Mevissen (eds), *South Asian archaeology 1991*. Stuttgart: Steiner, pp. 647-58

Zürcher, E.

1995 'Buddhist art in medieval China: the ecclesiastical view', in: K.R. van Kooij and H. van der Veere (eds) *Function and Meaning in Buddhist Art*. Groningen: Egbert Forsten (Gonda Indological Series. 3) pp. 1-20

Notes

1. Citralakṣaṇa, after the German translation by Laufer, 1913, p. 144.
2. Roth, 1990, p. 987 argues that the author of the Citralakṣaṇa must have lived before Varāhamihira.
3. Laufer, 1913, pp. 12-13. The name of the king in this story gave rise to a lot of speculation about this supposedly Jainistic or even Hellenistic origin of the work. However, the epithet Nagnajit, 'conqueror of the naked ones', that is the *pretas*, was given to the king because he had defeated Yama's army which consisted of these *pretas*, see Laufer, 1913, p. 138. It is true that the name Nagnajit is mentioned again as someone who transmitted the art of painting and as the author of the first chapter of the Citralakṣaṇa, and the story may be secondary. However, the intentions behind the story seem to be more illuminating than any possible historical core. The tale is about the first drawing of a portrait of a boy who died prematurely and thus became a *preta*. We surmise that *nagnajit* in this context means: someone who 'wins back' a *preta* by making a painting of him. This is what the king in his role of the primeval painter actually did. Otherwise: Roth, 1990, pp. 985-6.
4. The text was first translated into German by Berthold Laufer in 1913. The existing English translations are based on his translation, although they do not always follow it faithfully, e.g. Goswamy & Dahmen-Dallapiccola 1976. See the excellent study by Roth, 1990, pp. 979-1028. I summarize this story on the basis of Laufer's original German translation. A fresh translation of the complete text from the Tibetan text is urgently needed.
5. Disputes and fights with Yama about untimely death are not unknown in Indian literature and demonstrate that Indians were not always willing to accept death unconditionally. The Sāvitrī story is justly famous because the heroine defeats Yama in a dispute about the untimely death of her husband.
6. Laufer 1913, p. 135: 'Entsprechend seiner Gestalt und mittels der Farben sollst du den Sohn des Brahmanen in einer ihm ähnlichen Weise trefflich malen.'
7. Laufer 1913, p. 136: '..als einen Lebenden'.
8. Faithfulness in the case of 'portraits' was appreciated in classical Gupta India, see e.g. Sivaramamurti 1978, p. 118.

9. Suśruta Saṁhitā I 35, pp. 125-6 (text), pp. 144-5 (translation).
10. See Ruelius, 1974, p. 16.
11. Zürcher, 1995, p. 4.
12. *Ibid.*, p. 5: 'It is obvious that these traditions are utterly unhistorical.'
13. *Ibid.*, p. 4.
14. *Ibid.*, p. 6.
15. See e.g. Kālidāsa, Raghuvaṁśa 8, p. 92.
16. Clear proof of this intellectual freedom towards visual representations are the small fragments of ancient Buddhist theatre plays which were discovered in Central Asia, but hail from the city of Mathura in the second century AD. These fragments have been known ever since Lüders' critical edition in 1911. Yet, their value for art historical interpretation has never been done proper justice. What has survived of these plays, supplemented by indirect evidence of other plays which are lost, is clear testimony to ancient India's familiarity with *real* personifications, that is abstract concepts which are represented on stage by actors. 'Wisdom' (*buddhi*), 'Firmness' (*dhṛti*) and 'Glory' (*kīrti*) are personified and begin a conversation about the Buddha. The Buddha is called 'Light that bears the name of man', and 'Dharma in human form'. We will come back to these fragments in a later publication.
17. Sivaramamurti, 1978, p. 6.
18. Gail, 1989, p. 112 did not take this causative into account.
19. The new translations offered here are based upon the text of the critical edition of Shah, 1958.
20. We would prefer a somewhat different interpretation of this concise statement than C. Sivaramamurti has offered (1978, p. 9). In our view, it is not 'a close observation and reproduction of the world around us, in dance and painting alike' that is emphasized, but the dependence of painting on dance in terms of representation of reality. Dance and painting use the same 'choreographical' means of expression.
21. Shah (1961, pp. 4-5) prefers to call it 'interdependence of the arts', and rightly points to the standard education of a gentleman in ancient India, who ought to know about all the arts.
22. In the Indian tradition writing was never an art in itself. Only in some Buddhist manuscripts we come across some occasional attempts to calligraphy, for instance in a few Pāla manuscripts.
23. It is remarkable that Viṣṇu's dance is immediately associated with a violent act, viz. the killing of demons. The same association is made in the case of the dance of Śiva.
24. Cf. III 44,8: *padmapatradalāgrābhaṁ*, 'like the petals of the red lotus'.
25. Cf. III 44,5: *caturmukhaṁ*, 'with four faces'.
26. Vs. 15ab: The Bombay edition reads: *sarvatra* and *sthitaṁ*, ms C reads: *parvatam* and *sthiraṁ*. Cf. P. Shah (1958), critical apparatus. Shah (1961, p. 140), however, translates the readings of ms C, although her interpretation differs in a few respects. The whole line should run: *parvataṁ pārthivaṁ sthairyaṁ dhyānabandham ataḥ sthiraṁ*. Compare III 44,5 and III 44,9.

27. About the term 'branches of knowledge', *vidyāsthānāni* which may refer to fourteen or eighteen branches of knowledge, such as the four Vedas, Dharmaśāstra, History, the six Vedāṅgas, Mīmāṁsā and Nyāya, see Pollock, 1989, pp. 21-4.
28. Shah, 1961, p. 139.
29. Freedberg, 1989, p. 79.
30. It is difficult to establish how many Hindus would have known, or read, this part of the text at the time it was composed, or in later periods for that matter. To trace the impact of the symbolic explanations on the reception of Hindu art offered in this work would involve another type of iconographical and philological research.
31. Art historians in fields other than South or Southeast Asia developed this term to explain the phenomenon that kings usually choose strikingly similar iconographic means to create their own royal imagery.
32. Taddei, 1993, p. 649.

XVIII

WHAT LIES AT THE BASIS OF INDIAN PHILOSOPHY

*Sengaku Mayeda**

1. PREFACE

In 1983 I had an opportunity to meet His Holiness Abhinava Vidyātīrtha (1917-89) at the Shringeri monastery, the headquarters of the Śaṅkara School. On that occasion I casually noted that in Japan there were only a few scholars who were engaged in research on Indian philosophy although there were many researchers of Buddhism. Then His Holiness said as follows:

> Buddhism and our Non-dualism are exceedingly similar. The only point of difference lies in the fact that we, Advaitins, recognize the existence of a substantial entity but Buddhism does not do so.

From this remark I was fortunate enough to obtain a glimpse of His Holiness' view of Buddhism.

His statement was agreeable to me. It was, however, contrary to my expectations. The reason is that when we call to mind Śaṅkara's violent criticism of Buddhism as revealed in his commentary on the Brahma Sūtra and his other books, we could scarcely expect a statement of that nature from

* The substance of this paper is based upon my lecture in Japanese 'Indo-shiso no Kontei ni arumono' delivered at the University of Tokyo on January 25, 1991 which was published in the Shunju, No. 328, 1991, pp. 5-13 and this paper was read at the East-West Philosophers' Conference held at Massey University, Palmerston North, New Zealand from August 12 to 15, 1994. I would like to express my thanks to Professors Eli Franco and Anindita N. Balslev who kindly read through my original manuscript and gave me valuable suggestions. Thanks are also due to Mr. Richard A. Paulson who kindly took trouble to revise and correct my English.

a person like His Holiness Vidyātīrtha, who himself is none other than the successor of Śaṅkara, a great exponent of Advaita Vedānta in the eighth century, and the representative of his school.

2. ŚAṄKARA'S VIEW OF BUDDHISM

Now, it might be necessary for me to explain Śaṅkara's view of Buddhism. In order to know his view of Buddhism, I would believe that there is no better source material than his commentary on the Brahma Sūtra.[1]

Here in his commentary on the Brahma Sūtra, just prior to his criticism of Buddhism, he criticizes the Vaiśeṣika school and characterizes it as 'semi-nihilism' (*ardhavaināśika*). In this connection he defines Buddhism as a whole as 'the doctrine which asserts the nihilism of everything' (*sarvavaina-śikarāddhānta*). Then, as I have pointed out elsewhere, he makes an attempt at 'critical classification of the Buddhist doctrines', on the criterion of what is asserted as real or, in other words, from his ontological point of view, and criticizes the following three different types of Buddhist propounders: (1) 'the Buddhist who asserts the real existence of everything' (*Sarvāstitvavādin*), (2) 'the Buddhist who asserts the real existence of Vijñāna, i.e. Consciousness' (*Vijñanāstitvavādin*) and (3) 'the Buddhist who asserts the emptiness of everything' (*Sarvaśūnyatvavādin*), not to mention 'the Buddhist who asserts the emptiness of everything', 'the Buddhist who asserts the real existence of consciousness', as well as 'the Buddhist who asserts the real existence of everything' commonly advocate 'the doctrine of entities having but momentary existence' (*kṣanabhaṅga*).[2] Therefore, in Śaṅkara's understanding, Buddhism as a whole is a doctrine which asserts total nihilism.

For Śaṅkara, even that Buddhist doctrine which asserts that consciousness alone is real is nihilistic and without a shred of rationality (*upapatti*). Śaṅkara adds that it is not possible to lead an ordinary life on the basis of such a nihilistic view.

And finally he concludes his criticism against the Buddha and Buddhism with the following words:

> No further special discussion is in fact required. From whatever new points of view the Bauddha system is tested with reference to its probability, it gives way on all sides, like the walls of a well dug in sandy soil. It has, in fact, no foundation whatever to rest upon, and hence the attempts to use it as a guide in the practical concerns of life are mere folly.
> Moreover, by propounding the three mutually contradictory systems,

> teaching respectively the reality of the external world, the reality of consciousness only, and total nihilism, Buddha has himself made it clear either that he was a man given to making incoherent assertions, or else that hatred of all beings induced him to propound absurd doctrines by whose acceptance they would become thoroughly confused … . Buddha's doctrine has to be entirely disregarded by all those who have a regard for their own happiness.[3]

3. ŚAṄKARA'S OPPONENTS' VIEW OF ADVAITA

It is, however, well known that Śaṅkara was criticized by his opponents as 'a Buddhist in disguise' (*pracchannabauddha*) and his philosophy as 'the doctrine affirming the world to be an illusion' (*māyāvāda*)[4] which is but crypto-Buddhism.

Among the Vedāntins, Bhāskara (750-800) was probably one of the earliest critics of Śaṅkara. He called the Māyāvādin 'one who depends on the doctrine of the Buddhist' (*Bauddhamatāvalambin*), and said that this position has been negated by the author of the Brahma Sūtra.[5] Afterwards, Yāmuna (918-1038), Rāmānuja (-1137), Madhva (1197-1276), Vallabha (1473-1531) and other Vedāntins severely criticized the Advaita Vedānta, pointing out that it is in essence nothing but a Buddhist doctrine.[6]

This is true of modern Indian scholars. For example, in his *A History of Indian Philosophy*, S. Dasgupta remarks:

> Śaṅkara and his followers borrowed much of their dialectic form of criticism from the Buddhists. His Brahman was very much like the śūnya of Nāgārjuna … . I am led to think that Śaṅkara's philosophy is largely a compound of Vijñānavāda and Śūnyavāda Buddhism with the Upaniṣad notion of the permanence of self superadded.[7]

Even S. Radhakrishnan who asserts that 'there is no doubt that Śaṅkara develops his whole system from the Upaniṣads and the Vedānta Sūtra without reference to Buddhism',[8] says as follows:

> We need not say that the Advaita Vedānta philosophy has been very much influenced by the Mādhyamika doctrine … the Nirguṇa Brahman of Śaṅkara and Nāgārjuna's śūnya have much in common.[9]

4. ŚAṄKARA AND VIJÑĀNAVĀDINS

Then, what possibly could have been Śaṅkara's own reaction to the charges that his doctrine resembled that of Buddhism?

First of all, we must admit that, judging from his writings, there is no room to doubt the fact that his knowledge of Buddhism was of a high standard and accurate. Moreover, it is very probable that those critics against Śaṅkara appeared not only after his death but also already during his lifetime. In fact, Śaṅkara himself explicitly admits in his Māṇḍūkyopaniṣatkārikābhāṣya that Buddhism is said to be in close similarity to the Advaita of his paramaguru Gauḍapāda.

However, Śaṅkara was by no means easily amenable to accepting such a similarity between Non-dualism and Vijñānavāda. On the contrary he stressed in his Māṇḍūkyopaniṣatkārikābhāṣya[10] the Advaitins' originality and difference from Buddhism in the following two points: First, the Advaita philosophy expounds the non-dual, Ultimate Reality which, unlike the Vijñānavādins' Vijñāna consisting of three components,[11] i.e. knowledge, object of knowledge, and knower, does not possess any components like those, and secondly the Advaita philosophy is based on the absolute authority of the Upaniṣads.

In this connection the Ālayavijñāna of the Vijñānavādins may come into question. It is the substratum of transmigratory existence assumed by them and corresponds to the unchangeable substratum, that is, the Ātman of the fourth stage whose existence is strongly asserted by the Śaṅkara school. Śaṅkara's critique against this Ālayavijñāna is found in his Brahma Sūtrabhāṣya where he denies the possibility of Ālayavijñāna's being the substratum of residual impressions for the Vijñānavādins accept its momentariness, although they recognize the Ālayavijñāna as the substratum. In Śaṅkara's view our daily activities like memory and recognition would presuppose the existence of a continuous entity (*anvayin*) that persists in all the three times, past, present and future, or else, an unchangeable, permanent (*kūṭasthanitya*), quiescent Ātman. Therefore, this Ātman is essentially different from the Ālayavijñāna.

Even if Śaṅkara's Ātman and the Vijñānavādins' Ālayavijñāna differ in essence, they show mutual similarity in the function that they exercise within each system, and as both of the two systems equally assert the non-reality of the phenomenal world, they both belong to a similar monistic standpoint.

In this context it is quite interesting to note the following fact: Buddhist philosophers, Śāntarakṣita (725-788) and Kamalaśīla (740-796), probably Śaṅkara's contemporaries in the eighth century recognize even the fact that Non-dualism is as valid as Vijñānavāda although they also point out that there is 'a slight mistake' (*alpāparādha*) in the Advaita which asserts the eternal nature of the Ātman.

5. TWO TRADITIONS IN THE HISTORY OF INDIAN PHILOSOPHY

In his work entitled *The Central Philosophy of Buddhism*,[12] T.V.R. Murti once asserted:

> There are two main currents of Indian philosophy - one having its source in the ātma-doctrine of the Upaniṣads and the other in the anātma-doctrine of Buddha.

I would think that nobody can deny his assertion. Then, how has the Vedānta school, which is representative of the orthodox (*āstika*) brahmanical traditions, come to assume close similarities to Buddhism, which is representative of the non-orthodox (*nāstika*) traditions?

One major reason for this is the fact that during their long history, the two traditions have exerted mutual influence upon each other. They have gradually approached one another and eventually almost fused with one another. This is a fact that one might reasonably assume. Actually the process whereby these two traditions approached each other can be explained to a certain degree. In my opinion, the process in which these two philosophical traditions approached one another was more or less as follows.

One of the fundamental standpoints of early Buddhism is said to have been the suspension of judgement with respect to metaphysical problems (*avyākata*). Judging from this basic standpoint of Buddhism, I would think that it is first of all contrary to Buddhism even to consider the question as to whether the Ātman exists or does not exist. According to some able scholars of early Buddhism, the concept of non-Self as revealed in the Suttanipāta (a text which is reputed to belong to the earliest group of Pāli Buddhist scriptures), by no means denies the existence of the Ātman, rather it stresses 'the rejection of attachment, particularly attachment to ego', 'the rejection of an undue adherence to something' and 'the refusal to be enslaved by anything'. I personally believe that the theory of non-Ātman as revealed in early Buddhism, is not to be construed as denying the existence of the Ātman as revealed in the Upaniṣads. As observed in the well known metaphor of the poisoned arrow of Māluṅkyaputta, the major concern of early Buddhism was final release from pain and sorrow (*duḥkha*) and attainment of nirvāṇa. With regard, however, to the question of the subject who experiences that pain and sorrow or the subject of transmigration, all judgement was suspended. Accordingly, it seems to me that the doctrine of non-Ātman as revealed in early Buddhism, does not conflict with that of Ātman.

However, in the period of the Abhidharma the theory of *anātman* i.e. non-Self could not escape changes in its connotation, and eventually came to mean 'the non-existence of Ātman'. In order, however, to prove the non-existence of Ātman the existence of many substantial entities called *dharmas*, i.e. existential elements was emphasized. When Mahāyāna Buddhism arose, these existential elements were severely criticized by Nāgārjuna, and accordingly the concept of *śūnyatā*, i.e. emptiness was propounded as the basis of Mahāyāna Buddhism.

However, just as Gotama Buddha was silent or suspended his judgement concerning the subject of pain and sorrow, Nāgārjuna was also silent or suspended his judgement concerning the subject who observes that everything is *śūnyatā*. As a result of this, I would suppose that the Consciousness-only theory of the Yogācāra school and the theory of Buddha-nature were propounded in order to make clear, from the standpoint of *śūnyatā*, what Nāgārjuna did not try to do. In other words Buddhism finally came to take up such problems as the subject of pain and sorrow and that of cognition concerning which any judgement was suspended in early Buddhism. These are actually nothing but problems of Ātman and Brahman in the Vedānta Philosophy.

It was in the period of the Gupta dynasty that the Yogācāra school came to cope with those problems; in this period the social dynamism of Buddhism was on the decline and Hinduism which received the support of the Gupta dynasty was on the upswing. Against such a background there arose the 'vedāntinization' of Buddhism, in other words, 'sanskritization' of Buddhism. To explain briefly, Buddhism was transformed from the pluralism of the Abhidharma into the monism of Ālayavijñāna of the Yogācāra school through the negation of *dharmas*, i.e. substantial entities of Abhidharma by Nāgārjuna's theory of emptiness.

As is well known, in the Laṅkāvatāra Sūtra, which was probably composed around the year AD 400, *tathāgatagarbha* or the matrix-embryo of the *tathāgata* was not only at times identified with Ālayavijñāna,[13] but the definition of this *tathāgatagarbha* was also very similar to the definition of Brahman in the Vedānta. Besides, Brahman, Viṣṇu, and Īśvara were also used as synonyms of *tathāgata* and the highest Brahman came to be regarded as the ultimate state.[14] In the same sūtra, Mahāprajña Bodhisattva raises the question to the Buddha as to whether the doctrine of the *tathāgatagarbha* is not the same as the heretical Ātman doctrine (*tīrthakarātmavāda*).[15]

On the other hand, the Vedānta school was formed with the Brahma Sūtra as its foundation in the fifth century, comparatively later than other philosophical schools including Buddhist schools; the Vedānta school established a system of Brahmanic monism, based upon the Upaniṣads, and especially in reacting to the then rather influential dualistic Sāṅkhya school.

However, this realistic monism of the Brahma Sūtra was of a character of difference-and-non-difference (*bhedābheda*), i.e. a dualistic monism, as it were. In denying such a standpoint and in the process of developing towards an absolute monism (*advaita*), that is, the upaniṣadic thought of the identity of Brahman and Ātman, the realistic monism of the Brahma Sūtra was gradually transformed and moved closer and closer to the Buddhism which had a more advanced theoretical system than the Vedānta. While doing so, this Vedānta philosophy came to be 'buddhisticized' considerably.

The gradual process of 'buddhisticization' of the Vedānta school is reflected in the four chapters of the Māṇḍūkyakārikā of Gauḍapāda (640-90). This 'buddhisticization' of the Vedānta school reached its highest point in the fourth chapter of this work composed before Śaṅkara.[16]

What Śaṅkara, born in South India in the eighth century, found before him were: first, the Mīmāṁsā school which was flourishing with excellent theoreticians like Kumārila (650-700), Prabhākara (ca. 700), Maṇḍanamiśra (680-720) and others; second, a tide of popular Hinduism which penetrated in leaps and bounds through the masses, and third, a weakening, esoteric-oriented Buddhism which was the target of strong criticisms of Kumārila and others; fourth, a thoroughly 'buddhisticized' Vedānta school.

Just as Dayananda Sarasvati (1828-1883) severely criticized the downfall of Hinduism in the nineteenth century with the cry 'Return to the Vedas', in the same way Śaṅkara in his days, beholding the prosperity of the Mīmāṁsā school and the deplorable situation of his own Vedānta school, must have proclaimed 'Return to the Vedas'. This must be the reason why there is no one who has left us so many commentaries of Upaniṣads as Śaṅkara.

With the purpose of the revival of the Vedānta school from the standpoint of orthodox Brahmanism, Śaṅkara, while using his profound knowledge of Buddhism, transmuted Buddhist doctrines in the Māṇḍūkyakārikā into Advaitism and achieved the re-'vedāntinization' of the 'buddhisticized' Vedāntic tradition.[17]

In this way, Buddhist doctrines, which had been absorbed into the pre-Śaṅkara Vedānta philosophy and constituted an integral part of the Māṇḍūkyakārikā, were preserved without being removed and were vedāntinized by Śaṅkara. As a result, the realistic monism of the Brahma Sūtra was transformed and developed into an illusionistic non-dualism which, as he himself recognized, closely resembled the Buddhist doctrine.[18] For Śaṅkara, Buddhism is an important target to overcome. For this reason he attacks it very severely here and there in his works. Śaṅkara was an epoch-making reformer in the history of Vedānta who turned the extremely buddhisticized tradition of Vedānta towards the Vedāntic Vedānta or Upaniṣadic Vedānta.[19]

Śaṅkara was aware of the fact that non-dualism was said to be similar to Buddhism, but he personally refused to accept this similarity. His reason for insisting that the highest truth is revealed only in the Upaniṣads, is probably a conviction arising from his desire to purify Vedāntic philosophy of its Buddhistic influences, and bring to light the original Vedāntic philosophy, which had developed as an interpretative science of the Upaniṣads.

By the process explained above the two traditions seem to have approached and resembled each other in spite of Śaṅkara's denial.

6. WHAT LIES AT THE BASIS OF INDIAN PHILOSOPHY

In such a way the two traditions finally attained a high degree of similarity. However, the essential difference between the two systems, in view of Śaṅkara and Śāntarakṣita, lies in the concept of Ātman and Ālayavijñāna, and we may further add that their difference lies in their source of knowledge as well.

Now, Ātman is everlasting while Ālayavijñāna is momentary. But both of them are in their essence pure consciousness. Therefore, their difference appears trifling. In my opinion, however, their difference is essential and deep-rooted.

After examining the two main currents of thought carefully and even recognizing the similarity between the Absolute of Vedānta and that of the Mādhyamika or Vijñānavāda, T.V.R. Murti concluded that 'there could not be acceptance of any doctrinal content by either side from the other, as each had a totally different background of tradition and conception of reality'. However, he did not clarify just what is 'a totally different background and conception of reality'. I would like to discuss what lies in the background of the two main different currents of Indian philosophy.

If I were to express my conclusion in advance, I would say that the conflict between the concepts of the Ātman and Ālayavijñāna takes its source from that of Sat (Being) and Asat (non-Being) which can be traced back to the creation myths of the Ṛgveda. Since I would believe that this conflict between Sat and Asat, which has a history of many centuries, lies at the basis of Indian philosophy, I would like to set aside a little time to deal with this question.

7. THE CONFLICT BETWEEN SAT AND ASAT

In his inspiring paper entitled 'Early Philosophical Speculation in the Rig Veda' the late professor W. Norman Brown (1892-1975) says as follows:

Vedic man did ... consider that there were in the cosmos two kinds of beings more powerful than humans whom he could influence. These were the gods (*deva*) and the demons (*rakṣas*). The gods were a kind of bureaucracy administering and enforcing the rta. They and human beings were held to occupy the surface of the earth, the vault of heaven, and the mid-air ... between those two. This part of the cosmos was known as the Sat 'Existent'. It had light, warmth, and moisture, and in it the *ṛta* prevailed. The demons occupied the Asat 'Non-Existent', a fearsome region situated below the Sat, from which it could be reached by descending through a great chasm. The Asat was without light, warmth, and moisture, and in it the *ṛta* did not prevail. Rather, the Asat was without the *ṛta* (*anṛta*).

> ... The demons were relentless enemies of the *ṛta* and of all beings living in the *ṛta*, that is, gods and humans. The gods and the demons were in an unending war with each other ... . Men and gods should be allied against the demons. To effect such an alliance was the basic purpose of the Vedic sacrifice.[20]

Furthermore, Vedic man tried to understand the operation of the universe and for this purpose it was essential to penetrate to the origin of the universe, particularly the origin of the Sat and Asat. In the Ṛgveda there are a number of theories in answer to this inquiry. According to Brown the myth of the hero god Indra and the demon Vṛtra presumably demonstrates the earliest example of these theories.[21]

This Indra-Vṛtra myth began with the precreation state of chaos. In this state all the elements of the Sat were in existence but were unorganized and were concealed in the Asat, where they were held in restraint under a huge cover or obstruction by the demon Vṛtra.

On the other hand the gods, who could not cope with Vṛtra, probably arranged for the birth of Indra. As soon as Indra was born, he drank *soma* and finally slew Vṛtra. Then the cosmic Waters came out. The Sat was then differentiated from the Asat, and the Sat contained the Waters and the Sun, thus being provided with warmth, light and moisture, and the *ṛta* was made to prevail there with the Sat, and there the demons continued to breed and endanger the beings living in the Sat.

It did not take long for Vedic man to be dissatisfied with the myth which was too naïve to remain convincing. Vedic man continued to search for the origin of the Sat and the Asat. Finally Vedic man's efforts resulted in the Nasadīya hymn (10.129) of the Ṛgveda which starts by taking us back before the differentiation of the Sat and the Asat, even before their very existence.[22] There did appear neither Indra nor Vṛtra in this hymn but 'That One' (*tad ekam*) alone breathed, though uninspired by breath, by its own potentiality in a primeval state where there was neither Sat nor Asat. It was

That One that brought about the creation of the universe. That is to say, It brought about the separation of Sat from Asat. The Nasadīya hymn states as follows:

> The sages by pious insight into their heart (that is, by introspection) found the relation (*bandhu*) of the Sat with the Asat.[23]

According to another hymn of the Ṛgveda 'The Sat was born from the Asat in the primitive ages of the gods.'[24] In the Atharvaveda[25] the Skambha, the support of the universe, is regarded as containing both Sat and Asat. Brahman is praised as 'the Womb of both Sat and Asat' and it is said in other hymns of the Atharvaveda[26] that the Sat is established in the Asat while the living beings rest securely in the Sat and that the gods who were born from the Asat were indeed great.

Besides, in the Brāhmaṇas Asat is also regarded as a kind of fundamental principle and there are many creation myths wherein Asat was at the beginning of the universe. In the Upaniṣads this idea was transmitted and we come across such passages as 'In the beginning this (universe) was Asat. There from, verily Sat was produced.'[27] and 'In the beginning this (universe) was Asat. It became Sat.'[28] Judging from the above observation, the idea that Asat is the origin of Sat seems to have prevailed since the Ṛgveda.

It was, however, the famous Upaniṣadic philosopher Uddālaka Āruṇi who denied this idea completely. In imparting the philosophy of Sat to his son Śvetaketu, Āruṇi insisted that it was Sat alone that existed in the beginning and strongly rejected the view that Sat was produced from Asat.[29] I would suppose that this philosophy of Sat stressed by Āruṇi came to be regarded as the legitimate theory of the orthodox Brahmans while that of Asat probably vanished from the limelight of the orthodox philosophical world.

In my opinion, however, although it disappeared from the orthodox philosophical world, the tradition of Asat was not totally destroyed but did continue to develop under such different forms as non-orthodox systems (*Nāstika*) of philosophy like Materialism, Jainism and Buddhism. Furthermore, even orthodox systems (*Āstika*) of philosophy such as the Nyāya and Vaiśeṣika which assert the combination theory (*ārambhavāda*) based upon the causation theory that the effect is not existent in the cause fall within this stream of Asat. Even the Ālayavijñāna of the Yogācāra school seems to belong to this philosophical tradition of Asat.

In the beginning of the universe the Sat and the Asat as fundamental principles of the universe are undifferentiated from each other as seen in the case of That One in the hymns of Ṛgveda. As in the Atharvaveda, the Sat and the Asat are 'twins', rather they could be better described as 'Siamese twins' which have Brahman as their common womb. Indeed, they are like the two sides of a single sheet of paper. The orthodox Brahman philosophers look at the obverse side of a paper while the Buddhists look at the reverse side of the same paper. Both Śaṅkara and Vasubandhu observed the same sheet of paper, but they did so from opposite directions.

8. CONCLUSION

At the basis of Indian philosophy there were the two main currents of thought: one accepted the Sat and the other accepted the Asat as the fundamental cause of the universe and human beings; both of them were orthodox brahmanical traditions of thought. In Upaniṣads, however, Uddālaka Āruṇi denied the philosophy of Asat and advocated that of Sat. It seems to me that this was the turning point wherein the orthodox brahmanical line adopted the standpoint of Sat and, against this. Buddhism and the other unorthodox schools accepted the standpoint of Asat.

Here I do hope that you will remember Murti's assertion that there are two main currents of Indian philosophy, each of which had a totally different background of tradition and conception of reality. However, he did not make clear what is 'totally different background and conception of reality'. If we were to suppose that it is the firm tradition of Sat and Asat, it is easily understandable why Vedānta and Buddhism, by extension Hindu philosophy and Buddhist philosophy, even though extremely similar to each other, are still possessed of a difference in content that cannot be ignored. It is, I would think, the two currents of Sat and Asat that lie at the basis of Indian philosophy.

The concepts of Sat and Asat are entirely contradictory to each other from a logical point of view. However, if we were to retrace our steps to their original root cause, we would arrive at that unique and undifferentiated That One. It should not be forgotten that Vedic man searched for neither being nor non-being but for their origin, That One. The two traditions of Indian philosophy equally aim at the Absolute, the ultimate state of liberation (*mokṣa* or *nirvāṇa*), which transcends the world of normal human logic, the world of subject and object, the world of nescience (*avidyā*), the world of frivolous talk (*prapañca*), and the world of being and non-being.

Notes

1. Śaṅkara ad Brahma Sūtra II,2,18,22.
2. *Ibid.*, II,2,31.
3. *Ibid.*, II,2,32.

4. In fact it is wrong that Śaṅkara's Advaita philosophy is called 'māyāvāda'. Cf. *A Thousand Teachings: The Upadeśasāhasrī of Śaṅkara, tr. with Introduction and Notes* by S. Mayeda. Tokyo: University of Tokyo Press, 1979, pp. 78-9.
5. Bhāskara ad Brahma Sūtra II,2,29.
6. For a detailed introduction, cf. H. Nakamura, *Shoki no Vedānta Tetsugaku* (Early Vedānta Philosophy). Tokyo: Iwanami, 1950, pp. 159-61; Yoshiaki Kodate, 'Vedānta Tetsugaku to Bukkyo-Kosho no Ichidanmen' ('An Aspect of the Vedānta-Buddhism Relations'), *Nihonbukkyogaku Nenpo* 23, 1957, pp. 257-75.
7. Vol. I, Cambridge: Cambridge University Press, 1951, pp. 493-4.
8. *Indian Philosophy*, vol. II. New York: Macmillan Co., 1958, p. 472.
9. *Ibid.*, vol. I. p. 668.
10. Maṇḍūkyopaniṣatkārikābhāṣya IV,99 p. 182 (*śrīśaṁkara granthāvaliḥ, samputah* 4. *Śrīvānī vilāsa mudrāyantrālayaḥ*, n.d.). There is a problem concerning the authenticity of this commentary. Cf. S. Mayeda, 'On the Author of the *Maṇḍūkyopaniṣad-* and the *Gauḍapādīya-Bhāṣya*', *The Adyar Library Bulletin*, vols. 31-32, 1967-1968, pp. 73-94; T. Vetter, 'Zur Bedeutung des Illusionismus bei Śaṅkara', *Festschrift für Erich Frauwallner, WZKS(O)* 12/13, 1968, pp. 407-9; P. Hacker, 'Notes on the *Maṇḍūkyopaniṣad* and Śaṅkara's *Āgamaśāstravivaraṇa*', P. Hacker *Kleine Schriften*. Wiesbaden: Franz Steiner, 1978, pp. 252-4, note 2. Previously, K. Kumada, taking the hint from this *Kārikā*, IV, 99, wrote 'The Fundamental Difference between Buddhistic and Vedāntic Philosophies', *Journal of Indian and Buddhist Studies* IX, no. 1, 1961, pp. 403-410. There are different controversies concerning the interpretation of IV, 99 of the *Kārikā*, but I deal only with Śaṅkara's commentary in this article.
11. The expression *jñānājñeyajñātṛbhedarahitam* likely presupposes *grāhyagrāhakasaṁvittibhedavān* seen in Dharmakīrti's Pramāṇavārttika II,354. Incidentally, this famous verse is quoted in Śaṅkara's Upadeṣasāhasrī II,18,142. Cf. my article 'Śaṅkara to Bukkyo', ('Śaṅkara and Buddhism'), *Bukkyo Kenkyu* 3, 1973, pp. 104-89.
12. London: Allen and Unwin, 1974, pp. 10-35.
13. Laṅkāvatāra Sūtra p. 221, line 12; p. 222, lines 6, 9; p. 223, lines 2, 6, 11; p. 235, lines 7, 16.
14. Laṅkāvatāra Sūtra, p. 153, line 9.
15. *Ibid.*, p. 77, line 13; p. 79, line 9. Cf. H. Nakamura, *op. cit.*, p. 199-204.
16. Cf. my article 'Śaṅkara to Bukkyo', ('Śaṅkara and Buddhism'), *Bukkyo Kenkyu* 3, 1973, p. 104.
17. Cf. my article 'On the Author of the *Māṇḍūkyopaniṣad-* and the *Gauḍapādīya-Bhāṣya*', *The Adyar Library Bulletin*, vols. 31-32, 1968, pp. 88-93.
18. Cf. my article 'Śaṅkara to Bukkyo', *op. cit.*, p. 103.
19. This may be one of the reasons that Śaṅkara was more rigid and stricter with regard to the problem of caste than the author of the Brahma Sūtra as well as his followers.
20. *India and Indology: Selected Articles by W. Norman Brown*, ed. by Rosane Rocher. Delhi: Motilal Banarsidass, 1978, pp. 79-80.
21. *Ibid.*, p. 80.

22. *Ibid.*, p. 82.
23. Ṛgveda 10, 129, 4.
24. *Ibid.*, 10,72,2; 3.
25. Atharvaveda 10,7,10.
26. *Ibid.*, 4,1,1; 5,6 1.
27. Taittirīya Upaniṣad II,7.
28. Chandogya Upaniṣad III,19,1.
29. *Ibid.*, VI, 2,1-2.

XIX

THE STORY OF JARATKĀRU ON A BALINESE *ULUN-ULUN*

*Dick van der Meij**

Introduction

In Bali, traditional polychrome paintings have been produced in the villages of Kamasan, province of Klungkung, East Bali, and also in Tabanan, West Bali; Gianyar, Central Bali; and Karangasem in the North-East. Now, most of these traditions have died out, but painting in Kamasan is still very much alive.[1] These traditional paintings used to be made for traditional furnishings in palaces and temples but nowadays are produced for other purposes as well. There is a booming industry for tourists, consisting of not only paintings but also painted containers for papers, bags, and greeting cards. Paintings depicting astrological calenders (*palalintangan*) are especially popular.

Traditional paintings are used for the decoration of temples and palaces during rituals and festivals. There are five classes of paintings (Forge, 1978, pp. 7-8): *tabing*,[2] *langse* (painted curtains), *ider-ider* (long hangings which are tied just under the eaves of pavilions in temples and palaces), ceiling paintings and pennants. An *ulun-ulun*, which is the subject of this essay is roughly square and put up against the wooden head of the raised bed which is the centre of all household rituals - forming a backdrop to the offerings laid out on such occasions (Forge 1978, p. 8). They are also used in a similar way in temple pavilions.

The paintings show scenes from traditional epics and stories such as the Mahābhārata, Arjuna Wiwāha, Rāmayāṇa, Malat, Bima Swarga, Tantri and many others (for details, see Marrison, 1995). *Tabing*, which derive their

---

* I would like to thank Prof. Karel van Kooij and Prof. De Casparis for reading and commenting on earlier drafts of this article. Thanks go to Rosemary Robson for the correction of my English.

inspiration from the Mahābhārata, show scenes such as the churning of the milky ocean, Kala Rauh, and scenes from the Bhāratayuddha. Usually one painting shows various scenes from a particular story. No individuality is expected from the painters and most paintings depicting the same story look alike. The style of these paintings is often called '*wayang*-style', since the traditional way people, gods, demons and other figures are depicted seems to be similar to the *wayang* puppets used in the shadow theatre. I would prefer not to use this term as it refers only roughly to the figures and creatures mentioned above - and even then their depiction is not quite the same! - and not to a number of other features of the paintings such as the material, the choice of the subject matter, and the fact that various scenes from stories are usually depicted in one painting.

There are quite a few collections of these paintings all over the world, but they tend to be concentrated in Australia, Bali, and the Netherlands. The best way to get an idea of these paintings is by browsing through the catalogue published by Anthony Forge which includes many pictures from the Forge collection in The Australian Museum in Sydney (Forge, 1978). As most paintings are rather cliché-like, most of them have been identified and studies of individual paintings or collections of paintings belonging together have been made by people such as Galestin (1939, on Tantri; 1943, on Smaradahana; 1969, on Lubdhaka; 1948-1949, 1956, on Malat); Goedheer (1939, Ādiparwa); Grader (1939, Brayut), Hinzler (1981, Bima Swarga); Vickers (1984, Malat); Worsley (1984, Rāmāyaṇa; 1988, Arjuna Wiwāha), to name but a few.

There are naturally exceptions which prove the rule and some paintings have continued to puzzle scholars for a long time. The *ulun-ulun* which is the theme of this essay is one such. In this essay I will describe the painting and offer two interpretations at different levels.

JARATKĀRU ON AN *ULUN-ULUN*

This particular *ulun-ulun* was discovered in 1985, adorning a wall in a bar next to the princely palace in Ubud, Central Bali. I was attracted to it to such a point that I even had it taken down from the wall and put outside in the light so that I could photograph it. Two years later I found myself in the same bar, the painting was still there and I liked it just as much as before. Over a glass of beer I decided to buy it, even though its quality had deteriorated during those two years. This decision caused some alarm as such orders were rare, but finally a price was agreed upon and the *ulun-ulun* travelled with me to Holland.

Illustration

The *tabing*, Klungkung, Bali, 1950s. 150 x 108 cm, painted by I Wayan Suweca, Banjar Sangging, Kamasan (photo by Wim Vreeburg)

The *ulun-ulun* depicts a scene from the Āstīkacarita from the Ādiparwa,[3] the first book of the Old Javanese Mahābhārata. It shows the journey of Jaratkāru to Swarga.[4] This is mentioned in Balinese in the colophon in the bottom right corner of the painting:

> *Ulun-ulun puniki pakariyan I Wayan Suweca saking Banjar Sangging Klungkung, Kamasan, Bali, lelampahanipun daweg sang Jaraṭkaru lunge ka swargeloki. puput*
>
> This *ulun-ulun* was made by I Wayan Suweca from Banjar Sangging, Klungkung, Bali. The story is Jaratkaru's journey to Swarga. End.

In short, the Old Javanese version of the story of Jaratkāru's journey to *Swarga* recounts the story of Jaratkāru who is an exemplary brahmin. He does not want to be involved with women and thus will not have children. In *Swarga* he is told by the spirit of an ancestor that he should have offspring, for should he not do so, all the merits which accrued during his (ancestor's) life will vanish and he will not be able to enter the abode of the ancestors. Jaratkāru promises he will marry, but only to a girl who also bears the same name, Jaratkāru, as himself. The only girl of that name is the younger sister of the serpent Bāsuki, and Jaratkāru marries her. They have a son, Āstīka. The Mahābhārata is recited to King Janamejaya to console him because the serpent sacrifice meant to kill the serpent Takṣaka was a failure circumvented by the interference of Āstīka who managed to persuade King Janamejaya to have the ritual stopped just before Takṣaka is about to fall into the ritual fire.[5]

This painting is not the only one of its kind. Three more similar *ulun-ulun* have come to my attention whereas other paintings showing various scenes from Swarga are found in great numbers (for instance in Delft, Terwen-de Loos 1964, ills. 91, and Hooykaas 1973, ills. y 1).

The first painting which is almost identical to the one shown here is kept in the collection of The Australian Museum in Sydney (A.M. E74161). What this particular *ulun-ulun* had to tell remained something of a mystery for a long time, even though it was clear that it showed scenes from *Swarga*. The theme had presented problems to Anthony Forge, who reproduced a picture of this painting in his catalogue (Forge, 1978: nr. 28) remarking: 'There are various stories of personages who visit Swarga and witness the torments of the damned, but it has so far proved impossible to identify which particular story this is.'

The second painting was identified by Hunter and forms part of the collection of William and Mary Basket of Cincinnati, Ohio. The picture is almost the same as the one shown here, but instead of presenting the story of Jaratkāru in the top right hand side, it depicts scenes from the Rāmāyaṇa (Hunter, 1988, p. 79) and features Hanuman instead of the serpent Bāsuki (see below).

The third painting forms part of the collection of the Royal Tropical Institute in Amsterdam. This painting is not really the same as the one discussed here. It shows the familiar scenes of *Swarga*, but the overall layout is somewhat different. However, in the scene the serpent Bāsuki is presented in our painting, in the Amsterdam painting we see Droṇa and the Paṇḍawas facing a god.[6]

We do not know how old the painting is. The former owner of the painting who told me in 1987 that it was about twenty-five years old. This is consistent with the statement of Mangku Mure and I Nyoman Kondra who told me in 1989 that the story of Jaratkāru used to be popular with painters during the 1950s and earlier, but at present is no longer depicted (oral communication from Mangku Mure and I Nyoman Kondra, Kamasan, 1989). Since the catalogue mentioned above showing the Amsterdam *tabing* dates from 1948 the painting was painted earlier. The other two *tabing* contain no useful dates. The *tabing* Forge has described bears a date corresponding to AD 1381 which is impossible. Nevertheless, on the grounds of style he dates it to the nineteenth century.

INTERPRETATION OF THE *ULUN-ULUN*

In Balinese traditional paintings, individual scenes are separated by strings of rocklike formations. Bearing this fact in mind, we are able to distinguish seven different scenes.

Scene one

Starting at the top left hand side we see five creatures who have the body of a human being, but the head of an animal, and one demon-like creature, all clad in waist cloths. They may be identified as (starting from left top to right below) a bull, a billy goat, a horse, a demon, a dog, and a boar. Their attitudes suggest moving. What these creatures are or what they represent is unclear.

Scene two

Continuing to the right we find Garuḍa - standing on a bull and a billy goat -

facing a tiered construction (*meru*), which is faced on the right hand side by a snake I identify as Takṣaka. Takṣaka is standing under the sun, depicted directly above him. Behind Takṣaka stand three feline creatures, perhaps two panthers and a tiger. They are standing under a tree. To their right we find two two-tiered pavilions adorned with cloths suspended under the eaves. In the pavilion on the left is a seated lady (Jaratkāru), attended by her servants who are engaged in conversation. In the right pavilion is seated a man (Jaratkāru) likewise attended by his servants. The servants of the man are engaged in conversation with the servants of the woman. The man sits facing towards a man who is talking to him. Two priests - one a man and the other a woman (recognizable because they wear priestly garb and the distinct *ketu* headdress) - are looking on, while young bamboo shoots extend over their heads.

Scene three

Starting from under the five animal creatures and their demonic companion we see a large scene of the torments in Yama's abode. The centre is taken up by a dagger tree, being shaken by Jogormanik who is sitting in the tree. The daggers fall on the wretched souls under the tree. Other souls are being tormented by the birds of hell. The souls are shown making gestures of trying to ward off these evils.

Scene four

This scene shows demon-like creatures trying to contain a bull. To the right we see the brahmin, Jaratkāru, and a servant, talking to Bāsuki and his sister Jaratkāru. We can tell they are engaged in conversation by the way in which they are holding their hands. The girl Jaratkāru, is standing submissively with her head lowered.

Scene five

A naked-breasted female soul is being tormented by a demon who tries to hit her with his club. Two children's souls are depicted below him. To the right is the 'quaking bridge' under which a fire is burning. A female soul is being taken by the arm by a demon and helped across it.

Scene six

Souls are confronted by demons. One demon is threatening to strike a soul with a *kris*, the other soul is threatened by a demon brandishing a club. A tiger and a leopard are biting at the poor souls.

Scene seven

Souls of children and men are threatened by a bull, a boar, and by demons. One soul is hoisted into the air by an elephant, another soul is being roasted over a fire. A two-headed bull cauldron contains five souls who are being cooked, while two demons attend the fire under the cauldron, putting logs of wood on it and fanning it. Another demon is carrying wood to the fire. The majestic figure of Bhatara Yama, encircled by his halo, surveys these activities. Two more female souls are being tortured. The genitals of one of them are being sawn by a demon while another demon holds a caterpillar to the breast of the other female soul. Souls of children are being harassed as well.

Scene eight

A squatting female demon, perhaps Bhatari Durga.

Scene nine

People in boats are surrounded by a turtle and other sea creatures while on the right a fisherman is pulling a fish out of the water on his fishing rod.

Scene ten

Contains the colophon, see above.

The scenes from *Swarga* depicted in the *ulun-ulun* are not the ones described in the Āstīkacarita , but are inspired by the story Bima Swarga. In this story Bima goes to *Swarga* to retrieve the souls of Pandu and Madri, who were sent to Yama's abode after their deaths. Scenes from this story are found in many traditional paintings and temple walls (see Hinzler, 1981), and also adorn the ceiling of the Kertha Gosa, the traditional hall of justice in Klungkung, East Bali (see Pucci, 1992). This hall was devastated in 1908 during the Dutch conquest, restored in the 1930s and restored again in 1960 (Hinzler, 1981, p. 189). Many of the torments seen in our *ulun-ulun* are also depicted on this ceiling. The figure of Jogormanik, shaking the tree and releasing hundreds of daggers may have been especially inspired by the illustration found in the Kertha Gosa (Pucci, 1992, pp. 62-3) with which most painters in the Klungkung area are familiar.[7]

INTERPRETING A BALINESE PAINTING

The interpretation of a Balinese painting is a hazardous undertaking. The interpreter has to know the literature which is found in Bali both in Old Javanese and in Balinese. He or she should also know how to 'read' the painting, which sometimes is not at all easy. Where should the reading start? and, should it start at the same place all the time?

Most scenes depicted are usually not so difficult to identify, but there are always some scenes which remain a conundrum. The interpretation of this painting - and the paintings connected to it - is especialy difficult. Why, for instance, is this painting clearly concerned with Jaratkāru - at least in the eyes of the Balinese - and yet the depiction of the actual Jaratkāru scene is so obscure? And why do similiar paintings not show Jaratkāru but scenes from the Rāmāyaṇa or other scenes from the Mahābhārata?

In an attempt to resolve these mysteries I will offer two distinct interpretations at different levels of abstraction.

THE FIRST INTERPRETATION

In Bali, the Old Javanese Ādiparwa is available in a plethora of palm-leaf (*lontar*) manuscripts. Any royal or aristocratic household should have a copy and mostly they do. The story of Jaratkāru is read during cremation rituals and as such has a clear connection with death, the afterlife, heaven and hell and the consequences that our conduct in life have after we are dead and eventually reborn. In Bali, the corpse of a royal or aristocratic person is not buried or cremated immediately. For the time needed to complete the preparations for the cremation he/she is put to lie in state on a wooden bed in one of the open air pavilions in the palace grounds. The bedstead is decorated with a *tabing*, in some cases relating the story which is our theme.

While the soul of the deceased is waiting for his corps to be cremated, his/her soul wanders about, being irritable because he/she has not yet reached the final destination after death. The *padanda* (Balinese Brahmana Priest) has to give offerings to the soul and to redeem it from hell.[8] During this time the soul of the deceased stays in a sort of limbo. It cannot go to its definite abode and also cannot incarnate back to earth.

Seen in this light, this offers a feasible explanation of why the painting shows scenes of torments in the afterlife. The depictions of these torments show what happens to the poor soul while he/she is waiting, as a result of his/her conduct on earth. They are also a reminder of the tasks a person has to fulfil on earth. This link therefore seems fairly obvious. Moreover, according to the text itself, the story of the Āstīkacarita itself 'will purify your soul and bring long life (*dīrghāyuṣa*)' (Phalgunadi 1990, p. 59).

Intriguingly, the actual Jaratkāru part of the illustration is only small. It is depicted on the right, second scene from the top (see illus.) In fact, only the part where Jaratkāru is presented to Bāsuki and his younger sister who bears the same name as himself stands out clearly (at least, since we know from the colophon that it is this story). The scene on top of this is unclear, but we do recognize the same figures as those from the scene below it. Apparently the figures are also Jaratkāru (husband and wife) shown with servants, facing two priests.

The torments in *Swarga* shown have a relation to childbearing and the raising of children. The female soul who is having her private parts assaulted by a saw is not found in the Kertha Gosa, but the reference to a - perhaps too enthusiastic but unproductive - sex life may be assumed. The female soul whose breast is being suckled by a caterpillar is found on Kertha Gosa (Pucci, 1992, p. 69). She is the ghost of a mother of an only child who died and thus she is being punished for being childless. The female soul trying to cross the quacking bridge is being punished for having committed abortion. Her aborted children wait for her at the bridge (Terwen-de Loos, 1964, item 91). Many other punishments are shown as well. The scene in the left hand corner, depicting all sorts of animals, and the role of Garuḍa are an enigma to me.

As life in general is inextricably bound up with death, it seems immaterial to which particular story the hell scenes are attached. It may be Jaratkāru, it may be the Rāmāyaṇa, it may be other stories from the Mahābhārata. All who live must die, and whatever deeds people have committed during their lives will inexorably affect what will happen to them after they have passed away. The circle of life has to be maintained at any cost. Therefore, people have to have children, and naturally, people should behave well. Most hell scenes depicted show what will happen to people who do not have children, or fail to behave according to the rules laid down by society.

THE SECOND INTERPRETATION

An interpretation at a somewhat more elevated level might be construed as follows. The snake in the top centre of the picture is most important here. The picture reminds us of the reason the Mahābhārata came into being. The reasoning is as follows:

The most prominent figure in this picture is the serpent in the top centre, represented directly under the sun. In my view this is Takṣaka,[9] the serpent responsible for the death of Mahārāja Parīkṣit, whose son, Mahārāja Janamejaya, ordered a serpent-sacrifice (*sarpayajña*) to be performed to avenge his father's death. Despite his best intentions, this sacrifice was not successful, a failure which can be traced directly to the actions of Jaratkāru. Jaratkāru married the sister of Bāsuki and had a son by her. His son is thus half Brahmin and half serpent. He managed to convince Maharaja Janamejaya to stop the serpent sacrifice just as Takṣaka was about to fall into the burning fire. The serpent is therefore not killed and thus he is the actual focus of the whole picture, which explains why he is shown under the sun in the top centre of the picture. The Old Javanese Mahābhārata, composed by Bhagavan Byāsa was narrated to Mahārāja Janamejaya by Bhagavān Vaiśampāyana to soothe his feelings when the serpent-sacrifice proved a failure. Therefore, in essence, because of Jaratkāru we have the Mahābhārata.

CONCLUSION

Balinese paintings may be interpreted at various levels. The level of interpretation may vary according to the part of the painting taken as starting point of the interpretation. As shown above, the larger part of a painting may yield an interpretation at a face value level whereas an interpretation based on a detail may yield an interpretation at a level of higher abstraction. Both are valid and make Balinese paintings more intriguing.

Scenes of Heaven and Hell are found on many different Balinese paintings. Regardless of the story they relate the same message: behave well and provide the world with children. Then you will be able to enter the abode of the ancestors and will be able to return to earth in a new incarnation.

The snake in my painting is depicted quite big and prominent, resulting in my second interpretation. Interestingly, the painting described by Forge also shows the snake, but there it is depicted much smaller and far less prominent. Had I known only the Forge painting I mignt not have thought of the second interpretation. It may very well be that individual painters do leave a personal mark on the painting to express their specific ideas.

REFERENCES

Forge, Anthony

1978 *Balinese Traditional Paintings; A Selection from the Forge Collection of The Australian Museum, Sydney*, Sydney: The Australian Museum

Galestin, T.P.

1939 'Tantri illustraties op een Balische doek', *Cultureel Indië* 1, pp. 129-36

1943 'Eenige Balische illlustraties bij het Oudjavaansche gedicht Smaradahana', *Cultureel Indië* 5, pp. 76-87
1948-9 'Illustraties van een oud Balineesch verhaal', *Indonesië* 2, pp. 486-520
1969 'Four Balinese Illustrations of the Tale of Lubdhaka', in: A. Teeuw *et al. Śiwarātrikalpa of Mpu Tanakuṅ. An Old Javanese poem, its Indian source and Balinese illustrations.* The Hague: Nijhoff. (Bibliotheca Indonesica. 3)

Galestin, T.P., L. Langewis and R. Bolland
1956 *Lamak and Malat in Bali, and a Sumba loom.* Amsterdam: Royal Tropical Institute

Goedheer, A.J.
1939 'De strijd om de onsterfelijkheid op een Balische doek', *Cultureel Indië* 1, pp. 344-46

Grader, C.J.
1939 'Brajoet: de geschiedenis van een Balisch gezin', *Djåwå* 19, pp. 260-75

Hinzler, H.I.R.
1981 *Bima Swarga in Balinese Wayang*, 's-Gravenhage: Nijhoff. (VKI. 90)

Hooykaas, C.
1973 *Religion in Bali.* Leiden: Brill. (Iconography of Religions. XIII/10)

Hunter, T.M.
1988 'Crime and Punishment in Bali: Paintings from a Balinese Hall of Justice', *Review of Indonesian and Malaysian Affairs* 22, pp. 62-113

Juynboll, H.H.
1906 *Âdiparwa. Oudjavaansch prozageschrift uitgegeven door..* 's-Gravenhage: Nijhoff

Kamus Bali-Indonesia
1990 Dinas Pendidikan Dasar Propinsi Dati I Bali

Marrison, G.E.
1995 'Balinese classical painting: its literary and artistic themes', *Indonesia Circle* 65, pp. 1-21

Phalgunadi, I Gusti Putu
1990 *Indonesian Mahābhārata; Ādiparwa - The first book*, New Delhi: International Academy of Indian Culture and Aditya Prakashan

Pucci, Idanna
1992 *Bhima Swarga; The Balinese Journey of the Soul*, Boston etc.: Little, Brown and Company

Stutley, James and Margaret
1977 *A Dictionary of Hinduism. Its Mythology, Folklore and Development 1500 B.C.-A.D. 1500.* London and Henley: Routledge & Kegan Paul

Terwen-de Loos, J.

1964 *Indonesische kunst uit eigen bezit*. Delft: Ethnografisch Museum

Vickers, Adrian

1984 'Ritual and representation in nineteenth-century Bali', *Review of Indonesian and Malaysian Affairs* 18, pp. 1-35

Worsley, Peter

1984 'E 74168', *Review of Indonesian and Malaysian Affairs* 18, pp. 65-109

1988 'Three Balinese paintings of the narrative Arjunawiwaha', *Archipel* 35, pp. 129-56

Notes

1. For details about the technicalities of the production of these paintings, see Forge 1978.
2. In the literature two terms are used for a cloth measuring roughly 100 x 150 cm.: *tabing*, and *ulun-ulun* (or *ulon-ulon*). I shall refer to this painting as an *ulun-ulun*, whereas in discussing a painting described by another author the term he uses will be retained. Hinzler (1981, p. 192) argues an *ulon-ulon* should be almost square, measuring either 108 x 110 cm or 145 x 135 cm. Our *ulun-ulun* measures 108 x 150 cm, which makes it by no means square. The word *tabing* is either used for a wooden partition (Hinzler, *ibid.*), or for the painting to be hung on the partition (Forge, Hunter, Marrison).
3. The Old Javanese Ādiparwa was first published by Juynboll in 1906. The latest 'edition' was published by I Gusti Putu Phalgunadi in 1990 together with an English adaptation. The Old Javanese version of the Ādiparwa is not the same as versions of the same story found in India.
4. The Balinese believe that *Swarga*, which commonly means 'heaven', is any place where gods reside. Hell is where the god Yama resides and so, even though we would call this place 'hell', for the Balinese it is also a *Swarga* (Hinzler, 1981, p. 203). It is in essence a place where the soul of a deceased person stays before he is reborn into a new existence on earth. In this essay the Balinese term *Swarga* will be retained.
5. Interestingly, in India there is a legend which tells that Janamejaya, after having committed brahminicide was condemned to listen to a recital of the entire Mahābhārata by Vaiśampāyana (Stutley and Stutley, 1977, p. 126) which would seem to make more sense to me.
6. A picture of this painting is found in the catalogue: *Indonesian Art. A loan exhibition from the Royal Indies Institute Amsterdam, the Netherlands. October 31 to December 31, 1948*. New York: The Asia Institute, p. 200, and in Hooykaas, 1973, illustration y 2.
7. Actually, it may be the other way around. Maybe the paintings of the Kertha Gosa were inspired by paintings already existing.
8. See Hooykaas, 1973, p. 22.
9. This snake has been identified by Hunter as either Anantabhoga or Bāsuki (Hunter, 1988, p. 91).

XX

WO LAG DER ĀSTĀVA?

Klaus Mylius

1. DIE FRAGESTELLUNG

Der Jubilar dieser Festschrift, Frits Staal, hat uns gerade anhand seiner ritualwissenschaftlichen Studien in mustergültiger Weise gezeigt, daß jede noch so umfassende Erkenntnis aus Einzelbausteinen besteht. Auch die angestrebte vollständige Ausdeutung des vedischen Rituals ist ohne Kenntnis der Details nicht möglich. Um also das Ritual erklären zu können, muß man wissen, wie es sich konkret abgespielt hat. Daß trotz aller Erfolge der Forschung unsere Kenntnisse immer noch bemerkenswerte Lücken aufweisen, sollen die folgenden Betrachtungen zeigen. Auch sie können nicht beanspruchen, unwiderlegbare Resultate erbracht zu haben, doch sollen sie die Aufmerksamkeit darauf lenken, daß das altindische Opferritual einer vollständigen wissenschaftlichen Erschließung immer noch harrt.

Im Somaritual - das gehört zum Grundwissen des Vedaforschers - finden an jedem Preßtag drei tageszeitlich gebundene Hauptrituale statt: die Morgen-, Mittag- und Abendpressung (*prātaḥsavana*, *mādhyaṁdina savana*, *tṛtīya savana*). Zu allen drei Pressungen gehören neben anderen Kulthandlungen auch Lobgesänge (*stotra*). Das jeweils erste dieser *stotra*s wird als 'Reinigungs-stotra' (*pavamānastotra*) bezeichnet, da es während der Klärung des *soma* gesungen wird.

Unterschiedlich ist jedoch der Ort, an dem diese Lobgesänge ausgeführt werden. Denn abweichend von den *stotra*s der Mittagpressung (*mādhyaṁdinapavamānastotra*) wie auch der Abendpressung (*ārbhavapavamānastotra*) erfolgt das *pavamānastotra* der Morgenpressung 'außerhalb', 'draußen' und heißt somit *bahiṣpavamānastotra*.

Die Stätte, an welcher dieses *bahiṣpavamānastotra* ausgeführt wird, heißt *āstāva*. So stellt Baudhāyana Śrauta Sūtra XXV, 31 fest:[1]

āditaścāntataś ca bahiṣpavamānam āstāve stuvīran

Am ersten und letzten (Preßtag) sollen sie das *bahiṣpavamāna* beim *āstāva* singen

Wo aber ist im Gesamtbereich der Opferstätte der *āstāva* angesiedelt gewesen? Diese Frage ist durchaus nicht müßig, denn nach Auskunft der Primärquellen muß dem *āstāva* im Ritual eine hohe Wertschätzung zugemessen worden sein. So bekundet[2] Pañcaviṁśa Brāhmaṇa (PB) VI,7,9:

bahiṣpavamānaṁ sarpanti svargam eva tal lokaṁ sarpanti

Zum *bahiṣpavamāna* kriechen sie; zur Himmelswelt eben kriechen sie

Aussagekräftig ist auch[3] Jaiminīya Brāhmaṇa III,116:

eṣa ha vā udgātuś ca yajamānasya ca dhiṣṇyo yad āstāvaḥ. āyatanam āstāvaḥ

Dieser *āstāva* ist ja des *udgātṛ* und des Opferveranstalters Wohnsitz. Eine (feste) Stätte ist der *āstāva*

Im Rahmen der Besprechung des besonders hoch gewürdigten Roßopfers (*aśvamedha*) äußert sich[4] Śatapatha Brāhmaṇa XIII,5,1,16:

stute bahiṣpavamāne 'śvam āstāvam ākramayanti sa yady ava vā jighred vi vā varteta samṛddho me yajña iti ha vidyāt

Wenn das *bahiṣpavamāna(stotra)* gesungen worden ist, lassen (die Priester) das Roß den *āstāva* beschreiten. Wenn es (dabei) entweder schnaubt oder sich wegwendet, möge (der Opferveranstalter) wissen: 'Erfolgreich ist mein Opfer!'

Vgl. zu diesem Vorgang im *aśvamedha* auch[5] Kātyāyana Śrauta Sūtra XX,5,7. Nach Jaiminīya Brāhmaṇa I,76; II,269 setzt man sich beim *āstāva* nieder, um einen Feind zu schädigen. Am dritten Tag der *upasad* redet der *adhvaryu*-Priester den *āstāva* mit einem Spruch aus[6] Taittirīya Saṁhitā I,3,3 k an (vgl. CH I, S. 106):

pariṣadyo 'si pavamānaḥ

Umgeben (von Priestern, welche) sitzen müssen, bist du der Reinigende

Über die konkrete Lage des *āstāva* aber gibt es erhebliche Unstimmigkeiten; nicht selten wird die Frage nach ihr gar nicht erst gestellt. Eine kurzgefaßte Übersicht der bisher geäußerten Auffassungen soll der folgende Abschnitt bringen.

2. BISHERIGE BETRACHTUNGSWEISEN

Naturgemäß empfiehlt es sich, zunächst die von verschiedenen Autoren vorgelegten Pläne vedischer Opferstätten im Hinblick auf die Lage des *āstāva* zu prüfen. Entsprechend seiner Thematik brauchte Alfred Hillebrandt[7] auf die *mahāvedi* nicht einzugehen, sondern konnte sich auf die Wiedergabe des *prācīnavaṁśa* beschränken. Gleiches gilt für Hertha Krick[8], die sich ebenfalls mit der *śālā* begnügen durfte. Erst Julius Eggeling[9] hatte in seinem Plan die *mahāvedi* zu berücksichtigen, doch ist der *āstāva* weder innerhalb noch außerhalb derselben verzeichnet. Selbst in dem berühmten, grundlegenden Werk von Willem Caland und Victor Henry[10] über den *agniṣṭoma* sucht man auf dem Opferstättenplan vergebens nach der Lage des *āstāva*. Dasselbe Defizit weist der Plan von Mukunda Jha Bakshi[11] auf. Asko Parpola[12] dürfte der erste gewesen sein, der den *āstāva* auf einem Opferstättenplan verzeichnet hat. Er lokalisiert ihn mit der Bemerkung 'location uncertain' innerhalb der *mahāvedi* und südsüdwestlich von der *cātvāla* genannten Grube, die außerhalb der *mahāvedi* liegt. Chitrabhanu Sen[13] wiederum verzeichnet auf demjenigen Plan, der das Somaritual zum Gegenstand hat, keinen *āstāva*; Gleiches gilt für das von Frits Staal[14] herausgegebene Monumentalwerk über das 1975 in Kerala stattgefundene *atirātra-agnicayana*.

Es nimmt nicht wunder, daß die bisherigen verbalen Angaben zur Lage des *āstāva* gleichermaßen divergent bzw. unscharf sind. Besonders augenfällig sind jedoch bloße Statements, die von keinem Versuch einer Beweisführung begleitet werden. So definieren Caland und Henry den *āstāva* folgendermaßen:

> *Emplacement situé à l'intérieur de la mahāvedi, au sud du cātvāla et jonché de gazon, où plus tard on chantera les hymnes du bahiṣpavamāna.*[15]

An anderer Stelle äußern sich die beiden genannten Autoren: der *āstāva*

> *est situé dans la partie septentrionale du vihāra, au sud du cātvāla*[16]

Man kann sich des Eindrucks nicht erwehren, daß manche späteren Vedisten die Ansichten von Caland und Henry lediglich festgeschrieben haben. So schreibt Arthur Berriedale Keith in seiner Übersetzung der Taittirīya Saṁhitā über

> *The Āstāva (within the Mahāvedi, to the south of the Cātvāla)*[17]

Ganz ähnlich beschreibt Louis Renou in seinem Ritualwörterbuch den *āstāva* als

> *lieu situé hors du sadas, dans la partie N. de la mahāvedi, au S. du cātvāla, où les chantres assis chantent le bahiṣpavamāna*[18]

Chitrabhanu Sen, ebenfalls Verfasser eines Ritualwörterbuches, beruft sich bei seiner Beschreibung der Lage des *āstāva*

> *situated outside the sadas, to the south of the cātvāla*[19]

ausdrücklich auf Caland und Henry. Und Frits Staal definiert den *āstāva* als

> *place for chanting the outdoor chant of the purified Soma (bahiṣpavamānastuti), northwest of the Agni altar*[20]

Fest steht lediglich, daß das *mādhyaṁdinapavamānastotra* innerhalb des *sadas* ausgeführt wird;[21] für das *ārbhavapavamānastotra* gilt das Gleiche.

Mehr der Vollständigkeit wegen soll kurz auf die Etymologie von *āstāva* eingegangen werden. Das Wort beruht natürlich auf der Wz. *stu* + *ā*; als Verb ist diese Kombination jedoch nicht belegt. Immerhin deutet das *ā* eine auf ein Ziel gerichtete Bewegung an. Diese findet in der Tat statt und führt die technische Bezeichnung *sarpaṇa*, da sie in gebückter Haltung auszuführen ist. So interessant die dafür von den Primärquellen gegebene Begründung ist - das Opfer wird mit einem Hirsch verglichen, den man anschleichen muß -, so ist sie für unsere Frage nach der Lage des *āstāva* kaum ergiebig.

J. Eggeling hat das Verdienst, auf den Kernpunkt unseres Problems aufmerksam gemacht zu haben.[22] Kurz formuliert, lautet er so: Wurde das *bahiṣpavamānastotra* außerhalb der Opferstätte (so will es der Kommentar zu Pañcaviṁśa Brāhmaṇa VI,8,10-11) oder lediglich außerhalb des *sadas* (so der Standpunkt Sāyaṇas) gesungen? Unbestritten bleibt, daß der Gesang jedenfalls außerhalb des *sadas* stattfand.

Verfolgt man den Gang der das *bahiṣpavamānastotra* vorbereitenden Handlungen, so geht man am besten auf die *vaipruṣa*-Libationen zurück. Diese werden auf dem neuen *āhavanīya*, also auf der *uttaravedi* dargebracht. Von hier begibt man sich zum *āstāva*.

Welche Richtung wird dabei eingeschlagen? Die meisten Autoren sind sich darin einig, daß es die nördliche war. So äußert sich u.a. F. Staal:

They then move to a place to the north of the altar, the āstāva, where the three chanters sing their first chant, 'the outdoor chant for the purified Soma' (Bahiṣpavamānastotra).[23]

In bezug auf die Richtung (Norden) weicht Staal an einer anderen Stelle[24] allerdings etwas ab:

Starting from the uparava holes ... adhvaryu, the three chanters (prastotā, udgātā, pratihartā), yajamāna, brahman, and pratiprasthātā crawl on the altar ... They crawl off the altar and move toward the āstāva spot, northwest of the altar ... They sit down, the prastotā facing west, the udgātā north, and the pratihartā south. There they sing the Outdoor Chant for the Purified Soma (Bahiṣpavamānastotra).[25]

W. Caland ergänzt bei der Übersetzung des *sarpaṇa*, daß *adhvaryu*, *pratiprasthātṛ*, *prastotṛ*, *udgātṛ*, *pratihartṛ*, *brahman* und *yajamāna* aus der *havirdhāna*-Hütte kommen.[26]

Angesichts der unbefriedigenden Lage in der Sekundärliteratur bleibt also nur die Möglichkeit, die Primärquellen daraufhin zu prüfen, ob ihnen eine mehr oder minder klare Auskunft abzugewinnen ist. Leider ist die Zahl der in Betracht kommenden Belegstellen sehr gering. Etliche Stellen sind zudem irrelevant. So kommt der *āstāva* in Śatapatha Brāhmaṇa II,2,4,11 nur allgemein als Stätte des Lobgesangs im *agnihotra* vor. Chāndogya Upaniṣad I,10,8 sagt von Uṣasti Cākrāyaṇa lediglich:

tatrodgātṝn āstāve stoṣyamāṇān upopaviveśa

Auch die ausführlichere (*upasad*-)Stelle Kātyāyana Śrauta Sūtra VIII,6,23 bringt kaum etwas Substantielles:

sadodvāraṁ pūrveṇa tiṣṭhann anudiśaty āhavanīyabahiṣpavamānadeśacātvālaśāmitraudumbaribrahmāsanaśālādvāryaprājahitānt samrāḍ asīti pratimantram

Ostwärts der (östlichen) Tür des *sadas* stehend, richtet er jeweils an das *āhavanīya*(-Feuer), die Stätte des *bahiṣpavamāna*, die *cātvāla*(-Grube) das *śāmitra*(-Feuer), den *audumbarī* (-Pfosten), den Sitz des *brahman*, das *śālādvārya*(-Feuer und) das *prājahita*(-Feuer) den (mit) "Ein Großkönig bist du" (beginnenden) Spruch.

Es handelt sich also um einen Rundumblick, der über die spezielle Richtung und Lage des *āstāva* nichts aussagt. Um diese zu erkunden, müssen weitere Quellen herangezogen und geprüft werden.

3. VERSUCH EINER LÖSUNG

Hier bietet sich zunächst das Āpastamba Śrauta Sūtra an.[27] Während ĀpŚS XII, 19, 7 nur noch einmal den *āstāva* als Ort des *bahiṣpavamānastotra* ausweist, wird ĀpŚS XII, 16, 17 ein wenig genauer:

udañcaḥ prahvā bahiṣpavamānāya pañcartvijaḥ samanvārabdhāḥ sarpanti

Es kriechen die fünf Priester nach Norden in gebückter Haltung zum *bahiṣpavamāna*, wobei sie einander von hinten berühren.

Die Bewegung nach Norden wird hier also erneut gefordert. Wie schon oben erwähnt, wird auch bei ĀpŚS XII,17,3-4 das Opfer mit einer zu beschleichenden Gazelle (bzw. Hirsch) verglichen. Die gebückte Haltung erwähnt auch Kātyāyana Śrauta Sūtra IX,6,33; die Richtung nach Norden kommt ebenfalls hier vor:

prahvā udañco gacchanti

Der Kommentator dieser Stelle will wissen, daß nur morgens gekrochen, mittags dagegen gebückt und abends aufrecht gegangen wurde.

Obwohl für unsere Fragestellung nicht direkt relevant, soll noch kurz auf die Art eingegangen werden, in welcher die Gesangspriester am *āstāva* Platz nehmen. Die hierzu bei Caland und Henry aufgeworfenen Fragen[28] dürften anhand von Lāṭyāyana Śrauta Sūtra I,11,19-21 von A. Parpola dahingehend geklärt worden sein,[29] daß im Süden der *udgātṛ*, im Osten der *prastotṛ*, im Westen der *pratihartṛ* saß, die nach Norden bzw. Westen bzw. Südosten schauten. Die drei Sänger bildeten also einen Halbkreis; vgl. LŚS II,6,4.

So interessant diese Stellen sind, können sie zur Lösung der eingangs gestellten Frage Entscheidendes nicht beitragen. Einen Schritt nach vorn bedeutet dagegen ĀpŚS XII,17,5, wo es heißt:

cātvālam avekṣamāṇāḥ stuvate

Sie singen das *stotra*, indem sie auf die *cātvāla*(-Grube) blicken.

Hier wird also ein deutlicher Zusammenhang - auch in räumlicher Hinsicht -

zwischen *āstāva* und *cātvāla* hergestellt. Wenn der Kommentator bemerkt:

antarvedyāsīnāś cātvālam avekṣamāṇāḥ stuvate

so muß bemerkt werden, daß das *antarvedi* im Śrauta Sūtra selbst keine Stütze findet. Anders sieht es dagegen im Mānava Śrauta Sūtra[30] aus; hier bemerkt MŚS II,2,4,6:

dakṣiṇataś cātvālasyāntarvedy āstāvāya saṁstṛṇāti

Südlich des *cātvāla* innerhalb der *vedi* streut er für den *āstāva*.

Ähnlich äußert sich der Kommentator zu Kātyāyana Śrauta Sūtra IX,6,33:

te adhvaryvādayaḥ ... udaṅmukhāḥ cātvālasya dakṣiṇato vedimadhye vartamānaṁ bahiṣpavamānadeśam (gacchanti)

Der *adhvaryu* usw. gehen mit dem Gesicht nach Norden zu der südlich vom *cātvāla* innerhalb der *vedi* befindlichen Stätte des *bahiṣpavamāna*.

Man könnte annehmen, daß das Problem damit gelöst ist. Die oben mitgeteilten Zitate haben vermutlich auch A. Parpola dazu geführt, den *āstāva* versuchsweise innerhalb der *vedi* anzusiedeln.

Aber wenn man es recht betrachtet, so sind MŚS II,2,4,6 und W. Caland (nebst seinen Anhängern)[31] die einzigen Autoritäten, die für die Existenz des *āstāva* innerhalb der *vedi* eintreten, mithin *bahis* nur auf das *sadas* beziehen. Es ist aber jedenfalls zu prüfen, ob *bahis* hier nicht doch eine weitergehende Bedeutung hat. Und dafür scheinen ĀpŚS und MŚS als Ādhvaryava Sūtras nicht die geeignete Quelle zu sein. Vielmehr ist zu bedenken, daß das *bahiṣpavamāna* zum *audgātra*-Ritual gehört. Die entscheidenden Aussagen sind daher aus den hier kompetenten Quellen zu gewinnen.

Für das an das *bahiṣpavamānastotra* anschließende Ritual trifft Lāṭyāyana Śrauta Sūtra II,1,8 folgende wesentliche Feststellung:

bahir vedy udañco 'yuñji padāny utkrāmeyuḥ prāg daśamyaḥ

Außerhalb der *vedi* sollen sie nordwärts eine ungerade Zahl von Schritten, weniger als zehn, machen.

Hier ist also erstmals von einem Prozeß außerhalb der *vedi* die Rede; es wird darauf noch zurückzukommen sein.

Auf die Bedeutung des Zusammenhangs zwischen *āstāva* und *cātvāla* wurde bereits oben hingewiesen. Dieser Gesichtspunkt soll hier noch ein wenig vertieft werden. So heißt es bei Vaitāna(śrauta) Sūtra XVII,1:

cātvālād dakṣiṇata upaviśanti

Sie setzen sich südlich vom *cātvāla* nieder.

A. Hillebrandt[32] drückt es noch etwas deutlicher aus:

das (*pavamānastotra*) des Prātaḥsavana heißt *bahiṣpavamāna*, weil es außerhalb des Sadas draußen am Cātvāla stattfindet.

Vor dem *bahiṣpavamānastotra* geschieht[33] Folgendes:

En silence les chantres regardent le cātvāla et une cruche d'eau;

danach:

Les chantres vident sur le cātvāla la cruche à eau qu'ils ont regardée avant le stotra[34]

Aus Āśvalāyana Śrauta Sūtra V,3,16 geht hervor, daß *śāmitra*-Feuer, *ūvadhyagoha*, *cātvāla*, *utkara* und *āstāva* alle in gemeinsamer Richtung liegen. Auch Śāṅkhāyana Śrauta Sūtra VI,12,3-10 stellt einen inhaltlichen und damit offenkundig auch räumlichen Zusammenhang her zwischen (neuem) *āhavanīya*, *āstāva*, *cātvāla*, *śāmitra* und *utkara*. Lāṭyāyana Śrauta Sūtra schreibt vor, daß man sich nach der Verehrung des (neuen) *āhavanīya* nordwärts zur Verehrung von *āstāva*, *cātvāla* und *śāmitra* wenden solle.

Noch zwingender ist die Vorschrift, die LŚS I,11,18 gibt:

cātvāladeśaṁ prāpyādhvaryāv upaviṣṭe tasmāt pratyag upaviśeyuḥ

Wenn (die Priester) die Stätte des *cātvāla* erreicht haben (und) der *adhvaryu* sich gesetzt hat, sollen sie sich westlich von ihm niederlassen.

Diese Stelle indiziert, daß *āstāva* und *cātvāla* nicht weit voneinander entfernt gewesen sein können.

Ferner ist an die oben aus PB VI,7,9 zitierte Stelle zu erinnern. Wenn die Bewegung der Priester in Richtung des *bahiṣpavamāna* (also des *āstāva*) mit einem Gang in die Himmelswelt verglichen wird, deutet dies eher auf ein Ritual außerhalb als innerhalb der *vedi*.

Ganz in diesem Sinne ist auch die kurze Bemerkung in PB VI,8,10:

apariśṛte stuvanti

im nicht umfriedeten (Raum) lobsingen sie.

Es soll nicht als zwingender Beweis gewertet, aber doch immerhin erwähnt werden, daß Sāyaṇa *apariśṛte* mit *bahirvedideśe* kommentiert. Bedeutsamer ist jedoch, daß PB VI,8,10 wie folgt fortfährt:

tasmād aparigṛhītā āraṇyāḥ paśavaḥ

Daher (leben) die wilden Tiere, ohne eingeschlossen zu sein.

Der *āstāva* wird damit gleichsam in den Bereich der wilden Tiere verlegt. Ein solcher Gedankengang ist jedoch mit der Vorstellung, der *āstāva* könne innerhalb der *mahāvedi* gelegen haben, unvereinbar.

Eine weitere Hilfe bietet eine Stelle aus JB I,84; sie lautet folgendermaßen:

eṣā vai yajñasya dvār yad antar āgnīdhraṁ ca cātvālaṁ ca.tayābhya-vayāt. tayodeyāt

Hendrik W. Bodewitz, einer der besten Kenner des JB, übersetzt und interpretiert diese Stelle so:[35]

The room between the Āgnīdhra and the Cātvāla is the door of the sacrifice. Through this he enters upon (the Mahāvedi), through this (same) he goes out (for the Bahiṣpavamāna).

Diese JB-Stelle und Bodewitz' Interpretation derselben weisen also ebenfalls deutlich auf den Umstand, daß der *bahiṣpavamānadeśa*, also der *āstāva*, außerhalb der *mahāvedi* angesiedelt gewesen sein muß.

Hinzu kommt aber noch etwas anderes. JB I,84 verwendet nämlich zur Schilderung des Betretens der *mahāvedi* das Verb *abhy-ava-i*. Es handelt sich also um ein Abwärtsgehen. Dazu aber stellt LŚS I,1,14 unmißverständlich folgende Forderung:

prāgudakpravaṇaṁ devayajanaṁ lomaśam avṛkṣaṁ samam

Die Opferstätte (soll) nach Nordosten geneigt, grasig, baumlos (und) eben (sein).

Das Verb *abhy-ava-i* paßt also ausgezeichnet zum Betreten einer nach Nordosten geneigten, ansonsten aber ebenen Fläche von Norden her.

Umgekehrt hat es sich von der *mahāvedi* her um einen Anstieg zum *āstāva* gehandelt. Als erhöhter, der Himmelswelt gleichgestellter Platz bot er für einen eindrucksvollen Gesang bessere Voraussetzungen als eine Stätte innerhalb der *mahāvedi*.

Die beigebrachten zahlreichen Indizien lassen es mithin zu, *bahis* nicht nur als *bahiḥsadas*, sondern als *bahirvedi* aufzufassen. Der *āstāva* als Stätte des *bahiṣpavamānastotra* ist nördlich von der *uttaravedi* außerhalb der *mahāvedi*, und zwar unweit südlich des *cātvāla*, zu lokalisieren.[36]

Notes

1. W. Caland (ed.), *Baudhāyana Śrauta Sūtra*. Bibliotheca Indica, Calcutta: Asiatic Society of Bengal, 1904-24.
2. Chinnaswami Śastri (ed.), *The Tāṇḍyamahabrāhmaṇa (sic) with the commentary of Sāyānachārya* (sic). , Benares: Chowkamba Sanskrit Series Office, 1935. (Kashi Sanskrit Series. 105)
3. Raghu Vira und Lokesh Chandra (ed.), *Jaiminīya-Brāhmaṇa of the Sāmaveda*. Nagpur: Chandra, 1954.
4. A. Weber (ed.), *The Çatapatha-Brāhmaṇa in the Mādhyandina-Çākhā with extracts from the commentaries of Sāyaṇa, Harisvāmin and Dvivedaganga*. 2. Aufl., Varanasi, 1964. (Chowkhamba Sanskrit Series. 96)
5. A. Weber (ed.), *The Śrauta-Sūtra of Kātyāyana with extracts from the commentaries of Karka and Yājñikadeva*. Neudruck Varanasi, 1972.
6. N.S. Sontakke und T.N. Dharmadhikari (eds), *Taittirīya-Saṁhitā with the padapāṭha and the commentaries of Bhaṭṭa Bhāskara Miśra and Sāyaṇācārya*, vol. I. Poona: Vaidika Saṁśodhana Maṇḍala, 1970.
7. A. Hillebrandt, *Das altindische Neu- und Vollmondsopfer in seiner einfachsten Form*. Jena: Fischer, 1879. Der erwähnte Plan befindet sich auf S. 19.
8. H. Krick, *Das Ritual der Feuergründung*, hrsg. von Gerhard Oberhammer, *Veröffentlichungen der Kommission für Sprachen und Kulturen Südasiens*. Heft 16. Wien, 1982. (Österreichische Akademie der Wissenschaften, Philosophisch-Historische Klasse, Sitzungsberichte, 399. Bd.) Der erwähnte Plan befindet sich auf S. 66.
9. J. Eggeling, *The Śatapatha-Brāhmaṇa according to the text of the Mādhyandina school translated*, part II, ed. by F. Max Müller, 1885. Second ed. of the first reprint, Delhi: Motilal Banarsidass, 1966. (Sacred Books of the East. 26) Der erwähnte Plan befindet sich auf S. 475.
10. W. Caland und V. Henry, *L'Agniṣṭoma, description complète de la forme normale du sacrifice de Soma dans le culte védique*. Paris: Leroux, 1906-7. Der erwähnte Plan ist no. 4, in tome I.
11. M.J. Bakshi (ed.), *The Śrautasūtra of Lātyāyana (sic), ending with agniṣṭōma (sic) chapter*. Benares: Chowkamba Sanskrit Series Office, 1932. (Kāshi Sanskrit Series. 97) Der erwähnte Plan befindet sich im Anhang.

12. A. Parpola, *The Śrautasūtras of Lāṭyāyana and Drāhyāyana and their commentaries. An English translation and study. Vol. I: 2 The Agniṣṭoma.* Helsinki, 1969. (Commentationes Humanarum Litterarum, Societas Scientiarum Fennica. 43, no. 2) Der erwähnte Plan befindet sich auf S. 17-18.
13. Ch. Sen, *A dictionary of the Vedic Rituals based on the Śrauta and Gṛhya Sūtras.* Delhi: Concept Publication Company, 1978. Der erwähnte Plan ist Nr. 3.
14. Frits Staal (ed.), *Agni. The Vedic Ritual of the Fire Altar.* 2 vols., Berkeley: Asian Humanities Press, 1983. Der erwähnte Plan bildet das Vorsatzblatt in vol. II.
15. Vgl. Caland und Henry, *op. cit.*, vol. I, S. 106, Anm. 1.
16. *Ibid.*, vol. I, S. 172.
17. A.B. Keith, *The Veda of the Black Yajus School, entitled Taittirīya-Saṁhitā.* Delhi, 1967, (HOS. 18 - 19, reprint). Zitat in vol. I, S. 38, Anm. 1.
18. L. Renou, *Vocabulaire du rituel védique.* Paris, 1954, S. 32.
19. Sen, *op. cit.*, S. 48.
20. Staal (ed.), *op. cit.*, vol. II, S. 773.
21. A. Hillebrandt, *Ritual-Litteratur. Vedische Opfer und Zauber.* Grundriß der Indo-Arischen Philologie und Altertumskunde III/2. Straßburg: Truebner, 1897, S. 131.
22. Eggeling, *op. cit.*, SBE vol. XXVI, S. 310, Anm. 1.
23. Staal (ed.), *op. cit.*, vol. I, S. 58.
24. *Ibid.*, vol. I, S. 602.
25. Die beteiligten Personen nach KŚS und Vait; die anderen *Śrautasūtras* weichen etwas ab.
26. W. Caland, *Das Vaitānasūtra des Atharvaveda* = VKAW Afdeeling Letterkunde, Nieuwe Reeks, Deel XI, No. 2, Neudruck Wiesbaden, 1968.
27. R. Garbe (ed.), *The Śrautasūtra of Āpastamba with the commentary of Rudradatta.* Calcutta 1882-1903, second ed., New Delhi, 1983. (Bibliotheca Indica. 2)
28. Vol. I, S. 172-3.
29. Parpola, *op. cit.*, S. 155, Anm. 3.
30. J.M. van Gelder (ed., tr.), *The Mānava Śrautasūtra belonging to the Maitrāyaṇī Saṁhitā.* Delhi, 1985. (Sri Garib Dass Oriental Series. 31, reprint).
31. Die Sūtra-Kommentatoren schlechthin als Autoritäten zu bezeichnen, wäre doch etwas gewagt.
32. Hillebrandt, *op. cit.*, S. 129.
33. Caland und Henry, *op. cit.*, vol. I, S. 175.
34. *Ibid.*, vol. I, S. 181.
35. H.W. Bodewitz, *The Jyotiṣṭoma Ritual. Jaiminīya Brāhmaṇa I, 66-364. Introduction, Translation, and Commentary.* Leiden: Brill, 1990, S. 48. (Orientalia Rheno-Traiectina. 34, ed. by Jan Gonda)
36. Vgl. den Opferstättenplan bei K. Mylius, *Wörterbuch des altindischen Rituals.* Wichtrach: Institut für Indologie, 1995.

XXI

THE TANTRIC TRANSFORMATION OF PŪJĀ: INTERPRETATION AND STRUCTURE IN THE STUDY OF RITUAL

Richard K. Payne[*]

INTRODUCTION[1]

In his deservedly famous essay 'The Meaninglessness of Ritual' Frits Staal presented among other things an understanding of ritual as 'the precise execution of rules',[2] and suggested that rituals could be analysed in terms of these rules, i.e., that there is a syntax of ritual.[3] These ideas found further development in later publications.[4]

This essay attempts to accomplish three goals. First to exemplify some syntactic patterns in addition to those which Staal has already identified. Second, to call attention to the Shingon ritual corpus which, being large, elaborate and well-documented, is comparable to the Vedic ritual corpus as an important resource for the study of ritual.[5] Third, to assert that syntactic analysis needs to be as common in the study of ritual as it is in the study of language. For example, it can provide a common reference point for evaluating competing interpretations of ritual which narrative descriptions cannot.

We will begin with the third point by giving an example of an instance in which, because of the absence of a syntactic analysis, there is little possibility of choosing between two contradictory interpretations. This will be followed by a syntactic analysis of a Shingon ritual. This ritual exemplifies both some

---

[*] An earlier version of this paper was presented under the title 'The Enacted Feast: A Tantric Transformation of Pūjā' at the annual conference of the American Academy of Religion/Western Region, College of Holy Names, Oakland, California, March, 1990.

of the syntactic patterns which Staal has identified and some additional patterns as well. The background discussion which introduces the ritual will place it within the larger Shingon ritual corpus, and thus also introduces this body of material which can be a valuable resource to the further study of ritual. In concluding, two areas of future work - ritual semantics and historical ritual studies - will be commented on.

COMPETING EXPLANATIONS: AN EXAMPLE

In his important study of Śaivasiddhanta funerary practices[6] Richard H. Davis has described what might be called the tantric transformation of the Vedic funerary rites. 'Śaivas maintain in their ritual the basic format of Vedic cremation'.[7] This includes the purification and decoration of the corpse, conveyance to the cremation grounds, ritual establishment of the site, protection of the boundaries, placing the corpse on a bier of sacrificial wood, placing the large and small ladles in the corpse's hands, lighting the fire, reciting mantras including the invocation of Agni, circumambulation of the site in reverse order by the officiant and kinsmen, their departure and own purification.[8]

Davis identifies several ways in which the Śaiva rite differs from the Vedic, however. These include having the guru of the deceased act as the officiant, replacing the Ṛg mantras with mantras invoking Śiva, the use of only one fire instead of three, etc. However, Davis sees the most important aspect of the transformation to be the embedding of a ritual element in which the deceased is identified with Śiva. Such ritual identification between the deceased and the deity evoked in the course of the ritual is considered by some to be the minimal defining characteristic of tantric ritual practice,[9] being the central and unique soteriological concept of tantra. In the case of the Śaiva funerary rite, this ritual identification is a repetition of the primary Śaiva initiation, the *nirvāṇadīkṣā* ('initiation into liberation'), 'right in the middle of the funerary rite'.[10] Davis explains the centrality of the *nirvāṇadīkṣā* for the Śaivasiddhanta tradition:

> true liberation is impossible without the performance of initiation. Only those seeking *mokṣa* and those aiming to be temple priests or adepts undergo this complicated and powerful two-day ritual, which affects the initiate ... as a blazing fire affects a heap of cotton ... . The ritual annihilates the soul's fetters, literally burning them up in a sacrificial fire.[11]

It is this rite from the Śaivasiddhanta tradition itself which is embedded into the Vedic funeral ceremony. The sacrificial fire which consumes the fetters of the initiate is homologized with the cremation fires.

Davis claims that this transformation of the Vedic funerary practice constitutes a radical change of the Vedic rite: 'The embedded initiation becomes the dominant rite of cremation and reduces the Vedic portions to a subordinate role. In other words, it shifts the entire focus of the ritual, and in so doing brings cremation into accord with the Śaiva notion of death as the last door to liberation'.[12] Davis also claims that 'transforming Vedic ritual forms offered a means for the Śaivas to highlight their own claims to superior efficacy and comprehensiveness, to articulate not so much their continuity with Vedic precedents as their own distinctive departures from the Vedic system'.[13]

If we examine the rite structurally, however, such an explanation may seem problematic. Hypothetically, a structural analysis of this transformation might look like this:

V1 V2 V3 N V4 V5 V6

where the Vedic elements are represented by 'V' and the *nirvāṇadīkṣā* is represented by 'N'. Unfortunately, this analysis is only hypothetical, since Davis does not supply enough details of the ritual to create an adequate syntactic analysis. In contrast to Davis' interpretation that this is a comprehensive 'shift' of the ritual, the embedding of a single complex of ritual actions into an existing structure may be interpreted as a rather minimal change in the Vedic rite. From this perspective one might explain the Śaiva 'transformation' as a purposeful appropriation of the Vedic rite, perhaps with the intent of either concealing the changes implied by Śaiva soteriology or borrowing legitimacy from the Vedic tradition. Michel Strickmann has noted that the tantric traditions 'embody a conscious antithesis to Vedic rites and precepts ... Their antithetical stance need not represent a true break in continuity, however; explicit opposition may as often as not prove to be a rationale justifying pragmatic assimilation and continuance of ancient practices under altered social conditions.'[14]

It is not my intention here to establish that any of these alternative interpretations is in fact the correct one.[15] Rather, it is simply to establish that, in the absence of a detailed syntactic analysis which can provide a shared basis for discussion, competing interpretations are much more difficult to evaluate.

THE RITUAL CORPUS OF SHINGON

Staal has suggested that the Vedic ritual corpus is of primary importance for ritual studies on the grounds that it is large, elaborate, and well-documented,[16] and that a large body of secondary materials relevant to the study of Vedic rituals, e.g., 'texts, translations, monographs, surveys, dictionaries and encyclopedias'[17] is available. The scale and elaboration of the Shingon ritual corpus can be indicated by a summary of just one of the modern studies of the Shingon ritual corpus which is currently available - part of the extensive secondary material available in Japanese.

Shōun Toganoo (1881 to 1953) is one of the most famous scholars of the Shingon tradition. In his *Himitsu Jisō no kenkyū*[18] (1935) he describes the ritual system of Shingon. The rituals Toganoo discusses are first, the four training rituals: *Jūhachidō*, *Kongōkai*, *Taizōkai*, and *Goma*. The *Jūhachidō* establishes the practitioner in his relation to *Dainichi Nyorai* (*Mahāvairocana Tathāgata*), who has two forms. Each of the next two rituals is oriented specifically on one of these two forms. The *Kongōkai* is oriented toward the form of Dainichi found in the *kongokai mandara* (*vajradhātu maṇḍala*), while the *Taizokai* is oriented toward the form found in the *taizo mandara* (*garbhakośa maṇḍala*) The integration of these two *maṇḍalas* and the rituals which accompany them makes the Shingon tradition unique within Buddhist tantric traditions. The final ritual of the training sequence is the *Goma* (*homa*), a fire offering primarily directed to Fudō Myōō (*Acalanātha Vidyārāja*[19]) for protection of the practitioner. While in the training sequence only one type of *goma*, the *Fudō Myōō Soku Sai Goma*,[20] is performed, there are many different *goma* rituals known to the tradition. A recently published collection of *goma* manuals[21] includes sixty-five different *gomas*, comprising four different kinds of *goma* devoted to a variety of different deities. This gives some indication of the elaboration to be found in the Shingon ritual corpus.

Toganoo follows his discussion of the four training rituals with a discussion of the rites of initiation, *kanjō* (*abhiṣeka*), of which, although more are known to the tradition, three are still performed today. These represent different stages of progress, from the initial *kechien kanjō* (establishing a karmic bond) to the *dembo ajari kanjō* ('ordination' as an *ācārya*). Prior to receiving the *kanjō*, one must have taken the *samaya* precepts, the next ritual Toganoo discusses.

Once having become an *ajari*, the practitioner can perform a wide variety of rites. Toganoo discusses rituals for the discovery of treasures, for the attainment of *ṛddhi*, a set of eight 'major ceremonies' carried out for the protection or benefit of the state,[22] a set of fifteen 'secret rituals',[23] a set of sixty-two 'ordinary rituals'.[24] In addition there are a variety of meditative practices which are also ritualized. Toganoo selects just three of these to discuss, the *Gachirin kan* (Visualization of the Moon Cakra), *A ji kan* (Visualization of the Syllable A), and *Gosō-jōshin kan* (Visualization of One's Own Body Attaining the Five Aspects of Buddhahood).

This sequence of rituals studied by Toganoo also demonstrates a similarity between the Vedic and Shingon ritual systems: both are hierarchical in character. Staal says of the Vedic rituals, 'There is increasing complexity. A person is in general only eligible to perform a later ritual in the sequence, if he has already performed the earlier ones'.[25] The hierarchical character of the Shingon rituals is clear in the training of a Shingon priest. The requirements for performing/having later rituals are two-fold - the performed/received prior rituals and initiations in which the practitioner is the 'object' of the ritual. In both cases, the rituals become more complex as the practitioner progresses through the training.[26]

The *Jūhachidō* is a key for the study of the Shingon rituals. As the first full ritual which a Shingon priest in training learns to perform, it acts as a paradigm for all of the other rituals he/she will learn. The *Jūhachidō* is also paradigmatic for the other rituals of the training sequence since the latter three can be understood as expansions and transformations of the basic syntactic structure of the *Jūhachidō*.

STRUCTURAL ANALYSIS OF A TANTRIC *PŪJĀ*: THE SHINGON *JŪHACHIDŌ*

The name *Jūhachidō* literally means 'the Eighteen Ways'. Although originally this ritual was composed of eighteen mantras and mudrās, it is today more complicated.[27] Turning to its Indian origins, the *Jūhachidō* may be considered to be an example of *pūjā*. Although *pūjās* may be elaborated in a seemingly limitless fashion, Lawrence A. Babb has identified the offering of food to a deity as the central and defining ritual action of *pūjā*:

> without a food offering of some kind the ritual would simply not be *puja* in the conventional sense of the term. The type of food given may vary widely; anything men eat the gods eat too, although the superior deities tend to prefer vegetarian fare. But always, in *puja* some kind of food is given … . The food offering is … the central and indispensable act, the core around which all else is elaboration and overlay.[28]

The offering of food is, however, located within the metaphor of feasting an honoured guest. Ākos Östör has written:

> The most common account of pūjā given by the people [of Vishnupur, West Bengal] is based on an analogy between the service of a deity and the treatment of a guest. The guest is to be honoured above everyone else. The host invites his guest, goes part of the way to meet him, and welcomes him. In the house the guest is received with joy and respect, given refreshments, bathed to be clean from the dust of the road, rubbed with scented oils and perfumes, given new cloth to relax in, fed and entertained, and finally bad good-night and allowed to rest. When the time comes for him to leave the host bids him farewell and reminds him to come again. The ideal way to treat a guest is the way to treat the gods: guests are like deities, and gods are guests among men.[29]

The symbolic offering of food is a central part of the *Jūhachidō* ritual and, although not all of the specifics identified by Östör are to be found in the *Jūhachidō*, the general semantics of the ritual are those of feasting an honoured guest. Toganoo Shoun describes the *Jūhachidō* as 'a ritual which is centred around inviting an honoured guest to come to the altar and to there feast on a great banquet'.[30] At the same time, in the Shingon training the performance of a *pūjā* is also a *sādhana*, i.e., ritual is engaged in as a form of meditative practice. This is why these rituals lack the triadic relation between deity, officiant and sponsor as analysed by Östör,[31] having instead a dyadic relation between the deity and the practitioner.

As given in the *Shō Nyoirin Nenju shidai* ritual manual[32] compiled by Gengō (911 [var. 914] to 995 [var. 988])[33] the sequence of actions of the *Jūhachidō* are as follows:

- enter the shrine
- universal reverence: prostrations
- take the seat of practice
- arrange the *pūjā* offerings
- universal reverence
- rub the hands with powdered incense
- visualize the three mysteries
- purification of the three kinds of actions
- samaya of the Buddha-kula: purification of bodily karma
- samaya of the Padma-kula: purification of speech karma
- samaya of the Vajra-kula: purification of mental karma
- putting on armour: protecting the body
- empowerment of the *argha* water: purifying the practice hall
- empowerment of the offerings
- visualize the syllable RAṀ
- purification of the ground

- visualize the Buddhas
- arising of the *vajra*
- universal reverence
- declaration
- to the *kami*, and aspiration
- five vows of Samantabhadra Bodhisattva
- generating bodhicitta
- samaya
- vows
- five great vows of a bodhisattva
- universal offering, and three powers
- Mahā-Vajracakra: establishing the earth boundary
- vajra poles: binding the four corners
- vajra walls: enclosing the ritual space on the four sides
- visualize the seat of enlightenment
- mudrā of Mahā Ākāśagarbha Bodhisattva
- smaller vajracakra
- sending off the jewelled carriage
- return of the jewelled carriage
- inviting the chief deity
- four wisdoms
- clap the hands
- Hayagrīva-vidyārāja
- vajra net: enclosing the ritual space above
- vajra fire enclosure: fires at the four corners for protection
- great samaya
- offering of *argha* water
- lotus throne
- ringing the vajra bell
- five *pūjā* offerings, offered symbolically, i.e., mudrā and mantra: powdered incense, flower garland, burning incense, food and drink, and lights
- five *pūjā* offerings, offered materially: powdered incense, flower garland, burning incense, food and drink, and lights
- hymn in praise of the four wisdoms, and clap the hands
- hymn to the chief deity
- vast *maṇi* offering to the three powers
- worship of the buddhas
- visualization of ritual identity
- three mudrās and mantras of the chief deity
- initial recitation
- three mudrās and mantras of the chief deity

- visualization of the *akṣaracakra* (wheel of syllables)
- three mudrās and mantras of the chief deity
- subsequent recitation
- five *pūjā* offerings, offered symbolically
- five *pūjā* offerings, offered materially
- *argha* water
- latter ringing of the vajra bell
- praises
- universal offering, and three powers
- worship of the buddhas
- transfer of merits
- five vows and sincere transfer of merits
- dissolving the boundaries: great *samaya*, fire enclosure, Hayagrīva, vajra walls, earth element
- releasing the deities
- three classes
- remove the armour
- universal reverence: prostrations
- depart the shrine

Such a linear presentation of the steps of the ritual does not, however, display the way in which the ritual is organized, which elements go together with one another to create larger ritual units. As Staal has said, 'A *linear* representation of this type is not only extremely cumbersome, but it obscures all the elements of structure we have been so eager to detect'.[34]

Looking at the basic syntactic structure of the *Jūhachidō* ritual, the pattern of offering a feast to deities emerges quite clearly.

The fundamental activities of the ritual are

A) preparing to feast the deities, which corresponds to preparing oneself and one's dwelling for guests,

B) inviting and receiving the deities, which corresponds to inviting and receiving one's guests,

C) offering food and other items to the deities, which corresponds to the feast *per se*,

B') taking leave of the deities and sending them off, which is symmetrical to the inviting and receiving of guests, and

A') closing the ritual, which corresponds to cleaning up after the feast.

In more symbolic form then, the basic structure of the *Jūhachidō* ritual is symmetrical:

A B C B' A'

This is the basic structure of what may be referred to as votive *pūjā*. The *Jūhachidō*, however, is a tantric ritual and incorporates an additional ritual activity at its core. This is ritual identification, the visualized identity of the practitioner with the deity evoked - the same as discussed by Davis in relation to the Śaiva funerary ritual. In the process of adding ritual identification, D, symmetry of the ritual seems to have led the to making offerings after ritual identification as well as before. Hence, the structure of tantric *pūjā* is:

A B C D C' B' A'

Further analysis of the ritual structure can be facilitated by examining the tradition's own nine-part division of the ritual:

1. adorning the practitioner
2. practising Samantabhadra's vows
3. binding the earth
4. adorning the seat of enlightenment
5. requesting the deities
6. sealing for protection
7. *pūjā* offerings
8. recitation: ritual identification
9. second *pūjā* offerings

While symbolically symmetrical around the main act of ritual identification, the traditional divisions reflect the process which I have called 'terminal abbreviation', i.e., actions in the second half of the ritual are reduced in number and simplified. In terms of the meanings of the actions taken in the last division, 9. second *pūjā* offerings, these replicate the actions in the first seven divisions. This designation of all of the ritual following the central action of ritual identification indicates an implicit recognition of terminal abbreviation by the tradition itself. Combining the seven major stages of a tantric *pūjā* as analysed above with the nine traditional divisions, and taking terminal abbreviation into account, creates the following structure:

A. preparing the ritual site
  A. 1. adorning the practitioner
  A. 2. practice Samantabhadra's vows
  A. 3. binding the earth
  A. 4. adorning the seat of enlightenment
B. inviting and receiving the deities
  B. 5 requesting the deities

B. 6 sealing for protection
C. feasting the deities
C. 7 *pūjā* offerings
D. ritual identification
D. 8 recitation
C'. completion of the feast
C'.7. 'latter *pūjā* offerings'
A'.2. 'latter practice of Samantabhadra's vows'
B'. leave taking and sending off
B'.6. 'opening the ritual enclosure'
A'.3. 'unbinding the earth element'
B'.5. 'sending off the deities'
A'. completion of the ritual
A'.1. 'practitioner's departure'

The complete analysis of the ritual would then be:

A. preparing the ritual site
A.1. adorning the practitioner
A.1.a. enter the shrine
A.1.b. universal reverence: prostrations
A.1.c. take the seat of practice
A.1.d. arrange the *pūjā* offerings
A.1.e. universal reverence
A.1.f. rub the hands with powdered incense
A.1.g. visualize the three mysteries
A.1.h. purification of the three kinds of actions
A.1.h.i. purify the five places of the body
A.1.h.ii. samaya of the Buddha-kula: purify bodily karma
A.1.h.iii. samaya of the Padma-kula: purify speech karma
A.1.h.iv. samaya of the Vajra-kula: purify mental karma
A.1.i. putting on armour: protecting the body
A.2 practice of Samantabhadra's vows
A.2.a. empower the *argha* water: purify the practice hall
A.2.b. empower the offerings
A.2.c. visualize the syllable RAṀ
A.2.d. purify the ground
A.2.e. visualize the Buddhas
A.2.f. arising of the *vajra*
A.2.g. universal reverence
A.2.h. declaration
A.2.i. to the *kami*, and aspiration

A.2.j. five vows of Samantabhadra Bodhisattva
A.2.j.i. take refuge in the Three Jewels
A.2.j.ii. repent all wrongdoings
A.2.j.iii. rejoice in merits
A.2.j.iv. beseech the Lord Mahāvairocana
A.2.j.v. transfer merits
A.2.k. evoking bodhicitta
A.2.l. samaya
A.2.m. vows
A.2.n. five great vows of a bodhisattva
A.2.o. universal offering, and three powers
A.3. binding the earth element
A.3.a. Mahā-Vajracakra: establishing the earth boundary
A.3.b. vajra poles: binding the earth element
A.3.c. vajra walls: binding the four corners
A.4. adorning the seat of enlightenment
A.4.a. visualize the seat of enlightenment
A.4.b. mudrā of Great Ākāśagarbha Bodhisattva
A.4.c. smaller vajracakra
B. inviting and receiving the deities
B.5. requesting the deities
B.5.a. sending off the jewelled carriage
B.5.b. return of the jewelled carriage
B.5.c. inviting the chief deity
B.5.d. four wisdoms
B.5.e. clap the hands
B.6. sealing for protection
B.6.a. Hayagriva-vidyarāja
B.6.b. vajra net
B.6.c. vajra fire enclosure
B.6.d. great samaya
C. feasting the deities
C.7. *pūjā* offerings
C.7.a. offering of *argha* water
C.7.b. lotus throne
C.7.c. ringing the vajra bell
C.7.d. five *pūjā* offerings, offered symbolically
C.7.d.i. powdered incense
C.7.d.ii. flower garlands
C.7.d.iii. burning incense
C.7.d.iv. food and drink
C.7.d.v. lights

C.7.e. five *pūjā* offerings, offered materially
- C.7.e.i powdered incense
- C.7.e.ii. flower garlands
- C.7.e.iii. burning incense
- C.7.e.iv. food and drink
- C.7.e.v. lights

C.7.f. hymn in praise of the four wisdoms, and clap the hands
C.7.g. hymn to the chief deity
C.7.h. vast *maṇi* offering to the three powers
C.7.i. worship of the buddhas

D. ritual identification
D.8. recitation
D.8.a. identity of body
- D.8.a.i. visualization of ritual identity
- D.8.a.ii. three mudrās and mantras of the chief deity

D.8.b. identity of speech
- D.8.b.i. initial recitation
- D.8.b.ii. three mudrās and mantras of the chief deity

D.8.c. identity of mind
- D.8.c.i. visualization of the *akṣaracakra*
- D.8.c.ii. three mudrās and mantras of the chief deity

D.8.d. subsequent recitation
- D.8.d.i. Buddhalocanā
- D.8.d.ii. Mahāvairocana of the Garbhakośa
- D.8.d.iii. Mahāvairocana of the Vajradhātu
- D.8.d.iv. Akṣobhya
- D.8.d.v. Ratnasambhava
- D.8.d.vi. Amitāyus
- D.8.d.vii. Amoghasiddhi
- D.8.d.viii. Vajrasattva
- D.8.d.ix. Trailokyavijaya
- D.8.d.x. Mahāvajracakra
- D.8.d.xi. One Syllable Golden Cakra
- D.8.d.xii. Buddhalocanā

C'. completion of the feast
C'.7. latter *pūjā* offerings
C'.7. d. five *pūjā* offerings, offered symbolically
- C'.7.d.i powdered incense
- C'.7.d.ii. flower garlands
- C'.7.d.iii. burning incense
- C'.7.d.iv. food and drink
- C'.7.d.v. lights

- C'.7.e. five *pūjā* offerings, offered materially
    - C'.7.e.i. powdered incense
    - C'.7.e.ii. flower garlands
    - C'.7.e.iii. burning incense
    - C'.7.e.iv. food and drink
    - C'.7.e.v. lights
- C'.7.a. *argha* water
- C'.7.c. latter ringing of the vajra bell
- C'.7.f. praises
- C'.7.h. universal offering, and three powers
- C'.7.i. worship of the buddhas

A'.2. latter practice of Samantabhadra's vows
- A'.2.k. transfer of merits
- A'.2.j. five vows and sincere transfer of merits

B'. leave taking and sending off
- B'.6. opening the ritual enclosure
- B'.6.d. great *samaya*
- B'.6.c. fire enclosure
- B'.6.a. Hayagrīva
- A'.3. unbinding the earth element
- A'.3.c. vajra walls
- A'.3.b. earth element
- B'.5. sending off the deities (this one single action corresponds to all five of the actions in B.5. requesting the deities)

A'. completion of the ritual
- A'.1. practitioner's departure
    - A'.1.h. three classes
        - A'.1.h.ii. samaya of the Buddha-kula: purify bodily karma
        - A'.1.h.iii. samaya of the Padma-kula: purify speech karma
        - A'.1.h.iv. samaya of the Vajra-kula: purify mental karma
    - A'.1.i. remove the armour
    - A'.1.b. universal reverence: prostrations
    - A'.1.a. depart the shrine

A tree diagram of this structure is appended to this essay.

In addition to terminal abbreviation, the *Jūhachidō* also shows the embedding of rites into a ritual, a syntactic pattern which has been extensively discussed by Staal.[35] Most prominently, the transformation of a votive *pūjā* into a tantric *pūjā* by the addition of ritual identification exemplifies embedding.[36]

Initially, the presence of the transfer of merits between C'.7. and B'.6.

appeared to be another instance of embedding. The transfer of merits is, of course, one of the hallmarks of Mahāyāna Buddhism, and is found during the conclusion of almost all Mahāyāna practices.[37] On this interpretation, the transfer of merits serves as an embedded rite which in this case transforms the ritual from what might be called a 'Hindu' *pūjā* into a Mahāyāna Buddhist one, just as the embedding of ritual identification transformed it from a votive into a tantric *pūjā*.

However, upon closer examination, this appears to be an instance of interruption, which Staal has described, saying 'an embedded ritual may be interrupted, once or several times, by the ritual in which it is embedded, to be continued or completed afterwards'.[38] Since the transfer of merits is a bodhisattva action, it corresponds to the actions found in A.2, i.e., evoking bodhicitta and taking the five vows of Samantabhadra bodhisattva.[39]

In addition to the structural issues, however, there is an additional consideration regarding the transfer of merits. The eighteen mudrās referred to in the name of the ritual would appear to constitute the historically oldest strata of the *Jūhachidō*. None of the mudrās found in the 'practising Samantabhadra's vows' is among the traditional eighteen. This would suggest that both 'practising Samantabhadra's vows' and the transfer of merits were embedded in the process of transforming a Hindu *pūjā* into a Buddhist one.[40]

Not surprisingly, the *Jūhachidō* also shows the syntactic pattern of symmetry. There are, however, two different kinds of symmetry: mirror image symmetry and repetitive symmetry. Mirror image symmetry involves the reversal of the order of actions when they are performed in the latter part of the ritual. This is found, for example, in B'.6. opening the ritual enclosure, where the order is B'.6.d, B'.6.c. and B'.6.a. (The absence of B'.6.b. is an instance of terminal abbreviation.) Repetitive symmetry is where the actions are performed in the same order in the second half of the ritual as in the first. This is exemplified quite clearly in C'.7., latter *pūjā* offerings, where not only are the symbolic and material offerings made in the same order (C'.7.d. and C'.7.e.), but the individual offerings within each of these two are made in the same order (C'.7.d.i., C'.7.d.ii., C'.7.d.iii., C'.7.d.iv., and C'.7.d.v.). The idea that there are two different kinds of symmetry is at variance with the assumption Staal seems to make that there is only mirror image symmetry when he describes the order of the *agniṣṭoma* as a 'not quite successful' attempt 'to establish a regular "mirror-image" pattern'.[41]

CONCLUSION: AREAS FOR FURTHER STUDY

RITUAL SEMANTICS

One of the problematic issues which Staal has raised is that of ritual semantics. Although ritual is 'meaningless' (in the sense that it is performed for its own sake[42]), it does seem to have some aspects which are correlates of semantics. Three of these aspects are the organizing metaphor, the changes which occur when there is a change of chief deity evoked, and the way in which a ritual system is organized.

As described above, *pūjās* are organized around the metaphor of feasting an honoured guest. This is not, however, the only organizing metaphor. Strickmann has pointed out that Taoist ritual is organized around the metaphor of submitting a petition to a government bureau. Doubtless additional kinds of organizing metaphors will be discovered as work in ritual studies progresses.

While this study has examined a single ritual, comparison with other rituals in the Shingon corpus indicates a kind of semantic shift when the chief deity changes. For example, the *Ajikan* is a meditative ritual in which the chief deity is the syllable A. As the syllable is not anthropomorphic, offerings are not made to it - a sort of 'intransitive' form of *pūjā*.

The organization of a ritual system includes the hierarchical relations discussed above, i.e., the fact that a person must have completed the simpler rituals in the system or have received the proper initiations in order to perform a subsequent ritual. Thus, part of the 'meaning' of performing the *Jūhachidō* is that the practitioner has established a karmic bond with one of the deities of the Shingon *maṇḍalas*, received a Buddhist name, taken the three sets of precepts, performed the proper number of prostrations, practised the full moon meditation, and practised the visualization of the syllable A.[43]

HISTORICAL RITUAL STUDIES

Staal chose to adopt a synchronic perspective for the study of ritual syntax, since such a perspective has proven 'fruitful in the linguistic study of syntax'.[44] Synchronic studies will certainly continue to be necessary as they provide an understanding of the workings of a particular ritual system at a particular time. Studies which are not so chronologically delimited risk mixing together aspects of a ritual system which are not to be found together at any one time - the danger of creating anachronisms. Having taken such precautions, however, it would seem to be the case that historical studies of ritual, its change over time and across cultures, can be established in much the same fashion as historical linguistics now studies changes in language.

The Shingon corpus can be a valuable resource for such historical ritual studies. Many of the rituals have roots in the Vedic ritual system, underwent transformation when they became tantric and Buddhist rituals, and were then transmitted to China and eventually to Japan. A related set of rituals were also transmitted to Tibet where they have undergone development separately from the developments in East Asia. The historical and cultural complexities of the lineages of ritual practice will be a rich resource for the establishment of a history of ritual and religious praxis comparable to the histories of doctrine which have tended to dominate the academic study of religion up to the present time.

CHARACTER GLOSSARY

A ji kan 阿字観

dembô 伝法

Fudô Myôô 不動明王

Gachirin kan 月輪観

goma 護摩

Gosô jôshin kan 五相成身観

Himitsu Jisô no kenkyû 秘密事相の研究

Jûhachidô 十八道

kanjô 灌頂

kechien 桔縁

Kongôkai 金剛界

samaya 三摩耶

Taizôkai 胎蔵界

Structure of the Shingon *Jūhachidō*

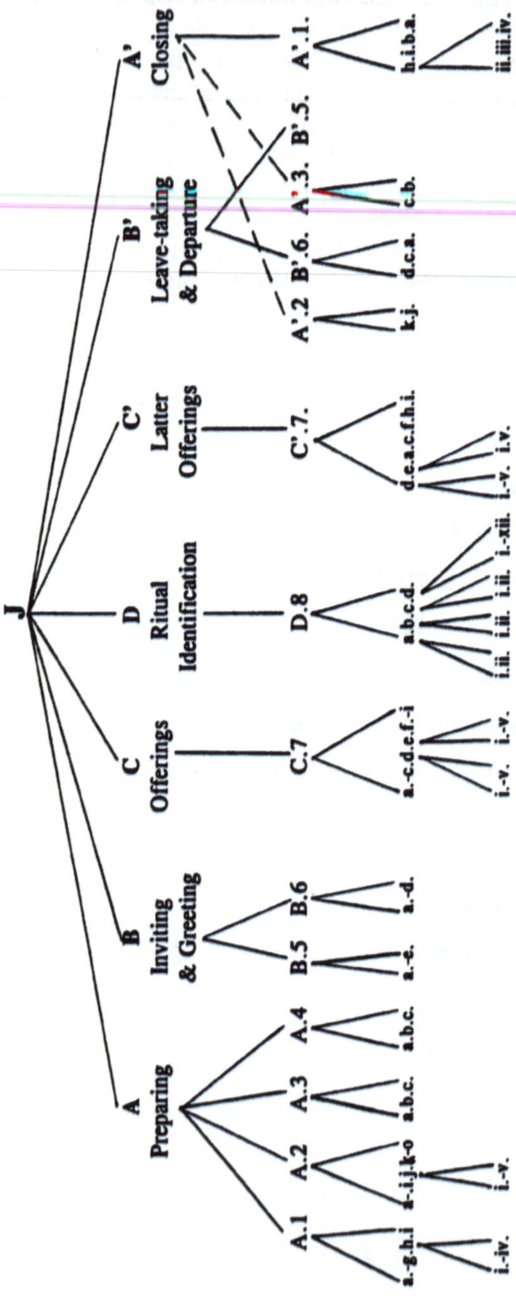

Notes

1. In 1981, when Frits Staal agreed to be a member of my dissertation committee, he gave me a copy of his essay 'The Meaninglessness of Ritual', *Numen, International Review for the History of Religions*, 1979. Reprinted in part in chapter 13C of his *Rules Without Meaning*. Toronto Studies in Religion, vol. 4. New York: Lang, 1989, pp. 131-40. The argument for the syntactic analysis of ritual which he laid out in that essay struck me as being the single most important methodological contribution to the study of ritual. It shaped my study of the *goma* [Ph.D. dissertation, 'Feeding the Gods: The Shingon Fire Ritual', Graduate Theological Union, 1985. Since published in an expanded version as *The Tantric Ritual of Japan: Feeding the Gods*. Sata Pitaka Series, no. 365. Delhi: International Academy of Indian Studies, 1989] and all of my study of the Shingon ritual corpus since. This essay is an attempt to repay the kindness Professor Staal has shown me over the years by adding my own arguments for the syntactic study of ritual to his own.
2. Staal, 'Meaninglessness', p. 15.
3. *Ibid.*, p. 17.
4. Frits Staal, 'Ritual Syntax', in: *Sanskrit and Indian Studies, Essays in Honour of Daniel H.H. Ingalls*; 'Ritual Structure', in: Frits Staal (ed.), *Agni, The Vedic Ritual of the Fire Altar*, vol. II; *The Science of Ritual*. Post-graduate and Research Department Series, no. 15. Professor P.D. Gune Memorial Lectures, first series. Poona: Bhandarkar Oriental Research Institute, 1982, and *Rules Without Meaning*.
5. Cf. Staal, 'Meaninglessness', p. 2.
6. Richard H. Davis, 'Cremation and Liberation: The Revision of a Hindu Ritual', *History of Religions*, 1988, pp. 37-53.
7. *Ibid.*, p. 46.
8. *Ibid.*, pp. 46-7.
9. See Michel Strickmann, 'Homa in East Asia', in: Staal (ed.), *Agni*, II. 418-55, p. 418. For a critique of such a monothetic view and the development of a polythetic view of tantra, see Douglas Renfrew Brooks, *The Secret of the Three Cities: An Introduction to Hindu Sakta Tantrism*. Chicago: University of Chicago Press, 1990, pp. 52-3.
10. Davis, *op. cit.*, p. 47. For a full description of the *nirvāṇadīkṣā*, see Richard H. Davis, *Ritual in an Oscillating Universe: Worshipping Śiva in Medieval India*. Princeton: Princeton University Press, 1991, pp. 89-100.
11. Davis, 'Cremation and Liberation', p. 43.
12. *Ibid.*, p. 48.
13. *Ibid.*, p. 40.
14. Strickmann, *op. cit.*, p. 418.
15. While I am not in fact asserting either of these explanations, at the very least after having considered the syntactic aspects of the ritual, one might be led to a more qualified conclusion about the relation between the Śaiva ritual and the Vedic. Davis' explanation seems to be based on the Śaivas' own understanding of what has occurred. However, the Śaivas from their own perspective may see

this transformation of the Vedic rite differently than would someone from further outside their range of concerns. What may seem to them to be a radical transformation, may look to someone else as a very minimal, very conservative reworking of an existing ritual practice. Davis' interpretation of the Śaiva funerary rite seems to implicitly accept the perspective of the Śaiva tradition itself. This is clearer in his very important and more extended *Ritual in an Oscillating Universe*. What I am attempting to establish here is not that there is anything wrong with taking such a perspective as the basis for evaluative judgements such as those he makes concerning the radical character of the Śaiva transformation, much less attempting to assert that there is some purely objective perspective or that a syntactic analysis provides such an objective perspective. Rather, I am asserting that the bases upon which evaluative judgements are made should also be made explicit in an author's presentation. If what Davis is saying is that from the viewpoint of the Śaiva tradition itself the transformation made constituted a radical critique of the inherited Vedic tradition, then I think there would be no reason to question his evaluation. Lacking the qualification, however, his evaluation is weakened since other perspectives do not support his evaluation. The potential value of syntactic analysis for ritual studies is that it can provide a common basis for discussing and comparing interpretive and evaluative judgements.

16. Staal, 'Meaninglessness', p. 2.
17. Staal, 'Ritual Syntax', p. 123.
18. Toganoo Shoun, *A Study of Esoteric Sādhanas* (tr. Leo Pruden. Xerographic copy, n.d.).
19. Also known as Caṇḍamahāroṣaṇa. See my 'Firmly Rooted: On Fudō Myōō's Origins', *Pacific World, The Journal of the Institute of Buddhist Studies*, 1988.
20. See either my *Tantric Ritual of Japan*, pp. 95-142, or my 'Ritual Directions for the Śāntika Homa Offered to Acala', in: Dale Todaro (ed.), *Handbook on the Four Stages of Prayoga, Chūin Branch of Shingon Tradition*. Koyasan: Department of Koyasan Shingon Foreign Mission, 1988, for a translation of the manual used in the performance of this ritual.
21. *Goma zenshū*. Koyasan: Senshugakuin, 1982.
22. *Ninnō-kyō hō, Shūjgō-kyō hō, Kujaku-kyō hō, Seiukyō-hō, Go-shichinichi hō, Taigen hō, Fugen Emmyō hō*, and *Hokke-kyō hō*. Different lineages within the Shingon tradition disagree as to the exact number of these 'major ceremonies', some adding, others deleting.
23. *Niken Kannon hō, Ōsashi hyō hō, Nyoihōju hō, Hija hō, Misoka go-nenju, Koya-nenju hō, Shari hō, Ryōbu-gōgyō hō, Zuigu hō, Nyohō Aizen hō, Nyohō Shōson hō, Go-dai Kokuzō hō, Go-himitsu hō, Kōmyō Shingon hō*, and *Rishu-kyō ho*. Again, there are variations between different Shingon lineages.
24. For the list of these, see Toganoo, *op. cit.*, pp. 280-3.
25. Staal, 'Ritual Syntax', p. 125.
26. A complete description of this system is to be found in my *The Tantric Ritual of Japan*, pp. 66-72.

27. Taisen Miyata identifies twenty-five mudrās. *A Study of the Ritual Mudrās in the Shingon Tradition: A Phenomenological Study on the Eighteen Ways of Esoteric Recitation (Jūhachidō Nenju Kubi Shidai: Chūin-ryū) in the Koyasan Tradition*. Sacramento: Northern California Koyasan Temple, 1984. Seventy-four mudrās are given in M. Horiou Toki, *Si-Do-In-Dzou: gestes de l'officiant dans les cérémonies mystiques des sectes Tendai et Singon*. Paris: Leroux, 1899, pp. 148-66, though this presentation shows every mudrā formed in the course of the ritual, many of which are duplicates of one another since the same mudrā is used for several ritual actions.
28. Lawrence A. Babb, *The Divine Hierarchy: Popular Hinduism in Central India*. New York: Columbia University Press, 1975, p. 54.
29. Ākos Östör, *The Play of the Gods: Locality, Ideology, Structure, and Time in the Festivals of a Bengali Town*. Chicago and London: University of Chicago Press, 1980, p. 50. For a description of *pūjā* in the Jain tradition, see Caroline Humphrey and James Laidlaw, *The Archetypal Actions of Ritual: A Theory of Ritual Illustrated by the Jain Rite of Worship*. Oxford: Oxford University Press, 1994, pp. 25-31.
30. Toganoo, *op. cit.*, p. 73.
31. Östör, *op. cit.*, pp. 152-3.
32. Toganoo, *Himitsu Jiso*, p. 80-3. Slightly variant versions of *Juhachidō shidais* have developed within the Shingon lineages since the rite was introduced by Kūkai. For the version edited by Yūkai (1345 or 1348 to 1416) cf. Taisen Miyata, tr., 'The Jūhachidō Nenju Kubi-Shidai: Chūin-ryū Text with Commentary', in: Miyata, *op. cit.*, p. 113-62.
33. Paul Demiéville, Hubert Durt and Anna Seidel (eds), *Répertoire du Canon Bouddhique Sino-Japonais*, 2nd ed. Paris: Libraire d'Amérique et d'Orient, and Tokyo: Maison Franco-Japonaise, 1978, p. 250.
34. Staal, 'Ritual Syntax', p. 127.
35. Staal, 'Ritual Structure', II, pp. 127-34.
36. This is a simple embedding, rather than an instance of a repeatable, self-embedding which Staal identifies as an instance of a recursive rule, Staal, *Rules Without Meaning*, p. 88.
37. See Yuichi Kajiyama, 'Transfer and Transformation of Merits in Relation to Emptiness', in: Katsumi Mimaki, *et al.* (eds), *Y. Kajiyama, Studies in Buddhist Philosophy (Selected Papers)*. Kyoto: Rinsen Book Co., 1989, pp. 1-20. Also, Gadjin M. Nagao, 'Usages and Meanings of Pariṇāmanā', in: Gadjin Nagao, *Mādhyamika and Yogācāra: A Study of Mahāyāna Philosophies*. Ch. 8, ed. and tr., Leslie S. Kawamura. Albany: State University of New York Press, 1991, pp. 83-90.
38. Staal, *Rules without Meaning*, p. 108.
39. One may speculate that the rationale for this interruption is the desire to make the transfer of merits while still in the presence of the deities.
40. Additionally in Yūkai's version the wording of the 'sincere transfer of merits' is the same as the fifth of the 'five vows of Samantabhadra'. See Miyata, *op. cit.*, pp. 157-8.
41. Staal, 'Ritual Syntax', p. 134.

42. Staal, *Rules Without Meaning*, pp. 131-2. Staal also asserts that ritual activity is distinct from ordinary activity in that ritual activity is 'useless'. Yet Staal identifies a kind of efficacy of ritual. For example, eligibility rules indicate that there is a distinction between those who have completed the required preceding rituals and those who have not. (See for example, Staal, *Agni*, I, p. 193.) Thus, the preceding rituals have the effect of qualifying the practitioner. Similarly, two people who have undergone the wedding ritual are considered married. Of course, such ritual efficacy results from the ritual system itself or from the nesting of the ritual system within the social system, rather than from having the pragmatic efficacy of ordinary activities. One is reminded here of Dan Sperber's discussion of how he identifies what is symbolic. Dan Sperber, *Rethinking Symbolism*. Cambridge: Cambridge University Press, 1975, pp. 2-4.
43. This aspect of a ritual system is what Lawson and McCauley seem to be delineating in their diagramming of rituals. E. Thomas Lawson and Robert N. McCauley, *Rethinking Religion: Connecting Cognition and Culture*. Cambridge: Cambridge University Press, 1990. Lawson and McCauley's approach seems to provide an important basis for describing ritual systems. Their structural analysis, however, seems to be an analysis of a narrative description of a ritual, rather than of the ritual itself.
44. Staal, 'Ritual Syntax', p. 122.

XXII

BHARTṚHARI'S PHILOSOPHY OF LANGUAGE SPHOṬAVĀDA AND ŚABDABRAHMAVĀDA: ARE THEY INTERRELATED?

K. Kunjunni Raja

Bhartṛhari, who flourished in the beginning of the fifth century, was an eminent Sanskrit grammarian and a deep and original philosopher of language. He is well known for enunciating two theories: the *Sphoṭavāda* and the *Śabdabrahmavāda*. They created a storm among thinkers; orthodox schools attacked them severely; modern thinkers of the present century are attracted by the novelty and significance of Bhartṛhari and are trying to understand these views in their proper perspective. Some scholars, like Biardeau, believed that the two theories are interrelated and cannot be explained in isolation. In my book *Indian Theories of Meaning*, I held that even if they are interrelated, each can be studied individually also. I shall try to give a brief summary of the two theories.

Bhartṛhari begins his Vākyapadīya with a clear statement that the Absolute Brahman is identical with the Speech Principle, and that the whole phenomenon of material existence is only a manifestation of this Śabda Brahman of which *śabda* and *artha*, the symbol and the meaning, are only two aspects. This speech essence, which is identical with consciousness (*cid*), has no beginning or end, and appears in the form of various symbols or expressions on the one hand, and the meanings in the form of ideas on the other hand, and thus constitutes the entire phenomenal world:

anādinidhanam brahma śabdatattvaṁ yad akṣaram
vivartate'rthabhāvena prakriyā jagato yataḥ.

Symbols and meanings, *śabda* and *artha*, *nāma* and *rūpa* are only two aspects of the same principle. Though this Ultimate Reality, this Speech Essence, is one, it appears different on the basis of *śaktis* like time and space, which are not different from itself; thus the world appears as evolutionary and pluralistic.

In language behaviour, the idea and the expression are unitary in the mind of the speaker and the listener. Actual expression or manifestation of the sentence can be effected only through time or space, the efficient cause by which Brahman controls the universe. Just as the wirepuller in a puppet play has full control of the puppets, time has full control over the running of the world. It is the pregnant forces within Śabdabrahman that form the first cause of the bursting forth of worldly phenomena.

Though Bhartṛhari uses the term *vivarta* for the manifestation of the world from the *śabdatattva*, in his theory the phenomenal world never loses its direct ontological identity with Brahman. Śaṅkara's rope-snake identity theory of superimposition (*adhyāsa*) does not explain Bhartṛhari's theory correctly. According to Bhartṛhari, the *śaktis* like time and space are not different from Brahman; they have the capacity to bring out its potentiality into activity. The question whether time and space are subjective or objective is solved by taking *śaktis* as real, which bring forth from a real Śabdabrahman a real world that is not *māyā* as in Śaṅkara's Advaita.

Bhartṛhari has no need to have a nirguṇa Brahman, indescribable and transcendent. Revelation through intuitive power, an instantaneous flash of insight (*pratibhā*), can penetrate reality well beyond the empirical level.

According to Bhartṛhari, the integral, indivisible *śabda*, which is the meaning bearing aspect (*vācaka*) of language, is called *sphoṭa*; its integral, indivisible meaning (*artha*) is also a flash of insight (*pratibhā*). There are different levels in the manifestation of speech. Just as the phonemes (*varṇas*) are unreal abstractions from the word, so also the words are abstractions from the sentence. The utterance (*uccāraṇa*) can be considered as the *vākya-sphoṭa*, the whole utterance being taken as an integral and indivisible unit (*vākyam eva sphoṭaḥ*), or as *padasphoṭa* or *varṇasphoṭa*. Bhartṛhari's view is that the whole utterance, *vākyasphoṭa*, should be taken as the unit of communication.

When we listen to the utterance of a person, what is it that we hear? Is it the whole utterance (*vākya*) of the words in a sequence, or the phonemes in a sequence? If we do not know the language at all, we may be hearing only some sound-bits which we cannot understand. Even the identity of the phonemes may not be clear. If we know the language very well, as in the case of the mother tongue, we actually hear the sentence as a whole, and immediately grasp its meaning also as a whole in a flash of insight. The fact that it is communicated in a temporal or spatial series in sequence is only due to the imperfection of the medium, but that is the only way of effective communication. Depending on our command of the language, we may hear the sentence as a whole, or as a series of words or phonemes. Bhartṛhari and his follower Maṇḍana Miśra (in his Sphoṭasiddhi) are definite that the *sphoṭa* is perceived through the ear when we listen (*śrotrapratyakṣa*). In actual life situations, what we hear is only the sentence-unit and from that we get the integral sentence meaning. In the case of a new sentence we have never heard before we may hesitate, but on the basis of the memory of the word meanings acquired through experience (word-meaning relationship), we understand the sentence and its meaning.

Whether Bhartṛhari accepted only three stages in the manifestation of *Vāc* or *śabda*, or four, continues to be discussed. Bhartṛhari has stated that the speech principle (*śabdatattva*) has three levels in the course of its manifestation, namely, *Paśyantī*, *Madhyamā* and *Vaikharī* (VP 1.143):

vaikharyā madhyamāyāś ca paśyantyāś caitad adbhutam
anekatīrthabhedayāḥ trayyā vācaḥ paraṁ padam.

When we speak of *vākyasphoṭa* as the revealer of meaning (*vācaka*) and *pratibhā* as its meaning, the two seem to be different; but actually they are only two aspects of he same entity. Whether Bhartṛhari accepted a fourth stage higher than the *Paśyantī* has been debated. Bhartṛhari's statement about the three levels of *Vāc* does not preclude the acceptance of a higher level about which we cannot say anything.

According to some Kashmir Śaiva scholars, *Paśyantī* has two aspects of which one is *parā-Paśyantī*. Utpaladeva states in his Iśvarapratyabhijñākārikā (IsMEO, Roma, 1994) that Bhartṛhari accepts the highest level as *parāpaśyantī*:

citiḥ pratyavamarśātmā parā vāk svarasoditā
svātantryam etan mukhyaṁ tadaiśvaryam,

'consciousness has as its essential nature reflective awareness (*pratyavamarśa*); it is the supreme word (*parā vāc*) that arises freely. It is freedom in the absolute sense, the sovereignty (*aiśvarya*) of the supreme self' (p.120). The *vṛtti* does not describe the different levels, but only the first. In the Kashmir Śaiva tradition, Bhartṛhari is criticized for giving a threefold division of *Vāc*.

According to Utpaladeva, *Parā-vāc*, the highest level of *Vāc*, is so close to silence that the distinction between the Supreme *Vāc* and its power or energy is non-existent and belongs only to theory. It is identified with Brahman by Bhartṛhari. Earlier, Somānanda in his Śivadṛṣṭi attacked Bhartṛhari's theory assuming that only three levels are assumed by him, and added (Iśvarapratya-bhijñāhṛdayakārikā, p. xx): 'Why can't the grammarians confine their attention to grammar; why should they intrude into philosophy?'

XXIII

À PROPOS DE RAPPORTS ENTRE RASAŚĀSTRA ET TANTRA: ÉTUDE SUR UN FRAGMENT DU RASENDRACŪḌĀMAṆI

Arion Roşu[*]

En Occident, les penseurs de la Renaissance ont diversement situé l'alchimie dans l'ensemble des savoirs de cette époque du XVe au XVIIe siècle, considérée comme l'âge d'or de l'art hermétique. Classée comme art opératoire ou discipline théorique, l'alchimie comporte aussi une incertitude en ce qui concerne sa finalité, à savoir la fabrication de l'or (aurifaction) ou la vocation thérapeutique.[1] En Asie, l'ancienne chimie indienne (*rasaśāstra*) comprend dans sa partie positive, classique, les arts chimiques (pharmacie, métallurgie), alors que par ses aspirations spéculatives, baroques, elle cultive le bithématisme récurrent dans l'histoire de l'alchimie: transmutation des métaux imparfaits (*dhātuvāda*) et recherche de la médecine universelle, qui est à la fois panacée (*sarvārha*) et élixir de longue vie (*rasāyana*). Auxiliaire de l'Āyurveda comme pharmacie empirique aux temps anciens, le Rasaśāstra devient à l'époque médiévale une 'iatrochimie', antérieure à la 'chimiatrie' paracelsienne de la Renaissance.[2]

Les incertitudes qui entourent les origines de l'alchimie indienne - supposées être marquées par certains apports chinois[3] - et le développement post-gupta de ses pratiques rendent malaisée la recherche historique en ce domaine, comme pour tant d'autres savoirs indiens. De ce fait, tout filon

* Au terme de cette étude, nous sommes heureux de témoigner à Madame Hélène Diserens (Paris) notre vive reconnaissance de son dévouement au traitement informatique du texte. Messieurs Michel Angot (Pontault-Combault) et Dick van der Meij (International Institute for Asian Studies, Leyde) ainsi que Madame Hanneke 't Hart (Instituut Kern, Leyde) y ont épisodiquement prêté leur concours, dont nous les remercions.

culturel doit être exploité dans la mesure où il est susceptible d'éclairer l'histoire lacunaire du Rasaśāstra. A ce titre, méritent attention, en raison de leurs nombreuses notations alchimiques, les légendes concernant le lieu saint de Śrīśailam (Śrīgiri, Śrīparvata) et les portes-sanctuaires de son approche, notamment celle d'Alampur.[4]

Dans cette région de pèlerinages du pays Āndhra, privilégiée pour les śivaïtes, certains textes du Rasaśāstra (ĀK, RRĀ) décrivent des pratiques alchimiques, alors qu'aux XIVe et XVe siècles, les littératures dravidiennes, principalement les sources telugu, abondent en éléments relatifs à l'imaginaire de l'art transmutatoire. Ces données reflètent les enseignements des mouvements *siddha*[5] et *nātha(yoga)*[6], qui se manifestent à l'époque en tant que composantes du tantrisme en Āndhra et ailleurs au Dekkan, et même au-delà. En effet, le phénomène est observable aussi au Gujarat, dans l'Inde occidentale, marquée par le *rasésvara-darśana* et où le recueil jaina *Vividha-tīrthakalpa* du XIVe siècle donne du mont Girnār l'image d'un lieu saint alchimique, dont la description affabulatrice enrichit l'imaginaire indien.[7]

Evoqué en Āndhra au culte quotidien, lors de la formule du *deśakāloccāraṇa / saṁkalpa*, prononcée par les dévots hindous, le lieu saint de Śrīśailam est aussi magnifié pour les mérites acquis aux pèlerins qui s'y rendent pieusement. A partir des épopées on retrouve ce *tīrtha* śivaïte mentionné dans les littératures indiennes, qui laissent parfois percer un climat tantrique à Śrīśailam, marqué par la présence d'ascètes *kāpālika* (*Mālatīmādhava*).[8] Ce centre de pèlerinage, où Nāgārjuna aurait, selon la tradition, fondé un laboratoire alchimique (*Rasendramaṅgala*), est aussi une source de récits invraisemblables (*Kādambarī*) et un lieu où fleurissent des procédés magiques (*Ratnāvalī*).

Réputé comme *siddhikṣetra* favorisant l'acquisition de pouvoirs surnaturels (*Agnipurāṇa*), le lieu saint dekkanais est connu traditionnellement comme un endroit idéal tant pour les ascètes que pour les magiciens et les *siddha*. Ces êtres 'parfaits', réels ou légendaires, sont censés posséder des capacités surhumaines d'ordre spirituel ou matériel, et notamment des procédés pour assurer la santé ou prolonger la vie (formules macrobiotiques, opérations mercurielles). Certains archéologues indiens n'hésiteraient pas à les considérer même comme bâtisseurs de plusieurs temples consacrés à Śiva Siddheśvara dans la région de Śrīśailam. Centrée sur le temple de Śiva Mallikārjuna et délimitée par huit portes-sanctuaires, en particulier celle d'Alampur, cette zone sacrée se trouve placée sous le signe de la rêverie alchimique.[9] Le répertoire de cet imaginaire nourrit les descriptions, proches par le contenu et l'étendue, que présentent les traités médiévaux ĀK 1.12 (*Śrīśaile siddhilābhaḥ*) et RRĀ 4.8 (*Śrīparvatasādhanam*).

Si la qualité d'alchimiste est généralement acquise aux méditants *siddha*, en particulier dans l'Inde du Sud, cette qualification s'impose apparemment moins pour les renonçants *kāpālika*. Attestés dans les sources indiennes dès le début de notre ère,[10] ces ascètes porteurs de crâne (*puruṣa-śiras*) sont célèbres plutôt par leur aspect et leur comportement répulsifs, caractérisés avec vivacité dans le troisième acte du drame allégorique *Prabodhacandrodaya* du XIe siècle. Vers la même époque, une autre œuvre dramatique, le *Caṇḍakauśika*, d'inspiration purāṇique, met en scène, au quatrième acte, le dieu Dharma sous les traits d'un *kāpālika*, qui s'est rendu maître des arts magiques (*guṭikā*, *añjana*, *pādalepa*, etc.) et alchimiques (*rasāyana*, *dhātuvāda*). En possession d'un grand trésor de substances merveilleuses, il fait l'éloge du mercure, qui chasse la mort, car l'usage de ce métal 'surparfait' (*saṁsiddha*) conduit les éminents *siddha* au monde des immortels (*amaraloka*), où ils vivent heureux sur les sommets du Meru. D'autre part, la tradition du Rasaśāstra connaît des maîtres qui répondent aux noms théophores de Kāpālin (RRS 1.2), de Kāpāli (SDS, p. 81, 1.8) ou encore de Kāpālika (RRS 1.4), attesté aussi comme auteur d'un texte sur la pathologie oculaire (*New Catalogus catalogorum*, vol. III, p. 343: *netraroganidāna*).

C'est la littérature technique sanskrite qu'il convient d'interroger pour d'autres allusions éventuelles à des pratiques du *rasavāda* en milieu *kāpālika*. Selon certains auteurs, les adeptes de ce mouvement śivaïte, diversement appelés (*kāpāla*, *kapālin/kāpālin*, *soma*, *mahāvratadhara*, etc.) n'ont pas produit un corpus canonique propre ni d'autres écrits.[11] En conséquence, le śivaïsme *kāpālika* a adopté comme autorité les Tantra de Bhairava (*bhairavāgama*), ainsi dénommés du fait que ces textes sont structurés par l'enseignement de Śiva donné à la Déesse, qui questionne en disciple. Cette tradition scripturaire a retenti en Rasaśāstra, qui cultive aussi Bhairava.[12] Ce Śiva, dieu de la transgression par excellence, qui a décapité Brahma, est le brahmanicide archétypique, dont les *kāpālika* suivent rituellement l'exemple. Cette tradition de la sacralité transgressive s'est prolongée avec les renonçants *aghorī*, qui ont grandement contribué à la diffusion du culte de Bhairava, comme en témoignent certains observateurs modernes.[13]

La recherche d'une littérature *kāpālika* nous conduit à citer quelques textes alchimiques inédits, dont les titres sont susceptibles d'évoquer ces *tāntrika* hétérodoxes: le *Kāpālikatantra,* dont le manuscrit est conservé à l'Université de Trivandrum,[14] le *Kāpālīsiddhānta* de la collection du célèbre lettré Kavīndrācārya, qui remonte vers le milieu du XVIIe siècle,[15] et le *Kāpālikatantra*, écrit en *nandināgarī* sur feuilles de palmier et conservé à l'Université de Mysore.[16] Dans son commentaire sur l'*Āpastambīyadharmasūtra* (1.10. 29.1), Haradatta mentionne un *Kāpālikātantra* à propos du sceptre magique *khaṭvāṅga*, insigne du brahmanicide pénitent (*mahāvratin*) et auquel sont associés les śivaïtes porteurs de crâne. Selon toute vraisemblance, le terme de *tantra* signifie ici 'école, doctrine', et non 'texte'.[17]

Le manuscrit de Mysore commence et s'achève par une invocation à la déesse Lakṣmī. Début: *śrīmahālakṣmyai namaḥ / avighnam astu // namas tasmai rasendrāya jarāmaraṇahāriṇe / daridragajasiṁhāya jagattimira-bhānave* // (folio 1r, ligne 1) ... / Fin: *catuṣṣaṣṭipuṭe bhāge śulvavedhaṁ pradāpayet / jāyate kanakaṁ divyaṁ devābharaṇabhūṣanam // ... / sa hi bhavati sahasravedhī tāre tāmre bhujaṅge ca // - //* Colophon: *iti kāpālikānantraṁ* (corr.*-tantraṁ*) *samāptam // - // śrīmanmahālakṣmyai namaḥ // - // śrīgopījanavallabhārpaṇam astu //* (folio 20v, lignes 6-8). Le contenu de ce texte inédit est incontestablement alchimique, mais son lien avec les ascètes porteurs de crâne est loin d'être évident. Le vocable *kāpālikā* du titre pourrait s'appliquer à des femmes ascètes. En effet, les thèmes sanskrits en *a-* présentent au féminin la tendance vers *-ā* dans les adjectifs, et vers *-ī* dans les substantifs et les formes substantivées (*kāpālikī*), sans exclure des variations,[18] telles que le théonyme Kāpālikā (Kapāla-Bhairavī).[19] Comme étymologie, le mot *kāpālika/kāpālikā* dérive de *kapāla* 'coupe, bol', qui désigne également, selon Suśruta (*Nidāna* 5.7-8), une dermatose grave (*mahākuṣṭha*). Cette maladie de la peau se manifeste par des taches, dont le noir ou bleu foncé rappelle la couleur d'un bol en terre cuite: *kṛṣṇa-kapālikā-prakāśāni kapāla-kuṣṭhāni*. La *Cikitsā* de Suśruta atteste aussi la forme *kapālikā* 'bol, récipient en terre cuite' (1.96a) ou 'affection dentaire' (22.38a). Le suffixe *-ikā* (au féminin) est associé à des termes sanskrits qui trahissent une langue tardive.[20]

A la suite de ces indications āyurvédiques, on pourra essayer de comprendre la signification du titre *Kāpālikātantra* dans l'éclairage du terme technique *kāpālikā* 'impureté du mercure' en Rasaśāstra. Les défauts (*doṣa*) du vif-argent sont classés en trois catégories, dont la dernière s'appelle *aupādhika*, synonyme de *kañcuka*, 'couche, enveloppe', qui recouvre la surface du métal au contact de l'oxygène. Absorbé sous forme de compositions médicinales, un tel mercure entraîne chez l'homme des effets toxiques ou pathologiques. Parmi les listes, on en connaît une de sept défauts *kañcuka*, classés d'après l'origine de ces *doṣa*, notamment ceux qui proviennent des minerais avec lesquels le mercure fut en contact avant l'extraction (plomb, étain, cuivre, fer):[21] *nāgaja* définit les impuretés *śyāmā* et *kāpālikā*, alors que *vaṅgaja* concerne les défauts *kapālī* et *kālikā*.[22] A ce propos, il est utile de mentionner, avec Somadeva, le groupe des substances dénommé *kāpālikā-gaṇa-dhvaṁsin*, censé éliminer les impuretés des métaux et du mercure fixé (RCM 9.27).

On doit maintenant étendre l'enquête lexicale à d'autres expressions, tirées principalement du *Rasārṇava*, dedate controversée (XIe siècle?) et que certains attribuent à un auteur *kāpālika*. Ces expressions, pour être tout aussi incertaines, ne se recommandent pas moins à notre attention. Elles se rapportent à des recettes (*yoga*) ou méthodes (*krama*) qui concernent la teinture (*rañjana*) des métaux[23] - connue déjà du monde méditerranéen ancien[24] - ainsi que d'autres opérations chimiques[25] et certains appareils.[26] La terminologie n'excelle pas en clarté, car les mots présentent une orthographe douteuse, révélatrice de textes corrompus. Nous y avons relevé de nombreux passages, où sont attestés les termes techniques suivants: *kapālin, kapālī, kāpāli, kāpālika, kāpālikā* et *kāpālī* (supra p. 398-9).

Selon le savant bengali P.C. Rây, il y a lieu de croire que certaines expressions, comme *kāpāliko yogaḥ* (RA 6.84ab) et *ravi-nāga-kapālī* (RA 16.51ab), seraient autant d'allusions aux *kāpālika* (index RA 1910, pp. 10, 14, 62). L'attestation de telles opérations en milieu tantrique renvoie au drame *Caṇḍakauśika* du Xe siècle, où l'on voit un ascète *kāpālika* qui cultive l'alchimie et les arts magiques (*supra*, p. 397). Vers la même époque, la comédie prakrite *Karpūramañjarī* présente le magicien Bhairavānanda (acte 2, 6.22), qui se réclame de la voie tantrique *kulamārga* (acte 1, str. 22-23), alors que la *Rucikaraṭīkā*, commentaire du *Prabodhacandrodaya*, considère, à tort ou à raison, ce personnage comme un adepte du *somasiddhānta*, la doctrine des *kāpālika*.[27] D'autre part, on notera que le *kāpālika* du *Prabodhacandrodaya* (*supra*, p. 397) est ironiquement qualifié de *kulācārya* par son interlocuteur jaina (acte 3).

Quoi qu'il en soit, les recherches récentes sur le tantrisme kaśmīrien, qui doivent beaucoup à Alexis Sanderson, rattachent le *kula / kaula* à la tradition *kāpālika*,[28] illustrée par la culture macabre des lieux de crémation.[29] Certains auteurs soulignent cependant les rapports étroits des *kāpālika* avec les *nāthasiddha*, alors que les liens, disent-ils, ne sont qu'indirects entre les *kaula* et les ascètes porteurs de crâne, bien que leurs mouvements śivaïtes relèvent tous les deux du tantrisme extrême de la main gauche (*vāmācāra*).[30] Par référence à cette tradition originelle *kula* ou *kaula* - termes qui ne sont pas toujours équivalents - se situent les 'transmissions' (*āmnāya*) du śivaïsme tantrique kaśmīrien. Ces écoles du Nord présentent une terminologie à plusieurs registres issue d'un fond commun, dont témoigne le terme de *kula*, avec une constellation de significations que sous-tend la doctrine du système. Au sens courant, le mot *kula* se traduit par 'groupe, famille, maison'[31], mais il désigne aussi le corps humain et son homologue macrocosmique, l'univers, somme toute, l'ensemble de la manifestation. A ces valeurs s'ajoutent d'autres significations plus abstraites ou métaphysiques, comme *ātman*, conscience, réalité ultime, distincte cependant de Śiva en tant que *akula*.[32]

Le terme de *kula* 'corps' se rencontre justement dans le fragment suivant du *Rasendracūḍāmaṇi*, qui temoigne par ailleurs d'une certaine perméabilité des *kaula* aux idées du *rasavāda*. En ce qui concerne ce passage relatif à l'idéologie et aux pratiques *kaulika* (RCM 1.6-14), on a l'impression que Somadeva fait en quelque sorte allusion au kaulisme dès la première strophe du chapitre initial, très élaborée et teintée d'un certain raffinement poétique.

En effet, on observera le mot *kula* sous plusieurs formes, dont une récurrence phonétique (*yamaka*) en fin de vers: *vīkṣaṁ vīkṣam anukṣaṇaṁ mṛti-kulair nṛṇāṁ kulaṁ saṁkulam* (RCM 1.1).

taṁpāradaṁsarva-gadābdhi-pāra-daṁdivyāṣṭa-siddhi-prada-kaulikeśvaram/
kalpāyur-ārogya-vidhāna-dakṣinaṁ sa-deha-mukti-pradam ekam ādriye //6//
go-māṁsa-bhakṣāmara-sīdhu-pānān vidhvasta-pāpān atimukta-tāpān /
tān kaulikān naumi sa-deha-muktān videha-muktān hasataḥ sadaiva // 7 //
go-śabdenoditā jihvā tat-praveśo hi tāluni /
go-māṁsa-bhakṣaṇaṁ tat tu mahāpātaka-nāśanam // 8 //
jihvā-praveśa-saṁbhuta-vahninotpāditaḥ khalu /
cāndraḥ sravati yaḥ sāraḥ sa syād amara-vāruṇī // 9 //
tat pānaṁ dvādaśābdena[33] *deha-siddhiṁ karoti hi* /
eṣaiva khecarī mudrā cirābhyāsena sidhyati // 10 //
prakṛty-ādi-dharānto yaś caturviṁśatiko gaṇaḥ /
tat kulaṁ tena dīpyeta yo jīvaḥ sa hi kaulikaḥ // 11 //
rasair vābhyāsa-yogena kalpa-sthāyī kulena hi /
sarvaiśvarya-guṇopetaḥ kauliko 'sau maheśvaraḥ // 12 //
pañcabhūta-samūhātmā nāstiko veda-nindakaḥ /
pañcatva-mukti-vādī ca sa hi pākhaṇḍa-kaulikaḥ // 13 //
tasmāt pākhaṇḍam utsṛjya śivoktāṁ kaulikīṁ kriyām /
saṁsevya sādhayen martyo jīvan-muktiṁ parāt parām // 14 //

TRADUCTION[34]

6. Je respecte le mercure (*pārada*) unique, qui donne l'autre rive (*pāra-da*) de l'océan de toutes les maladies,[35] en seigneur des *kaulika*,[36] il offre les huit pouvoirs merveilleux,[37] il est apte a procures la santé et une vie longue d'un *kalpa*,[38] il assure la délivrance avec le corps.[39]
7. Je salue les *kaulika* qui boivent la liqueur des immortels en se nourrissant de la viande 'bovine'[40] et dont les péchés sont détruits.[41] Ils échappent aux souffrances,[42] sont délivrés avec et sans corps,[43] et rient en permanence.[44]
8. La langue est désignée par le mot 'bovin' (*go*). Manger de la viande bovine[45] c'est faire entrer la langue dans l'arrière-palais. Cela détruit les grands péchés.[46]
9. L'essence lunaire qui coule, produite par le feu né de l'entrée de la langue, est la liqueur des immortels.[47]
10. Cette boisson donne en douze ans[48] la perfection du corps (*deha-siddhi*).[49] Cette position (*mudrā*) appelée *khecarī* réussit par une longue pratique.[50]

11. L'ensemble des vingt-quatre réalités élémentaires, qui commencent par la nature (*prakṛti*) et finissent par la terre (*dharā*),[51] constitue le corps (*kula*).[52] Le soi vivant (*jīva*),[53] qui peut briller par ce corps (*kula*), est *kaulika*.[54]
12. Ou bien celui qui se maintient durant un *kalpa*[55] avec son corps (*kula*), grâce au mercure et au *yoga* sans cesse pratiqué, qui est pourvu de toutes les qualités de la souveraineté (*aiśvarya*)[56], celui-là est le *kaulika* Maheśvara.
13. Le mécréant, pour lequel l'*ātman* représente l'agrégat des cinq grands éléments, qui dénigre le Veda,[57] qui dit que la délivrance [n']est [que] le retour aux cinq grands éléments, celui-là est le *kaulika* hérétique.[58]
14. Donc, en rejetant [la conduite] hérétique et en pratiquant constamment l'action *kaulika*, enseignée par Śiva, on réalisera la délivrance dès cette vie (*jīvan-mukti*), qui transcende le transcendant (*parāt parām*).[59]

Réputés pour leur comportement anomique, les śivaïtes *kaula* et *kāpālika* étaient connus jusqu'à présent au travers de textes littéraires et surtout tantriques. Maintenant certaines sources du Rasaśāstra s'avèrent donc susceptibles de mettre en lumière l'aspect alchimique, ignoré jusqu'à présent, des pratiques propres à ces adeptes du tantrisme extrême. Mais l'histoire de l'ascétisme indien connaît, outre les *siddha*, d'autres exemples de *yogin* ou de *sādhu* qui se prétendent faiseurs d'or.[60]

Formé premièrement à la pensée mathématique et philosophique occidentale, le Professeur Frits Staal s'est ensuite tourné vers l'Inde, attiré en particulier par les *śāstra* majeurs tels que la grammaire ou la linguistique et la logique, sans oublier le *gaṇita*, comme en témoignent ses derniers travaux sur le sanskrit scientifique.[61] Dans une toute récente contribution à l'histoire comparée des sciences en Europe et en Asie, il évoque la figure de Newton comme alchimiste, adepte de la *prisca sapientia* à la recherche d'une nouvelle méthode du savoir.[62] La leçon de cet immense savant du Trinity College nous fait espérer voir ces parcimonieuses pages de *nugae chemicae* lues avec un certain intérêt par le dédicataire du présent volume d'hommage. Au fait, il serait bien difficile de trouver un canton indologique sur lequel ne s'est pas exercée de quelque manière la curiosité de Frits Staal. En effet, l'éventail de ses travaux s'étend de la récitation védique (1961) aux techniques psychosomatiques des méditants (1975) et ensuite à la culture physique de l'Inde traditionnelle (1993).[63] Et, si nos souvenirs sont bons, le premier échange épistolaire entre nous deux a eu justement lieu au sujet des arts martiaux indiens, il y a plus de trois lustres.[64]

Autant que la *mallavidyā*, le Rasaśāstra aussi valorise, comme le Yoga, le corps, qui, selon Kālidāsa, représente le premier moyen de gagner du mérite en vue du salut.[65] Dans la quête de la délivrance, le corps, considéré comme une faveur divine (*prasādaodya*), occupe une place privilégiée, comme l'affirme avec force un maître *vīraśaiva* du XVIIe ou XVIIIe siècle.[66] Cette valorisation du physique est dans la droite ligne des idées indiennes depuis l'Āyurveda classique.[67] Ce *veda* de la [pleine] durée de vie (*āyus*), que désigne l'appellation sanskrite, vaut à la fois pour l'art de guérir et pour celui de prolonger la vie de l'homme en santé jusqu'à cent ans au moins.

En effet, la doctrine classique de la médecine indienne ne concerne pas seulement les malades mais aussi et surtout les bien portants, car l'équilibre physique et moral ainsi que la longévité sont des facteurs indispensables au dépassement de la métensomatose (*saṁsāra*) en vue d'atteindre à l'immortalité. Selon les gens avertis en la matière, affirme Caraka (*Sūtra* 1.43), l'Āyurveda est considéré comme le meilleur des savoirs (*tasyāyuṣaḥ puṇyatamo vedaḥ*), dans la mesure où son enseignement, destiné à assurer la santé (*ārogya*) et une longue vie (*āyus*), ainsi que le bon-ordre (*dharma*), préconise pour les humains le bien en ce monde et aussi dans l'autre (*lokayor ubhayor hitam*). Le commentaire de Cakrapāṇi, qui justifie le superlatif carakien *puṇyatama*, surprend par le fait que la médecine est jugée supérieure même au Veda.[68]

La longévité et l'immortalité sont des thèmes récurrents dans l'histoire des idées indiennes, depuis les hymnes védiques jusqu'aux sources tardives.[69] Les conceptions et les exploits des gymnosophistes indiens ont attiré l'attention des Grecs à l'époque d'Alexandre. Non rares furent ensuite les visiteurs asiatiques qui sont venus chercher en Inde la 'plante d'immortalité'.[70] Dans son livre en arabe sur l'Inde, terminé en 1031, le savant iranien al-Bīrūnī attribue aux Indiens le savoir du *rasāyana*, qui est 'un art faisant intervenir certaines opérations, drogues et compositions médicinales, dont la plupart sont d'origine végétale. Ses principes rétablissent la santé des incurables et rendent la jeunesse aux vieillards décrépits'.

La double orientation de la médecine āyurvédique se reflète dans la classification même que donne Caraka des remèdes (*bheṣaja*) suivant leur action, à savoir thérapeutique en cas de maladie ou roborative destinée à la santé des bien portants. On y décèle ainsi un prolongement de l' ancienne répartition ātharvaṇique en formules de guérison (*bhaiṣajya*) et formules de longue vie (*āyuṣya*).[71]

D'autre part, Suśruta distingue quatre catégories de maladies (*vyādhi*), dont la dernière se rapporte aux naturelles (*svābhāvika*), inhérentes à la nature humaine, à savoir la faim, la soif, la vieillesse, la mort et le sommeil (*Sūtra* 1.23-25). Selon Caraka, ceux qui suivent une cure de rajeunissement en observant les règles (*vidhivat*), non seulement vivront longtemps en ce monde, mais ils auront part à la voie heureuse, fréquentée par les divins *ṛṣi*, et parviendront au *brahman* impérissable, à la délivrance (*Cikitsā* 1.1.80).

La conception d'un corps incorruptible, merveilleux (*divya-deha*), nourrit à titres divers tant l'Āyurveda et le Yoga que le Rasaśāstra et les Tantra.[72]

Aux premiers siècles de son histoire, le système mercuriel indien présente deux orientations d'intérêt bien marquées mais inséparables, l'une alchimique (manipulations métalliques, aurifaction), l'autre, thérapeutique ou pharmaceutique, déjà présente dans la médecine classique, où le vif-argent est cependant quasi absent.[73] Les deux orientations sont indissociables, car le refaçonnement de l'individualité physique (*kāya-kalpa*) rejoint le perfectionnement des métaux, la transsubstantiation du corps (*deha-vedha*) étant calquée sur la transmutation des métaux vils (*loha-vedha*).[74] La fabrication de l'or s'associe à l'optimisme macrobiotique, dans la mesure où les manipulations métalliques rejoignent les procédés de rajeunissement (*rasāyana*). Constante dans les idées indiennes (*supra*, p. 410), la valorisation du corps se retrouve raffermie dans les sources sanskrites du Rasaśāstra tantrique (X^{e} - XIIIe siècle), qui se situe, dans le temps, entre la chimie āyurvédique de l'époque classique et l'iatrochimie indienne du XIIIe ou XIVe siècle à l'époque moderne.[75] Un des premiers traités sanskrits d'alchimie, diversement datés du Xe au XIIe siècle, rappelle que la délivrance vient de la connaissance libératrice, acquise par l'exercice ou l'étude, que conditionne finalement un physique robuste (*sthire dehe*).[76]

Dans le chapitre initial (*jīvasthitikathana*), une importante source *śākta* (KAT), de date incertaine (1000-1400), décrit la condition humaine placée sous le signe du cycle des renaissances. Nourri par des formules proverbiales, le thème est servi par un texte qui concorde étroitement avec un chapitre purāṇique,[77] où l'on met en valeur l'instrument physique de la délivrance (KAT 1.20-46). On assimile aussi le corps à un temple (*deho devālayaḥ*) et le soi ou l'âme (*jīva*) au dieu Sadāśiva (KAT 9.41). La maladie, le vieillissement, la mort exigent de l'être humain des efforts soutenus pour préserver son corps des maux (*roga*) tels que les dermatoses (*kuṣṭha*) et d'autres maladies (KAT 1.21).

Le souci du bien-être conjugue, dans le même *kāya-sādhana*, aussi bien le tantrisme et le *haṭhayoga* que la médecine et l'iatrochimie.[78] En effet, la culture du corps sert apparemment de toile de fond sur laquelle se rencontrent ces diverses traditions de savoir et de pratiques, dont les littératures relèvent du tantrisme protéiforme. Le terme sanskrit de *tantra* (interchangeable avec celui de *āgama* en milieu śivaïte) s'applique en premier à des exposés développés sur la tradition religieuse, autre que le Veda institué (*nigama*) depuis toujours, en tant qu'enseignements divins - base doctrinale et pratique - à l'intention des croyants. La même appellation désigne cependant aussi des textes qui traitent des sujets techniques, aussi particuliers que la médecine, l'(al)chimie, la magie ou encore l'astrologie et la démonologie.[79]

Si les origines du Rasaśāstra sont controversables, on doit cependant admettre que les conceptions védiques relatives au rajeunissement et à l'immortalité ont préfiguré le *rasāyana* āyurvédique ou alchimique.[80]

D'autre part, outre des allusions alchimiques dans les belles-lettres sanskrites, on observe à la même époque classique plusieurs textes bouddhiques, dont certains sont traduits en chinois, qui se réfèrent à la transmutation métallique.[81] A ce propos, il convient d'évoquer les relations culturelles sino-indiennes établies par la diffusion du bouddhisme et développées à la faveur des voyages entrepris au long des siècles par des religieux itinérants. Ainsi est-il permis de penser que ces contacts ont servi non seulement à l'édification spirituelle des moines, mais aussi à un échange d'idées en matière scientifique et technique, notamment thérapeutique ou (al)chimique.[82]

L'idée de la transmutation du corps (*deha-vedha*) - en symétrie avec la pratique de la transmutation des métaux vils (*loha-vedha*) - et la philosophie macrobiotique rattacheraient le système mercuriel indien plutôt au tantrisme, dont les enseignements ont imprégné les premiers documents significatifs du Rasaśāstra, vers l'an mil ou au début du deuxième millénaire. Dans l'état actuel de la recherche, désavantagée aussi par le manque de repères chronologiques de la littérature technique, on peut seulement observer une première phase à forte dominante tantrique, illustrée notamment par plusieurs textes (*MBhT*,[83] RAK[84]), et une seconde, appelée iatrochimique, plus thérapeutique que magique. Selon certains auteurs, à cette dernière appartient en fait la littérature technique proprement dite de ce savoir aussi fascinant que déconcertant, et ignoré jusqu'à présent des indianistes.

Si l'on reconnaît des idées et des pratiques alchimiques dans les sources tantriques (aurifaction, opérations mercurielles, élixirs de longue vie)[85], on est sensiblement plus reservé en ce qui concerne les rapports, difficiles à établir historiquement, entre tantrisme et médecine / iatrochimie traditionnelle. Quoi qu'il en soit, le savoir des *vaidya* diffère notablement de celui des *tāntrika*, bien que les premiers aient pu se montrer perméables à quelques apports venus des seconds:[86] l'examen du pouls, appelé *nāḍī-parīkṣā* (supposé d'influence chinoise), des procédés médico-magiques, dont certains remontent aux Veda (*mantra*, *yantra*, *maṇḍala*), et des drogues, qu'elles soient minérales, narcotiques[87] ou vénéneuses.[88] On connaît aussi des textes de démonologie qui ont pénétré dans les traités āyurvédiques, comme en témoigne le *Rāvaṇakumāratantra* relatif à la possession infantile.[89] On a relevé des parallèles parmi d'autres documents tantriques, à savoir un chapitre de *l'Uḍḍīśatantra* et la *Khaḍgarāvaṇabālacikitsā*.[90]

Si ces apports tantriques restent ponctuels et d'un emploi limité dans l'Āyurveda savant, il n'est pas moins vrai que les pratiques curatives et les procédés (al)chimiques ont intéressé les milieux ritualistes en Inde. L'Appendice du présent article (*infra*, pp. 416-18) en apporte une preuve formelle dans un fragment du manuel śivaïte *Nityādisaṁgraha(-paddhati)* de Rājānaka Takṣakavarta, postérieur à Abhinavagupta, actif vers 975-1025.[91]

Cette *paddhati* est contenue dans un manuscrit de la collection Mark Aurel Stein conservée à l'Indian Institute Library, annexe de la Bodléienne.[92] L'inédit oxonien est en fait une copie, dont l'original se trouve à l'Institut Bhandarkar de Poona. Classé par erreur, au siècle dernier, comme texte purāṇique et répertorié sous le nom de *Bhṛṅgeśasaṁhitā* (n° 76/1875-76),[93] ce document śivaïte ne figure pas dans le catalogue plus récent des manuscrits tantriques de l'Institut Bhandarkar.[94]

La *Nityādisaṁgrahapaddhati* est un abrégé de textes āgamiques disposés de manière à éclairer le culte quotidien (*nitya*) et occasionnel (*naimittika*) de Svacchandabhairava. Le manuel reproduit un chapitre pris à la *Śrīkaṇṭhī*, un āgama perdu, lequel chapitre contient la liste ancienne des textes canoniques des cinq 'courants' (*srotas*).[95] Pour le contenu des Écritures śivaïtes, cet ouvrage, appelé aussi *Śrīkaṇṭha-* ou *Śrīkaṇṭhīyasaṁhitā*,[96] faisait autorité parmi les maîtres de l'école *trika* du Kaśmīr. Il est mentionné à cet effet par Kṣemarāja (vers 1000-1050) dans son commentaire sur le *Svacchandatantra* (vol. IV, p. 19) et longuement cité dans le *Tantrālokaviveka* (vol. I, pp. 42-43) de Jayaratha (vers 1225-1275) pour la liste des 64 Bhairavāgama (*Sādāśivaṁ cakram*), qui se réfèrent à la tradition *kaula*.

Au 'courant' occidental (*paścima*) appartiennent 20 Bhūtatantra,[97] qui se rapportent à l'exorcisme[98] et concernent les *kāpālika*. Ces textes démonologiques, dont traite le chapitre *srotobheda*(*-paṭala*) et qui rappellent le *Kumāratantra* de Rāvaṇa,[99] sont énumérés dans les trois premiers *śloka* de l'Appendice. Les suivants donnent des indications sommaires sur le contenu de ces tantra démonologiques, qui présentent tous, sauf les deux derniers (*Yamaghaṇṭa* et *Ghaṭotkaca*), des préoccupations d'ordre médical, ou plutôt médico-magique, et (al)chimique. Les sujets sont très variés: fièvres (11a, 20a), panacées (11c), rajeunissement (27ab), compositions médicinales à base végétale (28a), thérapeutique chirurgicale des maladies des yeux (8ab), procédés médico-magiques tels que *mantra* et *yantra* (*passim*), opérations mercurielles et transmutations métalliques (9ab). Cet inventaire de sujets tantriques établis à la lumière du manuscrit inédit d'Oxford confirme les rapports entre tantrisme et arts de santé.[100] A ces mêmes rapports renvoie l'étude de la littérature alchimique/iatrochimique sanskrite en général et du *Rasendracūḍāmaṇi* en particulier.[101]

APPENDICE

Fragment du texte tantrique inédit *Nityādisaṁgrahapaddhati* de Rājānaka Takṣakavarta.

Manuscrit sanskrit en caractères *śāradā* conservé à la Bodleian Library (MS

Stein Or. d. 43, folio 13r[4] sqq.) et généreusement communiqué en transcription, avec apparat critique, il y a plus de dix ans, par Monsieur Alexis Sanderson, alors 'lecturer' en sanskrit à l'Université d'Oxford (lettre du 1er août 1985).

paścime bhūtatantrākhyaṁ śivabhedagataṁ bhavet[102] /
tatra helā hayagrīvaṁ kaṭaṅkaṁ katakāmayam / (1)
karoṭī muṇḍamālākhyaṁ kārkoṭaṁ khaḍgarāvaṇam /
caṇḍaṁ caṇḍāsidhārākhyaṁ vikaṭaṁ ṭakamaṇḍalam[103] / (2)
bhūtatrāsaṁ[104] *śikhārāvaṁ ghargharaṁ siṁhakoṭaram* /
ghorāṭṭahasam ucchiṣṭaṁ[105] *yamaghaṇṭaṁ ghaṭotkacam* / (3)
... (lacune indiquée dans le ms.) ... *bhūtānām anukampayā* /
sarvavyādhivināśārtham ... (lacune indiquée dans le ms.) ... / (4)
niḥsṛtaṁ raudramantroktaṁ mālāmantrais tu saṁkulam /
agamṛtyumantraiḥ pūrṇam ... (lacune indiquée dans le ms.) ... / (5)
helākhyaṁ prathamaṁ yat tu hy āyurvedādilakṣaṇam /
dvitīyaṁ bālarakṣārthaṁ hayagrīveti viśrutam / (6)
kaṭaṅkaṁ rūpikānāṁ ca nāśanārtham[106] *udāhṛtam* /
ratikāmanivṛttyarthaṁ caturthaṁ kaṭakāhvayam / (7)
netrāmayavināśārthaṁ śalyakarmapravartakam /
nītiśāstrādisaṁyuktaṁ karoṭī pañcamaṁ smṛtam / (8)
rasavādādisiddhyarthaṁ dhātuvādādilakṣaṇam /
tadarthaṁ muṇḍamālākhyaṁ ṣaṣṭhaṁ tantraṁ pradarśitam / (9)
khanyavādādikaṁ sarvaṁ nidhānākṛṣṭilakṣaṇam /
sarpākarṣaṇayuktaṁ ca kārkoṭaṁ saptamaṁ smṛtam / (10)
grahajvaraviṣādīnāṁ mālāmantraiḥ samākulam /
sarvavyādhipraśamanam aṣṭamaṁ khaḍgarāvaṇam / (11)
caṇḍaṁ rakṣapiśācānāṁ nāśāya navamaṁ smṛtam /
caṇḍāsidhāraṁ daśamaṁ sārdhalakṣatrayaṁ matam (12)
samastauṣadhaprāptyarthaṁ † *daśanāma* † *pratiṣṭhitam*[107] /
lakṣaṇam oṣadhīnāṁ tu pṛthakkalpādibhedataḥ / (13)
mantrarājasamūhena vikaṭaikādaśaṁ smṛtam /
parasainyavināśārthaṁ ca ṭakādiprasādhanam / (14)
bhūtānāṁ balibhedāś ca amaraṁ ṭakamaṇḍalam /
dvādaśaṁ ca mahādevi yena jñātena mantrabhāk / (15)
sarve bhūtā vinaśyanti yoginī śākinī tathā /
yathārthanāmam uddiṣṭaṁ bhūtatrāsaṁ trayodaśam / (16)
rāvoktaś[108] *cāṣṭadhā yatra bhūtānāṁ bhayavardhanaḥ* / (17)
raudrābhedāni sarvāṇi yoginīnāṁ ca lakṣaṇam /
tathā kumbhādibhedaṁ ca tac chikhārāvakīrtitam / (18)
tantraṁ caturdaśaṁ devi guhyamantrārthasaṁkulam /
ekabhedaṁ tu bhūtānāṁ tathā vyādhyādilakṣaṇam / (19)

jvarāvataraṇaṁ yac ca tannivāraṇam eva ca /
bhūtayoniṣu sarvāsu tadaṁśāni vibhagaśaḥ / (20)
jñāyante yena tantreṇa ghargharaṁ tat tripañcamam /
navalakṣapravistīrṇaṁ bahubhedais tu saṁkulam / (21)
mantravādaṁ tu yan mṛtyuṁ tad uktaṁ siṁhakoṭaram /
ghorāṭṭahāsaṁ deveśi lūtāpiṭakanāśanam / (22)
bhūmikājñānasaṁyuktam ṛddhyaṣṭakasamanvitam /
yantraprayogasāmānyam ṛddhinaṣṭādilakṣaṇam[109] / (23)
indrajālādisaṁkīrṇaṁ ṣoḍaśaṁ parikīrtitam /
devatānāṁ pramāṇaṁ ca yajanam ṛddhikarmiṇām (24)
raktayāgaṁ guhyasūtraṁ vaśyoccāṭanalakṣaṇam[110] /
kuhukarmādikaṁ yat tu raudramantrapariṣkṛtam[111] / (25)
ucchiṣṭaṁ nāma taj jñeyaṁ tantraṁ saptadaśaṁ smṛtam /
dhāraṇājñānabhedaṁ[112] *tu plutyādisamalaṅkṛtam* / (26)
sthiratvaṁ dehapiṇḍasya jarāvyādhivināśanam /
paracikīrṣājñatvaṁ[113] *ca parakāya*[114]*-praveśanam* / (27)
oṣadhīnāṁ tu saṁyogaṁ sthitatvaṁ † tadra † retasaḥ /
varṇitaṁ yatra devena sarvabhūtahitaiṣiṇā[115] / (28)
ucchiṣṭaṁ tad[116] *varārohe tantram aṣṭādaśaṁ smṛtam /*

BIBLIOGRAPHIE

ĀK *Ānandakanda*, éd. S.V. Radhakrishna Sastri. Tanjore: Saraswathi Mahal Library, 1952

Caṇḍakauśika de Kṣemīśvara, éd. Jīvānanda Vidyāsāgara. Calcutta: 1884

Carakasaṁhitā avec le commentaire de Cakrapāṇidatta, éd. Jādavjī Trikamjī Āchārya. Bombay: Nirṇaya Sāgar Press, Bombay, [3]1941

GSBh *Gorakṣasaṃhitā*, éd. Janārdana Pāṇḍeya II: *Bhūtiprakaraṇa*. Varanasi: Sampurnand Sanskrit Vishvavidyalaya, 1977

GŚ 1958 *Gorakṣaśataka*, éd. critique Swami Kuvalayānanda et S.A. Shukla. Lonavla: Kaivalyadhāma, 1958 (réimpr. 1974)

GŚ 1976 *Gorakṣaśataka*, éd. critique et trad. Fausta Nowotny. Köln: Nowotny, 1976

HP *Haṭhapradīpikā* de Svātmārāma, éd. Swami Digambarji et R. Kokaje. Lonavla: Kaivalyadhāma, 1970

HYP 1972 *Haṭhayogapradīpikā* de Svātmārāma, avec le commentaire *Jyotsnā* de Brahmānanda, éd. et trad. Radha Burnier *et al.* Adyar: Adyar Library and Research Centre, 1972 (réimpr. 1975)

HYP 1974 *Haṭhayogapradīpikā* de Svātmārāma, trad. et notes Tara Michaël. Paris: Fayard, 1974

Karpūramañjarī
de Rājaśekhara, éd. Sten Konow et trad. annotée Charles Rockwell Lanman. Cambridge: Harvard University, 1901.

KJN *Kaulajñānanirṇaya and some minor texts of the school of Matsyendranātha*, éd. Prabodh Chandra Bagchi. Calcutta: Metropolitan, 1934

KAT *Kulārṇavatantra*, éd. et trad. Ram Kumar Rai. Varanasi: Prachya Prakashan, 1983

Kumārasaṁbhava
éd. et trad. en latin Adolf Friedrich Stenzler. Berlin-London: Oriental Translation Fund, 1838

Kumārasambhava
(chap. IV et V), avec le commentaire de Mallinātha, éd. et trad. V.G. Paranjpe. Poona: Deccan Book Stall, [2]1941

Maitri Upaniṣad
éd. et trad. Anne-Marie Esnoul. Paris: Adrien-Maisonneuve, 1952

Mālatīmādhava
de Bhavabhūti, avec le commentaire de Jagaddhara, éd. et trad. annotée M.R. Kale. Delhi [etc.]: Motilal Banarsidass, [3]1967

Manusmṛti
avec le commentaire de Kullūka, éd. Gopāla Śāstrī Nene. Varanasi: Chowkhamba Sanskrit Series Office, 1970

MBhT *Mātṛkābhedatantra*, éd. Ram Kumar Rai. Varanasi: Prachya Prakashan, 1983

Mṛgendrāgama (Kriyāpāda et *Caryāpāda)*
avec le commentaire de Bhaṭṭa Nārāyaṇakaṇṭha, éd. critique Niddodi Rāmacandra Bhatt. Pondichéry: Institut français d'indologie, 1962

Mṛgendrāgama (*Kriyāpāda* et *Caryāpāda*)
avec le commentaire de Bhatta Narayanakantha, trad. annotée Hélène Brunner-Lachaux. Pondichéry: Institut français d'indologie, 1985

Prabodhacandrodaya
de Kṛṣṇamiśra, texte et trad. Armelle Pédraglio. Paris: Institut de civilisation indienne, 1974

Rājanighaṇṭu
de Narahari (*Varga* XIII), éd. critique et trad. annotée Richard Garbe: *Die indischen Mineralien*. Leipzig: Hirzel, 1882

RHT 1958 *Rasahṛdayatantra* de Govinda Bhagavatpāda, avec commentaire sanskrit et glose en hindi par Caturbhuja Miśra. Ajmer: Kṛṣṇa Gopāl Āyurved Bhavan, 1958

RHT *Rasahṛdayatantra* de Govinda Bhagavatpāda, éd. et trad. B.V.

Subbarayappa *et al.* Bangalore: Indian Institute of World Culture. (Manuscrit aimablement communiqué en photocopie par monsieur B.V. Subbarayappa)

RKDh *Rasakāmadhenu*, éd. Jādavjī Trikamjī Āchārya. Bombay: Bhagavān Lāl Tribhuvan, 1925

RPS *Rasaprakāśasudhākara* de Yaśodhara, éd. JādavjI5 Trikamjī Āchārya. Bombay: Nirṇaya Sāgar, 1911

RRĀ *Rasaratnākara* (section IV: *Rasāyanakhaṇḍa*) de Nityanātha Siddha, éd. Jādavjī Trikamjī Āchārya. Banaras: Caukhambā Saṁskṛta Pustakālaya, 1939

Rasaratnasamuccaya
éd., avec explications en hindi, Dharmānanda Śarmā. Delhi [etc.]: Motilal Banarsidass, [2]1962 (réimpr. 1977)

RRS *Rasaratnasamuccaya* (chap. I-XI), texte et trad. Damodar Joshi, 2 vol. New Delhi: Indian National Science Academy, 1991-1992

RA/RA 1910 *Rasārṇava*, éd. critique Praphulla Chandra Rây et Hariśchandra Kaviratna. Calcutta: Asiatic Society of Bengal, 1910

RA 1978 *Rasārṇava*, avec commentaire en hindi, éd. Indradeo Tripathi. Varanasi: Chowkhamba Sanskrit Series Office, 1978

RAK *Rasārṇavakalpa*, éd. et trad. Mira Roy et B.V. Subbarayappa. New Delhi: Indian National Science Academy, 1976

RCM *Rasendracūḍāmaṇi* de Somadeva, éd. Jādavjī Trikamjī Āchārya. Lahore: Motilal Banarsidass, 1932

Rasendracūḍāmaṇi
de Somadeva, texte et trad. explicative en hindi Siddhinandan Mishra. Varanasi-Delhi: Chaukhambha Orientalia, 1984

Le florilège de la doctrine śivaïte
Śaivāgamaparibhāṣāmañjarī de Vedajñāna, éd. critique et trad. annotée Bruno Dagens. Pondichéry: Institut français d'indologie, 1979

ŚDhS *Śārṅgadharasaṁhitā*, avec les commentaires *Dīpikā* et *Gūḍhārtha-dīpikā*, éd. Paraśurāma Śāstrī. Bombay: Nirṇaya Sāgar Press, [2]1931

SDS *Sarvadarśanasaṁgraha* de Mādhava, éd. par des lettrés de l'Ānandāśrama. Poona: Ānandāśrama, 1906 (réimpr. 1977)

Siddhāntakaumudī
de Bhaṭṭoji Dīkṣita, éd. et trad. Śrīśa Chandra Vasu, 2 vol. Delhi [etc.]: Motilal Banarsidass, 1906 (réimpr. sans date)

Śivasūtra et *Vimarśinī*
de Kṣemarāja, trad. Lilian Silburn. Paris: Institut de civilisation indienne, 1980

Somaśambhupaddhati
texte et trad. annotée Hélène Brunner-Lachaux, 3 vol. parus.

Pondichéry: Institut français d'indologie, 1963-77

Suśrutasaṁhitā
avec le commentaire de Ḍalhaṇa, éd. Jādavjī Trikamjī Āchārya et Nārāyaṇ Rām Āchārya. Bombay: Nirṇaya Sāgar Press, [3]1938

Luce delle sacre scritture (*Tantrāloka*)
a cura di Raniero Gnoli. Torino: Unione tipografico-editrice torinese, 1971

Vātūlanāthasūtra
avec le commentaire d'Anantaśaktipāda, trad. Lilian Silburn. Paris: Institut de civilisation indienne, 1959

YS *Yogasūtra*, avec les commentaires *Yogabhāṣya, Tattvavaiśāradī* et *Yogavārttika*, éd. Śrīnārāyaṇa Miśra. Varanasi: Bhāratīya Vidyā Prakāśana, 1971

Yogasūtra
avec les commentaires *Yogabhāṣya* et *Tattvavaiśāradī*, trad. James Haughton Woods. Delhi [etc.]: Motilal Banarsidass, 1914 (réimpr. 1977)

The Yoga-Upaniṣads
with the commentary of Śrī Upaniṣad-Brahmayogin, ed. A. Mahadeva Sastri. Adyar: Adyar Library and Research Centre, 1920 (réimpr. 1968)

YV *Yogavārttika* de Vijñānabhikṣu, texte et trad. annotée Trichur Subramaniam Rukmani, 4 vol. Delhi: Munshiram Manoharlal, 1981-89

Yogavāsiṣṭha
de Vālmīki, avec commentaire, éd. Vāsudeva Lakṣmaṇa Śāstrī Paṇśikar, 2 vol. New Delhi: Munshiram Manoharlal, 1918 (réimpr. 1981)

YH Le cœur de la Yoginī. *Yoginīhṛdaya,* avec le commentaire *Dīpikā* d'Amṛtānanda. Texte sanskrit traduit et annoté par André Padoux. Paris: Collège de France, 1994

Notes

1. Voir à ce sujet Jean-Marc Mandosio, 'La place de l'alchimie dans les classifications des sciences et des arts à la Renaissance', *Chrysopœia* 4, 1990-91, pp. 199-282. Cet article, paru en 1993, est une version développée de la communication présentée au colloque international *Alchimie et philosophie à la Renaissance* (Tours, 4-7 décembre 1991), dont les Actes sont publiés sous la direction de Jean-Claude Margolin et Sylvain Matton. Paris: Vrin, 1992, pp. 11-41.

2. Wolfgang Schneider, 'Chemiatry and iatrochemistry', dans: Allen G. Debus (éd.), *Science, medicine and society in the Renaissance. Essays to honor Walter Pagel* I. London: Heinemann Educational Books, 1972, pp. 141-50.
3. B.V. Subbarayappa, 'Chemical practices and alchemy', dans: D.M. Bose, S.N. Sen et B.V. Subbarayappa (éds), *A concise history of science in India*. New Delhi: Indian National Science Academy, 1971, pp. 316-18 (Possible origin of Indian alchemy).
4. Voir notre article à paraître sous le titre 'Alchimie et géographie sacrée dans l'Inde médiévale', *JA* 285, 1997. Résumé en anglais dans le *Journal of the European Āyurvedic Society* 2, 1992, pp. 151-7.
5. Cf. Sanjukta Gupta, Dirk Jan Hoens et Teun Goudriaan, *Hindu tantrism*. Leiden: Brill, 1979, p. 23. Sur les mouvements *siddha* et *nātha*, lire les contributions de Prabodh Chandra Bagchi et de Sukumar Sen à l'ouvrage *The cultural heritage of India* IV. Calcutta: The Ramakrishna Mission Institute of Culture, 1956, pp. 273-9 et 280-90.
6. Sur le nāthisme, voir Kalyani Mallik, *Siddha-siddhānta-paddhati and other works of the Nātha Yogīs*. Poona: Oriental Book House, 1954, pp. 1-28. Cf. Shashibhusan Dasgupta, *Obscure religious cults*. Calcutta: K.L. Mukhopadhyay, 1969, pp. 192-5 et 251-2.
7. Christine Chojnacki, 'L'imaginaire alchimique dans un texte jaina médiéval', à paraître comme appendice de notre article du *JA* 285, 1997 (*supra*, n. 4).
8. A.P. Singh, '*Mālatīmādhava*: A play with tantric design', dans: *VIII*th *World Sanskrit Conference. Abstracts*. Vienna: Institut für Indologie, 1990, non paginé. Cf. Vidya Dehejia, *Yoginī cult and temples: A tantric tradition*. New Delhi: National Museum, 1986, p. 83.
9. Cf. Sanjeeva Rao, 'Rasasiddhas of Alampur', *Bulletin of the Indian Institute of History of Medicine* 13, 1983, pp. 38-44.
10. Sur les ascètes porteurs de crâne selon les documents littéraires et épigraphiques, voir David N. Lorenzen, *The Kāpālikas and Kālāmukhas. Two lost Śaivite sects*. Delhi: Motilal Banarsidass, 1991, pp. 1-95, 215-23. Cf. le compte rendu de Minoru Hara dans *IIJ* 17, 1975, pp. 253-61. Cf. aussi Apurba Chandra Barthakuria, *The Kāpālikas. A critical study of the religion, philosophy and literature of a tantric sect*. Calcutta: Sanskrit Pustak Bhandar, 1984.
11. Cf. Mark S.G. Dyczkowski, *The Canon of the Śaivāgama and the Kubjikā Tantras of the Western Kaula tradition*. Albany: State University of New York Press, 1988, pp. 27, 29.
12. Cf. Alexis Sanderson, 'Śaivism and the tantric traditions', dans: Stewart Sutherland *et al.* (éds), *The world's religions*. London: Routledge, 1988, pp. 668-9. Sur le culte *kāpālika* de la forme terrifiante de Śiva, cf. Amal Sarkar, 'Bhairava images and the Kāpālikas', *Indian Museum Bulletin* 1/1, 1966, pp. 46-50.
13. Voir Elizabeth Chalier Visuvalingam, 'Bhairava's royal brahmanicide: the problem of the Mahābrāhmaṇa', dans: Alf Hiltebeitel (éd.), *Criminal gods and demon devotees*. Albany: State University of New York Press, 1989, pp. 157, 211 (n. 3). Cf. Mircea Eliade, *Le Yoga: immortalité et liberté*. Paris: Payot, 1954, pp. 294-304. Sur un rapproche entre les *kāpālika* du passé et les ascètes

aghorī de notre époque, voir David N. Lorenzen, *op. cit.*, (*supra*, n. 10) p. 53. Cf. Jonathan Parry, 'Sacrificial death and the necrophagous ascetic', dans: Maurice Bloch et Jonathan Parry (éds), *Death and the regeneration of life*, Cambridge [etc.]: Cambridge University Press, 1982, p. 87.

14. Suranand Kunjan Pillai (éd.), *Alphabetical index of the Sanskrit manuscripts in the University Manuscripts Library Trivandrum* I. Trivandrum: University of Kerala, 1957, p. 130: n° d'ordre 3291 (ms. n° 7475): *Kāpālikatantra* avec commentaire, 500 *grantha*, texte complet en *devanāgarī* sur le *rasavāda*. Sur certains noms d'alchimistes ou de *nāthayogin*, tels Kapāla, Kapālī et Kāpālikā, mentionnés dans plusieurs listes de *siddha*, lire David G. White, *op. cit.*, (*infra*, n. 101), pp. 80-6, 127, 189-92 (nr. 11-30), 417 (n. 21: lire *Āyurved*, p. 460, non 490).

15. R. Ananta Krishna Sastry, *Kavindracharya list*. Baroda: Central Library, 1921, p. 17, n° 979: *Kāpālīsiddhānta*, sous la rubrique *rasāyanasiddhānta*. Cf. *Kāpālikamata* sous la rubrique 'doctrine' (*matagrantha*), p. 34, n° 2184. La date supposée de cette liste (sūcī), nous fait remarquer M. Gerdi Gerschheimer, ne peut servir de *terminus ad quem* (1650) pour tous les manuscrits inventoriés, car il est avéré que certains documents remontent au XVIIIe et même au XIXe siècle. Cf. P.K. Gode, 'The Kavīndrācārya-sūcī - is it a dependable means for the reconstruction of literary chronology?', *New Indian Antiquary* 6/2, 1943, pp. 41-2.

16. H.P. Malledevaru (éd.), *Descriptive catalogue of Sanskrit manuscripts* XIII: *Vaidya*. Mysore: Oriental Research Institute, University of Mysore, 1986, pp. 48-9, 66-7 (appendice n° 94), n° d'ordre 41485 (ms. n° 3005/3): *Kāpālikātantra* contenu dans un recueil factice, qui comprend un ms. du *Rasahṛdayatantra* (n° 3005/2), également en écriture *nandināgarī*, et un autre l'*Upadeśasārasarvasvasaṁgraha* en écriture *grantha* (n° 3005/1). Les documents. alchimiques, incomplets, n° 2 avec 19 fol. (22 x 3, 5 cm, 9 l., 28 lettres) et n° 3 avec 21 fol. (22 x 3, 5 cm, 9 l., 30 lettres) sont foliotés séparément (10-38 et 1-21) et non à la suite, comme le laisseraient entendre B.V. Subbarayappa *et al.* (p. 9).
Nous remercions le Prof. P.S. Filliozat, qui a bien voulu consulter pour nous à Mysore le *Kāpālikātantra*, dont il nous a communiqué la description et une photocopie (lettre du 6 novembre 1995). D'autre part, le Dr. D. Wujastyk du Wellcome Institute (Londres) nous a fait aimablement parvenir plusieurs pages en photocopie relatives aux deux manuscrits alchimiques répertoriés dans le Catalogue de Mysore. Les délais de publication du présent article ne nous ont pas permis une étude plus attentive du *Kāpālikātantra*, sur lequel nous espérons pouvoir revenir plus tard.

17. Mark S.G. Dyczkowski, *op. cit.*, (*supra*, n. 11), p. 146, n. 138.

18. Voir Louis Renou, *Grammaire sanscrite*. Paris: Adrien-Maisonnneuve, 1961, pp. 279-82 (§ 211). Dans l'ouvrage, déjà cité (*supra*, n. 11), de Mark S.G. Dyczkowski (p. 18), la forme féminine *kāpālikā* du *Vāmanapurāṇa* (6.87) est une coquille. D'autre part, le même auteur (p. 26), en suivant David N. Lorenzen, *op. cit.*, (*supra*, n. 10), p. 13, se réfère par erreur à une renonçan *kāpālikā*, en fait *kāpālikī* selon la *chāyā* (*Saptaśatī*, str. 408).

19. Cf. David N. Lorenzen, *op. cit.*, (*supra*, n. 10), p. 30.
20. Louis Renou, *op. cit.*, (*supra*, n. 18), pp. 247-8 (§ 195).
21. Listes et explications par Damodar Joshi, 'Mercure in Indian medicine', *Studies in History of Medicine* 3, 1979, pp. 244-6 et du même auteur, *Rasaśāstra*. Trivandrum: Āyurveda College, 1986, pp. 78-82. Cf. Siddhinandan Mishra, *Āyurvedīya Rasaśāstra*. Varanasi-Delhi: Chaukhambha Orientalia, 1981, pp. 191-4 et Bhagwan Dash, *Alchemy and metallic medicines in Āyurveda*. New Delhi: Concept Publishing Company, 1986, pp. 44-5.
22. RKDh 3. 1.62cd (p. 321): *kapālī kālikā vaṅge nāge śyāmā kapālikā //*. Cf. RCM 15.23-25. Sur l'impureté *kālikā*, cf. RA 12.80, dont le texte se retrouve avec des variantes dans RAK 140.
23. RA 1978, 14.64ab: *taṁ khotaṁ rañjayet paścāt vaṅgābhraka-kapālinā /*; 14,85ab: *taṁ khotaṁ rañjayet paścāt kāpālikramayogataḥ/*; 16.28ab: *evaṁ jīrṇasya sūtasya śṛṇu kāpāli-rañjanam /*; 16.34ab: *esaḥ kāpāliko yogaḥ sarva-lohāni rañjayet /*; 16.43: *nāga-śulvaṁ tathā tīkṣṇaṁ kāpāli-kramam uttamam / tenaiva rañjayet tāraṁ sapta-vāraṁ punaḥ punaḥ //*; 16.45ab: *vaṅga-tīkṣṇa-kapālī* (var. éd. 1910: *-tīkṣṇaṁ kapālī*) *ca śulvaṁ tāraṁ tu rañjayet /* 16.46: *vaikrānta-nāga-kāpālī śuddha-tāraṁ tu rañjayet / rañjayet saha hemnā tu bhavet kuṅkuma-saṁnibham //* 16.47cd: *āra-kāpāli-cūrṇaṁ tu śuddha-tāraṁ tu rañjayet //*; 16.48: *vimalena ca nāgena kāpālī parameśvarī / rañjayet sarva-lohāni tāraṁ hema viśeṣataḥ //*; 16.51: *ravi-nāga-kapālī-nāgaṁ ko tu śuddha-tāraṁ tu rañjayet / rañjayet trīṇi vārāṇi tārāriṣṭaṁ tu jāyate / tenaiva rañjayed dhema sapta-vārāṇi parvati //*. RA 16.29-59ab reproduit avec des variantes dans ĀK, pp. 698-700, 127-155ab: *kapālī-yogāḥ* (manipulation des métaux). Cf. RKDh, pp. 95-8: *atha vajra-bandhādi-rañjanārthaṁ kapālī-yogāḥ / muṣāyāṁ bhasmī-karaṇaṁ tad rakta-varṇaṁ kapālī-yogaḥ / rasārṇave* (suit un fragment inspiré ou tiré du RA). La compilation RKDh, datable du XVIe ou XVIIe siècle, est précieuse, dans la mesure où elle conserve correctement des extraits de textes alchimiques plus anciens. Cf. RA 1910, p. 3.
24. Sur la manipulation des métaux, notamment leur teinture superficielle ou plus profonde (*diplōsis*), voir Marcelin Berthelot, *Introduction à l'étude de la chimie des anciens et du moyen âge*. Paris: Steinheil, 1888, pp. 53-62.
25. RA 6.84ab: *eṣaḥ kāpāliko yogo vajra-māraṇa uttamaḥ /*; 16.62cd: *etat kāpālikā-yogāt cūrṇam amlena mardayet //*. Sur l'auteur, supposé *kāpālika*, du RA, cf. Siddhinandan Mishra, *op. cit.*, (*supra*, n. 21), p. 33. Voir aussi David G. White, *op. cit.*, (*infra*, n. 101), p. 148.
26. ĀK 1. 26.124-125ab (p. 508): *sthālyāṁ sūtādikān kṣiptvā ... / ... kṛtvā mṛdv-agninā pākas tv etat kāpāli-yantrakam //*.
27. Voir le commentaire *Rucikaraṭīkā* cité par Giuseppe Tucci, 'Animadversiones Indicae', *Journal and Proceedings of the Asiatic Society of Bengal* 26, 1930, p. 131. Cf. David N. Lorenzen, *op. cit.*, (*supra*, n. 10), p. 49 et Mark S.G. Dyczkowski, *op. cit.*, (*supra*, n. 11), p. 147, n. 146.
28. Voir Alexis Sanderson, *op. cit.*, (*supra*, n. 12), pp. 661-704. L'enseignement et les travaux philologiques de cet auteur ont été profitables aux recherches de Mark S.G. Dyczkowski, *op. cit.*, (*supra*, n. 11), pp. 26-38 (The Kāpālikas) et 59-92 (The Kaula Tantras). Sur le *kula*, cf. André Padoux, YH, pp. 34-40.

29. Voir Alexis Sanderson, 'Purity and power among the Brahmans of Kashmir', dans: Michael Carrithers *et al.* (éds), *The category of the person. Anthropology, philosophy, history*. Cambridge [etc.]: Cambridge University Press, 1985, pp. 200-2.
30. Voir Krishna Kanta Handiqui, *Yaśastilaka and Indian culture*. Sholapur: Jaina Saṁskṛti Saṁrakshaka Saṅgha, 1949, pp. 354-9 (Śaivism: *vāmamārga*). Cf. David N. Lorenzen, *op. cit.*, (*supra*, n. 10), p. 35.
31. Cf. RA 12.71-73 (avec des variantes dans RAK 133cd-135), 12.80-82 et 15.139. Le composé *kulauṣadhi/ī*, synonyme de *divyauṣadhi/ī*, désigne un groupe de 64 plantes appellées 'divines', en raison de leur efficacité en alchimie. Elles font l'objet du sixième chapitre du RCM. Pour d'autres ensembles de 64 entités dans les sources indiennes, voir Gerdi Gerschheimer, *La théorie de la signification chez Gadādhāra* I. Paris: Collège de France, 1996, p. 9. Voir aussi les espèces d'arbres *kulavṛkṣa*, énumérées par Suresh Chandra Banerji, *Tantra in Bengal. A study in its origin, development and influence*. Calcutta: Naya Prokash, 1978, p. LVIII (appendices). Selon le même auteur (p. 120), le manuel de culte tantrique *Rahasyapūjāpaddhati* cite l'usage rituel de "*kula* flowers". D'après la classification des plantes alchimiques attribuée à (Śiva)Bhairava, sur les quatre groupes ainsi établis (rasa-, mahā-, siddha- et divyauṣadhī) le dernier est réputé le meilleur (RCM 6.3-4). Attestées depuis le RA (*supra*) et le RAK (131, 133-5, etc.), mentionnées par la suite dans le RRS (8.77), ces plantes merveilleusses sont énumérées et décrites dans le GSBh (ch. 7: *divyauṣadhi-lakṣaṇa-karma-vidhānaḥ*) et le RCM (ch. 6: *divyauṣadhi-nirupaṇa*) d'où sont tirées les listes du RPS (ch. 9) et du RKDh (1.3.55-123). Parmi les synonymes sanskrits de *manahśila* "réalgar" figure aussi le terme de *divyauṣadhi*, comme l'atteste le lexique de Madanapāla (4.29). Sur une interférence terminologique entre levégétal et le minéral en alchimie occidentale, cf. Marcelin Berthelot, *op. cit.*, (*supra*, n. 24), pp. 286-7.
32. Sur la polysémie du vocable *kula*, voir les observations de Kanti Chandra Pandey, *Abhinavagupta. An historical and philosophical study*. Banaras: Chowkhamba Sanskrit Series Office, 1963, pp. 594-7. Cf. André Padoux, YH, p. 187 (n. 37).
33. La leçon *dvāraśabdena*, que présente l'édition princeps de 1932, est peu sûre, bien qu'elle soit reproduite par S.N. Mishra dans l'édition plus récente du même texte, publié avec une traduction explicative en hindi (1984). Mais l'auteur se garde bien d'expliquer le composé douteux *dvāraśabdena*, auquel on pourrait préférer la conjecture *dvādaśābdena*, aimablement suggérée par le Pandit N.R. Bhatt (Institut français d'indologie de Pondichéry). Cf. *Yogatattvopaniṣad* 21cd: *mātṛkādi-yutaṁ mantraṁ dvādaśābdaṁ tu yo japet* //.
34. Version tirée de notre édition critique du *Rasendracūḍāmaṇi*, qui sera publiée à Londres par Kegan Paul International, avec traduction annotée et introduction, dans la collection 'Sir Henry Wellcome Asian Series'.
35. De la catégorie du *śleṣa*, la métaphore intervient souvent dans le discours alchimique, qui utilise tant la synonymie que la polysémie. Le sanskrit dispose de soixante vocables pour désigner le mercure, parmi lesquels figure le métaphorique *pārada*, dont le sens étymologique évoque la fonction salvatrice

du vif-argent, fondement de toute doctrine alchimique. Voir l'explication donnée en ce sens par Madhava dans le neuvième chapitre sur la doctrine mercurielle du SDS (p. 80, l.5): *saṁsārasya paraṁ pāraṁ datte 'sau pāradaḥ smṛtaḥ* (le vif-argent fait atteindre l'autre rive de la transmigration).

Sur la terminologie sanskrite du mercure, voir ĀK 1.23.7-10 et *Dīpikā* sur *ŚDhS* 2.12.2. Cf. S.N. Mishra, *op. cit.*, (*supra*, n. 21), pp. 161-2. Au sujet du vocabulaire de l'art hermétique, voir Robert Halleux, 'Problèmes de lexicographie alchimiste', dans: *La lexicographie du latin médiéval et ses rapports avec les recherches actuelles sur la civilisation du moyen âge*. Paris: CNRS, 1981, pp. 355-65. Cf. Michel Butor, 'L'alchimie et son langage', *Critique* 9, 1953, pp. 884-91.

36. Les *kaulika* sont les adeptes de la voie *kula* ou *kaula*.

37. Les sources indiennes présentent de multiples listes de pouvoirs extraordinaires (*siddhi*), que sont censés posséder les êtres appelés 'parfaits' (*siddha*). Ces listes varient peu ou prou selon que les textes sont anciens ou médiévaux, brahmaniques ou tantriques. Cf. Ferdinand D. Lessing et Alex Wayman, *Introduction to the Buddhist Tantrica systems*. Delhi [etc.]: M. Banarsidass, 1978, pp. 216 (n. 4), 220-1 (n. 13).

Outre les qualités merveilleuses associées aux pratiques visant à la délivrance, on observe des *siddhi* exploitées dans un but mondain, pour accomplir des actes de magie noire ou blanche, que l'on retrouve dans les classifications āgamiques des *siddhi*. Voir Hélène Brunner, 'Le *sādhaka*, personnage oublié du śivaïsme du Sud', *JA* 263/3-4, 1975, pp. 432-4. Afin d'acquérir ces pouvoirs 'magiques', on procède à des rites dits 'purs' (*śuci*) ou 'cruels' (*raudra, krūra*), que supposent les actions de magie blanche ou noire, dont fait mention André Padoux, 'Contributions à l'étude du *mantraśāstra*', *BEFEO* 76, 1987, p. 137. Cf. Mark S.G. Dyczkowski, *op. cit.*, (*supra*, n. 11), p. 37. A ce propos, il convient de mentionner l'équivalence *kaula* = *krūra-karman*, que signale Hemacandra dans *Ādīśvaracaritra* (1.410), analysé par Luigi Suali dans *Studi italiani di filologia indo-iranica* 7/1, 1909, p. 6.

Le qualificatif *siddha* est appliqué à des êtres mythiques circulant entre ciel et terre, car l'espace est la 'voie des parfaits' (*siddha-mārga*), d'après le *Mahābhārata* 3.43.37 (éd. critique). Sont également ainsi désignés des dieux tels que Dhanvantari, des *muni* comme Vyāsa et Vālmīki, ou encore des auteurs éminents de la tradition āyurvédique. En fait les authentiques 'parfaits' sont ceux qui ont acquis les huit pouvoirs merveilleux (*aṣṭasiddhi*), comme le souligne Jean Filliozat, *Les sciences et les techniques dans l'Inde dravidienne*, p. 12 (manuscrit). La liste des *mahāsiddhi*, à laquelle se réfère la compilation tardive (XVIe s.) *Śaivāgamaparibhāṣāmañjarī* (8.86-93), se retrouve, à quelques variantes près, dans la littérature philosophique et médicale. Les huit pouvoirs supérieurs concernent la capacité de se rendre petit comme un atome (*aṇiman*), immense (*mahiman*), lourd (*gariman*), léger (*laghiman*), d'atteindre tout (*prāpti*), d'avoir la volonté irrésistible (*prākāmya*), la souveraineté sur tout (*īśitva, īśitṛtva*) et la toute-puissance (*vaśitva*). Voir à ce sujet notre article intitulé 'Considérations sur une technique du *rasāyana* āyurvédique', *IIJ*, 17/1-2, 1975, p. 14, n. 38 (*ubi alia*). Cf. Ram Shankar Bhattacharya, 'Is it justified to read *garimā* in the list of the eight *siddhi*-s?', *Adyar Library Bulletin* 42, 1978, pp. 131-41.

38. Période mythique, le *kalpa* représente un jour dans la vie du dieu Brahmā et correspond à l'espace de temps entre la création et la dissolution d'un monde. Données chiffrées par S. Srinivasan, *Mensuration in ancient India*. Delhi: Ajanta Publications, 1979, pp. 147-8.

39. Le composé *sa-deha-mukti* ou *-muktatā* désigne, comme *jīvan-mukti* ou *-muktatā*, la délivrance du vivant de l'homme, promise aux *kaulika* dans les strophes 6 et 14. Cf. Surendranath Dasgupta, *A history of Indian philosophy* II. Delhi [etc.]: M. Banarsidass, 1932 (réimpr. 1975), p. 245.
Somadeva s'explique à ce sujet, un peu plus loin (str. 16), où il donne la définition suivante: le *sadeha-mukta* est celui qui est délivré de la multitude des couples [de contraires], tels le désir et l'aversion, etc. (*kāmādibhir dvandva-gaṇair vimuktaḥ*), qui éprouve la jouissance des dieux jusqu'à la fin d'un *kalpa* (*ākalpam ākalpita-divya-bhogaḥ*).
Selon la tradition alchimique indienne (SDS, p. 81, ll.5-10), parmi les délivrés en cette vie on reconnaît certains dieux comme Śiva, des démons (*daitya*), des sages (*muni*), des rois (*nṛpa*), des autorités en matière de Rasaśāstra, tels Govinda Bhagavatpāda (auteur du RHT), Kāpāli, Vyāli / Vyāḍi, etc., ainsi que de nombreux autres *siddha*, ayant tous acquis un corps incorruptible grâce au mercure (*tanuṁ rasamayīm āpya*). Sur l'alchimiste Vyāḍi ou Vyāli / Vyālācārya, voir Jean Filliozat, 'Al-Bīrūnī et l'alchimie indienne', dans: *Al-Biruni Commemoration volume*. Calcutta: Iran Society, 1951, pp. 102-4.
Cette transmutation du corps physique est le moyen préconisé par les *rasavādin* afin d'accéder à l'état idéal de délivré-vivant, que recherchent également les méditants *nātha*. Mais, dans cette 'culture du corps' (*kāya-sādhana*), les premiers font surtout appel aux substances minérales, notamment au vif-argent, alors que les seconds s'appuient sur le *haṭhayoga* et valorisent le *soma-rasa* comme élixir d'immortalité, comparable au *rasa* des alchimistes. Ce liquide lunaire coule dans le *cakra* du lotus aux mille pétales (*sahasrāra-padma*), au sommet du crâne, où, hors de l'espace, la *śakti* s'unit à Śiva dans un bonheur suprême, nirvāṇisant. Voir Shashibhusan Dasgupta, *op. cit.*, (*supra*, n. 6), pp. 192-4, 251-4. Cf. Kalyani Mallik, *op. cit.*, (*supra*, n. 6), pp. 21-2. Ce dernier auteur fait état simplement de références au Rasaśāstra dans la littérature *nātha*, alors que David G. White dans un ouvrage informé, qui vient de paraître (*infra*, n. 101), s'emploie, documents à l'appui, à relever des contacts et des interactions entre alchimistes et *nātha-yogin* dans le contexte medieval des traditions *siddha*, dont l'analyse est prolongée justqu'a l'époque contemporaine.

40. Propre au tantrisme, le langage figuré ou conventionnel (*sandhā-bhāṣā*) est valorisé aussi en alchimie (*supra*, n. 35). L'énigmatique *śloka* 7 devient parfaitement explicite à la lumière des trois strophes suivantes, qui décrivent la *khecarī*. Réputée comme unique (*ekā*) ou essentielle (*mukhyā*, chez Brahmānanda sur HYP 3.54), cette *mudrā* est hautement estimée dans la littérature de tradition *nātha*, influencée par le tantrisme et dont traite Christian Bouy, *Les nātha-yogin et les Upaniṣads*. Paris: Collège de France, 1994, pp. 9-28. Une édition critique de l'opuscule médiéval *Khecarīvidyā*, attribué à

Ādinātha, est en préparation, à titre de thèse, par James Mallinson, sous la direction du Professeur Alexis Sanderson de Université d'Oxford. Pour d'autres détails sur ce texte, voir Christian Bouy, *op. cit.*, (*supra*), pp. 12 (n. 15), 41 (n. 157), 73-5 et David G. White, *op. cit.*, (*infra*, n. 101), pp. 146, 169-70, 253, 324.

Sublime (*devatātmikā*), selon Abhinavagupta dans *Tantrāloka* 32, 4-5 (d'après *Vātūlanāthasūtra*, pp. 76-7), cette attitude, à la fois corporelle et spirituelle, est assimilée à l'état de Śiva par Kṣemarāja (*Śivasūtravimarśinī*, p. 147). La *khecarī* engage en particulier l'organe de la langue, qui, pour être retournée dans le cavum, doit subir au préalable un sectionnement progressif du frein ainsi que des manœuvres d'élongation avec un étirement du voile du palais, comme le précise Jean Filliozat, 'La nature du yoga dans sa tradition', dans: Thérèse Brosse, *Études instrumentales des techniques du yoga. Expérimentation psychosomatique*. Paris: École française d'Extrême-Orient, 1963, p. XIX.

Répété assidûment, ce geste est, dit-on, un moyen approprié (*sukaraṇa*) pour obvier aux misères inhérentes à la condition humaine, notamment les chagrins (*śoka*), les maux (*roga*) et la mort (*kāla*), et au-delà pour faire parvenir le yogin à réaliser l'Absolu (*brahman*) par la raison (*tarkeṇa*), comme l'affirme la *Maitri Upaniṣad* 6.20. Révérée par tous les méditants 'parfaits' (GŚ 1976 67: *sarvasiddhair namaskṛtā*), la *khecarī* conduit à la 'perfection du corps' (*kāya-siddhi*) (HYP 3.52 / HP 3.51, var. *kārya-siddhi*). Sur les détails psychophysiologiques concernant la mise en œuvre de cette position caractéristique du *haṭlayoga*, voir GŚ 1976 64-70 et HY / HYP 1.43; 3.6, 32-53/54; 4.38, 43-49. Cf. GŚ 1958 34, 62-6.

41. En l'occurrence, HY 3.47 / HYP 3.48 précise qu'il s'agit des grands crimes ou péchés (*mahāpātaka*), dont fait mention les *Lois de Manu* 9.235: meurtre d'un brahmane adultère avec la femme du maître, etc.

42. Cf. HY 3.49 / HYP 3.50: dans la *khecarīmudrā*, la langue, renversée dans le pharynx, recueille la sécrétion pharyngienne, assimilée ici au nectar lunaire (*soma*) aux saveurs multiples. Employée pour arrêter la chute de l'ambroisie dans le feu digestif, cette technique du *haṭhayoga* est censée enlever les maladies (*vyādhīnāṁ haraṇam*), mettre fin au vieillissement (*jarānta-karaṇa*) et faire acquérir au méditant les huit grandes *siddhi* ainsi que l'immortalité (*amaratva*). Voir aussi HY 3.38 / HYP 3.39 et HY 3.39 / HYP 3.40.

43. La double perspective sotériologique de l'homme est une conception partagée par les écoles philosophiques Vedānta et Sāṁkhya-Yoga, comme en témoigne en particulier le *Yogavāsiṣṭha* 5.75.49: *dvividhā muktatā loke vidyate deha-dhāriṇām / sadehaikā videhānyā vibhāgo 'yaṁ tayoḥ*
śṛṇu //. Cf.YV, vol. IV, pp. 40 (n. 1), 90 (n. 1).

Selon ce dernier ouvrage (YV, vol. II, pp. 8 (n. 6), 174-5), la délivrance en cette vie même correspond à la conviction de la discrimination (*viveka-khyāti*), le dernier stade du *samādhi* conscient (*saṁprajñāta*), alors que l'isolement (*kaivalya*) libérateur est atteint, après l'abandon du corps, dans le *samādhi* inconscient (*asaṁprajñāta*). L'expérience ou la jouissance du délivré-vivant (*jīvan-mukta*) est semblable à celle qu'éprouve le Seigneur (*īśvara*), avec la différence que celui-ci ne ressent jamais la douleur (*duhkha*), dont l'autre est touché avant d'accéder à l'état qui est le sien: *jīvanmuktasyāpīśvara-sadṛśa eva bhogo duḥkha-bhoga-mātram īśvarād vilakṣaṇam iti* (YV, vol. I, pp. 131-2). Pour le renonçant (*saṁnyāsin*), désireux de se détacher du phénoménal, les actes (*karman*) sont sans valeur pour une nouvelle naissance, du fait que les affects sont détruits (*kṣīṇakleśa*), comme l'affirme Vyāsa (YS 4.7). En effet, selon Patañjali, le dépôt des actes, qui a pour base les affects, est à ressentir dans une naissance actuelle ou future: *kleśa-mūlaḥ karmāśayo dṛṣṭādṛṣṭa-janma-vedanīyaḥ* (YS 2.12). Cf. YV, vol. II, pp. 176-7 et IV, pp. 123-4, 124-5, 142. Sur la doctrine de l'état de délivré-vivant dans l'hindouisme, voir la monographie de Gerhard Oberhammer, *La délivrance, dès cette vie (jīvanmukti)*. Paris: Collège de France, 1994.

D'après Somadeva (RCM 1.17), la délivrance sans corps est incomparable (*nirupama*), car elle comporte une félicité sans couples [de contraires] (*nirdvandvānanda-saṁyukta*) et se trouve débarrassée de toute condition limitative (*sarvopādhi-vivarjita*). Les délivrés 'sans corps' (*videha*) sont de purs esprits, que Vyāsa considère comme des dieux (YS 1.19). Bien que privés du 'corps grossier', dans leur état dit *vaidehya*, les 'désincarnés' éprouvent la même jouissance que les dieux, en raison du 'corps subtil' (*sūkṣma-*, ou *liṅga-śarīra*): *sthūla-deha-virahe 'pi liṅga-śarīreṇaiva yeṣāṁ devānāṁ bhogas te videhās tad-rūpatā ca vaidehyam* (YV, vol. I, p. 97). Ce substrat permanent de l'être psychique est formé des constructions du vécu (*saṁskāra*), qui restent actives et déterminent la métensomatose. Ainsi, la condition de *deva* n'est pas acquise pour toujours, car tout reliquat d'actes non encore soldés fait revenir l'âme dans la ronde des existences et l'engage dans un nouveau corps.

44. Selon la glose (*ṭīkā*) des *Gaṇakārikā*, le rire (*hasita*) est une des six sortes d'offrande (*upahāra*) que les śivaïtes *pāśupata* apportent au dieu Maheśvara. Voir texte, traduction et analyse par Minoru Hara, 'Nakulīśa-pāśupata-darśanam', *IIJ* 2, 1958, pp. 8-12, 26, avec les notes 122 et 123. D'après l'indianiste japonais, le *hasita* évoque ici le rire attribué a Śiva lui-même, tout comme la danse (*nṛtya*), autre forme d'offrande *pāśupata*, rappelle le dieu de la danse. Nous remercions Monsieur Gerdi Gerschheimer, qui a attiré notre attention sur la doctrine *pāśupata* décrite dans la glose des *Gaṇakārikā*.

Diversement apprécié dans le brahmanisme et le bouddhisme, le rire manifeste en général vitalité et joie de vivre, qui pourraient être interprétées comme l'expression du bonheur dont jouissent les *kaulika*. Voir la thèse de Volker M. Tschannerl, *Das Lachen in der altindischen Literatur*, Frankfurt am Main [etc.]: Lang, 1992, pp. 36-49 (Lachen und Geschlechtlichkeit) et pp. 141-5 (conclusions). Cet ouvrage nous a été aimablement signalé par Mademoiselle Nalini Balbir.

Enfin, dans ce contexte, il convient de mentionner aussi une nouvelle technique de méditation à base de rires et de pleurs enseignée par Rajneesh Chandra Mohar (1931-90), devenu Bhagwan Shree Rajneesh et, dernièrement, Osho. Après avoir créé à Poona un ashram renommé, ce *guru* fonda un nouveau mouvement religieux syncrétisant, qui s'est répandu également en Occident. Sur la religion du rajneeshisme, voir Jean Vernette avec la collaboration de Claire Moncelon, *Dictionnaire des groupes religieux aujourd'hui*. Paris: Presses universitaires de France, 1996, pp. 52-4, 166, 227. Cf. Harry Aveling, *The laughing swamis: Australian sannyasin disciples of Swami Satyananda Saraswati and Osho Rajneesh*, Delhi, 1996 (réimpr.).

45. Sur l'alimentation carnée en Inde d'après la littérature normative du *dharma*, voir Ludwig Alsdorf, *Beiträge zur Geschichte von Vegetarismus und Rinderverehrung in Indien*. Wiesbaden: Franz Steiner, 1962. Ce mémoire fait état aussi de l'Āyurveda, où la viande, bovine notamment, est valorisée du point de vue thérapeutique, comme le souligne Francis Zimmermann, *La jungle et le fumet des viandes*. Gallimard-Le Seuil, 1982, pp. 199-213 (ch. VII: Note sur le végétarisme et la non-violence), spécialement, pp. 205-8. Données textuelles sur l'alimentation carnée dans Om Prakash, *Food and drinks in ancient India*. Delhi: Munshiram Manoharlal, 1961, pp. 105-11, 164, 175-8, 209-16.

    Il est à noter que les normes brahmaniques concernant l'alimentation carnée sont assouplies en cas de détresse (*āpad-dharma*), à savoir famine ou maladie, comme en témoignent aussi certaines prescriptions médicales. Si, en l'occurrence, on accepte des situations d'exception, dans d'autres cas, il s'agit d'une transgression caractérisée des normes brahmaniques: le culte tantrique, notamment chez les *kāpālika*, comporte la consommation rituelle de viande et d'alcool, ainsi que de sécrétions coïtales. Voir André Padoux, 'Les cultes *śākta* de la main gauche', *Studia Orientalia* 70, 1993, p. 92. A ce propos, il est permis de rappeler que les pratiques sexuelles tantriques, en particulier *kāpālika*, sont parfois invoquées à l'appui d'une interprétation du répertoire érotique que présente la sculpture architecturale de certains temples de l'Inde médiévale, comme ceux de Khajurāho. Voir Jeannine Auboyer, *Introduction à l'étude de l'art de l'Inde*. Roma: Istituto italiano per il Medio ed Estremo Oriente, 1965, pp. 92-4 (*ubi alia*).

46. Voir *supra*, n. 41.

47. Les strophes 8 et 9 se retrouvent dans HY 3.47 / HYP 3.48 et HY 3.48 / HYP 3.49, avec deux légères variantes au dernier hémistiche: *candrāt sravati yaḥ sāraḥ sā syād amara-vāruṇī*. On doit rappeler que le traité de Svātmārāma est une anthologie du XVe siècle - d'après Christian Bouy, *op. cit.*, (*supra*, n. 40), p. 85 - et que ce florilège tardif ainsi que le traité iatrochimique RCM du XIIe-XIIIe siècle ont probablement puisé les deux strophes à une source commune antérieure.

    D'autre part, certains ont cru pouvoir reconnaître un adepte de la voie *kaula* dans le *kulin* qui pratique la *khecarī* régulièrement, dont fait mention HY 3.46 / HYP 3.47. Voir Tara Michaël, HYP 1974, p. 182.

48. L'instrumental *dvādaśābdena* présente une valeur temporelle, qui marque ici le résultat heureux obtenu au terme d'une action menée sans discontinuité dans le temps, suivant l'enseignement des grammairiens: *Siddhāntakaumudī* 563 sur Pāṇini 2.3.6.

49. Le terme de *siddhi* est ambigu, car il désigne aussi bien des pouvoirs extraordinaires que la réussite, le succès d'une action, comme le précise Hélène Brunner, 'Un tantra du Nord: le *Netra Tantra*', *BEFEO* 61, 1974, p. 133, n. 6.

50. Cf. HY 3.46 / HYP 3.47: la *khecarī* pratiquée sans discontinuer (*nityam*). Pour des résultats spectaculaires obtenus à court terme par une pratique qui ressemble à celle de la *khecarī*, voir KJN 6.18-19ab: ... *māsena jitayen mṛtyum* ... et KJN 6.19cd-20ab: ... *ṣaṇ-māsād abhyased devi mahā-rogaiḥ pramucyate /*.
51. Somadeva se réfère ici à l'enseignement du Sāṁkhya, qui distingue dans la structure de l'univers, outre l'être spirituel Puruṣa, vingt-quatre réalités élémentaires (*tattva*), depuis la Nature fondamentale (*prakṛti*) jusqu'aux cinq *mahābhūta*, dont la terre est le dernier élément.
52. Sur la polysémie du mot *kula*, voir *supra*, p. 401-2.
53. Le sixième chapitre du KNJ décrit la nature du *jīva*, diversement identifié avec le *manas*, la *buddhi*, le *prāṇa* ou le *vāyu* organique. Le soi est dit vivant tant qu'il anime un corps. En effet, lorsque celui-ci devient inanimé, le soi l'abandonne et alors son nom n'est plus *jīva* mais Śiva (KJN 6.7): *deha-sthas tiṣṭhate yāvat tāvaj jīvo 'pi gīyate / sa deha-tyakta-mātreṇa paraṁ śivo nigadyate //*. Sur Śiva incorporé (*dehin*), cf. le *Gahvaratantra* cité par Abhinavagupta dans le *Tantrāloka* 28, 224cd-227ab (trad. Raniero Gnoli, p. 654).
54. L'adjectif *kaulika* s'applique aux différents points de doctrine propres aux *kaula*; en l'occurrence, il s'agit de la conception du *jīva*. Cf. KJN 6, 9.
55. Sur *kalpa*, voir *supra*, n. 38.
56. La puissance suprême (*aiśvarya*) comporte l'octuple pouvoir surnaturel (*mahāsiddhi*) du yogin, comme le confirment les commentaires de Gauḍapāda (*Sāṁkhyakārika* 45) et de Vyāsa (YS 3.45: *aṣṭāv-aiśvaryāṇi*). Selon le śivaïsme *pāśupata*, l'expression *māheśvaram aiśvaryam* désigne la souveraineté propre à Maheśvara et cette puissance divine est susceptible d'être acquise par l'ascète parfaitement accompli (*siddha*). Voir Gerhard Oberhammer, *op. cit.*, (*supra*, n. 43), pp. 88-9.
    D'autre part, le *Mṛgendratantra*, célèbre texte de la littérature āgamique, distingue deux sortes d'initiation (*dīkṣā*), suivant le but poursuivi, à savoir jouissances (*bhukti*) ou libération (*mukti*). A ce propos, la *Rurusaṁhitā*, citée dans le commentaire du texte, déclare que les disciples purifiés par l'initiation 'ouissent longtemps de plaisirs en compagnie d'une troupe de femmes immortelles (*amara-strī-nikāya*); puis, leurs désirs tombés, ils ont part à la souveraineté suprême (*paraiśvarya*) dans le royaume de Śiva'. Voir la section des rites du *Mṛgendrāgama*, traduit par Hélène Brunner-Lachaux, pp. 198-9 et 203-4.
57. Cf. *Manusmṛti* 3.161: *veda-nindaka*. On doit cependant observer que les Tantra, notamment les Āgama, dans leur grande majorité, ne s'insurgent pas contre le Veda, qu'ils visent seulement à dépasser par la dévotion (*bhakti*) en vue de la délivrance (*mukti*), conception ignorée du védisme. Sans contester donc la validité du Veda en matière de jouissance des biens de ce monde, les Tantra relèvent leur avantage, dans la mesure où ils valent à la fois pour la *bhukti* et la *mukti*. Voir Jean Filliozat, 'Sur quelques désignations de textes sanskrits', dans: *Ludwik Sternbach Felicitation volume*. Lucknow: Akhila Bharatiya Sanskrit Parishad, 1979, p. 258.

58. Après d'autres écoles philosophiques, Somaveda évoque les matérialistes, appelés ici *nāstika* et connus aussi sous les noms de *lokāyata* ou *cārvāka*. Dans *Bhāvasaṁgraha* (str. 172-176), l'auteur jaina Devasena du Xe siècle attribue l'enseignement matérialiste à un maître *kaulācārya*. On doit cependant remarquer avec Krishna Kanta Handiqui, *op. cit.*, (*supra*, n. 30), p. 355, que les deux mouvements sont différents, bien que proches par certains points de doctrine, ce qui pourrait expliquer la confusion.
D'autre part, plusieurs sources sanskrites, comme la *Nyāyasiddhāntamālā* du XVIIe siécle (éd. Mangal Deva Shastri, Benares, 1927-28, p. 175), rapprochent les doctrines *cārvāka* et *somasiddhānta*, assimilable au *kāpālikadarśana*. A ce sujet, sont utiles les observations de Krishna Kanta Handiqui, dans *Naiṣadha-carita of Śrīharṣa*. Poona: Deccan College, 1956, pp. 640-5. Cet auteur appuie son analyse sur des textes aussi nombreux que divers, puisés en particulier dans le théâtre allégorique sanskrit du XIe ou du XVIIe siècle: *Prabodhacandrodaya* (*supra*, pp. 397, 400), *Amṛtodaya* et surtout *Vidyāpariṇayana* (éd. Śivadatta et K.P. Parab, Bombay: Nirṇaya Sāgar, 1893). Ānandarāya Makhī, l'auteur de ce dernier drame, met en évidence les idées religieuses ou philosophiques des *kāpālika*, pour lesquels le ciel (*svarga*) est le lieu de toute satisfaction et de plaisirs charnels sans limites (*samabhilaṣitasarvārthasiddhibhir aniyantritāḥ kāmopabhogā iti*, *op. cit.*, p. 42). Pour les ascètes porteurs de crâne, le but suprême de l'homme est de réaliser l'isolement libérateur avec le corps: *sadeha-kaivalyam idam eva naḥ paramapuruṣārthaḥ*, (*op. cit.*, p. 42). Voir aussi à ce sujet l'article, bien que dépourvu d'esprit critique, de Dakshina Ranjan Shastri, 'The Lokāyatikas and the Kāpālikas', *Indian Historical Quarterly* 7/1, 1931, pp. 125-37. Cf. l'étude de Gerdi Gerschheimer sur 'Les catégories (*padārtha*) selon Murāri Miśra', à paraître prochainement dans le *Bulletin d'études indiennes* 13-14, 1995-96.
En ce qui concerne les conceptions matérialistes, les *nāstika* ne croient ni à la réalité d'un autre monde ni à la validité absolue du Veda, comme le souligne Louis Renou, *Études védiques et pāṇinéennes* VI. Paris: Boccard, 1960, pp. 3 (n. 1), 41. Ils reconnaissent en général quatre grands éléments et acceptent parfois le cinquième, à savoir l'espace (*ākāśa*), selon Guṇaratna, cité par Dakshina Ranjan Shastri, *op. cit.*, (*supra*), pp. 127-8. Pour une analyse de la philosophie matérialiste, voir Surendranath Dasgupta, *A history of Indian Philosophy* III. Delhi [etc.]: Motilal Banarsidass, 1940 (réimpr. 1975), pp. 512-50.

59. La répétition est un fait de style parmi les plus frappants en sanskrit. Cf. Louis Renou, *La grammaire de la langue védique*. Lyon-Paris: IAC, 1952, p. 394.

60. Mircea Eliade, *op. cit.*, (*supra*, n. 13), pp. 275-7. Cf. George W. Briggs, *Gorakhnāth and the Kānphaṭa Yogīs*. Delhi [etc.]: Motilal Banarsidass, 1938 (réimpr. 1973), p. 23.

61. Frits Staal, 'The Sanskrit of science', *Journal of Indian Philosophy* 23, 1995, pp. 73-127, spécialement, pp. 88-101 (Mathemathical Sanskrit).

62. Frits Staal, 'Concepts of science in Europe and Asia', *Interdisciplinary Science Reviews* 20/1, 1995, pp. 7-19, spécialement, pp. 9-10 (Newton abstracts alchemy).

63. Frits Staal, 'Indian bodies', dans: Thomas P. Kasulis *et al.* (eds), *Self as body in Asian theory and practice*. Albany: State University of New York Press, 1993, pp. 59-102, spécialement, pp. 77-88 (Kalarippayattu: A martial art in Southwest India).
64. Arion Roşu, 'Les *marman* et les arts martiaux indiens', *JA* 269/3-4, 1981, p. 450, n. 84.
65. *Kumārasaṁbhava* 5.33d: *śarīram ādyaṁ khalu dharma-sādhanam*, rendu en latin par A.F. Stenzler: 'corpus enim primum virtutis subsidium'. Cf. le commentaire de Mallinātha *ad loc.*: *sati dehe dharmārtha-kāma-mokṣa-lakṣaṇāś caturvargāḥ sādhyante / ata eva 'satatam ātmānam eva gopayīta' iti śrutiḥ //.*
66. Cennabasavaṇṇa dans *Śūnyasaṁpādane*, texte kannaḍa édité avec traduction et notes par S.S. Bhoosnurmath et A. Menezes, vol. III. Dharwar: Karnatak University, 1969, p. 8.
67. Cf. Caraka, *Sūtra*, 1.15cd: *dharmārtha-kāma-mokṣāṇām ārogyaṁ mūlam uttamam //.*
68. Voir Arion Roşu, *Les conceptions psychologiques dans les textes médicaux indiens*. Paris: Institut de civilisation indienne, 1978, pp. 80-1. Cf. notre article 'Le *trivarga* dans l'Āyurveda', *Indologica Taurinensia* 6, 1978, p. 256.
69. Cf. Helmuth von Glasenapp, *Unsterblichkeit und Erlösung in den indischen Religionen*. Halle (Saale): Niemeyer, 1938, pp. 1-3.
70. Arion Roşu, *op. cit.*, 1975, (*supra*, n. 37), pp. 1-2.
71. Priyadaranjan Rây (éd.), *History of chemistry in ancient and medieval India*, incorporating the 'History of Hindu chemistry' by Prafulla Chandra Rây. Calcutta: Indian Chemical Society, 1956, pp. 37, 63.
72. Voir Arion Roşu, 'Yoga et alchimie', *ZDMG* 132/2, 1982, pp. 363-79.
73. Sur la pharmacie dans l'Inde ancienne, voir le chapitre de Prem Vrat Sharma et A.V. Sharma, dans: Priya Vrat Sharma (éd.), *History of medicine in India, from Antiquity to 1000* AD. New Delhi: Indian National Science Academy, 1992, pp. 399-417.
74. RA 1978 17. 165ab: *yathā lohe tathā dehe kartavyaḥ sūtakaḥ sadā /.* Repris dans SDS, p. 82, l.12 (lire *sadā*, non *satā*).
75. Succession des périodes établie par Prafulla Chandra Rây, le premier historien de la chimie indienne, dans l'ouvrage: *A history of Hindu chemistry*, 2 vol. London-Calcutta, 1902-9. Pour les idées renouvelées sur l'histoire de la littérature alchimique sanskrite, voir David G. White, *op. cit.*, (*infra*, n. 101), pp. 123-70 (chap. five: Tantric and Siddha Alchemical Literature).
76. Cf. RHT 1958 1.10: *iti dhana-śarīra-bhogān matvānityān sadaiva yatanīyam / muktis tasya jñānāt tac cābhyāsāt sa ca sthire dehe //.* Texte repris avec des variantes dans SDS, p. 80, ll. 17-8.
77. Le chapitre 16 du *Garuḍapurāṇa-sāroddhāra*. Pour le contenu du KAT, voir Teun Goudriaan et Sanjukta Gupta, *Hindu tantric and śākta literature*. Wiesbaden: Harrassowitz, 1981, pp. 93-6.
78. Voir Mircea Eliade, *op. cit.*, (*supra*, n. 13), pp. 277-82 et du même auteur, *Forgerons et alchimistes*. Paris: Flammarion, 1977, pp. 108-18 et 170-2. Cf. notre article 'Yoga et alchimie' (*supra*, n. 72).

79. Voir Jean Filliozat, *op. cit.*, (*supra*, n. 57), pp. 256-7. Cf. Teun Goudriaan et Sanjukta Gupta, *op. cit.*, (*supra*, n. 77), pp. 7-9 (définition du terme de *tantra*) et le chapitre VII (pp. 112-29), intitulé 'Tantras of magic', qui traite de textes techniques bien particuliers.
80. Voir notre article sur le *rasāyana* (*supra*, n. 37), pp. 1-29. Cf. Jean Filliozat, dans *L'Inde classique* II. Paris-Hanoï: Imprimerie nationale et École française d'Extrême-Orient, 1953, pp. 167-8.
81. Mircea Eliade, *op. cit.*, 1977 (*supra*, n. 78), pp. 110-12.
82. Outre certains apports supposés chinois aux origines du Rasaśāstra, on a étudié le développement des rapports sino-indiens en matière d'(al)chimie. Voir les travaux comparatifs de Vijaya Deshpande: 'Medieval transmission of alchemical and chemical ideas between India and China', *Indian Journal of History of Science* 22/1, 1987, pp. 15-28; 'Transmutation of base-metals into gold as described in the text *Rasārṇavakalpa* and its comparison with the parallel Chinese methods', *Indian Journal of History of Science* 19/2, 1984, pp. 186-92; 'Medieval Chinese and Indian alchemy: A note on comparative study of parallels in the Chinese text *Chun Zhu Ji Wen* (Records of things heard at Spring Island) and contemporary Sanskrit texts', *Indian Journal of Asian Studies* 2/2, 1990, pp. 50-61 (en collaboration avec Kamal Sheel).
83. Voir B.V. Subbarayappa et Mira Roy, '*Mātṛkabhedatantram* and its alchemical ideas', *Indian Journal of History of Science* 3/1, 1968, pp. 42-9.
84. Analyse par Mira Roy, '*Rasārṇavakalpa* of *Rudrayāmala* Tantra', *Indian Journal of History of Science* 2/2, 1967, pp. 137-42. Cf. B.V. Subbarayappa, Introduction à l'ouvrage RAK, pp. 3-6.
85. Le Lauhaśāstra, qui traite de remèdes métalliques, surtout à base de fer, est présenté, dans certains textes médicaux, sous un jour plus favorable que le Rasaśāstra: les préparations mercurielles sont moins bien tolérées que les médicaments contenant du fer. Voir Priya Vrat Sharma, 'A fragment of the *Lauhaśāstra* of Nāgārjuna', *Indian Journal of History of Science* 28/1, 1993, pp. 35-50.
86. Cf. Priya Vrat Sharma, 'Tantrik influence on Śārṅgadhara', *Ancient Science of Life* 3/3, 1983-84, pp. 129-31. On ne saurait toujours suivre l'auteur dans son analyse, qui appelle parfois certaines réserves.
87. Cf. G. Jan Meulenbeld, 'The search for clues to the chronology of Sanskrit medical texts, as illustrated by the history of *bhaṅgā* (*Cannabis sativa* Linn.)', *Studien zur Indologie und Iranistik* 15, 1989, pp. 59-70, spécialement pp. 66-7. A la terminologie sanskrite du cannabis, on ajoutera *saṁvid* 'conscience', qui signifie le chanvre indien au septième chapitre du texte tardif *Mahācīnācāra-tantra*, comme le signale André Padoux dans un compte rendu du *Bulletin d'études indiennes* 7-8, 1989-90, p. 367. D'autres textes *kaula* mentionnent le cannabis (KAT 5.42), que désigne *saṁvidā* dans le manuel de culte tantrique *Rahasya-pūjā-paddhati*, selon Suresh Chandra Banerji, *op. cit.*, (*supra*, n. 31), p. 120, n. 1.
88. Voir la conclusion de notre article '*Mantra* et *yantra* dans la médecine et l'alchimie indiennes', *JA* 274/3-4, 1986, pp. 263-4.

89. Jean Filliozat, *Le Kumāratantra de Rāvaṇa et les textes parallèles indiens, tibétains, chinois, cambodgien et arabe*. Paris: Imprimerie nationale, 1937, pp. 1-82, spécialement, pp. 67-8.
90. Teun Goudriaan, 'Khaḍga-Rāvaṇa and his worship in Balinese and Indian tantric sources', *WZKS* 21, 1977, pp. 143-69.
91. Avant de nous faire parvenir le texte inédit, accompagné d'une notice philologique, Alexis Sanderson s'y est référé lors d'une Table ronde de juin 1984, au cours de la discussion qui a suivi notre communication intitulée '*Mantra* et *yantra* dans la médecine et l'alchimie indiennes', dans: André Padoux (éd.), *Mantras et diagrammes rituels dans l'hindouisme*. Paris: CNRS, 1986, pp. 117-21 (résumé de l'exposé) et, pp. 121-6, spécialement 123-5 (discussion).
92. Gerard L. M. Clauson, 'Catalogue of the Stein Collection of Sanskrit MSS. from Kashmir', *Journal of the Royal Asiatic Society*, 1912, pp. 626-7 (ms. n° 207).
93. Shridhar R. Bhandarkar, *A catalogue of the collections of manuscripts deposited in the Deccan College*. Bombay: Government Central Press, 1888, p. 76 (n° 76/1875-76).
94. Har Datta Sharma, *Descriptive catalogue of the Government collections of manuscripts deposited at the Bhandarkar Oriental Research Institute*, vol. XVI2: *Tantra*. Poona: Bhandarkar Oriental Research Institute, 1976.
95. En l'occurence, André Padoux mentionne une 'liste des tantras des quatre *srotas*' (YH, p. 41, n. 60).
96. Cf. Teun Goudriaan et Sanjukta Gupta, *op. cit.*, (*supra*, n. 77), p. 14.
97. Le *Pratiṣṭhālakṣaṇasārasamuccaya* présente une autre liste des 20 Bhūtatantra, où l'on retrouve 5 titres figurant dans le texte de l'Appendice (*infra*, pp. 416-18). Cf. Mark S.G. Dyczkowski, *op. cit.*, (*supra*, n. 11), pp. 34-5.
98. Cf. Teun Goudriaan et Sanjukta Gupta, *op. cit.*, (*supra*, n. 77), pp. 7, 10, 16-17.
99. Cf. l'Appendice du présent article, *pāda* 6c: *bālarakṣārtham*.
100. A ce sujet, il convient de citer un récent recuil d'enseignements thérapeutiques puisés dans de nombreux Tantra (et d'autres sources), original sanskrit et traduction en bengali: Kṛṣṇacaitanya Ṭhākur, *Cikitsā bidhāne tantraśāstra*, 3 vol., Calcutta, 1987-92. Compte rendu de Rahul Peter Das, dans le *Journal of the European Āyurvedic Society* 4, 1995, pp. 229-34.
    En ce qui concerne les rapports entre Tantra et Rasaśāstra, il serait bien difficile de suivre S. Mahdihassan dans toutes ses spéculations, qui lui sont chères de longue date et qu'il reprend une fois de plus dans un article dont le titre est cependant prometteur: 'The origin of alchemy and of the tantric cult - an etymological approach', *Hamdard medicus* 29/1-2, 1986, pp. 7-21.
101. Des rapports entre tantrisme et alchimie traite aussi un tout récent ouvrage intitulé *The alchemical body: Siddha traditions in medieval India*, Chicago-London: The University of Chicago Press, 1996, auquel nous avons fait allusion plus haut (n. 39). L'auteur, David Gordon White, professeur associé d'études religieuses à l'Université de Californie (Santa Barbara), eut l'obligeance de nous faire parvenir un exemplaire à titre d'hommage, dont nous le remercions

vivement.

Ce travail de longue haleine représente une importante contribution à l'étude en profendeur des rapports entre les traditions alchimiques iatrochimiques, tantriques et haṭhayogiques de l'Inde médiévale. Pour ce système de vases communicants de la culture indienne, les recherches pluridisciplinaires de David G. White confirment et complètent nos investigations sur plusieurs points, notamment sur les ascètes *kāpālika* vus dans la perspective du Rasaśāstra (*supra*, pp. 397-401).

102. Cité dans le *Tantrālokaviveka*, vol. I, p. 44, ligne 2.

103. Ms.: *kaṭimaṇḍalam*. La forme *ṭakamaṇḍalam* se retrouve dans le *Jayadrathayāmala*, *ṣaṭka* 3, *paṭala* 24 (National Archives of Nepal, ms. 5/1975), folio 170v6.

104. Ms.: *bhūtrāsaṁ ca śikhārāvaṁ*. Cf. *pāda* (16)d.

105. Ms.: *uddiṣṭam*. Cf. *pāda* (26)a. Un *Ucchiṣṭatantra* est mentionné parmi les Bhūtatantras dans le *Pratiṣṭhālakṣaṇasārasamuccaya*: voir V.V. Dviveda, *Luptāgamasaṁgraha*, part II, Introduction (*Upodghāta*), p. 91.

106. Ms.: *nāśanānam*.

107. Ms.: *-aḥ*.

108. = *rāva uktaś*.

109. Ms.: *ṛddhyaṣṭanaṣṭādilakṣaṇam*. Ignorant du sens de ce texte, nous ne nouvons être sûr de notre diagnostic de ce *pāda* hypermètre.

110. Ms.: *vāśyoccāṭana-*.

111. Ms.: *parivṛtam* incompatible avec la cadence.

112. Ms.: *dhāraṇājjñāna-*.

113. Ms.: *paracikīrṣajjñatvam*.

114. Ms.: *parakāśa-*.

115. Ms.: *-hiteṣiṇā*; mais peut-être que ce solécisme est originel.

116. Ms.: *taṁ*.

XXIV

HIERARCHICAL IDEALISM PLOTINUS/PROCLUS, BHARTṚHARI

Ben-Ami Scharfstein

I have long admired Frits Staal's unique combination of breadth and detail of knowledge - of logic, linguistics, and much else - and his easy ability to move between what are, from the standpoint of scholarship, different intellectual worlds. Some years ago, I wrote an account of the philosophy of Bhartṛhari, the Indian grammarian and philosopher. Since I knew no one else who had so good a grasp of Bhartṛhari and could also be sympathetic to my purpose in describing his thought, I turned to Frits Staal with the request that he review and, if necessary, correct my account. Staal generously gave me the help I needed, for which I remain grateful to him. My contribution to the present Festschrift contains a more developed version of what Staal reviewed, this time in the context of a comparative history of philosophy to be called *From Uddālaka to Kant*. Except for the omission of the chapter number, I have made no changes. Should any reader give me suggestions for improvement, I will be grateful - the book will remain in the writing for some time.

Theme: The Atomists' Tactics Are Reversed to Create Structures of Principles that Culminate in Being-as-Such

The two philosophies dealt with in this chapter are called hierarchical because they argue that the everyday world depends on a large number of vital principles, which depend on fewer principles, which depend on fewer still, up to the vital principle on the top of the pyramid, being-as-such, on which everything else depends. I have said 'the top of the pyramid' because as the principles grow fewer they grow higher in the degree of their being; but because everything low in reality rests on what is higher and, eventually, on the highest, the pyramids can be seen as inverted. The two philosophies are idealistic in the philosophical sense because the principles on which each of their structures depend have the nature of consciousness. The philosophies are also mystical in that the Being (or One) and the Word (or Brahman) in which they respectively culminate cannot be understood by reason or described in words and is knowable only by means of the wordless experience that is an intimacy or identity like that of one's experience of oneself.

The two philosophies, the one Greek and the other Indian, are near mirror images of atomism in attitude and intellectual tactics. This is certainly not intentional and stems not from borrowing in either direction but from the philosophical environment of each of the two traditions and, of course, from the similarity of the problem involved - the search for reality by means of division as against the search for it by means of unification. In the spirit of division, the atomists begin with what they see as the relatively few, relatively complex and illusory things of the everyday world, which they divide repeatedly until, at the end of the process, there are only an innumerable number of material or other (Buddhist) atoms, all fully real, and all with the fewest characteristics possible. The descent to atoms shows that the plurality of beings and qualities of the visible world is an illusion produced by the limitations of human perception. In the contrary spirit, of unification, the hierarchical idealists reverse the atomists' reasoning and begin with what they see as the very numerous, relatively simple and illusory things of the everyday world, which they repeatedly reduce to a smaller and smaller number of immaterial principles until, at the end of the process, there is left only the single immaterial, fully real, absolutely indivisible atom, the absolute Being, the One or the Word that possesses all the infinitely many qualities of reality in an inseparable, indescribable union. This rise to unity shows that the perceived plurality of beings and qualities of Being-as-such is an illusion produced by the limitations of human perception and understanding.

In all these characteristics, the Greek and Indian forms of hierarchical idealism are alike. Whereas the atomists' atoms are completely devoid of consciousness and accessible only to analytic reason, the hierarchical idealists' Being-as-such, the One or Word, is completely devoid of matter and therefore accessible only to the self-intimacy found in consciousness. To the Indian philosophers, as we know, the highest reality had long been associated with self-intimacy, but among the Greek and Roman philosophers the unique nature of consciousness had hardly been explored until Plotinus, who characteristically insisted that 'the perceiving part of the soul must turn inwards and must be made to attend there'.[1]

Of course, the philosophical contexts and terminologies of the two systems are different. The most striking abstract difference is that Proclus - far more than Plotinus - uses a mathematical model of reality, and Bhartṛhari, a grammatical one. Proclus' model, like that of the Greek atomists, depends on the mathematical relationship between individual units and infinite quantities, which is to say, the kind of relationship that holds between individual numbers and the total of all numbers, which is uncountable because there is no last number among them. Plotinus uses recurrent mathematical imagery that is compatible with this position but does not make use of its quasi-mathematical logic; and his references to mathematics can be elusively quick.[2] The method of proof adopted by Proclus is an open though not inherently close imitation of the method of Euclid's axiom-based geometry. In contrast, the logic used by Bhartṛhari is based on grammar, which the Indians were the first to develop to a sophisticated level.

Plotinus (AD 205-270), Proclus (AD 410?-485):[3] 'Goodness Is Unification, and Unification, Goodness'

The only biography of Plotinus, written by his disciple Porphyry, opens with the words, 'Plotinus, the philosopher of our times, seemed ashamed of being in the body. As a result of this state of mind he could never bear to talk about his race or his parents or his native country.'[4] Plotinus was twenty-eight years old when he went to Alexandria to study philosophy. After eleven years with his teacher Ammonius, Plotinus had acquired so complete a training in philosophy, says Porphyry, that he grew eager to learn Persian and Indian philosophy.[5] For this purpose, Plotinus joined the army prepared to invade Persia, but the Emperor, who led the army, was killed and Plotinus, escaping with difficulty, settled in Rome, where he spent the rest of his life.

Plotinus could bear no more than a single rereading of anything he had written, and his eyesight made even a single reading difficult. After going mentally through a train of thought, he wrote it out unhesitatingly, as if copying from a book.[6] Yet his writings are filled, says Porphyry, with concealed Stoic and Aristotelian doctrines. He was, in fact, widely learned, with an excellent knowledge of geometry, arithmetic, mechanics, optics, and music.

Plotinus had many enthusiastic hearers. A gentle, helpful man, he was willing to become the guardian of the children brought to him by parents who felt themselves close to death. Knowing of a ruined philosophers' city, he formed the ambition of living there with his companions, according to Plato's laws. He therefore asked the Emperor and Empress, who venerated him, to revive it as the city of Platonopolis; but court intrigues prevented the city's establishment. However, Plotinus succeeded in his most profound ambition, to experience union with God. 'Four times while I was with him,' reports Prophyry, 'he attained that goal, in an unspeakable actuality and not in potency only.'[7] In a unique autobiographical passage, Plotinus himself writes:

> Many times it happened: lifted out of the body into myself; becoming external to all other things and self-encentred; beholding a marvellous beauty; then, more than ever, assured of community with the loftiest order ... After that sojourn in the divine, I ask myself how it happens that I can now be descending, and how did the Soul ever enter into my body ... (*Enneads* 8.1)[8]

Plotinus' great distant mentor is Plato. Plato, he believes, is the only philosopher who gathered the secrets of ancient wisdom, which he refused to reveal. Referring to his own sequence of metaphysical principles - the Good (or One), the Intellect, and the Soul - he says (after misquoting Plato) that this sequence, although made newly explicit, is not new, as is shown 'by the evidence of the writings of Plato himself'.[9] Understood correctly, Plato must always be right, he implies; but as he must be aware, he uses Plato selectively to stimulate or verify the ideas he, not Plato, wants to express. These ideas, together with similar ones of later philosophers, make a distinctive enough whole to merit the nineteenth-century name, Neoplatonism, by which it is now distinguished.

We know the life of Proclus mainly from the biography written by his disciple Marinus.[10] Sent by his father, a lawyer, to study in Alexandria, he was influenced by a revelation of the goddess Athena to study philosophy. In the then accepted way, he prepared for initiation in the mysteries of Platonism by first studying mathematics and Aristotle. Then he moved to Athens to study in the Platonic Academy. Eventually, he was chosen as the Academy's head. When lecturing to his many disciples, it is reported, his eyes glistened remarkably and his face glowed with 'divine brilliance'.[11]

In an empire that had become officially Christian, Proclus remained a pagan religiously free enough to observe the sacred days of both Egyptians and Greeks. He wrote on mathematics, astronomy, philosophy, literature, religion, mythology, and other subjects. Since he practised theurgy (etymologically, *divine work*), a magical form of salvation, it was necessary for him to know how to distinguish the different grades of gods, gods to whom he spoke in prayer and who sometimes answered him in his dreams, he believed.[12] Read now, his philosophy is fantastic and, in keeping with his period, very credulous. But it also reflects the exactness of the commentator on Euclid who gives clear, able explanations of the differences between geometrical definitions, hypotheses, postulates, axioms, and propositions. His writing is often garrulous and concerned with enumerations of abstract gods and powers, but his *Elements of Theology* is a succinctly authoritative systematization of Neoplatonic philosophy.

Proclus' ability to join credulousness and precision are expressed in two equally original and complementary aspects of his thought. One aspect is the conviction that Plato's authentic dialogues are to be regarded as divinely inspired. The dialogues should be studied to recover their clearly or obscurely expressed but always luminous, more than natural truths. These were transmitted in secret from most ancient times, hidden by the leader of the Academy, Arcesilaus, and then revealed again by Plotinus.[13] As will later be seen, Alfarabi and Maimonides read Plato in a similar light.

The second original aspect of Proclus' thought is his application of the Euclidean method to philosophy - Euclid, to him, is a member of the Platonic school. Like the Pythagoreans but in a more extensive, dialectical way, he fuses mathematics with ontology and for the first time demonstrates the universal harmony in the only language (he believes) that can express the harmony while proving its truth.[14]

In considering Plotinus' philosophy, we should keep in mind that the usual tension of his thought leads him to make unfamiliar uses of familiar themes and to multiply images - the inadequacy of which he stresses - in order to draw the reader deeper into his poetic yet reasoned thought.[15] Often, what he says is only a temporary fixing of his subtle, not very stable combination of many elements: Platonic Ideas (they are alive, as Plato would never agree), Platonic themes and images drawn un-Platonically together, the Aristotelian theory of potentiality and actuality, Stoic doctrines, and much else. As one sees in the essays that constitute his *Enneads*, Plotinus thinks systematically, but his system has to be pieced together to be seen as a whole. Like system makers generally, he is sure he is right, and yet, in the tradition of Socrates, he is able to tolerate objections and discussion.[16]

For what seem to me good reasons - clarity along with faithfulness to Plotinus' way of thought - my exposition will proceed along the following circuitous path:

1. A sketch of Plotinus' metaphysical hierarchy.
2. An explanation of the starting point of the hierarchy in terms of Plotinus' emotional and logical contrast between the evil nothingness of matter and the total goodness of the One.
3. An explanation of Plotinus' attempt to show how the One becomes apparently other than itself by 'shining forth' into Intellect and plurality generally.
4. An exposition of the Neoplatonic proof that existence is by nature the same as oneness/goodness.
5. An explanation of why the One is ineffable, but why negation is more successful than affirmation in hinting at it.

To present Plotinus' thought in the form of a neat hierarchy is to be faithful to its logic, but at the cost of misrepresenting its development and its open, shifting quality. The advantages of a hierarchical presentation are the clear perception of the whole it gives and its convenient approximation to what later thinkers make of Neoplatonism.

Plotinus' hierarchy reflects the attempts of Plato and of later Platonists to decide the order of the Ideas' importance, that is, of their reality or degree of *being*. Plotinus, who gives his levels of *being* the name of *hypostases* (meaning *what stands under, what supports, what is real*), begins his hierarchy, as I have said, with the One (*hen*), which he considers identical with the Good (*agathon*). In the universe of Plotinus, the One dominates everything and gives its nature and direction to everything, like Aristotle's god, not by any command but by virtue of its nature, which is that of complete actuality. Again like Aristotle's God, the One is affected only by itself in relation to itself. No other 'relation' is possible, because the One is everything there is. Everything that seems to be other than the One recognizes the relationship by desiring to return to what, in a sense, it has never left.[17] This process of going out while remaining identical is like a circle, says Plotinus, that expands and reflects in its circumference the nature of the pure, dimensionless point at its centre.[18]

To continue, for the moment still descriptively and dogmatically, the *One* by its shining forth, its radiation (*eklampsis*), creates all the universe without in any way growing diminished. First it radiates Intellect or Intelligence (*nous*), which differs from the absolute unity of *One* by being unity-in-plurality. Intellect remains one within the many that it includes in the same way as every mathematical theorem - and, therefore, the whole of mathematics - is contained in every mathematical theorem.[19] Intellect requires and is constituted by Forms or Ideas, which are not Plato's static pattern-Ideas, but have an inherent force (*dynamis*). This is the intuitive force of Intellect by means of which they shine or radiate forth their existence and, with the help of ideal matter, give rise to the world. As Socrates hesitantly suggested, individuals, too, have their Forms.

The hierarchy continues to develop in a logical sense; the relationships are timeless. By means of its spontaneous radiation, Intelligence comes to be Soul, which is unity and, as well, plurality (understood as the failure of unity to be evident).[20] Provided with Forms and ideal matter and radiating its force of life, Soul comes to be the World Soul (*psyche tou pantos*). This, the World Soul, produces the sensible world as a hierarchy of diminishing unity, power, and goodness. The diminution occurs, as will be explained, because the matter that pluralizes is 'truly not being'.[21]

The lower beings on the scale of existence need bodies (out of which they transmigrate into other bodies). To help themselves to survive and to recall what, as Intellect, they once knew and still know unconsciously, they also need sense organs. Unfortunately, bodies and sense organs are hindrances to higher understanding.[22] When the soul returns to intuitive contemplation, it no longer needs its lower faculties and the distinction between subject and object vanishes or nearly vanishes. Inspired by a final caution, Plotinus often says or implies that souls, although contained in the One and not really different from it, remain individual. Souls may, so to speak, see the light, or touch reality, or blend with reality, but the One remains transcendent.

So far, I have spoken for Plotinus dogmatically. Now, the dogmatic outline completed, we can begin in a philosophically more natural way. The beginning is that Plotinus dislikes matter and loves consciousness. He so opposes materialism that he is unwilling to think analytically about the atomists' arguments in its favour. Instead, he points out how heavy, helpless, and cruel matter always is. The weakness of matter is such that the heavier it is, the less able it is to lift itself up when it falls; simply material bodies are the most unpleasant to fall against and hurt more than living ones, which have natural sympathy.[23]

> 'This, then, is our argument,' Plotinus goes on, 'against those who place real beings in the class of bodies and find their guarantee of truth in the pushings and strikings and the apparitions which come by way of sense perception; they act like people dreaming, who think that the things they see as real actually exist, when they are only dreams.' (*Enneads* 3.6)[24]

Why, apart from emotion, does Plotinus speak of dreaming here? The philosophical answer is that he accepts Aristotle's definition of matter as in itself nothing but potentiality. To Plotinus, therefore, matter is *non-being*, 'a ghostly image of bulk, a tendency toward substantial existence ... invisible in itself' that needs something in addition to itself in order to produce body, for its being is only apparent and 'a sort of fleeting frivolity'.[25]

Since is it impossible for Plotinus to identify anything real with matter, he has to build his hierarchy on the immaterial principles he is familiar with, which are universals in the sense of Platonic Ideas. He follows clues from Plato - the clue in the *Republic* to the supremacy of the Good, the clues in the *Parmenides* to the existence of the One, and so on. Plotinus cannot help seeing reality as concentrated in oneness. He asks:

> What could anything be if it was not one? For if things are deprived of the one which is predicated of them they are not those things. For an army does not exist if it is not one, nor a chorus or a flock if they do not have their one, since the house is one and so is the ship, and if they lose

> it the house is no longer a house nor the ship a ship.[26] (*Enneads* 6.9.1)

Plotinus often speaks of the One with strong emotion and, helped by qualifying phrases such as 'as if', brings out the perfection of the One/Good's existence as superlatively self-centred love and superlatively self-sufficient introspection:

> He [the Good] penetrates, as it were, into his own interior as if in love with himself, the 'pure radiance', being himself this which he loves; this is to say that he brings himself into existence, since he is an abiding actuality ... He in a way holds fast to himself and in a way looks towards himself, and he is not as he chanced to be but as he will, and this will is not random or as it happened, for the will directed to the best is not random. (*Enneads* 6.8.16)[27]

It is easy to see that the description of the indescribable One taxes the resources of Plotinus' logic and language. So does the problem of the creative transition from absolute oneness to plurality, which is first revealed in the principle of Intellect. The Intellect comes second in Plotinus' hierarchy because he cannot accept the Aristotelian view that the supreme principle is thought thinking itself. But how, then, does thought, Intellect, come into being? The answer is that the One by its activity, so to speak, generates the Intellect. It generates because 'everything has an activity belonging to its substance'. For example, the substance of fire is heat, and 'when fire exercises the activity innate to its substance in remaining fire', another fire comes into being from the first. Similarly, the One remains what it is, but its perfection generates activity, which is the productive power of all things. 'For being is not a corpse, nor is it without life or intellection; Intellect and being are in fact the same.'[28]

The difficulty that Plotinus is facing is one that has come up before: Thinking implies a subject and an object, and the Intellect's thought would be an impossible, empty activity if it did not grasp some content:

> Intellection, which sees the intelligible object and turns toward it and is, in a way, being completed by it, is itself indefinite like vision, but it is defined by the object of its intellection ... Therefore, Intellect is not simple but many, and manifests a composition, though of course an intelligible one, and already sees many things. (*Enneads* 5.4.2)[29]

All such struggles by Plotinus to describe the indescribable and the beginnings of the creation of a plural world do little if anything to justify the metaphysical priority of oneness. Yet the Neoplatonists have a developed logical justification, which follows. Here I draw on the help of Proclus. I include in the proof only what is common to both Plotinus and Proclus, but to make the proof consecutive, I set the principles of the one in the quasi-mathematical order of proof of the other:[30]

> If you try to think of a world in which the quality of oneness or unity is absent, you see that the world is made possible by the metaphysical One - the One that is by nature absolutely and indivisibly single. The One does this by making it possible for any and all single things to exist. Every distinguishable thing is a *one* for itself, its oneness accounting for its unity and identity. Oneness is the same as existence. That is, no one thing can exist unless it is held together by being one rather than many. So if unity or oneness did not exist, nothing would exist, not even plurality or multiplicity, which is made up of ones.
>
> Suppose the opposite. Suppose that not *one* but multiplicity was basic. Then each of the ones or units this multiplicity contained would itself contain a multiplicity. Then the whole would be a multiplicity of multiplicities. And if, to go on consistently, each unit of each of this multiplicity of multiplicities contained a further multiplicity, the result would be that an infinite (uncountable large) number would contain an infinite number of infinities; and each of these infinities would contain an infinite number of infinities, and so on *ad infinitum*. Granting all this, we arrive at the following paradox: Although infinity, being large beyond the possibility of counting, is the greatest quantity that can be or be thought of, the infinity we have arrived at contains within itself an infinity of other infinities and is therefore infinitely smaller in quantity than the infinity of infinities it contains.
>
> Consider further: If the basic reality were not *one* but plurality, numbers could not exist. The reason is that all other numbers depend on *one*, the principle from which they are made by successive addition. If the basic reality were plural, it would be impossible to think about or know anything because every instance of knowledge is an instance of the identity of knower and known. Therefore, without unity, no such identity would be possible and all beings would be unknowable.
>
> Therefore, as has just been proved, *one* is the basic reality, which is shared in or 'participated in' by every being. This *one* of every being must be something purely one, with no admixture of anything more or different. If it were essentially both *one* and not-*one*, its unity would have to be imported into itself from another *one*, which would require the existence of another *one*, and so on *ad infinitum*. If such was the case, there would be no self-sufficient source of unity.[31]

> By this proof, we have in effect established not only that the One - capitalized again for its now established preeminence - is the basic reality, but also that plurality depends on unity and not vice versa. Because the One is one in more than a merely numerical sense, 'it does not come within the range of number' and is 'therefore not limited in relation to itself or anything else: since if it was it would be two'.[32] In theological terminology, the One is transcendent and yet immanent in whatever exists, singular or plural. As such, the One is the source of everything else:
>
> It is because there is nothing in it that all things come from it: in order that being may exist, the One is not being, but the generator of being ... The One, perfect because it seeks nothing, has nothing, and needs nothing, overflows, as it were, and its superabundance makes something other than itself. (*Enneads* 5.2)[33]

To grasp the hierarchy of Plotinus and Proclus, it is still necessary to prove that the One is identical with the Good. For this proof, all that is needed is to point out that everything strives to maintain its own being, whether by its physical resistance to destruction or by other means. Therefore *good* is defined as *remaining one*: 'Goodness, then, is unification, and unification goodness; the Good is one, and the One is primal good.'[34]

The world that results - the apparent world - reflects the real though diminished beauty of the One from which it is descended and to which it reverts. Therefore, says Plotinus, to the sensitive person the world is the sounds of all voices in a universal melody. Plotinus favours the metaphor of the sound of a voice over that of a sight because sound does not divide experience into parts but remains everywhere the same in the air, while sight, a pluralizing instrument, breaks things up and differentiates them. Yet although Plotinus makes use of the voice-metaphor, he at least once denies that in the fully real, intelligible world there can be any words or speech.[35]

Proclus alters, systematizes, complicates, and fills out Plotinus' sketch of the world. One alteration is in the conception of matter - potentiality or nonbeing - which Proclus is not able to identify with evil. Instead, evil is regarded as the absence of good, or the inability to receive good as purely as it is bestowed, or a perversion of the universal aspiration upward; but matter is not evil and evil is never real.[36]

Proclus' systematizing of Neoplatonism is heavily influenced, as I have said, by mathematics. In Neoplatonism generally, mathematical beings are intermediate between the highest beings - pure forms - and merely empirical beings. Unlike pure forms, mathematical beings are divisible and extend into series and geometric figures, but, unlike empirical beings, they are exact, perfectly ordered, immutable, and can be established and related to one another by irrefutable propositions.

According to Neoplatonism, sense perception draws the mind's attention to mathematical beings and so guides us to a higher, better unified understanding.[37] In his *Elements of Theology*, Proclus states, 'proves', and links his propositions like geometrical theorems, the purpose being to construct a system of exactly ordered levels of being. His propositions march along in their brave concatenation, from the first: 'Every manifold in some way participates unity', up to the last, the two hundred and eleventh, 'Every particular soul, when it descends into temporal process, descends entire: there is no part of it which remains above and a part which descends.'[38] The universal rationality that is represented in these propositions implies that everything has its exact level in the hierarchy of being and goodness. No level or place is empty or redundant, everything fits perfectly between what is above and below it, accidental beings or positions are ruled out. This 'axiomatics of perfection' has obviously strong echoes in the philosophies of Spinoza, Leibniz, and Hegel.[39]

Perhaps Proclus' most interesting complication of Neoplatonism is his development of the idea that at each level of reality there is a movement from cause to effect and back again. This movement is the reflection of the process by which *being* 'proceeds' creatively and becomes life, and life, by returning to *being*, becomes intelligence. Put in other, parallel terms, the movement is from permanence (eternal sameness) to manyness (eternal difference) and back to unity (the eternal limit in which difference is overcome).[40] But for our purposes it is best to avoid such complications and the technical justifications, which have not been mentioned, often based on Aristotle. Instead, we go on to the essential Neoplatonic proof that the One/Good is by nature indescribable.

The argument is that words can describe only a world of differences and limitations. In this world, thought is possible only if the thinker is split from the object of his thought. As the Neoplatonists see it, when a person thinks of something other than himself and becomes identified with this other, the person is more nearly split into two than when thinking of himself alone. Even thinking of himself alone, the person 'becomes a pair ... while remaining one'.[41] It is therefore inconceivable that the One or Good thinks at all - thinking is generated by desire and is a movement toward some good, so the Good does not think or need to think, 'for the Good is not other than itself'.[42]

Evidently, nothing can be predicated of the One because predication implies the distinction between subject and object and between what is and is not predicated. The consequence is that every description we give of the One is inaccurate and must be understood as though modified by the words *as if*.[43] Negation, however, is less inaccurate than the use of positive words. The reason is that whereas assertions cut up reality, negations tend to simplify things by making them less distinct, less well defined, more indefinite. It is only by such means as negation that it is possible 'to reveal the power of the One, which is incomprehensible and ungraspable and unknowable by particular intellects'.[44]

Bhartṛhari (c.450-510):[45]
'Grammar ... Is the Door to Salvation'

The power of language is one of the oldest themes of Indian thought. In Vedic India, speech is a great goddess and the sound of ritual chanting is essential to human welfare. In the Upaniṣads, Brahman, in the sense of absolute reality, is equated with language.[46] And among the orthodox, there is a universal belief in the absolute truth of Vedic words, taken to be heard and transmitted by the first sages. Given such beliefs, it is not surprising that language plays a considerable role in philosophy, as it does in Mīmāṁsā, Nyāya, and Buddhism, each of which sees it in the light of its characteristic interests. In Bhartṛhari, a grammatical view of language becomes quite central to philosophical speculation.[47]

Since nothing is really known about Bhartṛhari as a person and his dates, too, are uncertain, all one can do is repeat the old story that identifies him with a court poet of the same name. This poet wrote passionately regretful poems on separation from his mistress, gnomic verses on worldly wisdom, and very ambivalent poems on renunciation. The poet Bhartṛhari says, there are two worlds worth a man's devotion: the youth of beautiful women and the ascetic's forest retreat; between them, he cannot choose. While it seems psychologically improbable that the graceful, sensuous poet is the same person as the careful, not to say, pedantic philosopher, the two meet in their devotion to Brahman, unqualified reality. It has been remarked that the Chinese pilgrim I-Tsing knows of a Buddhist grammarian who seven times took a monk's vows and seven times returned to the laity. However, neither the poet nor the philosopher Bhartṛhari was a Buddhist. Besides, I-Tsing reports that the grammarian died about AD 650. This is later than it seems plausible to situate Bhartṛhari[48] the grammarian-philosopher, who was apparently known to the early sixth century Buddhist philosopher Dignāga.[49]

Though immodest about the powers of grammar-philosophy, Bhartṛhari takes little credit for his own perhaps great contribution to it. Of himself, he says merely, 'This summary of the science (of grammar) was composed by my teacher [Vasurata] after learning the various other systems and our own system.'[50] What at least appears to be new in his thought is the attempt to embody the grammatical point of view in a fully-fledged Indian philosophy, a *darśana*, a view of reality.

Bhartṛhari's chief work, *On Sentence and Word (Vākyapadīya)*, consists, Indian style, of some two thousand succinct verses distributed in three parts or chapters. It is accompanied by a commentary traditionally, though not certainly, by Bhartṛhari himself. Whoever was its author, it is often as obscure as the main text, sometimes because of its involved, ambiguous sentences, and sometimes because of its extreme terseness.[51] There are also three later commentaries. Because my aims are general, I will not point out the differences in point of view between the text and its commentaries, and I may sometimes include their interpretations without noting the fact. Bhartṛhari is more nuanced and more preoccupied with grammar than is evident from my account of him.

Here, there would be no point in more than mentioning the purely dogmatic characteristics of Bhartṛhari's thought. I mean by this his conviction that the Vedas are revealed truth, inseparable from the Sanskrit, the sacred primal language, in which they are embodied; that the whole, true knowledge is concentrated in a single word, the mantra *oṁ*, 'the creator of the worlds'; that the correctness of a grammatical form depends on its presence in scripture; and so on. Bhartṛhari is simply a believer, though by his own grammarian's standards, who trusts that there is a Yogic ability to enter others' minds and that certain persons have what to us appears to us miraculous powers of intuition.[52] Above all, he believes that 'even if the different doctrines disappeared and there were no authors to compose more, humankind would not turn away from the religious law that comes from revelation and tradition' (*revelation* translates the Sanskrit *heard*, applied to directly authoritative texts; *tradition*, in contrast, applies to texts whose authority is indirect because dependent upon memory).[53]

If we omit Bhartṛhari's more detailed linguistic analysis and concentrate on what remains in him of general philosophical interest, we find a group of interrelated ideas on the nature and importance of words and their organization into language:[54]

1. The 'word' - a term used by Bhartṛhari for language in general - has an inward meaning and an outward expression.
2. The 'word' cannot be an unrelated fragment, such as an unrelated word, but only the expression of a unitary meaning.
3. Regardless of how many words (in the usual sense) are used to express a thought, its meaning - its inward 'word' - is grasped as a whole by means of an intuitive act of understanding. Only after the whole is

grasped can its parts, its individual words, be understood individually and their relationships established by analysis.

4. All thought - awareness, perception, reasoning, judgement - and therefore all knowledge are infused with language and so cannot be separated from it.
5. The world as a whole is nothing other than a structure of interrelated meanings as grasped in 'words' (sentences), thoughts, and thought in itself. Essentially, therefore, the world is the same as the language by which it is known.
6. The relations between words express and are essentially the same as those between levels of reality. Because the structural relations between words are established by grammar, grammar is the science by which reality can be best understood.
7. The process by which our changing, material world was evolved from the unchanging, spiritual reality must have begun in and followed the process by which language was evolved.
8. In appearance, the power of the 'word' - language - is differentiated into that of space and of time, which is the more important. Time is the regulator of the order of empirical existence, the cause of every sequence and every instance of origin, existence, or destruction.
9. Even though the world is nothing other than a structure of potential, latent, and expressed meanings, its indwelling principle, that of the 'word' - language - is utterly beyond differentiation and expression in words.
10. Both knowledge and salvation come by way of grammar.

To go through these ideas, informally but more or less in the same order, we may begin by distinguishing between the inward sense and the outward meaning of the word, in the sense of language. Word and meaning are in origin the same. To be expressed, they must take on an outward form. The outward form is, of course, the sound, the word as audible to others. Among the experts in phonetics, says Bhartṛhari, there are various speculations on the cause of the audible word. Some believe that it results from the movement of the air that is set into motion by the speaker's effort and strikes the organs of articulation. Others believe that a word is made of the atoms that separate and transform themselves into shadow, light, darkness, and speech. Still other thinkers, among them Bhartṛhari, are more concerned with the transition from thought to sound. They believe that a spoken word is the transformation of the subtle form that is identical with the inner knower, who transforms himself in order to reveal his form. Becoming mind (*manas*) and 'ripened' by bodily heat, the knower enters the breath-air and 'colours' it with the mind's qualities. Splitting its densities into different sounds, the breath then merges with and manifests the sounds to speech, the phonemes.[55]

Bhartṛhari's conception of the process by which thought is turned into speech is, at once, psychological, physiological, and metaphysical. By his conception, the word/meaning is at first in the mind alone. The word/meaning is itself a power (*śakti*), but to be externalized it needs the help of the inward principle of activity, which is the inward breath. When this breath strikes the organs of articulation, they externalize the word, which then becomes the expressive, audible word. When the word's meaning is grasped by the hearer, the circuit of communication is completed - from one mind and inmost word to another mind and inmost word.[56]

If this explanation is accepted, it is evident that the words that are actually expressed have two elements. One is their meaning, their essential cause, and the other is the sequence of sounds by which the meaning is externalized. In spite of the sequence, the meaning remains whole and enters the listener's mind whole:

> From the differentiated, the undifferentiated word is born and it expresses the meaning. The word assumes the form of the meaning and enters into relation with it.[57]

Bhartṛhari uses an analogy to clarify his conception: When two sticks are rubbed together, their latent fire kindles visible fire and makes both itself and other things visible. In the same way, the word in the mind causes the audibly differentiated words to manifest themselves. That is, the inner word completes its germination, is impelled by the speech organs, manifests itself in different successive sounds, and makes both itself and other things known.[58] Put otherwise, the mind searches for a pronounceable form of the word and relates the form to what the speaker intends to say. The word then seems to change into the pronounceable form that is projected onto it. But the change of the meaning into something that is divided into sequential parts is only apparent. For the truth is, says the commentator, that the word in itself is neither sequential nor simultaneous but absolutely one; it appears to have parts only because of the sounds that are associated with it. In an image, the relation between the word in itself and the sound in which it is expressed is like that between the moon and the substratum of water that reflects it, the reflection being of something immobile but perceived as if joined to the water and moving with its waves. So, too, the word in itself takes on the properties of the sound that externalizes its meaning and appears to move with the sound, whether quickly, with moderate speed, or slowly.[59]

To use a Bhartṛharian image, the energy called the word is like the yolk in a peahen's egg: when the bird emerges from the egg, the essential uniformity of the yolk takes the form of a rich diversity. In still another image, when a painter wants to paint something made up of parts, such as the figure of a man, he scans it as a sequence of parts and then paints it sequentially, as he must. Likewise, a spoken word is at first perceived as a sequence and then, the sequence suppressed, as a partless mental unity superimposed on the sequential appearance. Just as the speaker first concentrates on the form of the word by isolating it from among other words, the hearer first tries to grasp the word's form in all the details that accompany it. Then the hearer makes out the meaning this particular spoken word has among the indefinitely many meanings that it can be given.[60]

The word has been used with caution, at first in quotation marks, to indicate that it is taken in the larger grammatical and metaphysical sense of the language in which a certain thought is expressed. This caution is the result of Bhartṛhari's contention that isolated words cannot be the basic units of meaning. To him, as is by now obvious, the true, unspoken word - the *sphoṭa*, as it is called - is a whole thought and is therefore more or less equivalent to a sentence, the least linguistic form that is complete.[61] It is the essence or fixed meaning that underlies the articulated sounds that reveal it. Opponents of Bhartṛhari claim that each word has its separate meaning, and that words or word-meanings become mutually dependent in the sentence and therefore convey a meaning the words do not have when taken separately. Put more technically, the separate words are universals that, when joined with other words, are restricted and made particular.[62]

In disagreement, Bhartṛhari begins with the whole sentence.[63] There are, he acknowledges, practical ways of establishing the meanings of individual words. A sentence can be analysed into components with artificial, secondary meanings. This is possible, he says, because a sentence is primarily an expression of some action or process, so that in analysing it we can at first isolate its main word - its main verb - to which we assign the meaning it has in that particular sentence.[64] All words that are not verbs are accessory and are assigned derivative meanings in order to help clarify the action or process.[65] But isolated words are indefinite, appear in many different forms, and have no fixed meaning, in the absence of which they are not only unclear but useless. Only words used together have fixed meanings; and the smallest clear, fixed meaning is that conveyed by a sentence: 'The nature of all word-meaning is dependent on the meaning of the sentence.'[66] It is true, Bhartṛhari acknowledges, that even before a sentence is finished, one grasps a vague meaning, not yet fit for communication; and as the later words appear, the first, vague meaning is abandoned or refined, up to the moment when the sentence ends and its full meaning can be grasped: the speaking out of the meaning and its grasping is itself a continuous process of clarification.[67]

Single words, says Bhartṛhari, can be compressed sentences, clear enough in the context of their use. But the simple addition of individual words each with its supposed independent meaning, or the analysis of sentences into such words, must fail.[68] The sentence is hardly more separate from its words than its words are from the sounds, the phonemes, that make them up. 'In the word there are no phonemes and in the phonemes there are no parts. Words have no existence separate from the sentence.'[69] If they try, those who are drinking sherbet can taste the flavour of each separate ingredient and understand what it contributes to the taste of the whole; yet the taste is indivisible. When the indivisible sentence cannot be understood all at once, the separate and unreal word-meanings are abstracted. As soon as the meaning of the sentence is grasped, the unreal meanings disappear.[70]

In order that language should fulfil its natural purpose, the conveying of meaning, it must be subjected to certain restrictions. Self-reference, for one, must be restricted. To give an example: When doubt is cast on something particular and then someone casts doubt on this doubt, the doubt loses its original form. Just so, when a cognition is determining the nature of its object, 'it cannot become the object of another cognition. If it did, it would lose its proper nature and become the "object" of another cognition.'[71] The purpose of cognition is to determine the nature of the external object, and while it is making the determination, it cannot itself be made the object of knowledge.

> The sentence 'all that I am saying is wrong' is not literally meant. If what it says is wrong, the point in question would not be conveyed. What is expressive cannot at the same time be the expressed. (VP 3.25-26)[72]

All such self-reference is illegitimate, says Bhartṛhari, because it nullifies the nature - the purpose and usefulness - of language, which it prevents from expressing anything. The sentence, 'All that I am saying is wrong' is not meant to be taken literally. If taken literally, it becomes impossible for the person who uses it to convey the point that he intends, that what he said just before is wrong - his language is not allowed to be a means of communication. What is used to convey a certain content cannot be turned back on itself as if it is fulfilling an intention other than its own. A function ought not take on another, reflexive one that leads to self-contradiction or infinite regress.[73]

This is all, in Bhartṛhari's eyes, part of an explanation of how language functions. The explanation, he is sure, is that given by the school of grammar, the tradition of which has been kept by an unbroken line of learned men - to give up tradition and rely on inference alone is like running ahead on an uneven path with one's hand extended but one's eyes quite blind.[74] As a partisan of the grammatical tradition, Bhartṛhari points out how fallible ordinary speech is and how limited the standard Indian instruments of knowledge are. The meanings expressed in words are understood differently at different times and by different persons. Much the same is true of perception: different persons perceive the same things differently and the same persons perceive them differently at different times. 'Therefore, both the comprehension and report of people who have not seen the truth (about things) are defective, unreliable, and perpetually inconsistent.'[75]

However, even the authority of sages is not enough. They perceive the true nature of things, but their perception cannot be transferred into words or put to use in ordinary life. Perception as a means of knowledge is obviously fallible, for each of us is subject to sensory illusions: We all see the sky as a surface even though logic teaches us that there is no surface there. So perception alone is inconsistent and unreliable. But logic, too, brings no certainty: Clever logicians are contradicted by still cleverer ones. And when merely pragmatic persons try to understand what is beyond ordinary words, they fail (as the science of grammar does not).[76]

Yet human beings do have a spontaneous, reliable kind of knowledge. This is intuition, *pratibhā*. Having the connotation of light and shining forth, it is the understanding that flashes lightlike on the mind.[77] Though essential to life, it is inexplicable:

> Having been formed from the function of one's inner self, its nature is not known even to the person. It effects the fusion of the (individual) word-meanings, without itself being logically thought out ... In the matter of the knowledge of what to do, no one transgresses it ... The whole world looks upon it as authority (for their conduct). Even in animals the knowledge of the beginning of behaviour dawns by virtue of it ... Who alters the note of the cuckoo in the spring? By whom are creatures taught to make nests and so on? Who directs animals and birds in functions like eating, loving, hating and leaping ...? (VP 2.144-47, 149-50)[78]

What is translated as *intuition* evidently includes instinct. The rather miscellaneous inclusiveness of Bhartṛhari's concept is evident in his enumeration of the six various kinds of *pratibhā* 'as obtained (1) by nature (2) by action (3) by practice (4) by meditation and (6) as handed down by the wise'.[79]

Some of the examples given by the earliest commentary may not seem fitting. Such an example, of intuition 'by nature', is the intrinsic tendency of primordial matter to evolve. As examples of intuition 'by practice', Bhartṛhari cites the ability of experts to know where to dig for water or how to distinguish what is genuine and false in coins or diamonds. Being indefinable, this intuitive ability to assess cannot be communicated to others. It is certainly no inference; if it was, it would have an inference's ordinary grounds.[80]

I will give no further examples, because what is more immediately relevant to our subject is the intuitive ability to understand language. It is intuition that enables us to grasp a whole meant object from a word that refers to part of it or to grasp the meaning of words with missing letters.[81] But above all, intuition is the ability to make one sense of all the words of a sentence, and to do this even before the sentence is finished. A single heard letter may yield the sentence's whole meaning. Intuition also gives us the ability to recognize that certain kinds of sentences, such as those meant to praise or blame, have a meaning different from that indicated by their analysis into separate words.[82]

If it is conceded that thoughts are always, to begin with, latent words, then all thoughts and all forms of knowledge are inseparable from words and from Word:

> All knowledge of what is to be done in this world depends upon the word ... All knowledge is, as it were, intertwined with the word ... It is this that is the basis of all the sciences, crafts and arts. Whatever is created due to this can be analysed (and communicated) ... Thanks to it, whatever is produced can be classified ... The consciousness of all living beings ... is of the nature of the word; it exists within and without. (VP 1.121, 123, 125, 126)[83]

Some explanation is in order. Knowing of every kind requires the ability to identify and remember. This ability is inseparable from words because it is inseparable from cognition, which is inseparable from words. Such is most obviously the case when technical distinctions are made by means of specially invented terms. The technical vocabularies distinguish what is necessary for action and might not otherwise be grasped. To understand music one must learn its science; merely listening to the different types of music notes is not enough.[84] Even that which in some weak sense exists 'is as good as non-existent as long as it does not come within the range of verbal usage'.[85] Even fantasies and imaginary objects and logical impossibilities are endowed with what measure of reality they have when they are brought to mind by words. And even children understand what they do because they have understanding in a vague sense, the result of 'the residual traces of words from their former births'.[86]

Without words, Bhartṛhari is saying, knowledge is indeterminate. For example, when, walking quickly, one steps on grass and clods of earth, one has an unverbalized awareness or sensation of them. The grass is no more than tinged with awareness, the 'word-seed' is only latent, only ready to sprout. Simply as awareness, this not yet verbalized awareness is speech, but only in the inarticulate form that cannot denote, and the sensation has little if any effect.[87] Without determination by words, awareness is too weak to be merit the name. To Bhartṛhari, therefore, what has not been verbally identified is to that extent unknown or almost unknown, near to unknowable, and incommunicable and unremembered. Nothing can be grasped when there is no language. In its absence, there are no latent words, no spiritual activities, no sciences, no arts, no crafts, no communication, no consciousness, no memory, and no self.[88]

If so, the word is essentially the consciousness and self-consciousness of all beings. We can therefore dare to say that the world is indistinguishable from the words by which it is known. And if we take one more metaphysical step along with Bhartṛhari, we conclude that the world is nothing other than a structure of meanings as fixed in language. Therefore

> the power which creates and regulates this universe rests on words ... 'It is the word which sees the object, it is the word which reveals the object which was lying hidden, it is on the word that this multiple world rests ...' (VP 1.118 and comm.)[89]

In an adventurous but reasoned sequence, it is concluded that the world is made of words, that expressed words are made of latent words or meanings, and that meanings are made of consciousness. This conclusion is the burden of the verse with which Bhartṛhari's book begins:

> Brahman without beginning or end, Word-Principle (*śabdatattva*), Immutable Phoneme (*akṣara*), who appears as ... the objects, from whom the world proceeds. (VP 1.1)[90]

The commentary goes on to say that in all its apparently different manifestations, the original material of the word remains. This is because we identify objects with the words by which they are named. And because reality, Brahman, must create the world in its own nature, and because the world's nature is words, Brahman itself must be the Word of words. And because Brahman creates words in their audible form, Brahman is the phoneme as well:

> The Brahman is called phoneme (*akṣara*) because it is the cause of the phonemes ... What is meant by 'it appears as the objects' is this - what is called appearance ... is the assumption by the One, without losing its one-ness, through apparent diversity, of the unreal forms of others. It is like the appearances in a dream. (VP 1.1. comm.)[91]

The understanding of language in this metaphysical sense teaches us that latent speech is the permanent possibility of cognition and cognition is the permanent possibility of the existence of the world. This is simpler to understand if we analyse how we know anything at all. The first source of possible knowledge is the highest universal, the Word, of which the lower universals are manifestations.[92] To know any particular thing is to know its indwelling, lower universal (*jāti*). To know this universal is first to know the equivalent word and, by its means, the universal that dwells in whatever we are perceiving.

Take an undramatic Indian example. When we know, that is, cognize a jar and say 'This is a jar', the word *jar* in the sentence is identified with the cognition and with the jar that is the cognition's content. And because the reality of the cognition and its object, the jar, is their indwelling universal - the jar itself is unreal - it makes sense to say that the word is the same as the jar. So the jar is derived from the word, and not the opposite - jars of clay come from verbal *jars*.[93] Conveying its form to an object, the word pervades and dyes it with its particular meaning, its variety of powers to function - in the case of a pot, the powers necessary to carry water.[94]

This raises a question: By what means does the power of the Word come to expression in the powers of the world of human experience? The answer is that time or, rather, Time is the causal agent.[95] It must be kept in mind that the many forms of the different objects of human experience do not affect the unity of the cognition that encompasses them all. They are differentiated, that is, projected as the contents of the empirical world, by Time's creative power, *kālaśakti*, which is that of the Word.[96] By its power, Time regulates the Word's transformations and, by doing so, is the cause of all origins, existences, and destructions. Time pulls the wires of the world automaton (the world as a robot, or perhaps as a puppet show). The action of Time is by means of its two eternal aspects, the one, the giving of permission for things to appear, and the other, the withholding of the permission. If not for Time's aspects, things would all be born at once and chaos result. Together with its associates, the heavenly bodies, Time enables the conventional fixing of seasons, months, days, and hours. And because all things happen by their association with different times, Time is the efficient cause of all particular effects. As such, it draws real objects out of potentiality as regularly as a waterwheel draws water from a river. In a metaphor, Time brings about changes in things just as a river places and displaces grass, leaves, and creepers on its banks.[97]

Such metaphors apart, objects come into being when universals trying to find themselves an 'abode' - an object to inhere in - guide potential causes into producing particular effects, which are the different particular objects. Having 'prompted' the appearance of these objects, the universals manifest themselves in them 'like reflections in clear water'. To all appearances, they are identical with the objects and are designated by the words that identify them.[98] Bhartṛhari sums up by saying, 'Time is the very soul of the universe. Hence it is identified with activity itself.'[99]

Brahman itself is, of course, atemporal.[100] Neither unity nor multiplicity can be attributed to it. Neither could be explained without the other, and since they are mutually dependent, disproving one would disprove the other. Clearly conceived, unity and difference are the same. The One Reality has no before or after even though 'it shines with the divisions of time'.[101]

This is the claim according to which everything in the world consists of outward appearances in which there resides the one source of all transformations.[102] Therefore it is the Word, with its temporal power and transformations into words, that makes the world. Therefore, all the sciences depend upon the correct, grammatical formation of language, so that grammar is their universal basis. Corrupted forms of language have no fixed meaning and are the cause of sin.[103] 'It is the word, language, that is the sole teacher.'[104] The very ability to conduct a reasonable discussion depends on the immutable power of words to convey a meaning based on their primary sense and contexts and grammatical use.[105]

Assuming that the world and language always go together, to discover how the world evolved, we turn to grammar to see how language evolved. In the beginning, as we have learned, there was, as there is and will always be, the Word. The Word, which is Brahman,

> is endowed with all powers, which are neither identical with nor different from it; and it has two aspects, that of unity and that of [apparent] diversity ... It is, in all states, unaffected by beginning and end, even though the manifestations appear in worldly transaction in a temporal and spatial sequence. (VP 1.1 comm.)[106]

There is a riddle here: Although words produce opposite effects, they cannot be opposed to one another because they all exist simultaneously in the same one Word or Brahman that constitutes them completely. The answer to the riddle is that words can have no existence apart from the Word, and their paradoxically separate-non-separate existence-non-existence is a differentiation into forms that are unreal as dreams are unreal.[107]

The science of grammar can lead us to see how the Word (Language), the metaphysical principle of the everything, leads us to salvation.[108] Salvation is an ascent toward the Word that reverses the direction of the evolution of language and of the world - in language, the ascent goes from articulated speech to the potential speech in the mind, to the total of word forms tinged with successiveness but going beyond it, and, finally, to the pure principle of language. Whoever aspires to this grammatical salvation must learn to see the constant Word without distinction. Such seeing requires intuition in its fullest sense, free of all differences.[109] To get to the Word in itself, one must learn to suppress sequential thinking. As has been hinted, to do this requires perseverance in the use of grammatically correct forms. The incorrect formation and sequencing of words obscures the light of the Word and the Vedic truth. Correct speech - Word-yoga - allows the light, the truth, to shine through, allows the sequences of words and thoughts to be suppressed, allows the knots of ego to be cut and union with the Word to be consummated.[110]

DISCUSSION

If Plotinus had gone to India, as he wanted, he would have found descendants of Uddālaka and Yajñavalkya with whom, interpreters permitting, he would have been able to find a common philosophical language. It is not likely, however, that the grammarian-predecessors of Bhartṛhari would have been among these discussants of the ultimacies of *being*. I say this because the grammarians' initially linguistic approach might have made them seem alien to Plotinus. The difference would have been sharpened if Proclus were substituted for Plotinus (I have not imagined Plotinus and Proclus together - psychological considerations make me doubt if Plotinus could suffer what Proclus made of his thought). Bhartṛhari and Proclus would no doubt exchange accounts of their respective ancient, holy traditions, submerged for a time and then recovered. In the debate between them, it would be the semantic, grammatical order of the Indian against the Greek's mathematical translation of semantic order. The Indian, struggling with the obdurate complexities of linguistic use, would often be tentative and allow for different possibilities. The Greek, putting each statement to a 'mathematical' true-false test, would claim unambiguous success at every step - in principle, his metaphysical mathematics rules out ambiguity. I do not know if the Indian would pit his mantras against the Greek's theurgic magic. But for the two, as for us, the philosophic crux might well be the possibility of casting linguistically conveyed meanings into forms with purely mathematical modes of proof.

It is therefore all the more interesting to see how closely the two very different lines of hierarchical idealism converge. They converge, first, in the logic of hierarchy, which, for Plotinus and Proclus, is that of the hierarchical reciprocity between Ideas and, for Bhartṛhari, that, ultimately, of the hierarchical reciprocity between meanings. Lines of thought also converge in the timeless *being* - Word, Brahman, One, Good - at the head of the hierarchy. To both sides, *being* is unlimited consciousness in an infinitely simple unity-in-variety beyond human conceiving and expression. There is also convergence in the idea of the creative overflow of *being* that projects and differentiates the world that, despite appearances, is identical with *being*.

How great a difference is there between saying, for Plotinus and Proclus, that reality is a structure of Ideas that proceed from and revert to their source, with which understanding humans should try to unite, and between saying, as does Bhartṛhari, that reality is a structure of meanings that proceed from meaning as such, with which understanding humans should try to unite? I pose the question without answering it. I have no doubt, however, that in both philosophies there is the same belief in creative emanation from *being*, which, as best we can understand it, is a consciousness that, in false appearance, grows increasingly distant from itself as its creations grow increasingly material. Nor do I doubt the comparison of the two doctrines of causality that say that everything is related to everything by the force exerted by conscious meaning.

In the end, what is most striking in hierarchical idealism is the intellectual and emotional need to identify being-as-such, the source of everything, with the quintessence of individual consciousness, our nucleus as human individuals. The moral that I would draw - differently from the two philosophers - is that what we most yearn to get close to is an inconceivably great version of ourselves.[111] This is, of course, the *ātman*-Brahman theme we have met before and will meet again, in different guises.

Notes

1. Plotinus, *Enneads* 4.8.8, 5.1.12.3-14. See M. Atkinson, *Plotinus: Enneads V.1*. Oxford: Oxford University Press, p. lxvi.
2. See the analytically clear expansion of *Enneads* 5.1.5.16-19 - an obscure use of Plato's unknown number theory - in Atkinson, *Plotinus: Ennead V.1*, pp. 20-1 and 109-23.
3. 'History and Intellectual Background of Philosophy', in: A.H. Armstrong (ed.), *The Cambridge History of Later Greek and Early Medieval Philosophy*. Cambridge: Cambridge University Press, 1967. A. Dihle, *Greek and Latin*

Literature of the Roman Empire. London: Routledge, 1994. E.R. Dodds, *The Greeks and the Irrational*. Berkeley: University of California Press, 1952. R.T. Wallis, *Neo-Platonism*. London: Duckworth, 1972.
Biography: Porphyry, 'On the Life of Plotinus and the Order of His Books', in: Plotinus, *Enneads*, tr. A.H. Armstrong, vol. 1; Porphyrius, 'Über Plotins Leben', vol. 5 of *Plotins Schriften*, tr. R. Harder. Hamburg: Meiner, 1958 (annotated).
Monographs: H.J. Blumenthal and A.C. Lloyd (eds), *Soul and the Structure of Being in Late Neoplatonism*. Liverpool: Liverpool University Press, 1982 (contains a chapter on Proclus). A. Charles-Saget, *L'architecture du divin: mathématique et philosophie chez Plotin et Proclus*. Paris: Les Belles Lettres, 1982. P. Hadot, *Plotinus or the Simplicity of Vision*. Chicago: University of Chicago Press, 1989. A.C. Lloyd, *Anatomy of Neoplatonism*. Oxford: Oxford University Press, 1990. J.M. Rist, *Plotinus*. Cambridge, Cambridge University Press, 1967.
Translations: Plotinus: *Enneads*, tr. A.H. Armstrong, 7 vols. Cambridge: Harvard University Press, 1966-88. *Enneads*, tr. S. MacKenna, abridged J.M. Dillon. London: Penguin, 1991.
Proclus: *A Commentary on the First Book of Euclid's Elements*, tr. G.R. Morrow. Princeton: Princeton University Press, 1970. *The Elements of Theology*, tr. E.R. Dodds, 2nd ed., London: Oxford University Press, 1963. *Proclus' Commentary on Plato's* Parmenides, tr. G.R. Morrow and J. Dillon. Princeton: Princeton University Press, 1987. *Théologie platonicienne*, tr. H.D. Saffrey and L.G. Westerink, 4 Vols. Paris: Les Belles Lettres, 1968-81.
Annotated Partial Translations: M. Atkinson, *op. cit.* J. Bussanich, *The One and Its Relation to Intellect in Plotinus*. Leiden: Brill, 1988.

4. Porphyry, 'On the Life of Plotinus and the Order of His Books' 1; in: *Plotinus*, tr. Armstrong, vol. 1, p. 3.
5. *Ibid.*, 3; tr. p. 9.
6. *Ibid.*, 8; tr. p. 29.
7. *Ibid.*, 23; tr. p. 71.
8. Plotinus, *Enneads*, tr. MacKenna, p. 334.
9. Plotinus, *Enneads* 5.1.8, tr. Atkinson, *op. cit.*, p. lxii and notes, pp. 186-92. On Plotinus' faithfulness and faithlessness to Plato see also Dodds, 'Tradition and Personal Achievement in the Philosophy of Plotinus', *Journal of Roman Studies* l, 1960, pp. 1-7; and Rist, *op. cit.*, pp. 179-87.
10. There are accounts of Proclus' life in the respective introductions to Morrow's translation of Proclus' commentary on Euclid and to Saffrey and Westrink's translation of the *Théologie platonicienne*, which also describes the Platonic, that is, Neoplatonic school of Athens, which Proclus came to head.
11. The life of Proclus 'presented the great teacher's progress towards moral perfection via exactly those stages which Neo-Platonist ethics had postulated since Plotinus. At that time moral perfection was seen as liberation of the spirit from all restrictions of corporeal existence in the empirical world. This liberation was to manifest itself in a life of intense prayer, in experiences of

mystic vision which transcend discursive thought, and in the ability, created by the close link with purely spiritual beings, to perform actions which seem impossible according to the laws of the material world...In this late period of Classical culture the philosopher occupies that position which is given to the saint in the Christian community.' (Dihle, *op. cit.*, p. 488).

12. Plotinus, who never uses the term *theurgy*, credits contemplation with the ability to defend the philosopher against magical attacks. In relative terms, he is a rationalist and denies that magic can affect the higher souls of either men or gods. *Ennead* 2.9, 'Against the Gnostics', is a notable attack against Gnostic doctrines and magical practices. The Gnostics, he says, pervert Plato and hate the material world and the great beauty it embodies. The Gnostics, he protests, refuse the traditional method of salvation through virtue and slowly acquired wisdom and turn instead to absurd magical procedures. But the Neoplatonic partisans of theurgy, who include Proclus, consider that Plotinian *theoria* is only contemplative, that is, passive, while theurgic acts allow its practitioners to come into contact with divine forces. According to Proclus, theurgy, which is known by revelation, is a power higher than all human wisdom. His essay *On the Hieratic Art* expounds the not un-Indian idea that by the principle of correspondence every part of universe mirrors every other. According to the theurgists, the manipulation the appropriate objects brings them into contact with the gods the objects represent (Dodds, *op. cit.*, p. 283-311; Wallis, *op. cit.*, pp. 70-2, 106-20).
13. Proclus, *Théologie platonicienne*, vol. 2, tr. Saffrey and Westerink, pp. clxxxvii-xix; and the text of the *Théologie* 1.1-5, pp. 5-26, with the corresponding notes, pp. 130-9. The only dialogue explicitly rejected by Proclus is the *Epinomis* - see note 3 (for p. 23), p. 138.
14. Charles-Saget, *op. cit.*, pp. 189-90, 209. Proclus, *A Commentary on the First Book of Euclid's Elements*, tr. Morrow, pp. xxvi-ii.
15. Bussanich, *op. cit.*, p. 1, 3-6.
16. Armstrong, *op. cit.*, pp. 211-12.
17. Some of the following description of the philosophy of Plotinus and Proclus draws from B.-A. Scharfstein (ed.), *Philosophy East/Philosophy West*. New York: Oxford University Press and Oxford: Blackwell, 1978, pp. 212-13.
18. Plotinus, *Enneads* 6.9.8.1-7.
19. *Ibid.*, 5.9.8.
20. *Ibid.*, 3.8.11.
21. *Ibid.*, 3.6.7; tr. Armstrong, vol. 3, p. 241.
22. *Ibid.*, 4.3.18-19, 4.4.1-2.
23. *Ibid.*, 3.6; tr. Armstrong, vol. 3, p. 237.
24. *Ibid.*, tr. Armstrong, vol. 3, p. 239.
25. *Ibid.*, 3.6; tr. Armstrong, vol. 3, pp. 241-3.
26. *Ibid.*, tr. Armstrong, vol. 7, p. 303.
27. Tr. Bussanich, *op. cit.*, p. 202.
28. *Ennead* 4.7.2; tr. Bussanich, *op. cit.*, pp. 8-9.
29. *Ibid.*, p. 8. See comm. p. 9.

30. The following account of justification of the priority of oneness in the philosophy of Plotinus and Proclus is mostly drawn from B.-A. Scharfstein, *Ineffability*. Buffalo: State University of New York Press, 1993, pp. 149-53. For the sources in Proclus, see the first six propositions in *The Elements of Theology*, tr. Dodds, pp. 3-7; and Saffrey and Westerink, *op. cit.*, vol. 2, book 2, chap. 1, pp. 3-14.
31. Proclus, *The Elements of Theology*, tr. Dodds, p. 5.
32. *Enneads*, 5.5.11; tr. Armstrong, vol. 5, p. 17.
33. *Ibid.*, tr. Armstrong, vol. 5, p. 59.
34. Proclus, *op. cit.*, tr. Dodds, p. 17.
35. Bussanich, *op. cit.*, pp. 97-8, citing *Enneads* 2.2.17.64-75; 5.1; 6.4.12; 4.3.18.13-20.
36. Wallis, *op. cit.*, p. 157.
37. Proclus, *A Commentary on the Book of Euclid's Elements*, tr. Morrow, p. xxxiii.
38. Proclus, *The Elements of Theology*, tr. Dodds, pp. 3, 185.
39. Charles-Saget, *op. cit.*, pp. 231-4. The phrase 'axiomatics of perfection' is from p. 253.
40. Proclus, *op. cit.*, props. 35-51, with Dodds' notes. See also his notes to prop. 65.
41. *Enneads* 5.6.1.
42. *Ibid.*, 5.6.5.
43. *Ibid.* 6.8.13.
44. Proclus, *A Commentary on Plato's Parmenides*, tr. Morrow and Dillon, p. 427.
45. *Indian philosophy of language*: M. Biardeau, *Théorie de la connaissance et philosophie de la parole dans le brahmanisme classique*. Paris: Mouton/Ecole pratique de hautes Etudes, 1964. H. G. Coward, *Sphoṭa Theory of Language*. Delhi: Motilal Banarsidass, 1980. H.G. Coward and K.K. Raja, (eds), *Encyclopedia of Indian Philosophies*, vol. 5, *The Philosophy of the Indian Grammarians*. Princeton: Princeton University Press, 1990 (with an extensive introduction, detailed paraphrases, and many blank pages containing names of members of the school about whom nothing but their names is known). B.K. Matilal, *Epistemology, Logic, and Grammar in Indian Philosophical Analysis*. The Hague: Mouton, 1971; *The Word and the World*. Delhi: Oxford University Press, 1990 (a lucid general introduction). D.S. Ruegg, *Contributions à l'histoire de la philosophie linguistique indienne*. Paris: Boccard, 1959.
Monographs: R. Herzberger, *Bhartṛhari and the Buddhists*. Dordrecht: Reidel, 1986. K.A.S. Iyer, *Bhartṛhari*, Poona: Deccan College, 1969; *The Vākyapadīya: Some Problems*. Poona: Bhandarkar Research Institute, 1982. G.N. Sastri, *The Philosophy of Word and Meaning*. Calcutta: Calcutta Sanskrit College, 1959.
Translations: Bhartṛhari, *The Vākyapadīya*, cantos (chaps.) 1, 2, tr. K.R. Pillai, Delhi: Motilal Banarsidass, 1971 (Since Iyer [see just below] omits verses 108 through 115 in Bhartṛhari's first chapter, the numbering in Pillai's equivalent translation differs from 108 and on). Bhartṛhari, *Vākyapadīya brahmakaṇḍa*, tr.

M. Biardeau. Paris: Boccard, 1964. *The Vākyapadīya of Bhartṛhari*, chap. I, tr. K.A.S. Iyer. Poona; Deccan College, 1965; *The Vākyapadīya of Bhartṛhari*, chap. 3, part 1, tr. Iyer. Delhi: Motilal Banarsidass, 1971; *The Vākyapadīya of Bhartṛhari*, chap. 3, part 2. Delhi: Motilal Banarsidass, 1973; *The Vākyapadīya of Bhartṛhari, kaṇḍa* (chap.) 2. Delhi: Motilal Banarsidass, 1977 (this volume has not been available to me). Passages from Bhartṛhari that appear here have sometimes been modified by me from one or more of the above translations. Aklujkar, 'An Introduction ...' (see just below) says that because Iyer's translation - the only complete one - is meant to convey the general sense of the original, it often does not attempt to get at the exact meaning of a knotty Sanskrit passage. Biardeau's French translation of the first book is more helpful in this regard, Aklujkar says (pp. 19-20).

Articles: A. Aklujkar, 'An Introduction to the Study of Bhartr-Hari', in: Bhate and Bronkhorst, eds, *Bhartṛhari*. S. Bhate and J. Bronkhorst (eds), *Bhartṛhari: Philosopher and Grammarian*, Delhi: Motilal Banarsidass, 1994 (First published Bern: Lang, 1993). J.E.M. Houben, 'Bhartṛhari's *Samaya* / Helārāja's Saṁketā', *Journal of Indian Philosophy* June 1992. F. Staal, 'Sanskrit Philosophy of Language', in *Current Trends in Linguistics* 5, 1969. F. Tola and C. Dragonetti, 'Some Remarks on Bhartṛhari's Concept of Pratibhā', *Journal of Indian Philosophy*, June 1990.

46. E.g. in the Bṛhadāraṇyaka Upaniṣad 4.1.2.
47. For a historical sketch of the place of language in Indian philosophy, see Coward and Raja (eds), *op. cit.*, pp. 3-32. See also Ruegg, *op. cit.*
48. The dating and sequence of philosophers proposed by Frauwallner and accepted by Iyer are as follows: Vasurata, Bhartṛhari's guru, AD 430-90; Bhartṛhari, 450-510; Dignāga, 480-540.

    On the poet Bhartṛhari and his possible identity with the grammarian, see: Iyer, *Bhartṛhari*, p. 2. D.H.H. Ingalls, *An Anthology of Sanskrit Court Poetry*. Cambridge: Harvard University Press, 1965, pp. 40-3. B.S. Miller, *The Hermit and the Love-Thief: Sanskrit Poems of Bhartṛhari and Bilhaṇa*. New York: Columbia University Press, 1978. A.K. Warder, *Indian Kāvya Literature*, vol. 4. Delhi: Motilal Banarsidass, 1983, pp. 121-2. M. Winternitz, *A History of Indian Literature*, vol. 3. Delhi: Motilal Banarsidass, 1963 [1922], pp. 256-8.
49. In defence of the identification of the poet with the grammarian, it is argued: 'The poetry of Bhartṛhari shares with the grammatical philosophy of Bhartṛhari, as expounded in the *Vākyapadīya*, ideas with terminology drawn from traditional systems of Vedānta and Sāṅkhya metaphysics, as well as from classical Yoga psychology. Also common to the poetry and the philosophy is a critical interest in the nature of time. In Bhartṛhari's philosophy, time is a creative power that is responsible for the birth, continuity, and destruction of everything in the universe. Much of the poetry shows a pessimistic preoccupation with the beginning and end of things. The inevitability with which time is said to ravage the life of man may conceivably represent the poetic expression of the futility and dejection upon a philosopher profoundly impressed with the power of time. Good arguments are put forth to date Bhartṛhari the philosopher to the fifth century AD. The core of the collected poems attributed to Bhartṛhari

also probably dates to this period.' (Miller, *op. cit.*, p. 23)
In India, it was fairly usual for the philosopher to be the poet as well. Śaṅkara, for example, wrote poetry, though all on the side of renunciation. The twelfth century sceptic, Śrīharṣa, who will be discussed later, wrote both his austerely philosophical text and an ornate, sexually explicit epic of love, which has (of course) been given an allegorical interpretation.

50. VP 2.482; tr. Pillai, p. 146.
51. Iyer, *The Vākyapadīya of Bhartṛhari*, chap. 1, pp. xii-xv.
52. *Ibid.*, pp. 84, 90.
53. VP 1.133. For Bhartṛhari's attitude toward revelation and tradition, see Iyer, *Bhartṛhari*, pp. 92-7.
54. Some of the following account of Bhartṛhari's philosophy is drawn from B.-A. Scharfstein, *Ineffability*. Buffalo: State University of New York Press, 1993. The principles of Bhartṛhari's philosophy are given a particularly brief, clear, orderly summary, to which there is no equivalent in Bhartṛhari himself, by Ashok Aklujkar in Coward and Raja, *op. cit.*, pp. 122-6. Aklujkar's summary makes Bhartṛhari sound more organized and modern than he is. His following, detailed summary brings out the linguistic side of Bhartṛhari's philosophy with particular clarity. The account I give below omits many of the points of even Aklujkar's brief summary, as it does of Bhartṛhari's own grammatical observations, but it remains relatively close to the way in which Bhartṛhari and his commentators expressed themselves.
55. VP 1.107-15.
56. Iyer, *Bhartṛhari*, pp. 66-7. Iyer, *Bhartṛhari*, p. 153; Iyer, *The Vākyapadīya: Some Problems*, pp. 15-17.
57. Commentary to VP 1.44; tr. Iyer, chap. 1 (i.e., vol. 1), p. 53).
58. VP 1.44-46.
59. VP 1.47-49.
60. VP 1.51-53, with comm. Iyer, Bhartṛhari, pp. 153-4.
61. The literal meaning of *sphuṭ, from which sphoṭa* is derived, is *to burst forth. Sphoṭa* is the idea that bursts forth in or, put in calmer words, manifests itself to the mind when one hears someone speaking. The successively larger units of spoken language are the phoneme or *varṇa* (the shortest articulated sound), the (ordinary) word or *pada*, and the sentence or *vākya*. Bhartṛhari believes that the *sphoṭa* can be at least vaguely revealed in any of these, but his basic contention is that the unrelated phoneme has no meaning at all (Iyer's explanation, VP 3.1, p. 2; and VP 2.210). If a single word is meaningful, he says, it is because it is the equivalent of a sentence. Therefore, the sentence should be seen as the basic linguistic unit or *śabda*. *Śabda* is has a variety of meanings, including that of the uttered *sound*, to which Bhartṛhari assigns the name *dhvani*. (Coward and Raja (eds), *The Philosophy of the Grammarians*, pp. 5-6; Iyer, *Bhartṛhari*, pp. 155-61; Iyer, *The Vākyapadīya: Some Problems*, pp. 24-9)
62. VP 2.208-9 (the opponents in question are the exponents of Mīmāṁsā). Iyer, *Bhartṛhari*, pp. 143-4, 183; and Staal, 'Sanskrit Philosophy of Language', pp. 509-14. The Neo-Vedantist philosopher Śaṅkara was also opposed to the *sphoṭa*

doctrine. See Chakrabarti, 'Sentence-Holism, Context-Principle and Connected-Designation *Anvitābhidhāna*', *Journal of Indian Philosophy*, March 1989, where the Mīmāṁsā and Nyāya positions are compared with the sentence-holism of Bhartṛhari, Quine, and others. For the word-sentence controversy in Indian thought, see the 'Introduction to the Philosophy of the Grammarians', Coward and Raja (eds), *op. cit.*, pp. 63-97; and Matilal, *The Word and the World*, pp. 106-19.

63. VP 2.421, 438-9.
64. VP 3.1, comm. and 2.414.
65. Iyer, *Bhartṛhari*, p. 62. In saying this, Bhartṛhari is taking a position in the old argument between, among others, the grammarians and the adherents of Nyāya. The argument is over the primacy of verbs, whether even nouns (or nominal stems) are derived from verbs (or verbal roots) - many Sanskrit sentences, especially those in the 'nominal style' are without verbs. Of course, Bhartṛhari is on the side of the grammarians and the verbs (Matilal, *op. cit.*, pp. 8-10, 55-61; and Staal, 'Sanskrit Philosophy of Grammar', p. 508).
The long third *kāṇḍa* or chapter of the VP makes detailed grammatical analyses to show that a sentence expresses a particular action or process, which is directly denoted by its main word, a verb. Bhartṛhari thinks that the function of most nouns is to show what means or accessories the action or process requires. Analysis of a sentence is artificial, he insists, but helps to explain the indivisible word that comes to expression in the sentence. Examined with care, no two of the infinite number of sentences are alike. Individual words abstracted from a sentence by analysis of their meaning are unreal, as unreal as the stem and suffix similarly abstracted from an individual word (*The Vākyapadīya of Bhartṛhari*, chap. 3, part 1; tr. Iyer, pp. 1-2).
66. VP 2.324; tr. Pillai, p. 110.
67. Iyer, *Bhartṛhari*, pp. 192-3; Iyer, *The Vākyapadīya: Some Problems*, pp. 25-6.
68. VP 2.327-282, 345.
69. VP 1.73.
70. VP 2.7, 8 and 3.4.1-2 (including Iyer's explanation, pp. 121-2). Also Iyer, *Bhartṛhari*, p. 86.
71. Iyer's explanation of VP 3.3.24. In *The Vākyapadīya of Bhartṛhari*, chap. 3, p. 91.
72. Tr. Iyer, p. 90.
73. VP part 1: 3.3.27 comm. and 3.3.28; tr. Iyer, p. 92.
There is a clear likeness between Bhartṛhari's ruling and that of Bertrand Russell in his (ramified) theory of types. By Russell's theory, too, a string of elements of a language is regarded as ill formed - rather than false - if it does not observe certain restrictions of level or form. The object is, of course, to outlaw paradoxical self-reference so that the language continues to be consistent. The preference of some of Russell's critics for a linguistic or semantic over a logical cure for problems of self-reference returns us to Bhartṛhari.
In contrast, John Buridan (1300-1358) argues that the liar's paradox - *What I am saying is false* - is false. As the paradox goes: if it is true it is false, if it is false it is true, if it is true it is both true and false, if it is false it is both true

and false, and, finally, it is both true and false. 'I maintain, briefly,' Buridan responds, 'that the sophism is false, because from it and a proposition expressing the case there follows something false, and yet the proposition expressing the case is taken to be true. (The 'something false' that follows is that the sophism is both true and false at once.) Now any proposition which, together with something true entails something false, is itself false.' (G.E. Hughes, *John Buridan on Self-Reference*. Cambridge: Cambridge University Press, 1982, p. 58).

74. VP 1.131, 142. The image of the blind man running is from 1.41. The line of grammarians, with its temporary eclipse by 'the followers of dry logic', is described at the end of chapter 2 (2.476-81).
75. VP 2.138; tr. Pillai, p. 69.
76. For the discussion on the adequacy of words and of the instruments of knowledge, see VP 2. 134-42. The clever logicians are from elsewhere, VP 2.34. See Iyer, *Bhartṛhari*, chap. 4; and Iyer, *The Vākyapadīya: Some Problems*, lecture 3. That grammar can realize true, indescribable knowledge: VP 2.234.
77. Tola and Dragonetti, 'Some Remarks on Bhartṛhari's Concept of Pratibhā', pp. 109-10. The knowledge of children is based on the residual traces of the words they used in previous lives (VP 1.121).
78. Tr. Pillai, pp. 71-2.
79. VP 2.151; tr. Pillai, p. 72.
80. VP, 1.35. Iyer, *Bhartṛhari*, pp. 88-92; *The Vākyapadīya: Some Problems*, pp. 56-8.
81. VP 2.156-58, 214-15.
82. VP 2.46 (on the single letter), 2.247 (on praise and blame).
83. Tr. Iyer, pp. 109-10, 112-13.
84. VP 1.119, with comm.
85. VP 1.121 and comm; tr. Iyer, p. 109. In Pillai, verse no. 122.
86. VP 1.121, comm.; tr. Iyer, p. 109.
87. Matilal, *op. cit.*, p. 134, interpreting difficult texts from the earliest commentary to the VP 1.124 and trying to clarify what is unclear in Iyer's translation, pp. 111-12.
88. VP 1.123-5, comm.; tr. Iyer, pp. 110-12.
89. Tr. Iyer, pp. 105-6.
90. My rendition. The word *akṣara* ordinarily means *imperishable* or *immutable*, but in the commentary to the *Vākyapadīya* it is given the sense of *(immutable) phoneme* (Iyer's note 5, p. 3 of his translation). The term *śabdatattva*, Word-Principle or Word-Essence, emphasizes that all words are transformations of the Word, while *śabdabrahman*, Brahman-Word or Supreme Word, emphasizes that the Word is the being of everything and is subject to no limitation.
91. Tr. Iyer, pp. 1-2. See Iyer, *Bhartṛhari*, p. 101.
92. Sastri, *op. cit.*, pp. 27-8.
93. Iyer, *Bhartṛhari*, p. 100.

94. Ruegg, *op. cit.*, p. 67, drawing on Helārāja, Bhartṛhari's commentator. According to Helārāja, Bhartṛhari's tenth-century commentator, the essential nature of objects consists in their particular powers to exert effects. Their essential nature is therefore equivalent to the nature of sentences, because 'a normal sentence expresses a complex meaning of which the central meaning is some action, to which the other elements contribute' (K.K. Raja, in Coward and Raja (eds), *op. cit.*, pp. 194-5).
95. VP chap. 3, part 2, tr. Iyer, section 9 (pp. 36-74). See Iyer, *Bhartṛhari*, pp. 111-30; 'Introduction to the Philosophy of the Grammarians', in: Coward and Raja, *op. cit.*, pp. 38-44; Sastri, *op. cit.*, chap. 2.
96. VP 1.2 and 1.3, both with comm.
97. VP chap. 3, part 2: 9.1.3,4,5,8,9,14,41.
98. *Ibid.*, 9.18, 19 (quoted). See Iyer, *Bhartṛhari*, p. 101 and the exposition in Coward and Raja (eds), *The Philosophy of the Grammarians*, p. 154.
99. VP, chap. 3, part 2: 9.1.12.
100. *Ibid.*, 3.9.62, with comm.
101. *Ibid.*, 3.6.26, 28; 3.7.40, 42 (quoted); trans Iyer, pp. 146, 172.
102. VP 1.130, comm.
103. VP 1.15.
104. VP 1.137 comm.
105. 'The meanings of words are determined from (their) syntactical connection (in the sentence), situation-context, the meaning of another word, property, place, and time, and not from their mere form. [Meanings of words are determined by] (constant) association (of two things), (their) dissociation, company, and hostility, the meaning (of another word), situation-context, evidence from another sentence, and the proximity of another word' (VP 2.314-15; tr. Pillai, p. 108).
106. Tr. Iyer, p. 1.
107. VP 1.4.
108. VP 1.142. 'Introduction to the Philosophy of the Grammarians', in: Coward and Raja (eds), *op. cit.*, pp. 44-50.
109. VP 1.142, 1.5.
110. VP 1.31, 142, 51, all with comm. Coward and Raja (eds), *op. cit.*, pp. 48-50, 138-53 (paraphrase of VP, chap. 2). Pillai, *op. cit.*, pp. 36-146.
111. See B.-A. Scharfstein, *op. cit.*, chaps. 4-5.

XXV

A PLAY ABOUT RITUAL: THE 'RITES OF TRANSMISSION OF OFFICE' OF THE TAOIST MASTERS OF GUIZHOU (SOUTH WEST CHINA)

Kristofer Schipper

PREAMBLE

Among the many contributions of Frits Staal to science, his study of the Vedic *Agnicayana* ritual, published in the splendid two volumes of *Agni, the Vedic Ritual of the Fire Altar*, has the status of a landmark.[1] It is an achievement which marks the culmination of the scholarship of generations of Sanskrit scholars, including that of Frits Staal himself. At the same time, the publication has marshalled an increased interest in ritual studies and has directly contributed to their present renewal. At the heart of this scientific revolution lies the encounter between the study of texts - such as the *Veda* - and living traditions of ritual performance. For a long time the study of these two approaches were separated. Text scholars - philologists and historians - did not do fieldwork, whereas field workers such as ethnologists and anthropologists did not read texts. Yet in many civilizations great ritual traditions are based on both textual and oral transmission. In these cases, it often takes a scholar steeped in classical learning to recognize the importance of the living practice. That this has taken so long may be because, as the Chinese proverb puts it: 'The horse that can journey ten thousand leagues is less difficult to find than the person who can recognize such a horse.' But once the proof had been established, the significance of the breakthrough was recognized, not only by scholars the world over, but also by the Indian government. Frits Staal was honoured by President Indira Gandhi, to whom he personally presented a copy of *Agni*.

China too has an abundance of ritual traditions, which, like those of the

Nambudiri brahmins, are in danger of disappearing. It is most encouraging that in recent years more attention is being paid to this vital aspect of Chinese culture and that efforts are being made to collect and study first hand field materials. The most important contribution to date is the research program on Chinese ritual theatre directed by Professor Wang Ch'iu-kuei of the National Ch'ing-hua University in Taiwan.[2] The program, which started in 1991, conducted a number of investigations and collected materials, including field descriptions, photographs and video recordings. These are published in what is already an impressive collection of monographs. Forty volumes have already been published, and twenty are scheduled to appear in the near future![3] The field report on the 'rites of transmission of office' of the Taoist Masters of Cengong county in Guizhou province, on which the present article is directly based, has been published as a volume in this series.[4]

Nowadays authentic Taoist ritual mainly survives in the more remote rural areas of China and especially in those places marked as 'autonomous regions' for so-called 'minority peoples'.[5] It is therefore understandable that most of the material was collected in these regions.

Another vast collection devoted entirely to Buddhist and Taoist rituals, this time also coming from more central locations, is also due to appear in the near future.[6]

The publication of all these materials is matched by scholarly articles in which the material is studied. These articles generally appear in the journal *Minsu quyi* (*Chinese Folklore: Reports and Studies*).[7] The above-mentioned series, as well as the journal, fill important gaps in our knowledge. We may say that they revolutionize our understanding of Chinese society and Chinese culture in general. It is therefore surprising that, relatively speaking, little attention has been paid to these publications in the scholarly world. Luckily, in April 1996, Professor Wang presented the results of the program in a series of lectures at Leiden University, thus making these materials known to a wider public.

RITUAL OR THEATRE?

Taoist ritual is generally enhanced by performances of a theatrical nature. Whereas the rituals themselves are as a rule executed by priests, the theatricals are usually played by laymen. Ritual and drama are, however, intimately entwined. The priest, called *shi*, 'master', or more politely *shigong*, 'sir master' - may be a true professional, but more often than not he (sometimes she) is a specialist whose main occupation is profane: peasant, shopkeeper, etc.[8] The actors are often laymen, but in some cases professionals, especially in the case of puppet or shadow play. Frequently there will be ritual parts performed by the master in the drama performance, and dramatic parts done by laymen may be embedded in the ritual.

In the case of the masked dance theatre of South West China (provinces of Guizhou, Guangxi and Yunnan) which is under consideration here, the dramatic action is commonly called *tiaoshen*, i.e. 'dancing gods'. This term is widely used throughout China for 'spirit medium'. Indeed, as is clearly shown in the texts of the masked theatre as published in our Monograph here, the spirits of the deities impersonated by the actors are considered to descend on them and take possession of them. To name an instance: in a play performance called 'Uniting the Emblems' (*Hebiao*), the text reads in the beginning:

- The Master Ancestor of Uniting the Emblems descends on me!
On my body he comes, the Master Ancestor!
- With my disciple, I will now unite the emblems!
The sun goes down over the mountains; twilight sets in.
- The Master Ancestor who Cleans the Body descends on me!
On my body he comes, the Master Ancestor!
- With my disciple, I will now unite the emblems!
The sun goes down over the mountains; twilight sets in.
Etc. (page 288)

It is a general custom in China that the spirit medium, once possessed by a deity or a saint, is then provided with the costume and attributes of this deity or this saint.[9] For the facial make up, either cosmetics or masks (in the tradition under study) are used. It should be noted that the '*nuo*' masks mentioned in this paper are consecrated through the 'opening of light' (*kaiguang*) rite just as is done with statues of gods, ancestral tablets or string puppets.[10] Trance - in more or less pronounced forms - is also experienced.[11] The ritual performance is thus enhanced by theatre and we may say that the significance of the rites is 'given to see' through the accompanying dramatic play.

To patronize a ritual performance - including formal rites and dramatic performances - is called in these regions of South and South-West China: *huanyuan*: 'to repay a vow', inasmuch it is assumed that these services are rendered as an offering and a thanksgiving for divine protection and assistance. The occasion can be most general and periodical - as in the case of seasonal services - or very specific and as a result of a vow for recovery from an illness or relief from a crisis.[12]

The notion of *huanyuan* applies to the service as a whole, rites as well as theatre. And indeed, as much as the drama is ritualized, so is the ritual dramatized. True enough, the priest does not put on a disguise making him -

or her - the image of a god. But the priestly robes themselves are divine vestments. At certain moments of the ritual, the priests represent or personify important gods. For instance, during the central transmission rites, the initiating master is considered representing The Old Lord (Laojun, the divinized sage Lao Zi) himself.

To sum up: it is possible and legitimate to study Taoist ritual in the context of Chinese drama, as it is equally possible to study Chinese theatre as a subsidiary activity to Taoist ritual. For two reasons the first of the alternatives prevails and Taoist ritual is mostly studied from the angle of theatre: (1) because of the religious background of Chinese scholars, who in general remain very Confucian in their outlook and (2) because of the fact that most Taoist ritual continues to be classified in the People's Republic of China as 'feodal superstitious activities' (*fengdian mixin huodong*), and therefore to be forbidden.

However, there are several theoretical frameworks which allow contemporary Chinese scholars to study so-called 'regional drama' (*difang xi*; this term was coined in contrast to the more metropolitan forms of theatre such as Peking Opera) and still do justice to Taoist ritual. One of them is to consider that both theatre and Taoism originated in shamanism and continue to be closely linked to that form of 'primitive religion'. Shamanism thus becomes the umbrella under which both can be studied. The other theory takes its point of departure from the fact that many forms of regional theatre - especially in South and South Western China - uses masks. This specific element is reminiscent of *nuo*, the exorcism performed by masked persons in ancient China. *Nuo* is something ancient Confucians approved of and which therefore is considered worthy of study bý present day Chinese scholars. The general idea is that this 'classical' form of exorcism lived on in primitive surroundings, such as peasant communities of the deep South, ethnic minorities, etc., and that over time it developed into theatre. Another postulate is that *nuo* theatre was at some time in history 'influenced by Taoism'. The question is rendered even more confusing by the fact that one of the terms used by the guizhou priests for exorcistic rites is *chongnuo*, an expression in which *chong* should mean 'to drive away' and *nuo* for evil spirits. If one thinks of *nuo* as masked theatre, the link with the masked exorcistic dances of ancient China becomes easy to make. These savant reconstructions are pushed further by taking into account legendary materials where one of the patron saints of *nuo* rituals should be no other that the famous Confucian thinker, Dong Zhongshu,[13] and genealogies concerning the lineages of the local Taoist masters tracing their ancestry as far back as the third century AD are simply taken at face value.[14]

THE RITUAL TRADITION OF THE 'RUSTIC MASTERS' (TULAOSHI) OF GUIZHOU

I will no longer elaborate on these abstract issues. They may be useful here only inasmuch as they explain why the book concerning the Taoist ordination ritual, which forms the main subject of the present article, carries the title: 'Report on the Investigation on the Rites of Transmission of Office of the Nuo [Theatre] Altar of the Gelao Ethnic Group of Pingzhuang County, Cengong District in Guizhou Province' (*Guizhousheng Cengongxian Pingzhuangxiang Gelaozhu nuotan guozhi yishi diaocha baogao*). This report - which I will henceforth call 'the Monograph' - was prepared by a group of researchers from the People's Republic of China, on the basis of their direct observation of a major ordination ritual in February 1992. Cengong district is in Eastern Guizhou. The district was formerly called Sizhou or Xixian. It is considered the oldest region colonized by Chinese coming from the north. The entire region of east Guizhou and west Hunan is also called Qianzhong, and it appears that this form of Taoist liturgy is prevalent throughout the region.[15]

The above-mentioned Monograph emphasizes the 'theatre' aspect of the ritual although the other aspects have not been neglected. Thus it does provide a fairly detailed description of the ritual proceedings and also reproduces part of the corresponding texts transcribed from the manuscripts transmitted by the priests. Some transcripts from recordings made at the occasion have been equally added, giving precious elements from the oral tradition which forms an important part of the ritual tradition. A videotape was made, excerpts of which have been made available by Professor Wang Ch'iu-kuei for research purposes. One of the co-directors of the project, Mr.Tuo Xiuming, published a separate article on the ordination ceremony with additional information. Approximately a year later another ritual by a troupe in the same region, this time prearranged as a demonstration of ritual and theatre performance took place. This was also reported on by the same scholar and his assistants. The monograph produced from this occasion contains also some new material.[16] Thanks to these materials, even someone who did not witness the ritual can get an accurate idea of the performance. It also provides the student with such a wealth of information, with so many new elements and directions to be explored, as to make any attempt at in-depth research and analysis impossible within the scope of this article. Consequently, the remarks that follow are purely exploratory in nature.

The ordination service for the 'transmission of office' - also vulgarly called: 'displaying the tokens' (*paopai*) - was held the at the home village of the ordinee at Pingzhuang, Cengong County. The 'Tokens', which occupies such an essential role in this ordination - or should we say 'investiture' ? -

of a Taoist priest, comprise a number of documents (see below). These are lists of deities as well as titles of office, which the officiant keeps in a long wooden box, sewn closed and decorated with thirty-six ribbons. This box with its ribbons is called 'the token-seal' (*paiyin*) or 'the token-ribbons' (*paidai*) and is indeed an instrument of authority used to summon and direct the gods. It is carried on the officiants left shoulder, where the long ribbons give it the aspect of a kind of short cape or shoulder ornament. When in use, the officiant takes it in his right hand, making sweeping movements with it in front of him. The use of this characteristic implement is not restricted to the priests of this particular area, but can also be found in the north-eastern Guizhou district of Dejiang. Two studies, one Chinese and one Japanese, on the ritual and theatre of this area have been published (based on field observation with one and the same Taoist master!), which confirm this.[17] A more complete study than the present one should take into account all the complementary material offered by these reports. Here it must suffice to say that the Chinese survey of the Dejiang ritual tradition offers a somewhat more detailed description of the 'token box' and the beliefs that are connected with it.[18]

The ordination service consisted of two perfectly distinct parts. First, during three days, a general service of thanksgiving (*huanyuan*) took place. Because the written petitions, which would carry the 'intention' (*yi*) of the ritual, had not been reproduced in the Monograph, it is difficult to tell what the purpose of this general service was considered to be. In any case, it is also customary elsewhere that a specialized service for ordination, expelling epidemics, etc. is preceded by such a general liturgy. The *huanyuan* service was composed of a series of major rituals: an invitation to the gods, a presentation of memorials, here called *Tiandishui yangke* (Rites for the Benefit of the Living of Heaven, Earth and Water), a general offering to all the gods (curiously called 'Exorcising the Baleful Stars' (*Rangxing ke*) and a great sacrifice of cooked meats (*Shangshu*). These four solemn rituals were performed by the old initiating master alone, as if he were the only one who still knew how to perform these.

In addition to the solemn rites, there were dramatic performances with songs and dances. These took place during the evenings and the nights. Some lasted up to three hours and were performed by three, four or five actors. Their contents were invariably closely linked to the thanksgiving service.

The altars for the ritual were set up at the ancestral home of the ordinee. The 'inner altar' (*neitan*) was constructed in one of the lateral buildings, whereas the 'outer altar' (*waitan*) for the 'army of the Five Camps' (*wuying*) of spirits soldiers was established at some distance in an open field. The ritual lasted four days, from March 5 to March 9, 1992. In the Chinese calendar, these dates correspond, in the Chinese calendar, to the third to the seventh day of the second moon. That means the rites started on the day after the festival of the Earth God. All the inhabitants of the village and many others participated either directly or by giving money and presents or by doing both. The performance went on without much interruption, with most of the Taoist rituals taking place during daytime and the dramatic performances during the evening and night.[19] The rites on the third day and last day of the general thanksgiving service started at 9.30 A.M. continuing, without interruption until 5.40 A.M. the next day. After a short rest, at 9.45 A.M. they resumed activities. The actual rites of ordination, transmission and passage of office, took place at the end of that very morning and lasted for two hours. We will examine them in some detail below.

Tuo Xiuming counts a total for the two services of twenty-seven distinct Taoist rituals, eight theatre pieces, and seven demonstrations of magical skills by the disciple to be ordained. This count is open to question as the exact nature and organization of the different items is not always clear.[20]

The entire performance was public. The altar area was easily accessible to onlookers. The theatre performances and the demonstrations of magical skills drew huge crowds. Ordinations of this kind were said to have been interrupted for many years. Only in 1983 did the old Masters take on disciples again. The initiating Master in this ritual was born in 1920 and consequently seventy-two at the moment of this transmission of his office. His disciple was born in 1964 and only twenty-eight. He entered into the old man's service as an apprentice in 1983.

The district of Cengong where the ordination took place is situated in the east of Guizhou near the Hunan border in an 'autonomous region' of the Miao and the Dong ethnic groups. In fact the region is inhabited by a number of different minorities, among others the Gelao, who are rather numerous in the rural county of Pingzhuang where the transmission service took place. The Gelao have not been extensively studied. Originally they may have been Lao people.[21] However, although some fragments of their language were recorded in the first half of this century, they appear to no longer know the old language as Tuo Xiuming reports that they speak Mandarin. Both the Initiating Master and the ordinee are Gelao. The group of Taoist priests to whom they belong is headed by the Initiating Master, and also comprises several members of the Miao and Dong minorities.

As noted in the texts published in the Monograph, the ritual itself is in perfect colloquial Chinese, as far as I can tell almost without any dialectical expressions. Consequently it is safe to assume that the Taoist tradition which is represented by the Masters of this particular area of Guizhou province is not linked to any specific minority group, be they Gelao, Miao or Dong, and that it belongs to Chinese religion in general, like the masked theatre performances and the mythical characters it puts on stage.

This does not imply that the tradition we find here is devoid of any regional flavour. The priests are called not just *fashi* ('ritual master') as in other parts of China,[22] but also *duangong* (orthodox sire), *zhangtanshi* (altar manager) and, apparently most frequently in this particular area, *tulaoshi* (litt. 'earth master', although in this context with the connotation of 'rural', 'agrarian', or 'rustic' Taoist master).[23] Duangong is a specific title for Taoist masters in South West China, especially among the so-called minorities.[24] The term *duangong* appears to have been coined in contrast to *xieshi*, 'Unorthodox Master', a term given to those who practice black magic or to priests of outlawed sectarian movements. It may be noted that the 'nuo theatre' in this case, which authors of the Monograph call *nuotangxi*, 'theatre of the *nuo* hall', is also, more authentically, called *duangongxi*, 'duangong theatre'.[25] When discussing the *duangong* of the Miao minority, D.C. Graham gives another valuable indication stating that:

> He uses an assistant who is called in Chinese a *ma-chüeh* (*majue*) or horse's hoof... . This man is put under a spell, after which he leaps about with a hatchet in each hand, striking in all directions ... .[26]

Here the *majue* does so in order to expel demons. The same kind of assistant is documented, discretely, in our Monograph where it is stated that at the beginning of the transmission ritual that the ordinating master

> uses his magical skills to put a man called 'horse's hoof' into trance and thereafter lets him chase away the demons and pernicious influences in the *nuo* hall, so as to cleanse it ... (page 106)

In conclusion it appears to me that these rituals most probably did not originate in the region of Guizhou, nor should the area of distribution be limited to that province. The *duangong* of this region in Guizhou claim to belong to an otherwise unknown Jade Emperor (Yuhuang) school which has as its centre the neighbouring province of Hunan (see below). Everything suggests that what we have is a local or regional Taoist order in some ways comparable to the Lüshan or Sannai school of Fujian and adjacent areas, with its 'fashi' or 'red head' priests of that school. In both cases the 'masters' are only part-time practitioners, whose main occupation is husbandry.[27]

Let me now try to list a few elements concerning the liturgical tradition, the pantheon and the ritual repertoire of these *Tulaoshi* as they can be found in the manuscripts. As said above, a number of these have been reproduced as an appendix to the Monograph.[28] There are manuscripts with the texts of the Taoist rituals - such as the one examined below as well as those of the theatrical performances. The latter are generally in the characteristic vernacular Chinese ballad form of a seven character rhymed verse. They stand between the written and oral tradition.[29] There are purely orally transmitted texts as well. These are the 'secret formulas', *mijue*, which a disciple can only obtain directly from his or her teacher. These are composed of recited spells (*zhou*) and hand gestures (*jue*) which accompany the secret spells, pronounced inwardly, without any sound. The secret spells have sometimes been noted down and from these examples we can see that they are in classical language, but not unintelligible.[30] In addition, all the rules and conventions concerning the ritual performance i.e. the music, the offerings, as well as the lore of the deities, are all transmitted orally.

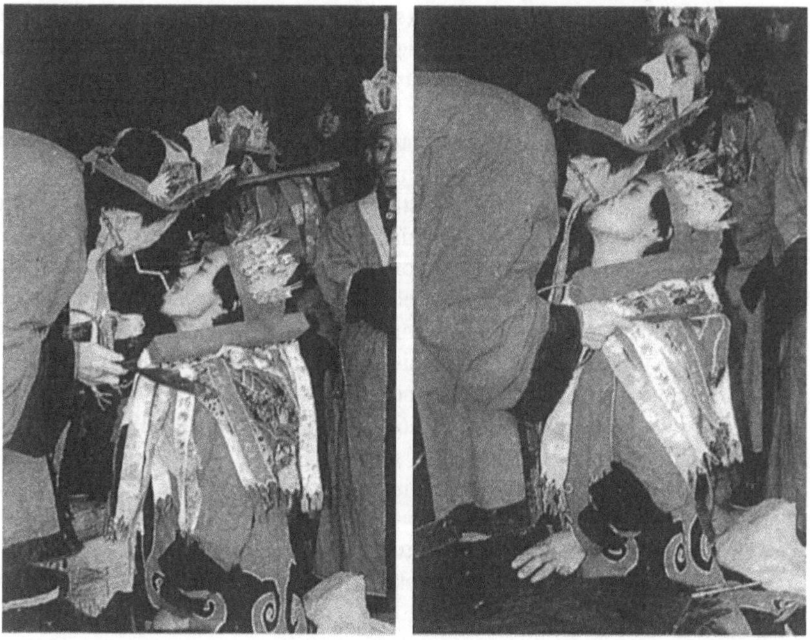

Illustration

The oral transmission of the 'secret formulas'. Photos reproduced from: *Guizhousheng Cengongxian Pingzhuangxiang Gelaozhu nuotan guozhi yishi diaocha baogao* (Report on the Investigation on the Rites of Transmission of Office of the Nuo [Theatre] Altar of the Gelao Ethnic Group of Pingzhuang County, Cengong District in Guizhou Province). Tuo Xiuming, Yang Qixiao, Wang Ch'iu-kuei (eds), *Minsu quyi congshu*, Taipei, Shi Ho-cheng Folk Culture Foundation, 1994, p. 423.

At the occasion of the ordination (the 'transmission' ritual) different rites are concerned with the initiation and confirmation of the disciple in both traditions. The written tradition is all wrapped up with the box which contains the Tokens (*pai*), while the oral tradition (*kouchuan*) is expressed in terms of feeding. But before getting into this differentiation, let us take a closer look at the manuscripts.

The age of the manuscripts is uncertain. Almost none of the copies seem to have been dated by their copyists.[31] An important volume which contains models for memorials for different services and which carries the title of *Tiandishui yangke* (Rites for the Benefit of the Living of Heaven, Earth and Water) was apparently copied during the Republican period (Minguo) uses the ancient place name of Sizhou (see page 238 and page 240). The corresponding volume for Exorcising Baleful Stars (*Rangxing ke*; pages 252 to 278) is based on a Qing dynasty manuscript copy from Guiyang, the provincial capital (page 259: Daqing tianxia Guizhou shoufu ...). A complete analysis of this text should yield much important information on the exact nature of this tradition. In this paper I can only mention a few points.

This particular ritual is of a distinct synchretistic nature. Petitions are addressed to the 'saints of Confucianism, Taoism and Buddhism' (Ru-, Dao-, Fojiao san shengren; page 278). It contains a number of Buddhist hymns, but even a more important number of elements taken from the classical Taoist liturgy. For example,we find the famous Formula for Purification called the 'Divine Spell of the Central Mountain' (*Zhongshan shenzhou*), on page 259. We also have the Laud of Jade Emperor (*Yuhuang hao*), the foremost deity of the Taoist pantheon on page 256. It is noteworthy that this ritual also contains some 'Tantric' elements, such as the hybrid half Chinese, half 'Sanskrit' mantra which immediately precedes the above-mentioned purification spell on page 271.

The synchretistic nature of this liturgical tradition is borne out by the list of principal deities as they are invoked, for instance, at one of the incense offerings. They are:

1. Ziwei, the Pole star;
2. The stars of the Ursa Major, the 'Dipper' constellation;
3. The constellation presiding over the present year;
4. Pangu, as 'Lord of the Ritual of the Three Origins';[32]
5. The Taoist Three Officials (Sanguan);
6. Guanyin, the Boddhisattva of Merci
7. Xuantian shangdi, the Taoist Lord of the North
8. Wenchang dijun, the God of Literature.[33]

But the ordination title (usually called *fapai*) of the performing priest as given in the manuscript is purely Taoist: it reads:

> Disciple of the Jade Emperor of the three Pure Ones (this is a general title used throughout the ritual),
> Initiated in the Orthodox Religion of the Heavenly Immortal with the Three Thousand Seven Hundred Dragon-and-Phoenix Golden Degrees of rank of [the school of] Huainan, Libationer who Protects the Walls, Heralding Minister[34] in Charge of the Law (this would correspond to the *register of rank* which is held by the priest),
> Disciplinary Official of the Northern Apex, Controller of the Department of Wind and Rain, Lightning and Thunder, Judge of the Empire and Executioner of the Perverse Gods and Demons of the Three Worlds, etc.[35] (here we have the definition of the 'office' of the priest),
> The common and ignorant servant, the disciple so-and-so ...

Like similar and better known ordination titles,[36] the long appellation is composed of two main parts denoting respectively, the 'register of rank' and the 'office'. As it stands here it must be a complete summing up of all the different variants a new master can chose from at his ordination (the normal procedure of selection being divination). It is most noteworthy that his or her rank in term of 'register' still carries the title of 'Libationer' (*jijiu*), a most ancient and honoured appellation of the priests of the communal Taoist movements of the Han dynasty.[37]

As to the 'school' to which the priests belong, it is given as that of the Jade Emperor, Yuhuang, the head of the Taoist pantheon. In the ordination certificates which are bestowed on the occasion of 'transmission of office' rites, this affiliation is indicated by the fact that the initiating authority is called 'disciple of Yuhuang' (Yuhuang menxia). Tuo Xiuming indicates that the origin of this school could be at Mayang in Hunan province (page 970). This is a place at some hundred miles east of Cengong, just across the Hunan border.[38]

At several places in the manuscripts it is specified that this Yuhuang school is also defined as the 'Orthodox Religion of the Heavenly Immortal of Huainan'. Huainan, the region 'south of the Huai river' is the traditional name of a early civilized area in the present province of Anhui. Both this name and that of Heavenly Immortal, which corresponds to the title given to the great mother goddess Bixia yuanjun,[39] suggest many other relationships to be explored, but it is impossible to go into these hypotheses now as I have no other material on this school at my disposal.

At the end of the ordination title, the priest inserts his or her own name. It is composed of the usual family (clan) name and the personal religious name. At ordination, the ordinee does receive a religious name of two characters, the first of which is always, in this instance, the character '*fa*', meaning 'rite', 'magic' and 'law'. There is no distinction related to different generations, as in other Taoist schools.[40]

So much for the subject of the school to which the *Tulaoshi* of Cengong belong. As to the kind of liturgy they are performing. For those more classical in style and which incorporate hymns and spells from the Taoist Canon, the kind of liturgy they perform clearly indicates that they are considered to belong to that of the 'One Orthodox School', that is, the Zhengyi order of the Heavenly Master (see pages 239 and 241).[41] There is at present no way to verify this claim.

In the eyes of the officiants, the ritual of the *huanyuan* service defines itself and enumerates the main items which it contains. These are:

- three sticks of incense (to be burned at a time);
- nine kinds of offerings, filling the hall;
- clear tea, freshly brewed;
- well-sealed clear wine;
- a whole pig;
- a chicken and a fish, which (with the preceding) make the Triple Sacrifice;
- twenty-four red emblems;
- seven condensed (?) plays;
- the music of gongs, horns and drums;
- rhythmic and respectful dances;
- songs and hymns;
- three valedictory representations.

All these items are presented 'in thanksgiving' (*huan*). Let us examine at which position in the ritual these ingredients are utilized. After the invitation of the gods with incense, tea and wine, a sacrifice of three kinds of animals (*sanxing*) is offered. This sacrifice here is cooked meat and it is presented during the night of the third day, at the conclusion of the *huanyuan* (redemption of vows) service which precedes the ordination rites. This is followed immediately by two theatrical pieces and a final address (*huixiang*) which close the service. These are presumably the 'three representations' mentioned above. This *huanyuan* then is a general liturgical service preceding the distinct service for the 'transmission of office'. It should be noted that this ordination is not a 'redemption of vows' ritual, as it is absent from the long and apparently exhaustive list of such services published in the Monograph. Let us now turn to the real topic of the present article.

DISPLAYING THE TOKENS

During these first three days (or, more precisely, two and a half days and one entire night at the end) during which the *huanyuan* service takes place, the ordinee has prepared himself - also in the eyes of his colleagues and the public - by performing a number of magic feats which are the main attraction of the occasion.

On a platform surrounded by a crowd of onlookers on the evening of the first day, with his bare hands he takes a red-hot steel tripod and puts it on his bare head. Although blue fumes rise up on the contact of his hair and head with the glowing implement, not a single hair is singed. He then proceeds to make a somersault with the burning tripod still on his head.

On the morning of the next day, he washes his hands in a wok with boiling oil which has just served to deep fry some shrimp crackers. Then he throws his keys in the wok and calmly fishes them out again. In the afternoon, something less scary but equally amazing: two volunteers are called from the crowd who are placed face to face at a certain distance. Two bamboos about ten feet long and of a certain thickness are brought in. Their extremities are tucked under the armpits of the volunteers. The distance between the two men, who stand immobile is bridged by two ten-feet bamboos, and these are kept at chest-wide interval from each other. The ordinee now starts reciting and calls out: 'Get together!' Soon his calls are accompanied by drum and gong beating. Slowly but surely the two bamboos, kept apart by the men, start to move in the middle towards each other and even intertwine.

Later the same night, the disciple walks, dances and somersaults over a heap of fierce burning charcoal. A final feat, that of climbing a ladder of twenty-four sharp knives, is not performed at this preliminary stage, but, as we shall see, at the very end of the ordination service and with a distinct ritual function.

These feats are not just 'tricks' but have each a ritual function.[42] Here they are performed to prove that the ordinee is mature enough to become a master. At several occasions, he already wears the priestly crown and gown. He also makes use of the priestly *pai*, 'tokens' in the wooden box with the ribbons although he does not yet hold the rank of master. For the moment he can only manipulate this instrument of authority as a delegation of the power of the master until he receives his own power.[43]

Almost at dawn the morning of the fourth day when the *huanyuan* service is finished, just a few hours rest bridge the gap before the beginning of the ordination service. Like the great *huanyuan* service it begins with an invocation of all the gods (*qingshen*). But now this 'invitation' ritual is completed by the installation of the Five Camps for the spirit soldiers. At this 'outer altar' (see above) the knife ladder is also set up. After completing the ordination, it will be from the top of this ladder that the ordinee will announce his new status to the demons and gods alike.

It is in the Five Camps that the ordinee will obtain his army of spirit soldiers which are the divine helpers of his priestly office. No details are offered concerning their actual recruitment.[44] At the end of the service, the soldiers are rewarded for their help and given leave. Then also the gods are sent home. Between these two antithetic rites, the actual transmission (*chuanfa*) takes place. This gives:

A	B	A'
invitation; setting up the Five Camps	transmission/ordination	sending off; demolishing the Five Camps

During the rites of transmission and ordination, nothing specific in the way of instruction is added, but the capacity of master is confirmed by a series of highly significant symbolical actions. Ritual takes the form of a mystery play about the nature and the meaning of ritual and priestly office, more than that of an actual initiation.

During the major part of the proceedings the ordinating master assumes the role of the 'Old Lord' Laojun, i.e. the deified form of Lao Zi. He engages in a series of dialogues and actions with the disciple. He is assisted by a number of his colleagues who occupy the functions of Assessor (zhengming shi), Guarantor (baoju shi), Presentator (yindu shi), Transmitter (jiefa shi), etc.

Once the Five Camps have been installed, the actual ordinationwealthy begins (page 106) with the ordinating master ordering the disciple and the assisting masters 'to run to the Five Camps and have a look'. Once they are back, he asks: 'What did you see?' 'Nothing' is the answer, as is normal, inasmuch the ordination which should enable the disciple to see his spirit helpers has not yet taken place. This is repeated three times, with the same result. Thereupon the master asks the disciple:

> You who have come here to establish an altar and receive the rites, are you of one heart (pure and sincere intentions) or double-hearted (dubious and insincere intentions)?

The disciple swears that he is sincere, lifting his finger up to heaven saying:

> 'If I am insincere, that with broken bones I may repay (my debt) to

Heaven.' The ordinating master then asks again, if something was seen or heard at the Five Camps, and again, all the assisting masters answer in unison that nothing was perceived, implying therewith that the young disciple should be initiated. This is then repeated for a third time.

From this point on the ordinating master assumes the role of 'Old Master' Laojun, in Heaven. Addressing himself to the Presentator, he asks:
- I ask you Presentator, who is there outside blowing his horn, disturbing the peace at my Heavenly Gate?
- The one who, the other end of the bridge (that spans the distance between heaven and earth, represented here by two benches put between Laojun and the disciple), has blown the horn three times, who has disturbed you Old Master Jade Emperor,[45] and respectfully begs his master-father on the altar, that is the disciple whom I have come to present here, as he wants to establish an altar and receive the rites.

  The disciple is asked again to swear that he is sincere, whereupon Laojun asks again the Presentator:
- The new disciple, from where did he come, where to does he go?
- He has come by the east road, and will leave by the west road.
- Who is in charge of the Gate of Heaven, of the Abode of the Earth, of the Road of Demons, of the People of Human Beings?
- I know it: The Jade Emperor is in charge of the Gate of Heaven, Yanluo (Yama) of the Abode of the Earth, the August Ancestor Fuxi presides over the People of Human Beings, and the Master of the Valley of Demons Guigu xiansheng rules over the Road of Demons. (...)
- What is the name of Laojun, of the Great God of Purple Tenuity; how many surnames are there?
- Laojun's name is Li, his personal name is Er. The God of Purple Tenuity is called Ji. That makes two surnames.
- How many ten-thousands, how many thousands of horsemen of the Nine Yi [tribes] of the East, how many of the Eight Man [tribes] of the South, of the Six Rong [tribes] of the West, of the Five Di [tribes] of the North, of the Three Qin [tribes] of the Centre, should be transferred to the disciple?
- Ninety-nine thousand horsemen of the Nine Yi, eighty-eight thousand horsemen of the Eight Man ... (etc.) should be transferred to the disciple.[46]

  (...)

The questions and answers concerning mythology and cosmology continue for some time. Then once more, for the third time the disciple has to swear that he is sincere.

Now follows the real 'transmission'. It consists of two parts. The first part concerns the transmission of the 'written tradition', the second one concerns the 'oral tradition'. The written tradition is represented by the box with the Tokens, *pai* or more fully: *paiyin*, literally: 'the seal of the tokens'. 'Seal' has to be understood here as 'instrument of authority, octroi, token', etc. In fact, the priests also have ordinary seals with which they stamp their liturgical documents.

The box holds several documents in addition to a number of potent medical herbs and also seven ribbons with valedictory sentences made by young virgins. The documents have not been transcribed in the Monograph, and I can only guess their contents from some rather vague photographs (ill. 77 to 83). There is a *certificate* (*die*) attesting the capability to perform exorcisms, a *petition* (*shen*) addressed to the Three Officials (Sanguan) to obtain their permission to officiate in their name and several *registers* (*lu* or *fulu*) with names of gods and demons together with their sacred names (in talismanic writing) over whom the priest has authority.

On the day before at a preparatory ceremony, the box of the master is opened and some of the contents taken out to be put in the box of the disciple. At the same time, some of the contents prepared for the box of the latter are then put in his master's box. This mixing of contents establishes a bond through which 'in you there is something from me, in me there is something from you'. (pages 101-3). The possession of the box enables the disciple to summon the gods and demons (spirit soldiers) of his *register*. While performing, he carries the box, with the thirty-six ribbons attached to it on his left shoulder or waves it in front of him while pronouncing the summons. Together with the box, the disciple also receives the other insignia, especially his seal of office, and sacred implements, as well as his priestly vestments. After the transmission of these objects, the ceremony culminates with the 'oral transmission' (*kouchuan*; see below).

Standing before his end of the 'bridge', the disciple, clad in a dark western-style suit, with white shirt and tie, holds before him a nylon travel bag. He bows. Sitting behind the altar table the master now addresses himself directly to the disciple and asks him:

- What kinds of things did you learn? What did you not learn?
- I learned nothing at all. I am incapable of anything!
- You point on the ten directions (in your hand), then you get the twelve time divisions of the day, yes! Once you point, its efficacious, yeah! And so you get Metal, Wood, Water, Fire and Earth (the Five Elements), yes!

Then he sings:

- The Gate of Ritual (*famen*, also the 'Gate of the Dharma' in Buddhism) of the Thirty-three Heavens opens, ah! the Heavens transmit the rites, ah! The master brings his soldiers to transmit them down below, ah! There is Earth, there is Heaven, there is Water, ah! The transmitting Master has rites to transmit, has rites to transmit, ah! The disciple has money to drop in the water basin, ah!

During the singing, the master shows the secret hand movements of this 'pointing' (thus activating) the energies of the universe. Whenever such a 'secret formula' has been performed, the Receiver (*jiefa shi*), who stands to the right of the master, mimics receiving it from the master and holding it in his cupped hands. He then carefully carries them over to the disciple and opens his hands above the travel bag, indicating that they have been put in there. At the same time the Guarantor (*zhengming shi*) takes a handful of cooked beans and gives these to the disciple, who starts vigorously chewing and swallowing them. These beans represent the spirit soldiers and the act is said to symbolically express the idea of 'sowing beans to harvest soldiers', a reference to a famous episode in the war of the Three Kingdoms where the magician Zhuge Liang performed this very feat. Finally the disciple throws a few coins in a basin that stands below the 'bridge'.

- Have you got the formula? Asks the master.
- I have got it. But I still do not have the method for sweeping the house (rites of purification).
- Kindly allow me (to show you). You call upon those Generals of the Three Principles. Make a mess of it (i.e. do as you like to do it), and that's just fine!

The master sings the same text as before while acting out the secret hand movements. These are again 'captured' by the assessor to his right and 'dumped' into the disciple's travel bag, whereas the other assessor gives once more a handful of beans, which the disciples chews up; and, again, a few coins in the basin.

This play continues for the following empowerments: intertwining bamboos, exorcising with masked exorcists (*chongnuo*),[47] the trick of the boiling oil, expelling epidemics of plague and cholera and healing child diseases. Here, in the case of these healing techniques, the transmission of the secret formulas is accompanied by the vow: 'When you heal a man, may that man find a mate, when you heal a woman, may that woman find a match.'

When again the master has assured the disciple that he can do 'just as you like and that's just fine!' and has asked him if he has learned it all, this part of the ceremony comes to an end. The disciple thanks the master and offers him a complete set of clothes (presumably ordinary clothes, as the Monograph specifies that there is a pair of trousers among them).

The first to come and congratulate the disciple on his new status are the old ladies of the village (the wealthy and healthy amongst them). They escort a young virgin who, in the name of the Three Mother Goddesses, is now to sew the red cloth wrapper around the Tokens box. The box is now forever closed, until the newly ordained disciple shares some of its contents with his own pupil. The box is handed over to the ordinee, together with all the other objects and vestments. He then puts on his new clothes, headdress and Tokens box.

The final 'oral' - mouth to mouth - transmission (*kouchuan*) is the most dramatic. The master stands on the 'bridge', and invokes with his Tokens box, his own spirit helpers.The disciple, stands below and turns his head upwards towards his master.

First the master spits in the mouth of the disciple. Then he takes a gulp of water which he passes into the disciple's mouth. This is then repeated. No words are pronounced.

After the water, comes the transmission of fire. This is done by means of three pieces of twisted paper which have been lighted at both ends. The master takes one (burning) end in his mouth and the disciple the other. Both bite off their part of the paper, and swallow it.

Then follows the transmission of wine. Three mouthfuls of wine pass from the mouth of the master into that of his disciple.

Finally, the transmission of meat; three pieces of cooked pork meat are passed over in the same way as the wine. The disciple chews and swallows everything carefully.

The Master asks:

- Disciple, did you get everything?
- Yes, everything.
- If you have taken it all, then you should carry me on your back from this bridge!

And that is what happens. Like a filial son carrying his old father, the new priest takes his master on his back and carries him out of the sacred area.

AFTERTHOUGHTS

If we consider that the dramatic ceremony of transmission we just witnessed expresses something about the way the Taoist *duangong* think about their ritual, the first thought that comes to mind is: 'exchange'. As David Holm[48] has shown, the notion of exchange is fundamental in the *huanyuan* rites (Holm, more precisely than my 'thanksgiving' here, translates *huanyuan* as 'redemption of vows'). In the transmission and ordination ceremony, the 'elaborate gift exchange' is ever more present and dramatically expressed. Moreover, in this exchange, we can distinguish different kinds.

1. The most important 'Tokens' of authority, kept in the long wooden *paiyin* box are not only exchanged but mixed. Tuo reports that to the *duangong*, this mixing of contents establishes a bond through which 'in you there is something from me, in me there is something from you'. This is a form of swapping.

2. When receiving the secret hand formulas (*mijue*) which are demonstrated by the master, captured and transferred to the disciple's travel bag, and rendered visible in the form of cooked yellow beans which the disciple eats on the spot, the counterpart is expressed in the payment of coins, thrown in a water basin. This is a form of trading.

3. The priestly vestments and ritual instruments are a gift of the master. In exchange, the disciple offers him a full suit of ordinary clothes. This is a form of substitution. Henceforth, the disciple is the priest, successor to the master who becomes an ordinary citizen. This is of course not true in practice, but the exchange of clothes expresses clearly the notion of succession.

4. The 'oral transmission' is even more significant as an expression of transmission. Water and fire are the two fundamental elements and the two main agents of sacrifice. Wine and meat are the two most fundamental oblations. This mouth to mouth feeding as a final gift by the master recalls the succession of kings in ancient Africa where the person who captures the dying sovereign's last breath is designated as the lawful heir to the throne.[49] This exchange is a form of communion.

Having given all, not only the rites of sacrifice, but also their oblation, the master has rendered his forces and empowerment to the next generation. The former disciple has to carry him over the bridge out of the sacred area.

It is significant that the 'oral transmission' is closely linked to the *koujue*, the 'oral formulas', the core of the secret instructions which truly empower the disciple as an exorcist and a healer. One is tempted to make a comparison with the 'New Testament' through the communion of wine and bread ('this is my body') and 'the word that has become flesh'.[50]

Another vein to explore would be even more speculative. The way the final rite is performed - with the master standing on the 'bridge' above the disciple who turns his head upwards to him; both with their 'cape' of ribbons and their headdress which from afar looks like an Indian crown of feathers, the pair looks like feeding birds. Taoists are of course 'feathered gentlemen' (*yushi*), their robes are 'feather vestments' (*yuyi*), and immortals can fly and have sometimes wings. But in the context of feeding, the admittedly far-fetched comparison that comes to mind is with *Zhuangzi* XII: 'The True Sage is a quail at rest, a little fledgling at its meal, a bird in flight who leaves no trail'. Whatever may be the value of these comparisons, they do show some kind of universality in the way people think about ritual, in the way ritual spontaneously expresses itself. In the play on transmission and tradition of our Rustic Masters, spontaneity is the very essence of ritual. There is nothing to learn, nor is there anything to know: 'Just do it, and that's fine!'

Notes

1. Frits Staal, *Agni, The Vedic Ritual of the Fire Altar*, I-II, Berkeley, 1983.
2. The project started in 1991 with a grant of the Chiang Ching-kuo Foundation for International Scholarly Exchange and was conducted through the cooperation of an international group of scholars from Taiwan, Hong Kong, the United States and Europe. The monographs published hitherto have, by and large, been written by scholars from the Chinese mainland.
3. The series is called *Minsu quyi congshu* (Chinese Folklore Reports and Studies Collection) and is published by the Shi Ho-cheng Folk Culture Foundation in Taiwan. The volumes, containing a single distinct monograph each, are not numbered.
4. *Guizhousheng Cengongxian Pingzhuangxiang Gelaozhu nuotan guozhi yishi diaocha baogao* (Report on the Investigation on the Rites of Transmission of Office of the Nuo [Theatre] Altar of the Gelao Ethnic Group of Pingzhuang County, Cengong District in Guizhou Province). Tuo Xiuming, Yang Qixiao, Wang Ch'iu-kuei (eds), *Minsu quyi congshu*, Taipei, Shi Ho-cheng Folk Culture Foundation, 1994. 426 p. (including 100 photographs). The report was written by a group of scholars under the auspices of the Centre for Research on Nuo Culture of the Academy of Ethnology of Guizhou (Guizhou Minzu xueyuan Nuowenhua yanjiu zhongxin).
5. On the question of the definition of chinese 'minority peoples', see C. Qiao and N. Trapp, *Ethnicity and Ethnic Groups in China*. New Asia Academic Bulletin, vol. VIII, Hong Kong, 1989. Minorities enjoy greater religious freedom, inasmuch they are supposed to have attained a lesser stage of historical evolution compared to the majority 'Han' population.
6. The title has been defined as 'Collection of Traditional Chinese Ritual texts'. The collection is to be published by the Hsin Wen-feng Publishing House in Taipei.
7. Published by the Shi Ho-cheng Folk Culture Foundation in Taiwan since 1979.
8. The Taoist masters we are studying here are peasants, or as our Monograph (page 121) puts it: 'half peasants, half shamans'. There are also female masters in Guizhou, who have exactly the same function as the male (see 'Guizhou di Nuoxi yu Mianju' in *Minsu (Folklore)*, Beijing, 1989.5, pp. 2-6.
9. See A.J.A. Elliott, *Chinese Spirit-Medium Cults in Singapore*. London: SOAS, 1955, plate VI.

10. See the Monograph, page 87. A more complete description is given in the monograph on the so-called Xinuoshen ritual (see below note 13) page 131. Compare my 'The Divine Jester: Some remarks on the Gods of the Chinese Marionette Theatre', *Bulletin of the Institute of Ethnology Academia Sinica* 21. Taipei, 1966, pp. 81-95.
11. This is a personal communication made to me by Professor Wang Ch'iu-kuei on the experiences of the Guizhou masked 'Nuo' dancers of the so-called 'Dixi' type.
12. The Monograph, pages 12 to 17, gives a list of the more than fifty different services that can be performed by the present Taoist masters.
13. See *Guizhousheng Cengongxian Zhuxixiang Cenwangcun Laowuqi xinuoshen diaocha baogao*, Wang Ch'iu-kuei, Tuo Xiuming (eds), *Minsu quyi congshu*. Taipei, 1995, p. 18.
14. *Ibid.*, pp. 17-18 and Monograph, pp. 11-12.
15. See Tanaka Issei, *Chugoku Fukei Engeki Kenkyū*. Tokyo: Institute of Oriental Culture, University of Tokyo, 1993, p. 1061.
16. Tuo Xiuming, 'Nuotan chuancheng di shenmixing yu xiquxing', *Minsu quyi* 92 (volume 3 of the collected papers of a conference on sacrificial ritual and ritual theatre). Taipei, 1994, pp. 967-1012; and the monograph quoted above, note 13.
17. *Guizhousheng Dejiangxian Wenpingxiang Huangtucun Tujiazhu chongshounuo diaocha baogao* (hereafter, the 'Dejiangxian monograph'). Wang Ch'iu-kuei, Tuo Xiuming (eds), *Minsu chuyi congshu*. Taipei, 1994; and References of Dejiangxian monograph and Tanaka Issei, *op. cit.*
18. See the Dejiangxian monograph quoted above, pp. 132-5.
19. For the detailed program, see Monograph, pp. 70-2.
20. For instance, this is the case for the rite of climbing a ladder of knives. As in other parts of China, the rite is performed at the very end of the ordination ceremony; for the announcement of the accomplishment of the ordination to the Heavenly authorities. This fact is clearly shown here in the Monograph on page 117. But, at the same time, the rite is given among the 'nuo tricks' (*nuoji*) on page 100, and, as if this was not confusing enough it is given in the general program of the rites (page 71) as happening and the day *before* the ordination!
21. See Inez de Beauclair, 'The Keh Lao of Kueichow and their History', *Studia Serica* 5, 1946, pp. 1-44 (reprinted in *Tribal Cultures of Southwest China*, by Inez de Beauclair, Asian Folklore and Social Life Monographs, vol. II, Taipei, 1970).
22. See Monograph, p. 165.
23. See Dejianxian monograph, p. 4 and compare our Monograph, p. 347.
24. This term is frequently given by D.C. Graham, *Folk Religion in Southwest China*. Washington: Smithsonian Institution 1961. His descriptions are rather vague.
25. See *Zhongguo shaoshu minzu cidian*. Peking: Minzhu chubanshe, 1991, s.v. 'nuoxi'.
26. Graham, *op. cit.*, p. 71.
27. See Monograph, p. 121.

28. It should be noted that the texts of some rituals and theatre pieces performed are not given, whereas other, which were not performed, are reproduced.
29. See my 'Vernacular and Classical Ritual in Taoism', *Journal of Asian Studies* XLV.1, November 1986, pp. 21-57.
30. Some of these spells, which belong to the 'secret' tradition have been reproduced in the Monograph, pp. 357-61.
31. An exception is the ms. for the transmission ritual, on page 355, dated 1904 (or 1964?), and copied by the *daoshi* Chen Faling.
32. Pangu is the mythical creator of the universe, here 'founder' of the liturgical tradition, and called 'Lord of the Ritual of the Three Origins'.
33. Monograph, p. 254; I have simplified their titles somewhat.
34. Reading *qing* for *xiang*.
35. Reading *deng* for *zheng*.
36. See my 'Quelques remarques sur la fonction de l'Inspecteur des Mérites', *Dôkyô no Zôhôteki kenkyû*. Sakai Tadao, ed. Kokushu kankyokai, Tokyo, 1977, pp. 252-90 (in Japanese).
37. See R.A. Stein, 'Remarques sur les mouvements politico-religieux au IIe siècle ap. J.C.', *T'oung Pao* 50, 1963.
38. See Monograph, p. 418 (illustration no. 83); this document here dates from 1981, but is copied after a Qing model. On Mayang and its school, see also the other Cengong monograph, p. 35.
39. And looks like it too: the 'Nuo' female deity worshipped on the altar has a head dress with the symbol of a bird, just as Bixia yuanjun.
40. Most other schools indicate generations by choosing a distinct character from a given poem, etc. See Koyanagi Shigeta, *Haku'unkan* (Monograph on the Baiyunguan in Peking). Tokyo, 1934, who gives lists of genealogy poems for different Taoist schools.
41. See my *Corps taoïste*, pp. 86-99. The Heavenly Master is not mentioned among the patriarchs listed in the rituals.
42. Each of these so-called 'tricks' is in fact a ritual and part of a general service. The intertwining bamboos are a form of divination in case of severe illnesses; others are used in purification or exorcisms.
43. Such a practice, documented in the video-tape is not current elsewhere, where the ritual implements of the ordained master cannot be touched by an uninitiated disciple.
44. *Corps taoïste*, p. 72.
45. Here Laojun is assimilated with Yuhuang, the Jade Emperor. In ancient Taoist rituals of the fourth and fifth centuries, Yuhuang is indeed an epithet of Laojun.
46. These spirit soldiers of the 'Five Barbarian Tribes' are frequently mentioned in early medieval ritual texts.
47. This is how I understand the term *chongnuo*, frequently used by the masters of Guizhou, but not clearly explained by the researchers.
48. David Holm, 'The Redemption of Vows in Shanglin', *Minsu quyi* 92, 1994, p. 896.

49. J.G. Frazer, *The Golden Bough (Abridged Edition)*, chapter 27, 'Succession of the Soul', Wordsworth Reference Reprint, 1994, p. 294. This comparison was suggested to me by Mrs. Yuan Bingling of Leiden University.
50. Following the translation by Burton Watson, *The Complete Works of Chuang Tzu*. New York: Columbia University Press, 1970, p. 130.

GLOSSARY

baojushi	保舉師
Bixia yuanjun	碧霞元君
Cengong	岑鞏
Chen Faling	陳法靈
chongnuo	冲儺
chuanfa	傳法
deng	燈
daoshi	道士
Daqing tianxia Guizhou shoufu	大清天下貴州首府
Dejiang	德江
Di	帝
die	牒
difang xi	地方戲
duangong	端公
duangongxi	端公戲
fa	法
famen	法門
fapai	法派
fashi	法師
fengjian mixin huodong	封建迷信活動
fulu	符錄
Fuxi	伏羲
Gelao	仡老
Guanyin	觀音
Guigu xiansheng	鬼谷先生
Guiyang	貴陽
Guizhousheng Cengongxian Pingzhuangxiang Gelaozu nuotan guozhi yishi diaocha baogao	貴州省岑貢縣平莊鄉仡老族儺壇過職儀式調查報告
Guizhousheng Cengongxian Zhuxixiang Cenwangcun Laowuqi xinuoshen diaocha baogao	貴州省岑貢縣注溪鄉岑王村老屋基喜儺神調查報告
Guizhousheng Dejiangxian Wenpingxiang Huangtucun Tujiazhu chongshou nuo diaocha baogao	貴州省德江縣穩坪鄉黃土村土家族沖壽儺調查報告
hebiao	和標
Huainan	淮南

huan	還
huanyuan	還願
huixiang	回鄉
jiefa shi	接法師
jijiu	祭酒
jue	訣
kaiguang	開光
koujue	口訣
Laojun	老君
Li Er	李耳
Man	蠻
Mayang	麻陽
mijue	密訣
Minguo	民國
Minsu quyi	民俗曲藝
Minsu quyi congshu	民俗曲藝叢書
neitan	內壇
nuo	儺
nuoji	儺祭
nuotangxi	儺堂戲
pai	牌
paidai	牌帶
paiyin	牌印
Pangu	盤古
paopai	拋牌
Pingzhuang	坪莊
Qin	秦
qingshen	請神
Rangxing ke	禳星科
Rong	戎
Ru-, Dao-, Fojiao san shengren	儒道佛教三聖人
Sanguan	三官
sanxing	三星
shangshu	上書
shen	神
shi	師
shigong	師公
Tiandishui yangke	天地水陽科
tiaoshen	跳神
tulaoshi	土老師

Tuo Xiuming 庹修明
waitan 外壇
Wang Ch'iu-kuai 王秋桂
Wenchang dijun 文昌帝君
wuying 五營
xieshi 邪師
Xinuoshen 喜儺神
Xuantian shangdi 玄天上帝
Yang Qixiao 楊啓孝
Yanluo 閻羅
Yi 夷
Yindushi 引度師
Yuhuang 玉皇
Yuhuang hao 玉皇號
Yuhuang menxia 玉皇門下
yushi 羽師
yuyi 羽衣
zhengming shi 證明師
Zhengyi 正一
Zhongguo shaoshu minzu cidian 中國少數民族辭典
Zhongshan shenzhou 中山神咒
zhou 咒
Zhuangzi 莊子
Ziwei 紫微

XXVI

HOMELESSNESS AND HOMECOMING NIETZSCHE, HEIDEGGER, HÖLDERLIN

Hans Sluga

1. 'Among Europeans today there is no lack of those who are entitled to call themselves homeless in a distinctive and honourable sense', Nietzsche wrote in 1882. He himself was more or less homeless at the time, living in boarding houses and cheap hotels in the years of his greatest productivity between 1879 and 1889. But he was speaking of homelessness here above all as a metaphor for the condition of modern man. 'We children of the future,' he wrote, 'how *could* we be at home in this today? We feel disfavour for all ideals that might lead one to feel at home even in this fragile, broken time of transition.' In declaring himself homeless in this sense, he took above all distance from the political ideals of his time, from its conservatism as well as its liberalism, from its humanitarianism and its nationalism. Of the latter - the 'national scabies' as he called it - he wrote: 'For that we are too open minded, too malicious, too spoiled, also too well informed, too "travelled"'.[1]

A century after Nietzsche homelessness is still with us, both in its literal and its metaphorical sense. There also still exists a nostalgic longing for home, for a place that we can truly call our own, for the safety of community and nation. In consequence, we find it hard to accept Nietzsche's optimistic claim that 'there is great advantage to be gained in distantly estranging ourselves from our age'.[2] We are less confident than he that this condition alone prepares us for the redemptive gift of a 'philosophy of the morning', the gift 'of all those free spirits who are at home in mountain, wood, and solitude ... . Born out of the mysteries of dawn, they ponder on how, between the tenth an the eleventh stroke of the clock, the day could present a face so pure, so light-filled, so cheerful and transfigured.'[3]

2. Among those struggling with these issues a generation or so after Nietzsche was Martin Heidegger for whom the question of homelessness and the longing for community became intertwined with the political tribulations

of his own time.

The themes of homelessness and community make their first appearance in *Being and Time* where Heidegger speaks of authentic existence as an unhomely and homeless condition:

> Anxiety brings Dasein face to face with its Being-free for the authenticity of its Being … . In anxiety one feels 'uncanny' (*unheimlich*). This refers, first of all, to the peculiar indefiniteness of that with respect to which Dasein finds itself in anxiety: the 'nothing and nowhere'. But here 'uncanniness' also means 'not-being-at-home' (*Nicht-zuhause-sein*).[4]

Being-at-home, he argues, is merely the tranquillized reassurance of an average everydayness of Dasein. On the other hand, as Dasein falls, anxiety brings it back from its absorption in the 'world'. Everyday familiarity collapses. Dasein has been individualized, but individualized *as* Being-in-the-world. Being-in enters into the existential 'mode' of the 'not-at-home' (p. 189). Being-not-at-home is, then, for the Heidegger of *Being and Time* an existential mode of Dasein in its fallen, authentic condition, a mode which Dasein reaches when it has freed itself from the everyday publicness of the One (*das Man*) and has attained an individualized existence in the world.

What was familiar in its everydayness is now revealed to provide no home for Dasein. What once seemed home, reveals itself to be just as uncanny as all that we find outside the original home. For the truly homeless, there is no turning back home. Heidegger writes that for an anxious, authentic existence the world has 'sunk into insignificance' and shows itself 'in its empty mercilessness' (p. 343). In anxiety 'Dasein is made to realize that it cannot derive its projects from the possibilities of the world. Then, "*Dasein*" is taken all the way back to its naked uncanniness, to its naked not-being at home, and becomes numbed by it ("von ihr benommen"). The condition takes "*Dasein*" back from its "*worldly*" possibilities, but at the same time *gives* it the possibility of an *authentic* potential-for-Being' (p. 344).

Yet in Division II of *Being and Time* Heidegger appears to abandon the radical - we might say Nietzschean - implications of these words. For he now declares that authentic Dasein must turn back from its condition of homelessness in order to find a hold. He announces a turning back home, a *Heimkehr*, whose exact nature remains, however, unspecified till much later. But he takes a first step to such a turning back in the puzzling section 74 of *Being and Time* in which a 'more concrete working out of temporality' is promised as an interpretation of Dasein's historicality. 'In the existential analysis we cannot, in principle, discuss what Dasein *factically* resolves in any particular case', Heidegger writes there. 'Nevertheless, we must ask whence, *in general*, Dasein can draw those possibilities upon which it factically projects itself' (p. 383). As thrown, Dasein is delivered over to itself but always as something in the world. This world, Heidegger declares now, discloses to Dasein possibilities of authentic existing and does so in two distinctive ways - as heritage and as destiny.

He writes: 'The resoluteness in which Dasein comes back to itself, discloses current factical possibilities of authentic existing, and discloses them *in terms of the heritage* which that resoluteness, as thrown, *takes over* … . If everything 'good' is a heritage, and the character of 'goodness' lies in making authentic existence possible, then the handing down of a heritage constitutes itself in resoluteness' (p. 383). As being in the world, Dasein is for Heidegger moreover a being with others and shares as such in a common destiny. 'This is how we designate the happening of the community, of the *Volk*, the people'. It is a destiny shared with one's generation. 'Dasein's fateful destiny in and with its 'generation' goes to make up the full authentic happening of Dasein' (p. 384-5).

We may ask: how can a Dasein reduced to its naked not-being-at-home have a heritage which it must take over and hand down in order to be authentic? What is the heritage of the homeless? From where does it come to the homeless and to whom do they owe it? Who is the *Volk* to which Heidegger refers Dasein in this passage and what constitutes a generation? What does it mean to share a destiny with this *Volk* and this generation? How can such shared destiny be significant to a Dasein in its naked uncanniness?

Heidegger's remarks in section 74 of *Being and Time* remain peculiarly abstract, despite their promise of a concrete working out of Dasein's temporality. It is only in his subsequent political involvement, that his words take on a determinate meaning. For in 1933 Heidegger will say in his rectorial address that authentic Dasein is called to a 'true and common rootedness' in the 'destiny of the German people'.[5] And later, in November of 1933, he will foreswear 'the idolization of a groundless and powerless thinking', claiming that the origin of science and knowledge lies in the will to truth of a people. Knowledge, so Heidegger, is 'bound into the necessity of a self-reliant folkish 'Dasein''.[6] Authentic Dasein has, so it seems, been finally rescued from its not-being-at-home. It has found its home in the shared German heritage, in the ground and destiny of the German people. Homeless Dasein has turned into folkish existence.

3. But how are we to understand the destiny and folkish existence of the German people? It was in pursuit of that question that Heidegger turned to Nietzsche and Hölderlin in the years between 1935 and 1945.

He had by then already passed through that rash phase of political enthusiasm which had made him identify so unconditionally with Hitler's revolution in 1933. Wary of many aspects of the actually existing National

Socialism, he had, however, still not broken openly with the system and still hoped to find an inner truth in the Nazi movement (the 'inner truth and greatness of National Socialism' of which he spoke in his 1935 lectures on metaphysics). Under these circumstance, Nietzsche and Hölderlin became for him indispensable guides to the question what it meant to be truly German, what it meant to be at home in Germany.

He began his engagement with them in an interpretation of Hölderlin's hymns 'Germania' and 'The Rhine' and he continued it with a critical reading of Nietzsche's metaphysics. Heidegger's Nietzsche lectures took up the years between 1936 and 1941. They were meant to be concluded in the Winter of 1941-2 with a course on the five basic concepts of Nietzsche's metaphysics, intended as a summary of the whole preceding series. But at the last moment Heidegger cancelled the course and embarked on a new series of lectures on Hölderlin's hymns.[7]

In the period between 1936 and 1945 Nietzsche and Hölderlin were, thus, never far apart in Heidegger's mind. The two together defined for him the essence of what it meant to be German. Insisting that Germany was still the nation of thinkers and poets ('we are not only called this but also are this') he spoke of a historically necessary interdependence of poetry and thinking that was exemplified 'in Nietzsche, who as a thinker is a poet, and in Hölderlin, who as a poet is a thinker'.[8]

The two played, however, very different and, indeed, antagonistic roles for Heidegger. While Nietzsche presented him with the sharpest formulation of the problem of modernity, Hölderlin suggested to him the inevitable step beyond that problematic. For that reason he resisted any attempts to equate the two. 'The recent fashion which puts Hölderlin and Nietzsche side by side is completely misleading', he told his students. The two were, in fact, separated by an abyss. But: 'Abysmally different, the two together, nevertheless, still determine the nearest and the farthest future of the Germans and the West.'[9]

The relation between the two was finally to be made clear in lectures scheduled for the Winter semester of 1944-5, but because the war was now coming speedily to an end, that course was never actually given. Heidegger managed only to deliver the first lecture before he was drafted into the home defence.[10] His notes for that course, though incomplete, are, however, indispensable for understanding how he finally saw the relation between Hölderlin and Nietzsche.

Heidegger had previously characterized Nietzsche as the philosopher of a technological understanding of Being and he had quietly woven into his critique of Nietzsche's metaphysics a critique of important aspects of National Socialism: of its expansionist ambitions, its Social Darwinist tendencies, its obsession with bureaucratic organization, and its power politics. He had also already announced the end of the epoch of the metaphysics of the will to power and of metaphysical thinking as a whole. By confronting Nietzsche with Hölderlin he meant, above all, to contrast a thinker still committed to a form of metaphysical thinking and a poet hinting at its overcoming.

In the lectures scheduled for the Winter of 1944-5 Heidegger had planned to add to the earlier critique of Nietzsche the charge that he was the philosopher of the godlessness and worldlessness of modern man. 'Nietzsche thought that the gods and all things are 'products' of creative man gives voice to a destiny of the essential history of Western man', he wrote in his notes. And: 'In the absence of the gods and in the decay of the world homelessness (*Heimatlosigkeit*) is specifically assigned to modern-historical man.'[11] Unlike Nietzsche, Heidegger did not take this condition as providing the chance for a new beginning, for a 'philosophy of the morning'. He saw it more bleakly as a threat and drew in support on Nietzsche's well-known lines:

> The crows are screaming,
> flying quick to town
> soon snow will fall,
> woe to the one who has no home.

He also, surprisingly, maintained that Nietzsche had anticipated the possibility of a possible homecoming. In a perverse turn that can be explained only by assuming that Nietzsche's name stands here for Nazi politics Heidegger ascribed to him, in fact, the idea that such homecoming was to be achieved through wilful conquest. However we understand this, it is clear that Heidegger himself by this time was nostalgically looking for an escape from the homelessness of the modern condition, that he did not himself believe this to be possible through wilful conquest, and that he was looking for such an escape instead along the path of Hölderlin's poetry. Heidegger might have done well to recall at this point Nietzsche's remark:

We have left the land and have embarked. We have burned our bridges behind us - indeed, we have gone farther and have destroyed the land behind us ... . Woe, when you feel homesick for the land as if it had offered more *freedom* - and there is no longer any 'land'.[12]

4. It was in his lectures on Hölderlin's hymn '*Der Ister*' in the Summer of 1942 where Heidegger spoke most eloquently of Hölderlin as the poet of homecoming.

The river of which '*Der Ister*' speaks, the Danube - the stream that crosses both Hölderlin's and Heidegger's homeland - becomes for Heidegger at once a symbol of the human condition itself. He writes: 'The stream "is" the location which rules through the dwelling of men and destines him into that into which he belongs and where he is at home (*heimisch*).'[13] The stream is both a place of dwelling and journeying, of *Wohnen* and *Wanderschaft* in a 'primordial unity'. It is 'the journeying of the journey in which coming to be at home has its essence' (p. 51).

Hölderlin, the poet of streams, is for Heidegger also the one to illuminate for us what it is to be journeying, to be far from home and homeless, and what it means to return home, to find home again, and to be able to say what home is. It is to the theme of departure and return that Heidegger devotes much of his discussion. Speaking at a moment when German troops were departing to all parts of Europe, when they had conquered Poland and France, had moved deep into Russia to the gates of Stalingrad, and when policies had just been set in motion to bring about the final solution of the Jewish question, Heidegger declared with the greatest emphasis that 'we must from now on think more German in relation to ourselves than any other Germans before us' (p. 100). What drew him to Hölderlin was the thought that the poet had spoken 'out of a poetical concern for the homecoming of the historical and Western humanity of the Germans' (p. 84). Heidegger was convinced that 'coming to be at home in one's own is the only concern of Hölderlin's poetry when it enters the form of the hymn' (p. 60). And this own with which he saw Hölderlin concerned was nothing less than 'the fatherlandish for the Germans' (*das Vaterländische des Deutschen*).

But what was fatherlandish for the Germans? That might not be easy to say. In his lectures Heidegger distanced himself from the popular, political definitions of 'the German mission'. But this distancing was not meant to achieve a distance from National Socialism as a whole. He rejected, instead, the common political rhetoric only in order to affirm what he called 'the historical uniqueness of National Socialism' (p. 98). He was, however, sure that there was no short cut for determining that uniqueness. He thought that it required instead a long, roundabout journey. Leaving familiar assumptions behind, we had to turn to what might seem at first foreign to the political needs of the moment. Coming back from these meandering and far-reaching journeys, and only then, would it be possible to say what for the Germans was their own.

> One's own, the finding of one's own, and the appropriation of what one has found as one's own are not the most self-understood and not the easiest, but they remain the most difficult and as such a most difficult they are placed into the poet's care (p. 60).

The need to find home meant that man was to begin with and for a long time not at home, Heidegger said. And this not being at home was due to the fact that man misunderstands, denies, and even is bound to flee his home. To find home meant always to pass first through what was strange, alien, or foreign (*das Fremde*).

In his account of '*Der Ister*' Heidegger speaks of poetry as a homecoming, as something in the nature of a return from abroad. Hölderlin can, as Heidegger puts it, find his home only through coming back. A year later, in an address delivered at the centenary of the poet's death, Heidegger took the opportunity to elaborate further on this theme of homecoming. Commenting on Hölderlin's poem 'Homecoming' (*Heimkehr*), a description of the poet's return from Switzerland to his Swabian homeland, he treats it as a metaphor for a more profound kind of homecoming. Travelling across Lake Constance in the early morning light, the poet finds himself come back to where everything appears familiar, where even the most passing salute seems a greeting from friends, and where every face seems related. 'Naturally, it is the land of birth, the soil of the homeland, what you are seeking is already near and is meeting you.'

The poet, as Heidegger reads these lines, has travelled to the Alps out of faithfulness to his own homeland in order to return to it as his own. Coming home is then a 'return to the nearness of the origin'. But this nearness can be attained only by passing through that which is not home, which is other and foreign. Those who merely live on the soil of their land of birth, are not those who have come home into the own of their homeland. The true origin is found only through a return home. Hölderlin's, the poet's calling is that homecoming through which alone the homeland becomes the land of the nearness to the origin. We, too, are called to share in the poet's concern. And if we do, 'then there is kinship with the poet. Then there is homecoming. And this homecoming is the future of the historical essence of the German'.[14]

Hölderlin's deepest insight is, indeed, the recognition of 'the law of being at home' according to which historical man is not familiar with his home at

> the beginning of his history, that he must even become not at home for this home, in order to learn from the other, by departing to it, the appropriation of his own, and that he can come to be at home only in the return from this other.[15]

According to Heidegger, Hölderlin had given voice to this 'law of the return' when he had written to his friend Böhlendorf in 1801 that 'nothing is more difficult to learn to use freely than the national' (p. 168).

These terms offer us, at the same time, an account of Heidegger's own preoccupation with Hölderlin. In interpreting the poet's words he wants to determine what being at home, what being German might mean for himself and his contemporaries in 1942. This home can reveal itself as its own, however, only in so far as we first pass through something other. Hölderlin's poem serves the philosopher as that other through which we must pass and from which we must return in order to discover what is truly our's. 'All interpreting is a translating', Heidegger says in his lectures. (*Alles Auslegen ist übersetzen*), but as such it is also always a setting over (an *über-setzen*).

> In so far as we are forced to interpret the poetical and thoughtful works of our own language, it is shown that every historical language is in and for itself in need of translation and not only in relation to some other. This, in turn, shows, that a historical people is not in itself and without effort at home in its own language. For that reason it can be that we speak 'German', but talk at the same time 'American' (p. 79-80).

Hölderlin's return into his own, his concern with what is fatherlandish for the German, is, thus, necessarily mediated through a passage through another. That other was for Hölderlin, as Heidegger argues, the world of the ancient Greeks. In undertaking his translation of Pindar's poetry and of Sophocles' *Antigone*, Hölderlin passed through the 'foreign and ancient land of the Greeks' and he did so in order to discover his own fatherland. That was no arbitrary passage. For the Germans as a whole can find home, according to Heidegger, only by passing through the foreign and ancient land of the Greeks. Greek poetry and Greek thinking alone can reveal to them what their home is. But they must remain conscious at all times that they are separated from the Greeks, that the Greeks have to be for them the primordial other. Hölderlin had understood this. For him Greek culture had not been an object of nostalgic longing. He had seen the Greeks neither in classical, nor in metaphysical or romantic term. The important thing was, as Heidegger puts it, to understand that 'Greekness is not equal to and not the same as Germanness' (p. 67). Setting himself against contemporary tendencies to view National Socialism as the rebirth of ancient Greek culture, Heidegger says contemptuously:

> The Greeks appear in most 'research reports' as pure National Socialists. In their zeal the scholars seem unaware that such 'results' do no service to National Socialism and its historical uniqueness and that National Socialism is not in need of them (p. 98).

5. The Greeks were, in Heidegger's views, rather that farthest point to which one has to move in order to return to the land of one's fathers and recognize it as one's own. In his lectures on '*Der Ister*' he exemplifies this double move by turning from the examination of Hölderlin's hymn to a discussion of the words of Sophocles's *Antigone* and then back again to Hölderlin's

poem. The structure of the lecture course itself corresponds, thus, to the law of return previously announced. The same law is exhibited in the course of Heidegger's thinking between 1935 and 1945. Beginning with lectures on Hölderlin, the poet with whom he identifies, he turns to Nietzsche as Hölderlin's 'abysmally different' other and from there back again to the poet. This three-fold movement is, indeed, for Heidegger the movement of all genuine philosophical thinking. For philosophy is always initially a loss of home which, passing through the other of philosophical and poetical texts and reflections, becomes finally able to escape its bloodless and groundless condition into a being at home in a newly appropriated fatherland.

More than half of the text of Heidegger's lectures on '*Der Ister*' is taken up by his examination of the second chorus of Sophocles' *Antigone*. Heidegger had previously discussed this text in his *Introduction to Metaphysics* of 1935. But where he had earlier characterized the polis as the place of the political, he now insists that it is not at all political in character but rather metaphysical. He begins his discussion with the famous first lines of the Sophoclean chorus:

pollà ta deinà koudèn anthrôpou deinôteron pèlei.

According to Heidegger have not yet been adequately translated, not even by Hölderlin himself in his rendering of *Antigone*. He argues that only his own philosophical language can reveal the full meaning of the Sophoclean words. For Heidegger it is evident that they can only be rendered as: 'Manifold the uncanny, but nothing more uncanny stirs towering above man.' By translating '*deinon*' as 'uncanny' (*unheimlich*), he argues, we expose a connection 'which reaches presumably beyond the consonance of the words "*unheimlich*" and "*unheimisch*", to the deep link between the "uncanny" and the "not-being-at-home"' (p. 84). He adds:

> We mean the uncanny in the sense of that which is not at home - which is not at home in its own home. Only because of that, can the 'not-being-at-home' be in consequence also the 'uncanny', meaning that which is strange, anxiety-provoking, and 'terrible'. The word of Sophocles that man is the most uncanny being says then, that man is in a unique sense not at home and that coming to be at home is his concern (p. 87).

While the Sophoclean chorus seems at first sight pre-occupied with the homelessness of man, Heidegger insists that it hints in effect at the possibility of a homecoming. Man is the uncanniest being, but his uncanniness, his not-being-at-home is possible only in so far as he relates himself to that which is as such, because he has an understanding of Being itself. The chorus that begins by speaking of the uncanniness of man, ends therefore by singing of the hearth from which this uncanny being is banned.

Let him not become familiar with the hearth
Who does not share my thought and my efforts.

In Heidegger's reading that hearth is nothing less than 'the original harmony of what is, the unifying one, and centre of the sphere'. It is 'the centre of whatever is to which all that is remains forever primordially directed because and in so far as it is what it is. The hearth like centre of what is is Being. Being is the hearth' (p. 140). If finding home means to find back to one's hearth and to recognize that hearth as one's own, then homecoming is to enter into the truth of Being. And in so far as poetry, that of Sophocles and of Hölderlin, is a coming home, it is a 'speaking and finding of Being' itself (p. 149).

We can say then that Being speaks to philosophy through the mouth of the poet and that philosophy is a being-spoken-to by Being itself. But this being-spoken-to requires an asking. The philosopher must ask about the words of the poet and the poet must ask for the truth of Being. Finding the truth of Being, hearing it speak to us calls then for a journey that seems, at first, to take us far from what is our ultimate concern. Being is at once our home and that which we can reach only after a long journey. Being at home is possible for us only as a homecoming and hence after a not-being-at-home.

6. With this dramatic announcement terminates the second part of Heidegger's lecture course on '*Der Ister*'. It remains for him now only to return to the poet of his own homeland and to appropriate the home that the passage through the Greeks is said to have made possible. Heidegger, in fact, never delivered the third and completing part of his course since the semester had come to an end. Fully composed, the text of the lecture remained in his drawer until it was finally published in 1984. The homecoming that Heidegger was envisaging remained, thus, curiously incomplete at the time of the lecture.

What was left unsaid in this way was the coming together of the truth of Being and the truth of a people as Heidegger saw it at the time. For Hölderlin was for him not only the poet who could bring the truth of Being to speak to the philosopher, but also the one who could help the Germans to think in a more German way, the poet of the fatherlandish idea of the Germans.

By examining the words of Sophocles, Heidegger says in the final, unspoken part of his lecture course, we discover what we still need to learn from Hölderlin's words. Hölderlin's river poems, he tells us, and Sophocles's chorus say, in fact, poetically the same. The Hölderlin of the hymns and the Sophocles of *Antigone* are tied together in a poetical and historical dialogue. It is not that Hölderlin identified himself with the Greeks; on the contrary, he fully recognized the difference between these two historical embodiments of humanity. But the two different embodiments were not alien to each other or antagonistic, they were complementary and hence necessary for each other. For what was for the Greeks their own is, at once, the other, the alien (*das Fremde*) for the Germans and what was for the Greeks the other is for Germans their own. The two nations belong together then, each needing the other for its own completion.

It remains to be discovered what Hölderlin brought back from this journey to Greece. The answer, Heidegger says, is to be found, once again, in his letter to Böhlendorf from 1801. Hölderlin writes there: 'As I see it, clarity of presentation is original to us and just as natural as the fire from the sky is for the Greeks.' The sentence is, so Heidegger, the key for what each nation must learn from the other to find back to itself and its own hearth. Clarity of presentation is indeed, the Germans' own, native endowment. 'The making of frames and compartments, dividing and organizing propels them' (p. 169). What they lack and what must come to them from outside is the 'fire from the sky', the inspiration of the divine. The Greeks, in contrast, had possession of that fire, but still require German organization and order to bring it under control. In this difference and mutual dependence lies for Heidegger the truth of the Greek and the German people.

> The Germans must be hit by the fire from the sky in order to use their own freely. For that reason the departure into southern lands is inevitable. For that reason the northeast is the promise of a poetical destiny (p. 170).

7. We may wonder what kind of conception of Germanness Heidegger expressed in these words. How much distance did he want to create between his own understanding of the German homeland and that advanced by the National Socialists? We know that he effected no radical break with National Socialism, but thought to discover its unique truth elsewhere than where general opinion sought it. Yet the question is, how much of an alternative to the existing National Socialism Heidegger was offering.

While we may grant to Heidegger the uniqueness of National Socialism, we see that uniqueness now in an altogether more sombre light. We will certainly not identify 'the truth of Being' with the unique truth of National Socialism. Heidegger's talk of such sameness strikes us now as perverse and repulsive. We are also struck now by the arbitrariness with which he links these thoughts to Hölderlin's poetry. To us that poetry speaks of other, more universal, human concerns. Such strictures must not blind us to the fact that Heidegger's concern with Hölderlin has led him to ask some questions of the highest significance. They concern what is our own and how we can come to recognize it; they concern what being at home and what not being at home is; they concern, above all, the homelessness that we can now see residing at the heart of philosophy.

In trying to listen to Hölderlin's words Heidegger has confronted the question what it is for him to think from a particular place and a particular time. In the uncompleted lectures of the Winter semester 1944 he said that philosophy must be concerned with the question 'what is now?'

> When man thinks what is, ... then he thinks and has always already thought the question what has been and what is going to come. Thinking in this way, he moves everywhere in that form of thinking which has since ancient times been called 'philosophy'.[16]

And this brought him to the further question in what he saw the current meaning of his preoccupation with Nietzsche and Hölderlin. Why should he be speaking of those two rather than of Kant and Goethe? And his answer was that 'Nietzsche is that thinker who thinks what is now. Hölderlin is that poet who expresses poetically what is now.'

Admittedly, Heidegger's attempt to understand what is now is characterized by a peculiarly narrow focus on the world. He possessed neither Hölderlin's appreciation for the achievements of the French revolution nor Nietzsche's vision of a new European politics. This narrowness is evident in his peculiarly negative assessment of the Anglo-Saxon world. 'We know today', he said in his lectures on '*Der Ister*',

> that the Anglo-Saxon world of Americanism is determined to destroy Europe, that is the home, and that is the beginning of the West. What is primordial is, however, indestructible. The entry of America into this planetary war [which had occurred half a year before Heidegger's lecture] is already the final act of the American loss of history and self-destruction ... . The hidden spirit of the beginning in the West has nothing but contempt for this process of self-destruction.[17]

Heidegger spoke crudely in these lectures of an America pre-occupied with the quantitative and hence given to a lack of all measure and moderation. 'This is the principle we call Americanism', he said. 'Bolshevism is only a variety of Americanism. This is the truly dangerous form of being without measure.'[18]

There lay in these words a lack of estimation of the power and resilience of the American continent which Heidegger shared with Hitler. Such words speak an ability to see that the Anglo-Saxon world, that America, too, has roots in the Greek tradition, that they also participate in the hidden spirit of the beginning. For Greece has produced many concepts and these, surely, include the concepts of liberty and democracy, of science and quantitative measurement as well as those of a poetical thinking of being and nothingness. Many fatherlands have a right to draw on these origins. The origin does not dwell in a single place: in Swabia, for instance, near the source of the Danube, or in Germany; it has many embodiments and resides in many places. Nietzsche saw it right when he wrote:

> German philosophy as a whole ... is the most fundamental form of *romanticism* and homesickness there has ever been: the longing for the best that ever existed. One is no longer at home anywhere; at last one longs back for that place in which alone one can be at home, because it is the only place in which one would want to be at home: the *Greek* world! But it is in precisely that direction that all bridges are broken - except the rainbow-bridges of concepts! And these lead everywhere, into all the homes and 'fatherlands' that existed for Greek souls![19]

We know now that the wind blows where it wants to and that it is not confined to this or that continent. We understand that there is a homelessness to all human thinking, at the heart of all consciousness. In its current and heightened form we call this homelessness the modern condition. Terrifying as the condition may be at times, we also know that there is no return for us to the warm assurance of animal nature. Human beings are, after all, what Heidegger deplores, adventurers who find their home in that which is not home. 'For the adventurous heart', Heidegger said in a deprecating tone, 'the difference between being at home and not being at home is lost. The wilderness becomes an absolute in itself and serves as the 'fullness of being.'[20] To this we may readily assent; it is in the wilderness that we construct our shelters. We are for that reason, in the end, not like Heidegger's homecoming philosopher but like Nietzsche's free spirits,

> [at] home, or at least having been guests, in many countries of the spirit; having escaped again and again from the musty agreeable nooks into which preference and prejudice, youth, origin, the accidents of people and books or even exhaustion from wandering seem to have banished us ...[21]

Like Nietzsche we say that 'he who has attained to the least degree of freedom of mind cannot feel other than a wanderer on the earth - though not as a traveller *to* a final destination: for this destination does not exist'.[22]

Notes

1. Friedrich Nietzsche, *The Gay Science*, tr. W. Kaufmann, New York: Vintage, 1974, sect. 377.
2. Nietzsche, *Human All Too Human*, tr. R.J. Hollingdale. Cambridge: Cambridge University Press, 1986, vol. 1, sect. 616.
3. *Ibid.*, sect. 638.
4. Heidegger, *Being and Time*, tr. by J. Macquarrie and E. Robinson. New York: Harper & Row, 1962, p. 188. Page references are to the standard German pagination reproduced in the margins of the English text. Throughout this essay I have altered translations where that seemed desirable.
5. Heidegger, 'The Self-Assertion of the German University', in: Gunther Neske and Emil Kettering (eds), *Martin Heidegger and National Socialism*. New York: Paragon House, 1990, p. 5.
6. Heidegger, 'Bekenntnis zu Adolf Hitler und dem nationalsozialistischen Staat', in: Guido Schneeberger (ed.), *Nachlese zu Heidegger*. Bern, no publisher, 1962, p. 149.
7. Heidegger, *Nietzsches Metaphysik*, Gesamtausgabe, vol. 50. Frankfurt, Vittorio Klostermann, 1990, p. 161, and *Hölderlins Hymne 'Andenken'*, Gesamtausgabe, vol. 52. Frankfurt, Vittorio Klostermann, 1982, p. 1.
8. Heidegger, *Einleitung in die Philosophie. Denken und Dichten*, Gesamtausgabe, vol. 50. Frankfurt: Vittorio Klostermann, 1990, pp. 95-6.
9. *Hölderlins Hymne 'Andenken'*, p. 78.
10. The next time Heidegger lectured at Freiburg was well after the end of the Nazi period when he had been restored to the status of emeritus professor.
11. *Einleitung in die Philosophie, Denken und Dichten*, pp. 114 and 116.
12. *The Gay Science*, sect. 124.
13. *Hölderlin's Hymne 'Der Ister'*, Gesamtausgabe, vol. 53. Frankfurt: Vittorio Klostermann, 1984 p. 23. The following page references are to the same text.
14. 'Heimkunft/ An die Verwandten', in: *Erläuterungen zu Hölderlins Dichtung*, Gesamtausgabe, vol. 4. Frankfurt: Vittorio Klostermann, 1981, p. 30.
15. *Hölderlins Hymne 'Der Ister'*, p. 156. The following references are to the same text.
16. *Einleitung in die Philosophie*, pp. 90-1.
17. *Hölderlin's Hymne 'Der Ister'*, p. 68.
18. *Ibid.*, p. 86.
19. *The Will to Power*, tr. W. Kaufmann and R.J. Hollingdale. New York: Vintage, 1966, sect. 419.
20. *Einleitung in die Philosophie*, p. 91.
21. *Beyond Good and Evil*, tr. W. Kaufmann. New York: Vintage, 1968, sec. 44.
22. *Human All Too Human*, vol. 1, sect. 638. I have written this essay for Frits Staal not because he is interested in Nietzsche, Heidegger, or Hölderlin (which he may not be), but because he is, like me, 'a wanderer on the earth'.

XXVII

THE SOCIAL AND INTELLECTUAL ORIGINS OF HUBERT AND MAUSS'S THEORY OF RITUAL SACRIFICE

Ivan Strenski

'THE' DURKHEIMIAN THEORY OF SACRIFICE

In 1898, Henri Hubert and Marcel Mauss wrote a classic in the study of religion, '*Essai sur la nature et la fonction du sacrifice*' (hereafter *Sacrifice: Its Nature and Function*. It first appeared at the head of the second volume of *L'Année sociologique* as the second of the two 'Mémoires originaux' published there; the other was Durkheim's *Concerning the Definition of Religious Phenomena*. Remarkably, this slim volume of barely one hundred pages (less notes) has proven to be the most frequently cited theoretical work on sacrifice ever written. Indeed, Staal refers to it as 'the best starting point from which sophisticated and possibly adequate theories of ritual can be derived'.[1] Durkheim's confused and derivative view of sacrifice is by contrast now almost totally forgotten. In any event, Durkheim eventually fell in behind Hubert and Mauss and in *The Elementary Forms* says that of the two essential elements of sacrifice, communion and oblation, it is oblation that is 'even more permanent than communion'.[2] For this reason alone, Hubert and Mauss's Sacrifice can fairly be said to have formulated *the* Durkheimian theory of sacrifice.

WHAT IS THE THEORY ABOUT? PROBLEMS WITH 'PLAIN MEANING'

Knowing that Hubert and Mauss's *Sacrifice* is *the* Durkheimian theory of sacrifice, how can we tell what it might mean?

In posing the question this way I am not asking - at least primarily - about the so-called 'plain' meaning of the theory. Anyone sufficiently diligent can, with text in hand, decipher much of what the Durkheimians say about sacrifice. No great mystery lurks in the 'plain' meaning of the theory, although it would be tedious and involve a rehearsal of all of it here. But one can illustrate something about the 'plain' meaning of the theory. We know, for example, that Hubert and Mauss wanted to put two points at the centre of any definition of sacrifice: consecration and the victim:

> Sacrifice is a religious act which, through the consecration of a victim, modifies the condition of the moral person who accomplishes it or that of certain objects with which it is concerned.[3]

Thus, unlike both Robertson Smith and Durkheim who made communion central to sacrifice, what is essential for Hubert and Mauss is the *process* of something's *becoming* (or ceasing to become) sacred, and less, or sometimes not at all, what happens before or after.

But there are complications about what Hubert and Mauss may mean from place to place even at the level of 'plain meaning'. For although it seems obvious that Hubert and Mauss's *Sacrifice* is about *ritual* sacrifice, the book ends with a passionate discourse on *civic* duty! This raises questions as to how much of what they had already said about ritual sacrifice was informed and rhetorically conditioned by their political commitments and patriotic feelings about civic sacrifice? Conversely, how much of what Hubert and Mauss urge us to do in regard to our civic duties is grounded in the precedents they see in their analyses of primitive rituals? Not so differently today, contemporary thinking about ritual sacrifice casts a broad shadow of concern about violence in our culture. As the debate swirling round recent books on sacrifice like Girard's *Violence and the Sacred* or Burkert's *Homo Necans* indicates, we continue to mirror our concerns about public policy in subtle ways against what we believe about the nature of ritual sacrifice. Similarly, and perhaps notoriously in a thinker like Girard, how much of what he says about ritual sacrifice projects his own *moralist* and pacifist politics? What should we then take the 'plain meaning' of the theory of sacrifice of the equally political and *moralist* Hubert and Mauss to be, given that they too encompassed both ritual sacrifice and civic sacrifice within one analytic and rhetorical effort?

SALIENCE, CONTEXT AND AUTHORIAL INTENTION

Thus, if this degree of mystery lies in the 'plain meaning' of what Hubert and Mauss wrote, even more mystery lies in the *salience* of their theory - in their intentions and in the possibility of their being understood to be fulfilling certain intentions within a certain context. This is thus to ask what Hubert and Mauss *originally* meant in making certain claims, *why* they bothered to say these things at all, why, when a whole variety of things *might* have been said, and were being said about sacrifice, they said some things and not others?

But to these questions of our authors are silent. Virtually no direct evidence of the intentions of Hubert and Mauss in writing *Sacrifice* exists. A few letters have turned up; a few lines in short autobiographical pieces by the authors point in certain directions; the introduction to their *Mélanges d'histoire des religions* is perhaps the greatest help.[4] But summing up we have to admit that we have very little direct evidence to help us answer questions about Hubert and Mauss's intentions or the salience of their work. What then is to be done?

Lacking such evidence of authorial intent, the task of understanding the particular *historical salience* of *Sacrifice* requires understanding the social *context* in which the Durkheimians created their theory of sacrifice. What set of common understandings or collective assumptions would have been taken for granted by our authors and by potential members of their audience about sacrifice? How can what Hubert and Mauss wrote be understood as reflecting certain choices, and thus reflecting certain choices, reflecting certain intentions?[5] This approach (pioneered by Skinner and Jones) yields in our case a series of questions: what kind of *theory* of sacrifice would one create, if, like Hubert and Mauss, one were a patriotic, dreyfusard, humanist, philosemitic, socialist, rationalist, somewhat anticlerical, iconoclastic scientific student of religion? Likewise, what kind of *theory* of sacrifice would Hubert and Mauss be expected to write in a context of right-wing nationalist hysteria, emergent antisemitism, where the political rhetoric of the duty of individuals to 'sacrifice' themselves for the fatherland merged with the theological rhetoric of sacrifice of Roman Catholic religious revival, often articulated in language perfected by figures such as Joseph De Maistre? What kind of theory would one write in light of the desire of the Durkheimians to enter and command that citadel of the scientific study of religion, École Pratique des Hautes Études, Fifth Section, then dominated by cryptically theologizing extreme liberal Protestants, such as Albert Réville? What kind of theory as well would one write given that Mauss tells us of the congenial thought of Sylvain Lévi, and the great Jewish Indologist's book on sacrifice of the same year?[6]

RELIGION, THE STUDY OF RELIGION AND LOCATING SACRIFICE: ITS NATURE AND FUNCTION

Ever since the beginnings of the Durkheim revival of the early 1970s, the history of Durkheimian thought has been written primarily by persons professionally committed to philosophy (Lukes), history (Jones, LaCapra, Vogt), or sociology itself (Besnard, Karady, Nisbet). Here, when not viewing Durkheim as fathering a new science of sociology, he is rightly cast as something the French call a '*moraliste*'. In the process, and without, perhaps, intending it, the main exceptions here are Robert Alun Jones and Pickering, both of whom recognize the theological context of their description of Durkheim as 'moralist' has tended to exclude the 'religious' dimension of his thought and its context in any serious sense of the term.[7] In general both Jones and Pickering have sought to begin understanding the work of Durkheim in terms of his engagement in the larger questions of the nature of religion and the study of religion in *fin-de-siècle*. Thus in Pickering's careful tour through the thickets of Durkheim's thinking about religion, he takes on all the thorny issues, including such matters as Durkheim's supposed Jewishness, the sacred-profane relation, his relation to religious humanism, the sanctity of the individual, as well as the relation of Durkheim's work to the one time member of the *équipe*, the Protestant, Gaston Richard. Jones as well has shed light on intriguing places in Durkheim's work on religion, especially on Durkheim's debt to William Robertson Smith and even as far afield as the German neokantian liberal theologian, Albrecht Ritchl.

Building on what Jones and Pickering have done so well, I have tried to locate Durkheimian theory of sacrifice within the context of the history of France's three great religious traditions, within the academic discipline of the study of religion in France, and within the larger public concerns of the time, such as individual sacrifice in warfare, or more limited public policies of civic sacrifice in behalf of the nation.

In Durkheim's day, sacrifice was classified as part of religion. In our day, it has come to be classified within the general study of ritual - primarily to understand the general process of ritualization. The founders, however, assumed that the study of sacrifice was important because it was a privileged way to understand the 'real' nature of *religion*. They likewise assumed religion itself was important, because of the part religion has played in France's peculiar history, and doubtless because very few of them were indifferent to religion itself. They may have blown 'hot', or just as often 'cold', but few of the founders of the study of religion in France were 'lukewarm' about religion.

Hubert and Mauss thus take their places in the company of the great theorists and founders of the study of religion, such as Robertson Smith, Frazer, Tylor, and Durkheim. But an examination of theoretical books written on sacrifice in the period when Hubert and Mauss's *Sacrifice* was written reveals three major works by the academics in Hubert and Mauss's circle in Paris, especially at the École Pratique des Hautes Études, Fifth Section. First among books by Roman Catholics was Alfred Loisy's, *Essai historique sur le sacrifice*, which although published in 1920 represents the culmination of work he had undertaken since the 1890s. Among French Protestants, there was no theoretical work, (a significant fact itself), not even something like William Robertson Smith's *Lectures on the Religion of the Semites*. We do however have Albert Réville's Collège de France Lectures of 1884, later published as *Prolegomena of the History of Religions*, which, as we will see, would give one all the reason one would want for avoiding the subject of sacrifice completely. As for Jewish scholarship, standing out above all is Sylvain Lévi's *La Doctrine du sacrifice dans les Brâhmaṇas*, published in the same year as Hubert and Mauss's *Sacrifice*, 1898. It would seem reasonable procedure in trying to understand, at least the academic context religious scholarship of Hubert and Mauss's *Sacrifice* to attend to these key works in the theory of sacrifice, produced after all by scholars, who for the most part, commanded the production of knowledge about religion, (and thus sacrifice) in the premier venue of the study of religion in France, (and perhaps the world), the École Pratique des Hautes Études, Fifth Section.

Second, the religious commitments of the founders came strongly into play in what they said about sacrifice. Since the founders assumed that the nature of sacrifice held the key to the nature of religion, the status of their own particular religious communities were therefore necessarily in question. Further, the identity of each major religious tradition of the West was heavily invested in the issue of sacrifice - whether as centre of its eucharistic theology (Roman Catholics), or as antithesis of its reformation view of the holiness of the individual, (Protestants) or as a source of stock-in-trade antisemitic blood libels (Jews). As critics and scholars of traditional religions, the Durkheimians well understood the theological history of sacrifice in France. What Hubert and Mauss wrote about sacrifice engaged these three religious histories with all their wider political and cultural associations. Likewise as part of a broad attempt to redefine the nature of 'real' religion, and usurp its functions what the Durkheimians proposed about the nature of sacrifice could also be read along religious - or at least quasi-religious - lines as well.

Third, by singling out the subject 'religion', I have played down their intimate interrelation in the France of the *fin-de-siècle*. Let me correct this. I believe that reading *Sacrifice: Its Nature and Function* historically requires locating Hubert and Mauss's text within contexts where religion and politics merged imperceptibly. Here I believe the influence of intransigent Roman Catholicism in both political and religious venues was informed by a particular theology of sacrifice, as worked out in a particular post-Tridentine theology of the eucharist, then hammered out on twin anvils of '*réaction*' to the Reformation and '*révocation*' of the Edict of Nantes. In the crisis of renewed nationalism which gave rise to extensive appeals for civic sacrifice and heroism in advance of the First War, the Church and this theology of sacrifice played a large part.

In the same spirit, I think we can locate the scholarly concerns of theorists of sacrifice mentioned earlier within the pertinent social and political contexts of the political and religious history of the three major religious traditions of France.

This approach then differs from what Jones and Pickering have undertaken in that I am more persuaded that the so-called 'external' dimension of scholarly work may give scholarly inquiry often as much salience as the 'internal' dictates of the agendas of the academic profession. But in emphasizing the 'external' constraints upon scholarship, I wish to make it clear that this is not a piece of ideology as far as I am concerned. It represents a judgement about the empirical nature of things in the France we study when we study the Durkheimians - the France where academic matters were often as political and social as they were anything purely intellectual. I probably do not have to remind my audience of the crucial political role of critical history writing in the Third Republic, of the rise of such periodicals as the *Revue historique* under the leadership of Protestant republican, Gabriel Monod. Whatever can be said about Monod and the *Revue historique* can be said with equal force about Protestant Maurice Vernes and his *Revue d'histoire des religions*. Indeed Vernes intended this great scholarly journal of the study of religions to play an explicitly political role in the France of his day. As François La Planche recently noted:

> Vernes's project with the *Revue d'histoire des religions* was perfectly situated within the context of the vast republican cultural enterprise - which naturally enough was not distinct from a desire for political transformation.[8]

THE DURKHEIMIANS, NOT JUST DURKHEIM

But having raised the issue of the importance of the study of religion in the *équipe*, it would be well to set the record straight about the collective nature of the study of religion there. At one point Pickering says that it was

'Durkheim rather than his disciples who gave prominent place to religion and who was willing to write about it in such positive terms ... who was unique in his enthusiasm for things religious.'[9] This seems to me most certainly mistaken - at the same time Pickering is correct to indicate Durkheim's unique concern with religion in contemporary society and its future, or with his imaginative explorations of the religious dimensions of humanism and individualism, as well as his passion for what sociologists since Bellah might today call 'civil religion'.

But Hubert and Mauss contributed as much, if not more, to the study of religion in the *équipe*. Hubert and Mauss after all directed the *Année* sections on '*sociologie religieuse*'; Durkheim never did. Moreover they seem to have been concerned with religion even prior to their association with Durkheim. Before the work of *L'Année sociologique* had begun, Hubert and Mauss had trained extensively in history of religions, and identified themselves professionally in terms of the study of religion; Durkheim did not. Hubert studied with critical church historian Abbé Louis Duchesne, an influential figure in the rise of Roman Catholic modernism, with Israel Lévi, the chief French rabbinic scholar of his day, holder of the first chair in Talmudic Judaism in France, and one day to become grand rabbi of Paris. Mauss alone wrote 206 reviews on religious topics in the *Année*'s first series; Durkheim managed 50. While Durkheim published virtually nothing substantial on religion between his 1899 'De la définition phénomènes religieux' and the 1912 *Elementary Forms*, Hubert and Mauss, singly or in collaboration, produced studies on magic, myth, the papal states, prayer, sacrifice, and sacred time. Both Hubert and Mauss likewise won chairs in the prestigious Fifth Section, (religious sciences) of the École Pratique des Hautes Études, and taught history of religions throughout their careers. Taking their teaching and publishing careers together, then an entire program of Durkheimian comparative history of world civilization was worked out by a radical division of labour between Mauss and Hubert: Mauss was Africa, Australia, the Americas, India and Oceania; Hubert was the Celts, Greeks, Romans together with their Mesopotamian and Near Eastern counterparts. Mauss was Buddhism, Hinduism and the 'primitive' religions; Hubert was Druidism, Gnosticism, Judaism, Christianity and the 'mystery' religions. In 1904 Hubert wrote his 'Introduction à la traduction française' to P.D. Chantepie de la Saussaye's *Manuel d'histoire des religions*, in its time recognized by Henri Berr, (the founder of the journal *Revue de synthèse historique*, a kind of forerunner of the *Annales*) as a 'manifesto' for the Durkheimian program of religious studies. While Hubert taught 12 of his 17 courses between 1901 and 1912 on religion and Mauss 20 of 23 in roughly the same period, Durkheim taught only 1! Even when we consider the creation of the *Elementary Forms*, we should recall Mauss's long-standing interest in ethnography and 'primitive religions'. This was perhaps due to the appropriation of the work of anthropologists like Quatrefages de Bréaux by the chief of Mauss's neokantian philosophy mentors at Bordeaux, Octave Hamelin.[10] Durkheim seems to have reoriented his interests toward religion and the 'primitive' after 1895, thanks to Robertson Smith he tells us, but incidentally also only after the arrival of Mauss in Bordeaux in the same year. Mauss collaborated with Durkheim in *Primitive Classification*, a work generally acknowledged as formulating ideas later developed by *Elementary Forms*; Condominas also claims Mauss's inaugural lecture at Hautes Etudes not only retraces elements of his introduction to *Sacrifice*, but that Durkheim repeats 'prèsque textuellement' Mauss's position in the *Elementary Forms*. Durkheim was thus far from being 'unique' in the *équipe* in his devotion to religious studies, nor is it clear that he was in many areas its prime mover.[11] Thus, when we seek to understand a 'Durkheimian' study of sacrifice, we need to keep in mind the privileged place of Hubert and Mauss in the Durkheimian study of religion, along with the larger context of French religious practice and academic study in which theorizing about sacrifice was carried out in their day. Let us look more closely at these religious contexts.

FRENCH ROMAN CATHOLICISM OF 'LA RÉACTION'

In the *fin-de-siècle*, this means understanding so-called 'intransigent' Roman Catholicism and its central theology of eucharistic sacrifice established by the Council of Trent, put into play by the *réaction* to the Reformation and revived by the anti-modernist campaign of the late nineteenth and early twentieth centuries. The terms of this theology, regnant in Hubert and Mauss's time, derived from the Oratorian fathers of the counter-reformation, 'révocation', and the religious campaigns waged against the Protestants from the sixteenth and seventeenth centuries.

The Oratorians founded a theological and practical religious movement which became known as the 'école française de spiritualité'. Essentially a movement promoting eucharistic piety, nothing was more central to the Oratorian conception of the eucharist itself than sacrifice.

Sacrifice was here conceived to have four characteristics. First, the true model for every sort of sacrifice, Christian or not, past, present and to come, was Jesus's death on Calvary. In effect, the theologians of the 'école française de spiritualité' asserted that Jesus's death on Calvary was a cosmic event, of such potency that all other sacrifices are but typically unconscious mystical imitations or anticipations of it. Because of this primacy, it becomes something of an 'elementary form' of sacrifice. Second, because Jesus is seen as both victim and priest, his death is *self*-sacrificial. Recalling Robertson

Smith, this is in effect to assert the theological and ontological primacy of a particular kind of sacrifice - namely a *sacrifice of the god*, in this case, Jesus. Third, Jesus's self-sacrificial death is in turn celebrated as a total giving of an individual to the deity - an utter *annihilation of the self* in which nothing is held back. Fourth, this total giving up of self by Jesus on Calvary effects a perfect *expiation* of human sin before the demands of deity.

Although the Oratorian tradition may not be widely known today, its sacrificial interpretation of the eucharist passed into popular piety in France. It later emerged conspicuously in the political and religious thought of Joseph de Maistre, where sacrificial immolation of the individual - even an innocent like Jesus - becomes an ideal of human behaviour. Then again in the nineteenth and twentieth century literature of the Roman Catholic revival, such figures as Balzac, Huysman, Bernanos, (and even the Graham Greene of *The Power and the Glory*) carried on the traditions of expiatory, annihilating self-sacrifice. Finally, this annihilationist ideology of sacrifice achieved perhaps its highest level of erudition by a French Roman Catholic author in Alfred Loisy's, *Essai historique sur le sacrifice*. Later, after his excommunication and writing as a Free Thinker, Loisy viewed the common religion of France as patriotism: accordingly, sacrificial death for the nation became the paramount religious act, even given a radical turn. Loisy speaks of sacrifice here as pure altruism, as complete 'giving up' of oneself as the highest form of religious action a person can take. In fact, total self-sacrifice made in a spirit of 'love' is the 'soul of sacrifice'; it alone makes an act religious: 'the sacrifice of self remains the religious act *par excellence* - all the more religious to the extent that it is more consciously and voluntarily consented to'. Even though he no longer could abide in Roman Catholicism, Loisy could not really break with its counter-reformation eucharistic spirituality. Perhaps the strongest testimony to the hold of this demanding spirituality comes in what Loisy wrote about sacrifice after the war. His absolute commitment to the norm of 'giving up' oneself to the nation seems not at all shaken by the grotesque waste of human life in the now classically senseless engagements of the First World War. Loisy ends his 1923 *La Morale humaine* with what can only be called a hymn to total self-sacrifice:

> Sacrifice, which seems to be the negation of happiness, is on the contrary, its condition and reality ... because, in one way or another, the sacrifice of individuals is required for the equilibrium and preservation of society, for the common good, for the establishment and maintenance of concord.[12]

Now, it is clear from the popular religious press, pastoral letters, and preaching of the time in question that these attitudes toward sacrifice were simply taken for granted by many if not most French Roman Catholics. Sacrifice was thus a moral notion much employed in the heated nationalist politics of a France readying itself ideologically for the First World War. With anti-clerical feeling running high, Roman Catholics, for instance, saw opportunities to re-establish their legitimacy as patriots by urging heroism and national military sacrifices. From its inception at the dawn of Bourbon absolutism, through the reactionary revivals in the nineteenth century, and well into the twentieth century, this complex of Oratorian and Maistrean sacrificial theology had shaped the personal piety, and thus civic life, of many of the faithful. It is, therefore, inconceivable that two men such as Hubert and Mauss, equally sensitive and knowledgeable about both religion and politics, would not have taken this syndrome of religion and politics into account in writing *Sacrifice: Its Nature and Function*.

DURKHEIMIANS AND ROMAN CATHOLICS ON SACRIFICE

In *Sacrifice: Its Nature and Function*, Hubert and Mauss conceived a relationship of the individual to the group which directly opposed the legacy of the Oratory and De Maistre. While approving of civic sacrifice Hubert and Mauss were clear about qualifying their approval. In brief, sacrifices typically assume limits. As if proposing a theory of sacrifice tailor-made for members of the liberal bourgeoisie they were, Hubert and Mauss say:

> In any sacrifice there is an act of abnegation since the sacrificer deprives himself and gives ... But this abnegation and submission are not without their selfish aspect. The sacrificer gives up something of himself but does not give up himself. Prudently, he sets himself aside. This is because if he gives, it is partly to receive. Thus sacrifice shows itself in a dual light; it is a useful act and it is an obligation. Disinterestedness is mingled with self-interest.[13]

Further in the theological domain, Hubert and Mauss also directly reject the view that the death of the god-man, Jesus on Calvary is anything of a universal and elementary form of sacrifice.[14] For one thing, Roman Catholics have reversed the correct order of sanctity in sacrifice. It is not the pre-existent divine *logos* who appears in history in order to broadcast holiness; it is the *rite* of sacrifice itself which makes beings into gods, and/or sustains the gods in their divine existence.[15] Rather than looking back to some such 'saving event', Hubert and Mauss propose that they have discovered an elementary mechanism of sacrifice which makes things holy.[16]

Thus sacrifice is a religious offering which is in effect a consecration of a victim, which in turn also modifies the condition of the sacrificer or the objects of sacrifice.

This structure underlies not only sacrifice among the primitives, but also the Roman eucharist itself: it, like all the rest, has built upon ancient pagan 'models',[17] and participates in the same elementary sacrificial mechanism found in religions around the world.[18] Second, for Hubert and Mauss, annihilating self-sacrifice is exceptional, and certainly not an elementary form of the institution. The sacrifice of the god theme likewise occurs infrequently in the texts, and when it does, there is no evidence that the text reports an actual occurrence of a real sacrifice. Rather, the sacrifice of the god is a mythologically projected ideal of what perfect sacrifice might be - sacrifice completely free of self-interest or calculation typical of human efforts.[19] More typical, as we have noted in calling attention to the bourgeois character of their theory of sacrifice, was to see sacrifice requiring a victim to intervene to *protect* the sacrificer, to save the, (bourgoise) individuality of the person giving '*of* themselves', but not giving '*up* themselves'. Then, Hubert and Mauss undercut the reactionary Roman Catholic-Maistrean view of the prestige of expiation in sacrifice. 'Expiation' does not mark, say Hubert and Mauss, a 'real type of sacrifice',[20] since it is not clearly separable from other forms of sacrifice. It 'cannot thus form the basis for a general and rigorous classification of sacrifices'.[21] To be sure expiation may be a factor in sacrifice, but it is not for Hubert and Mauss its centre.

WHY ALBERT RÉVILLE REJECTED SACRIFICE

If Roman Catholic idolization of sacrifice in political and religious senses seemed to push the Durkheimians to soften their theory accordingly, the Protestant dismissal of the sacrificial traditions of the religions did just the opposite. Prominent above all other French Protestant scholars of religion, was Albert Réville. Professor at the Collège de France, *titulaire* of the Fifth Section of the École Pratique des Hautes Études, Hibbert lecturer, moving spirit behind the first international congress of the history of religions in Paris 1900, active and well known in French politics, Réville was the major voice of authority for Protestant attitudes toward sacrifice. For Réville, sacrifice was not only a religious abomination, it also was a social and political one as well. Réville, for example, believed there were such things as 'sacrificial' societies - human groupings in which the spirit of ritual sacrifice determined key values, such as the attitude toward the sacredness of the individual. It was no accident that Aztec society both widely practised human and other sorts of sacrifice, and reduced the value of the individual before the almighty state. The monarchist France that the intransigent Roman Catholics sought to re-establish fell into the same pattern - a sacrificial eucharistic theology fitting hand in glove with anti-republican social formations, only to be relieved by Leo XIII and the first *ralliement*.

Beyond this complex of religious and political reasons for rejecting sacrifice, Réville had ample academic reasons, (also resonant and congenial with his political views) for being critical of ritual sacrifice's role in the history of religion. These number five.

First was a theological commitment to the ideal of progressive but *discontinuous evolution*. Replaying Reformation polemics against the Roman Catholics, Réville gladly observes that as one stage of evolution succeeds another, it leaves it behind. 'What has been, is not; what is, will not be', says Albert Réville.[22] The present thus does not 'participate' in the past, but represents a radical break with it. Albert Réville's son, Jean, noted simply that with regard to the past we need feel no gratitude.'[23] It was likewise in religious history,

> each new degree of spiritual development negates that preceding it ... Monotheism only acquires its value and consciousness of itself in disengaging from its polytheistic context ... In this sense, to abolish is to fulfil.[24]

Thus between the pagan or Roman Catholic sacrificial 'them' and the enlightened non-sacrificial 'us' there was no real relationship.

Second, one reason ritual sacrifice was judged so low on the scale of religious evolution was because it involved gift. As gift, sacrifice would then be something which could oblige God to act in behalf of the gift giver, (also Robertson Smith's view).[25] Further, while it may have been true, says Réville, that 'all religions' considered sacrifice 'as pre-eminently the means of realizing the union of man with the divinity',[26] in the eyes of the Reformation, the union thus created would be fundamentally corrupted by the failure of human confidence in divine salvation.

Third, recalling his indictment of the Aztecs, Réville believed that ritual sacrifice culminated inevitably in ritual *human* sacrifice. Since sacrificial ritual necessarily involves renunciation, and since for human beings, the greatest act of renunciation would be to sacrifice one of their own kind, the practice tends at once toward suicide or human sacrifice.[27] In its civic sense, it is interesting to note, Réville was an early opponent of the 'sacrifice' of Dreyfus. There, he felt, one could see how civic sacrifice developed into symbolic human sacrifice - into expiating scapegoating the beloved of the followers of De Maistre. The indignity heaped on Dreyfus can only be explained, believes Réville, in terms of the long historical formation of human minds effected by the sacrificial eucharistic theology of the Roman Catholic church in France, especially so, since the days of royal absolutism and the '*révocation*'.[28]

The fourth reason Réville rejected sacrifice was his belief that 'real' religion was internal, a matter of 'spirit and truth'. Typical of the spirituality of the Church of the Desert, and also the Methodist purity of heart taken up by the French Protestants, real religion not something material or 'embodied'.[29] Alternately, real religion was also about morality and defined by beliefs. As such, tangible, social and cultural dimensions of religion were mere 'externals', and thus of no particular importance.[30]

The fourth reason Réville rejected sacrifice was his belief in natural religion. By definition however, such an essential religion, a 'religion-as-such' was good. It constitutes a bond which a human being 'delights in feeling';[31] it is a 'natural property and tendency, and consequently an innate need of the human spirit'.[32] Given its brutality and violence, ritual sacrifice could not possibly be part of such an essentially good thing as religion.

It was thus this commitment to this natural religion, together with what was compatible with it in the Reformation tradition, which 'overdetermined' Réville's rejection of sacrificial ritual. Besides the evil aspects of sacrificial religion, sacrifice is a kind of bribery, as we saw, a pathetic and paradigmatic attempt at salvation by works. Sacrificial religions are at best only 'fallen' or greatly diminished approximations of what religion-as-such really is. 'Religions constitute an evolution - the evolution of religion; but religion, under the most various forms, abides and will abide', says Réville.[33]

DURKHEIMIANS REJECT ALBERT RÉVILLE ON SACRIFICE

Part of the reason Hubert and Mauss do not treat sacrifice like Albert Réville is, of course, that they sought to do *science* rather than theology or a psychology. Thus, they rejected an approach to religion which was based on appeals to an *a priori* natural religiousness, the essential goodness of this religion, or to Réville's private realms of inner feeling to which no public access was permitted. This desire for good facts, for public data was one of the reasons the Durkheimians were at least *methodological* ritualists - to understand the religion, understand the ritual. As Durkheim himself claimed, ritual is 'religion made visible and tangible'.[34]

Moreover, as would be scientists, the Durkheimians wanted to explain religion, and this required discovering its causes. We will shortly see how the view that rituals *cause* religion developed in conjunction with Sylvain Lévi. But before doing so, let me turn to gift.

As it turned out, ritual satisfied this requirement, not only because it could be seen as an empirical cause, but also because it was *social*. In the Durkheimian world, this brings us to gift. We know that despite their criticism of Tylor's gift theory of sacrifice, Hubert and Mauss did not reject a gift theory of sacrifice in its entirety. Sacrifice is a gift which is also essentially a consecration, thus requiring the destruction of the offering in part or in whole.[35] Gift must be maintained because, as Mauss's *The Gift* shows, gift was essential to society, and as Durkheim puts it, religion, (under its ritual aspect) was nothing more than a special mode of exchange between gods and humans.[36] This is to say, along with Hubert and Mauss in *Sacrifice*, that there is 'no sacrifice without society',[37] and no society without sacrifice.[38] Thus, because religion is for them essentially social, Hubert and Mauss could never agree with Albert Réville that sacrifice had no part in the essential definition of religion.

DURKHEIM'S EVOLUTIONISM AFFIRMS THE SACRIFICIAL PAST

Less obviously theological than Réville's objections to sacrifice as gift and ritual was his framing of the question of sacrifice within an evolutionary perspective. Both Réville and the Durkheimians were evolutionists, but while Réville stressed the discontinuity between past and present, the Durkheimians stood for continuity. Against Albert Réville's view that, 'each new degree of spiritual development negates that preceding it' and that 'to abolish is to fulfil',[39] Durkheim dissolves the present into the past: 'the present ... is by itself nothing; it is no more than an extrapolation of the past, from which it cannot be severed without losing the greater part of its significance'.[40] The Durkheimians believed thus that the present *participated* in the past: evolution was continuous and cumulative, rather than marked by the clean, sharp and radical breaks between past and present. Consider again how differently, for example, Durkheim approaches the problem of understanding such a highly evolved creature as 'modern man'.

> Indeed what do we even mean when we talk of contemporary man, the man of our times? ... in each one of us, in differing degrees, is contained the person we were yesterday, and indeed in the nature of things it is even true that our past *personae* predominate, since the present is necessarily insignificant when compared with the long period of the past

> because of which we have emerged in the form we have today. It is just that we don't directly feel the influence of these past selves precisely because they are so deeply rooted within us.[41]

In effect, Durkheim explicitly lays out a theoretical position on the relation of past and present implied in Hubert and Mauss's *Sacrifice*. When they say that the 'Christian imagination has built on ancient models',[42] they in effect argue that the sacrificial structures established in the past seem indeed to inform and condition the present - which itself only continues the past. Therefore even for the sake of understanding the present, we need to comprehend the causality and persistence of past events which produced it. In this, the Durkheimians ironically approximated the Roman Catholic sensibility about sacraments such as the Roman eucharist, while challenging the primacy claimed for it. But there is more.

Roughly at about the same time Durkheim was writing these words, a patron of the Durkheimians and a great student of religion in his own right would write these:

> The present is no longer a spontaneous creation - autonomous and independent of the past. It prolongs the past; it continues it and condenses it. It is only the past itself on the way to being transformed.[43]

This man was Sylvain Lévi.

His idea of an eternal past has hardly been explored, but it was apparently deeply rooted - at least in liberal Jewish rejection of personal bodily resurrection, while holding fast to a belief about their collective survival in a continuous ever present past. In an address to an audience of Jewish scholars, Lévi related how Jews of old would escape the spiritual turmoil and mental stress of persecution by retreating into their studies. There they discovered a 'dream' world of 'eternal calm'. These scholars of former ages are 'the men who made us', Lévi reminds his fellows. But the dead do not leave us alone, separated in time and space, with little but reverent thoughts to connect us to them. In some perhaps strange sense, if Lévi be taken in earnest, the old scholars go on actually living with us:

> If we lend an ear to those secret voices in our minds [*consciences*], if we pay heed to those things we do without explicitly wanting to, if we get to the bottom of some of the subtle sources of the inchoate thoughts which play in our brains - then, we would hear *them*. We would take them by surprise. While we may believe that by means of history, we live in them, it is actually they who live in us. They continue their history through ours.[44]

Does this kind of piety explain why Mauss could say of the Durkheimian *équipe* fallen in the First War: 'This team lives and has ever been reborn.'[45] Does it also shed light on how Hubert and Mauss could remark triumphantly again their view just recalled that the contemporary Roman Catholic mass actually built upon archaic models. I think it does suggest that Hubert and Mauss's *Sacrifice* may owe a possible debt to Jews like Sylvain Lévi. And thus we come to the relation of Hubert and Mauss's theory of sacrifice and the Jewish context, and as promised, to the work of Sylvain Lévi.

SYLVAIN LÉVI AND MAUSS

For Jews in the *fin-de-siècle*, sacrifice produced vexation. The rise of antisemitism, culminating perhaps in the Dreyfus Affair, brought pressures to bear on the entire Jewish community from many sources. In the academic community, ancient Jewish traditions of ritual sacrifice served as evidence to libel modern Jews with renewed charges of deicide. The liberal Jewish academics of the time, some even patrons of the careers of Hubert and Mauss, such as James Darmesteter and Salomon Reinach, reacted as religious liberals might be expected to react. They 'spiritualized' sacrifice; they minimized its place in 'essential' Jewish religion, and tended to diminish its place in the nature of 'real' religion, just as Albert Réville had.

In terms of civic sacrifice, Jews were caught in a nasty 'double bind'. On the one side, Jews were accused of eagerness to 'sacrifice' the lives of others in pursuit of community self-interest. In the heat of the Dreyfus Affair, even the philosemitic Charles Péguy had charged that the French Jewish leadership was willing to 'sacrifice' Dreyfus to his accusers rather than sponsor his defence and risk retaliation. On the other side, Jews witnessing the onset of the First War, and under a cloud of suspicious antisemitism, felt pressures to sacrifice themselves for the country to *prove* their patriotism. A social commentator of the time observed that the Jewish Durkheimian, Robert Hertz, had had the presentiment

> that he would not return [from the war]. But this presentiment changed into a determination to become a sacrifice. Jewish by origin and French by all the thoughts of his mind and strivings of his moral being, he reckoned that the blood of the men of his race and of his own conscience would be usefully shed to liberate their children from all reproaches of egoism, particularist interest and indifference in the eyes of a suspicious France.[46]

But at the same time, more traditional Jewish thinkers such as Sylvain Lévi, resisted the spirit of their times. Lévi saw sacrifice as *relatively* valuable in humanity's rise to 'higher' religious consciousness. I am prepared to argue that it was indeed Lévi's role as theoretician of sacrifice at the time he was Mauss's teacher in Paris which accounts for much of the same positive attitude to sacrifice found in Hubert and Mauss's *Sacrifice: Its Nature and Function*.

Questions about the relation of Judaism to Durkheimian sociology tend to start naturally enough with Durkheim himself. To put it bluntly, Jews and Durkheim cared about each other's work because Durkheim was Jewish. This viewpoint is understandable, given his principal place in the formation of the Durkheimian research project. But it is misdirected for at least two reasons. First, focusing on Durkheim alone blinds us to the absolutely crucial role performed by his confederates in the *équipe* which produced the *L'Année sociologique* - in particular Marcel Mauss. Thus even though one may want to begin with Durkheim, I have already moved the discussion more and more to Hubert and Mauss as the focus of the inquiry draws closer to the Jewish social origins of *Sacrifice*.

Second, focusing on Durkheim and his Jewishness alone fails because Durkheim apparently did all he could to suppress his Jewish roots. David Emile Durkheim was in this sense David in name only - witnessing that ethnicity, like many other cultural things, is as much a matter of nurture as nature.

I am thus arguing that when trying to understand Durkheim in relation to being Jewish and Judaism, scholars make at least two great mistakes: first, they search for his 'essential', typically hidden, Jewishness, rooted in Durkheim's Jewish ethnicity and early nurture; second, they likewise conceive the problem of the relation of Durkheim to Judaism as one involving his personal or strictly individual or personal piety or attitudes. Why do I call these research aims 'mistaken'?

First, to seek Durkheim's 'essential' Jewishness would seem at least a low research priority. What is this 'essential Jewishness' in Durkheim's thought? A concern for group over individual, ritual over myth, and so on? If so, why not include those 'essential' features of Jewish identity that Durkheim's thought does not manifest, e.g. theism, Chosen People, Zionism, dietary interests? Further, the 'essentials' of a French nationalist or Roman Catholic world view might answer the call in the France of Durkheim's day just as easily as Judaism. They too value collectivity and ritual.

A second tack is also possible. Durkheim's Jewish identity has also not been found simply because in many senses it is not there. Durkheim actively dissolved it in assimilating to French national identity. Attempts to identify French Jewish identity with reverence from the French revolution and the

Declaration of the Rights of Man has been well sketched out at least since the days of Bernard Lazare. On this view Durkheim was an *Israëlite* rather than a *Juif* - assimilated rationalist, positivist and modernist not mystical, mysterious and symbolist. Durkheim objected to the idea of the Jews as 'a people apart'. He not only predicted total assimilation of French Jews within a generation or two, he *welcomed* it. That was the kind of person he wanted to be.

Further, to the extent such an essential Jewishness is said to be hidden or covert, we are in a realm of speculation, but lacking controls on our inquiry. If Durkheim was a 'secret' Jew, perhaps he was also a 'secret' Protestant or secret Muslim. There is no end to these possibilities, and no way to decide between them. The kind of circumstantial enquiry alluded to above where Durkheim's collectivism and ritualism are matched with features of the Jewish tradition is always a possible line of investigation. But then Judaism would need to be as likely a candidate as Roman Catholicism, and good reasons would need to be found for such links.

But beyond these negative reasons for abandoning the quest for Durkheim's essential, (hidden or not) Jewishness, there are positive reasons for seeking another approach to this question. Even if we discovered that Durkheim was secretly Jewish, (whatever this means), and that his scholarly agenda was informed by Jewish ideas, we would still have to evaluate the significance of such an ethnic contribution to Durkheim's thought. Is it significant or merely incidental? What research strategy does emphasis on this fact serve?

To the degree that seeking to uncover the Jewish roots of Durkheim's thoughts focuses on Durkheim's past, this research strategy should be challenged. It is unlikely that new evidence will come to light about Durkheim's childhood or Jewish home life. It is therefore unlikely that we will ever be able to make further progress in the quest for Durkheim's roots or for the vague notion of his essential Jewishness - even if we should be able to overcome the problems already raised above.

The same restrictions do not afflict inquiry into Durkheim's intellectual relations with his Jewish contemporaries, with Jewish learning, with issues pertaining to Jews alive in his time. Durkheim was, for example, a member of the board of directors of the Society for the Aid of Jewish Victims of the War in Russia, although not a member of the Alliance Israëlite Universelle or Société des Études Juives. Here there is much to know, and relatively little known. Part of choosing a research project involves making pertinent choices about investigating matters that *can* be illuminated, and which *would* have consequences for a deeper understanding of Durkheim's thought. This would entail abandoning the quest for Durkheim's Jewish roots in favour of inquiry into the nature of Durkheim's place in the network of relations and intellectual issues involving him in the Jewish intellectual community and public issues of his time.

There are at least three compelling reasons to seek the origins of Hubert and Mauss's *Sacrifice* in the social context of Sylvain Lévi's Judaism: First, Mauss tells us explicitly that *Sacrifice* owed much to Sylvain Lévi's *La Doctrine du sacrifice dans les Brâhmaṇas* 1898. Second, *Sacrifice* itself is in fact largely based on data drawn from Hindu and Jewish traditions and formed by Jewish scholars like Sylvain Lévi. Third, both authors of *Sacrifice*, studied under Sylvain Lévi and were nurtured intellectually within a context in which his influence was prominent.

Critical for us is the fact that Hubert and Mauss conclude this discussion by saying that 'the ultimate aim of our researches [is] the sacred', and in the same breath go on to say that it was also the 'highest reward of our work on sacrifice'.[47] Thus, the conclusions Hubert and Mauss reached about sacrifice are for them internally related to those reached about the new positive sacred which characterizes Durkheimian thought after 1895 or so. They are part of this same theoretical breakthrough. Thus, if we could understand the origins of one, we might very well be in a position to understand the origins of the other.

Briefly, this is what is at stake. Against the anti-ritualist position of Jewish and Protestant liberals, Hubert and Mauss adopt a *causal* ritualism: Religion *is* its rituals, not just its beliefs or even morality. Ritual is the locus of the positive power of the sacred that injects effervescence, energy and power into people, and because of which people are religious at all. Thus, sacrifice is for them what *makes*, (things) *sacred*, as the root meaning of 'sacrificium' testifies. It even creates the gods. Sacrifice performs a positive function of creating the religious life of people. They say all this, of course, because as we noted in relation to the Durkheimian critique of the anti-ritualism of Albert Réville ritual is religion in social form, and thus 'religion made visible and tangible'.

The 'origin' of Hubert and Mauss's conviction, I shall argue, is to be found in the thought of Sylvain Lévi, and behind that in the life he led.

The position on the sacred developed by Hubert and Mauss seems impossible to identify with the views of typically anti-ritualist 'iconoclast' Jewish modernists such as Salomon Reinach. But seen from the position of a more modest sort of Jewish liberal, such as the 'progressive' Sylvain Lévi, we can see the appropriate context of Jewish modernism from which *Sacrifice* arose. What is more, as mentor, patron and personal friend to both Hubert and Mauss at the École Pratique des Hautes Études, Fifth Section, Sylvain Lévi was perfectly positioned to understand what the Durkheimians wanted to do. Sylvain Lévi had maintained very close personal relations with Marcel Mauss from at least the last years of the nineteenth century until his own death in 1935. The great Indologist was thus well placed to influence the course of the Durkheimian revolution in our thinking about religion in ways that, for example, William Robertson Smith's books never could. Most importantly, Sylvain Lévi taught Mauss a course on sacrifice especially designed for him, and produced an important book, *La Doctrine du sacrifice dans les Brâhmaṇas* 1898, based upon this course of lectures. Marcel Mauss noted in his review of Sylvain Lévi's *La Doctrine du sacrifice* that this work has the 'greatest interest for the sociologist', and that 'we have greatly drawn upon it'.[48] *La Doctrine du sacrifice dans les Brâhmaṇas* is duly cited numerous times in Hubert and Mauss's *Sacrifice*. Although passing reference has been made by scholars to the possible influence of Sylvain Lévi on Mauss, no one has ever tried to read Hubert and Mauss's *Sacrifice: Its Nature and Functions* in light of Sylvain Lévi, his work and social milieu. Now this is precisely what I intend to do.

SYLVAIN LÉVI: MAUSS'S 'SECOND UNCLE'

What do we know of Sylvain Lévi and of his relation to the Durkheimians?[49] Sylvain Lévi (1863-1935) was one of the most distinguished Indologists of his day. From early in his career as a student in Paris, Lévi had resolved to do oriental studies, even though he could not decide about a speciality. So Sylvain Lévi sought the advice of Ernest Renan, himself a former *élève* of the Indologist Eugene Burnouf. Renan confided in Lévi that Abel Bergaigne, the incumbent in the chair of Indian religions at the École Pratique des Hautes Études, Fifth Section, had no *élèves* under his direction. Sylvain Lévi took up his studies with Bergaigne, and thus began his scholarly life.[50]

After his first teaching post in 1889 at the rabbinic school in Paris,[51] he eventually succeeded Bergaigne in the chair of Sanskrit at the École Pratique des Hautes Études, Fifth Section. By 1894, he was elevated to the Collège de France in Sanskrit Language and Literature, where he finished out his illustrious career.

Sylvain Lévi's scholarly interests were broad: he wrote books on the history of Nepal, the first systematic study of classic Indian theatre, studies of classic Buddhist idealist philosophy and among other things his masterpiece, *La Doctrine du sacrifice dans les Brâhmaṇas* (1898). Lévi also pursued cultural influences across Asia, and between Asia and the West even into modern times. Later in his life, he acted upon these instincts by taking a leading role in establishing cultural ties between France and Asian nations, such as India, Japan, Nepal and Vietnam.

As for Sylvain Lévi and Mauss, we have Mauss's testimony that he was 'my second uncle'.[52] He was also one of the great patrons of the *équipe*, especially in the all important École Pratique des Hautes Études, Fifth Section.

> He did more for us and all of you who have followed us [than one might imagine]. He never separated the advancement of a concern for science and for our careers from the advancement of our progress and work. One of his great 'merits' was that he thought about each one of us in material, fatherly, and fraternal ways ... He evaluated us and placed us all exactly as we ought to have so been - for another task even greater, in relation to our studies, teaching, and science - but also for ourselves and for him...[53]

Mauss claims Sylvain Lévi was responsible not only for his own election there, but for that of another Durkheimian student of his as well: this was none other than Henri Hubert. As we will see later, this testimony is borne out along with its deeper implications in the records of the meetings of the Fifth Section held to decide such matters. Thus Sylvain Lévi's relation to the nucleus of the Durkheimian *équipe* could hardly have been greater.

TOWARD A THEORY OF SACRIFICE: WHAT SYLVAIN LÉVI TAUGHT

After this conversion to a kind of methodological ritualism, the next and decisive step toward the formation of Hubert and Mauss's *Sacrifice* came in 1896. As director of Mauss's studies, Sylvain Lévi selected the topic of sacrifice for special study by Mauss. In doing this, he also directed the course Hubert and Mauss would take in their study of sacrifice later to appear in the *L'Année sociologique*. Mauss tells us explicitly of Sylvain Lévi's care and influence in this matter:

> His course on the *Brāhmaṇas* was personally destined for me. His *Idea of Sacrifice in the Brâhmaṇas* - his chief work - had been made for me. From its first words, it delighted me with a decisive discovery: 'the entry into the world of the gods'; there, right under our noses, was the starting point of the labours which Hubert and I realized in *Sacrifice*. We were only bearing witness.[54]

Thus what Sylvain Lévi meant to the nucleus of the Durkheimian *équipe*, Hubert and Mauss, was nothing less than a positive source of influence urging them on to serious study of ritual and sacrifice that came to fruition in *Sacrifice: Its Nature and Functions*.

Sylvain Lévi's *La Doctrine du sacrifice dans les Brâhmaṇas* (1898) is a substantial work, and one I shall not try to summarize here. But we can at least bring out its salience for the work of Hubert and Mauss.

First, although the Durkheimians seem to have been more nuanced, (or confused) than Sylvain Lévi about causal ritualism, or the priority of ritual to myth and word, But also note that Sylvain Lévi allows a role for the spoken word:

> Sacrifice is a learned and complicated combination of ritual acts and sacred speech, or rather it is the impalpable and irresistible power which is released from their reconciliation, like electricity is born from elements put into contact.[55]

This differs from the point made by William Robertson Smith that to *understand* primitive religion, we should begin by trying to understand ritual life. Here the issue is the nature and origins of religion itself, not how to go about studying it. Confirming this, Sylvain Lévi says that the *nature* of the religion revealed in the *Brāhmaṇas* is constituted by sacrificial ritual. Thus sacrifice 'is God and God *par excellence*'. Further, sacrifice 'is the master, the indeterminate god, the infinite, the spirit from which everything comes, dying and being born without cease'.[56] So potent is the sacrifice, that even if gods are relevant, those very gods are 'born' from sacrifice, are 'products' of it.[57] Sylvain Lévi in effect argues for what I have earlier termed causal ritualism. Renou refers specifically to sacrificial ritual[58] - a point duly recorded by Mauss in his review of *La Doctrine du sacrifice dans les Brâhmaṇas*.[59]

Second, because the gods themselves were the causal consequences of sacrificial ritual, the *definition* of religion could be separated from a belief in the existence or even the idea of God. Instead of the gods as defining religion, the notion of a sacred power, behind the gods and empowering them, took over. It requires little imagination, of course, to see here the later *sacré* of the Durkheimians. For Sylvain Lévi, this power - the *brahman* of Indian thought - was a property of sacrificial ritual itself - a 'deep seated energy' which is an 'impalpable and irresistible power which is released ... like electricity'.[60]

Thus Sylvain Lévi, in effect, provides the Durkheimians with the materials for their idea of the sacred. It is significant that when Durkheim defines religion in terms of the sacred, and rejects a definition by reference to god in *The Elementary Forms*, he virtually cites Sylvain Lévi by citing the great man's teacher, Abel Bergaigne.[61] Further, he employs Sylvain Lévi's domain of expertise, Buddhism, to demonstrate his claim that a science of religion must therefore be methodologically atheistic.[62]

Together with what we have seen in connection with what he taught about the priority of ritual to theism, the idea of religion as a source of energy, and the perennial power of ritual action, we can see that Sylvain Lévi was one of the key figures influencing the development of Durkheimian thinking about religion, and thus about sacrifice.

CONCLUSION

One question remains: What accounts for Sylvain Lévi's departure from the liberal norm common to his Jewish colleagues and to Protestant liberals in the Fifth Section, like Albert Réville? To give it a name, I am pressed to say something like 'particularism'- the view that religion is nothing, (or at least destined to become mere philosophy) if it is not embodied in some way. Thus, Sylvain Lévi observes that the reason Buddhism becomes a mere philosophy when it moved to China from the Indian subcontinent was its loss of institutional and cultural bases. In its Mahayana developments, Buddhism even cheerfully dissolved its ties with India as an historical and geographic entity. The 'new Buddhas', says Lévi, 'had no sacred geography'. Interestingly enough for Lévi, this was tantamount to becoming a religion in the sense that our religious liberals such as Albert Réville imagined 'religion as such' to be - a religion without ritual, a religion devoid of 'any concern with social formation'. In Lévi's eyes, this disembodied 'religion' was a poor thing indeed.

Confirming this vision, Sylvain Lévi's little known Zionism decisively confirms this judgement against disembodied religions. Like other liberal Jews of his generation, the Dreyfus Affair apparently startled him into a new appreciation of the difficulties of being Jewish. Despite national policies decreeing that Judaism was just a 'religion' alongside other 'religions', historical events seemed to turn Lévi toward the view that this could not in practice ever be so: Jews were a 'people', even if they also sought to be wholeheartedly French. By 1918, Lévi thinks more like Herzl than Salomon Reinach. He applauds plans to settle Palestine with Jewish 'colonies'. He thrills at the renewal of Hebrew, because 'verbal effervescence responds to a boiling up of ideas and doctrines on their way to being realized'. He marvels at Jewish enterprise in all areas of commerce and agriculture, and especially how intellectuals work the land as a way of recovering their Jewish identity. Thus for Sylvain Lévi, the ideal of an embodied religion, in touch with its own 'soil' informed his scholarly judgement about the status of Buddhism set adrift from its home and his feeling for what the Hindu tradition was able to achieve in contact with its land of origins. The 'law' of his thought was thus that as much as his universal aspirations favoured Buddhism, the Enlightenment and modernist, prophetic Judaism, the reality of a life of embodiment put him more on the side of particularity - traditional India and Judaism.

We have only to recall Marcel Mauss's words about the significance of Sylvain Lévi's influence to confirm the link.

After having worked with Durkheim in Bordeaux, Mauss moved to Paris to do his doctorate with Sylvain Lévi who, says Mauss, gave him nothing less than a 'new direction to my career'. This I believe is to cite the role Sylvain Lévi in moving Durkheimian thought, in several stages, toward the study of ritual and sacrifice.

First, Mauss had to be converted from philosophy to history. Put otherwise, Sylvain Lévi gave Mauss a taste for particulars. He was made to reconcile the particulars of historical data with his philosophical tendencies toward the universal. Mauss tells us: Sylvain Lévi 'made me plunge into a sea of facts ...', and after two years of immersion in facts, Mauss confesses in exasperation, 'I kept on collecting and sifting facts ...'. Second, Sylvain Lévi seems to have taught Mauss the knack of reading, (especially Indian) texts as indicators of rituals, rather than philosophical arguments in reaction to Max Müller's mythological reading of the ancient Indian texts as having only 'incidental dramatic value'.[63]

The social significance of this methodological move should not be underestimated. The methodological ritualism shared by Bergaigne and Sylvain Lévi, and passed on to the Durkheimians was as deeply rooted in the larger moral, political and religious context of its day, and in part played its role in the creation of Hubert and Mauss's *Sacrifice*.

Notes

1. Frits Staal, *Rules without Meaning*. New York: Lang, 1989, p. 150.
2. Emile Durkheim, *The Elementary Forms of the Religious Life*, tr., Joseph W. Swain. New York: Free Press, 1915, p. 385.
3. Henri Hubert and Marcel Mauss, *Sacrifice: Its Nature and Function* [1899], W.D. Halls, tr. Chicago: University of Chicago Press, 1964, p. 13.
4. Henri Hubert and Marcel Mauss, 'Introduction à l'analyse de quelques phénomènes religieux', in: Victor Karady (ed.), *Marcel Mauss, Oeuvres. Volume 1. Les Fonctions sociales du sacré*. Paris: Minuit, 1968, originally in *Mélanges d'histoire des religions*. Paris: Travaux de l'Année sociologique, 1909.
5. Quentin Skinner, 'Meaning and Understanding in the History of Ideas', *History and Theory*. Vol. 8, 1969, p. 49.

6. Sylvian Lévi, *La Doctrine du sacrifice dans les Brâhmaṇas*. Paris: Leroux, 1898.
7. The chief exceptions to this rule have in some real sense inspired my own work. Here, of course, I must single out W.S.F. Pickering's *Durkheim's Sociology of Religion*. London: Routledge and Kegan Paul, 1984, and several seminal articles; Robert Alun Jones, 'Robertson Smith, Durkheim, and Sacrifice: An Historical Context for the Elementary Forms of the Religious Life', *Journal for the History of the Behavioral Sciences* 17, 1981, pp. 184-205. Robert Alun Jones and W. Paul Vogt, 'Durkheim's Defense of *Les formes Élémentaires de la vie religieuse*', in: Henrika Kuklick and Elizabeth Long (eds), *Knowledge and Society: Studies in the Sociology of Culture, Past and Present*. Vol. 5, Greenwich: JAI Press, 1984, pp. 45-62. Robert Alun Jones, 'Demythologizing Durkheim', in: Henrika Kuklick and Elizabeth Long (eds), *op. cit.*, 1984, pp. 63-83. Émile Durkheim, 'The Problem of Religion and the Duality of Human Nature', (1913), Robert Alun Jones and W. Paul Vogt, tr., *Knowledge and Society: Studies in the Sociology of Culture Past and Present*. Vol. 5, Greenwich: JAI Press, 1984, pp. 1-44. Recently W.S.F. Pickering has argued that a 'deep concern for matters religious is ... at the heart of Durkheim's theory', (W.S.F. Pickering, *Durkheim's Sociology of Religion*. London: Routledge and Kegan Paul, 1983, p. 23).
8. François La Planche, 'La méthode historique et l'histoire des religions: les orientations de la *Revue d'histoire des religions*', in: Michel Despland (ed.), *La tradition française en sciences religieuses: pages d'histoire*. Les cahiers de recherche en sciences de la religion 10. Québec: Université Laval, 1991.
9. W.S.F. Pickering, *op. cit.*, p. 508.
10. Ivan Strenski, 'Durkheim, Hamelin and the "French Hegel"', *Historical Reflections/Réflexions Historiques* 16, 1989, p. 158.
11. Georges Condominas, 'Marcel Mauss, père d'ethnographie française', *Critique* 297, 1970, pp. 130 f.
12. Alfred Loisy, *La Morale humaine*. Paris: Nourry, 1923, p. 290.
13. Henri Hubert and Marcel Mauss, *Sacrifice*, p. 100.
14. Henri Hubert and Marcel Mauss, 'Introduction', p. 15.
15. Henri Hubert and Marcel Mauss, *Sacrifice*, pp. 80, 88, 91.
16. *Ibid.*, p. 13.
17. *Ibid.*, p. 94.
18. *Op. cit..*, 1964, pp. 93-4.
19. *Ibid.*, p. 101.
20. *Ibid.*, p. 14.
21. *Ibid.*, p. 17.
22. Albert Réville, 'Evolution in Religion, and Its Results', *Theological Review* 12, 1875, p. 235.
23. Jean Réville, *Liberal Christianity*. New York: G.P. Putnam's Sons, 1903, pp. 198 f.
24. Albert Réville, 'De la renaissance des études religieuses en France', [1859] *Essais de critique religieuse*. Paris: Cherbuliez, 1860, p. 388.

25. William Robertson Smith, *The Religion of the Semites*, Revised edition. London: A & C Black, 1923, p. 394.
26. Albert Réville, *Prolegomena of the History of Religions*, English tr., A.S. Squire, London, (Williams and Norgate), 1884, p. 128.
27. *Ibid.*, p. 132.
28. Albert Réville, 'The Dreyfus Affair', *The New World* 8, 1899, p. 621.
29. Albert Réville, *Prolegomena*, pp. 66, 72.
30. *Ibid.*, p. 126.
31. *Ibid.*, p. 25.
32. Albert Réville, *Lectures on the Origin and Growth of Religion*, [1884] 2nd ed., P.H. Wicksteed, tr. London: Williams and Norgate, 1905, p. 6.
33. Albert Réville, 'Evolution in Religion, and Its Results', *Theological Review* 12, 1875, p. 246.
34. Émile Durkheim, review of Guyau, *L'Irreligion de l'avenir* [1887] in *Durkheim on Religion*, W.S.F. Pickering, ed. London: Routledge and Kegan Paul, 1975, p. 26. Also cited in W.S.F. Pickering, *Durkheim's Sociology*, 1984, p. 326.
35. Henri Hubert and Marcel Mauss, *Sacrifice*, 1964, pp. 11-12.
36. Émile Durkheim, review of Guyau, *L'Irreligion*, 1975, pp. 26 f.
37. Henri Hubert and Marcel Mauss, 'Introduction', 1968, p. 16.
38. Henri Hubert and Marcel Mauss, *Sacrifice*, 1964, p. 102.
39. Albert Reville, 'De la renaissance', 1860, p. 388.
40. Emile Durkheim, *The Evolution of Educational Thought, (Lectures on the Formation and Secondary Education in France*, 2nd ed., [1938] Peter Collins, tr. London: Routledge and Kegan Paul, 1977, p. 15.
41. *Ibid.*, p. 11.
42. Henri Hubert and Marcel Mauss, *Sacrifice*, 1964, p. 94.
43. Marcel Royannez, 'L'eucharistie chez les évangéliques et les premiers réformés français, 1522-1546', *Bulletin de la Société de l'Histoire du Protestantisme Français* 125, 1979, p. 554.
44. Sylvain Lévi, Allocution to the General Assembly of the *Société des Études Juives*, 24 November 1904, *Revue des Études Juives* 66, 1913, p. iii.
45. Marcel Mauss, 'An Intellectual Self-Portrait', *The Sociological Domain*, Cambridge: Cambridge University Press, 1973, p. 140.
46. Hubert Bourgin, *De Jaurès à Léon Blum*. Paris: Artheme Fayard, p. 484.
47. Henri Hubert and Marcel Mauss, 'Introduction', 1968, p. 17.
48. Marcel Mauss, review of *La Doctrine du sacrifice dans les Brâhmaṇas*, by Sylvain Levi, *Oeuvres. Volume 1. Les Fonctions sociales du sacre*, ed., Victor Karady. Paris: Minuit, 1968, p. 352 (First published in *L'Année sociologique* 3, 1900, pp. 293-5).
49. Major sources for the life and works of Sylvain Lévi are Marcel Mauss, 'Sylvain Lévi', [1935] in: Victor Karady (ed.), *Marcel Mauss, Oeuvres. Volume 3. Cohesion sociale et divisions de la sociologie*. Paris: Minuit, 1969, pp. 535-47, and Jacques Bacot (ed.), *Mémorial Sylvain Lévi*. Paris: Paul Hartmann, 1937. See also the *nécrologie* by Isidore Lévy, 'Sylvain Lévi 1863-1935', *Revue des Études juives* 100, 1935, pp. 1-3. See as well Staal's grasp of this relationship, Frits Staal, *Rules without Meaning*.

50. Jean Filliozat, 'Diversité d'oeuvre de Sylvain Lévi', in: Luciano Petech (ed.), *Hommage à Sylvain Lévi: pour le centenaire de sa naissance 1963)*. Paris: Bocard, 1964, p. 53.
51. Marcel Mauss, 'Sylvain Lévi', 1969, p. 542.
52. *Ibid.*, p. 537.
53. *Ibid.*, p. 539.
54. *Ibid.*, p. 538.
55. Lévi certainly taught the Durkheimians that ritual, not the idea of gods was the key to the origins of religion. This is confirmed by Louis Renou, who says that for Lévi 'ritual dominates mythology'. ('Sylvain Lévi et son oeuvre scientifique', Jacques Bacot, *op. cit.*, p. xxiii.)
56. Sylvain Levi, *La Doctrine du sacrifice dans les Brâḥmaṇas*, 1898, Ch. 2. Noted as well by Mauss in his review of *La Doctrine du sacrifice* in *L'Année sociologique* 3, 1900, pp. 293-5, (Victor Karady (ed.), *Oeuvres. Volume 1. Les Fonctions sociales du sacre*. Paris: Minuit, 1968, p. 353).
57. See note 56.
58. Louis Renou, 'Preface', to *La Doctrine du sacrifice dans les Brâhmaṇas* [1898], 2d ed., by Sylvain Levi. Paris: Presses Universitaires de France, 1966, p. viii.
59. Marcel Mauss, *op. cit.*, 1968, p. 353. Originally published in *L'Année sociologique* 3, 1900, pp. 293-5.
60. Sylvain Lévi, *La Doctrine du sacrifice*, p. 77.
61. *Ibid.*, pp. 10-11. Renou even notes that *La Doctrine du sacrifice* serves as a 'counterpart' for Abel Bergaigne's *Religion vedique*, (Louis Renou, 'Sylvain Levi et son oeuvre scientifique', in: *Memorial Sylvain Levi*. Paris: Hartmann, 1937, p. xxiii).
62. Durkheim however does not explicitly refer here to the authority of Sylvain Lévi, but to Abel Bergaigne - Sylvain Levi's own teacher and also to Eugene Burnouf and Hermann Oldenberg, (*The Elementary Forms of the Religious Life*. New York: Free Press, 1965, pp. 45-50).
63. Sylvain Lévi himself had learned this from his teacher the Indologist Abel Bergaigne, (cited in *The Elementary Forms*) Jean Réville, *Liberal Christianity*. New York: Putnam, 1903, pp. 198 f.

XXVIII

PARTICIPATION IN, AND OBJECTIFICATION OF, THE CHARISMA OF SAINTS

Stanley J. Tambiah

In his *Rules Without Meaning*,[1] Frits Staal makes reference at several points to my writings on ritual - he is generous while not suppressing his critical sense - and in Chapter 25 he discusses some features of my book *The Buddhist Saints of the Forest*.[2] What Staal found most interesting in this work was the notion of the 'objectification of charisma in objects and fetishes' which I deployed in the description and interpretation of the cult of amulets associated with the forest monk saints of Thailand who were prominent at the time of my fieldwork in the seventies.

Staal labelled my approach 'object semantics' and made these remarks about it which I will take as my text for further commentary and elaboration in this essay written in his honour.

'Though Weber's theory allowed for the "routinization of charisma" in social institutions and social positions', Tambiah notes that 'he was blind to the symbolism of objectification of charisma in objects and fetishes'. Staal is taken with the processes that cause the relative scarcity and ensuing commercial value of the Thai amulets, and the developments by which amulets become 'private possessions of laymen who expect to use the amulets' potency to manipulate, overpower, seduce, and control their fellow men and women in an ongoing drama of social transactions'. 'Tambiah applies to his amulets Mauss' expression of a "magico-religious guarantee of rank and prosperity". Actually, if one replaces ... "amulets" by "dollars" one obtains an interesting theory of the origination of capitalism.'[3] This seemingly throw-away remark on the origination (and expansion) of capitalism is in fact suggestive as we shall see in the concluding part of this essay, which alludes to the processes of fragmentation, multiple reproduction and expanding circulation of relics and images as integral to the spread of Buddhism and Christianity as 'universalizing' and missionizing religions.[4]

This essay is divided into four parts:

1) I shall first outline two interrelated processes which I label 'participation in charisma' and 'objectification of charisma' that appear to me as discernible in certain selected religious traditions, such as early and medieval Christianity, many branches of Sufi Islam, much of Theravada and Tibetan Buddhism, in which the veneration of 'saints' (and their counterparts in non-Christian traditions) and associated cults of relics, amulets, images, icons, and tomb shrines are found in *different combinations*. It seems to me a plausible hypothesis that in these religious traditions where sainthood is recognized and institutionalized, there exists a set of shared conceptions about the profiles of their highest practitioners, as for example the *saint* in Christianity, the *wali* or *sayid* in Sufi Islam, the *arahant* in Theravada Buddhism, the great *lama* of Tibetan Buddhism. (We might extend the list to include the *tsaddik* in Hasidism, the *guru sahib* in Sikhism, the *jina* in Jainism, and so on.)[5] I do not propose to substantiate this comparative hypothesis here, but shall proceed as if it is plausible.

2) Next, I shall give a brief summary of the cult of amulets focused on the forest saints in Thailand in the form I studied it in the 1970s.

3) Thereafter, I describe in greater detail, the subsequent development of the relic and *stupa* cults after the first crop of forest saints (Acharn Man and his contemporaries and first generation of disciples) passed away. This sequel to my own narrative is provided in the work of another scholar, J.L. Taylor, who followed me in time, and whose documentation is worthy of recognition.

4) In the fourth part, I shall return to the general theme of the processes of participation in, and objectification of charisma, and offer some conceptual and substantive elaborations of them, concluding with a critique and revision of Walter Benjamin's famous essay on 'mechanical reproduction' as a special technical capacity and propensity only of the modern industrial secular civilization.

Readers, please note that I have left out diacritical marks in my text, except when I am quoting from other authors.

PART 1. THE PROCESSES OF PARTICIPATION IN, AND OBJECTIVATION OF, SAINTLY CHARISMA

Comparative study suggests that saints have an impact on this world which they strive to transcend, and that in special ways they are both conduits to the divine or supramundane realm and creators of community. The world renouncer, to use a term that has coinage in Buddhist and Hindu contemplative asceticism, has in the past and present exerted a special kind of influence on life in the world, precisely because having emancipated himself (or herself) from worldly involvements, and gained a priority in relation to it by burning away all defilements, and by virtue of his detached and impartial yet compassionate love for all beings, he can, as Heesterman puts it in his explication of the renunciatory ideology, 're-enter into relation with the world, where he now enjoys unequalled prestige ... He is no longer a party to the affairs of the world because he is independent from it.'[6] Much the same logic seems to inform the manner in which many Christian and Islamic (particularly Sufi) saints are seen as radiating their virtues to their lay devotees, without losing their world-transcending sanctity.

Another focus, closely related to that adumbrated above, directs our attention to a distinctive patterned relation between a saint and his or her followers which we may call (borrowing from Troeltsch) the 'circulation of grace' in regard to Christianity, and for Buddhism the 'circulation of merit'. Troeltsch characterized the way in which the medieval Christian church redistributed the supernumerary good works of its ascetics and saints as a process of 'vicarious oblation' and 'circulation of grace'. These terms aptly illuminate the way in which certain religious traditions see the manner in which the saint's virtues are radiated to his or her devotees. Writes Troeltsch:

'The idea of vicarious repentance and achievement is really a living category of religious thought; the vicarious offering of Christ both as a punishment and as a source of merit is only a special instance of a general conception ... Thus the duty of those who live 'in the world' towards the whole is that of preserving and procreating the race, a task in which ascetics cannot share, while they for their part have the duty of showing forth the ideal in an intensified form, and of rendering service through intercession, penitence and the acquisition of merit. This is the reason for the enormous gifts and endowments to monasteries; men wanted to make certain of their own part in the oblation offered by monasticism.'[7]

Troeltsch's notion of the circulation of grace can be compared with the Buddhist idiom without excessive distortion. While there is of course no sense of a vicarious offering of the Buddha, there is a strong sense in which the members of the *sangha* (order of monks) by leading exemplary lives, and thereby occupying an elevated position in relation to the laity that surrounds them, keep the universe in moral balance. Thus the purity of the *sangha* and its adherence to the discipline is of crucial concern to the laity. The circulation of merit is as follows: the exemplary monk is the receiver of *dana* (material gifts) from lay householders. He in turn provides the field and context in which the layman makes merit (*punnak khetta* in Pali). The monk in return, not only by virtue of his disciplined conduct but also through preaching the *dhamma* (doctrine), and through ritual recitations, makes a return in the form of *dhamma dana*, a return which does not compromise his journey in search of liberation.

Though many saints, during their life were situated outside the 'establishment', and were marginal to the ecclesiastical and governmental centres of their societies, they were by another mode of counting, according to the doctrinal traditions of their religions, their greatest achievers. Typically the Thai forest monk saints, like the Hasidic *tsaddiqim* and the Christian ascetic saints referred to by Troeltsch, and the Islamic *sufi* saints, are 'illuminates'. More often than not, rather than creating doctrine, they bring doctrine to life through practice, especially for their lay communities, and are capable of circulating their grace and radiating their charisma as quintessential achievers of the highest values of their religions. In the end, the personal, interiorized, mystical illumination of the saints is seen in these religious traditions as flooding the vast spaces of the world with their cosmic love. Religion is embodied in and proceeds from them, just as they, as individuals, interiorize the whole religion. Frequently the saint as holy man devoted to practice is differentiated from the bookish legists, the *'ūlama'* of Islam, from the *ganthadhura* (those dedicated to the vocation of books) in Buddhism, and from the hierarchs like the bishops of the Christian church. But after their death these saints, in the form of their relics and their other material traces such as talismans and amulets *are incorporated* by an *established* religion and placed at the heart of institutionalized religion.

There are two modalities or dimensions that are present in the saint's circulation of grace and/or merit and creation of a community which I shall call 'objectification of charisma' and 'participation' in charisma. Objectification of charisma relates first to the process of transferring or transmitting charismatic energy or virtue from the living saint to an object (amulet, talisman, clothing, image, etc.), and secondly, to the object in turn subsequently embodying and radiating that energy and potency to its possessors. By participation I mean the interaction and fusion between the community of devotees and their radiant saint or *guru* at the centre: they are first of all in a relation of contiguity, and then they translate that relation into one of existential immediacy, contact and shared affinities. There is an 'indexical' transference, to use the jargon of Peircean semiotics, of energies and grace from the leader to follower. (Staal is on record as being allergic to my use of 'technical' words, and I might yet persuade him that these are merely disposable rafts used only to get to the other bank of the river.)

In the language of Hinduism, the same two modalities are conceptualized by the terms *prasad* and *darshan*, and both are contained within the larger conception of *bhakti* worship, and the even more general phenomenon of *miracles*.

THE CULT OF RELICS, STUPAS AND SHRINES

The 'objectification of the charisma' of great men (in Pali Buddhist terminology, *mahapurisa*) and the capacity of these exemplary persons to perform 'miracles' and marvels provides the links between living saints and the cult of relics, *stupas* and tomb shrines as a posthumous celebration and manipulation of their continuing presence (in Latin: *praesentia*) after their death. The parallels of 'miracle' in Christianity are *karamat* in Islam, and acts of *iddhi* in Buddhism and *siddhi* in Hinduism.

In late Antiquity, and in Medieval Christian, and in Sufi Islamic traditions, the saint's distinctive capacity to perform miracles while alive finds its posthumous extension and continuation in the veneration of the saint's relics and/or his or her tomb as the repositories and conductors of miracle making powers. In Buddhism the cult of relics, embedded in *stupas*, relates first and foremost to the bodily remains of the Buddha himself, though various Buddhist saints throughout the centuries have also been the foci of relic cults, including the forest monks of Thailand of our time.

PART II. THE FOREST SAINTS AND CULT OF AMULETS IN THAILAND

Let me now explicate processes of the 'objectification of charisma' and 'participation in charisma' by ethnographic reference to the cultic aspects surrounding the forest monks of contemporary Thailand.

In *Buddhist Saints of the Forest and the Cult of Amulets*,[8] I described how a wave of popular religiosity had become focused on a number of reclusive, ascetic and meditative monks residing in forest hermitages in the interior provinces at the periphery of Thailand. These monks acclaimed as *arahants* (saints) by the public at large are to a varying degree followers of the path of purification as set out in the *Visuddhimagga*, a classical text composed by Buddhaghosa in the fifth century AD.

Somewhat unexpectedly in our modern times which give prominence to politicians, generals and captains of industry, and to the acquisition and expenditure of power and wealth, it is the very holiness, saintliness and detachment of the virtuoso forest monks that have raised them into national prominence. First acclaimed by the common people at the grass-roots level in the rural hinterland and small provincial towns, they have later been lionized and patronized by the ruling and commercial elites in the capital of Bangkok, including the royalty. The King and Queen, the Crown Prince, the ruling generals and ministers, and the managers of the largest banking and business houses have all sought these *arahants* in their remote habitations and have bowed before them, hoping to be edified and strengthened by their visits to them. The centre of the polity and economy goes to the periphery to be reinvigorated, because it is in the periphery that the heart of the religion beats.

I have also described how the amulet itself represents a conjunction of two parties, the lay sponsor who organizes and finances the manufacture of the amulet-medallion and has his or her insignia imprinted on one side of it, and the holy man, the famous monk, who through chanting sacred words of protection and through meditation sacralizes the amulet-medallion and transfers his virtuous energies to it,[9] and whose image (head or bust) is imprinted on the other side. Thus the amulet-medallion is physically, iconically, and indexically a meeting and a linking of the lay sponsor and monk adept, and in itself becomes an 'animated' object of great anthropological interest.

Let me sum up the wider implications and involvements of the cult of saints and amulets in Thailand in this way. The cult which on the face of it is a religious phenomenon soon becomes implicated in the world of politics, commerce and influence.

1) The cult brings meditation masters in forest hermitages, their monk disciples and co-residents, and the outer circles of lay society in the rural periphery as well as urban centres, capped by the metropolitan capital, all these, into one religio-political space. All these distributions of persons and their networks define and participate in a field of power and merit.

2) A complementary and symbiotic dialectic operates in this field. The sponsorship of the amulet cult - the manufacture of amulets, their sacralization by the saints, and their distribution to the lay devotees and consumers at large - is largely in the hands of the powerful and affluent lay patrons, such as royalty, generals, bankers and higher bureaucrats. These powerful laity 'legitimate' and mark the religious achievements and charisma of the saintly monks at a society-wide or national level, just as they in turn are also indirectly empowered and legitimated by the saints.

3) The Thai attribution of charisma to their saints, and the ability of these saints to transfer their radiant presence and energies to amulets, which become concrete repositories of power is a form of fetishism constituted of two social cycles or loops in which saint and lay public are implicated. One cycle is the ideologically developed and transparent one; the other is more hidden and manipulative and rides on the former. The first cycle is to be understood in terms of the Buddhist path of salvation: According to its cosmography, the ascetic meditator attains progressively higher spiritual levels of consciousness, leaving the grosser material excrescences behind, and through the control of his sense doors and sensory states attains understanding wisdom and universal compassion. This, then, is a state of transcendence,

which also generates supranormal powers with which to affect the phenomenal world through detached action. The specifically Buddhist formulation, then, is that detached action can also become effective pragmatic action, because by being removed from the immediacy of desires and entanglements, it is all the more encompassing and creative. Now it is in the space of this first cycle that these virtuoso saints are approached by merit-making laymen, to whom the amulets are distributed as part of the saints' dispensation of blessings. At this level of exchange, the conventional Buddhist exegesis has some explanatory value: Amulets (like relics) act as 'reminders' of the virtues of the saint and the Buddha, and the saints act as 'fields of merit' in which laymen plough, sow, and harvest their donations. Moreover, in this frame the pious intentionality of the lay person's merit-making is stressed.

At the next remove, we have the cycle of transactions by which lay persons possess, accumulate, and secrete on their persons or otherwise employ these amulets to influence, control, seduce, dominate, and exploit fellow laymen for worldly purposes - in the corridors of politics, the stratagems of commerce, the intrigues of love, and the sycophancy of clientage. In this arena we perceive two developments: First, the iconic and indexical properties of the amulets are recognized not as mere reminders but as pragmatically efficacious. The emblems of the saints, the Buddha, and other sacred beings, come to embody the monk's virtue and power by existential contact with him, and by virtue of his impregnating them with sacred words, purifying them with sacral water, and other similar acts of transference. This objectification of the virtue of the saint also implies the descent of his spiritual and transcendental powers from the higher realms of spirituality and universality to the lower realms of material desires and limitations and particularities of space and time.

It is inevitable in the Thai case that this process of materialization, this process of gravitation, should have further consequences. One is that the amulet moves in time from a context of donation and love (*metta*) to a context of trade and profit: it is converted into a highly saleable good and enters the bazaar and marketplace. When it does so, it also stimulates the production of fakes and becomes a pawn in the usual publicity media of advertisements, collectors' catalogues, magazine articles, books, and the mythology of miracles. A second consequence is that the more amulets are produced, the more they are faked, and the more they are purchasable for money, the more they deteriorate in their mystical powers (despite the inflationary spiral of prices for those amulets regarded as rare antiques). This in turn ensures that new amulets come into fashion, and that many others already in circulation are condemned to be forgotten or less desired; moreover, the propensity for collectors to accumulate amulets increases, in the simple arithmetic calculation that the more you possess, the more clout you have. Thus the comparison of the relative virtues of amulets leads to mystical power itself, which at its source in the form of the saints' cosmic love was both limitless and rare, becoming graduated, differentiated, and quantified by the play of market forces, in short, commoditized.

PART III. THE RELICS OF FOREST MONK SAINTS OF THAILAND

Taylor's book, *Forest Monks and the Nation State*, published in 1993,[10] takes the account of the forest monk saints of Thailand, whom I dealt with in my *Buddhist Saints of the Forest* (1984), further along in time and in elaboration especially in terms of the cult of *stupas* built to commemorate famous forest saints and the associated cult of relics generated by their remains.

I am citing him here because he gives valuable new information about how many of the monks associated with the blessing and distribution of amulets during their life became after their death and cremation the source of sacred relics, and the posthumous focal points of monumental building activities such as *stupas* (*jedii*) and other structures (the equivalent of the Thai word *jedii* is *cetiya* in Pali).[11]

Taylor appropriately asserts that 'Normatively, amulets, which are in the same category as sacred relics, function as reminders of the pure sanctity of the virtuoso (whether 'town' or 'forest') and attributed with mystical powers.'[12]

That the ashes and bone fragments collected from the sites in which famous forest monks were cremated are sacred relics [*phrathaat* (Thai); *saririka-dhatu* (Pali)] preserved in urns and pots is a well-established practice in Thailand. In *Buddhist Saints of the Forest* I cited from a hagiography of Acharn Man how after the cremation of his body in 1950, his ashes were distributed to 'the bhikkhu delegates from various towns ... so that they could be enshrined in places to be specially built',[13] and to his various lay disciples and devotees also attending from various towns. On the very place where the cremation pyre stood was later built the convocation hall of Wat Suddhavas. The master's ashes in due course were reported by some collectors as crystallizing into jewel-like relics. In the words of the hagiography, they turned into 'smooth and glossy grains, sandlike in appearance, resembling relics of the Buddha and some other arahant disciples in ancient times'.[14] In the case of the deceased Acharn Man, the most famous lay collector and possessor of his relics was a certain lady, Mrs. Khamanamool.

Of Acharn Man's own teacher Acharn Sao it is reported that after his cremation at Wat Burapha in Ubon in 1941, his remains were distributed and ground down by some of his pupils and made 'into composite personal amulets'.[15] Acharn Man's hagiographer asks this critical question, 'Why do the ashes of arahants become relics?' and advances this impeccable answer:

> The body of an *arahant* and a worldling … have the same ingredients, with this difference: The mind of an arahant has been absolutely purified, whereas that of a worldling contains defilements. 'Body matter is then transformed in accordance with the condition and nature of the mind.' A purified mind can therefore purify the physical body to eliminate its 'toxic ingredients' by virtue of the mind's regular withdrawal into the most profound depths of concentration and hence the likelihood of his ashes turning into relics after his death. Noble Disciples can be divided into two groups, according to their history of attainment: The *dandābhiññā* is one who, having reached the level of Non-Returner, takes a long time before achieving *arahant*ship the *khippābhiññā* is one whose attainment is 'sudden', taking little time, and he does not live long after his attainment. It is the former whose prolonged and many meditative exercises have a chance to purify the body, and Acharn Man is a quintessential example of such a Noble Disciple.[16]

Taylor provides information of similar relic formation from the collected ashes of many other forest saints, contemporaries as well as disciples of Acharn Man (examples are Acharns Orn, Fan, and Jan Khemapatto).[17] One notable feature of this process of relic formation is that the crystallized bone fragments are alleged under certain conditions to be capable of 'miraculous multiplication', a characteristic of relics whose significance I shall comment on shortly. Taylor reports that the duration of time when the ashes turn to crystallized relics may vary from saint to saint, depending on whether they had made a wish (*athat-thaan*) before death that their bones turn to relics to convince people of the truth of the arahant's path and the fruits of meditative practice, or depending on when in their career they attained arahantship (the later the attainment the longer the interval before crystallization) and so on.

In the case of Acharn Phrom, Acharn Man's first saint disciple who died at the age of 79, 'less than one year after his cremation in 1971, his bone fragments were found to have crystallized'. In fact his prophetic wanderings and exemplary ascetic regimen were very similar to his master's, and this may explain the speed of crystallization.

> After ordaining Phrom undertook extensive wandering and went as far as Burma, into Laos, and various parts of the north and northeast. He

> eventually met up with Man in Chiang Mai, where Phrom - like his teacher before him - reputedly attained *arahantship*. He returned to Baandung district in Udornthaanii province not far from his home district to settle and died at the age of seventy-nine. Less than one year after his cremation in 1971, his bone fragments were found to have crystallized. Phrom, and the other monks mentioned above, were paradigmatic of the early *arahan* ideals; their lives were read as acts of courage and endurance leading to spiritual 'liberation' (*wimutti*).[18]

It seems that just as amulets blessed by forest monks find their way into the market place, a process that I have reported, so do the minute partially or fully crystallized bone fragments of forest saints. Thus, it is reported that the relic fragments of Acharn Fan Aajaro 'who died in 1977 (cremated the following year) currently [in the early nineties] have a market value of around 150,000 to 200,000 baht (US$ 5,769 - US$ 7,692)'.[19]

JEDIIS (STUPAS) AND RELIC REPOSITORIES AND PILGRIMAGES

Since my inquiries into the forest monk saints of Thailand in the late seventies, a subsequent development in the form of building *stupa*s or *jedii*s in which to store and enshrine 'relics' of famous *arahant*s constitutes a kind of final phase in the careers of forest saints, and arguably also, the inevitable waning of the 'presence of the saint' with the passage of time, as the *stupa*s and their relic chambers become less and less frequented as places of pilgrimage. This is a gradual process, and some *stupa*s and the temple complexes in which they are located are more durable than others depending on the continuing faith in the miraculous power (*ithirit*) of the dead saint combined with the organizational energies and the maintenance of disciplinary practices, especially meditation and ascetic practices, by the dead saint's successors and 'lineage'.

Once again I rely on Taylor to provide the remainder of the forest monk narrative where I left off. He reminds us that forest monks are popularly conceived as 'store houses' of merit and mystical powers, and are 'regarded (though rarely discussed as such directly) as national 'spiritual treasures'.

> The extent of such veneration can be appreciated by visiting the many grand relic-museums or repositories (*jediiphiphithaphan*) marking northeastern sites where Man's early pupils died in the 'terminal' settlement phase.

> Dry-season *jedii* tours are now a common feature and the focus of collective merit-making in a sense of 'communitas' (... though social hierarchies flow over from outside and persist during the merit-making tour ...). Noteworthy *jedii* include the master's [Man's] at Wat Paa Sutthaawaat in Sakon Nakhorn and his pupils such as Juan Kulachettho (1920-80), Fan Aajaro (1899-1977), Wan Uttamo (1922-80), and Khao Anaalayo (1888-1983). This latter-mentioned monk was particularly popular among the elite.[20]

Khao was supposedly the second pupil of Man to have attained *arahant*ship. He wandered extensively in the mountainous areas of Nongkhai province and secluded regions of Nakhorn Phanom province in the Northeast. He died at the age of ninety-five in May 1983; 'he was given a royal cremation the following year attended by royalty and national political elite'.[21]

There are many other cases of *jedii* cult, characterized by devotion of saints' relics, that emerged in relation to other disciples of Man.

Perhaps the most splendid of all *jedii* in the northeast is Ajaan Fan Aajaro (Wat Paa Udom Somphorn, Sakon Nakhorn), replete with hand-carved stone murals with scenes of the teacher's wandering life, like *Jataka* tales. The *jedii* cost well over half-a-million US dollars collected from donations coming largely from the capital. Because corruption was so rife, Mahaa Bua, as widely regarded lineage head, was asked to supervise the fund-raising as he had earlier with Man's *jedii* at Wat Paa Sutthaawaat in Sakon Nakhorn. In the latter instance, although Man had died in 1949, it was not until twenty-four years later that the first foundation stone was laid and, by that ceremony, marking the commencement of the modern '*jedii* cult' for forest monks throughout the northeast.

In Man's *jedii*, the original 975 donors (including business, military, and political elite) were listed along with their contributions (averaging over 3,000 baht [US$ 118] per person) in a widely distributed booklet. Many of the contributions were sent through Man's disciples, or prominent individuals, lay organizations, private and public corporations. The largest single donation came through Mahaa Bua's Wat Paa Baan Taat, largely because of Mahaa Bua's links with establishment elite in the provinces and the capital.[22]

Juan Kulachettho's newly completed *jedii* rising above the sparse Maekhong plain cost over 10 million baht (US$ 384,615), most of the support coming from fund-raising initiated by the Electricity Generating Authority of Thailand (EGAT) Buddhist Association led by its recently retired senior executive Mrs. Suriiphan Maniiwat ... Suriiphan has for a number of years organized annual merit-making tours to the monastery which take in a number of other well-known parent forest monasteries. At the time of writing the *jedii* was still not yet finished and because of increasing costs Suriiphan had to raise private funds to keep the work going. Juan was highly regarded by the king who had great 'faith' in the teacher and after his cremation was given most of the relics and ash which he took with him back to the palace (though promising to return them when the *jedii* was completed). According to informants, Suriiphan ... has been trying for many years to solicit royal approbation and personal favour for her prominent tutelar role among the northeastern forest *arahan*.

Perhaps one of the most important sponsors of *jedii* in the northeast and long-time patron of forest monks in the tradition of Ajaan Man is the Norngkhaai-based entrepreneur, Kimkai and his family. Kimkai, a local Member of Parliament, initially made his fortune wheeling and dealing across the Maekhong and presently has numerous business ventures spread across the country.[23]

Taylor gives further details of the enormous money and other material contributions made by Kimkai and his close kinsmen, for the construction of various monuments and buildings in many temple complexes, including the building and maintenance of *jedii* housing the relics of many forest saints,[24] and for the conspicuous merit-making festivities that are staged when completed *jedii*s are opened.

> At Wat Tham Klongphen and Wat Hin Maak Peng, Kimkai contributed a great deal of money to the construction of various monuments and buildings. He also provided material support to the *jedii* for Ajaans Jan Khemapatto, Orn Yaanasiri, and Bua Siripunno. A new *jedii* for Ajaan Singthorng Thammawaro (Wat Paa Kaew Chumphon) and nearby Ajaan Suphat (Wat Paa Baan Taai) is a current Kimkai project at the time of writing. Seemingly, Kimkai's family were supporters of Ajaan Juan prior to his untimely death in 1980 and were going to build the *jedii* but lost out to EGAT's Buddhist Association.
>
> In the past ten years Kimkai's family supported Ajaan Chorp (Wat Paa Khokmon), Ajaan Bunjan (Wat Paa Santikaawaat), Ajaan Thui Chanthakaro (Wat Paa Daan Wiwek), and the recently deceased Ajaan Lui Janthasaaro (Wat Tham Phaabing and at Wat Tham Phuukhaa in Sakon Nakhorn associated with a lesser-known pupil of Man, Ajaan Kuu, who died in 1953) until he was taken to Bangkok shortly after by another wealthy supporter. There were also a number of other pupils of Ajaan Man under the patronage of Kimkai.[25]

The following excerpt from Taylor confirms what I had asserted earlier in my discussion of the cult of amulets surrounding the Thai forest saints of the sixties and seventies: that their charisma is first and foremost appropriated by the sponsors of amulet making and sanctification - the royalty, the banking houses, the *nouveaux riches millionaires*, the generals and top bureaucrats. In Thailand, in other words, the moneymakers and powerwielders, both in the capital of Bangkok and in the provincial towns, are the primary donors and merit-makers towards holy men, once they have attained fame, and receivers of their legitimating blessings and generative powers.

This documentation by Taylor concisely says it all:

> During the official opening of Ajaan Khamdii Paphaaso's 7 million baht (US$ 269,230) *jedii* in Loei, Kimkai's relatives had arranged a free food stall for the many hundreds of weary lay visitors over the two days of the ceremony. The *jedii* was opened by the governor of Loei and one of the King's esteemed Privy Councillors (Ong Khamontrii), Dr. Chao na Siilawan ... In fact this ceremony held in December 1989 was the largest single gathering of northeastern Phra Kammathaan monks in many years. It was also well attended by high-ranking *pariyat* monks from Wat Noranaat and Wat Raatchabophit in Bangkok, as if to affirm the indubitable links with the centre. Well over 400 monks (twenty to thirty of these famous forest teachers from around the northeast) came to pay respects to Khamdii and participate in joint religious rituals.[26]

THE WANING OF THE JEDII CULT: THE LAST PHASE IN THE CYCLE

The process of objectification of charisma in amulets and later in the relics of forest monk saints, some of which are deposited in *jedii*, raises the issue of the duration of time when the presence of the dead saint and the powers associated with his or her sanctity prevails intact. It is part of the dynamics of these cults that they rise and decline over time at differential rates, to be replaced by the emergence of successor saints and their ritual activities and new followings of devotees.

Already in Thailand, the *jediis* and monumental halls (*saalaa*) built to commemorate and store the remains of dead saints are experiencing the waning of public support and the decline in pilgrimage attendance.

> After *jedii* have been built at forest monasteries during the terminal phase, hitherto prolific support starts to dwindle and to be diverted elsewhere, with many parent monasteries struggling to maintain their upkeep. At Wat Doi Maepang in Phrao district, Chiang Mai, two years after the death of Luang Puu Waen Sujinno (1888-1986), there were few

indicators of either its primitive origins or the vigour and height of its popularity in the late 1970s. Local sellers of photograph reproductions, amulets, flowers, and incense along the largely deserted rows of empty stalls are not so active these days. The abbot, Ajaan Nuu Sujitto, spent millions of baht on now-empty buildings in preparation for the nationally broadcast two-day ceremony attended by the royalty and some of the nation's highest-ranking dignitaries. The central *mondop* (Pali: *mandapa*, literally 'temporary hall') holding Waen's relics, life-like wax figure and personal monk-requisites was built by a wealthy Thai-Chinese physician from Bangkok, Dr. Amon Mahaphatthanaangkuun. The Railways Authority of Thailand supported Waen's expensive *kuti* situated opposite his original simple wooden shingle dwelling. The king donated funds for the beautiful *bot*, though these days there are often too few monks in residence for the performance of the fortnightly ritual 'acts of the *sangha*' (*sangkhakam*). The contrast is sharp compared with either the bustling, heady days when convoys of luxury motor vehicles were coming and going and a large community of wandering monks staying close to the teacher, and the beginning phase when Man and his northeastern pupils Ajaans Sim, Teur, Khao, and others encamped in the teak forest during their period of wandering in the north. But from this group of northeastern monks it was Waen, finding peace and solitude at Doi Maepang, who finally stayed on ...

At Wat Paa Udom Somphorn in Sakon Nakhorn, Ajaan Fan's *jedii* is giving the present abbot something of a headache to maintain, especially with its extensive landscaped gardens. Donation boxes (*klornglapborijaak*) have started to appear, something which would have been abhorred by the teacher fifteen years ago.

A similar problem of maintenance support was also felt at Ajaan Khao's former monastery, Wat Tham Klongphen, according to its abbot, Bunpheng Khemaaphirato. These above case-studies are some examples where prolific ritual merit-making has been drastically reduced, leaving a complex infrastructure which many of the early monks did not in any case particularly want, especially electricity and telephones. Besides donation boxes, the last resort seems to be selling cheap artifacts, 'blessed water', pictures of the deceased master, medallions, and the like. But then in a wider sense this has long been a feature of Buddhist Southeast Asia where the landscape is potted with *jedii*, indigenous Buddhalogical sacra. The establishment of *jedii* throughout the northeast has had the purpose of sanctifying and legitimating religious sites in the far provinces in a direct line to the centre and king as the guardian of *dhamma*, the *Thammaraachaa* ...

In *Buddhist Saints of the Forest*[27] I suggested that the motivating impulses and the dynamics of forest monk saints and their followings is such that, although the forest monk path is valued in the Buddhist scheme of vocational attainment for monks, these charismatic formations are found to be in the long run volatile and less enduring, in spite of their networking, when compared with the *establishment sangha*, centred on town and village monasteries focused on 'learning' and on everyday ritual transactions with the laity, and from whose base was erected the hierarchy of ecclesiastical offices and titles from which the forest monks were excluded.

In the Thai case, not only were the forest monk movements and their charismatic capacities appropriated and domesticated in the long run by the political authorities in place, but also they were under pressure to be encompassed and incorporated by a branch of the establishment *sangha* entrenched in Bangkok and long allied with royalty and the ruling military (I refer to the Thammayut Order). Thus while some of the quickly blossoming *jedii* complexes face the fading of their fragrance, other monastic residences began by forest monks might also in time become regularized village or town monasteries dispensing *pariyat* learning and training chanters of sacred words.

These statements by Taylor aptly signal the terminal phase of the forest monk upsurge in Thailand in the last few decades. The 'practising' residences of forest monks including those in the line of Acharn Man have in time become either institutional *pariyat* centres or *wat baan* ('village monasteries'), catering for the ritual needs of an agricultural community. At some of these now-domesticated monasteries, the resident novices and monks could not even remember the founding teacher or had little interest in the monastery's inchoation. The links are clearly outside of institutional form in the personalized face-to-face pupillary lineage, and once the teacher dies and pupils disperse, the monastery ceases to have any symbolic significance to forest monks.

In time even the teacher's *jedii*, that classic locus of sanctity, will likewise fade into obscurity (despite what the present devotees say). Some of the once-assiduous, active centres for forest monks are now either taken over by the domesticated *sangha* or left to the elements, gathering dust with fading lineage photographs eaten by termites in the communal hall (*saalaa*) now housing tarnished Buddha images, and in the monks' grounds (*sangkhaawaat*) overgrown forest tracks and once well-trodden meditation paths (*jongkram*; Pali: *cankama*) in receding circumjacent forest - all part of the changing scenario in the forest monastic tradition. The remaining ascetic monks have now gone deeper into the last frontier areas, to small pockets of forest further from the provincial towns and less spoilt *samnak* which themselves one day will doubtlessly go the same way as their parent monasteries, an emergent phase which bespeaks finality, and a '*commencement de la fin*'.[28]

In the light of the foregoing discussion I am left with a speculation: In Buddhist countries like Sri Lanka, Burma, and Thailand the vast distribution of ruins of monastic residences and more significantly *stupas* and *caityas* - if viewed as one vast configuration *synchronically*, this distribution of sites and remains might convey an illusory sense of a thriving religion catering to thickly populated habitation; but if viewed *diachronically* we might also see this map of ruined holy places as the traces of such short-lived forest monk communities, and the merit-making acts of waves of lay devotees who eagerly wished to be patrons and donors of their own special holy men.

I have no doubt that the module of the expansion of early Buddhism and its colonization of new areas at the periphery was to a large part effected by the twinning of monastic residence and the *stupas* as a duality. How valid was this process for later times?

PART IV. SOME SEMIOTIC ISSUES PERTAINING TO SACRA AND OTHER 'VALUED' POSSESSIONS

In the foregoing pages I have highlighted two related and interrelated complexes of processes.

1) The process of 'participation' of disciples, devotees, and followers in the charisma of living saints and gurus through distinctive communicational, interpersonal and intersubjective exchanges. A prominent feature of this participation is the living saints' capacity to perform *miracles* (acts of *karamat*, *iddhi*, *siddhi*, etc.).

2) The processes of 'objectification of charisma' by which objects which are initially considered to be 'reminders' of a person, or to have a similarity and contiguity with that person are seen and felt cognitively, emotively and experientially to embody the posthumous 'presence' of that person and to cause performative effects, such as healing illnesses, conferring reproductive fertility, and revitalizing life. The objects in question are relics, icons, amulets, talismans, etc. In my jargon, the absorbing issue is how the passage is made from a doctrine of reminder (an analogical, or metaphorical, and/or contiguity (metonymical) relation between the saint and an object) to a doctrine of *presence* (an *identity* relation between them), so that the object becomes potent and 'animated' in its own right having the capacity to evoke responses and create practical effects. This effectiveness of images, cult objects, and other sacra depends on the conflation and fusion of sign and signified. David Freedberg coming at this issue from a different perspective refers to this as a 'fusion and inherence' resulting in 'the circularity of

inherent divinity and the perception of liveliness'.[29]

Posing the issue of how images (paintings, icons, statues, etc.) evoke powerful responses in their viewers, Freedberg in his professional affiliation as an art historian has attempted to explicate this issue in terms of 'figuration' and the acts of 'consecration' that 'make images work'. I have independently dealt with the issues of images, relics, and amulets in Thailand in terms of what I have called 'the objectification of charisma', especially concentrating, while not ignoring figurations, on the ritual transfers by authorized officiants of energies and powers through 'installation ceremonies', including the rite of 'opening the eyes' of the Buddha image, which thereby enliven the images and ensure their 'radiance' and the radiation of their power.

While 'figuration', in Freedberg's sense, is important in the case of images, statues, icons, and amulets (Freedberg has misrepresented me on this matter),[30] and there are standard manuals on the iconographical and aesthetic requirements for making images in many religious traditions, yet it seems to me that the power of these objects cannot be fully comprehended within the scope of Art Theory and Art History alone. Inevitably they must be viewed within the larger framework of religion and ritual, the sources and manifestation of charisma, holiness and sacrality; the hagiographical conceptions and conventions relating to saints and holy women and men; and the cosmological postulates including the relations between this worldly and other-worldly (or supra-worldly) domains. Ultimately here lies the limitation of Freedberg's history and theory of response whose lengthy meanderings cannot break out of the circle of 'figuration'.

THE SENSORY MODALITIES OF PARTICIPATION IN CHARISMA

Participation through visualization

Participation in the charisma of gods and saints (and other holy persons) occurs in multiple sensory modalities. (A primary kind is the visual mode of cognition and affective experience involved in the explication of the power of images.)

In Hinduism and Buddhism the concept of *darshan* exemplifies all the nuances of 'viewing' and 'visualization'. Visiting a temple to view the 'god' or the Buddha involves, as Diana Eck has nicely explained, the double process by which the god or Buddha views the devotees as the devotees worshipfully gaze at the former. Hence the importance of the 'ceremony of opening the eyes' of an image or statue at its installation ceremony.

Christianity has parallel practices. In early Byzantine history there are references to pilgrims going to see and gaze upon the holy man in the desert.

Even more strikingly perhaps, in early Buddhism and early Christianity there was the notion that when the devotee views the 'relics' of the saint he sees the saint himself. Thus the *Mahavamsa*, the sixth century Sri Lankan chronicle reports Mahinda who brought Buddhism to Anuradhapura as asserting, 'When we behold the relics we behold the Conqueror', an episode repeated in other texts such as the *Thupavamsa*. Mahinda is credited with saying this in explanation of his wish to return to India because there were as yet no relics of the Buddha on the island.

Athanasius, the biographer of Antony the Anchorite, reports that 'simply by seeing his conduct many aspired to become imitators of his way of life'.[31] Again, describing his travel companion Paula's visit to Egyptian monks, Jerome writes: 'in each holy man she believed she was seeing Christ'.[32]

Participation through ingestion

Especially in Hinduism, the offerings (*prasad*) made to a deity and the deity's 'food leavings' are highly auspicious to consume. Quite common to Hindu and some Islamic traditions is the drinking of water in which the saint's or holy man's feet have been washed. The Roman Catholic Mass involves the ingestion of consecrated bread and wine which through 'transubstantiation' have become the flesh and blood of Christ.

Participation through touch (tactile act) and through other forms of physical contact (contiguity relation)

Quite common to Christian, Hindu, and Buddhist traditions are touching the robe or feet of the holy person, or touching his or her statue and image as a way of receiving grace and sacred power. Touching tombs or sleeping next to them is a special act of veneration among Sufi Islamic devotees. Burial '*ad sanctos*', that is adjacent to or near a saint's tomb, was much valued in early Christianity, and is a continuing tradition among certain Muslim sects from Cairo to Karachi. Touching or holding a relic is especially sought after by Buddhist and Christian worshippers. The laying of hands by the saint is similarly a powerful mode of transferring blessings. The King's touch in medieval France was a cure for dropsy.

> Thomas of Celano relates of St. Francis that, after a miracle of the swallows people, having seen the sign, 'were filled with great admiration and they hastened with deep reverence to kiss the hem of the saint's

garment, and they praised God, saying truly this man is a saint and a friend of the Most High!' On other occasions, people sought 'to touch him in their devotion ... [and] they laid hands on him, pulled his habit and even cut pieces from it so as to keep them as relics.[33]

Participation through hearing

This is our last example of a sensory modality in the transmission of charisma. In Buddhist settings in Thailand (and elsewhere in Southeast Asia) there is a marked emphasis on hearing the sermon and the chanting of sacred words, and gaining merit, and even liberation, through listening. The auditory channel is of course a major conduit for receiving religious instruction, moral edification and 'merit', but a matter that cannot be ignored is the sense that the sacred words as sound waves in themselves are powerful and impact on and penetrate the body, even if the words are not understood in the semantic and syntactical senses. At the extreme stands the case of sacred or 'magical' formulae compounded of syllables and words, borrowed and mixed up from 'other' languages or concocted from sounds with no morphemic sense. Malinowski coined the phrase 'the coefficient of weirdness' to label this mode of constructing powerful language.

The power attributed to the recitation of sacred texts can be illustrated from some examples drawn from Mahayana Buddhism. Gregory Schopen has remarked with regard to 'the cult of the book' in Mahayana that the protection of the recited text extends not only to the hearer of the words but also to the place or site where it was preached.[34] Again, Leon Hurvitz reports that the text in question asserts: 'If there is a person who shall read and recite the *Saddharmapuṇḍarīka* [the *Lotus Sūtra*] be it known that person of himself shall be adorned with the ornaments of the Buddha ... What is the reason? When this man preaches the Dharma with joy, anyone who hears it for a moment shall straightaway achieve ultimate *anuttarasamyaksaṁbodhi*.'[35] In these examples, it is certainly not the semantic meanings of the recitations that are being foregrounded for the effects produced.

It is important to give in this context a prominent place to Staal's submission that in the enactment of Vedic rites (of his discussions pride of place goes to his awesome documentation of the *Agnicayana* performed in Kerala in 1975), it is the *syntactic rules*[36] which generate and constitute the ritual recitations (*mantras* and chants) that are primarily relevant to consider, for in Vedic ritual, as in *mantra* meditation, the function of language is phonetic and syntactic, not semantic. The rituals are meaningless in a referential semantic sense, and his bold claim about 'the meaninglessness of ritual' has to be situated in relation to his argument that the logic of ritual language and its efficacy are locked within its syntactic impulsions and ordering. Whether this formulation can be a general theory of all ritual language is no doubt open to debate, but Staal has certainly foregrounded the necessity to look first at and foremost the syntactic structures and patterning of ritual sequences before tackling questions of semantic meaning.

A note on Walter Benjamin: On the reproduction of 'authentic' sacra as cult objects.

I would finally like to make a comment on Walter Benjamin's famous essay on 'The Work of Art in the Age of Mechanical Reproduction'.[37]

Benjamin deals in this essay with, among other things, the implications of the modern process of secularization of art combined with the technical capacity for 'mechanical reproduction'. He first develops the concept of 'aura' in relation to a work of art as a cult object in past tradition.[38]

> The uniqueness of a work of art is inseparable from its being imbedded in the fabric of tradition ... Originally, the contextual integration of art in tradition found its expression in the cult ...

Such objects were characterized by *aura*, which he defines as a 'unique phenomenon of distance however close it may be to the viewer'. This aura 'represents nothing but the cult value of the work of art ...; and its 'inapproachability' (distance being the opposite of closeness) is 'the quality of the cult image'.

Now 'with the secularization of art, authenticity displaces the cult value of the work, and mechanical reproduction is with 'literature and painting' an external condition for mass distribution'.[39]

> With the advent of the first truly means of reproduction, photography, simultaneously with the rise of socialism, art sensed the approaching crisis which has become evident a century later ... for the first time in world history, mechanical reproduction emancipates the work of art from its parasitical dependence on ritual ... Instead of being based on ritual, it begins to be based on another practice - politics.

Much of what I have discussed in relation to the cult of amulets in Thai Buddhism and of relics in both Christianity and Buddhism, goes somewhat counter to Benjamin, for in relation to their efficacy and wide distribution I would claim a necessary *dialectical connection* between the two modalities he separated the 'aura' and 'distance' of the unique work of art or cult image and the secularized closeness of mechanically reproduced objects of present day mass consumption. In the case of the Thai amulets, 'aura' and 'sanctity' are allegedly ensured by claims of 'original markings' associated with the authentic production of amulets and their 'likeness' to the 'original' (the saint's figure). At the same time there has been from way past the capacity to make a stock of numerous replicas and mechanical reproductions of the authentic amulet for distribution to a considerable (but not unlimited) number of recipients.

In the case of relics, we have seen how general and necessary was their virtually unlimited fragmentation and reproduction into smaller and smaller particles, none of which were claimed to have lost the aura and authentic presence and charisma of the original and holy person. There was an inherent logic to this mass production and distribution of cult objects tied to the massive spread of missionizing religions both *ritually* and *politically* involved in this process of mass consumption and religio-political control.

Moreover, these processes of multiple production of amulets and the progressive fragmentation of relics, and the potency attached to the 'authentic' images and relics inevitably encourages the processes of copying, replicating and reproducing the 'original' in the case of amulets and images, and the manufacture of relic fragments for sale in the marketplace by alleged antique dealers and experts. The production of fakes is integrally related to the commoditization of sacra and the desire of the people at large to possess them. Mass consumption and the motivation to acquire a larger and larger stock of sacra in the belief that one can materially aggregate and concentrate the power contained in them and use it as a capital fund of power by the possessors drives towards both the massive production of copies and the attendant weakening and depreciation of their efficacy.

That the phenomenon of multiplication and reproduction through copying from originals, copies of originals, copies of copies of originals and so on was a familiar occurrence in medieval Christianity has also been remarked on by David Freedberg.[40]

Take the instance of New Church at Regensburg, built in 1519 on the site of a Jewish synagogue and cemetery purposely razed to the ground. This church dedicated to the Virgin who had performed a miracle on the site, was adorned with a copy of the already miraculous picture of Madonna attributed to Saint Luke in the Alte Kapelle. It was set up on a marble altar. 'Albrecht Aetdorfer was swiftly commissioned to make further copies. A statue of the Virgin by Erhard Heydenreich, master of works at the cathedral, was also placed on a column outside. Already in 1519, over 50,000 people had come to visit the *Schöne Maria* at Regensburg for both images had begun to work miracles.'

'Although it seems clear that the Regensburg pilgrimage was largely a manipulated affair, initiated by an anti-Semitic priest, the fact remains that 'images were a rallying point and a confirmation of deeds done. As on innumerable other occasions, the pilgrimage has its origins in an image that works miracles, or it records gratitude for an apparently miraculous act. The image, whether itself miraculous or not, is placed in a more elevated and appropriate setting; it continues to work miracles; people passionately venerate both it and its copies; they make further copies and take them away with them, or set them up elsewhere; they invest more hope in them; and record their thanks for further favours by bringing still more images ... to the site of the shrine. In the first tumultuous year of the Regensburg pilgrimage, a contemporary noted that there were not enough tokens to take home, and many people cried and were tearful at having to go back without any. The following year, the surviving accounts testify to the manufacture of 109,108 clay pilgrimage badges, and 9,763 silver ones. Such figures are typical of the great shrines of Europe, especially in Bavaria and Italy.

Enshrinement, adornment, and multiplication: all these are illustrated by Michael Ostendorfer's print of 1519-21 showing the pilgrimage to and adoration of the *Schöne Maria*. Representation of the miraculous image feature at least four times in this scene: first, the actual image deep within the church; then the statue in the foreground; then on the huge banner suspended from the tower; and finally on the pennant fluttering by the left-hand side of the church. In the background and from the sides, a great crowd of pilgrims file into the wooden church, while in the foreground a scene of frenzied devotion and supplication takes place before the statue to which still more candles are attached. From the eaves of the church hang a profusion of votive objects, including limbs and implements. The whole scene gives a vivid sense of the extraordinary strength and the extent of the effectiveness of the image at the shrine, and the compulsion to make copies of it. Even though it is itself hidden in the mysterious depth of the shrine, we are left with no doubt of its powerful presence there: the copies alone clamorously testify to that.'[41]

Notes

1. Frits Staal, *Rules Without Meaning. Ritual, Mantras and the Human Sciences*. New York: Lang, 1989.
2. Stanley Jeyaraja Tambiah, *The Buddhist Saints of the Forest and the Cult of Amulets*. Cambridge: Cambridge University Press, 1984.
3. Staal, *op. cit.*, pp. 319-20.
4. In the spread of Sufi Islam a parallel role has been played by the proliferation of the tomb-shrines of saints.
5. I am of course by no means saying that the cult of saints is universal, nor that the traditions named share the same cultic components. In Sufi Islam amulets and tomb shires occur, but Islam by and large repudiates bodily relics and images. Bodily relics are rare in Hinduism, icons are a feature of Byzantine and Eastern Christianity. I am suggesting there is a family resemblance between the

practitioners I have named, so that the term 'saints' can serve as a general term to enable comparison.

6. J.C. Heesterman, *The Inner Conflict of Tradition. Essays in Indian Ritual, Kingship, and Society*. Chicago: The University of Chicago Press, 1985, p. 38. Similar ideas about the Buddhist saints are formulated by me in *The Buddhist Saints of the Forest and The Cult of Amulets* (*op. cit.*).
7. Ernest Troeltsch, *The Social Teachings of the Christian Churches*. Tr. Olive Wyon, vol. 1 New York: Free Press, 1949, pp. 239-40.
8. Tambiah, 1984, *op. cit.*
9. As Staal appositely remarks 'the forest monks are in fact given to "practice" which includes ritual as well as meditation, and meditation itself is closely related to both recitation and ritual' (Staal, *op. cit.*, 1989, p. 315).
10. J.L. Taylor, *Forest Monks and the Nation State: An Anthropological and Historical Study in Northeastern Thailand*. Singapore: Institute for Southeast Asian Studies, 1993, p. 188.
11. It is personally gratifying that another anthropologist (trained in Australia and not associated with me) confirms in *most* respects all I reported and interpreted concerning the cult of amulets in Thailand that centred on certain forest saints in the sixties and seventies.
12. Taylor, *op. cit.*, p. 188.
13. Tambiah, *op. cit.*, 1984, pp. 109-10.
14. *Ibid.*, p. 109.
15. Taylor, *op. cit.*, p. 178.
16. Tambiah, *op. cit.*, p. 109
17. Taylor, *op. cit.*, p. 176.
18. *Ibid.*, pp. 182-3.
19. *Ibid.*, p. 187.
20. *Ibid.*, p. 156.
21. *Ibid.*, pp. 156-7.
22. *Ibid.*, pp. 157-8.
23. *Ibid.*, pp. 159-60.
24. *Ibid.*, p. 160.
25. *Ibid.*, p. 160.
26. *Ibid.*, pp. 160-1. Phra Kammathaan refers to monks devoted to meditation, and *pariyat* monks are scholar monks who have passed ecclesiastical examinations.
27. Tambiah, *op. cit.*, ch. 21.
28. Taylor, *op. cit.*, p. 164.
29. David Freedberg, *The Power of Images, Studies in the History and Theory of Response*. Chicago and London: University of Chicago Press, 1989, pp. 31-2.
30. Freedberg, *op. cit.*, pp. 135-6 is primarily concerned with the role of figuration in making images effective, alleges that in my discussion of the cult of amulets, I 'almost entirely' pass over the visual components of effectiveness. Freedberg has given the barest attention to the early sections of Chapter 14 in my book which deals with the figurative and iconographic, conventions associated with the fashioning of Buddha images and which generate 'radiance' and 'fiery energy'.

31. R.C. Gregg (tr.), *The Life of Antony and the Letter of Marcellinus [by] Athanasius*. New York: Classics of Western Spirituality, 1980, p. 66.
32. NPNF. A Select Library of the Nicene and Post Nicene Fathers of the Christian Church. 28 vols. Repr. Grand Rapids: Eerdmans, p. 202.
33. Stephen Wilson (ed.), *Saints and Their Cults. Studies in Religious Sociology, Folklore and History*. Cambridge: Cambridge University Press, 1983, pp. 6-7. Wilson cites Karrer (ed.), *Saint Francis of Assisi* as the source.
34. Gregory Schopen 'The Phrase "*sa pṛthivīpradeśaś caityabhūto bhavet*" in the Vajracchedikā Notes on the Cult of the Book in Mahāyāna', *IIJ* 17, 1975, p. 147-81.
35. Leon Hurvitz (tr.). *Scripture of the Lotus Blossom of the Fine Dharma*. New York: Colombia University Press, 1976, pp. 175-6.
36. Staal, following Chomsky's analysis of syntactical rules and structures, uncovers the constitutive features of Vedic ritual in terms of phrase structure, transformational and selfembedding rules, recursivity, and so on. He also sees similarities between music and ritual, and highlights ritual structures in the form of 'refrain', 'cycle', 'palindrome', 'overlapping', 'threesomes', etc. The same structures and sequences may be employed in different contexts and can have multiple uses (functions).
37. Walter Benjamin 'The Work of Art in the Age of Mechanical Reproduction', in: Hannah Arendt (ed.), *Illuminations, Essays and Reflections*. New York: Schocken Books, 1968.
38. The following quotations were taken from, pp. 223 and 224 of the above essay.
39. By contrast, according to Benjamin, in film it is an 'inherent condition'.
40. David Freedberg, *op. cit.*, pp. 103-4.
41. Excerpts from: David Freedberg, *op. cit.*, pp. 103-4.

XXIX

LINEAR TIME IN HISTORICAL TEXTS OF EARLY INDIA

Romila Thapar

History as it has evolved in recent times is linked to the legitimation of nation-states. In pre-modern times there were perceptions of the past and reflections of historical consciousness. These took the form of chronicles and annals and occasionally narratives of selected events, as for instance a campaign. There were some histories of specific institutions such as a dynasty, or a religious institution or a religious sect, differentiated from other texts by some emphasis on chronology and sequential narrative, an emphasis which in Europe increased with the propagation of the Renaissance sense of the past.[1] Such texts were sought in India by the early Orientalist scholars, the officers of the East India company working in Bengal and Madras, and it became a search for a history of India. Failing to find these histories, the scholars were convinced that there was an absence of historical consciousness in India and that they therefore, would have to rediscover the Indian past. In their reconstruction of this past, the premises were the current intellectual preconceptions of Europe.[2]

The supposed absence of history was attributed, among other things, to the cyclic concept of time which was an important theme in the texts initially used in the rediscovery of the Indian past, particularly the Manu Dharma Śāstra and the Purāṇas. Neither of these, nor brahmanical eschatology, was thought to provide evidence of linear time which was believed to be necessary to the writing of history.[3] Other civilizations, such as that of the Greeks, also had a cyclic concept, but comparisons with these were not in order.[4] Linear time and the eschatology which began with the Garden of Eden and ended with Judgement Day was seen as essentially Judeo-Christian and therefore also absent in the Greek past. However it was thought that this did not prevent the Greeks from developing a sense of history. In other cases cyclic time was not only viewed as characteristic of non-monotheistic religions but also came to be used as a device to exclude such societies as ahistorical.

This argument has continued to be influential, as is evident in the writings of Mircea Eliade. According to him, time in early India was seen as a continuous, cyclic repetition and on such a vast scale that human activities are minimized. This disallows any event from being unique and there is inevitably a repetition of events in each new cycle. The construction of the cycle derives from an avalanche of figures which is co-related to the philosophic notion of the world being illusory.[5] By way of contrast, linear time had a beginning and an end, therefore events were not repeated and could be treated as unique. Linear time also implied a differentiation between history and myth, myth belonging to a prehistorical past. The direction of linear time was viewed as moving towards 'progress' as understood in its nineteenth-century sense and often equated with the achievements of modern Europe. The intention of this essay is to briefly refer to the construction of time - both cyclic and linear - and to suggest their varied functions, especially where historical consciousness is evident. These time concepts were not viewed in as dichotomous a manner as they are today, since there could be some over-lapping in their use.

A denial of history is also postulated on the argument that the authority of the Veda is timeless.[6] Mention is made of the *aitihāsikas* who attempted to explain the Vedas by drawing from events which are believed to have actually happened. Unfortunately their texts have not survived. There is the ambiguous inclusion of the Itihāsa-purāṇa as the fifth Veda.[7] Judging by the other subjects included in this category, the fifth Veda seems to have been applied to those branches of knowledge which were important but outside the focus of Vedic knowledge. If this was so then the timelessness of the authority of the Veda would not apply to Itihāsa-purāṇa, for, being a fifth Veda it would be beyond the purview of the recognized four. One is reminded here of the theory that the Itihāsa-purāṇa tradition was essentially a tradition of the *kṣatriyas* and therefore not centrally the concern of the *brāhmaṇas*, at least in the initial period.[8]

Time can be seen in diverse forms, as abstract, as a calendar, as cosmological time, as part of eschatology and as historical chronology. At an abstract level time was a creator begetting heaven and earth and all that has been and shall be.[9] Sometimes the simile of a five-spiked wheel is used with an emphasis on repetition and regular spacing.[10] In some of the Upaniṣads it is a possible ultimate cause and weaves the past, present and future across space like a warp and woof.[11] It is also equated with deity and can destroy the world although it remains imperishable.[12] Perhaps the most evocative image is that of time being the *sūtradhāra* of the universe.[13]

Time is mapped through calendars which become part of what might be called the technology of time, and are based on the two most visible planets, the sun and the moon, and on the *nakṣatras* or constellations, and these become part of what might be called the technology of time. Those who make the calendars control the reckoning of time and are themselves proficient in astronomy. Observations of the motion of planets and the seasons were used to demarcate units of time. These included the lunar day/*tithi* with its sub-divisions of *muhūrtas*, the day and the night/*ahorātra*, the fortnights of the waxing and waning moon/*pakṣa*, the lunar month/*māsa* and the seasonal cycle of the year generally starting with the spring equinox. The year was also divided into the *uttarāyana* and the *dakṣināyana*, the northern and southern course based on the solstice. Initially the *yuga* was a five-year cycle but was later expanded to a much longer period. These units in some cases were also developed into the calendar of ritual time. The year of three hundred and sixty days which constituted the calendar was also associated with the sacrifice.[14] Spatially time was visualized as the altar, the setting up of which was suggestive of the regeneration of time.[15] To this was added the notion of the periodic creation and destruction of the universe.[16] Ritual time tends to be cyclic and therefore different from concrete time. This is reinforced by the cyclic motion of lunar time and by the repetition of the cosmogony as symbolized in the sacrificial ritual.[17]

When such time concepts are later tied into astrology they acquire greater precision. The contiguity of Hellenistic culture both overland and through maritime trade at the turn of the Christian era, brought about an interaction which enhanced activity in astronomy. Yavanas from the west often came as visiting traders and some were associated with works on astronomy used in India.[18] It has been suggested that there was a radical change after circa AD 400 when the Siddhānta-jyotiṣa superseded the Vedāṅga-jyotiṣa and calculations earlier using stellar and lunar observations now preferred to incorporate planetary motions and solar reckonings.[19]

Cosmological time required the construction of immense cycles of time and there is some overlap between cosmology and the calculations of astronomy in the figures which were used. The largest unit was the *kalpa*, calculated by Indian astronomers as 4,320 million years. This figure may have been borrowed by astronomers from purāṇic sources, the astronomers requiring a long period of time for their calculations.[20] References to *yuga*s also occur in the Manu Dharma-Śāstra and the Mahābhārata. In the Purāṇas, the *kalpa* is the period which covers the creation and continues to the ultimate destruction of the universe. In Buddhist, Jaina and Ājīvika texts, there is often a spatial description of the *kalpa* such that in temporal terms it can hardly be measured. Thus we are told that if there is a mountain in the shape of a cube, measuring one *yojana*, and if every hundred years the mountain is brushed with a silk scarf, then the time taken for the mountain to be eroded would constitute a *kalpa*.[21] The Ājīvika description is more elaborate. If there is a river, 117,649 times the size of the Ganga and if every hundred years one grain of sand is removed from the bed of this river, the time required for the removal of all the sand would be one *sara* and 300,000 *saras* constituted one *mahākalpa*.[22] In most descriptions the theme of 'every hundred years' is consistent and introduces a recurring temporal dimension, but the spatial element overrides this and tends to negate measurement. At a literal level the silk scarf would have quickly disintegrated, and it is virtually impossible to remove a grain of sand from the bed of a flowing river.

The *kalpa* is the full stretch of time. Smaller cycles, although still immense in terms of years, made up the *mahāyugas* and these in turn were divided into four *yugas* of unequal length, frequently quoted as time periods. Manu gives a series of minuscule and maximal measurements.[23] In the description of the four *yugas* mention is made of the Kṛta lasting for four thousand years with a preceding and subsequent twilight period, each of four hundred years. The length declines by a thousand years from each of the later *yugas* - the Treta, Dvāpara and Kali, and the twilight periods also decrease, each by a hundred years. The total length of a *mahāyuga* of twelve thousand years makes up an age of the gods and this figure has to be multiplied by 360 to convert it to human terms, which makes it 4,320 million human years. Purāṇic sources tend to use the figures as calculated for human years. The Mahābhārata has a similar description of the *mahāyuga*, when at the end of the cycle, the Kṛta returns.[24]

But the Purāṇas also include the unit of the *manvantara* which is equal to seventy-one times the number of years in a *mahāyuga*, with some to spare. Clearly the arithmetic of the *manvantara* did not fit that of the *mahāyuga* and this has been interpreted as an imperfect synthesis of more than one independent doctrine deriving from different sources.[25] What is significant is that an attempt was being made to coalesce the two in purāṇic time-reckoning. This is particularly noticeable given that there was a concern with mathematical orderliness in the construction of time, as evident for example, in the shortening of the length of each *yuga* in descending arithmetical progression.

In all these texts there is an emphasis on a change in the character of the four *yugas* manifested in a gradual decline in *dharma*. This is indicated in the descriptions of utopian conditions which change from *yuga* to *yuga*. The initial utopia is eroded by a decrease in longevity, by the necessity of marriage and by the decrease in man's height from six miles to six feet.[26] The imagery of decline in such instances tends to be linear. Or *dharma* is likened to a bull which stands on four legs in the Kṛta age but drops one leg in each of the following.[27] The Kali is an age in which the world turns upside down, for rulers are *mleccha* or *vrātya-kṣatriyas*, the *śūdras* take on the function of the *brāhmaṇas* and the norms of the *varṇa* ordering of society are done away with.[28] The *yugas* are not identical although the form is similar and there is therefore no bar on new events taking place.[29] The awareness of a changing past is evident even if the change carries an element of inevitability.

The wheel of time as a symbol of eschatology is common to more than one view of time. The golden age occurs at the start of the cycle and in some it returns when the cycle is completed. In this context it is interesting that a saviour-figure is introduced in the last age when the decline is at its worst. Such a figure could be seen as representing a linear element in eschatology. The Buddha Maitreya, the Buddha to come, was mentioned briefly in the early texts, but gradually the legend grew incorporating detailed descriptions of Maitreya,[30] some of which were doubtless linked to the dialogue between Buddhism and other religions prevalent in Iran and Central Asia. It was believed that there would come a period when the Buddhist *dhamma* would decline and when anarchy would prevail and people would take to the forests and the hills. The coming of the Buddha Maitreya would bring them back from their hiding and the *dhamma* would once more be established. The Buddha Maitreya is associated with the upward movement of the wheel of time and heralds the return of the golden age. It becomes in effect a millenarian movement and not surprisingly the number of years associated with the coming of Maitreya are often 500, 1000, 1500, or 5000.

The elaboration on the coming of the Buddha Maitreya seems to coincide with another saviour-figure, though perhaps of a more messianic kind, that of Kalkin, the tenth *avatāra* of Viṣṇu, described at length in the early Purāṇas. With the decline of *dharma* in the Kali age there is again a fleeing into the hills and a return to subsistence living. Kalkin we are told, will be a *brāhmaṇa* and will restore the norms of *varṇa* society.[31] The concern here is also with the restoration of the *dharma*, but with the *dharma* as projected by brahmanical interests.

The occurrence of the saviour-figure is doubtless part of the inevitable decline of *dharma* at the end of the *mahāyuga*, yet it encapsulates a major change and a return to another golden age. There is an intervention in the human condition and the inevitability of decline is reversed. Curiously this intervention also coincides with a seemingly greater awareness of the possible separation between myth and history, and the introduction of an evident historical chronology with its centrality of linear time.

In the composition of the earliest Purāṇas, there is one manifestation of linear time which has not been fully recognized. This is the listing of generations in a genealogy. In the Vaṁśānucarita chapter of the Viṣṇu Purāṇa, purporting to narrate the succession of those who ruled, an attempt is made to provide a view of the past in the form of a continuous set of genealogies from the beginning of the *mahāyuga* to the then present - which was probably about the fourth century AD.[32] These genealogies also occur, perhaps earlier, in the Ādiparvan of the Mahābhārata and in the Bālakāṇḍa of the Rāmāyaṇa.[33] Here they are shorter than in the Purāṇa and more concisely related to the main families. In the epics the attempt is to provide an ancestry for the heroes, whereas in the Purāṇa, the attempt is to provide an overview of the past through genealogical patterns. These patterns can be divided into three distinct, although unequal sections.

The chapters prior to the Vaṁśānucarita in the Viṣṇu Purāṇa describe the genesis of the universe and provide the details of cosmography and of time cycles in the form of the two systems, that of the *mahāyugas* and that of the *manvantaras*. In the first section of the Vaṁśānucarita, the Manus are said to be the earliest rulers, each rule covering immense periods of time. There are no events or descent lists connected with this rather vague section. In the reign of the seventh Manu there occurs the great Flood from which Manu is saved by Viṣṇu in his *matsya-avatāra*, a story which is narrated at length in the Matsya Purāṇa. It has its analogies with the Mesopotamian Flood legend and the earliest version in Indian sources occurs in the Śatapatha Brāhmaṇa.[34] In the Vaṁśānucarita it acts as a time-marker with the obvious symbolism of water washing away the past, separating the period of the Manus from the subsequent rule of the *kṣatriya-rājās*.

The second section changes radically from the first. There is little or no association with cyclic time, since it is concerned with generational time through listing genealogies. The sequence of generations is for all practical purposes an exercise in linear time, even if within the overall framework of the *yugas*. Was this a mechanism of demarcating myth which was seen as occurring in cyclic time from what is projected as approximating history and which is set in another system of time-reckoning? The link with the earlier section is limited to the statement that the founders of the *kṣatriya* descent groups were the progeny of the seventh Manu. The keeping of genealogies and reckoning by generations - whether factual or fictive - are important to lineage-based societies where rank is determined by birth and clans with genealogies claim status.[35] Genealogical time can be manipulated either to stretch or to telescope genealogies in order to suggest status for some groups. The genealogies in the Purāṇas are distributed into two major lineages, that of the Sūryavaṁśa and that of the Candravaṁśa, where the symbolism of the sun and the moon touches on many facets of observance and belief. It is interesting that the choice was of the two planets central to the making of calendars. It also ties in with the epics since the Rāmāyaṇa is the narrative of the Sūryavaṁśa and the Mahābhārata draws together many of the descent groups of the Candravaṁśa.

There are significant differences in the construction of these genealogies which point to ways of both recording and perceiving the past, although the lists were evidently not intended necessarily as authentic historical records.

But genealogical patterns provide a perspective on social and political assumptions. The Sūryavaṁśa records descent from father to eldest son, suggestive of primogeniture and possibly therefore also of an awareness of the beginnings of monarchy - a reading which might be appropriate to the narrative of the Rāmāyaṇa. The Candravaṁśa on the other hand frequently lists more than a single line of descent among brothers and fans out in the process. It parallels more closely a segmentary lineage system, which would give it a very different form from that of the Sūryavaṁśa. The segmentary lineage system is generally associated with clans rather than with kingdoms and might therefore be said to conform more closely to the narrative of the clans as it unfolds in those sections of the Mahābhārata, which are thought to be prior to the intervention of the Bhṛgus.[36] The Sūryavaṁśa is traced back to Ikṣvāku, the eldest son of Manu, whereas the Candravaṁśa has an ancestress in Iḷā, the androgynous child of Manu. Social norms also differ and the Candravaṁśa records a larger variation, some of which would have been unacceptable to the prescriptions of the Dharma-Śāstras. The descent list of the Sūryavaṁśa gradually dies out, but that of the Candravaṁśa moves to the dramatic war described in the Mahābhārata, in which virtually all the clans are involved. The war becomes another major time-marker, for it marks the destruction of the age of the *kṣatriyas*. Its importance as a time-marker is heightened by the statement that the Kali age began after the war and that it was calculated with reference to a constellation, thus suggesting indirectly a point in time.[37] Astronomers calculated the date for the start of the Kali age as equivalent to 3102-1 BC which date was adopted in inscriptions and histories wherever a precise time reckoning was required.[38]

In the third section of the Vaṁśānucarita the narrative changes from the past tense to the future, claiming to be a prediction of events.[39] This evident change in the presentation of time is also indicated by a change in the pattern of the genealogies, for now the record is that of ruling dynasties. The genealogies are framed within each dynasty and in some Purāṇas they are accompanied by the mentioning of regnal years for each member of the dynasty. The figures may well be exaggerated in some cases, nevertheless the statement which is being made is that there was a change to a different kind of political authority. Listing by dynasties introduces a new pattern of measuring time and is evidence of linear time. Unlike the descent lists of the *kṣatriya* clans, dynasties were not related to each other and sometimes are even said to have been *bhṛtyas* or in the service of, the previous one. Since the age of the *kṣatriyas* had ended, the dynasties, it was said, would often be of *śūdra*, *vrātya-kṣatriya* or even *mleccha* status and others beyond the pale of caste society.[40] Clearly the worst of the Kali age had arrived.

In the seemingly seamless web of the Vaṁśānucarita, three distinct sections can be recognized. The use of time in each changes. Thus the period of the

Manus is set within cyclic time and the cycles continue to form the backdrop for later forms. These however, were more clearly drawing on linear time, initially through the genealogies of generational time, and later through the more measured regnal years of dynasties. This is an indication of the recognition of historical time as also the blurring of the dichotomy between cyclic and linear time. This can be seen in a new development which starts around the Christian era and is not entirely unconnected with the recording of regnal years. This was the invention and use of historical eras, an innovation immediately noticeable in inscriptions.

The recording of regnal years goes back to the third century BC in the inscriptions of the Mauryan emperor, Aśoka. Regnal years point to time reckoning being based on a precise point in time - the consecration of the king. The introduction of eras required that the point to be used as the basis for calculating time be widely known. It has been argued that the system of eras, reflected in the use of the term *saṁvatsara*, grew out of points of time associated with calculations in astronomy and were presumably adapted to historical purposes. This may be true of some, but certainly not of all eras. The Kṛta or Mālava era of 58 BC also known from the eighth century as the Vikrama era was used so widely in documents of a historical kind that it was thought to have originated in some historical event, but since there is little agreement on such an event it has been suggested that it was associated with astronomy.[41] Attempts to relate eras to historical events continue.[42] There is a similar ambiguity about the Śaka era of AD 78 which has for long been associated with the accession of Kaniṣka although the date of AD 78 for this event remains controversial. Associations of a more clearly historical kind are evident in eras such as the Gupta Era of AD 320 and various others initiated by rulers, as for example, the Harṣa Era of 606, the Kalacuri-Cedi Era of 248 or the Vikrama Cālukya Era of 1075, to mention just a few. That such a wide range of eras was in use, points to the recognition of linear time for functions associated with royal authority.

The extensive use of eras, tended to separate the cosmology of time from its historical function. Eras introduced more precise dating in official documents. Such documents are available largely in the form of inscriptions of various kinds and some are in the nature of annals. Most were issued by ruling families but also by others in authority. Some were records of votive offerings by traders, artisans and small-scale landowners. That there might have been an influence from Hellenistic astronomy in the use of eras may be one explanation, but it also has to do with the nature of monarchical states and the activities of their rulers from the first millennium AD onwards. Precision in time reckoning, especially in official documents, lent greater authority to the state. The significance of linear reckoning was in part tied to the legitimation of this authority. Inscriptions recording grants of land, generally given to religious donees by those in authority, were legal documents and required a precise and detailed date. This included the era or the regnal year of the ruler, the season, month, fortnight and the day. Where the donation was meant to negate the influence of something inauspicious, such as an eclipse, there the precise date would be tied to a time appropriate from astrological concerns. In either case it is the precision of the date which is essential and the reckoning is in a linear calendar.

Another possible influence on the importance of precise dating could have come from the parallel traditions of Buddhism where events were dated in relation to the *mahāparinirvāṇa*, which became the starting point of something akin to a Buddhist era. This was not a universally accepted date but the discrepancies were reasonably close, 486 and 483 BC.[43] Recently these dates have been re-examined and alternative dates of a few decades later have been suggested.[44] What is important however, is that within particular Buddhist traditions there was a stable date. Chronicles written by Buddhist monks display a sharp sense of time, and the need to keep records of breakaway sects and of monastic property would also have encouraged a greater reliance on history.

A sense of linear time is also evident in other categories of texts which would broadly conform to being regarded as historical. Biographies of kings, such as the Harṣacarita of Bāṇabhaṭṭa narrating the early life of Harṣavardhana or the Rāmacaritam of Sandhyākaranandin on the Pāla king, Rāmapāla, are narratives of the lives of a single ruler but include a short history of his dynasty. Often a chapter on the historical antecedents of the author is added to establish the qualifications of the author.[45] A more extended form of such records are the many *vaṁśāvalis* or chronicles of dynasties and regions, available from many parts of the country, such as the Mūṣakavaṁśam of the Ay dynasty of Kerala or the *vaṁśāvali* of the kingdom of Chamba in Himachal Pradesh.[46] Brief histories of dynasties also occur in a large number of inscriptions and these contain not only the origin myths of the dynasties but a sequencing of rulers as well. The Khajuraho inscription engraved on the orders of the Candella ruler Dhaṅga for his father Yaśovarman at the Lakṣmaṇa temple, starts with an origin myth in which the Candella ancestor is given a high status, being linked to the Candravaṁśa and the *ṛṣi* Atri. It then provides brief vignettes on all the earlier rulers in succession, mentioning political alliances, marital alliances and major campaigns.[47]

Inscriptions and historical biographies carried the official version of events as perceived from the perspective of the court of the king concerned. This is at times confirmed or else contradicted in similar sources from other dynasties, particularly in regard to the claiming of success in campaigns. Part of the recognition of authenticity of such sources is that they narrate events with a strong sense of the past and the present and in a linear format. The ambiguity of cyclic time functioned better as a metaphor, the actuality of power tending to be rooted in a form of linear time, although the distinction was by no means absolute.

To characterize societies as using either cyclic or linear time is questionable. Most societies use more than one category of time. The shape of time has a variety of functions and early Indian society used both cyclic and linear time differently. This also makes such a characterization an inadequate explanation for the centrality or otherwise of history. The incorporation of one category as in cosmology or eschatology, does not preclude the use of other forms for other purposes. In attempting to understand how a society views time, the particular use that is made of various categories is revealing, for their statements are not identical. Simultaneous and layered forms of time indicate historical complexity.

This also raises the question of whether it is valid to generalize for a society or culture as a whole, as for instance, to speak of an Indian concept of time. Do all social groups within a society visualize time in an identical way or are there particular uses of time which are linked to social differences, and do the same groups use different categories of time according to occasion? For example, brahmanical views are generally associated with cyclic time. When the *brāhmaṇas* supersede the bards by taking over the authorship of the *vaṁśāvalis*, they do not change the sequential, generational time of the genealogies and dynasties or the sequential narration of events, but in fact intensify them as is evident from a comparison of the *vaṁśas* in the epics and in the Purāṇas and by the narratives in the royal biographies and the chronicles, not to mention the brief histories in inscriptions.[48] Inscriptions of a later date were often composed either by *brāhmaṇas* or by *kāyasthas*, but even where the former are the authors, their use of linear time is quite apparent and established.

Indian civilization has been characterized as knowing only cyclic time, a view which was occasioned by a narrow reading of the texts - largely ritual and normative texts with limited functions. The inclusion of other genres of texts introduces other forms of time and it becomes apparent that the emphasis was not on cyclic time alone. If it is conceded that these other forms, and especially linear time, were familiar and functionally important, then the question of historical consciousness in early India cannot be dismissed as irrelevant, as it often has. The construction of history from this perspective becomes a new area of investigation.

Notes

1. P. Burke, *The Renaissance Sense of the Past*. London: Arnold, 1969.
2. Romila Thapar, *Interpreting Early India*. Delhi: Oxford University Press, 1993.
3. W. Jones, 'On the Chronology of the Hindus', *Asiatic Researches* I, 1789, p. 345 ff.; J. Bentley, 'Remarks on the Principal Eras and Dates of the ancient Hindus', *Asiatic Researches* V, 1808, p. 315 ff.
4. For a recent view of history and the Greeks, see, A. MacIntyre, *After Virtue*. Notre Dame: Notre Dame Press, 1984.
5. M. Eliade, *Cosmos and History: The Myth of the Eternal Return*. New York: Harper and Row, 1959; 'Time and Eternity in Indian Thought', *Man and Time*, Bollingen Series XXX.3. New Jersey: Princeton University Press.
6. S. Pollock, 'Mīmāṁsā and the Problem of History in Traditional India', *JAOS* 109/4, 1989, pp. 603-10.
7. Chāndogya Upaniṣad 7.1.2-4.
8. F.E. Pargiter, *The Ancient Indian Historical Tradition*. Oxford: Clarendon, 1922.
9. Atharvaveda 19.53.1.-12; 54.1-5.
10. Ṛgveda 1.164.13-14.
11. Śvetāśvatara Upaniṣad 1.2; Bṛhadāraṇyaka Upaniṣad 3.8.3-4.
12. Bhagavadgītā 11.32.
13. Bhartṛhari in Vākyapadīya 3.9.3-5. Quoted in W. Halbfass, *On Being and What there is*. New York: Suny, 1992, p. 205.
14. Aitareya Brāhmaṇa 7.7.2.
15. Śatapatha Brāhmaṇa 10.5.4.10.
16. Atharvaveda 10.8.39-40.
17. M. Eliade, *op. cit.*, p. 78.
18. D. Pingree, *Jyotiḥśāstra*. Wiesbaden: Harrassowitz, 1981.
19. V. Krishan, 'The Astronomical Revolution in India about AD 400 and its Implications', *Vishveshvarananda Indological Journal*, 15.2.1977, pp. 265-84.
20. D. Pingree, personal communication.
21. Saṁyutta Nikāya 2.180-181.
22. A.L. Basham, *History and Doctrine of the Ājīvikas*. London: Luzac, 1951, pp. 253-4.
23. 1.60-86.
24. Vanaparvan 186.107 ff.; Śāntiparvan 222.6 ff.
25. A.L. Basham, *The Wonder that was India*. London: Sidgwick and Jackson, 1954, p. 321.
26. Manu 1.83; Viṣṇu Purāṇa 6.1 and 2; Mahābhārata, Vanaparvan 148.10 ff. 186.23 ff. Śāntiparvan 200.35 ff. Aṅguttara Nikāya 4.156.
27. Manu 1.81-82; 6.16.
28. Mahābhārata, Vanaparvan 186.28 ff.; Viṣṇu Purāṇa 4.24.70 ff.
29. J.N. Mohanty, 'Philosophy of History and its Presuppositions', in: P. Bilimoria (ed.), *Essays on Indian Philosophy*. Delhi: Oxford University Press, 1993.

30. Dīgha Nikāya 3.75 ff. J. Legge, *The Travels of Fa-hien*. Oxford: Clarendon, 1886, p. 110. P.S. Jaini, 'Stages in the Bodhisattva Career of the Tathāgata Maitreya', in: A. Sponberg and H. Hardacre (eds), *Maitreya, the Future Buddha*. Cambridge: Cambridge University Press, 1991, pp. 54-90.
31. Viṣṇu Purāna 4.24.98 ff.
32. Viṣṇu Purāṇa 4. Romila Thapar, 'Genealogical Patterns as Perceptions of the Past', in: *Studies in History*. n.s. 7.1.1991, pp. 1-36
33. 1.53-56; 1.70.
34. 1.8.1.1 ff.
35. Romila Thapar, *From Lineage to State*. Delhi: Oxford University Press, 1984.
36. V.S. Sukthankar, 'The Bhṛgus and the Bhārata: A Text-Historical Study', *ABORI*, 18, 1937, pp. 1-76; *On the Meaning of the Mahābhārata*. Bombay: Asiatic Society, 1954; R.P. Goldman, *Gods, Priests and Warriors*. New York: Columbia University Press, 1977.
37. Viṣṇu Purāṇa 4.24.104-107, 113.
38. E.g. F. Kielhorn, 'Aihole inscription of Pulakeśin II', *Epigraphia Indica* 6. 1900-1901, p. 7.
39. Viṣṇu Purāṇa 4.21 ff.
40. Viṣṇu Purāṇa 4.24.70 ff.
41. D.C. Sircar, *Indian Epigraphy*. Delhi: Motilal Banarsidass, 1965, p. 251 ff.
42. G. Fussman, 'Nouvelles inscriptions śaka: ère d'Eucratide, ère d'Azès, ère vikrama, ère de Kaniṣka', *BEFEO* LXVII, 1980, pp. 1-43.
43. The chronicles of Sri Lanka used 544 BC, taking back the date by a cycle of sixty years, but this appears to have been a later recalculation to authenticate certain local events.
44. H. Bechert (ed.), *The Dating of the Historical Buddha*. 2 vols. Göttingen: Vandenhoeck and Ruprecht, 1991.
45. For references to the extensive *carita* literature see V.S. Pathak, *Ancient Historians of India*. Bombay: Asia Publishing House, 1966; A.K. Warder, *An Introduction to Indian Historiography*. Bombay: Popular Prakashan, 1973.
46. Gopinatha Rao, 'Extracts from the Mūṣakavaṃśam', *Travancore Archaeological Series* II, 1 No. 10, 1916, pp. 87-113. M.G.S. Narayanan, 'History from the Mūṣakavaṃśakāvya of Aṭula', *PAIOC*, Jadhavpur 1969.
47. F. Kielhorn, 'Inscriptions from Khajuraho', *Epigraphia Indica* 1. 1892, p. 122 ff.
48. Romila Thapar, 'Society and Historical Consciousness: the *itihāsa-purāṇa* tradition', in: *Interpreting Early India*. Delhi: Oxford University Press, 1992, pp. 137-73.

XXX

ON MANTRAS AND FRITS STAAL

George Thompson

In spite of Staal's arguments to the contrary, there are good reasons to retain the view that mantras are essentially a linguistic phenomenon, although this view may require the broadening of our conception of language and language use. Nevertheless, it is argued in this paper that Staal's recent attacks on conventional (i.e. symbolic) views of ritual and mantras have greatly advanced the treatment of both. In the case of mantras in particular, it is clear that Staal's challenge has forced Sanskritists occupied with mantras to re-think their arguments. Most of them as a result have resorted to speech-act theory and other current forms of the philosophy of language, with varying degrees of success. Finally, a few new ways of viewing mantras are suggested.

Over the years Frits Staal has performed the valuable service of clearing away a good number of unnecessary and misguided assumptions, as well as many basic errors, that have hindered our understanding of Vedic ritual, on the one hand, and mantras on the other. Besides the important field work in Vedic that Staal has done through the years, and the meticulous presentation of the facts of the monumental *agnicayana* ritual in *Agni* (Staal, 1983), an example of his conceptual ground-clearing is his treatment of the interpretations of various ritual phenomena in the *brāhmaṇa*s (e.g. in Staal, 1989, p. 117). That these largely *ad hoc* and inconsistent interpretations have little relevance to the study of the rituals themselves has slowly, though perhaps reluctantly, come to be conceded: Vedicists have been singularly slow to recognize the difference between ritual *per se* and ritual exegesis.[1] Staal has clarified matters by distinguishing these more or less mythological and hermeneutical texts from the *śrautasūtra*s, the ritual handbooks, which he has rightly emphasized are essentially descriptive in intent, *not* prescriptive, as was commonly assumed, and to that extent far more valuable as a resource to us, as students of ritual, than the *brāhmaṇa*s. We know that the Vedic ritual tradition placed greater emphasis on the technically correct performance of ritual procedure than on any presumed meaning of the performance. For this reason, too, it is appropriate to grant more weight to the ritual sūtras, which carefully delineate what one is to do, than to the *brāhmaṇas*, which only afterwards rationalize the doing, as Staal asserts. Beyond that, Staal's syntactic analyses of these *śrauta* rituals, the solemn, public rituals of the classical Vedic period, have undeniable force, because they highlight the formal, almost mathematical, structures of these remarkably abstract, very priestly, rituals. At the very least, Staal's focus on these structures has clearly revealed the remarkable complexity of the Vedic ritual system, in ways that no Vedicist before him had fully appreciated.

It has taken a while, certainly, since his first, iconoclastic venture into the 'meaninglessness of ritual' (in Staal, 1979), but many, though certainly not all, Vedicists have become more or less willing to consider the possibility that both ritual in general and mantras are in fact meaningless - at least in some strictly semantic sense. This is evident, for example, in the conceptual shift that seems to dominate the volume devoted to mantras, edited by Alper (1989). To a remarkable extent, there is in this collection a growing consensus that Staal's thesis must be contended with, even if ultimately rejected, for he has pointed out to us the unconsidered assumptions which have driven our own largely *ad hoc* interpretations not only of Vedic ritual but of mantras. I for one have found it very useful to consider mantras as - to summarize Staal's thesis - pre-linguistic, akin to music, and in structure more similar to the syntax of bird-song than to the syntax of human language. However, I would want to pause at this last term - human language - and would raise the question of whether or not mantras may be best viewed, after all, as examples rather of language - extraordinary or otherwise - if we first concede that ordinary language itself is still very little understood, especially by Sanskritists, who have been by and large rather hostile to theory of any sort. In my view, which in ways appears to approach the traditional view of most Sanskritists, as well as the native tradition itself, mantras may well reside firmly within the realm of language, even if, at times, at its very borders. But on this point I disagree not only with Staal, who insists that mantras are non-linguistic, but also with most Sanskritists who have speculated about mantras: under the influence of Staal's critique, they have tended to characterize mantras as extraordinary, a highly peculiar and problematic use of language. They have tended to agree, in short, that 'mantras do not abide entirely by the rules of language' (thus one of the most insightful students of mantras, Padoux, 1989, p. 301). I would propose a number of objections to this view.

It seems to me that the use of speech-act theory by Tambiah and Wheelock, among Indologists, has been fruitful, even if, as Staal argues, they are not entirely successful.[2] On the whole, Staal's treatment of these two scholars (e.g. at 1989, p. 161-3; 223-5; 238-40) seems a bit too cavalier, as he himself seems to acknowledge, when he softens his remarks on Tambiah 'with a grain of salt' (1989, p. 163). Staal is certainly right to challenge Wheelock's assertion that 'the Vedic mantra truthfully *describes* and thereby actualizes a *bandhu* (i.e. a connection) between ritual object and cosmic entity', and the claim that the Vedic mantra 'stands as a *means* to the ends of the sacrifice' (Wheelock, 1989, p. 118, also cited in Staal, 1989, p. 223), since there are many Vedic mantras which clearly do not explicitly describe anything, nor do they refer in any way to cosmic or ritual entities. But Wheelock's assertions are not simply 'inspired by the *brāhmaṇa* literature', as Staal insists (1989, p. 224), though they may be *partially* so. Wheelock has also attempted, I think in an interesting way, to characterize mantras, and ritual language in general, as 'situating speech', that is, as 'speech acts whose intention is to create and allow the participation in a known and repeatable situation' (besides Wheelock, 1989, p. 99, cf. in more detail Wheelock, 1982, p. 59 f.). Wheelock's examples of this sort of speech are worth noting. 'When the Vedic adhvaryu priest says, "I carry you (the bundle of grass) with Bṛhaspati's head" (BŚS) this mantra, coupled with the fact that he is presently carrying the grass bundle on his head, serves to establish his divine status in the ritual situation' (Wheelock, 1989, p. 100). According to Wheelock, this very typical passage shows how a mantra relates to the ritual action which it accompanies: it confers divine status upon the ritual practitioner and divine significance upon the ritual action.

In 1982, Wheelock uses a familiar Christian example of ritual speech, the words 'This is my body ... . This is my blood', which are used, it is generally assumed, to symbolically, or even *actually*, turn the bread into the body and the wine into the blood of Christ. These words are, Wheelock says, 'the affirmation of what one believes to be the identity of the bread and wine at this moment' (Wheelock, 1982, p. 58). The words are clearly not meant to be informative, at least not directly. They are, in Wheelock's view, a type of performative utterance that 'situates' the participants in the central event of their religion. To put it briefly, this is what, according to Wheelock, mantras in general are also supposed to do.

Besides dwelling on the 'situating' significance of such ritual equations (*bandhus*), Wheelock also makes interesting observations about the use of first and second person pronouns and a variety of verb forms used in Vedic mantras, classifying them according to Searle's taxonomy, as e.g. 'assertives, directives, commissives, and expressives' (cf. Wheelock, 1982, p. 62). In my view, Wheelock's use of speech act theory here needs to be more fully addressed, since in fact it points to conspicuous features even of some of Staal's own examples of mantras. In *Rules* (Staal, 1989, p. 269), Staal quotes a number of Vedic mantras in order to compare them with, among other things, the utterances of 'mental patients' and infants. This comparison itself is, I think, quite fruitful and may well suggest, as Staal claims, that such utterances and mantras are archaic or regressive, and thus capable of revealing to us certain features involved in the origins of language.[3] But in fact a surprisingly large number of Staal's preferred mantras here are manifest precisely in the form of second person equations (i.e. *bandhus*), precisely like those which Wheelock has called attention to: '*You are* the head of Makha ... . *You are* the two feet of the ritual ... . *You are* the unchanging direction ... . *You are* the waistband of Aditi ...'. And these identifications are also, as Wheelock observed, frequently accompanied by imperative or optative verbs, a fact which suggests that such commands are intended not merely to get the priest to perform a required ritual action, but also that such mantras are intended to magically confer divine status upon the ritual specialist.[4] In contrast, in the two passages quoted by Staal from mental patients there is a striking *absence* of the personal pronouns, except for 'the Fourth Prayer', which consists of the following cryptic, though undeniably 'meaningful', statement: 'to destroy my consciousness and my ego'. In a second, longer and more complicated passage, recorded from a hospital patient in Riga (Staal, 1989, p. 271), there are no pronouns whatsoever. As a result, the verb forms in this passage are markedly ambiguous - interpretable either as indicatives (with subjects unexpressed) or as commands. Regressive though such language may be, it is also very suggestive of *meaning*, because of its evident sophistication in the use of euphemism and indirection, and therefore deserving of more attention than it has been given to it by Staal. Certainly, to dismiss it as meaningless, as Staal seems to do, would seem to be a mistake.[5]

Likewise, the comparison of mantras with the babbling of infants (cf. Staal, 1989, p. 272-3) needs further attention. It seems to me that what motivates children to this sort of pre-linguistic behaviour is in no way regressive or archaic (*pace* Staal's suggestion that it illustrates 'the recapitulation of phylogeny by ontogeny', *ibid.*). It appears to be a commonplace in ritual studies (cf. the popularizing book by Driver, 1991, p. 28 f.) to suggest that such behaviour is a 'ritualization of speech production'. But it seems clear that infant babbling is also a fundamentally practical, exploratory, behaviour. Far more than is generally realized, children, and even infants, are obsessed with language (far more than most adults), and in their early babbling and in their pre-sleep monologues, alone in their cribs, they are clearly engaged in an effort to reconstruct the distinctive features, the phonological system, the syntactic patterns, and the semantic oppositions, of the language of the community into which they have only recently entered. This is active, self-initiated, language-learning, and in so far as it is this, babbling is *not* ritualization. Thus we should not be satisfied with comparisons between mantras and babble which overlook their underlying motivations, which I assume to be different. It seems to me that Staal, here, as elsewhere, has relied too much on merely formal similarities between phenomena that are, or at least appear to be, very differently motivated. As far as I can tell Staal has ignored the issue of motivation, perhaps because it appears to bring us into the sphere of 'mythology and hermeneutics', matters in which Staal has expressed no interest at all, except to dismiss (cf. Staal, 1989, p. 421 ff.). This is a strategy with which I have to disagree.

Furthermore, while it *may* be true that neither Tambiah nor Wheelock has adequately exploited speech act theory (as well as communication and information theory), the possibility remains that these theories can be profitably exploited in the study of mantras (or at least until Staal has shown otherwise!). Let me reiterate, then, that I do agree to an extent with Staal's criticisms of Wheelock's and Tambiah's use of speech act theory. I myself happen to find Jakobson's treatment of the functions of language far more useful than Austin's or Searle's, especially when it is modified to accommodate the insights of Austin and Searle into performatives. I think Jakobson's schema of the functions of language is relevant to Wheelock's views in particular, since it covers much more of the full range of language use than the taxonomies of Austin and Searle do. But speech-act theory in general has succeeded, at the very least, in stretching our conception of what we do with ordinary language, and in getting us to re-examine it. We of course do many, many things with it, and some of these things may have little to do with what Staal refers to as 'individual expression or communication',[6] which are too hastily assumed by Staal, as well as Wheelock and Tambiah, to be the primary functions of ordinary discourse. On the other hand, I think that in general the 'extra-ordinariness' of ritual language and of mantras has been very much exaggerated. In what follows I will try to indicate why I think so.

To take some familiar examples of very ordinary language: how much 'individual expression' is involved in a quick and efficient transaction between me and a bank teller whom I will meet once and once only? On the other hand, how much actual information is communicated between two people, perfect strangers, who are, perhaps uneasily, killing time with small talk, while waiting for the elevator to start up again? Speech-act theory invites us to look more closely at daily acts like these, and it would seem useful to remember that these two examples of ordinary speech-acts are also examples of what is called 'secular ritual' - that is, the 'interaction ritual' of sociologists like Erving Goffman. Many interesting things could be said about such speech-acts. For example, they are often highly stereotyped in form, highly repetitive, and even 'meaningless' occasionally, and in this sense perhaps not unlike mantras as they are characterized by Staal.

Consider the following very ordinary though still remarkable conversation which Jakobson has culled from the fiction of Dorothy Parker (Jakobson, 1987, p. 68-9: the line-breaks have been added for the sake of clarity and emphasis):

'Well!' the young man said.
'Well!' she said.
'Well, here we are,' he said.
'Here we are,' she said, 'Aren't we?'
'I should say we were,' he said, 'Eeyop! Here we are.'
'Well!' she said.
'Well!' he said, 'Well.'

This nervous conversation strikes me as *highly* ritualized - secular though it is, of course - and heavily laden with a mantra-like expletive, the much relied-upon 'Well!' It is also a perfect example of phatic language, to use Malinowski's term for language use that aims to establish and maintain *contact* with one's audience. Jakobson makes several interesting observations about such language use, for example that it 'may be displayed by a profuse exchange of ritualized formulas, by entire dialogues with the mere purport of prolonging communication' (Jakobson, 1987, p. 68). It also typically includes, in my view, mantra-like affirmatives like 'Um-hum' or 'Uh-huh'. Phatic communication is also something that humans share with, for example, talking birds, by the way, an observation of Jakobson's which may throw light on Staal's provocative comparison of bird song and mantras (Staal, 1989, p. 279 ff.) In any case, that there is something phatic about at least some mantras has been completely ignored, and needs to be addressed.

This brings us back to Staal's speculations about the origins and evolution of language out of the mantra-state. These speculations are provocative and suggestive, and are among Staal's most interesting, though controversial, contributions to the study of ritual. Non-Sanskritists should understand that the wrath that Staal sometimes inspires in Sanskritists is to some extent due to his baiting iconoclasm, but that a significant portion of that wrath is directed against Staal's greatest virtue: his freedom of thought, his willingness to speculate, his risk-taking. Sanskritists, Vedicists in particular, are a remarkably conservative group - 'philologers' rather than thinkers, as Staal likes to remind them. They tend to dismiss Staal's views as not *textual* enough. But in fact, his views are frequently beyond their competence to judge.

I do not wish to get involved in a lengthy discussion of Staal's speculations about the role of ritual and of mantras in the origins of language,[7] except to say that I have been convinced by him of ethology's relevance to ritual studies, and like him I assume that ritual is therefore older than language, older than human culture, and older than our species itself. For that reason, one point in Staal's discussion of bird song puzzles me, and so I will address it.

This is the concept of ritualization as used by ethologists, starting with Huxley.[8] Staal repeatedly refers to this concept in *Rules* (Staal, 1989),[9] primarily to argue that ritual is indeed older than humans and human language. As far as I understand it, however, the central point of this concept is that ritualized behaviour is *essentially* communicative. Consider the definition posed by Eibl-Eibesfeldt in his textbook of ethology (Eibl-Eibesfeldt, 1970, p. 97), a book which Staal himself frequently cites:

> Whenever it is of advantage for an animal that some of its incidental behaviour be understood by another, selection operates to transform the behaviour pattern in question into a conspicuous signal. This modification of a behaviour pattern to serve communicative function is called ritualization.

It is surprising how little attention Staal pays to the fact that for ethologists ritualization implies the re-direction of behaviour away from its primary or original function,[10] *toward* 'a new function, that of communication' (thus Burkert, 1983, p. 23). Staal notes (Staal, 1989, p. 135) that 'among animals, ritualization often implies that the goal of an activity has changed', and he proceeds to talk about the changing function of various ritual displays, like fighting, which he notes is sexually stimulating. He also suggests that the concept of ritualization 'is invoked when the original function or behavioral pattern is no longer visible or known' (1989, p. 285). Finally, he argues that the concept is a category into which is put all behaviour whose motivation cannot be deduced - i.e. behaviour which appears to be meaningless (just like objects in a museum whose function cannot be determined: they become viewed as, and designated as, 'ritual objects'). Ultimately, the notion of ritualization is used in opposition to the notion of function. Ritual activity is perceived to be the opposite of goal-oriented activity. It is activity for its own sake, like play - and also like skating, skiing, or dancing (all cited by Staal, *ibid.*) - or, more pointedly, like music and mantras (cf. 1989, p. 288). This discussion is all very suggestive and compelling, and at times reminds me not only of Huizinga's theory of the role of play in culture, but also of Bataille's wild but brilliant attack on the notion of 'utility', which he contrasts with free, non-productive - even wasteful and destructive - 'expenditure'.[11] To this extent I recognize in Staal's theories a certain, perhaps paradoxical, kinship with performative approaches to ritual studies, which tend to minimize the role of cognitive and emotive factors in ritual, emphasizing instead ritual as pure action.[12] But what about the notion that ritualization is essentially communicative? Staal does not seem to face it squarely.

There is a remarkable example of animal ritualization, which has been frequently discussed in standard ethological literature (e.g. Eibl-Eibesfeldt, 1970, p. 121; Lorenz, 1966, p. 65 f.; Sebeok, 1979, p. 18, and Sebeok, 1976, p. 89), and which illustrates vividly the point that ritualization is essentially communicative. The example is striking also because it involves the use of a comparative method not unlike that used by historical linguists.[13] Lorenz's lengthy summary is worth repeating (Lorenz, 1966, p. 65-6):

> In several species of so-called Empid Flies (in German very appropriately called *Tanzfliegen* - Dancing Flies), closely related to the fly-eating Asalid Flies, a rite has developed as pretty as it is expedient. In this rite the male presents the female, immediately before copulation, with a slaughtered insect of suitable size. While she is engaged in eating it, he can mate her without fear of being eaten by her himself, a risk apparently threatening the suitors of fly-eating flies, particularly as the male is smaller than the female. Without any doubt, this menace exerted the selection pressure that has caused the evolution of this remarkable behaviour. However, the ceremony has also been preserved in a species, the Hyperborean Empis, in which the female no longer eats flies except at her marriage feast. In a North American species, the male spins a pretty white balloon that attracts the female visually; it contains a few small insects which she eats during copulation. Similar conditions can be observed in the Southern Empid, Hilara maura, whose males spin little waving veils in which food is sometimes, but not always, interwoven. But in Hilara sartor, the Tailor Fly, found in Alpine regions and deserving more than all its relations the name of dancing fly, the males no longer catch flies but spin a lovely little veil, spanned during flight between the middle and hind legs, to which optical stimulus the female reacts. In the revised edition of Brehm's *Tierleben*, Heymons describes the collective courtship dance of these flies: 'Hundreds of these veil-carriers whirl through the air in their courtship dance, their tiny veils, about 2 mm. in size, glistening like opals in the sun'.

We can see here, across the behaviour of several species, the development of this originally *highly* functional appeasement or gift-giving behaviour among 'fly-eating flies'. The gift of a slaughtered insect is used as a substitute for the male himself, i.e. to distract the hungry, and therefore potentially dangerous, female. This gift is presented by a related species with a kind of gift-wrapping which functions to further preoccupy the female. In yet another related species the male no longer offers a gift of food - nor even an inedible substitute - but merely an 'empty box' with gift-wrapping, 'a lovely little veil'. The original gift of food has evolved, in this case, into an empty box which contains nothing any more - *except* for a message. It is generally observed in the literature that this 'pretty white balloon' has in fact become a *symbol*.

Now, this example of animal ritualization shows, among other things, that even animal ritual can be highly 'meaningful'. Among these flies, the male signals to the female that he wants to mate, using a gesture that, we can clearly see, has evolved from one of appeasement or gift-giving; to a stylized gift-giving including wrapping; to a gift-wrapping with a substitute 'gift'; and finally, to a gift-wrapping with no gift at all inside. This last 'gift' (that of the appropriately named species Hilara sartor), i.e. an empty balloon, or a 'lovely gift-wrapped box' with nothing at all inside, might best be seen as a pure gesture.

This sequence is a remarkable illustration, it seems to me, not only of ritualization as it is understood by ethologists, but also of the semiotic process (semiosis). It also serves as a striking little allegory of the nature of the sign: which turns out to be, in this light, an empty box into which one puts one's meaning. But for present purposes the essential point would seem to be that even if ritual is older than language, ritual can still have something to do with meaning, even if we insist with Staal on using the term in the strictest of senses. Ethology is fundamentally concerned with communication. This is to say that the spheres of ethology and semiotics overlap to a significant degree - to such an extent in fact that Sebeok (1972) has coined the term 'zoosemiotics' to cover this overlap. Semiotics is by nature (and, by the way, by etymology[14]) *semantic* - it is all about meaning-bearing elements, or signs.

Like rituals, semiosis, sign-behaviour, is older than language. Therefore meaning also must be older than language. It does not arise *only* with language, as Staal so adamantly insists. Rather, meaning arises with communication, a far older process than language (meaning is also generally acknowledged to be present, e.g. in the communicative dances of bees, as von Frisch has elegantly shown). Unlike the arrival of human beings and language, therefore, the arrival of meaning upon the primordial ritual scene may not have been very late at all. In other words, from the point of view of ethology (*pace* Staal), ritual and meaning seem to have a great deal to do with each other.

Before going on to address a few, much more specific, and briefer, issues raised in Staal's book (1989), I would like to call attention to some remarks made by Sebeok comparing communicative functions in various non-human species with the functions of human language.[15] Sebeok expresses certainty that emotive and phatic communication occurs among non-humans. He considers cognitive and conative communication probable in non-humans. 'But the remaining two - the poetic and the metalingual - seem to be exclusively human' (Sebeok, 1972, p. 17). However, in light of Staal's observations about the syntax of bird song, as well as those of the ornithologists whom he cites, Sebeok may well be wrong about the absence of the poetic function in non-humans. For, as Staal points out, a remarkable aesthetic sense has been many times observed, by ornithologists and philosophers, in birds. In my view this amounts to saying that 'a poetic sense' exists in birds. But it is also interesting to note that Sebeok here repeats Jakobson's assertion (mentioned above) that bird song is essentially phatic, regardless of its elaborate structure. It essentially serves as a means of intra-species contact. Even if this is true, however, I don't think that Staal's discussion of structure is any less important, insightful, or true. But I would suggest that Staal's *exclusive* attention to structure and syntax is a self-imposed, and unnecessary, handicap, even as it may have proven itself to be a positive, heuristic, research device. There seem to be other important features to bird song that it would be useful for us to take into further consideration.

The same applies to Staal's treatment of mantras. Putting aside the apparently meaningless bīja mantras of Tantrism,[16] let us consider Staal's treatment of the so-called Puranic mantras.[17] His main point seems to be that 'whereas they are literally meaningful, unlike the Tantric bīja mantras, they are treated as if they were devoid of meaning' (Staal, 1989, p. 233). That is, the distinguishing feature of these mantras (e.g. *namaḥ śivāya*, *oṁ namaḥ śivāya*, *oṁ namo nārāyaṇāya*, etc.) is *not* the deities to which they refer, nor their meanings, but rather their number of syllables. For this reason, Staal says, these mantras 'are not treated like utterances of language' (Staal, 1989, p. 234).

I would offer two points in response to this: syllable-counting is a well-known feature in Vedic and Sanskrit poetry in general, as in many if not most poetic traditions of the world. Staal must either exclude poetry from language, or he must distinguish mantra syllable-counting (i.e. not language and therefore meaningless) from poetic syllable-counting (i.e. language, and therefore at least potentially meaningful). Staal seems in fact to beg the whole question. He does not demonstrate that syllable-counting in mantras (or anywhere else) is a non-linguistic act. He merely asserts this. He asserts that mantras which are governed by syllable-counting are therefore to be excluded from language. But this, it seems to me, is precisely what he has to prove.

The next point, perhaps more telling, is that in poetry the focus is placed not on the emotive or cognitive or on any other social function of language - at least not *primarily* - but rather on formal patterns of various sorts - not just syllable-counting, of course, but also rhyme schemes, syntactic and other sorts of grammatical parallelisms, phonic echoes, etc. Of course, the fact that, in poetry, directly semantic elements are subordinated to such non-semantic or formal patterns (or constraints, in the case of strictly observed poetic convention) clearly does not mean that poetry is therefore non-semantic. Likewise, it would seem to me, the presence of non-semantic features in mantras does not necessarily mean that they are by that very fact not semantic, i.e. meaningless.

The point may be obvious, but it is worth remembering that language operates on several levels at once. That is, phonic, phonemic, morphemic, lexical, syntactic, and semantic features do operate simultaneously in any given utterance. Staal seems to exclude the last level - semantics - out of hand, perhaps because in his view it is the last to have developed, at least diachronically if not logically. I concede that Staal's focus on syntactic features in mantra sequences, and in ritual sequences, in general has been fruitful. But not so his insistence on excluding these other levels of analysis.

Let us return to syllable-counting in the case of puranic mantras. Consider the mantra *oṁ namaḥ śivāya*, for example: perhaps its *primary* feature (i.e. the one considered most important by the ritualists), is that it consists of six syllables (and is therefore appropriately called *ṣaḍakṣara*, i.e. 'consisting of six syllables', as Staal observes). But, even as a mantra, the utterance *oṁ namaḥ śivāya* may still operate, simultaneously, on these other levels - including the semantic level. And a good place to begin an analysis would be to consider the question whether this mantra is, and to what extent it is, cognitive, or emotive, or phatic, etc. (certainly, the role of the divine name is significant!). The point is that we need to overcome the habit of assuming that a given utterance must have *only* one function at a time. Likewise, in the case of mantras.

I'd like to turn again to Staal's suggestion that mantras are like songs: some have words in them, and some don't - that is, words aren't necessary, but the music is. What about poems, for example a dadaist poem by Hugo Ball or an animal-cry poem by Michael McClure? These have all your basic linguistic features in them, as well as meaningless words or sounds - but no music. Or is the difference between poetic structure (which they undeniably have) and musical structure, in the final analysis, negligible? If so, we may once again find ourselves, in the case of mantras, still very much within the sphere of language. For the patterns which we find in mantra sequences are not necessarily musical - that is, such patterns are not necessarily confined to music. We all know that highly elaborate structures have been found in literary works, both oral, like the Vedas, and written, ranging from Biblical Psalms to Homer to Dante to Shakespeare to Joyce, as well as to the Gathas of Zarathustra, as Schwartz's recent studies (cf. e.g. Schwartz, 1986) of them make clear. Such structures exist also in classical Sanskrit kāvya and to some extent also in Vedic poetry, where we frequently encounter, e.g. palindrome-like lexical and syntactic patterns extending over lines of verse and even whole stanzas (illustrative examples can be found abundantly in Gonda, 1959: 'Chap. V: Chiasmus'; cf. also the recent and important study of Elizarenkova, 1995). The point is that structuralism, whatever else it has done (cf. Staal's discussion of it, 1989, p. 441 f.) has taught us to detect such elaborate patterns not only in nature, not only in a wide variety of the arts of human culture, but in language as well, and there perhaps pre-eminently. As far as I can tell, Staal's arguments from structure do not establish that mantras are not linguistic phenomena.

To turn to a few other specific issues raised by Staal's theory of mantras. Mantras like OṀ are said to be invariant, virtually indestructible, and therefore unlike language, and in fact, much older than language (cf. Staal, 1989, p. 275). But the Sanskrit word *mātar*, 'mother', it seems to me, has also been 'virtually indestructible' over perhaps a much longer period than 3,000 years, and if we consider the putative IE form, **mater*, from which it is derived, we can probably double or even triple that age. Staal himself has referred to Jakobson's discussion of a child's 'first acquisition of conventional speech' (Jakobson, 1962, p. 541, cited by Staal, 1989, p. 275), and in particular of the consonant-vowel pattern in the reduplicated form *mama*, with its primitive contrast between oral opening and closure and re-opening. It may well be that with *mama* we are standing at the threshold to conventional speech, and that, as Staal proposes, with Sanskrit OṀ we are standing on the other, or perhaps the far side of, i.e. *before* language. But both OṀ and *svāhā* (another of Staal's examples of invariant mantras) are probably also *words*. That is, they both have reasonable *etymologies* (if we accept Parpola's view [Parpola, 1981] of the former, and Renou's [Renou, 1958] of the latter). And *as words*, they have a history, which it may be worth knowing about if we want to know their significance even as mantras (likewise, I would attribute at least some of the tenacity of the phonic shape of the Sanskrit term *matar* to the emotional charge of its reference). I realize, of course, that from Staal's point of view, in so far as they are mantras, none of this matters. The point is that in this case also, the features by means of which Staal distinguishes mantras from language can in fact exist within language itself.

Here's a perhaps typical example of mantra use, which may be useful as illustration. At the moment of his death, gunned down by an assassin's bullet, Gandhi is said to have cried 'Rām, Rām!' This is the name of course of a popular Hindu god. But besides being a name, 'Rām' is also a popular mantra, and in this particular use it seems much less 'akin to music' than, in fact, to language - intensely expressive language perhaps: in this case perhaps a cry for help. Or a kind of command. Or perhaps it is rather (or as well!) an example of phatic language. Doesn't it seem that in this, his final act, Gandhi sought, at the moment of death, that is with his dying breath, to make contact with his God, perhaps to unite himself with him? Of course, this is something that mantras are typically said to do, both by contemporary users of mantras and by the native Sanskrit tradition itself. Such a usage needs to be explained, not because it seems an extraordinary use of language, but rather because in this usage mantras are not entirely unlike a rather ordinary - a very commonplace - usage of language. In fact, it may well be, statistically speaking, that phatic language is far more frequent than any other usage that language is put to. Jakobson, by the way, also notes that phatic language 'is the first verbal function acquired by infants; they are prone to communicate before being able to send or receive informative communication' (Jakobson, 1987, p. 69). It seems to me, then, that there may be other grounds for agreeing with Staal about the priority of mantras over language, although I would speak more strictly of the priority of *phatic* language over *informative* language. Once again, I would make the claim that mantras may be significantly phatic. If so, the important question then becomes: in the case of mantras, *contact* is established between the reciter and what in particular? Since it has been emphasized by Staal and Wheelock, among others, that mantra-utterance frequently exhibits little need of an audience, we cannot assume that mantras are directly phatic: i.e. that they establish and/or maintain contact between a speaker and a hearer. It may be that mantras function instead to establish contact between the speaker and a divine figure, as in the case of Gandhi's final cry to Ram. Or it may be that mantras present to the reciter a desired or a traditional representation of self which the reciter must embrace as a paradigm, i.e. that a mantra-utterance may be usefully characterized, at least on occasion, as an act of self-representation.[18]

Also, if we can talk about silent mantras, as the native tradition itself does, in this silent form, are mantras really closer to music, as Staal insists (cf. Staal, 1989, p. 279 f.)? That is, are silent mantras more comparable to 'silent music' than to 'silent language'? It would seem to me that the proper analogy would be with the latter, or perhaps with inner speech, or thought, or even meditation. Recall *mantra's* etymology, after all: it is literally an 'instrument of thought' - and it is clearly perceived as such in the Ṛgveda and Avestan, where the term *mantra* first occurs.[19] In any case, the native tradition itself says - like Renou and other Sanskritists - that mantras dissolve *not* into music but into *silence*. In that case, beyond language, yes. But, it would seem clear, beyond music as well.

On the other hand, Staal emphatically denies that mantras are acts, citing the sūtra *ekamantrāṇi karmāṇi*, which he glosses as 'each act is accompanied by one mantra' (Staal, 1989, p. 193). But starting from the Ṛgveda we have instances of mantras like *svāhā*, *vaṣat*, and even OṀ, etc. occurring in compounds such as *svāhākṛti*, *vaṣatkāra*, *oṁkāra*, etc. These expressions can mean only one thing, it seems to me: 'the *performance*, which is to say, the *utterance*, of *svāhā*, *vaṣat*, OṀ'.[20] In fact, the well-known use of the term *-kāra* in the Sanskrit grammatical tradition to mark phonemes, e.g. *akāra* = 'the phoneme *a*', *cakāra* = 'the phoneme *c*', etc. which is derived from this Vedic usage, as Renou (Renou, 1941-2) has shown, indicates an early awareness of speech-utterance as an *act*. Similarly, the occurrence of mantras like *svāhā*, *vaṣat*, OṀ, etc. with forms derived from the verb *kṛ*-, 'to act', would seem to indicate an awareness that mantra-utterance is in some sense an *act*. Granted that the sūtra *ekamantrāṇi karmāṇi* distinguishes mantras from *karmāṇi*, i.e. from ritual acts; but it does not necessarily imply that mantras are *not* acts. It might rather indicate either that mantras are a sub-set of a larger category, *karmāṇi*, or (probably more likely) it indicates that *karmāṇi* and mantras are two different types of acts. These could well turn out to be, to use the terms of Marcel Mauss, *rites manuels* on the one hand and *rites oraux* on the other. Thus there seems to be strong evidence in Vedic itself, *pace* Staal, to suggest that mantras were perceived to be acts, and if so, couldn't they very simply be *speech*-acts, even if of a very special - perhaps ritual - type?

It may be useful to turn briefly to still another point emphasized by Staal, about the phonology of mantras (cf. Staal, 1989, p. 254 f.). They do not *always* conform to the phonology of the language in which they are inserted (*pace* Staal, *ibid.*). Consider the Tantric mantra KHPHREM, or its variants HSHPHREM or HSHRPREM or HSKHPHREM (all cited by Padoux, 1990, p. 423). While each individual sound or phoneme is permissible in Sanskrit, they are not all permitted to occur together as clusters, and in fact, as Padoux himself points out, these unpronounceable mantras are understood to be 'not so much an utterance as a mental representation of the mantra's phonic elements' (Padoux, 1990, *ibid.*). To such mantras, of course, are attributed various symbolic and mystical significances having to do with the gods and spiritual powers. The point is that these mantras seem to have less to do with music than with some sort of relatively obscure but nevertheless *symbolic* thinking (cf. again the verbal root *man*-: its etymological connection with our term mantra is very significant, I would insist, *pace* Staal). Once again, mantras would appear to have something significant to do with representation. In such cases, it would not be surprising that their communicative function would be minimized, or even entirely absent.[21]

All of this is to say, in conclusion, that in spite of Staal's daring and often breathtaking originality, the sweeping breadth of his knowledge, the valuable connections that he makes, and the coherence and clarity of his argumentation, I have strong reservations, in the end, about his view of mantras. But if I have come to reject Staal's view, it is not because I prefer to cling to an approach to mantras that has dominated Indology from its beginnings. Staal has, I think, convincingly dismantled the old approach, typified in both its strengths and in its weaknesses by, say, Gonda's still important and influential essay of 1963. This approach is essentially, and rather simply, philological; Gonda catalogues relevant passages from a wide range of sources, and then he interprets, relying on his no doubt deep familiarity with the texts. But for the most part Gonda brings no further methods to the discussion, and, it should be said, very little insight (cf. Gonda, 1963). In the same way Sanskritists, Vedicists in particular, with regard to mantras at least, have been singularly reluctant to cross into alien disciplines, and as a result they tend to accept without comment notions and images from 'their' texts which they vaguely feel to be absurd or bizarre.[22]

In spite of my disagreements with Staal, it would seem beyond dispute that the landscape in mantra interpretation has been significantly changed by Staal's theories, as well as by his criticisms of more conventional approaches, and particularly by his persistent attacks on the exclusively philological orientation of most Sanskritists. It is now taken for granted, as the Alper volume (Alper, 1989) devoted to mantras clearly shows, that one needs to apply insights from a number of other disciplines, the chief among them being linguistics, of course. But beyond relatively familiar disciplines like anthropology and comparative religion, Staal ventures into more exotic areas like logic and ethology and even ornithology, and from Chomsky and Wittgenstein to Schoenberg and Boulez. There is certainly a bit of recklessness in these ventures (Staal does not deny this), but his explorations, I think, remain fruitful, suggestive, stimulating, and above all challenging. I would want it to be very clear: I think that much of the hostility that Staal's work arouses is reactionary; my disagreement with Staal in no way implies a return to those old ways. Let them rest in peace. In my view the debate about mantras *must* begin on Staal's terms: if we Sanskritists are going to insist that mantras are a form of language, then we should know more about linguistics (and not just etymology). Staal has challenged us to do so, and I think that we are obliged to respond.

So then, while I would insist that mantras are, after all, linguistic, in spite of Staal's arguments to the contrary, the issue remains for me a problematic one, for mantras are, I have come to believe, a complex, multi-purpose, phenomenon, and I know of no wholly satisfactory treatment of them.[23] Other perspectives than the ones offered by Staal and those represented in Alper's collection need to be considered and evaluated. It has been suggested, for example, by Strenski (Strenski, 1993, p. 224, in a review of Staal, 1989) that mantras may actually be, rather than a pre-linguistic, archaic, and vestigial atavism, actually parasitic on language, and therefore 'post-linguistic' in some interesting sense.[24] There is in fact a great deal more to be said about such a view.

It would seem useful to continue, as others have done, including Staal himself, to compare the phonology and syntax and perhaps even the semantics of mantras to other pre- or non-linguistic phenomena - like bird-song. But we should also consider para-linguistic, or other fringe linguistic phenomena - like the recitation of prayers, the chanting of magical spells, or the ecstatic experience of speaking in tongues - with all of which, it seems to me, mantras share certain significant features. Staal himself has observed this parallelism on the level of phonology at least: as in glossolalia so in mantras there is a high degree of invariance and repetition; the use of unknown or meaningless sounds; a severely reduced phonology, or at least a preference for a small range of phonemes such as nasals or nasalized clusters, or sibilants or sibilant clusters. Among non-Sanskritists, Jakobson especially has made some very interesting observations about such phenomena.

Whereas glossolalia appears to be an involuntary by-product of ritually induced trance, a mantra seems to be a voluntary device (or voluntary *action*) that may be used, among other things, to induce trance, or, as our texts suggest, various altered states of consciousness, or perhaps heightened awareness of both internal or external phenomena (e.g. breathing, muscle tension, the stream of thought and other bodily sensation, etc.). The popular and perhaps facile view that there is a correlation between trance and music, trance and dance, trance and ritual, and finally between trance and mantras, is not necessarily wrong. Also, suggestions like the anthropologist Rodney Needham's (Needham, 1967), that there is a consistent correlation between percussion instruments of various sorts (including firecrackers at Chinese New Year's) and a variety of rites of transition may in fact be relevant to the study of mantras. Consider the mantra *phaṭ*, with its pair of truly *plosive* voiceless stops: it is not surprising that this mantra, with its strongly percussive sound quality, is considered, like many other mantras, a weapon. Likewise, there is a common correlation between various rites and non-linguistic objects such as masks (cf. Napier's interesting monograph on the role of masks in various rites of transition, Napier, 1986). This fact may also be of use to us, I think. For mantras, like masks, sometimes serve to induce a kind of dissociation from one's ordinary self, or from one's ordinary concerns and troubles (this may explain the use of mantras in healing). Needless to say, these and other suggestions about mantras and trance need to be explored further than is possible here.

And this brings me to my final point. It seems to me that recent discussions of mantras have not stressed enough the fact that mantras are significantly motivated by a folk linguistic idea that lies at the heart of much of Vedic speculation. Throughout Vedic literature, and to a remarkable extent throughout much of all Sanskrit literature, and Indic culture as a whole, there is an underlying conviction that the spoken word, more particularly the ritually, solemnly uttered word (or even a sound sequence without meaning), is a thing of great power. 'Les pouvoirs de la parole' (to use the phrase of the Sanskritist most responsible for our general recognition of this fact: Louis Renou, cf. Renou, 1955, p. 1-27) were a major preoccupation of Vedic brahmins, both in their role as poets as well as in their role as priests, and such powers of the word have been a major preoccupation of much of Indian intellectual life. India's remarkably sophisticated and remarkably old grammatical tradition has its roots in this folk linguistic attitude. Probably one of the most conspicuous features of Indian cultural (not just religious) life is the presence, even the pervasiveness, of mantras. In India, people of all sorts turn to mantras as a means of last resort in moments of crisis. Perhaps this is because India, more than other cultures, has fallen under the spell of language, the *power* of language, the magical belief that merely by saying OṀ, or some other comparably charged mantra, we can overcome any obstacle. But though this phenomenon may be most conspicuous in India, it is certainly a widespread, if not a universal one: humans everywhere seem prone to the belief that merely by saying so we can it make it so.[25]

In my view, then, mantras presuppose an *idea* about language. Any theory of mantras which ignores this fact cannot be an adequate one. For if mantras presuppose a folk speculation about the power of language, i.e. if they presuppose a philosophy of language, or a metalinguistics, they must first be linguistic. Of course, I have deliberately reversed the sequence of events as we find them in Staal's version of the evolution of mantras and language. For him, mantras came first, followed by meaning, and therefore followed by any possible ideas concerning them. But have we any evidence that his version is more viable than mine? Can we say with confidence that birds have mantras, or that our primate cousins do, or that Neanderthals knew how to use them, say, to heal, to contact deities or the dead, or to make themselves stronger in the face of this or that danger? Is there any real evidence, after all, to show that mantras are anything but a purely human achievement, or, if you will, a purely human foible, i.e. a product of the, perhaps incestuous,[26] union of human imagination and human language?

REFERENCES

Alper, Harvey (ed.)
1989 *Mantra*. Albany: State University of New York Press (SUNY Series in Religious Studies)

Bataille, Georges
1962 *Erotism: Death and Sensuality*. San Francisco: City Lights Books
1973 *Théorie de la religion*. (Texte établi et présenté par T. Klossowski). Paris: Gallimard
1985 *Visions of Excess: Selected Writings 1927-1939*. (Edited by A. Stoekl with C.R. Lovitt and D.M. Leslie, Jr.). Minneapolis: University of Minnesota Press
Borst, Arno
1957-63 *Der Turmbau von Babel: Geschichte der Meinungen über Ursprung und Vielfalt der Sprachen und Völker*. Stuttgart: Hiersemann
Brown, Roger
1958 *Words and Things*. New York: The Free Press
Burkert, Walter
1983 *Homo Necans: The Anthropology of Ancient Greek Sacrificial Ritual and Myth*. Berkeley: University of California Press
Deren, Maya
1953 *Divine Horsemen: The Living Gods of Haiti*. New York & London: Documentext, McPherson & Company
Driver, Tom F.
1991 *The Magic of Ritual: Our Need for Liberating Rites that Transform Our Lives and Our Communities*. San Francisco: Harper
Eibl-Eibesfeldt, Irenäus
1970 *Ethology: The Biology of Behavior*. New York: Holt, Rinehart and Winston
Elizarenkova, Tatyana
1995 *Language and Style of the Vedic Rsis*. Albany: State University of New York Press (SUNY Studies in Hindu Studies)
Findly, Ellison
1989 '*Mántra kavisastá*: Speech as Performative in the Rgveda', in: Alper, 1989
Foucault, Michel
1970 *The Order of Things*. New York: Random House
Girard, René
1977 *Violence and the Sacred*. Baltimore: Johns Hopkins University Press
Goffman, Erving
1959 *The Presentation of Self in Everyday Life*. New York: Doubleday
1967 *Interaction Ritual: Essays on Face to Face Behavior*. New York: Doubleday
Gonda, Jan
1959 *Stylistic Repetition in the Veda*. Amsterdam: Noord-Hollandsche Uitgeversmaatschappij (Verhandelingen der Koninklijke Nederlandse Akademie van Wetenschappen, Afd. Letterkunde LXV/3)

1963 'The Indian Mantra', *Oriens* 16 (Reprinted in *Selected Studies IV*, 1975) Leiden: Brill

Holdrege, Barbara

1994 'Veda in the Brāhmaṇas', in: Patton, 1994

Huizinga, J.

1970 *Homo Ludens: A Study of the Play Element in Culture*. New York: Routledge & Kegan Paul

Huxley, Julian (ed.)

1966 'A Discussion on Ritualization of Behavior in Animals and Man', *Philosophical Transactions of the Royal Society of London*, Series B, No. 772, 251, pp. 247-526

Jakobson, Roman

1987 *Language in Literature*. (Edited by K. Pomorska and S. Rudy). Cambridge: Harvard University Press

Jamison, Stephanie

1991 *The Ravenous Hyena and the Wounded Sun*. Ithaca and London: Cornell University Press

Lévi-Strauss, Claude

1981 *The Naked Man*. New York and London: Harper & Row (Introduction to a Science of Mythology. 4)

Lorenz, Konrad

1966 *On Aggression*. London and New York: Harcourt, Brace & World

Malinowski, Bronislaw

1923 'The Problem of Meaning in Primitive Languages', Appendix to: C.K. Ogden and I.A. Richards, *The Meaning of Meaning*. New York: Harcourt, Brace & World

Mauss, Marcel

1968 *Oeuvres*, Tome I. (Edited by V. Karady). Paris: Minuit

Napier, A. David

1986 *Masks, Transformation, and Paradox*. Berkeley: University of California Press

Needham, Rodney

1967 'Percussion and Transition', *Man*, n.s. 2, pp. 606-14. London: The Royal Anthropological Institute of Great Britain and Ireland

O'Flaherty, Wendy Doniger

1980 *Women, Androgynes, and Other Mythical Beasts*. Chicago & London: University of Chicago Press

Padoux, André

1989 'Mantras: What Are They?', in: Alper, 1989

1990 *Vāc: The Concept of the Word in Selected Hindu Tantras*. Albany: State University of New York Press (SUNY Series in the Shaiva Traditions of Kashmir)

Parpola, Asko
1981 'On the Primary Meaning and Etymology of the Sacred Syllable OM', *Proceedings of the Nordic South Asia Conference Held in Helsinki*, June 10-12, 1980
Patton, Laurie
1994 *Authority, Anxiety, and Canon: Essays in Vedic Interpretation*. Albany: State University of New York Press (SUNY Series in Hindu Studies)
Renou, Louis
1941-2 'Les connexions entre le rituel et la grammaire en sanskrit', *JA* 233, pp. 105-65. (Reprinted in Staal, 1973, pp. 434-469)
1955 'Les pouvoirs de la parole dans le Ṛgveda', in: *Études védiques et paninéennes, Tome 1*. Publications de l'institut de civilisation indienne. Paris: E. de Boccard
1958 *Études sur le vocabulaire du Ṛgveda*. Pondichéry: Institut français d'indologie
Révész, G.
1956 *The Origins and Prehistory of Language*. (Translated by J. Butler, no date given for the German edition). London: The Philosophical Library
Rocher, Ludo
1989 'Mantras in the Śivapurāṇa', in: Alper, 1989
Schechner, Richard and Willa Appel
1990 *By Means of Performance: Intercultural Studies of Theatre and Ritual*. Cambridge and New York: Cambridge University Press
Schwartz, Martin
1986 'Coded Sound Patterns, Acrostics, and Anagrams in Zoroaster's Oral Poetry', in: R. Schmitt & P.O. Skjaervø (eds), *Studia Grammatica Iranica, Festschrift für Helmut Humbach*. München: Kitzinger
Searle, John
1969 *Speech Acts: An Essay in the Philosophy of Language*. Cambridge and New York: Cambridge University Press
1979 *Expression and Meaning: Studies in the Theory of Speech Acts*. Cambridge and New York: Cambridge University Press
1983 *Intentionality: An Essay in the Philosophy of Mind*. Cambridge and New York: Cambridge University Press
Sebeok, Thomas
1972 *Perspectives in Zoosemiotics*. The Hague: Mouton (Janua Linguarum, Series Minor. 122)
1976 *Contributions to the Doctrine of Signs*. Bloomington and Lisse: Peter de Ridder Press (Studies in Semiotics. 5)
1979 *The Sign and Its Masters*. Austin: University of Texas Press

Sperber, Dan
1982 'Is Symbolic Thought Prerational?', (Originally published 1980, ed. M.L. Foster and S.H. Brandes in Symbol and Sense, New York), reprinted in 1982, M. Izard and P. Smith (eds), *Between Belief and Transgression: Structuralist Essays in Religion, History, and Myth*. Chicago: University of Chicago Press

Staal, Frits
1973 *A Reader on the Sanskrit Grammarians*. Cambridge: MIT Press
1979 'The Meaninglessness of Ritual', *Numen* 26, pp. 2-22
1983 *Agni: The Vedic Ritual of the Fire Altar*, I-II. Berkeley: Asian Humanities Press
1989 *Rules Without Meaning: Ritual Mantras and the Human Sciences*. New York: Lang (Toronto Studies in Religion. 4)

Strenski, Ivan
1991 'What's Rite? Evolution, Exchange and the Big Picture', *Religion* 21 (July 1991), pp. 219-25

Tambiah, S.J.
1979 'A Performative Approach to Ritual', in: *Proceedings of the British Academy* 65, pp. 113-69

Thompson, George
1995 'The Pursuit of Hidden Tracks in Vedic', *IIJ* 38, pp. 1-30

Todorov, Tzvetan
1978 'Le discours de la magie', in *Les genres du discours*, pp. 246-82, Paris: Seuil

Turner, Victor
1967 *The Forest of Symbols: Aspects of Ndembu Ritual*. Ithaca and London: Cornell University Press

Wheelock, Wade
1982 'The Problem of Ritual Language: From Information to Situation', *The Journal of the American Academy of Religion* 32, pp. 49-71
1989 'The Mantra in Vedic and Tantric Ritual', in Alper, 1989

Notes

1. Anthropologists have been far more sophisticated about such distinctions: they have read, for example, Turner's discussion (Turner, 1967, p. 50 f.) of the difference between exegetical, operational, and positional meaning. They have also had the benefit of Lévi-Strauss's important reflections in the Finale to his 'Introduction to a Science of Mythology' (Lévi-Strauss, 1981, p. 625 f.), where he points out the difference between glosses and commentaries on ritual - which amounts to mythology - and what he calls 'ritual proper'. Staal has demonstrated quite convincingly that the brahmanas have value only as 'mythology', not as ritual *per se*. The difference is fundamental. A different

view of Staal's position, considered as 'isolationist', can be found in Jamison, 1991, p. 4 f.

2. For example, Staal's criticism of their rather indiscriminate use of the term 'performative' seems to me to be perfectly valid. The following distinction made by Staal (1989, p. 163) seems to be particularly telling: 'To call an utterance performative is interesting because it calls attention to a special class of utterances; to call an act performative is trivial because most acts are performative'.
3. Staal's speculations concerning the role of mantras in the origins and development of language have been developed recently in a paper delivered at the Tenth Meeting of the Language Origins Society, July 1994, held in Berkeley, CA.
4. The combination of an identification or equivalence and an accompanying directive or conative verb is very frequent, not only in Vedic discourse but in magical discourse in general. It is a means of transferring the effects of an operation performed on one of the elements to the other, as when one performs an operation on the hair of a distant patient in order to influence the patient. For a general discussion, cf. the chapter on 'le discours de la magie' in Todorov, 1978.
5. Such utterances may appear at first glance to be an incoherent jumble, what psychologists like Brown (1958, p. 293) call a 'word salad', but in fact these word deletions tend to be systematic and the resulting utterance appears *telegraphic* or cryptic, rather than incoherent in the strict sense. There is an interesting discussion in Jakobson (1987, p. 135) concerning the poet Hölderlin, who in his debilitating madness attempted 'to eliminate his 'I' from conversations and, later, from his writing as well' (thus Jakobson, *ibid.*). And yet he continued to compose remarkably lucid poems that conformed gracefully and faithfully to the strictest of metrical forms. Such linguistic aberrations invite speculation not only about the autonomy of the various functions of language, but also about its presumed nature.
6. This quotation is taken from Staal's unpublished paper, 'Mantras: Sounds Beyond Language', delivered at a conference on 'Extra-ordinary Language and Religious Experience', at the University of California at Berkeley, in May 1993. My response to that paper, also delivered at the conference, forms the basis of the present paper. It should be noted that in this passage Staal is characterizing what mantras *are not*. It is thus only by inference that Staal can be grouped with Wheelock and Tambiah on this point.
7. I have presented a paper at the Language Origins Conference cited above in fn. 3, entitled 'Vedic Folk Linguistics and the Origins of Language', in which some of these issues are discussed.
8. Cf. the collection edited by Huxley, 1966; a good overview is also offered in Burkert, 1983, p. 22 ff.
9. Staal, 1989, p. 111 f.; 135 f.; 284 f.; as well as the concluding 434.
10. Consider, e.g., aggression, the centrality of which explains why discussions of ritual frequently turn into discussions of violence and sacrifice, as if ritual and sacrifice were indistinguishable. Cf. in particular Girard and Burkert.

11. The notion is specifically defined in Bataille, 1985, but clearly it is present as an undercurrent in his other works on religion, sexuality, and economics: cf. e.g. Bataille, 1962 and 1973.
12. A fairly representative sample of such an approach can be found in Schechner and Appel, 1990. Maya Deren's remarkable account of Haitian Voudoun also comes to mind (Deren, 1953). It should be pointed out, however, that Staal finds little of interest in performance theory (cf. for example his comments at 1989, p. 247), as is also suggested by the quotation cited above in fn 2.
13. This is a point noted already by Foucault, 1970, whom Staal discusses rather unsympathetically, and is important in so far as it suggests that the two sciences share similar goals and methods, as well as intellectual lineage.
14. I point out this etymology because of Staal's marked hostility to it (cf. for example Staal, 1989, p. 23). Such expressions of hostility have no doubt added fuel to the fire which philologically-oriented Sanskritists have directed against him.
15. Here, Sebeok relies to a great extent on Jakobson's model of language functions: cf. Sebeok, 1972, p. 16 f.
16. It should be noted, however, for the record, that Wheelock, 1982, has attempted a semantic interpretation of them - using ellipsis as an interpretive device; again, this attempt may not be entirely successful, but it is certainly worth considering.
17. Cf. Staal, 1989, p. 233-4; see also Rocher's useful survey (1989).
18. This view, which I suggest tentatively, seems close, perhaps, to Wheelock's notion of 'situating speech.' I am aware of the danger of confusing phatic language with other language functions, such as the representational or cognitive, by extending it here to contact between a speaker and imaginary, mythical, or spiritual entities. But this is similar to what occurs in prayer, with which mantras should be compared, not because they are identical, but because they may be allied uses of language. Need phatic language and phatic gestures be always social? On contact theories of language origins in general, one might consult the relevant chapters in Révész, which I think are still valuable. In any case, if mantras are in fact phatic in function, then it would make sense that they are primitive, and pre-linguistic: as Jakobson observed, contact must precede communication.
19. Findly's claim (1989, p. 15) that the semantic development of the term *mantra* toward its current sense is relatively late (i.e. 'a younger period of Ṛgvedic composition') is probably not right, in spite of the relative paucity of the term in the Ṛgveda. The fact that it is attested in Avestan with essentially the same meaning and usage strongly suggests that this development had already taken place before the Vedic period.
20. For a fuller discussion of this use of the verb *kṛ-*, cf. Thompson, 1995.
21. In general, it seems to me that an analysis of bīja mantras could well proceed along these lines.
22. A similar criticism has been made also by Jamison, 1991, p. 36.
23. This appears also to be the final point of view of Padoux, in his admirably lucid and well-balanced overview of the Alper collection on mantras, Padoux, 1989.

24. This approaches Sperber's discussion of symbolism in general as 'post-rational' rather than pre-rational (Sperber, 1982).
25. As I have argued in the paper cited above in fn. 7, a good deal of popular and even scholarly speculation about the power of language is similarly motivated. Notions of linguistic relativity, for example, are susceptible to this sort of analysis. In general cf. Borst's monumental study of the history of such ideas (Borst, 1957-63).
26. Here I am alluding to a well-known motif in Vedic cosmogony, which has many parallels, both within Indo-European and beyond, concerning which one might consult O'Flaherty, 1980. More specifically, on the coupling of *Manas*, Mind, and *Vāc*, Speech, cf. most recently Holdrege, 1994.

XXXI

TIBETAN EXPERTISE IN SANSKRIT GRAMMAR (3): ON THE CORRECT PRONUNCIATION OF THE INEFFABLE

Peter Verhagen

In my contribution to the present volume in honour of Professor Dr. Frits Staal I will touch on three topics that figure prominently in his academic work, the phenomenon of the *mantra*, the traditions of Sanskrit indigenous grammar, and the Veda. The main focus will be on the former two subjects, but a number of observations on Vedic matters will also be included.

In the various forms of later Buddhism, under the general designations Mahāyāna and Vajrayāna, particularly in the branches of the latter, an important place is occupied by the phenomenon of the *mantra*. In this context I translate *mantra* tentatively as 'esoteric formula', a string of terms and syllables, of evidently Indic or Sanskritic origin, to great extent untranslatable, yet always meaningful, i.e. it is attributed a very specific meaning and function in specific ritual and meditational contexts.[1] A further factor contributing to the untranslatability of these *mantras* is the fact that much if not all of the force or the efficacy of the formulas lies in the actual sound and form of the *mantras*.

The complex process of the dissemination of Buddhism in Tibet, from the mid seventh century CE onwards, involved the translation of a massive corpus of Sanskrit Buddhist literature. A great many forms of lore centred on such *mantras* were transmitted to Tibetan practitioners, and continued within Tibet, through the usual processes of personal instruction and initiation, from master on to pupil-adept. The literature pertaining to these esoteric traditions, both the basic texts generally termed *Tantras*, as well as the extensive technical and exegetical literature that developed around them, received due attention as well and a great many such texts were translated into Tibetan.[2]

In the Tibetan translations of such Tantric materials we find that the *mantras* are left untranslated.[3] Naturally this led to all kinds of problems for the Tibetan adepts, in particular with regard to the pronunciation, graphical notation etc. of these basically Indic formulas. Various forms of investigation into the practical and theoretical aspects of the *mantras* had already sprung up within the Tantric exegetical traditions in India before the introduction of this lore in Tibet. In addition to this, to meet the specific needs of the Tibetan apprentice, we find the development in Tibetan literature of a genre of treatises dealing with the pronunciation of *mantras*. Such *klog-thabs*, 'pronunciation manuals', var. *sṅags-kyi-klog-thabs* '*mantra* pronunciation manuals', have been written by several prominent specialists in the fields of linguistics and the esoteric. In my research thus far I have been able to locate twelve. They were produced throughout the history of Tibetan scholastics: the earliest that I have been able to trace dates from the twelfth century, the most recent from the nineteenth century. Investigation of these materials has proved to be extremely fruitful, leading to interesting results both from a Buddhological as well as from a linguistic point of view.[4]

It should be mentioned at this point that in Tibetan Buddhist scholastics a strong interest existed in several indigenous Indic linguistic traditions. In the field of grammar particularly the two most influential Buddhist systems of Sanskrit grammar, *Cāndra* and *Kātantra*, were studied extensively. Numerous texts belonging to these systems were available in Tibetan translation, most of them incorporated into the Buddhist canon,[5] and throughout the centuries of Tibetan Buddhist scholastic history, say from the eighth to the twentieth century, a considerable number of Tibetan scholars have contributed significant original writings to this literature as well, ranging from exposés on specific details to broad surveys and compendiums, and extensive commentaries.[6] Also other linguistic disciplines originating from the Sanskrit traditions such as lexicography, poetics, prosody and dramatic arts, found their way into the Tibetan Buddhist curriculum in the more scholastically oriented circles.[7]

I will now focus on a work of one Tibetan grammarian who can justifiably stand as the representative of his generation and more. I mean the scholar commonly known as Si-tu, or Si-tu Paṇ-chen (or Si-tu Mahāpaṇḍita), 'the Great Scholar Si-tu'. He was the eighth reincarnation in the (Ta'i) Si-tu lineage of hierarchs in the Karma Bka'-brgyud-pa tradition, the most commonly used of his ordination names being Chos-kyi-'byuṅ-gnas and Bstan-pa'i-ñin-byed Gtsug-lag Chos-kyi-snaṅ-ba (1699-1774). Although relatively late he was eminently versed in the whole range of Indo-Tibetan grammatical traditions, both earlier as well as contemporaneous, as becomes abundantly clear from his writings. He is generally considered to be one of the most important linguists in the history of Tibetan scholastics, if not the most important.[8]

The treatise we will presently be having a look at is not a *klog-thabs*, it is not an instruction manual on the pronunciation of Sanskrit *per se*, but it does deal - particularly in these segments we will investigate here - with a number of the theoretical and practical problems involved in maintaining purity and orthopraxy of the *mantra* lore in Tibet. It is in fact a collection of replies by Si-tu to questions on a variety of subjects. The collection, entitled *Rje-btsun-mchog-gi-sprul-pa'i-sku-dgyes-par-byed-pa'i-dri-lan-nor-bu'i-me-loṅ-zhes-bya-ba*, 'Answers to queries by the reincarnation of [or: i.e.] the highest venerable master, who is a source of joy, entitled 'Mirror of Jewels'', covers thirty-one folios in volume 8 of the xylograph edition of his collected works prepared in Sde-dge.[9]

The main part of the collection (ff. 1-29r1) contains the replies to questions that were put to Si-tu by one particular reincarnated Lama who has thus far remained unidentified. This main collection of queries is divided into two sections, (1) questions of a general nature, that is on matters that are not specifically Buddhist (ff. 1v2-13v5), and (2) questions of a non-general nature, i.e. dealing with topics within the realm of Buddhist dogmatics and philosophy (ff. 13v5-29r1). We find that almost all questions in section (1) pertain to grammar and linguistics in a broader sense.

In this fascinating little text we get a rare insight into the versatility of this master of grammar. An impressive range of subjects passes in review, for instance: etymologies of Indic and Tibetan terms, problems of pronunciation and phonological features of Indic speech-sounds (both individual phonemes and clusters), linguistic terminology, and information on authors and texts belonging to the indigenous traditions of Sanskrit grammar.

The colophon dates the collection to 1749 CE. By that time Si-tu was well established as a major authority on grammar: he had finished his magnum opus on Tibetan grammar, his extensive commentary on the two basic texts, in 1744, and he had commenced work on his three-volume commentary on the Sanskrit grammar *Cāndra* in 1747 (after an interruption of some years to be resumed in 1754 and finished in 1756); moreover, since 1734 he had been working on one of his life's major tasks, namely revising the dozens of translations of works on Indic linguistics contained in the Tibetan Buddhist canon.

In the following pages I will present a first investigation of a selection from the questions and Si-tu's replies in the section on secular topics in this 'Mirror of Jewels' collection. The Tibetan text of longer passages translated in this article is given in the endnotes.

Question 2 (f. 2r1-2r3):

Referring to a particular xylographic edition of a commentary on the Sanskrit grammar *Sārasvata-vyākaraṇa* by the famous Jo-naṅ-pa scholar Tāranātha Kun-dga'-sñiṅ-po (1575-?), this question is raised:

> On the opening page of the book [i.e. blockprint] of that same commentary, on the right hand side, there is a depiction of Sarasvatī, holding an attribute that looks like a two-topped staff with a bowl attached to it at the curving end. What is this? And, a similar [attribute] being required for her [i.e. Sarasvatī] is not taught, etc., in the Buddhist *Tantras*. What [is your explanation]? (Appendix 1)

Answer:

> That depiction of an attribute, is the well-known attribute of Sarasvatī, the *tambura*, which looks like the *sgra-sñan* [Tibetan string instrument], but is different.
> In countries such as India and Nepal, [many] types of such [instruments] are depicted; I myself have also seen [them; i.e. the instruments or the depictions ?] frequently. (Appendix 2)

The blockprint of this commentary available in the collected works of Tāranātha indeed displays on its opening page three illustrations, depicting Śākyamuni Buddha, Sarasvatī and Tāranātha. The depiction of the goddess, on the left for the viewer - note that in case of such depictions the Tibetans always determine left/right from the standpoint of the persona(e) depicted - shows her holding an object looking precisely as described in the question (see ill. 1). There can be no doubt that the present question is in fact speaking of the Phun-tshogs-gliṅ print which is still available (facs. ed. Sherab Gyaltsen, 1990).

Illustration 1.

Depiction of Sarasvatī, Tāranātha's commentary ad *Sārasvata*, Collected works, vol. 14, f. 1v1; facs. ed. *Collected works of Jo-naṅ Rje-btsun Tāranātha*, Leh 1987-14: 2.

A different print of this text, struck from the same blocks, and a print of a bilingual version of the *Sārasvata sūtra* text, also from Tāranātha's collected works (the latter unfortunately missing in the facs. ed.), have been made available through the Nepal-German Manuscript Preservation Project. The opening page of the latter texts shows depictions of the same figures, in an almost identical layout, in the same style, different only in some minute details (see ills. 2 & 3). Evidently the blocks for the two texts were produced as a set, in the larger context of the Phun-tshogs-gliṅ xylographic edition of Tāranātha's collected works. No exact date is available for the carving of the blocks;[10] my impression of the style of the depictions leads me to suppose that an early date is plausible, possibly the project was undertaken shortly after the author's death, for which we have no exact date either, ca. mid seventeenth century.

Illustration 2.

Depiction of Sarasvatī, Tāranātha's commentary ad *Sārasvata* Collected works, vol. 14, f. 1v1; facs. ed. *Collected works of Jo-naṅ Rje-btsun Tāranātha* microfilm Nepal-German Manuscript Preservation Project.

The depiction of Sarasvatī, including that of her attribute, is extremely similar to that in the commentary. Inspection of the various prints shows that, in its stylized depiction, the instrument has a pot-like appendage at both ends. Only the (for the viewer) right-hand end of the instrument has a distinct upward curve. In a glorious 17th-century scroll painting from eastern Tibet, the goddess is shown holding the same instrument, here with an upward ornamental flourish at both ends.[11] Si-tu observes, in his reply to this question, that Sarasvatī's attribute is the *Ṭanbura*, the Indian four- or five-stringed instrument, used typically to provide a drone in improvisational music. However, the attribute of the goddess is usually identified as a *vīṇā*, the more lute-like string instrument, in Tibetan iconography. With its typical form of usually seven strings on a long rounded board, with two gourds attached towards the ends, the identification as a stylized depiction of a *vīṇā* seems the more plausible here.

Illustration 3.

Depiction of Saravatī, Tāranātha's version of *Sārasvata* basic text, f. 1v1; microfilm Nepal-German Manuscript Preservation Project.

I should note here that the form of the Tibetan string instrument *sgra-sñan*,[12] to which Si-tu likens the attribute in his answer, is not completely similar to a *vīṇā* or a *ṭanbura*, but it is the closest correspondence to either in the limited repertoire of Tibetan musical instruments. And, indeed, we do find that the goddess Sarasvatī is also represented in Tibetan visual arts holding an instrument that appears to be more similar to a *sgra-sñan*.[13]

Question 13 (f. 5v2-6r1):

This question requests clarification, with regard to the pronunciation, of certain unusual phenomena typical for mantras, mainly irregular consonant clusters, and in particular the combinations with the so-called *upadhmānīya* and *jihvāmūlīya* variants of the *visarga*.[14] An extract from Si-tu's answer:

> As the clusters <*ḥ*>*k*<*ḥ*>*kā* etc., as found in the *Kālacakra(-tantra)*, are unusual utterances (*brda* ?), that are not known from the grammatical literature,
> […] when considering the literature [available on this topic], I would think no practical advice is available other than the general maxim:[15]

> When combining dissimilar consonants in a cluster, one should pronounce [in] quick [succession] the mere [i.e. brief ?] pronunciation of each [constituent] phoneme. (Appendix 3)

The Sanskrit final *ḥ*, a voiceless breathing, traditionally termed *visarga* or *visarjanīya*, has two conditioned variants: a velar fricative termed *jihvāmūlīya* occurring before voiceless velar stops, and a labial fricative known as *upadhmānīya* which occurs before voiceless labial stops. These allophonic variants are given their own graphic representation only in certain branches of Vedic literature, notably the Paippalāda recension of Atharva-saṁhitā.[16]

It here becomes evident that these signs, known from certain writings in the Vedic period, an awareness of which however had remained in the technical description they received in the grammatical traditions, are employed again in certain branches of Tantric literature. Possibly the tendency towards schematization and optimalization of existing schemata in Tantric literature (through accretion as well as reduction) may have played a role here: the inclusion of the two allophones leads to optimalization of the schema of phonemes, the category of fricatives now having a separate element for each of the five major phonological classes of velar, palatal, retroflex, dental and labial, viz. *jihvāmūlīya*, *ś*, *ṣ*, *s* and *upadhmānīya* (in the classical schema the same *visarga ḥ* occurring in the first and the fifth class).

The variants were most probably known to the Tibetan specialists through the *Kātantra* and *Cāndra* grammars. *Kātantra* introduces the allophones in *sūtras* 1.1.17-18, rules derived from the Vedic *Prātiśākhyas*, i.c. *Vājasaneyi Prātiśākhya* 8.19; in general *Kātantra* is known to rely heavily on the *Prātiśākhyas* for his phonological description. *Cāndra* introduces the two *visarga* variants in 6.4.31, following Pāṇini 8.3.37. That the Tibetan experts were well aware of these variants is shown by their treatment in a number of *klog-thabs* and related treatises.[17]

Question 14 (f. 6r1-6r6):

The question addresses the problem of the pronunciation of *mantras* graphically represented in various scripts, and/or stemming from various languages:

(a) 'As regards the pronunciation of [elements of *mantras*] that are known from *mantra-tantras* [written in / translated from] the symbol language of the *Ḍākinīs*,[18] the *Paiśācī* language, etc.: is their pronunciation similar to Sanskrit, or, if not, what is their particular pronunciation?

(b) Similarly, phrases such as *Oṁ-dar-du* [?] or *Ma-ma-sño-ta-sño-ta* etc., occurring in [texts] stemming from Skag [and ?] Zlog [?], and China:

what is their pronunciation? Is it similar to Sanskrit, Tibetan, or another [language]?

(c) In general, to what language should these and similar *mantras* stemming from China be classified?' (Appendix 4)

Answer:

(a) '[The pronunciation of phrases] such as *Ma-da-na-chaṅ-ba-la-śa*, occurring in the two chapters in the symbol language does not differ from the pronunciation of Sanskrit.
[The pronunciation of phrases] occurring in treatises on dramatic arts in the *Paiśācī* language, is also similar to that [scil. of Sanskrit].

(b) As regards some [phrases] occurring in *Rñiṅ-ma*[*-pa*] *Tantras*, as in general they are completely dissimilar to anything like the Indic or Tibetan [language / pronunciation ?], and as there are no specific details available, there is no [other] proper [course] than to pronounce them merely with respect [i.e. as 'literally' as possible ?].

(c) The *mantras* stemming from China are in fact *mantras* translated from an Indic [language into Chinese], which could not properly be represented in the Chinese script, nor could they be conveyed by the [Chinese] speech sounds.
Therefore, phrases such as at the beginning of the *mantra* of the *Prajñā* [*-pāramitā*] *Hṛdaya*, viz. *gyas-ti-gyas-ti-po-lo-gyas-ti* for [Sanskrit] **gate gate pāragate*, and *nan-bū-o-mi-tha'o-hwo* for [Sanskrit] *namo 'mitābhāya*, appear to be merely corrupt, and therefore it is not possible to decide on one particular manner of pronunciation for these mantras translated from [texts from] this country.
However, as it seems [likely] that the Tibetan translators of that era have maybe made the pronunciation and spelling correspond [lit.: made the pronunciation correspond to the spelling], I think that there is no objection if it is done [i.e. pronounced] according to Tibetan pronunciation.' (Appendix 5)

A number of aspects of both question and answer are problematic. For instance, I have not been able to trace the designations Skag and Zlog, apparently names of countries or peoples (or perhaps Skag-zlog is a bisyllabic name).

The reference to Chinese notations of two specific *mantras* is not without interest. The most common representation in Chinese of the phrase *gate gate pāragate* from the well-known *mantra* in the *Prajñāpāramitā-hṛdaya*[19] is *jie di jie di bo luo jie di* (Kumārajīva's translation, in pinyin), while the second *mantra*, *namo 'mitābhāya*, usually reads in Chinese, again in pinyin, *nan wu* *a mi tuo fo* (i.e. **namo 'mitābhāya buddhāya*); both readings quite comparable to those that Si-tu quotes here.

Si-tu's acquaintance with at least some basics of the Chinese language becomes evident elsewhere in this same text, where a number of etymologies are discussed (question 9, f. 3r3-4r2). In this question the interlocutor asks for the etymology of certain names of countries, inter alia *Bahir, supposedly a name for India, to which Si-tu replies:

> [The term] *bahir* is not the name of the [or: a] country of India, but it appears that the Indians that had come to Tibet called India *bahir*, i.e. 'outside', and Tibet [they called] *antar*, 'inside'. As did, for instance, the Chinese who had come to Tibet, who called China *wa'i-thu*, i.e. 'outside' and Tibet *li-thu*, 'inside'. (Appendix 6)

Here Si-tu's analysis involves two Sanskrit indeclinables, *bahir* 'out(side)' and its antonym *antar* 'in(side)',[20] and a parallelism with two Chinese terms, which can be identified as (pinyin) *waitou* and *litou*, very common words for 'outside' and 'inside'.[21] Moreover, in Si-tu's autobiography[22] we find reference to his attempts at translating Chinese works on astronomy and astrology.[23]

Question 16 (f. 6v6-7v3):

The question asks for clarification on the traditional Indic complex of variant forms of vowels, distinguishing the feautures of length, accent and nasality. In particular the two former features are inquired after. Si-tu's answer (omitting the final sentence, which is irrelevant for our purposes) runs as follows:

> The vowels *a*, *i*, *u*, *ṛ* and *ḷ* [each] have [three] variant forms, due to [the distinction in] the three 'short', 'long' and 'protracted' [Skt. *pluta*]. [However], as [the vowel] *ḷ* does not have a long [form], that [one] is subtracted, [giving the total of] fourteen. As these [each again] have [three] variant forms due to [the distinction in] the three '*udātta*' etc., [this gives a total of] forty-two. As regards the four *saṁdhy-akṣara* [vowel] phonemes [scil. *e*, *o*, *ai* and *au*], as they do not have a short [form], due to [the distinction in] the other two, [this gives a total of] eight variant forms, to which again [the distinction in] '*udātta*' etc. [is applied], [giving a total of] twenty-four. These two [totals] added up give [a total of] sixty-six. Due to [the distinction in] the two 'nasalized' and 'non-nasalized', these [sixty-six forms give a total] of one hundred and thirty-two different forms of vowel. This corresponds to the view of [grammars] such as *Cāndra* and *Sārasvata*.

> In this connection [a duration of] one, two and three morae is applied to the short, long and protracted vowels [respectively], while [a duration of] half a mora [is attributed to] consonants without a vowel.
> As regards the formation of the variant forms of '*udātta*' etc. for the short [vowels] etc.: When these vowels are pronounced in combination with consonants, when provided with a distinct 'strong' sound, they are *udātta*, i.e. 'with high pronunciation'; when provided with a 'weak' [sound], they are *anudātta*, i.e. 'with low pronunciation'; and, when provided with a 'middle' [sound], they are *svarita*, i.e. 'with middle pronunciation.[24]
> (Supralinear gloss: Also when the mere vowels are pronounced, they have variant forms depending on their 'strong', 'weak' or 'middle' sound [i.e. realization].)
> Such is the formation [of these variant forms]. Although there is a detailed exposé [on these matters] in the basic text on [lit.] vowels [i.e. phonology?] written by Rgyal-dbaṅ, this [text] has not reached Tibet. It is possible to find out certain traits of this lore from marginal remarks in the *Kāmadhenu* commentary on *Amarakoṣa*, and one can also get to understand mere aspects [of this] from the *Dhātu-sūtras* in Pāṇini's [tradition]. A detailed description, based on a basic text, a commentary, or the like, such a thing has not been translated into Tibetan thus far, so I am not able to give a proper exposé [on this matter]. (Appendix 7)

Evidently Si-tu's presentation of phonological and prosodic facts in this passage is primarily based on the traditional Indic descriptions that are usually found in separate sections of the various grammatical corpora, commonly termed *varṇa-sūtras*, 'basic statements on phonology' [or, perhaps more literally, 'basic statements on the phonemes']. These in their turn were ultimately modelled on the detailed phonological and phonetic statements contained in the Vedic *Prātiśākhya* literature, which contains precise and sophisticated descriptions of the phonology, but also of the suprasegmental features of the pitch accent of the language of the Vedas.

Most likely Si-tu's actual source here was the concise *varṇa-sūtra* belonging to the *Cāndra* system of grammar.[25] The order of treatment and the phrasing of the facts stated show close correspondences with this *Cāndra varṇa-sūtra*. In the latter part of his answer, Si-tu himself mentions three further sources for information specifically on the Vedic accent, a treatise on phonology by a Rgyal-dbaṅ (possible Sanskrit reconstructions *Jinendra or *Rājendra) which I have not been able to identify, the *Kāmadhenu* commentary on the Sanskrit lexicon *Amarakoṣa*, and the lexicon of verbal bases (*dhātupāṭha*) belonging to Pāṇini's grammar.

Question 18 (f. 7v5-8r1):

Although certain aspects of this question and its answer have remained problematic and opaque, I have decided on including it here as it contains interesting observations on the Vedic *anunāsika*:

> As regards the graphical form of the phoneme dealt with in [*sūtra*] *ṁ chandasi* in the basic text of *Sārasvata*, [GRAPH; see ill. 4], is this an actual Indian graph? If not, I would like also to know the graphical form [this phoneme] is given in Indian writing. In the commentary by Jo(-naṅ Tāranātha) it is stated that it should be pronounced as *sgyiṅ* or *sgya-yaṅ*. However, in [the works] of many experts on pronunciation [it is stated that it should be pronounced] as *gyiṅ*. (Appendix 8)

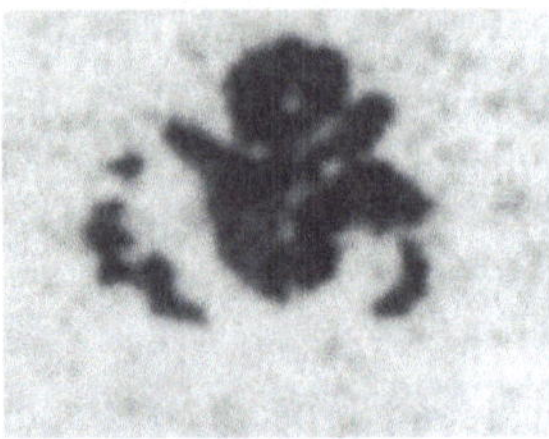

Illustration 4.

Anunāsika graph, Si-tu's *Dri-lan-nor-bu'i-me-loṅ*, Collected works vol. 8, f. 7v5; facs. ed. Sherab Gyaltsen (1990-8: 390 line 5).

The rule from the *Sārasvata* grammar referred to here is the last *sūtra* in chapter 1.4 on sandhi of consonants, *ṁ chandasi*, which describes, specifically for the language of the Veda, the optional substitution of *anusvāra* by *anunāsika* before fricatives, *r* and *h*.[26] In the Tibetan Buddhist canon *Bstan-'gyur* three translations of *Sārasvata* materials are available: two different translations of the basic *sūtra* text, and one of the *vṛtti*.[27] Tāranātha and Si-tu both have written commentaries on this grammar,[28] and in Si-tu's collected works we find his revision of the canonical translation of the *vṛtti* text.[29] This rule, and its interpretation in the Tibetan canonical translations and in the work of Tāranātha, has been the subject of an earlier article by Meisezahl.[30]

My tentative translation of Si-tu's answer:

> The graphical form *dzñīṁ* was determined in accordance with the Tibetan manner of pronunciation.
> Nevertheless, the proper manner of pronunciation of the *dzñ* phoneme

appeared not to be used in recitation[31] by the Tibetans.
Moreover, when in the commentary [on *Sārasvata*] by Jo[-naṅ Tāranātha], where the form *sgyiṅ* is [introduced] in the treatment of the examples, the traditions [scil. of pronunciation] of [this] Indian speech sound are investigated, this [treatment in the commentary] is quite elaborate (lit. not small, *cuṅ-min*), and this [pronunciation] can be inferred from the very words [in the text at hand], and it would seem I cannot elucidate aspects [of this matter] by means of [any further] examples.
Also other [quite] divergent forms of graph for this [phoneme] exist, for example [the graphical form] occurring in the **Nagara* [i.e. *Devanagarī*] script is: [GRAPH; see ill. 5]. (Appendix 9)

Illustration 5.

Anunāsika graph, Si-tu's *Dri-lan-nor-bu'i-me-loṅ*, Collected works vol. 8, f. 8r1; facs. ed. Sherab Gyaltsen (1990-8: 391 line 1).

The treatment of this *sūtra* in Tāranātha's commentary, and his discussion of the pronunciation of the Vedic *anunāsika*, to which the question alludes, and to which we are referred by Si-tu for further information, unfortunately is not as informative as Si-tu suggests. As regards pronunciation, it does not do much more than stating that the pronunciations *sgyiṅ* and *sgya-yaṅ* also occur [as already mentioned in the question], and that although the graphic notation *dzñīṁ* has been adopted by the Tibetans, it should not be regarded as a consonant cluster of *dz* (i.e. Sanskrit *j*) and *ñ* with vowel *ī*.[32] In his study of the treatment of the Vedic *anunāsika* in *Sārasvata*, both in the original and in the Tibetan versions of the canon and Tāranātha, Meisezahl (1965-6: 145) has already drawn attention to a significant aspect of the Tibetan notations for the Vedic *anunāsika*, namely their attempts to indicate the double nature of this allophone, both vocalic and consonantal.

Question 29 (f. 13r1-13v1):

Finally, the last question to be discussed here deals with the characterization of the four *Vedas*. It should be noted that the literature of the *Veda* was not unknown to the Tibetan Buddhist scholars, if not via direct access to this literature, in any case through the quite frequent references to Vedic and Brahmanical lore in the Buddhist literature. Testimony to this is the inclusion of the names of the four *Vedas* in the Sanskrit-Tibetan lexicon of Buddhist terminology known as *Mahāvyutpatti*, which served as the standard lexicon for Tibetan translators from the early ninth century onwards.[33] The question reads:

> In the *Sārasvata* commentary by Jo-naṅ [Tāranātha it is stated that]:
> 'A distinction is made between the *Veda* and the *Śāstra*, and no matter whether or not one holds that the *Veda* is self-arisen, the root texts [of the *Veda*] are primordial basic texts, but the *Śāstras* [contain] exposés on the meaning of these [*Veda* texts]. [In] what [sense] are [the terms] 'self-arisen' and 'primordial' intended here?
> Moreover, as regards the Vedas under consideration here, is it correct that they are four in number, viz. *Ṛg-*, *Yajur-*, *Sāma-* and *Atharva-*[*veda*]?
> Of these, what kind [of text] is the *Ṛg-*[*veda*]? And, as regards the *Yajur-*[*veda*], does it also contain [materials] harmful to living beings, or is it only beneficial?
> As regards the *Sāma-*[*veda*], as it is evident that [*Kāvya-*]*Ādarśa* etc. must be a *Śāstra* [i.e. poetics], is it so that of the *Vedas* this *Sāma-*[*veda*] in the literary tradition of the Brahmins is similar [?; scil. to the poetics described in *Kāvyādarśa*?]?
> I would like to request clarification on what is [said on these matters] in the basic texts. (Appendix 10)

The passage in Tāranātha's *Sārasvata* commentary which is mentioned in the question, can be found in the section on the *kārakas*, in the treatment of the dative case and the syntactic function of indirect object that this case indicates. In this connection Anubhūti's autocommentary on *Sārasvata* gives the examples *vedavide gāṁ dadāti*, 'To the man who knows the *Vedas* he gives a cow', and *chāttrāya kanyāṁ dadāti*, 'To the scholar he gives a daughter.' Anent these examples Tāranātha explains the traditional distinction between the Vedic and Śāstric specialists among the Brahmins, and it is in this passage that the phrase quoted here occurs.[34]

Si-tu's answer:

> As the Brahmins hold that the *Veda* was not made by man, and has always existed, in their tradition it is taken as the basis, being established as the primordial text. However, as the Buddhists etc. do not hold this [view], [they consider it as] established as a primordial text, in the sense that this

text came into existence in the far past.

What is known as 'the *Veda*' consists of the set of four, *Ṛg-* etc. Of these [the *Veda* of] the 'definite statement' [lit. trl. of *ṅes-brjod*], i.e. *Ṛg-veda*, appears to be [a text] teaching many methods of propitiating the gods, and [the *Veda* of] the 'sacrificial offering', i.e. *Yajur-veda*, is for the most part harmful to the living beings, through the five great sacrificial offerings etc., the most important of which is the **āhuti*[35] [i.e. the *homa* sacrifice[36]], and regarding [the *Veda* of] the 'pleasant words' / 'song', i.e. *Sāma-veda*, among many [scholars] known as the wise of Tibet, a misconception has arisen, namely that [this term *Sāma-veda*] means 'poetry';[37] this is not [the case]. If one asks, what [the *Sāma-veda*] is like, it is a basic text consisting of well-composed and pleasant verses on a variety of topics, like a praise of the habit of singing the six-*Oṁ*-syllable [*mantra*] [?], and an exposé of the true nature [?], etc., and many such [collections] appear to have existed. (Appendix 11)

Although the question refers to four *Vedas* and Si-tu mentions the number four, he enumerates only three, omitting the *Atharva-veda*. The interest of the Tibetan scholiasts in the form of their standpoint heterodox views of non-Buddhist traditions in India is shown by the attention given to these views in certain philosophical compendiums (termed *Grub-mtha'*). For example, Dbus-pa Blo-gsal (14th cent.) devotes almost one third of his famous *Grub-mtha'* to the 'heterodox' philosophies of the *Lokāyata*, *Sāṁkhya*, *Śaiva*, *Vaiṣṇava* and *Jaina* traditions.[38] It seems probable that Dbus-pa Blo-gsal did not consult any brahmanical or similar heterodox textual sources directly; he appears to quote such sources only second-handedly, taking his quotes from quotations incorporated in Buddhist *śāstras*.[39] I am not so sure about this with regard to Si-tu; he would seem to have had first-hand knowledge of a variety of non-Buddhist texts. As a case in point, note that Si-tu has translated a chapter from a 14th/15th-century Pāṇinian commentary, dealing specifically with grammatical peculiarities of Vedic Sanskrit. This translation, included in his collected works,[40] describes in considerable detail syntactic, morphological and morphophonemic idiosyncracies of the Vedic textual materials.

This brief investigation of some paragraphs from this fascinating collection of replies by the major Tibetan grammatical authority Si-tu is a mere first presentation of the wealth of materials contained therein and elsewhere in his collected works. I hope it has in any case given an impression of the wide range of linguistic disciplines belonging to the domain of the elite of classical Tibetan scholastics, exemplified here by the versatility and broad orientation of the eminent grammarian Si-tu Chos-kyi-'byuṅ-gnas. As such this article may hopefully serve as a fitting tribute to a similarly versatile scholar.

APPENDICES

1./yaṅ-/‘grel-pa-de-raṅ-gi-dpe’i-dbu-logs-g’yas-su-dbyaṅs-can-ma’i-sku-bris-pa’i-phyag-[2r2:]mtshan-dbyug-pa-rtse-mo-gñis-’khyog-po’i-ṅos-la-bum-pa-bciṅs-pa-’dra-ba-gaṅ-yin-daṅ-/de’i-dgos-pa-de-’dra-naṅ-pa’i-rgyud-las-bśad-min-sogs-ji-ltar / zhes-pheb-pa /

2. phyag-mtshan-gyi-gzugs-brñan [?] -de-ni-dbyaṅs-can-ma’i-phyag-mtshan-yoṅs-grags / tambu-ra’am / sgra-sñan-gyi-dbyibs-kyi-khyad-par- [2r3:] te / rgya-gar-daṅ-bal-po-sogs-su-’di-lta-bu’i-rnam-pa-bris-pa-kho-bos-kyaṅ-maṅ-du-mthoṅ-lags /

3. dus-’khor-gyi- <ḥ>k<ḥ>kā / sogs-ni-brda-sprod-pa’i-bstan-bcos-sogs-la-ma-grags-pa’i-thun-moṅ-min-pa’i-brdar-’dug-pas/...lam-tsam-gsuṅ-rgyun-la-dpag-na / [6r1:] / gsal-byed-mi-’dra-brtsegs-gyur-na // raṅ-sgra-thon-tsam-myur-bar-bklag(/)ces-bśad-pa’i-’ol-spyi-tsam-las-gzhan-bya-thabs-ma-mchis-sñam-lags /

4./yaṅ-/mkha’-’gro’i-brda-skad-daṅ-śa-za’i-skad-sogs-sṅags-rgyud-la-grags-pa-rnams-kyaṅ-bklag-tshul-legs- [6r2:] sbyar-daṅ-mthun-nam / mi-mthun-na-zur-du-bklag-tshul-ji-’dra-mchis-daṅ-de-bzhin-skag-zlog-rgya-nag-nas-byuṅ-bar / oṃ-dar-du [or: ṅu ?] -zhes-pa-daṅ- / ma-ma-sṅo-ta-sṅo-ta-sogs-legs-sbyar-ltar-ram/bod-daṅ-mthun-pa-sogs-’don-lugs-ji-ltar-dgos/spyir-’di-sogs-rgya-nag-nas-byuṅ-ba’i-sṅags- [6r3:] rnams-skad-rigs-gaṅ-’dra-zhig / ces-pheb-pa /

5. brda’i-skad-brtag-gñis-las-’byuṅ-ba’i / ma-da-na-chaṅ-ba-la-śa / zhes-pa-lta-bu-ni-legs-sbyar-gyi-bklag-tshul-ñid-las-ma-’das-śiṅ- / śa-za’i-skad-zlos-gar-gyi-bstan-bcos-su-’byuṅ-ba-sogs-kyaṅ-de-daṅ-mtshuṅs-śiṅ- / rñiṅ-ma’i-rgyud-las-’byuṅ-ba- [6r4:] ‘ga’-zhig-ni-’ol-spyi-tsam-du-rgya-bod-gaṅ-daṅ-mthun-par-’dug-pa-de-ga-ltar-bklag-ciṅ-gaṅ-du-yaṅ-mi-mthun-pa-der-ni-klog-tshul-dmigs-bsal-med-pa-rnams-la-mos-pa-tsam-gyis-klog-pa-las-’os-ma-mchis-sñam-zhiṅ- / rgya-nag-nas-byuṅ-ba’i-sṅags-rgya-gar-nas-bsgyur-ba-rnams-kyaṅ-rgya-nag-yi-ger-ji-bzhin-’god-ma- [6r5:] thub-ciṅ- / ṅag-gis-ma-’khyor-bas-śer-sñiṅ-gi-sṅags-la / dṅan [?] -gyas-ti-gyas-ti-po-lo-gyas-ti-zhes-’don-pa-daṅ- / na-mo-a-mi-ta-bhā-ya-la / nan-bū-o-mi-tha’o-hwo / / zhes-’don-pa-lta-bu-ma-dag-pa-kho-nar-snaṅ-bas-yul-de-nas-bsgyur-ba’i-sṅags-’di-rnams-la-’di-ltar-klog-dgos-zhes-kha-tshon- [6r6:] gcod-mi-nus-mod / ‘on-kyaṅ-de-skabs-bod-kyi-lo-tsā-ba-rnams-kyis-klog-tshul-yig-sdeb-daṅ-mthun-par-byas-pa-śas-che-’dra-bas-bod-klog-ltar-byas-kyaṅ-’gal-ba-med-dam-sñam-lags /

6.ba-hir.-ni-rgya-gar-gyi-yul-miṅ-ma-yin-yaṅ-/rgya-gar-pa-bod-du-phyin-pa-[3v5:] rnams-kyis-rgya-gar-la-ba-hir-te-phyi-rol-daṅ- / bod-la-antar.-te-naṅ-du-'bod-par-snaṅ-ba/dper-na/rgya-nag-gi-mi-bod-du-phyin-pa-dag-gis-rgya-nag-la-wa'i-thu-ste-phyi-daṅ- / bod-la-lī-thu-ste-naṅ-du-'bod-par-snaṅ-ba-bzhin-no /

7. dbyaṅs-a-i-u-ṛ-ḷ-rnams-la-thuṅ-ṅu-riṅ-po-plu-ta-gsum-gyis-bsgyur-ciṅ- / ḷ-la-riṅ-po-med-pas-de-phri-ba'i-bcu-bzhi / de-rnams-u-dātta-sogs-gsum-gyis-bsgyur-bas-bzhi-bcu-rtsa-gñis/mtshams-sbyor-gyi-yi-ge-bzhi-la-thuṅ-ṅu-med-pas-gzhan-gñis-kyis-bsgyur-ba'i-brgyad-la / u-dātta-sogs- [7r5:] kyis-bsgyur-bas-ñi-śu-rtsa-bzhi / de-dag-bsdoms-pas-drug-cu-rtsa-drug / de-rnams-sna-ldan-yin-min-gñis-kyis-bsgyur-bas-dbyaṅs-kyi-dbye-ba-brgya-so-gñis-su-'gyur-ba-ni-tsāndra-pa-daṅ- / dbyaṅs-can-ma-sogs-bzhed-pa-mtshuṅs-par-'dug-pas / skabs-de'i-gcig / gñis / gsum-gyi-yun-tshad-rnams-ni-thuṅ- [7r6:] riṅ-śin-tu-riṅ-po-rnams-daṅ-sbyar-zhiṅ- / yun-tshad-phyed-pa-ni / gsal-byed-dbyaṅs-med-rnams-so / thuṅ-ṅu-la-sogs-pa-rnams-u-dātta-sogs-su-'gyur-ba'i-tshul-ni / dbyaṅs-de-rnams-gsal-byed-daṅ-sbyar-nas-brjod-pa'i-tshe / gsal-byed-sgra-drag-pa-rnams-daṅ-phrad-pas-u-dātta-ste-mtho-bar-'don-pa-daṅ-/[7v1:]zhan-daṅ-phrad-pas-a-nu-dātta-ste-dma'-bar-'don-pa-daṅ- / bar-ma-daṅ-phradpas-swa-ri-ta-ste-bar-mar-'don-pa-rnams-su-'gyur-zhiṅ- / [supralinear gloss: dbyaṅs-kho-na-brjod-pa'i-tshe'aṅ-sgra-drag-zhan-bar-ma-las-'gyur-te /] de-ltar-'gyur-ba'aṅ- / rgyal-dbaṅ-gis-brtsams-pa'i-dbyaṅs-kyi-mdor-zhib-mor-bśad-yod-par-'dug-kyaṅ-bod-du-ma-byuṅ-la-de'i-luṅ-'dren-'ga'-zhig-'chi-med-mdzod-kyi-'grel-pa- [7v2:] 'dod-'jor-yod-pa'i-tshig-zur-las-śes-par-nus-śiṅ- / pā-ṇi-pa'i-byiṅs-kyi-mdos-kyaṅ-phyogs-tsam-go-mod / zhib-mo'i-rnam-gzhag-ni-de-ston-pa'i-gzhuṅ-daṅ-'grel-pa-sogs-la-rag-las-śiṅ-de-'dra-bod-du-sṅar-ma-'gyur-bas-ji-bzhin-'chad-par-mi-nus-so /

8. dbyaṅs-gzhuṅ-gi / dzñīṃ-sdeb-sbyor-la'o / / zhes-pa'i-skabs-kyi-yig-gzugs / [*anunāsika* graph] rgya-yig-ṅo-ma-lags-sam / min-na-rgya-yig-raṅ-du-byas-pa'i-dbyibs-kyaṅ-ma-'tshal-zhiṅ-/jo-'grel-du-sgyiṅ-zer-ba'am/sgya-yaṅ-zer-ba-ltar-bklag-dgos-par-gsuṅs-kyaṅ- / [7v6:] 'gyiṅ-zhes-pa-bzhin-klog-mkhan-maṅ-na-sṅa-ma-ltar-lags-sam /

9. dzñīṁ / zhes-bod-kyi-klog-tshul-la-bsams-nas-bris-par-'dug-ciṅ- / dzñā-yig-gi-klog-tshul-ji-lta-ba'aṅ-bod-pas-mi-thon-par-snaṅ-bas / jo-'grel-du-sgyiṅ-zhes-dper-brjod-pa-mdzad-la-rgya-gar-pa'i-ṅag-sgros [?] -la-brtags-tshe-da-duṅ-de'aṅ- [8r1:] cuṅ-min-pa-zhig-'dug-kyaṅ-ṅag-ñid-las-mkhyen-par-bya-dgos-kyi-dper-brjod-kyis-rnam-pa-gsal-mi-nus-par-snaṅ-ṅo-//de'i-yig-gzugs-kyaṅ-mi-mthun-pa-gzhan-yod-srid-kyaṅ- / na-ga-ra'i-yi-ger-'byuṅ-ba-ni / [four-part mark] 'di-lta-bu'o /

10. jo-naṅ-dbyaṅs-can-ma'i-'grel-par-rig-byed-daṅ-bstan-bcos-tha-dad-de / rig-byed-raṅ-byuṅ-'dod- [13r2:] pa'am / mi-'dod-kyaṅ-rtsa-ba-gdod-ma'i-gzhuṅ-yin-bstan-bcos-kyaṅ-de'i-don-'chad-pa-zhes-pa'i-raṅ-byuṅ-daṅ-/gdod-mar-'jog-tshul-gaṅ- / 'jog-pa'i-rig-byed-kya-ṅes-brjod / mchod-sbyin / sñan-tshig / srid-sruṅ-zhes-bzhir-bgraṅs-pa-ñid-yin-min / de'i-ṅes-brjod-ji-lta-bu-[13r3:] daṅ- / mchod-sbyin-yaṅ-sems-can-la-gnod-pa-daṅ-bcas-pa'am-dge-ba-can-kho-na-lags / sñan-tshig-la-me-loṅ-sogs-ni-bstan-bcos-yin-dgos-par-snaṅ-na-rig-byed kyi sñan-tshig-de-bram-ze-rnams-kyi-ṅag-rgyun-lta-bur-yod-dam / gzhuṅ-du-yod-pa'i-gsal-kha-yaṅ-zhu-lags /

11. bram-ze- [13r4:] rnams-kyi-rig-byed-ni-skyes-bus-ma-byas-pa / rtag-par-byuṅ-du-'dod-pas-lugs-de-la-gdod-ma'i-gzhuṅ-du-grub-pa-gzhir-bzhag-nas / naṅ-pa-sogs-kyis-de-ltar-mi-'dod-kyaṅ-dus-ches-sṅa-ba-zhig[or: sṅa-bzhig?]-nas-gzhuṅ-de-byuṅ-bas-gdod-ma'i-gzhuṅ-du-grub-ciṅ- / rig-byed-ces-pa'aṅ-ṅes-brjod-sogs-bzhi-po-ñid-du- [13r5:] ṅes-la / de'i-ṅes-brjod-ni / ṛgbe-da-ste-lha-mñes-par-byed-pa'i-thabs-du-ma-ston-pa-zhig-yin-par-snaṅ-zhiṅ-/mchod-sbyin-ni-yā-yurbe-da-ste-sreg-blugs-gtso-bor-gyur-pa'i-mchod-sbyin-chen-po-lṅa-sogs-sems-can-la-'tshe-ba-can-śas-che-ba-lags-śiṅ- / sñan-tshig-ni / sā-ma-be-da-zhes-bya-ba-ste / bod-kyi- [13r6:] mkhas-par-grags-pa-maṅ-por-yaṅ-de-sñan-ṅag-la-'dod-pa'i-'khrul-pa-byuṅ-'dug-pas / de-ni-ma-yin-la / 'o-na-ci-lta-bu-zhe-na / oṃ-yig-drug-glur-blaṅs-nas-'dren-dgos-pa'i-bstod-pa-daṅ- / de-kho-na-ñid-ston-pa-sogs-don-du-ma-can-tshigs-su-bcad-pa-sdebs-legs-śiṅ-sñan-pa'i-gzhuṅ-zhig-ste-'di- [13v1:] 'dra-du-ma-'byuṅ-bar-snaṅ-ṅo- /

REFERENCES

Allen, W.S.

1962 *Sandhi. The theoretical, phonetic and historical bases of word-junction in Sanskrit*. The Hague/Paris: Mouton. (^{2}1972) (Janua Linguarum, Series Minor. 17)

Beguin, G.

1977 *Dieux et démons de l'Himâlaya. Art du Bouddhisme lamaïque [Grand Palais 25 mars - 27 juin 1977]*. Paris: Editions des musées nationaux

Belvalkar, S.K.

1915 *An account of the different existing systems of Sanskrit grammar, being the Vishwanath Narayan Mandlik gold medal prize essay for 1909*. Poona

Causemann, M.

1994 'Der *'Dzam-taṅ Dkar-chag* der gesammelten Werke des Rje-btsun

Tāranātha', *ZAS* 24, pp. 79-112

Chandra, L. (ed.)

1968 *The Autobiography and Diaries of Si-tu paṇ-chen. With an introduction by E. Gene Smith*, New Delhi. (Śata-Piṭaka Series. 77)

Chatterji, K.C. (ed.)

1953-61 *Cāndravyākaraṇa of Candragomin*, part 1 (Chapters 1-3), Poona 1953, part 2 (Chapters 4-6). Poona 1961

Dargyay, E.M.

1977 *The Rise of Esoteric Buddhism in Tibet*. Delhi etc.: Motilal Banarsidass.

Karmay, S.G.

1988 *The Great Perfection (Rdzogs chen). A Philosophical and Meditative Teaching in Tibetan Buddhism*. Leiden etc.: Brill

Meisezahl, R.O.

1965-6 'Über *jñīṃ* in der tibetischen Version der Regel *ṁ chandasi* der *Sārasvata*-Grammatik', *IIJ* IX, pp. 139-46

Mimaki, K.

1982 *Blo gsal grub mtha'. Chapitres IX (Vaibhāṣika) et XI (Yogācāra) édités et Chapitre XII (Mādhyamika) édité et traduit*. Kyoto

Mimaki, K. & A. Akamatsu

1985 'La Philosophie des Śaiva vue par un auteur tibétain du 14e siècle', in: M. Strickmann (ed.) *Tantric and Taoist Studies in honour of R.A. Stein - volume 3*. Bruxelles, pp. 746-72. (Mélanges Chinois et Bouddhiques XXII)

Monier-Williams, M.

1899 *A Sanskrit-English Dictionary. Etymologically arranged with special reference to cognate Indo-European languages*. Oxford:

Renou, L.

1952 *Grammaire de la langue Védique*. Lyon-Paris: IAC. (Collection 'Les Langues du Monde')

Rhie, M.M. & R.A.F. Thurman

1991 *Wisdom and Compassion. The Sacred Art of Tibet*. New York: Abrams

1996 *Weisheit und Liebe. 1000 Jahre Kunst des tibetischen Buddhismus*. Bonn: Kunst- und Austellungshalle der Bundesrepublik Deutschland

Ruegg, D.S.

1995 *Ordre spirituel et ordre temporel dans la pensée Bouddhique de l'Inde et du Tibet. Quatre conférences au Collège de France*. Paris: Boccard. (Publications de l'Institut de Civilisation Indienne, Série in-8o, Fasc. 64)

Sherab Gyaltsen (ed.)

1990 *Collected Works of the Great Ta'i si tu pa kun mkhyen chos kyi byun gnas bstan pa'i nyin byed*. Sansal/Delhi: Palpung Sungrab Nyamso Khang Sherab-ling Institute of Buddhist Studies

Snellgrove, D.L.

1987 *Indo-Tibetan Buddhism - Indian Buddhists and Their Tibetan Successors*. London:

Tillemans, T.J.F. & D.D. Herforth

1989 *Agents and Actions in Classical Tibetan. The Indigenous Grammarians on Bdag and Gzan and Bya byed las gsum*. Wien. (Wiener Studien zur Tibetologie und Buddhismuskunde. 21)

Uhlig, H.

1995 *On the Path to Enlightenment. The Berti Aschmann Foundation of Tibetan Art at the Museum Rietberg Zürich*. Zürich.

Verhagen, P.C.

1993 '*Mantras* and Grammar. Observations on the study of the linguistical aspects of Buddhist "esoteric formulas" in Tibet', in: K.N. Mishra (ed.) *Aspects of Buddhist Sanskrit (Proceedings of the International Symposium on the Language of Sanskrit Buddhist Texts, Oct. 1-5, 1991)*. Sarnath: Central Institute of Higher Tibetan Studies, pp. 320-46. (Samyag-Vāk Series. 6)

1994 *A History of Sanskrit Grammatical Literature in Tibet. Volume 1: Transmission of the Canonical Literature*. Leiden etc.: Brill (Handbuch der Orientalistik Abt. 2 Bd. 8).

1995 'Studies in Tibetan Indigenous Grammar (2): Tibetan phonology and phonetics in the *Byis-pa-bde-blag-tu-'jug-pa* by Bsod-nams-rtse-mo (1142-1182)', *Asiatische Studien / Etudes Asiatiques*, XLIX.4, pp. 943-68

1996 'Tibetan Expertise in Sanskrit Grammar [2]: Ideology, Status and other Extra-linguistic Factors', in: Jan E.M. Houben, (ed.) *Ideology and Status of Sanskrit: Contributions to the History of the Sanskrit Language*. Leiden etc.: Brill, pp. 275-87. (Brill's Indological Library 13)

fc. 'Tibetan *Klog-thabs*, "Manuals of Pronunciation"', in: G.L. van Driem (ed.) *Himalayan Linguistics*. Berlin: Mouton de Gruyter, 1997 (Trends in Linguistics)

fc. '38. The influence of the Sanskrit tradition on Tibetan indigenous grammar.', in: S. Auroux, K. Koerner, H.J. Niederehe & K. Versteegh (eds) *Geschichte der Sprachwissenschaften. Ein Internationales Handbuch zur Geschichte der Sprachforschung von den Anfängen bis zur Gegenwart*, 3 vols. Berlin-New York: Walter de Gruyter, 1998?, (Handbücher zur Sprach- und Kommunikationswissenschaft)

Whitney, W.D.
[1]1879 *Sanskrit Grammar, including both the classical language and the older dialects of Veda and Brahmana*. Leipzig. ([2]1889) 1).

Notes

1. Cf. Snellgrove, 1987, pp. 122; 141-4.
2. On account of its special nature the translating of the esoteric Vajrayāna literature was more restricted and stood under stricter supervision than the rendering of the Mahāyāna literature, cf. e.g. Karmay, 1988, pp. 5-6; Verhagen, 1996, pp. 285-6.
3. Cf. e.g. Snellgrove, 1987, p. 143.
4. Work on these materials published or in press thus far: Verhagen, 1993; 1995, (forthcoming A), (forthcoming B). The main results of this investigation, however, will be included in the second volume of *A History of Sanskrit Grammatical Literature in Tibet* which I am currently preparing.
5. Verhagen, 1994, describes forty-seven such titles; sixteen (possibly twenty) of these belong the *Cāndra* system, and thirteen to *Kātantra*.
6. These materials will be described in the monograph I am currently preparing, under the projected title *A History of Sanskrit Grammatical Literature in Tibet. Volume 2: Assimilation in Indigenous Scholarship*, to appear at Brill's, in Handbuch der Orientalistik Abt. 2.
7. On the importance of certain secular sciences in the context of Mahāyāna Buddhism, specifically in the Tibetan traditions, cf. Ruegg, 1995, in particular the second part, pp. 93-148, under the title 'Sciences religieuse et sciences séculières en Inde et au Tibet: *Vidyāsthāna* Indo-bouddhiques et *Rig gnas* Indo-tibétains - Remarques sur la nature et les finalités des études Indo-tibétaines'.
8. On the historical figure Si-tu Chos-kyi-'byuṅ-gnas and his importance as a scholar, cf. Smith introd. Chandra, 1968, pp. 7-12; 15-17, Tillemans & Herforth, 1989, pp. 2-3; 8-10, Verhagen, 1994, pp. 192-3; 199; 215-6.
9. Facs. ed. Sherab Gyaltsen, 1990-1998; N.B. correct the order of pages in that edition to: pp. 377-84; 323-6; 389-94; 333-4; 397-436; 375/376.
10. The print was known to the redactors of an edition of Tāranātha's collected works in 'Dzam-thaṅ monastery, but, again, we do not know the date of that either; cf. Causemann, 1994, pp. 79-80.
11. Cf. Rhie & Thurman, 1991, p. 135, no. 27, and ibidem in the expanded German version (1996). This interesting painting is also discussed in David Jackson's review of this catalogue (*WZKS* 37, 1993, pp. 117-8), where he draws our attention to the association with grammatical and linguistic studies evident from the six smaller figures that are depicted. He tentatively suggests a possible identification of the most prominent, i.e. the top central figure as the famous Sanskrit grammarian Candragomin. I propose another candidate: could this perhaps be a depiction of Anubhūti-svarūpācārya, the author of Sārasvata-vyākaraṇa, Sarasvatī's eponymous grammar?
12. Cf. e.g. N.N. 1977, p. 267, no. 342.
13. Cf. e.g. Uhlig, 1995, p. 157, no. 102.

14. Note that <*ḥ*> is my notation for the *upadhmānīya* form of the *visarga*.
15. This quote possibly, with variation in the first two syllables only, stemming from a pronunciation-manual by Si-tu himself, Coll. Works vol. 10, *Legs-par-sbyar-ba'i-skad-kyi-klog-thabs-ñuṅ-ṅu-rnam-par-gsal-ba*, f.5v5: / *der-yaṅ-mi-'dra-brtsegs-gyur-na* / / *raṅ-sgra-thon-tsam-myur-bar-bklag* / (ed. Sherab Gyaltsen, 1990, p. 134).
Comparable statements in other *klog-thabs*:
(a) Tāranātha (1575-?), *Rgya-skad-du-bris-pa'i-yi-ge'i-klog-thabs-'phags-yul-mkhas-pa'i-lugs*, f.2r6-v1:*gsal-byed-'dra-daṅ-mi-'dra-gaṅ-brtsegs-kyaṅ-/phal-cher-steṅ-rnams-sgra* [?] *-thon-myur-bar-bklag* /
(b) A-mes-zhabs Ṅag-dbaṅ Kun-dga' Bsod-nams (1597-1659) [?], *Legs-sbyar-klog-tshul-gyi-bstan-bcos-blo-gsal-kun-dga'-ba-zhes-bya-ba*, f. 2v1 / 345v1: *brjod-du-mi-ruṅ-* [*ba-*] *maṅ-* [*po-*] *daṅ-ñuṅ-* [*du*] / *brtsegs-pa-* [*rnams-ni-yi-ge-raṅ-*] *raṅ-* [*gi-*] *sgra-dran-* [*pa-ste-don-*] *tsam-* [*du-*] *bklag* /
(c) Dṅul-chu Dharmabhadra (1772-1851), *Sṅags-kyi-yi-ge'i-klog-tshul-dṅos-bstan-pa*, f. 2r6: / *gsal-byed-mi-'dra-maṅ-ñuṅ-du-brtsegs-kyaṅ-* / / *so-sor-raṅ-sgra-thon-tsam-myur-bar-bklag*.
16. Cf. Whitney, 1889, par. 67, Renou, 1952, par. 13, 143, Allen, 1972 pp. 57; 78.
17. E.g. Bsod-nams-rtse-mo's *Byis-'jug* 3.58-59, Bu-ston's *bśad-thabs* ad *Kālacakra-tantra*, f. 8v2-6, A-mes-zhabs' *Blo-gsal-kun-dga'-ba* f. 2r3-5 = 345r3-5.
18. Cf. e.g. Dargyay, 1977, p. 88.
19. The full *mantra*: *Oṁ gate gate pāragate pārasaṁgate bodhi svāhā*.
20. Cf. for instance also *bāhīka* and *bāhya* occurring as names of peoples, Monier-Williams, 1899, pp. 730-1.
21. Cordial thanks are due to Dr. P. Harrison (Christchurch, New Zealand) for kindly supplying the information on the Chinese terminology.
22. Facs. ed.: Chandra, 1968.
23. Smith introduction to Chandra, 1968, p. 11.
24. Alternative translation: 'When these vowels are pronounced in combination with consonants, when combined with consonants with a 'strong' sound, they are *udātta*, i.e. 'with high pronunciation'; when combined [with consonants] with a 'weak' sound, they are *anudātta*, i.e. 'with low pronunciation'; and, when combined [with consonants] with a 'middle' sound, they are *svarita*, i.e. 'with middle pronunciation'. 'This alternative interpretation, taking the term *gsal-byed* in its usual meaning of 'consonant', could, for instance, be taken as pointing to a fundamental misunderstanding on the side of Si-tu, namely that the Vedic accent somehow was determined by specific phonological features of the initial consonant in the syllable.
25. Editions: Belvalkar, 1915, p. 117, Chatterji, 1953, part 2, pp. 394-5.
26. The Sanskrit as reconstructed from the bilingual canonical version and the Tibetan translation: *ṁ chandasi* / *chandasy anusvāro ṁkāram āpadyate* / *śa-ṣa-sa-ha-repheṣu parataḥ* / *vayaṁ somaḥ* / *vayaṁ somaḥ* / *dzññ̄ṁ-sdeb-sbyor-la'o* / *sdeb-sbyor-las* [emend *la*] *-rjes-su-ṅa-ro-dzññ̄ṁ-gi-yi-ger-* [add *'gyur-* ?] *te* / *śa-ṣa-sa-ha-re-pha-rnams-pha-rol-la'o* /, in Tāranātha's translation CG 31, Peking *Bstan-'gyur* vol. *no* f. 310r3-4); cf. also CG 43 Peking *Bstan-'gyur* vol. *pho* f.

198v6 & CG 44 Peking *Bstan-'gyur* vol. *pho* 213r1-2.

27. Designated as CG 31, 43 and 44 in Verhagen (1994).
28. *Dbyaṅs-can-brda'-sprod-kyi-'grel-pa-mchog-tu-gsal-ba-zhes-bya-ba* occupying all of volume 14 of Tāranātha's collected works, and *Mtsho-ldan-ma'i-brda-sprod-gzhuṅ-gi-'grel-pa-legs-bśad-ṅag-gi-'od-zer*, which is not contained in his collected works yet preserved separately (cf. Ruegg, 1995, p. 123; unfortunately the latter text is not accessible to me at the moment of writing.
29. *Mtsho-ldan-ma'i-brda-sprod-pa'i-rab-byed*, collected works vol. 7 ff. 1-91.
30. (1965-6); his investigation did not include the work of Si-tu.
31. *thon* = instead of, or a form of *'don* 'to utter'?
32. Tāranātha, *Sārasvata* comm., *Bka'-'bum* vol. 14 f. 34v1-2: *'di-ni-rig-byed-las-'byuṅ-ba'i-yi-ge-logs-pa-zhig-la-'bod-tshul-sgyi ṅ-zer-ba'am / sgya-yaṅ-zer-ba-'dra-bas / dzñīṅ-gi-gzugs-su-'bri-ba-yin-gyi / dza-ña-brtsegs-pa-la-i-zhugs-pa-min-no /*
33. *Mahāvyutpatti* 5047-5050 give *ṅes-brjod*, *mchod-sbyin*, *sñan-tshig* and *srid-sruṅ(s)* as translations for *ṛc*, *yajus*, *sāman* and *atharva* respectively.
34. Tāranātha, *Sārasvata* comm., *Bka'-'bum* vol. 14 f. 155r4-6: *rig-byed-daṅ-bstan-bcos-ces-tha-dad-de / rig-byed-raṅ-byuṅ-'dod-pa'i* [em.: *pa'aṁ*] */ mi-'dod-kyaṅ-rtsa-ba-gdod-ma'i-gzhuṅ-yin / bstan-bcos-de'i-don-'chad-pa'aṁ / min-kyaṅ-raṅ-blos-'god-pa'aṁ / thos-pa-rjes-su-sgrog-pa-yin-te / lta-ba-drug-daṅ- / yan-lag-drug-daṅ- / brda'-sprod-pa-la-sogs-pa-bstan-bcos-bcu-bzhir-grags-pa-la-sogs-pa'o / ces-'dod-do / da-lta-yaṅ-bram-ze-rig-byed-pa-daṅ- / bstan-bcos-pa-zhes-so-sor-grags-pa-yod-do /* 'A distinction is made between the *Veda* and the *Śāstra*, and no matter whether or not one holds that the *Veda* is self-arisen, the root texts [of the *Veda*] are primordial basic texts, but the *Śāstra*s [contain] exposés on the meaning of these [*Veda* texts], and, even if they do not, they are creations of the own intellect and instructions secundary to the *śruti*. [Such *Śāstras* are] inter alia the six *Darśanas*, the six [*Veda*-]*aṅgas*, and what is famed as the fourteen *Śāstras*: grammar, etc. Even today Brahmins are classified as "Vedic" or "Śāstric".'
35. Tib. *sreg-blugs*; cf. *Mahāvyutpatti* 4253 *āhuti-dravyam* = *(b)sreg-(b)lug / glug-gi-rdzas.*
36. Cf. *Mahāvyutpatti* 4245 *homaḥ* = *sbyin-sreg.*
37. Or 'poetics', Tib. *sñan-ṅag*, i.e. **kāvya.*
38. Folios 10v5-53r5 from a text totalling 127 folios, cf. Mimaki, 1982 p. 18.
39. Cf. Mimaki & Akamatsu, 1985, p. 750.
40. *Phā-ṇi-pa'i-rig-byed-sgra-sgrub*, [sub *Sdeb-sbyor-rnams-kyi-mtshan-ñid-thos-pas-chub-pa*], f. 9v1-13v6. coll. works vol. 7, ed. Sherab Gyaltsen, 1990-1997, pp. 482-90, a translation of the *Vaidika-prakriyā* chapter of Rāmacandra's *Prakriyākaumudī.*

XXXII

ON SYNTACTIC AND SEMANTIC CONSIDERATIONS IN THE STUDY OF RITUAL

Henk J. Verkuyl

INTRODUCTION

To my own surprise, this paper deals with Staal's work on ritual, in particular with Staal (1989) rather than e.g. his work on Indian logic and linguistics, as e.g. Staal (1988). I know virtually nothing about Indian rituals except for what I read about them in Staal (1983; 1989), so I will not say very much about them. However, one of Staal's leading hypotheses (roughly) says that rituals should be studied as rules without meaning. It is even a motto capturing a leading idea in his work on Indian rituals and mantras, namely that one should study them without appealing to meaning. This thesis presupposes a strict distinction between syntax and semantics. This strictness shows up in the possibility to study syntax without any appeal to semantics, whereas the rverse is not possible, of course, because the meaning of an expression is dependent on the expression.

My contribution to this Festschrift will try to take away some of the naturalness of the assumption that a strict distinction can be made between form and meaning, and thus between syntax and semantics. Not that I would like to blur the distinction in every situation in which we have a natural or formal language and a model in which it is interpreted, but recent developments in philosophical logic as appied to natural languages suggest that it seems possible to have less strict distinction which may effect a better understanding of the way language is a part of our cognitive organization. I have no idea how much of what I am going to say really bears on the main hypotheses of Staal's work on ritual. After all he underscores the point that he considers ritual as an activity rather than as a language. But here, I think, one should be careful: only if language is defined as a system of forms and their meanings, may one claim that rituals are *not* languages on the ground that they are just systems of meaningless forms. In that sense, it becomes natural to focus on rituals as systems of activity. Yet, the very fact that Staal attributes a syntax to ritual implies that it is possible to attach meanings to its expressions even if the meaning of the forms is reduced to their own form itself. One may study syntax without semantics, but that does not mean that there can be no meanings. The basic question becomes whether there are sufficiently developed theories of meaning to provide meanings to apparently meaningless forms, or whether the forms are assigned a sort of 'zero-meaning' as in Chomsky's work. Staal (1989, p. 137) says that 'The chief provider of meaning being religion, ritual became involved with religion and through this association, meaningful.' What I will suggest is that if Staal's thesis should be modified or rejected, the opposition should not come from those who locate meaning in the area of religion and belief, but rather in deeper lying principles of cognitive organization.

Although this position might surprise him - the study of logical systems is not normally associated with the study of cognition - it brings Staal's thesis to the very heart of semiotics, the study of formal languages, and it is from this point of view that I would like to approach his thesis.

TWO MODELS, TWO STRATEGIES

There are two well known models in terms of which the relation between a teacher and his pupil can be characterized: the Socrates/Plato-model (the pupil basically agrees and tries to explain what his master taught), and the Plato/Aristotle model (the pupil disagrees and tries to reject virtually everything his teacher said). Now, Frits Staal took my final examination in logic and the philosophy of language, but he did not teach me in these fields. He taught me as the leader of the famous work group - Staal in Amsterdam in the early sixties. He was the central stimulating person and having had a mathematical training he could explain many formal aspects of the transformational-generative syntax of that period to linguistic students. During the year 1965/1966 he taught me also extra-curricularly as one of the members of an illustrous pair of teachers, Frits Staal and Richard Montague, jointly and harmoniously chairing another working group in which generative grammar and Montague grammar (not yet so called) were systematically compared. The double loyalty of a pupil to a pair of teachers makes it virtually impossible to fully apply any of the two models to one of them, certainly because the teachers did not agree: Staal was on the generative side and Montague on his own side, and there was a huge abyss between these two approaches, certainly in that period.

It is in this context, that with regard to Staal's purely syntactic approach,

I would like to operate on the Plato/Aristotle-side, and on the Socrates/Plato-side with respect to Montague. Yet, with regard to Staal I shall operate on the Socrates/Plato side as well, simply because I work in a generative framework albeit combined with Montagovian model-theoretic tools. In other words, I agree basically with the sort of approach Staal advocates, but I will show that the picture of the relationship between syntax and semantics is more complicated than suggested by him and though I have no real alternative, it might be good for scholars working in his discipline to also have a more complex picture of the issues involved. As said, this is also the criticism one may have with respect to those forming theories about language just from the point of view of syntax. In this sense, my intention is to contribute to a discussion about what belongs to syntax proper and what to semantics proper. Perhaps there are certain areas in which it may be very hard or even impossible to draw a clear borderline. And if this convinces my generative teacher, he might even feel inclined to have a look at his main thesis from the point of view of this more complex picture of the relation between syntax and meaning. On the other hand, I am fully aware that he argues against those who start at the meaning side of a complex system and I agree with his strategy to first put things in the right semiotic format and making the syntactic point. My remarks are for those who are prepared to follow that move and then want to see whether or not it might be the whole story. My point will be that Staal's thesis might have to be given a more complex character, the more so because Staal employs basically the same strategy as Chomsky does in his study of natural language: to restrict oneself to a syntactic approach. If the Chomskyan strategy could be shown to fail to deliver a reasonably complete picture of the empirical domain to which his theory applies because it could be argued that some sort of semantic patterning must be drawn into the theory formation about natural language. it might also influence Staal's position.

Staal's thesis presupposes that ritual is considered as a language whose syntax is to be studied without an appeal to the notion of the meaning its forms have or can be said to have. This is not the thesis defended by Staal; we need to refine its formulation. Staal entered into the discussion about the nature of ritual seeing that people focused on their meaning: rituals were given an interpretation. The first step by Staal was to connect interpretation with form: he put his study of the ritual in the standard semiotic form which requires that there be no semantics without syntax. At that point, he asked himself whether or not it would be possible to study ritual without my appeal to meaning. This is also fairly standard in semiotics because many logicians study the properties of their formal languages by only looking at their uninterpreted forms. They set up some axioms and some rules of inference and then look at what can be derived by applying these rules: rules without meaning. This sort of approach is called proof theoretic. That Staal is on this side is clear by his remarks on the role of syntax in natural language. He adopts Chomsky's position (1989, pp. 52-60. but especially p. 138) by arguing that in natural language the relation between sound and meaning is mediated by an unnecessarily complex roundabout system: syntax. Like Chomsky, Staal says that language is not (only) for the sake of communication and he derives from this the view that syntax is 'a structured domain of specific rules which in fact makes language unlogical and inefficient. These specific rules, which are without rhyme or reason, must come from elsewhere. They look like a rudiment of something quite different. This supports the idea that the origin of syntax is ritual.'

However. there is a second sort of approach to formal languages in semiotics: one may be interested in truth assignment to formulas of the language and in this case a relation is presupposed between the forms of the language and some domain of interpretation. This is the model-theoretic approach: it characterizes the conditions under which expressions are true in a certain model. i.e. with respect to a domain of interpretation. It investigates the relations between the structure of the language and the structure of the domain. Montague is nowadays considered as the chief scholar who made model theory available to the study of natural language. His central thesis, developed in different papers collected in Montague (1974) is that a natural language is a formal typed language, which means among other things that the categories of natural language are treated as syntactic expressions systematically related to semantic objects that are part of an algebraic structure attributed to a domain of interpretation. I will demonstrate how this can be done shortly. Montague assumes a one-to-one correspondence between forms and meanings and in my view this provides a means to relate semantic structures to syntactic forms, though not by taking semantic structure as a point of departure but by assuming a genuine match.[1]

Whatever the differences between proof theory and model theory, both approaches make a principled distinction between syntax and semantics and the idea is clear: keep them strictly separated. I will operate on the model theoretic side trying to show that in the analysis of a genuinely syntactic pattern, the NP VP distinction, we may hit upon underlying principles which give away the structuring of models in which we interpret expressions of this form.

TWO TRADITIONS

There are two traditions in which the distinction between form and meaning is expressed explicitly. In linguistics, this insight is attributed to De Saussure (1965) who made a principled distinction between a sign and its denotation. This is used by Staal in order to strengthen his main claim: it is dependent on the split between form and meaning. Staal also underscores the importance of the Saussurean claim that the relation between a 'signe' and its 'signifiant' is conventionally determined. Of course, there are some peripheral structural ties between forms and their meanings, such as onomatopoeias, but these do not affect the claim that the relation between form and meaning is arbitrary in the sense that a different meaning could have been associated with any of the forms we have available in a language. It should be added that once the assignment of meaning has taken place, much of this conventionality is restricted by the presence of existing form-meaning pairs. But this is outside the realm for which the claim has been made.

The Saussurean tradition has merged with another tradition in philosophical logic which developed the notion of semiotics. In the semiotic tradition of the thirties the distinction between syntax and semantics (and pragmatics) was given its present standard form, as in Carnap 1958. The strict separation between the three corners of the semiotic triangle, in particular between expressions and their denotation is compatible with distinction between 'signe' and 'signifiant' as well as with the Saussurean claim of arbitrariness. This solved several problems that would otherwise burden theory formation in natural and formal languages. It enables us to develop a precise theory of inference but also a theory of reference (and hence of quantification), and in so doing one is in the process of developing a successful theory of meaning. As observed above, the distinction between syntax and semantics led to two research strategies in mathematical logic: proof theory and model theory. It is clear that Chomsky has always been siding with the proof theoreticians: he does not appeal to meaning in constructing his theories about natural language. In applying Chomsky's insights to the area of ritual Staal has retained this proof theoretic position. It also follows that my interest in Montague's work will lead to raising some problems.

SOME PROBLEMS

As indicated, Staal appeals to De Saussure for an important argument about the reference of linguistic expressions, in particular to his insight that the relation between a linguistic form and its reference is conventional. This insight has gained the status of one of the few axioms of linguistics. In fact, there is no way to avoid the position that the relation between linguistic form and the semantic object to which it refers is arbitrary. However, it is important to see that the Saussurean axiom is formulated with respect to basic, lexical forms, such as nouns, verbs and adjectives. As formulated, it holds only at the lexical level. As soon as we arrive at the level at which the principle of compositionality can be said to be operative, one is forced to rely on the functional nature of the rules combining basic forms into derived meanings. The most common assumption is that compositionality requires (mathematical) functions, the complex output, i.e. a structured meaning, being determined by its basic constituent members and hence the arbitrariness 'percolates' as it were to the top of the tree to which the meaning is assigned.

Let me put these things in a more precise format. Interpretation in the logical-semiotic tradition that currently has been extended so as to include linguistics, can be seen as a function mapping linguistic forms onto semantic objects. The interpretation function is defined as operating on a domain consisting of linguistic forms and as yielding semantic objects which are part of the domain of interpretation. It is standard to distinguish between the interpretation of basic forms (say, lexical items) and the interpretation of complex forms: the latter are constructed from the basic forms. So, the interpretation of language forms begins by considering any form as complex breaking it down into less complex units until one arrives at the levels at which the forms can be considered basic. I will briefly discuss these two modes of interpretation and see how we can extract some arguments from it pointing in the direction suggested above.

THE INTERPRETATION FUNCTIONS I AND $[\![\, . \,]\!]$

In model theory, interpretation takes place with respect to a model. A model M for a language L consists (basically) of a domain of interpretation D and an interpretation function I : $\mathbf{M} = \langle D, I \rangle$. The basic forms of L are interpreted by I which assigns to the constants c_α in a language L their value $I(c_\alpha)$ in the domain of interpretation D_α, where D_α is construed out of D:[2]

$$I : L_{con} \rightarrow D_\alpha$$

The subscript a stands for types (categories). For example, $John_e$ says that the proper noun *John* denotes an individual (*e* stands for entity); $walk_{et}$ says that walk is a predicate constant (*et* stands for a set), etc. In this way the categories of natural language are systematically mirrored by the types of a logical language. The domain of individuals D_e is taken here as the point of departure: all types are construed from individuals *e* and truth values *t*.

The domain D may be any domain we like. In (approaches based on) first order logic, however, D is standardly taken as a domain of individuals. Assuming that in sentences like (1) there are two basic forms *John* and *walk*, we obtain (2) as the result of applying the function I.

(1) John walks

(2) *a*. I (*John*) = John
b. I (*walk*) = W = $\{\chi : \chi \text{ move by foot}\}$

Here I (*John*) ϵD_e, that is, John is an element of the domain, and I (*walk*) $\subseteq D$. That is: W is a subset of D_e, the set of all individuals. As said, I could have been given different linguistic forms as its input and yet have given the same values.[3]

Full sentences are considered complex expressions that are reduced to more simple ones. In first order predicate logic (1) is standardly interpreted as in (3):[4]

(3) $[\![\textit{John walk}]\!] = 1$ iff $[\![\text{walk}]\!]([\![\textit{John}]\!]) = 1$ iff
$I(\textit{walk})(I(\textit{John})) = 1$ iff $I(\textit{John}) \in I(\textit{walk})$ iff John $\in$ W

The sentence is considered true if and only if John is an element of the set W: $I(\textit{John}) \in I(\textit{walk})$. The reduction to the I-interpretation applies to all sorts of complex forms, e.g. conjunctions like *John, Mary and Sue* and to *walk and talk*. Conjunctions are first taken as a complex form and then reduced to the basic forms which receive an I-interpretation. For example, $[\![\textit{walk and talk}]\!]$ is first reduced to $[\![\textit{walk}]\!] \sqcap [\![\textit{talk}]\!]$ and then to $I(\textit{walk}) \sqcap I(\textit{talk})$.[5] The same sort of pattern holds for disjunction, negation, quantifiers, etc.

SPLITTING THE DOMAIN

The set W in (2b) is the set of those individuals that walk in D, which means that the predicate *walk* is conceived of as a function partitioning the domain into a set W and its complement W′. This structures D. In fact, D is bipartitioned as many times as there are predicates in the language. Due to this property of predicates (a two-place predicate partitions D into a set of pairs standing in a certain relation and a set of pairs not standing in this relation, etc.) one may discern types of semantic objects in D. In modem type-logic, these structures have been dealt with extensively. The leading idea is that D is stratified by a set of functions so that different categories in the language correspond to different semantic types in the domain. It is now standard to say that $D = D_e$ i.e. the set of all individuals of type *e* (*e* stands for entity). The verb *walk* is considered as pertaining to a set W of type *et* (the notation represents a function sending entities *e* belonging to W to truth values *t*). It is treated as a predicate constant whose value is an element of D_{et}.

One of the things Montague (1974) taught us was that the proper name *John* is an individual constant whose value is not only an element of D_e, but also an element of the domain $D_{(et)t}$. This is the domain of functions sending sets, i.e. semantic objects of type *et*, to truth values *t*. For example, we saw that the set W was treated as a (characteristic) function sending all the entities that walk to the truth value 1 and all the entities *e* that do not walk to 0. Now, one may think of *D* as being structured into the set of all its subsets. This collection, being a set, called $\wp(D)$ may also be split by a function sending all the sets with a certain property to 1 and all the sets lacking this property to 0. In this way, *I*(*John*) can be seen as the collection of sets containing all the sets of which John is an element: there are two ways of looking at one syntactic element. It makes the relation between syntax and semantics more flexible in the sense that *John* may correspond to an individual or to a collection of sets, whereas semantically an equivalence holds. Yet there is some other constant relation between form and meaning: a mathematical function.

A UNIVERSAL FORMAT

An intriguing question arises: how do we explain the fact that our interpretation has the format of a function? And again, this amounts to asking whether it is a matter of syntax or of semantics. Let me work out this question in some detail, because we need more ingredients to get to the point. Take a sentence like (4).

(4) Three children walked

Here again one needs application of *I*, so we end up with something like:

(5) *a*. *I*(*three*) = 3
b. *I*(*children*) = C = $\{\chi : \chi$ is a child$\}$
c. *I*(*walk*) = W = $\{\chi : \chi$ move by foot$\}$

However, (5a) would not work as formulated: there is structure involved in (4): (4) is interpreted with the NP analysed as: Det N and the function $[\![.]\!]$ works, provisionally, as follows:

(6) $[\![\textit{three children walk}]\!] = 1 \Leftrightarrow$
$[\![\textit{three children}]\!]([\![\textit{walk}]\!]) = 1 \Leftrightarrow$
$([\![\textit{three}]\!]([\![\textit{children}]\!])([\![\textit{walk}]\!]) = 1 \Leftrightarrow$
$(I(\textit{three})(I(\textit{children})))(I(\textit{walk})) = 1 \Leftrightarrow$
$W \in I(\textit{three})(I(\textit{children}))$

That is, like *John* the NP *three children* is taken as a function splitting the domain $D_{((et)t)}$. In (6) $D_{((et)t)}$ is split into a collection of sets containing three children and into the complement of this collection. In (4), the function *I*(*three*(*children*)) takes the set W as its input and maps it to 1 iff W contains three children in which case the sentence is true, otherwise false. This sounds rather complicated because so much machinery seems to be involved, but as I will show shortly with the help of a diagram, the basic ideas are simple.

Montague (1974) laid the foundations for the theory of Generalized Quantification that developed in the eighties. It is one of the most successful semantic theories to date, because it added significantly the scope and the depth of the insight in quantifiers among which the numeral *three*. It is taken as a determiner. Nowadays, the determiner is considered a key element in the approach to quantification. It has become standard to interpret sentence (4) in terms of a subject-predicate combination of the form NP VP yielding the interpretation that three individuals in D_e are members both of the set of children and of the set of those who walk, as shown in Figure 1.

What the Det *three* does is to provide a format for the interpretation of the relation between two sets expressed by elements of the sentence: the head noun of the NP in the NP VP combination denotes a set (in this case C) intersecting with the denotation (in this case W) of the VP. This appears to be a universal format available for sentences in all languages.

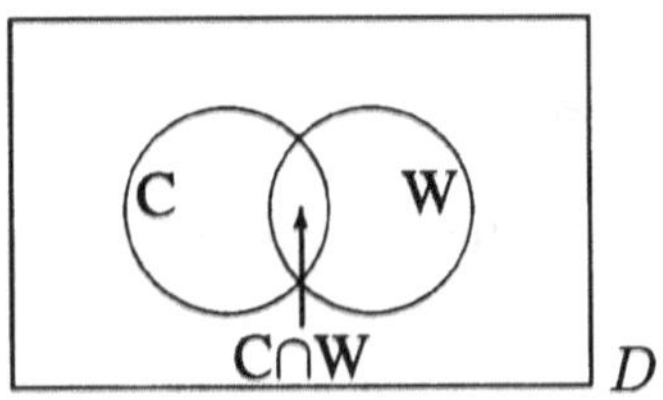

Figure 1

W is mapped to 1 if W contains three children in the intersection C ∩ W.

THREE

The English numeral *three* taken as a determiner expresses a relation between the head noun N and the VP. Yet, this format is not directly visible in the syntax of English: one of the arguments of the 'Three'-relation is part of the NP, the other argument is the VP. In other languages, the format may be expressed by different syntactic configurations. Yet, any sentence expressing that three children walked has the same format due to the universality of the meaning of the numeral. We therefore replace (5) by (7):

(7) $I(three) = \{<\chi_{et}, y_{et}> : |\chi \sqcap y| = 3\}$

That is, *three* is to be interpreted as a relation between two sets such that their intersection contains three individuals. Formally:[6]

(8) $[\![three\ children\ walked]\!] = 1$ iff $< [\![children]\!], [\![walk]\!] > \in [\![three]\!]$

In this way, the universal format of Figure 1 is encoded in a general lexical definition that is universally applicable. Another way of expressing this is by saying that once people master elementary operations over the set of natural numbers, they have universal semantic objects looking for a name. That this name is arbitrary is a trivial matter as compared with the fact that a structure is available reflecting something of the order we assign to the domains we are speaking about.

Of course, the number 3 could have been given a different name (e.g. *trois*) but the question is whether a numeral could have a different semantic format given the notion of a numeral, i.e. arguably given the notion of an individual. Quantificationally, the number 3 may be defined as in (7), so there is a semantics involved which escapes an all too trivial formulation of the Saussurean axiom of conventionality. Nothing would prevent another sort of relation between the NP and VP. Yet, one may argue that the relation between the NP and the VP expressed as $[[_{NP}$ Det N] VP] receives a fixed interpretation on the basis of the configuration in Figure 1 which does not seem so incidental or arbitary.[7]

THE LAW OF DISTRIBUTIVITY

Sentences like (9):

(9) *a*. John, Sue and Mary ate four sandwiches
b. Three children ate four sandwiches

strongly suggest that one may think about the NP VP relation in terms of a function giving to each of the members of the NP-denotation its own private VP, so to say. That is John ate four sandwiches, Sue ate four sandwiches and Mary ate four sandwiches. And due to the irreversible nature of eating this means that we are speaking about 12 sandwiches. These events may have occurred at the same time or at different times and places, but the main point is that (on this interpretation) each of the individuals of the subject NP have its own individual 'path' expressed by the predicate. Now, this is reminiscent of, or even similar to what we are used in arithmetical operations like:

(10) $3 \times 4 = 12$

We know that (lO) is equivalent to $(2 \times 4) + (1 \times 4)$ and to $(1 \times 4) + (1 \times 4) + (1 \times 4)$. This equivalence is known as the law of distributivity. It governs a class of mathematical operations among which the Boolean intersection in Figure 1 occurs. We break ⟦*John and Sue and Mary*⟧ x ⟦*Eat four sandwiches*⟧ down into ⟦*John*⟧ x ⟦*Eat four sandwiches*⟧ + ⟦*Sue*⟧ x ⟦*Eat four sandwiches*⟧ + ⟦Mary⟧ x ⟦*Eat four sandwiches*⟧.

The basic idea emerging here is that in offering the information conveyed by (9), the 'lumpsum'-information expressed by their subject NP is automatically broken down into information concerning the basic elements constituting the sets {John, Sue, Mary} in (9a) or the set of three children in (9b). This is what the law of distributivity is about: it guarantees that no information given at the level of 3 x 4 is lost at a lower level of organization.

Given the crucial role played by the law of distributivity this is the point where semantic considerations come in: the level at which information expressed by language is organized interpretively seems to be determined by principles of computation presumably more basic than syntax itself. In fact, there are two things that should be explained away before (re-)assigning primacy to syntax. Firstly, one must explain why it is that a basic computational principle like the law of distributivity available to us as part of our linguistic abilities should determine syntax in such a fundamental way as to determine the NP VP structure. Secondly, and even more important: one should explain why it is that the law of distributivity is only applied asymmetrically. It works one way, so to say, because we can derive from (9) that there were 12 sandwiches involved but not that there were twelve children involved. But this can only mean that syntax serves as a constraint on what otherwise would be the full application of the distributivity principle. Syntax uses just one part of a more general principle and is in this way dependent on it.

Let me recapitulate from another point of view. Suppose that in a syntactic structure of the form $[\alpha \oplus \beta]$, with $\oplus$ a syntactic operation joining α and β into a group $\alpha\beta$, and suppose that $\alpha\beta$ is always of type α', which would mean that $\alpha\beta$ is always an α-phrase, much in the way a VP is a V-phrase. And suppose that we are able to interpret $\oplus$ as a function which is similar to or is structurally related to (a principle determining) an operation in one of our number systems. Assume furthermore that our cognitive organization heavily depends on our use of number systems (discretization of individuals, order, measurements, etc.), then the very fact that our syntax in natural language is tied up with this system may be explained as so closely tied up with semantics that one is obliged to think in terms of interaction rather than of primacy. With respect to (10) a plausible hypothesis is that our cognitive organization of the world forces us into the intersection relation $\sqcap$ because it mirrors certain features of our computing capacity in which distributivity takes its natural place.[8] Note that this does not mean the return of religion as the semantic fosterplace of rituals, cause we are speaking about semantics at a very abstract level of cognitive organization. We are speaking about the semantics of our number systems involving our capacity to distinguish between individuals and mass and assume that these semantic principles cannot be abstracted from in the study of syntax, not even in the syntax of rituals.

NEGATION

How much syntax goes into (11)?

(11) John does not walk

The syntax of predicate logic tells us that *(does)* not is to be taken as an expression which, when applied to (1), yields the wellformed expression (11). But its semantics says that if (1) is true, (11) is not true. If (11) is true, (1) is not true: $[\![\neg\varphi]\!] = 1$ iff $[\![\varphi]\!] = 0$. Again the syntactic element bringing this about - *(does)not* - could have been different but the set-complement relation seems to be something which is brought about by the fact that we use a name for a constant, however arbitrarily, to split a domain. The very use of a name invokes the use of a splitting function characterizing the members of a set, and so inevitably there is the structuring of a domain and in this structuring semantic considerations must be given a place: any function having {1,0} as its co-domain can be aid to produce negation as corresponding to a complement. There is syntax in this but the fact that we apply these functions systematically in our use of language reveals that beneath the superficial fact that their input could have been differently, there is the question of why we pick them out among so many different functions that could have been employed otherwise. The point is again: as soon as we are able to systematically relate certain expressions in a language to algebraic structure, we cannot maintain the strict distinction between syntax and semantics and subsequently the abandonment of semantics. The next step is: as soon as we need algebraic structure in order to explain our cognitive organization, it is very hard to maintain the Chomskyan claim about the study of natural language and hence Staal's claim about the syntax of ritual. The two claims are logically independent, so it might turn out that Chomsky's claim may be rejected and Staal's claim may be maintained, but given the fact that Staal stresses the connection, the point on negation may apply to him as well.

PLURALITY AND SINGULARITY

This line of thought may be applied to the relationship between the stem *child-* and its plural form *-en*. If we analyse the internal structure of the plural form *children*, we need interpretively something like (12):

(12) a. $I(child\text{-}) = \mathrm{C} = \{\chi_e : \chi \text{ a child}\}$
b. $I(\text{-}en) = pl = \{<\chi_{et}, y_{((et)t)}> : y = \{z_{et} \subseteq x : |z| \geq 2$

In (12), χ ranges over sets and y over the collections of sets having two members. So, (12b) applied to the set C says that *pl* takes C and yields the set of all sets formed out of C minus the emptyset and the singletons. This means that C, as proposed by Jespersen (1924), is taken to be neutral as to singularity and plurality and that *sg* can be taken as an instruction to structure the set C into a set consisting of all the sets containing one child and that *pl* is as shown in (12): the instruction to form a collection of sets of pairs, triples, etc. out of the members of the set C.

Suppose that this analysis can be defended (which means an adaptation of the model in Figure 1 but not an essential one for the point made above), then again we see a division of a collection into two disjoint subsets.[9] And again we see that a fundamental semantic phenomenon is visible. In this case, there is an interesting fact: Japanese does not distinguish between singular and plural in the way English and many other languages do. So, rather than having (12b) and for *sg* the meaning $\{z_{et} \subseteq \chi : |z| = 1\}$, Japanese would have something like (13).

(13) $sg/pl = \{<\chi_{et}, y_{((et)t)}> : y = \{z_{et} \subseteq x : |z| \geq 1\}$

This means that Japanese would neutral as to how the set C is structured. It does not mean that C is not structured. On the contrary the mechanism to produce all the subsets of a certain set is something that can be argued to be available also in Japanese. Note that the fact that Japanese has a different strategy from English enhances the present argument rather than weakening it. Languages are bound to develop different strategies. The Japanese case suggests that this is a selection from possible choices from algebraic options. A difference between 1: many vs neutral. So, again the question arising is: how much syntax is actually going into this analysis and how much semantics? If combinatorial principles or mathematics underlie the ordering of sets and their structuring into sets of sets meeting certain universal conditions, then it might be argued that some of these universal conditions could have to do with the way we structure the world. We organize an unorganized set of individuals into a domain full of structure about which we can speak, but it is doubtful whether it would be revealing to call these principles syntactic only, unless we can show that the structuring has nothing to do with our cognitive organization. One may, of course, call any form of cognitive organization syntactic, but this would make the debate about a syntactic or semantic approach of strings a terminological debate rather than a real issue.

CONCLUDING REMARKS

My contribution to this collection of papers is an appeal to those who want to carry on with the discussion, to pay some attention to factors that I have been pointing at. What I have said might or might not be relevant but this can only be assessed on the basis of taking Staal's claim seriously and exploring it further.

The three cases that I discussed as an illustration of my remarks on the relation between syntax and semantics have in common that an of them may be reduced to fundamental constraints on set theoretical structures. An important question is whether any syntax can escape from a set-theoretically based ordering giving away fundamentally cognitive modelling of the world as we perceive it and cognize it. I do not want to push this point any further but I think that in the discussion about the relation between syntax and semantics, the issue of the structure of our cognition should be given a role. I accept without any problem the thesis that the proof-theoretical approach of generative linguistics, leading to the exclusion of semantic considerations, has been shown to be very fruitful for certain areas of language structure, though not all I should add. It has been shown to be fruitful for Staal's treatment of ritual, as far as I can judge, at least for a more profound discussion of the issues involved. But in linguistics it might turn out that in certain areas semantic considerations must interact with purely syntactic ones, certainly in view of computational principles in which the basic organization of our cognition is involved. I have tried to argue that with respect to formats as in Figure 1 these principles may be discovered, because basic patterns of predication may be related to principles determining our capacity to compute. An interesting consequence of our capacity to count, i.e. to distinguish entities, to keep measure, to make music, to dance, could be that all sorts of mathematical structure which is part of this capacity might be due to the fact that we relate our syntax to the world and get its structure back as meanings.

REFERENCES

Carnap, Rudolf

1958 *Introduction to Semantics and Formalization of Logic*. Harvard University Press. [1942;1943]

Chomsky, Noam

1981 *Lectures on Government and Binding*. Foris: Dordrecht.

Jespersen, Otto

1924 *The Philosophy of Grammar*. Allen & Unwin: London.

Montague, Richard

1974 *Formal Philosophy. Selected Papers by Richard Montague*. Edited and with an introduction by R.H. Thomason. Yale University Press.

Saussure, Ferdinand de

1965 *Cours de Linguistique Générale*. Publié par Charles Bally et Albert Sechehaye. Payot: Paris [1915].

Staal, Frits

1983 *Agni. The Vedic Ritual of the Fire Altar*. Berkeley.

1988 *Universals. Studies in Indian Logic and Linguistics*. The University of Chicago Press.

1989 *Rules Without Meaning. Ritual, Mantras and the Human Sciences*. Toronto Studies in Religion. Peter Lang: New York etc.

Notes

1. I refer here indirectly to the discussion in the generative framework that took place at the end of the sixties. Some proposed to take first order logic as the deep structure and they thought that they took semantics as primary. In fact, this turned out to be nothing but (an inferior sort of) syntax, because there was no genuine interpretation involved in the model-theoretic sense. Montague employs higher order techniques in order to overcome some of the shortcomings of first order languages and added the necessity to interpret expressions with respect to a model in which they are true or false.
2. I restrict myself here to the interpretation of constants as they appear in natural languages and shall not pay attention to the assignment of values to variables. All members of traditional lexical categories are treated as constants.
3. One of the things ignored in the claim of conventionality is that once the relation between a constant and its reference is made, it remains more or less fixed at the language side, apart from phonological changes to which morphemes of a language are subjected. In general, one could say that once they are paired there is a tendency to retain both the sound and the meaning as constant as possible. It opens up the question of whether the rigidity just mentioned is a fact of syntax or a semantic fact. I do not know how this point relates to Staal's discussion of rites that receive 'ambiguous' intepretations, i.e. the aspersion and the fecundity interpretations (1989, p. 115 ff., in particular p. 128).
4. I will not discuss tense, because this would not throw any light on the issue under discussion.
5. I simplify here the treatment of conjunction to some degree: for the present purpose it suffices to show the relation between $[\![.]\!]$ and *I*.
6. The formula in (8) is often given in a functional form with so-called λ-abstraction. In the case of (4) this means that one can see *three* as a function taking the set C (χ in (7)) and yielding a function that takes W (y in (7)) and has as its output the truth value 1 if and only if three individuals are both in C and W.
7. This bears on the discussion about configurationality, as e.g. in Chomksy (1981). The question is whether or not all language can be put into the NP VP format even though they do not have in on the surface. I will not enter into this discussion here, but if one rejects the position that all language have the NP VP format (at least at a 'deep structural' level, a reasonable position seems to be that non-configurational languages organize their predication into the format of Figure 1 with different linguistic means.
8. $\sqcap$ in a Boolean algebra is here the counterpart of the x in arithmetics.
9. A collection of sets is a set of sets, so if it splits up one obtains two sets of sets.

XXXIII

THIN, THINNER, THINNEST: SOME REMARKS ON JAIMINĪYA BRĀHMAṆA 1.144

A. Wezler

1. Ten years ago K. Klaus published a slightly revised version of his doctoral dissertation, *Die altindische Kosmologie. Nach den Brāhmaṇas dargestellt.*[1] This book exudes the spirit of Klaus' supervisor W. Rau and in fact continues and supplements Rau's pioneering work on state and society in ancient India.[2] It is equally based on a comprehensive and thorough study of the Brāhmaṇas, a clear awareness of the philological problems these texts involve and with a full knowledge of the secondary literature. The material is presented in a lucid manner, following a clear systematic plan;[3] in brief, a highly informative and useful monograph, albeit perhaps a bit too much influenced by the painful aftermath of positivism and because it falls short of becoming exactly enthralling it does not go beyond mere fact-finding[4] nor cross the boundaries of traditional Indological or Vedological methods. Not just an inventory, it is an indispensable tool of research. It is hoped that it is widely and regularly consulted in spite of the fact that it is written in German. Klaus also prepossesses his readers in his favour by his intellectual honesty, which nowadays is to be regarded as a remarkable virtue, but used to be the natural attitude of many of our academic ancestors. This is shown e.g. by explicitly stating that he fails to understand a particular phrase or expression in the texts he utilizes. In my view such frankness is not just a laudable trait of character, but has rather a significant methodical advantage in that it is, if not intended, then in any case, destined to directly further later research: gaps in one's knowledge once publicly admitted and clearly stated[5] will sooner or later be filled by a fellow scholar, or oneself as they provoke particularly intense and persistent intellectual curiosity, a sportsmenlike competition so to say.

2. In this paper I would like to draw attention to one case of an admitted failure of understanding in Klaus' book. In Chapter 3 which deals with 'the shape of the universe',[6] or more precisely, with 'the statements about the shape of the universe which vary in accordance to the perspective in which it is seen',[7] Klaus discusses the passage Jaiminīya Brāhmaṇa (JB) 1.144.[8] Including his corrections[9] it reads as follows:

> *tad āhuḥ: prādeśamātrād vā etad imaṁ lokaṁ na spṛśati, prādeśamātrād amuṁ neti. atho āhur: yāvad eva gos sūtāyā ulbaṁ tāvataivemaṁ lokaṁ na spṛśati, tāvatāmuṁ neti. atho āhur: yāvad eva śakṛty ulbaṁ*[10] *tāvataivemaṁ lokaṁ na spṛśati, tāvatāmuṁ neti .... tad u vā āhur: yathā vā akṣeṇa cakrau viṣṭabdhāv, evam etenemau lokau viṣṭabdhau. nīvemaṁ lokaṁ spṛśati, nīvāmum iti.*

Klaus' German translation[11] can thus be rendered into English:

> With regard to this [some] say: 'It is the measure of a span by which that (i.e. the intermediate world [between heaven and earth]) does not touch this sphere here, the measure of a span by which [it does not touch] yonder [sphere].' And further [some others] say: 'Only by this much as is the amnion of a cow who has calved does it not touch this sphere here, by this much [does it not touch] yonder [sphere].' And further [some others] say: 'Only by this much as is ... does it not touch this sphere here, only by this much [does it not touch] yonder [sphere].' ... And with regard to this [some] forsooth say: 'Verily just as two wheels are kept apart by the axle, so the two spheres here are kept apart by that. It touches this sphere here directly, so to say, yonder [sphere] directly, so to say.'

At two places a series of full stops marks passages deliberately not translated by Klaus. The second is a merely Sāmaveda technical remark: it was left out because it does not seem to be relevant for the question at issue, viz. various 'profane' views on the size of the distance between the intermediate world and the earth and the intermediate world and heaven. Yet in a footnote referring to the first lacuna Klaus informs us that 'he does not understand *śakṛty ulbaṁ* or *udvaṁ*'. And in fact one cannot but wonder what in the name of God an 'amnion on/in faeces' could be. No help can be expected from G. Ehlers as his attention is focused on emendations, and so far only that part of the JB has been published which covers the second book (*kāṇḍa*).[12]

3. The whole passage quoted just now has been translated by H.W. Bodewitz in his latest book on this Brāhmaṇa.[13] He renders it as follows: 'Now they

say: "This (intermediate world) is within a span of touching[14] this world, within a span of reaching yonder world." And they also say: "It is within a new-born calf's membrane (i.e. within an ace) of touching this world and reaching yonder world." And they also say: "It is within a pellicle (found) in the faeces of touching this world and reaching yonder world." ...[15] And they also say: "As two wheels are firmly fixed by an axle, thus these two worlds are firmly fixed by this (Vāmadevya). He softly touches (i.e. fondles) as it were this world and yonder world."'

There are other points too where Bodewitz differs from Klaus, whose work he knows,[16] and these should be inspected first for the sake of convenience. Bodewitz takes *gauḥ sūtā* to mean 'a newborn calf', whereas Klaus, perhaps under the impression of what is found in the Large Petrograd Dictionary,[17] takes the participle to have an active meaning. Indeed at Manusmṛti 8.242,[18] to which Böhtlingk-Roth refers, exactly the same expression is used and it quite clearly[19] denotes 'a cow who has calved'. But what is true of classical Sanskrit need not be true of Vedic; certainly, the problem is first of all a linguistic, grammatical one, and therefore cannot be solved by biological considerations, viz. does the 'membrane' really belong to the cow or the calf. According to the Vaidika Padānukrama-Koṣa[20], JB 1.144 is the only place in the whole of the Brāhmaṇa literature where the feminine form of the participle[21] occurs! Yet there is at least circumstantial evidence that Klaus is right in this case if we consider the even older expression *sū́tikā*, 'woman in childbed, i.e. a woman who *has* recently given birth', and its derivation as explained by Debrunner.[22] Besides one could not but wonder why a calf should be referred to by the very general expression *go* and, especially, why the chorion of a *female* calf should be spoken of here, i.e. in a context where the sex of the calf is of no importance at all.

Bodewitz justifies his translation of the phrase *yathā vā akṣeṇa cakrau viṣṭabdhau* by referring[23] to J(aiminīya) U(paniṣad) B(rāhmaṇa) 1.20.3 and Oertel's translation[24] of this passage a'... or two wheels by means of an axle, so these two worlds are propped apart by means of this atmosphere'. It is of minor importance that the predicate used there is *viṣkabdhau*, not only because this is one of the variants of *viṣṭabdhau* at JB 1.144, but also because the semantic difference is very small, almost negligible.[25] Yet what is important here are Klaus' and Bodewitz' differing interpretations of the finite verb *ni spṛśati*, and especially of its subject(s) and, in connection with this latter question, of the reference of *etena* in the sentence preceding the concluding one: Klaus starts from the assumption that the intermediate world is also referred to here, while Bodewitz takes it for granted that at this point the author has already resumed his main topic, the *sāman* called Vāmadevya.[26] Neither of them seems to recognize a problem here, at least neither of them addresses a problem but both refer to parallels in other Vedic texts as supporting their interpretation. Bodewitz adds in a note[27] the remark: 'The softness is explained with a comparison by PB (= Pañcaviṁśa Brāhmaṇa) 7,9,11', viz. that of the cat who takes her young ones between her teeth without hurting them by biting, or that of the wind which blows gently over the water.[28] 'There it should protect the animals or the cattle (see PB 7,9,9); here it is connected with the soft contact between the intermediate world and the other two worlds.' JB 1.144, on the one hand, and PB 7.9.11, and 7.9.9, on the other, can appear to be related to each other only to an interpreter whose mind is already preoccupied with the idea of 'softness'. That is to say, the exegetical problem is directly connected with a semantical one: what is precisely denoted and meant by *ni-spṛś*? 'To touch softly' or 'to touch directly'? It is quite probable that Bodewitz thought of the occurrences of *ni-spṛś* in the RV which Monier-Williamson, on the basis of the Large Petrograd Dictionary,[29] refers to as meaning 'to touch softly, caress, fondle'.[30] In two cases (RV 8.96.11 and 10.91.13) one is indeed easily given the idea that 'to touch' in the sense of 'to move' (German 'anrühren') is the right meaning, but at 10.95.9, i.e. in the famous *ākhyāna*-hymn of Purūravas and Urvaśī,[31] what is meant is clearly a concrete, and (highly?) erotic 'touching'.[32] Yet it is not so much this latter case, but the passage in the JB under discussion in connection with the general function of the preverb *ni-* that suggests a meaning 'to touch intentionally/continuously'[33] or even 'to touch by penetrating'. For, after all, an axle, i.e. 'the pin, bar, shaft, or the like, on which or by means of which a wheel or pair of wheels rotates'[34] runs through the hole in the nave of the wheel. In this respect the comparison 'as two wheels are firmly fixed by an axle, thus these two wheels are firmly fixed by that', is not entirely appropriate if 'that' refers to the intermediate world, and is for this very reason rectified in the subsequent sentence by adding *iva* to *ni spṛśati*.[35] In contradiction to a real axle, the intermediate world does not run through this world and yonder world but it only 'as if' touches them (by penetrating them)! What the author evidently means to say is that it is in direct contact with the other two worlds, i.e. that there is not the slightest gap between them. The reference of *etena* is entirely a matter of interpretation, especially since *etad* is already used for 'the intermediate world' in the first sentence of the passage as quoted above. Bodewitz, however, would most probably argue that it is the Vāmadevya chant which is the subject referred to by the verb in the immediately preceding sentence not translated by Klaus, and that it should therefore form the subject of this sentence and the subsequent one as well. But the initial *tad u vā āhur* of this sentence is, among other things,[36] a comparatively clear indication that on the contrary the author returns to the topic of the distance between the three worlds. Klaus has a rather strong argument in favour of his interpretation of *etena*. He adds to his translation the remark:[37] 'The opinion expressed last was evidently the most widely held,' and lists in a footnote not less than eight references[38] to parallels in other Brāhmaṇas, not all of which are of the same importance.[39] I agree with him in substance, but cannot help adding that the fact that a particular view is 'the most common' does not, of course, necessarily warrant the conclusion that it is also found at a particular point in a text.

The expression *śakṛty ulbaṁ*, which Klaus leaves out, is rendered by Bodewitz by 'a pellicle (found) in the faeces'; since no additional explanation is given by him, one wonders whether this is just an attempt at a literal translation or whether Bodewitz had in mind a particular 'pellicle', and, if so, what kind of 'pellicle', because he adds 'found' within brackets and uses the definite article.

4. It would be tempting now to enter into a examination of all the passages in the Sāṁhitās and Brāhmaṇas where the term *úlba-* m. and *jarā́yu-* m. occur; but time and space being limited it suffices to emphasize that the Vedic Indians as of old clearly[40] distinguished between the 'amnion' and the 'chorion' and knew that the embryo is covered (*āvṛta*) by it,[41] and that the chorion is the outer (*uttara*) embryonic membrane. Procreation of man and cattle was too important and too much an everyday experience for them not to gain that kind of knowledge, which in the modern world has become rather a physician's or veterinary's prerogative. It can safely be assumed that the author of JB 1.144 had a clear idea about the thickness of an *úlba*, of e.g. a human foetus, or a bovine foetus, viz. that it certainly measures less than a millimetre. And it is equally obvious that when using the expression with reference to 'faeces', he was aware of the metaphor involved.

If the structure of the exposition of the various views on the distance between the intermediate and this world and yonder world is taken into consideration, one cannot fail to observe that these views are arranged in a systematic manner, according to the size of the distance, from one *prādeśa* (ca. 16 cm)[42] to the 'amnion' of a calf (less than 2 mm), from that to the *śakṛty ulbaṁ*, and from that to nil. Even if *śakṛty ulba* were an instantaneous coinage of a term, we would nevertheless be justified in looking for a membrane somehow related to faeces that is still, perhaps even markedly, thinner than a mammal's amnion.

5. There is in fact such a type of membrane or pellicle which meets the requirements. Many species of songbirds show a special manner of defecation. After feeding, the nestling turns around and secretes a small ball of faeces wrapped by a mucous pellicle, and this is grasped by the parent bird with his beak and eaten or carried away from the nest.[43] This membrane is further stated[44] to be 'very resistant, formed by glutinous

mucus that is secreted in the final part of the intestinal tract' of the nestling. Zoologists do not seem to have been particularly interested in that 'Kothäutchen', fecal sac, so no data are available about its chemical composition, or the thickness of the membrane.[45] Nevertheless the Indologist does not, in this case,[46] have to carry out the 'fieldwork' himself and wait for spring to come because he has seen this very scene of the care of the brood so often in TV films in the course of the last 10 years or so. He does not take any serious risk when he maintains that this particular membrane is indeed thinner than the amnion of mammals, almost invisibly fine. Consequently this could be the JB's 'pellicle on[47] faeces'!

Or could it not be the solution? To the best of my knowledge there is no second occurrence of the phrase *śakṛty ulba* in the whole of the vast Sanskrit literature.[48] Nor am I able to point out another passage that could at least be evidence of the ornithological fact as such and its having been observed in India. Strictly speaking, it is just a proposal that I am making here, not more than a possibility that the Vedic Indians knew the fecal sac. Nevertheless, some additional support is rendered by the following considerations and observations:

a. Vedic knowledge about animal, and vegetable life is of such a nature that this particular element cannot by any means be regarded as absolutely extraordinary and therefore simply falling outside the scope of what was accessible to people whose means of observation of natural phenomena was just seeing with the naked eye.

b. Songbirds are very common, and many of them can practically be considered companions of man in his settlements; in any case they already are a particularly conspicuous part of the fauna of the *vana*, not to mention the *araṇya*.[49]

c. At BĀU 3.3.1[50] it is stated that the extension of the *samudra* is twice that of *pṛthivī*, which in its turn, is twice as large as 'this world'; subsequently the *ākāśa*, the 'space' separating the latter two or all three of them from each other is said to be 'as large as the edge of a razor or the wing of a fly' (... *tád yắvatī kṣurásya dhắrā yāvad vắ mákṣikāyāḥ páttraṁ tắvān ántareṇākắśás* ...). That is to say, in a thematically very similar passage we find two phrases one after the other meant to describe something very thin, the second also being taken from the animal kingdom.

Admittedly all the arguments put forward by me cannot turn a possibility into a certainty, but taken together they lend the solution proposed a high degree of probability, nay verisimilitude.

6. But even if this interpretation of *śakṛty ulbam* could be shown to be intenable,[51] this would not affect the main point I want to make with regard to the JB passage under discussion, at least as long as there is agreement that the expression denotes something that is thinner than amnion, and to all appearances this agreement can be safely presupposed. I purposely use 'the main point', because all I have said until now was intended to prepare the ground for a more substantial analysis of the passage, viz. an analysis in terms of its significance for the Indian history of ideas.[52]

The first question one cannot help asking is whether what is said in the text is the result of mere speculative thinking or somehow related to observation or also because of observation. Can the spheres where the intermediate world reaches this or yonder world be perceived at all? Certainly the former, one is tempted to answer. But how does one perceive it? By standing upright and looking down at one's feet or by lying flat on the ground? Above all, how can empty space[53] be perceived? Only on certain occasions, e.g. when dust or dry leaves are driven over the bare ground by a playful breeze or when deep hanging clouds quickly drift from one side of the horizon to the other? Or what about dawn and dusk, i.e. the horizon in the east and west observed during these times of the day (although it is rather this world and yonder world which then 'touch' each other)?

I am well aware of the fact that this is mere guesswork; yet in my opinion it is not absolutely useless because at least it shows that the possibility of a relation with particular observations cannot be entirely precluded.

In spite of this possibility, one also wonders whether the three first views (viz. that the distance is that of a span or of an amnion or of a membrane on faeces) are real in the sense of ever having been held by different groups of people, or whether they are but an invention meant to gain a background against which the fourth and only really existent view stands out more clearly and convincingly. This, too, is no doubt possible in principle, but not probable in view of the absence of parallels of similar fabrications.

Consequently there is great likelihood that each of the four views has had its advocate(s) or rather its creator. This then leads to the following observations regarding the passage as a whole:

a. Four views, or perhaps I should say, theories about one and the same 'cosmological'[54] problem are assembled.

b. As already stated earlier (cf. § 4), they are arranged in a systematic manner starting with the largest distance between the various worlds and ending with the absence of any distance at all.

c. This sequence is not just logical in itself, but as has been rightly observed by Klaus (see § 3), it is evidently caused by the fact that the author regards the last view as correct, no matter whether he was also aware of its being the most widely accepted one or not.

d. If one is of the opinion that the three worlds in question in fact touch each other directly, what most naturally suggests itself is to regard the remaining three deviant views as being more or less far removed from truth, as getting more or less close to the truth. This is exactly what the author of the JB passage has done, perhaps even with particular emphasis because he separated the fourth view from the three other ones by a technical remark (see above p. 5).

Thus this passage exhibits a number of remarkable features: Various, thematically directly connected views are brought together, and presented, or reported, systematically and apparently following the principle of gradual approximation of reality.

It would not, however, be possible to simply say that the author of JB 1.144 regards *only* the last view as correct. After all his main concern is to give an explanation of the ritual, the chanting of the Vāmadevya after the preceding Bṛhat and Rathamtara and before the subsequent Naudhasa (Sāman), i.e. the fact that the *Vāmadevya* must be chanted 'independently', 'on its own support', as is stated at PB 7.9.12 and 15, and that there is a time gap, however small, which separates it from the preceding and subsequent Sāmans. It is evidently for this reason that, in the sentence not translated by Klaus, the author uses *evam*, with reference to the third view, which he consequently cannot unconditionally reject. Quite remarkably he is not content with adding this remark (with the consideration of which the Vāmadevya should be chanted), but resumes the topic of the distance between the intermediate and the other two worlds in order to state yet another view according to which they are in direct contact with each other. To all appearances the cosmological question as such continues to captivate him, and he deems it advisable, or necessary, to add a fourth theory he was also familiar with. Yet, the main problem we are confronted with is what his opinion is regarding the validity of this latter theory. It is tempting to answer that he considers this view ultimately true because he places it at the very end of his exposition and not amongst those mentioned at the beginning, which are too far removed from reality. But one should not succumb to this temptation because it is not possible to decide whether the particular sequence in which the four theories are mentioned corresponds to the decreasing size of the distance or to the increasing degree of correctness, or whether it is perhaps caused by both these aspects which luckily coincide in the present case. But I think this much can be said: he does not consider the fourth view to be untenable in principle, at least not outside the context of the Vāmadevya Sāman and the Agniṣṭoma. Most probably he wants to intimate this by adding the view that it is rather to be preferred if the question of the distance between the three worlds is examined as such, i.e. independent from the special ritualistic connection in which discusses. It is this probability on which my subsequent considerations are based.

The author did not also deem it necessary to mention the name of the creator and advocate(s) of the individual views; the reason for not mentioning the names remains anybody's guess, but the fact as such deserves to be noted,[55] even though it cannot be decided whether this is an earlier stage or a different strand of 'reporting views'. It is equally important to observe that in each case the view obviously is presented apodictically, no reasons being adduced, and that none of them is discussed except for the internal gradation and the criticism of the preceding view(s) it implies. All four views are most probably formulated by the author of the JB passage himself:[56] what is of importance to him is the essential contents of the various views, not how they have been used or expressed by a particular person. Equally unclear is whether the creators of these views knew each the others' positions, and whether and how the views are historically related to each other - because the sequence in which they are 'reported' at JB 1.144 does not by any means also reflect historical development.

7. Not a few of the noticeable distinctive elements of the passage strongly remind one of the later Indian tradition of doxography,[57] even to such an extent that it at least seems legitimate to wonder whether this tradition might not have much deeper roots than has been recognized until now and that they are recognizable in early Jain and Buddhist literature. This is not what I want to examine here a bit more closely, but rather another aspect of its significance which is not as readily perceived. Of course, what I have in mind is the significance this passage has in the context of the origins of science ('Wissenschaft') so stimulatingly studied by Frits Staal.[58] For it can hardly be disputed that this passage is a relevant piece of evidence, and a remarkably early one, for an attitude of mind which may be called scientific or pre-scientific although the notion of science defies easy definition, and is controversial (even) among historians[59] and theoreticians of science.[60] Provisionally a simple concept of science may be taken as a basis of the following deliberations, according to which it consists in collecting, describing and classifying 'facts'; I need hardly add that among other things the decision about something being a 'fact' or not depends to a large extent on the historical period and the cultural tradition(s) of a country or group of

countries to which the corresponding science or constituent part of a science belongs.

As for the JB passage under discussion here, what has to be noted first (although it does not form a specific characteristic of it) is that the three worlds, the intermediate, this and yonder, are considered to be separate and distinct entities, of whichever shape they may be;[61] yet they are in contact with each other. To say that this view is but a natural continuation of the Rigvedic mytheme of Indra or Varuṇa fixing asunder heaven and earth could not be accepted as a valid counter-argument. First because this mytheme is by itself only another expression of the human 'Urerfahrung' that we live on the surface of earth and that the sky/heaven is far above our heads and that there is an intermediary space which alone makes life on earth possible. Secondly the passage at issue here is itself no longer mythological. The question which the author and his predecessors/colleagues[62] want to answer, the problem they address is quite clearly that of the precise nature of this contact: is it a direct one or not, and if not, how large is the distance between the worlds (respectively)? In our case collecting facts consists in the gathering of knowledge,[63] viz. the juxtaposition of four different answers to this very question. Yet this is not just (another) result of the 'transmission of knowledge from generation to generation'[64] and its growth in the course of this process; it is rather a deliberate compilation of *all* corresponding views which, as has been shown, are arranged in a very particular sequence. In its turn, this implies a particular form of classification of knowledge, and, even more important, testifies to a mental acquisitiveness which is both theoretical, critical in that it starts from the assumption that only one statement can be absolutely true (see above, § 6) and rational in the sense of believing in the capacity of the intellect to improve upon knowledge received and to find out the truth. An essential element of what is commonly called science, viz. reason(s) for the view(s) stated, seems to be lacking; but before jumping to this conclusion I should like to suggest that we should examine the question more thoroughly whether (the) reason(s) could not perhaps be implied in the views themselves as they are stated: *prādeśa* could indicate that otherwise a dangerous friction between the worlds might arise; *ulba* could be an indication that the worlds are regarded as something born,[65] the problem being only that of the thickness of the 'amnion'/membrane,[66] and the fourth view could imply that the three worlds are made/created.

Finally I should like to note that the JB passage deserves scholarly attention in yet another respect. N. Luhmann[67] reminds us of 'a peculiarity of the Jewish exegesis of law, viz. that it is important to raise dissent to an adequate level and to preserve it as a tradition'.[68] Quite evidently this is equally a peculiarity of the Indian tradition and to be sure since the Vedic period, a peculiarity which demands respect, not only because to a certain degree it compensates for the scarcity of historiography proper.[69] Instead of lamenting the lack of any information on the persons who created or held these views about the contact between the three worlds, about their dates, etc., Indologists should rather be grateful that these views as such have been preserved by a tradition which regarded disagreement among scholars and philosophers as (almost) equally important as truth itself, or significantly rather as *siddhānta* itself which by its very nature is not entirely independent of other people's ideas.[70]

Ceterum censeo terram Tibetanam
in veterem dignitatem ac libertatem
esse vindicandam.

Notes

1. Bonn, 1986. (Indica et Tibetica. 9)
2. *Staat und Gesellschaft im alten Indien nach den Brāhmaṇa-Texten dargestellt*, Wiesbaden, 1957.
3. Cf. the very detailed 'Table of Contents' ('Inhaltsverzeichnis'), p. 7-10.
4. In his review, published in *IIJ* 32, 1989, pp. 294-300, H.W. Bodewitz dwells at some length on what he regards as the main sore point of study, viz. the difficulty of distinguishing 'die profanen Vorstellungen' from other conceptions, i.e. of gleaning the profane ideas from texts which are by and large religious and ritualistic. Note that according to information I was kindly given by Klaus himself no other review did ever appear.
5. The various forms of deliberate camouflage or unconsciously wrong translations are not so easily recognized.
6. This is the expression used in the 'Table of Contents' ('Form des Weltalls').
7. Cf. the caption of chapter 3 on p. 29: 'Die Angaben über die Form des Weltalls variieren je nach der Perspektive, aus der es betrachtet wird.'
8. Klaus adds in brackets the reference to page and line of the edition by Raghu Vira and Lokesh Chandra, Nagpur, 1954. (Sarasvati-Vihara Series. 31)
9. Of his footnotes in which he records variants only one is retained by me; see fn. 10.
10. Klaus reports in a footnote the corresponding critical apparatus of the edition, viz. mss. read *udvaṁ* (sic!) Ra: *udbakaṁ*; La: *ḍakam*.
11. The German original reads as follows (p. 30 f.): 'Diesbezüglich sagen [einige]: "Um das Maß einer Spanne, fürwahr, berührt dasselbe (das Zwischenreich) den Bereich hier nicht, um das Maß einer Spanne den dort nicht". Und weiter sagen [einige]: "Bloß um soviel, wieviel die Schafhaut (Amnion) einer Kuh, die gekalbt hat, ist, berührt es den Bereich hier nicht, um soviel den dort nicht." Und

weiter sagen [einige]: "Bloß um soviel, wieviel ... 17) ist, berührt er den Bereich hier nicht, um soviel den dort nicht." ... Und diesbezüglich, fürwahr, sagen [einige]: "Wahrlich, wie zwei Räder durch die Achse auseinandergestemmt werden, so werden die beiden Bereiche hier durch dasselbe auseinandergestemmt. Es berührt den Bereich hier sozusagen direkt, den dort sozusagen direkt".'

12. Reference is to his book *Emendationen zum Jaiminīya-Brāhmaṇa (Zweites Buch)*. Bonn, 1988. (Indica et Tibetica. 14)
13. *The Jyotiṣṭoma Ritual. Jaiminīya-Brāhmaṇa 1,66-364.* Leiden etc.: Brill, 1990. (Orientalia Rheno-Traiectina. 34) The translation of the passage in question is found on p. 81 f.
14. In note 57, on p. 253, Bodewitz rightly asks: 'Are there any parallels for this construction of the ablative + *na* (or the instrumental + *na*)?', see also below fn. 56.
15. The passage was left out by Klaus, but not by Bodewitz, reads as follows: *tad evaṁ ivaiva manyamānena geyam*, 'therefore it should be chanted by someone who considers it like this as it were' (tr. Bodewitz).
16. Cf. fn. 4 above and the 'Bibliography' attached to his book (pp. 322-8, viz. p. 326).
17. Vol. III, column 1023.
18. *anirdaśāhaṁ gāṁ sūtāṁ*
vṛṣān devapaśūṁs tathā /
sapālān vā vipālān vā
na daṇḍyān manur abravīt //.
19. Only under this condition does the qualification 'within ten days' (*anirdaśāhā*) make sense (a cow is regarded as difficult to manage the first days after calving). Further, newborn calves drink milk and cannot do damage to crops by grazing.
20. *Brāhmaṇa Section*, Pt. II, Hoshiarpur, 1973, p. 1595.
21. The masculine forms cannot be expected to solve the problem.
22. J. Wackernagel, *Altindische Grammatik*. Band II,2 *Die Nominalsuffixe* von A. Debrunner, Göttingen: Vandenhoeck & Rubrecht, 1954, pp. 315-17.
23. Viz. in note 59 on p. 253.
24. H. Oertel, 'The Jaiminīya or Talavakāra Upaniṣad Brāhmaṇa: Text, Translation and Notes', *JAOS* 16, 1894, pp. 79-260.
25. Klaus' rendering ('auseinandergestemmt') is (in spite of RV 10.89.4 where, according to him, this topos is met with for the first time) a bit too literal.
26. On which see F. Staal, 'The Twelve Ritual Chants of the Nambudiri Agniṣṭoma', in: J.C. Heesterman, *et al.* (eds), *Pratidānam* (i.e. *Festschrift Kuiper*). The Hague-Paris: Mouton, 1968, pp. 409-29, especially 423 f., and *Agni*, Vol. I, 1983, p. 643.
27. Viz. no. 60 on p. 253.
28. Cf. W. Caland, *Pañcaviṁśa-Brāhmaṇa. The Brāhmaṇa of the Twenty Five Chapters*. Calcutta: Asiatic Society of Bengal, 1931, p. 157. (Bibliotheca Indica. 255)

29. Cf. s.v. *spṛś, ni-* and s.v. *nispṛś*. Note that in both entries the references to the RV contain mistakes, i.e. read 8.96.11, in the former, and 10.95.9, in the latter.
30. Bodewitz renders *ni-spṛś* by 'fondle' also in his translation of JB 1.113 (*op. cit.*, p. 65, [see fn. 13]).
31. In his yet unpublished article on this hymn P. Thieme, too, renders *nispṛś* by 'zärtlich berührend'.
32. Note that *ni-spṛś* is in all these cases connected with a locative.
33. A continuous touch e.g. of the body of another person by one's hand - which need not remain always at the same place - can, of course, express tender feeling, etc., but I doubt that such is also denoted by the verb *ni-spṛś*.
34. Quoted from *Webster's New Encyclopedic Dictionary*, Cologne, 1994.
35. I do not, of course, want to deny that the *iva* can be accounted for also if one considers *ni-spṛś* to have the meaning 'to touch softly, caress, fondle'.
36. Note also the *iva* by which the last of the three views referred to by *evam* is qualified.
37. *Op. cit.* [see fn. 1], p. 31 ('Die zuletzt geäußerte Ansicht ist offensichtlich die verbreiteste gewesen').
38. The reference to ŚB 8.2.1.7, however, seems to be due to an error. Note that Klaus adds three more references in fn. 1 on p. 80.
39. Particularly instructive are passages like ŚB 7.1.2.23 where the *antarikṣaloka* is said to be *anatarhita*, 'not separated at all, from this world', or PVB 15.4.8 where heaven and earth are stated to be *samante*, 'having a common border(line)', with the intermediate world.
40. This statement is not falsified by the fact that *ulba* is, as already noted in the *Large Petrograd Dictionary*, used (at least once in the *Vājasaneyī Saṁhitā*) with reference to the uterus, because this is most probably not caused by a mistake ('Verwechslung'), but just a metonymy. Nor can Medhātithi be regarded as a witness for the other side, for his explanation of *jarāyu*, of Manu 1.43, by *ulbam garbhaśayyās* does not by itself prove that he did not know of the difference between the amnion and the chorion.
41. Cf. also Mahābhārata 6.25.38c (= Bhagavadgītā 3.38.3c) (*yatholbenāvṛto garbhas*), the only passage in the whole of the Mahābhārata where *ulba* occurs - which I was able to find out with the help of Prof. Tokunaga's computerized text.
42. Cf. A. Michaels, *Beweisverfahren in der vedischen Sakralgeometrie. Ein Beitrag zur Entstehungsgeschichte von Wissenschaft*. Wiesbaden: Steiner, 1978, pp. 76 and 156 f.
43. Cf. E. Bezzel und R. Prinzinger, *Ornithologie*, 2., völlig neubearbeitete und erweiterte Auflage. Stuttgart, 1990, p. 248.
44. Translated from *Handbuch der Zoologie. Eine Naturgeschichte der Stämme des Tierreiches*, gegründet von W. Kükenthal ... hrsg. von T. Krumbach. 17. Bd., 2. Hälfte, Sanropsida: Aves. Bearbeitet von E. Stresemann. Berlin-Leipzig: De Gruyter, 1927 bis 1934, p. 399. My thanks are due to zoologist Jochen Martens for his kind help and information.
45. Only the weight of the small ball of faeces seems to have interested zoologists.

46. Cf. A. Wezler, 'Bemerkungen zu einigen von Naturbeobachtung zeugenden Textstellen und den Problemen ihrer Interpretation', in: *StII* 13/14 (*Festschrift W. Rau*), 1987, pp. 321-46, esp. p. 339 (in order to find out the anatomical structure of a stalk of the lotos plants I had to pluck one in a pond in my own garden and to cut it open).
47. That is to say, I take the ending of the locative to denote what is called *sāmīpyakam adhikaraṇam* (*Bṛhadvṛtti* on Hem. 2.2.30) by Hemachandra.
48. I need hardly clarify that I had to rely on the existing published dictionaries.
49. On the difference between the two expressions and the ideas connected with them, see the important studies of J.F. Sprockhoff, 'Āraṇyaka and Vānaprastha in der vedischen Literatur. Neue Erwägungen zu einer alten Legende und ihren Problemen', *WZKS* 25, 1981, pp. 19-90, *WZKS* 28, 1954, pp. 5-43 and *WZKS* 35, 1991, pp. 5-46.
50. = ŚB 14.5.1.2. It was my friend Michael Witzel who kindly drew my attention to this passage.
51. I do not think that the text of the JB has to be emended here. In any case, I do not find the suggestion to read *yakṛty ulbaṁ* instead of *śakṛty ulbaṁ* acceptable (a suggestion that was made in the course of a discussion after a lecture given at the University of Tokyo). It is true that Monier-Williams in his *Sanskrit-English Dictionary* lists a word *yakṛtkośa*, 'the cyst or membrane enveloping the liver', and that its being a real part of the Sanskrit lexicon cannot be denied even though no text reference is given by the author. But *ulba* is, as far as I can see, always used, even metaphorically, with reference to something that comes out of a body or an entity, i.e. is born or born, as it were.
52. In passing only I should like to add that I am aware of the necessity of a methodological clarification of this term. I don't use it as an equivalent of 'Geistesgeschichte'. Ideas could be defined as 'entities that exist only as contents of some mind' (cf. *The Oxford Companion to Philosophy*, ed. by T. Honderich, Oxford-New York: Oxford University Press, 1995), but what I have in view are those of these entities which are still accessible because they are expressed (not necessarily by words) in an element of the Indian culture heritage. Yet I do not, of course, want to study ideas alone, and disregard their general socio-cultural and historical context - if only this context can be determined in terms of relative or absolute chronology. Besides, it is almost impossible to do everything at the same time. That is why, in my work published until now, I have confined myself more or less to pointing out certain ideas which struck me as particularly interesting. See also below fn. 70.
53. Cf. e.g. Klaus (cf. fn. 1 above), p. 81 f.
54. It is rather for the sake of convenience that I stick to this term, and not because I regard it as the most appropriate.
55. Cf. A. Wezler, 'Über Form und Charakter der sogenannten 'Polemiken im Staatslehrbuch des Kauṭilya' (Untersuchungen zum 'Kauṭilīya' Arthaśāstra II)', *ZDMG* 143, 1993, pp. 106-34. This was kindly pointed out to me by Prof. Ryutaro Tsuchida from the University of Tokyo.

56. The rare, or even unique, construction of the ablative + *na* (see fn. 14 above) is not sufficient a piece of evidence to assume that the author of the JB passage quotes, or closely follows, the wording of his source(s).
57. On which cf. W. Halbfass, *India and Europe. An Essay in Understanding*. Albany: State University of New York Press, 1988, pp. 349-68.
58. Reference is to the booklet *The Fidelity of Oral Tradition and the Origin of Science*. Amsterdam etc., 1986. (MNKAW, afd. letterkunde, nwe reeks 49/8)
59. As for the former I was provided with very useful information by one of my colleagues at the University of Hamburg, Jost Weyer, whose field of specialization is the history of chemistry. Note that Staal in his booklet (fn. 56) does also not define the concept of science presupposed by him.
60. Cf. e.g. Michaels mentioned in fn. 42 above, pp. 1-20.
61. On which cf. Klaus, *op. cit.*, (fn. 1), p. 72 (earth), p. 80 (intermediate world) and p. 164 (sky/heaven), but also fn. 18 on p. 31 (the comparison between the three worlds and two wheels on an axle 'does not permit to draw any conclusion about the shape of the earth and the sky/heaven').
62. They could, of course, also have been his contemporaries.
63. According to Staal (*op. cit.*, cf. fn. 56 above) - who in his turn refers to Gilbert Ryle, *The Concept of Mind*. London, 1949 - it would be 'knowledge *how*'.
64. Quoted from Staal's booklet p. 251.
65. Already in the RV the idea of being wrapped by an/the *ulba* is transferred to mythological and non-biological entities. - The reason or argument would, of course, in this case be: 'No, the distance is not that of a span, but that of the thickness of the amnion because these worlds are beings/entities that are born'.
66. The physical nature of that which separates the intermediate world from this world, etc., apparently did not form a problem to the Vedic thinkers.
67. *Das Recht der Gesellschaft*. Frankfurt am Main: Suhrkamp, 1995, p. 8.
68. The German original is: '(Aber man sollte sich) an eine Eigenart der jüdischen Rechtsexegese erinnern: daß es wichtig ist, Dissense auf ein angemessenes Niveau zu bringen und als Tradition zu bewahren.' Preservation of various traditions is one of the most significant features of Indian culture, and therefore also of Indian texts.
69. A particularly impressive and convincing example is Bhartṛhari's Vākyapadīya, and the Vṛtti on the first two Kāṇḍas, in which much, most probably the essential parts of the earlier philosophy of language has been preserved, at least in substance.
70. This is one of the reasons why the study of ideas suggests itself so naturally, as it were, in Indology.

XXXIV

THEOLOGY AND THE ACADEMIC STUDY OF RELIGION IN THE UNITED STATES

Donald Wiebe

1. My concern in this essay is with the role of theology in the academic/scientific study of religion in the United States. More precisely, my concern will be with the question as to whether theology *rightfully* has a role to play in the scientific study of religion and religions in the publicly funded university, for it is only in light of that question of legitimacy that I think we can make any sense of the developments in the study of religion in the USA or to understand the debates regarding its current status within the American scientific community. The question of the legitimacy of theological and religious studies in US universities is, unfortunately, very complex and will not be easily treated, for it is, in the context of the publicly funded university, both a political and a scientific (or methodological) question. In the privately funded universities and colleges in the US, and, obviously, in the privately funded seminaries and divinity schools, the question of the teaching of religion and theology presents no political crisis of legitimacy at all. And whether it presents a methodological, and therefore scientific, problem depends, I think, on whether the teaching of religion and theology in those contexts is undertaken as an autonomous intellectual activity only, or is seen to be an aspect or form of *Religionswissenschaft*.[1]

The teaching of religion or theology in the publicly funded university, on the other hand, is, or seems to be, directly prohibited by the first amendment of the constitution of the United States which clearly demarcates the state from the church and prohibits the state from any activity that would involve, or appear to involve, the establishment of religion.[2] In this respect the study of religion in Europe is considerably different from its counterpart in the United States because theology is not excluded from the curriculum in European universities. The question as to whether theology, ought to be allowed a place in the curriculum of the publicly funded university in the United States, has however, been answered, indirectly at least, negatively.

But that has not prevented theologians and other theologically minded scholars from trying to find some indirect means by which to bring theology back into the curriculum of the public universities.[3] And without some understanding of that fact I do not think it possible fully to understand the character of what is called 'religious studies' in the United States. With that issue in mind, I shall proceed to a review of the historical development of the field of 'religious studies' in the United States.

2. I cannot, obviously, in the space of a single essay hope to provide a fully adequate history of the study of religion and religions in the colleges and universities of the United States. What I shall do, therefore, is to review briefly several recent treatments of that history with a special focus on the emergence of the academic study of religion as an element in the curriculum of the publicly funded university. In stating the matter in this way, it needs pointing out that history is predated, and very much influenced by, a scholarly but religious study of religion that constituted a fundamental element in the early, colonial, university in America. University education in religion was essentially training of the Protestant clergy, with advanced study of theology coming after acquiring a classical undergraduate education, including the study of the Christian religion, common to all students in the university. To speak of a Protestant hegemony here, especially in the early years of the Republic, is quite appropriate given that the first Roman Catholic school was not established in America until 1789. However, by the end of the eighteenth century explicit instruction in bible and theology had been removed from the curriculum of the Protestant universities and relegated to seminaries and divinity schools, with moral philosophy replacing it as the integrating factor of the undergraduate programme of studies in the arts. This displacement of the study of religion from the arts faculty, however, also meant that that study was not likely to develop in the same way as other subjects in the arts curriculum, for as D.G. Hart maintains, 'the upshot of these developments was to remove serious study of religion from the liberal arts curriculum' (1992, p. 200).[4] One sees in the early history of the study of religion in the United States, then, an assumption that religious experience and commitment are necessary preconditions for the study of religion, and that that assumption was essentially of a Protestant nature; as Charles Long put it in a recent essay (1992), one sees in the early history of the study of religion in America the foundations of a cultural Protestantism that becomes the foundation for future developments in the study of religion, or at least profoundly influences and affects those developments.[5]

The only book length treatment of the study of religion in the United States is Robert S. Shepard's recent *God's People in the Ivory Tower. Religion in the Early American University*. Shepard's project in this book is to trace the development of the early programmes of religious studies in six American universities: Boston University, Cornell University, New York University, University of Pennsylvania, University of Chicago, and the Divinity School at Harvard University. He focuses his attention, however, chiefly on the developments in the study of religion at the latter two institutions because in his opinion, William Rainey Harper of Chicago and Charles Elliott of Harvard, the presidents of their respective institutions, 'provided an academic structure for the scholarly treatment of religion ... [and] promulgated a new rationale, [and] new conception of the place of religious studies in the academic environment' (Shepard, 1991, p. 42). The 'educational landscape' with respect to the study of religion in the post-colonial period was rather bleak for those interested in the study of religions other than Christianity; the cause of this, as Shepard somewhat circularly puts it, was the 'irregularity and fragility of the American university's interests in the scientific study of religion' (Shepard, 1991, p. 9). However, Shepard is also aware of the importance of the Protestant hegemony in religious education in American colleges and universities and points out that despite the emergence of a liberal form of theological reflection that embraced science and the importance of scientific method in all learning, and despite the availability of scientific models for the study of religion in various European contexts, most religious studies in American institutions of higher learning in the late nineteenth century were still dominated by theological and ministerial concerns. This was the case, he insists, at Boston University even though the university appointed to its ranks a professor of comparative theology whose work was to be dedicated to the 'historic and scientific study of religions' (Shepard, 1991, p. 12) as early as 1873. That study of comparative theology did not become a distinct academic unit free to develop in terms of principles peculiar to itself but rather remained an element in Boston University's School of Theology. There was no redirection of the School's programme, claims Shepard, 'from its primary mission of ministerial education' (Shepard, 1991, p. 15). It was a welcome handmaiden to theological study, but only in so far as such study helped to broaden the perspectives of potential ministers and missionaries.

The comparative study of religion at New York University, on the other hand, was intended to be scientific in the sense of treating 'other religions' fairly, claims Shepard, but was nevertheless dominated by an apologetic impulse. 'There was no design,' he writes, 'to support a strong academic program in the emerging discipline of Religions Wissenschaft' (Shepard, 1991, p. 27). Although some of the members of the teaching staff there thought an objective and scientific approach to the study of religions more appropriate for their university - which claimed to be non-sectarian - there was no intention on the part of those doing the teaching 'to engage in serious research in the science of religion' (Shepard, 1991, p. 26).

Cornell University Shepard claims, had but a 'brief flirtation with the science of religion' (Shepard, 1991, p. 18) which failed to become established because of its subordination to the work of Cornell's School of Philosophy. That 'flirtation', however, was important for its influence on Morris Jastrow of the University of Pennsylvania who saw it as an important contribution to the emergence of a scientific study of religion in the university context and drew support from it for his own work at the University of Pennsylvania. Although no formal programme for the academic study of religion emerged there until the twentieth century, when it did, writes Shepard, 'there is little doubt that serious scholarship was the objective of the history of religions group [and that] ministerial training had no formal role to play' (Shepard, 1991, p. 36). Because it was originally founded as a secular institution and without affiliation of a divinity school of any kind, Jastrow was able to establish a study of religion that accepted the historical method as the *sine qua non* of the study of religion, although shortly after his death, George A. Barton moved the programme in a religious direction so that, as Shepard puts it, it 'became a haven for ministers and rabbis of proximate churches and synagogues to continue their studies' (Shepard, 1991, p. 38).

Although Shepard, as I have already noted, believes that Harper at Chicago and Elliott at Harvard provided a new conception of the place of religious studies in the context of the academy, it is closer in intent to the traditional Protestant religio-theological approaches found at Boston, New York, and Cornell than it is to Jastrow's European influenced conception of the study of religion at the University of Pennsylvania. Shepard nevertheless maintains that both men were convinced of the value of scientific learning and that both also saw the value of the study of religion in the curriculum of the university. Consequently 'the outcome in both cases', he insists, 'was a welcoming of Religions Wissenschaft', (Shepard, 1991, p. 59) although it was very unlike that of the European scholars who founded the discipline.[6] 'For both men,' Shepard continues, 'the university was the proper place to study religion and to study it - Christian theology included - scientifically. Moreover, the scientific study of religion was appropriate for minister and scholar alike' (Shepard, 1991, p. 59). And on further reading it becomes clear that even though Chicago established the 'first nonsectarian, graduate research program in religious studies in America' (Shepard, 1991, p. 51) it received the majority of its students from the divinity school and, as Shepard admits, never really became 'a significant training ground for scholars of religion' (Shepard, 1991, p. 68). As for Harvard Shepard also recognizes that '[it] was ... in the training and preparation of ministers that Harvard Divinity School carried out its fundamental mission' (Shepard, 1991, p. 68). From this it is quite apparent, I think, that the academic study of religion was not conceived by either Harper or Elliott along the scientific lines on which Jastrow developed his programme at the University of Pennsylvania, for at both Chicago and Harvard the task of the university was considered to be not only that of the mediator of a knowledge about religions but also the mediator of religious knowledge. It is true that their programmes were based upon an acceptance of the role of scientific method in higher learning which, as Shepard claims, at least implicitly 'contained a sympathetic and harmonious role for the study of Religions Wissenschaft' (Shepard, 1991, p. 124) but the influence of the seminary within the framework of the university had a major and negative impact on the development of *Religions Wissenschaft* as an independent and autonomous undertaking in the university setting. Those who supported *Religions Wissenschaft* in the university, that is, were also, and primarily, involved with religious instruction and with the professional training of clergy and missionaries and they made the emerging discipline of *Religions Wissenschaft*, at best, an ancillary subject in a theological curriculum (Shepard, 1991, pp. 107, 111, 113, 119-121). This meant that there was not a simple openness to and acceptance of *Religions Wissenschaft* but rather that the results of such scientific research depended to some extent on the assurance of the supremacy of Christianity among the religions of the world (Shepard, 1991, p. 125). Indeed, Shepard comes to very much the same conclusion when he writes: 'Whether apologetic or moralistic in purpose, the important fact was that Religions Wissenschaft programs arose within a Christian - mostly Protestant - ethos' (Shepard, 1991, p. 126). And this must be clearly recognized if one is to understand why the current debates over the role of theology in programmes of religious studies in the university's curriculum are what they are. Early *Religions Wissenschaft* in the American university context, therefore, was, in effect, a 'Christian *Religions Wissenschaft*'. This Christian *Religions Wissenschaftliche* influence in the nineteenth century university certainly helped improve the level of scholarship on religion in the university - essentially by helping it to transcend narrow ecclesiastical boundaries placed upon thought about religion and religions - but it also prevented the development of a genuine, full-blown *Religions Wissenschaft*. And Shepard does not himself altogether shy away from drawing such a conclusion, even though it undermines his earlier claim that Harper and Elliott produced a new rationale for the place of religious studies in the academic environment. 'As the theological school within the emerging university became the place to study religion,' he writes, 'Religions Wissenschaft became subject to the particular curricular structure, objectives, and ethos of the seminary' (Shepard, 1991, p. 128). *Religions Wissenschaft*, he continues, was unable to separate [itself] from the theological and professional concerns of the nascent university, particularly the rising seminary within the university. A theological agenda accompanied the entrance of comparative religion in American higher education despite the arguments, some rhetorical and some sincere, that the new discipline was objective, scientific, and appropriate as a liberal arts subject. Students valued the new science of religion for its professional utility; [but] few even considered its potential as a discipline coterminous with other humanities disciplines (Shepard, 1991, p. 129).

In effect, then, Shepard concludes that the study of religion failed to emerge as an independent academic enterprise that could attract its own identifiable group of scholars with a *Religions Wissenschaft* identity. Such an identity eluded the grasp of scholars in the field of religion, he insists, until after the second World War (Shepard, 1991, p. 129).

3. In coming to this conclusion Shepard adopts a view about the development of religious studies in the United States which he claims is first put forward by Joseph M. Kitagawa in his *The History of Religions in America* (1959) and *Humanistic and Theological History of Religion With Special Reference to the North American Scene* (1983). Roughly stated that amounts to the claim that the religious liberalism of the World's Parliament of Religions was the fundamental impetus that led to the establishment of the study of comparative religions in American universities, and kept the interest in that enterprise alive until the late 1920s but which thereafter declined with the demise of the liberal attitude caused by the dominance of neo-orthodox theology in American thought in the 1930s. For the most part Shepard's reading of Kitagawa is quite acceptable for it is true that Kitagawa does talk of a decline of interest in the scholarly study of religion in the 1930s, and of a renewed interest in that field 'after the end of World War II' (Kitagawa, 1959, p. 5; 1983, p. 556). But Kitagawa's claims in these essays are more complex, and more confused, than Shepard acknowledges. In the earlier essay, for example, Kitagawa notes that among the participants at the Parliament there were a number of scholarly figures, including students of religion, but emphasizes that 'they attended the parliament as representatives of their faiths or denominations and not of the discipline of the history of religions' (Kitagawa, 1959, p. 4). And in drawing attention to this fact Kitagawa seems to acknowledge that the academic study of religion in the United States has at least a two-fold character - indeed, that it is of two kinds. In *The History of Religions in America*, for example, he demarcates the History of Religions as a scientific enquiry from the 'theological history of religions', insisting that the former 'developed only at the dawn of the modern period, namely, during the Enlightenment' (Kitagawa, 1959, p. 16) and he further insists that such a scientific *Religions Wissenschaft* 'did not develop in America until a relatively recent date' (Kitagawa, 1959, p. 1). The reasons for the late development he claims, moreover, are to be found in the religious character

of the education provided by the American universities in the colonial period.

In his later essay on 'Humanistic and Theological History of Religions' Kitagawa argues that 'any attempt to assess the development of the History of Religions during the past 100 years, especially in North America, cannot ignore the impact of two of the major international assemblies on the discipline, namely the World's Parliament of Religions, held in Chicago in 1893, and the first Congress of the History of Religions, held in Paris in 1900' (Kitagawa, 1983, p. 553). He acknowledges, however, that it is the Parliament of Religions that becomes 'inseparably related to the aim of Comparative Religions or History of Religions' (Kitagawa, 1983, p. 553) in the United States but can only half-heartedly claim that the tone for later developments in the History of Religions in America was set by the 1900 Paris Congress (Kitagawa, 1983, p. 554). He reviews, very briefly, the discussion of the nature of the scientific study of religions among those scholars involved in the International Association for the History of Religions (IAHR) - the official body formed in 1950 to oversee the Congresses that followed the Paris meeting of 1900 - but nowhere shows *how* it influenced developments in the United States. Excluding the influence of a few European trained scholars like Morris Jastrow, for instance, it is really not possible to argue, as Kitagawa does, that the IAHR, or even the Paris Congress, has had much of an influence on developments of the field in the US. Although the American Society for the Study of Religion (ASSR), formed in 1959, is an affiliate member of the IAHR, its influence, late as it came in the post-war development of the field, has been negligible; its severely restricted membership has simply not allowed it to play a significant role in the development of the field.[7] The American Academy of Religion (AAR), the main professional association representing the majority of those who teach religion in institutions of higher education in the United States, not only has never had either a formal or an informal connection with the IAHR, it has deliberately kept itself aloof from, if not opposed to, their activities.[8]

This I think accounts for the rather ambiguous character of Kitagawa's understanding of the development of the field of religious studies in the United States. Although in his 1959 article he recognizes the emergence of a scientific study of religion in the Enlightenment (Kitagawa, 1959, p. 17) he insists that *Religions Wissenschaft*, (the History of Religions), is not simply scientific but rather religio-scientific because it must 'view the data 'religio-scientifically' (Kitagawa, 1959, p. 21). And in the 1983 essay he refers to the History of Religions as 'an autonomous discipline situated between normative studies ... and descriptive studies' (Kitagawa, 1983, p. 559). Unlike other social sciences, then, this discipline does not, he insists, simply seek descriptions or explanations of events and processes as do other social sciences, but rather is inquiry directed to the meaning of the religious data on which it focuses attention and is therefore a mode of inquiry that, in a sense, links descriptive with normative concerns (Kitagawa, 1983, p. 560). This (hybrid) enterprise he then refers to as the humanistic history of religions which he contrasts with a more explicitly theological history of religions that is more directly and overtly normative. The former, according to Kitagawa, derives from the work of scholars involved in the international congresses and, later, the IAHR and is what Kitagawa considers to be a *bona fide* scientific undertaking, but he provides no evidence upon which to base this claim. Kitagawa is entirely correct, of course, in his claim that such a humanistic history of religions comes to the rescue, so to speak, of the field of religious studies after World War II. However, to argue that such a view of the field derives from the work of those involved in the Paris and subsequent Congresses, or in the IAHR, is difficult to maintain since those scholars quite clearly saw themselves as undertaking a scientific enterprise similar in structure to any of the other sciences included in the curriculum of the modern university. Moreover, if the IAHR were the source and foundation of such a humanistic *Religions Wissenschaft*, as Kitagawa maintains, the lack of cooperation with, and even outright hostility to, the IAHR by the American Academy of Religion is a puzzle that becomes difficult if not impossible to solve.

That the academic study of religion in the United States experienced some kind of rejuvenation after the Second World War is, I think, fairly generally accepted. Whether Kitagawa has adequately characterized and accounted for it, however, is another matter. Whereas Kitagawa attributes this renewal of interest in religion to 'factors' such as a sudden interest in things Eastern, a growing fascination in the social sciences with myth and symbol, and an interest on the part of influential theologians in inter-religious dialogue (Kitagawa, 1983, p. 556), D.G. Hart maintains that religious studies became a growing concern in large part because of 'the cultural crisis generated by the two world wars' (D.G. Hart, 1992, p. 218). 'Science and technology,' he continues, 'seemed to be more of a threat than a salvation to social ills, and educators took steps to ensure that undergraduates received a proper training in values and morality' (D.G. Hart, 1992, p. 218).[9] He notes in this regard that in 1943 the American Council of Learned Societies (ACLS) 'recognized' as self-evident the value of religion to a liberal education, despite the dangers of indoctrination that it considered to accompany the teaching of religion; the benefit that accrued to the study of religions in the university from this widespread concern for a general liberal education of America's youth was immeasurable and had an obvious impact upon the growth of religious studies in the post-war years. As Hart points out, from the end of World War II until 1960 programmes in religious studies in

American universities doubled and student enrolments showed a dramatic increase (D.G. Hart, 1992, p. 211). But what one sees here is the coincidence of the concerns of the ACLS with those of American Protestantism and not with those who wish to establish a science of religion(s); the concern of liberal Protestants was not the objective study of religion but rather to show the ethical and spiritual relevance of religion for the modern, now fragmented, world. Hart writes:

> Calls for unity in higher education provided Protestants concerned about the study of religion with added leverage. Generated by the crisis of the Second World War, many books and articles by mainline Protestant leaders appeared that picked up on the sense of upheaval within American universities ... [lamenting] the increasing fragmentation of knowledge and the secularization of learning (D.G. Hart, 1992, p. 209).

Given this informal alliance between Protestant Christianity and the defenders of the Humanities, it is not surprising that programmes in religious studies departments simply resembled the programmes and curricula in Protestant seminaries and divinity schools.[10] And what is evident in this is that the rejuvenated *Religions Wissenschaft* of which Kitagawa speaks - the humanistic history of religions - is more a continuation of the programmes Shepard calls 'hristian *Religions Wissenschaft*' since it took on the character of the Christian - by which he means mostly Protestant - ethos within which it arose (Shepard, 1991, p. 126). Shepard is right to point out that after World War II this 'Christian *Religions Wissenschaft*' was less directly associated with the professional training of theologians and ministers and that it attempted, with some degree of success, to become a more scholarly and therefore academically acceptable enterprise, but he provides no indication that its more humanistically based identity made of it something fundamentally different from what it had been through the late Victorian and Edwardian periods of American history. That so-called new, humanistically based, *Religions Wissenschaft*, that is, far more nearly resembles the theologically oriented history of religions in the early American university of which Shepard writes than it does either the contemporary work of European scholars connected with the IAHR and its international congresses, or the new ideal of *Religions Wissenschaft* that is said to have emerged in the academic community in the 1960s. As Hart puts it, 'Proponents of religious studies, which often meant the ethos of liberal Protestantism, made common cause with humanistic scholars and resurrected the philosophical idealism of the old genteel tradition under the label of general education' (D.G. Hart, 1992, p. 218). Consequently, the period from 1945 to 1960 was not only no advance in the development of religious studies as an academic or scientific undertaking, it was a retrograde step in that, like the classics in the colonial curriculum, it defined itself only in terms of a reaffirmation of commitment to civilization that in effect put it into opposition with the sciences.

4. The 1960s, Hart maintains, saw a decline in the consensus between the university and the church - between the university and the dominant Protestant ethos, although he does not provide an explanation for this development. This was due in large part, I would argue, to the growing religious pluralism of Western civilization which made it impossible any longer to assume the superiority of Christian ideals and beliefs. It became obvious, therefore, that if the study of religion was to remain an element in the university curriculum it would have to find a more clearly scientific and objective justification. According to Hart, 'the formation of the American Academy of Religion signalled ... [that] reaction against the humanistic and liberal Protestant orientation of religious studies during the 1940s and 1950s' (D.G. Hart, 1992, p. 219), and provided the discipline with a new, radically altered structure and identity. Hart is right, I think, to claim that the formation of the American Academy of Religion was a reaction against the 'Christian *Religions Wissenschaft*' that had dominated the study of religion in the American university, but he is seriously off the mark, I suggest, in claiming that the Academy was successful in its bid to put religious studies on a more objective and scientifically acceptable basis. Without understanding this, it seems to me that it is impossible properly to understand the present state of *Religions Wissenschaft* in the American academic context. I shall, therefore, provide a brief analysis and critique of Hart's thesis as it is presented in his essay 'American Learning and the Problem of Religious Studies'.

The essential element in Hart's analysis, in my opinion, is the claim that the debates and discussions among students of religion in the early 1960s which led to the formation of the American Academy of Religion constitutes a watershed in the history of the study of religion in America. According to Hart it does so because the formation of the Academy involved acceptance of the scientific commitments generated by the Enlightenment in that it moved from seeking religious, theological, and humanistic explanations to seeking for scientific explanations of religious phenomena. As he puts it, '[t]his tradition abandoned the notion that religious guidance or belief was necessary for the study of religion and insisted instead that religious phenomena could be explained in naturalistic categories as well as any other artifact' (D.G. Hart, 1992, p. 197). Unfortunately, I think Hart fails to make his case. There is no doubt that the study of religion had a very uncertain status as an academic discipline prior to the 1960s because it was neither sufficiently 'classical' to be wholly accepted as one of the humanities nor scientific enough to be included among the sciences, as Hart suggests. As he points out, for example, it was not until 1979 that the ACLS even listed the AAR among its disciplines, and even today it does not appear that even the social sciences, let alone science more generally, recognize the work of the so-called student of religion as worthy of consideration.[11] The formation of the National Association of Biblical Instructors in 1909 and of the American Association of Colleges in 1915 he claims had some influence in this respect because they promoted standardization of religious instruction in both universities and ecclesiastical educational institutions, but they did not, in my opinion, make a significant impression upon the broader academic/scientific community; those teaching religion were still not readily accepted by the academy at large. The links between the study of religion and the humanities were strengthened, but for many in the university, this connection was considered to undermine the scientific value of the claims of their disciplines. Hart's further claim that many within the field of religious studies felt that their historical associations with ecclesiastical institutions constituted a problem and that the perception of bias and lack of objectivity in their work required some response is also on the mark. Whether their response to the problem - namely, undertaking a self-study of the National Association of Biblical Instructors which led to the birth of the American Academy of Religion by renaming the Association - was adequate, however, is another matter altogether.

According to Hart, the emergence of the AAR marked a radical change in the way religion was studied and taught in the United States. 'The new methods of studying religion advocated by the AAR,' he writes, 'signalled the demise of this Protestant dominance as professors of religion became increasingly uncomfortable with their religious identification … . By striving to make their discipline more scientific, religion scholars not only embraced the ideals of the academy but also freed themselves from the Protestant establishment' (D.G. Hart, 1992, p. 198). The National Association of Biblical Instructors, he argues, established the self-study committee to review the status of the field within the academic community and to see what the Association could do to improve the 'health' of the discipline - 'to make the study of religion more academically respectable' (D.G. Hart, 1992, p. 212). To that end he maintains that the members of the Association 'sought to establish their field on empirical and scientific grounds', and therefore not only rejected their Protestant associations and affiliations but also rejected the idea of religious studies as a humanistic discipline. Thus he claims that from the time of the self-study report, '[to] be religious was no longer as important for professors of religion as methodological sophistication and academic achievement' (D.G. Hart, 1992, p. 213).

Hart bases his argument for such a transformation of the study of religion in American universities in the 1960s essentially upon an analysis of Claude Welch's *Graduate Education in Religion: A Critical Appraisal*, which was the result of a major two year research project on graduate studies in religion as distinguished from professional theological education. As the president of the American Academy of Religion, Welch presented a preliminary report on the results of his research in his address to the Academy in 1970 under the title *Identity Crisis in the Study of Religion? A First Report from the ACLS Study*. Hart's claim that Welch's report 'helped to legitimate the discipline within American higher education' (D.G. Hart, 1992, p. 214) I am convinced is entirely correct. I am also convinced of his claim that Welch's report 'called for increased methodological sophistication within the profession' (D.G. Hart, 1992, p. 214) and that to some extent that was achieved. However, his further claim that the report 'marked a clear defeat for the older humanistic and Protestant orientation of religious studies ...' (D.G. Hart, 1992, p. 214) I think an overstatement at best. It is true that Welch called for an independence of religious studies from ecclesiastical interference and control, and that his study helped undermine the Protestant hegemony that had until recently wholly dominated religious education and the study of religion in America. But it did not, I would argue, persuade scholars to shift their loyalty from the church to the academy; it simply made the nature of the relationship of the scholar to the church and to the academy more 'sophisticated'. Part of the reason for Hart's faulty assessment here is due, I think, to the success of Welch's claims that as of 1970 students of religion had clearly recognized the distinction between theology and religious studies and had, formally at least, accepted the fact that it was religious studies alone that was the domain of the university teacher of religion. Welch's claims were more window-dressing than reality, however, for although a formal kind of recognition of the scientific character of the study of religion had been achieved, it was at best lip-service that was being paid to that notion while the reality on the ground was considerably different. Indeed, Welch himself seems to have been aware of this for he notes that despite the successes in programmes of religious studies finding acceptance in a number of public universities there were still many signs that provided a clear indication that the issue of identity had not yet been resolved; there were, that is, a number of serious unresolved matters which in Welch's consideration still left the field of religious studies, as he put it, at risk of 'falling back into the arms of confessional interests' (Welch, 1970, p. 12).

It is interesting to note that D.G. Hart himself is aware of precisely those matters that concerned Welch, but he does not seem to take them as seriously as in my opinion he ought. He admits, for example, that the change in name of this professional body from the National Association of Biblical Instructors to the American Academy of Religion did not settle the issue of advocacy and indoctrination - aspects characteristic of the teaching of theology and religion in seminary and divinity school contexts (D.G. Hart, 1992, p. 213). Scholars in the field, he notes, wanted an increase in academic respectability, but they also wanted to make students more humane, religious, and moral (D.G. Hart, 1992, p. 213). In this regard, moreover, he points out the irony in Holbrook's chairmanship of the Self-Study Committee that brought about the change of name of the National Association given the fact that Holbrook had argued strenuously for a view of religious studies as a humanizing and liberalizing discipline in his *Religion. A Humanistic Field*, published in the same year. Furthermore, though claiming that scholars shifted their loyalties from the church to the academy, Hart also acknowledges that this did not mean a shift in loyalties away from religion, despite having insisted that the new methods advocated in the study of religion by the AAR made it clear that to be religious was no longer important for professors of religion (D.G. Hart, 1992, pp. 213-15). Closer analysis of the presidential addresses to the AAR would, I think, have made it even more clear to Hart that his claims regarding the defeat of the 'older humanistic and Protestant orientation in religious studies' is an exaggeration at best, if not entirely false. And within two years of the presentation of Welch's first report of the ACLS study to the Academy the real ethos of, and state of affairs in, the field were laid bare in Robert Michaelson's presidential address entitled 'The Engaged Observer: Portrait of a Professor of Religion'. Michaelson acknowledged a trend within the Academy that sought a clear demarcation between religious instruction and the teaching about religion but noted that a clearly opposite tendency was re-emerging in the field, one that rejected objective scholarship and demanded 'involvement' (Michaelson, 1973, p. 422). In fact, so strong is that 'opposite tendency' that he talks of it as a process giving birth to a new form of religion which he calls 'classroom religion' (Michaelson, 1973, p. 422). Although Michaelson expresses some surprise with respect to the scholars involved in this development, he himself seems to buy into the concerns which they expressed, and therefore appears to contribute to their project of replacing the ideal of an objective and scientific study of religions with a more traditional model of religious studies. Michaelson is concerned in this essay, that is, with the risk to the soul presented by the study of religions, and claims therefore that the teacher of religion, even in the university context, has an obligation to help her or his students to achieve wholeness. Consequently he proposes a study of religion that, *per impossible* one might think, combines detachment with involvement because, he insists, it is not possible for the scholar to keep her or his scholarly concerns clearly distinct and separate from existential concerns. 'That much,' he writes, 'the age of involvement has taught us' (Michaelson, 1973, p. 423).[12] In view of these comments it is not surprising, then, that Eric Sharpe came to quite a different conclusion about the history of the development of the field in America. Contrary to Hart, Sharpe maintains that '... by the early 1960s the American schools were moving firmly into the acceptance of this wider role' (Sharpe, 1986, p. 280) for religious studies.

5. Although there is more one might say about the history of the growth of the academic study of religion in the United States it seems to me that a enough has been said about the matter to allow some significant conclusions to be drawn. I think it abundantly clear, for example, that although there has been much activity and apparent shifts of attitudes within the field, the field has not really advanced very much beyond what it was at its origins in the nineteenth century. Kitagawa's suggestion, followed by Shepard, that a new style of religious studies emerged in the immediate post-war years, and D.G. Hart's suggestion that with the birth of the American Academy of Religion in the 1960s a transformation occurred that catapulted the study of religion in American universities from the religious and humanistic realms into the scientific is simply not substantiated; as I have shown, the arguments provided on behalf of such claims are internally inconsistent and the evidence adduced in their favour either weak or non-existent. The study of religion in the United States is, for the most part, therefore, still conceived of in religious or quasi-religious terms. Two recent essays, both commissioned, so to speak, by the AAR make this abundantly clear: the first is Leo O'Donovan's overview of the 75th Anniversary meeting of the AAR held in Chicago in December of 1984, and the second is Ray L. Hart's recent study on religious and theological studies in American higher education.

Although O'Donovan talks of the programme of that special meeting of the Academy as having been so structured as to provide 'a general indication of the disciplinary challenges which have emerged for the *scientific* study of religion today' (O'Donovan, 1985, p. 560, emphasis added), it is clear that the relationship of advocacy learning, and theological and even ecclesiastical involvement in the study of religion were not among the items up for consideration. Indeed, it is also clear that O'Donovan and his colleagues in the Academy simply assume that such learning and involvement are necessary elements in any proper academic study of religion. O'Donovan has no hesitation, for example, in encouraging the Academy to involve itself in the reshaping of American religiosity which he sees as already under way. 'If a new appropriation of faith is under way in the United States,' he writes, 'we shall need guidance in understanding the experience and courage in learning how it may be socially constructive. The American Academy of Religion is, of course, the largest single professional association for students of religion who might contribute to meeting that need' (O'Donovan, 1985, p. 558). He places great emphasis in his review on the contributions of theologians, and particularly those who call for a contextual or 'relativized theology' because he thinks that in so doing the scholar can assist religion in use of its emancipatory potential for the betterment of society. Members of the Academy may expect, he says, that it will 'be more - or less - challenging to the society in which it exists and to be more - or less - concerned for the unity of purpose among members who are both citizens and believers' (O'Donovan, 1985, p. 564), and he urges the members of the Academy in this regard to consider seriously constructing their own versions of such religion-social agendas as may be found in various documents of 'the World Council of Churches and by Vatican II's Pastoral Constitution on the Church in the Modern World' (O'Donovan, 1985, p. 565). O'Donovan concludes his review with the following statement which, because of its revealing character, I quote at some length:

> Whether the academic study of religion can respond to this social situation will surely be a matter of concern for its largest organization when it meets again next year at the Anaheim Hilton just before Thanksgiving. Will the grand contemporary sculpture of Chicago - Calder and Picasso and Miró - be replaced in a dispiriting way by Disneyland? Next year we will again be searching for images that travel well, images that even bear transcendence. If our study of religion is genuinely concerned for its cultural context, then there should be time enough for a good-humoured visit to the Enchanted Kingdom. But if we pursue religion for our own sakes alone, then the Meeting will pass by emptily. Next fall in California, it seems, there will be another opportunity to see which face of religion will be uppermost, the timelessly trivial or the historically transcendent, the giddiness of self-centred human ceremony or the true joy that accompanies God's pleasure in creation (O'Donovan, 1985, p. 565).

Given this picture of the study of religion in the Academy one would be very hard pressed indeed to make out a case for a radical transformation having occurred in the academic study of religion due to the transformation of NABI into the AAR. A review of the substance of the scholarly contributions made to the AAR - as found in the programmes for the annual meetings of the Academy - moreover, will clearly show that the kinds of ethical, religion-social, and theological concerns that are close to O'Donovan are also the concerns of most of the members of the Academy. And that, of course, means that such concerns are also the dominant element of the programmes of study in the various college and university departments in the United States in which those scholars teach.[13]

It is, therefore, difficult, if not impossible, in light of O'Donovan's account of the work of the Academy and its members, to talk of new and radical developments in the scientific study of religion in American academic circles. The so-called new developments in the 1940s and the 1960s have simply proved not to be radical departures from the 'Christian *Religions Wissenschaft*' model first established in the field in the nineteenth century. There can be no doubt that the theological character of today's brand of *Religions Wissenschaft* is considerably less Christian and Protestant in character - i.e., that it is more pluralistic - and that it is often less explicitly and loudly religious and theological. But this is a far cry from showing that it is religiously and theologically neutral and, therefore, more scientific. The intention to make the academic study of religion, already existent in the university curriculum, more scientific no doubt constituted the hopes of some of those who aided in the formation of the American Academy of Religion, but it was an intention that was obviously short-lived or otherwise overwhelmed by other more pressing concerns. Nor has that scientific impulse with respect to the study of religion re-emerged in the Academy (or in very many departments of religious studies in American universities and colleges in the thirty years since the formation of the Academy, despite the rhetoric that is sometimes heard in this regard.[14] Although no full-scale study of the current state-of-the-art of the field of religious studies has been undertaken which is able to provide evidence in support of these claims, I think the results of the recent report of the AAR's Committee on Education and the Study of Religion, which has been billed as an update on the Welch report, go a long way toward substantiating them.

If any radical changes have taken place in the academic study of religion in the three decades of the existence of the AAR one would be justified, I would argue, in expecting to find some indication of that in Ray Hart's report on behalf of the Committee on Education and the Study of Religion entitled 'Religious and Theological Studies in American Higher Education'. In fact, however, the religious studies scene seems not to have changed at all over that period of time: the report, that is, shows rather clearly, I think, that the 'state-of-the-art' in religious studies is still essentially at the 'NABI stage', if we may call it that, and that it simply reflects, as did the substance and structure of the NABI, what the academic study of religion was all about at its first emergence on the American scene - an inchoate enterprise often indistinguishable from theology and laced with apologetic and moral concern. As Hart's report makes clear, the members of the Academy are still clearly divided along 'study of religion'/'practice of religion' lines, with a deep division between those concerned with knowledge about religion and those concerned for the truth of religion (Ray Hart, 1991, pp. 734, 778).[15] They therefore differ little from their NABI forebears. Moreover, it appears that by far the majority of the members of the Academy favour the style of scholarship that attempts to combine the two activities, or at the very least does not insist on a clear demarcation between them, for one of Hart's concerns is to try to resolve the tension between them without, as he puts it, dishonouring the virtues the field is claimed to inculcate: tolerance, pluralism, respect for opposing positions, etc.' (Ray Hart, 1991, p. 790). What appears to be at stake here is a loss of power in the Academy; either a loss of power in alienating a large contingent of members in the Academy and risking a split in it, or a loss of academic legitimation by being too closely associated with religious and theological matters. To fudge the issue is to maintain the status quo whereas to clarify the theology/study of religion tension is to risk radical alteration in the structures of control that presently characterizes the field. It is not wholly surprising, therefore, that Hart maintains that the conflict of approaches to the study of religion revealed by his report is a problem in the definition of the field that has 'not yet been sufficiently named to permit whatever is at stake to come to the fore' (Ray Hart, 1991, p. 731).

There are some hints in the report that seem to suggest some advance in the discussion of the nature of the field since the days of the NABI, but they do not amount to much. Although Hart notes the discomfort of many 'with the nomenclature that discriminates 'religious' and 'theological studies', he does admit that 'religious studies' is now generally used to refer 'to the scholarly neutral and non-advocative study of multiple religious traditions' (Ray Hart, 1991, p. 716). And he is also aware of the fact that the younger scholars, recently out of graduate school, find theology to be a problem in the study of religion in the university context (Ray Hart, 1991, p. 732). All he does with these 'discoveries' however is to claim, rather lamely, that the study undertaken was unable to get to the bottom of this opposition. Significantly, the 'opposition' is indistinguishable from that which plagued the academic study of religion in the American setting from the time of the emergence of the field in the late nineteenth century. And it seems to me, therefore, that one is still forced to the conclusion to which Shepard comes in his analysis of the early history of the field, namely, that academically/scientifically respectable identity still eludes the field of religious studies in the United States.

Three decades after the submission of the report of the NABI Self-Study Committee the Academy has been presented with a report from another, the AAR, Self-Study Committee. Although the number of goals or aims delineated in the reports has increased in the latter version of the Association's/Academy's Mission Statement, they are in all essential respects identical. The seventh goal, which requires of the Academy acceptance in its 'conversation' of 'the various voices in the field of religion', appears to be directed entirely to ensuring that the academic or scientific study of religion within the Academy will never exclude the presence of religion itself. Where this presented a problem in the early days of the AAR in its search for academic legitimation, the intention was muted; today it need not be because, as Barbara DeConcini, the current Executive Director of the Academy, puts it in an interview regarding the Mission Statement, it is no longer possible to assume 'a clear distinction separating scholarship from advocacy', and therefore 'all disciplined reflection on religion' must be endorsed by the Academy of Religion, and by implication, the academy in general (DeConcini *et al.*, 1993, p. 5). Indeed, she maintains that a number of the recent presidential addresses to the Academy have clearly shown 'that religionists have an important contribution to make to the entire conversation in the academy' - small 'a' - 'in this regard because of the particular insights and understandings of the religionists with respect to 'knowledge and power, evidence and advocacy' (DeConcini *et al.*, 1993, p. 5). In putting matters this way, however, it seems to be fairly clear that the AAR has succumbed to the temptation - a temptation which Welch thought the student of religion had transcended in the 'NABI to AAR' development - to look upon the study of religion as much as a religious undertaking as a scientific one.

It seems, then, given the dominant view of the nature of the study of religion within the academic religious studies community, that the teaching of religion and theology within the public university in the USA has become a reality and that this has come about as a result of the fact that the scientific study of religion has been so described as to make it virtually indistinguishable from theology. This constitutes a victory for the American theological community in a double sense, for not only does it evade the strictures of the constitution regarding the 'separation of church and state', it also provides a kind of scientific legitimacy to the undertaking it did not have when wholly ensconced in the seminary and divinity school setting. The price for this victory, which has not yet been paid, will, I think, be rather high because it seems to me that this is a development which when fully understood will not be found acceptable to the academy at large. My best guess is that the broader scientific community within the university will consider the Academy of Religion to have jettisoned the very thing they appear to have set out to gain in the process of transforming the National Association of Biblical Instructors into the American Academy of Religion - namely, a genuinely objective and scientific knowledge about religion and religions. For those in the field still concerned to achieve that ideal, there is then ground for deep disappointment in the history of the development of religious studies in the United States. There is, nevertheless, some hope to be found in the 'hints' of progress in the development of the field which I have suggested are to be found in Hart's report. Those hints say nothing of the current status of the field, but they may well be harbingers of a genuinely new and radically different future for the academic study of religion in America.[16]

REFERENCES

Alexander, Kathryn O.
1988 'Religious Studies in American Higher Education: a Bibliographic Essay', *Soundings* 71, pp. 389-412
Cady, Linell E.
1993 *Religion, Theology, and American Public Life*. Albany: State University of New York Press
DeConcini, Barbara, *et al.*
1993 'Barbara DeConcini on the AAR's Self-Study: An Interview', *Religious Studies News* 8/3, p. 5
Hart, D.G.
1992 'American Learning and the Problem of Religious Studies', in: G.M. Marsden and B.J. Longfield (eds), *The Secularization of the Academy*. New York [etc.]: Oxford University Press, pp. 195-233
Hart, Ray L.
1991 'Religious and Theological Studies in American Higher Education', *Journal of the American Academy of Religion* 69, pp. 715-827
Holbrook, C.A.
1963 *Religion. A Humanistic Field*. Englewood Cliffs: Prentice-Hall
1964 'Why the Academy of Religion?', *Journal of the American Academy of Religion* 32, pp. 97-105
Inglis, Fred
1993 *Cultural Studies*. Oxford [etc.]: Blackwell
Kitagawa, Joseph M.
1959 'The History of Religions in America', in: M. Eliade and J.M. Kitagawa (eds), *The History of Religions. Essays in Methodology*. Chicago: University of Chicago Press, pp. 1-30
1983 'Humanistic and Theological History of Religions with Special Reference to the North American Scene', in: Peter Slater and Donald Wiebe (eds), *Traditions in Contact and Change*. Waterloo: Wilfrid Laurier University Press, pp. 553-63
Lawson, E. Thomas
1967 'A Rationale for a Department of Religion in a Public University', in: Milton D. McLean (ed.), *Religious Studies in Public Universities*. Carbondale: Southern Illinois University Press, pp. 38-44
Long, Charles H.
1985 'The Study of Religion in the United States of America: Its Past and Its Future', *Religious Studies and Theology* 5, pp. 30-43
1992 'A Common Ancestor: Theology and Religious Studies', in: Joseph M. Kitagawa (ed.), *Religious Studies, Theological Studies, and the*

University Divinity School. Atlanta: Scholars Press, pp. 137-50

Michaelson, Robert S.

1967 'The Study of Religion: A Quiet Revolution in American Universities', in: Milton D. McLean (ed.), *Religious Studies in Public Universities*. Carbondale: Southern Illinois University Press, pp. 9-14

1973 'The Engaged Observer: Portrait of a Professor of Religion', *Journal of the American Academy of Religion* 41

Murphy, Murray G.

1989 'On the Scientific Study of Religion in the United States, 1870-1980', in: M.J. Lacey (ed.), *Religion in Twentieth-Century American Intellectual Life*. Cambridge: Cambridge University Press, pp. 136-71

O'Donovan, Leo J.

1985 'Coming and Going. The Agenda of an Anniversary', *Journal of the American Academy of Religion* 53, pp. 557-65

Prebish, Charles S.

1994 'The Academic Study of Buddhism in the United States: A Current Analysis', *Religion* 24, pp. 271-78

Robertson, Roland

1993 'Community, Society, Globality, and the Category of Religion', in: E. Barker, J.A. Beckford, and K. Dobbelaere (eds), *Secularization, Rationalism, and Sectarianism*. Oxford: Clarendon, pp. 1-17

Schüssler Fiorenza, Francis

1991 'Theological and Religious Studies. The Contest of the Faculties', in: Barbara G. Wheeler and Edward Farley (eds), *Shifting Boundaries: Contextual Approaches to the Structure of Theological Education*. Louisville: Westminister Press, pp. 119-49

1993 'Theology in the University', *Bulletin of the Council of Societies for the Study of Religion* 22/2, pp. 34-9

1994 'A Response to Donald Wiebe', *Bulletin of the Council of Societies for the Study of Religion* 23/1, pp. 6-10

Sharpe, Eric J.

1986 *Comparative Religion. A History*. end ed. London: Duckworth

Shepard, Robert S.

1991 *God's People in the Ivory Tower. Religion in the Early American University*. New York: Carlson

Welch, Claude

1970 'Identity Crisis in the Study of Religion? A First Report from the ACLS Study', *Journal of the American Academy of Religion* 39, pp. 3-18

1971 *Graduate Education in Religion. A Critical Appraisal*. Missoula:

University of Montana Press
Wiebe, Donald
1994 'On Theology and Religious Studies: A Response to Francis Schüssler Fiorenza', *Bulletin of the Council of Societies for the Study of Religion* 23/1, pp. 3-6
n.d. 'Does Theology Have a Place Within the Academic Study of Religion?' (unpublished mss)
n.d. 'Against Science in the Academic Study of Religion. On the Emergence and Development of the AAR', in: George Bond and Thomas Ryba (eds), *Perry Festschrift*, forthcoming

Notes

1. This has been the essential point of contention in a recent exchange of views between Professor Schüssler Fiorenza and me in the *Bulletin of the Council of Societies for the Study of Religion*; see Francis Schüssler Fiorenza, 1991, 1993, and 1994 and Donald Wiebe, 1994.
2. About the time of the transformation of the National Association of Biblical Instructors into the American Academy of Religion the decision in the 'Abington Township School District Vs Schemp' case (1963), also introduced a profound change in the broader social context of the academic study of religion. As Kathryn Alexander puts it, it 'altered the total picture of the study of religion in America by introducing the language which has governed its subsequent development in secular, public colleges and universities' (1988, p. 389). The distinction it introduced into the field is that between the teaching *of* religion and teaching *about* religion, with only the former being excluded from public institutions by the constitution, thus opening a 'window' for the inclusion of the academic study of religions in the curriculum of the modern public university. The claim that 'theology' is an essential aspect of the academic study of religion, of course, re-opens the debate about the (political) legitimacy of 'religious studies' in the state colleges and universities.
3. See here, for example, Schüssler Fiorenza in note 1 above, or Linell E. Cady (1993).
4. This point may appear somewhat paradoxical. However, it must be remembered that the displacement spoken of here refers to the removal of instruction in Bible and theology from the arts curriculum as elements of religious instruction and therefore represents an element of the secularization of the university's ethos. That, of course, would have created a more hospitable general ethos for the scientific study of religion except for the fact that other venues for the study and teaching of religion were created on most campuses - venues in which the religious intention predominated. Given those developments Hart is certainly correct in his assessment that the removal of the study of religion from the liberal arts community made it nearly impossible for that study to develop as did others in the university curriculum.

5. Long's ultimate intentions in his analysis of the development of religious studies on the American scene differs in a number of respects from mine. Although in his essay on the past and future of the study of religion in the United States (1985) he laments that too little attention is paid to the founding figures of the field of the history of religions (30-1), he nevertheless also cautions his readers against the assumption that the study of religion in the public university either is, or ought to be, a guarantee of its objectivity (32). He writes: 'Constraint by law, and most probably from conviction by the members of its faculties, enables these institutions to present their teachings on religion in a non proselytizing style; such institutions could then be declared the objective centres for the academic study of religion. This temptation must be resisted and denied because it does not portray an authentic appraisal of the meanings and nuances of the history of the study of religion prior to this innovation, nor does it pose in a correct manner the relationship and matrices of the meaning of American academic life as it is related to a serious concern for the understanding of religion' (32). The Enlightenment model of scientific study, he insists therefore, 'thwarted the development of a full range of methods for the study of [the] ultimate meaning of these expressions [i.e., religious expressions]' (39). As will become clear in this essay, it seems to me that Long's desire for a deeper probing of the 'essential meaning of "the religious"' (37), as he puts it, is really to make of the academic exercise a religious one. Despite these differences, on some matters we are in at least partial agreement, and that is especially so on our views of the meaning for the scientific study of religion for the emergence of the AAR in the early 1960s. As Long puts it in his 'A Common Ancestor: Theology and Religious Studies' (1992): '... the American Academy of Religion, in taking over from its predecessor, the National Association of Biblical Instructors, stayed within the limited arenas already defined by religious and secular Protestantism. When the change was made from NABI, very few of those involved in making the transition knew anything of the scholarly study of religion that had been inaugurated in the early part of this century by Professor Morris Jastrow, Jr. This was the tradition of *Religions Wissenschaft*. While this orientation might not have solved the American dilemma of the study of religion, it would have placed its problem and meaning within a wider historical and comparative context' (144). Moreover, I agree with Long that there were cultural reasons that account for the course of action taken by the Academy, having to do with the fundamentally Protestant character of its constituency. However, as will become clear, I do think the objective, scientific model for the study of religion represented by Jastrow is the only appropriate model for the study of religion in the public university and reject Long's claim that such study needs 'the critique of theology to save it from the illusions of objectivity and to point out the ethical and teleological nature of all our intellectual endeavours' (149). That kind of 'authentic Human science' (150) is not, I would argue, science.
6. See here, for example, my essay 'Religion and the Scientific Impulse in the Nineteenth Century: Friedrich Max Müller and the Birth of the Science of Religion' (1994).

7. It is important to note here, moreover, that the understanding of *Religions Wissenschaft's* task was radically at odds with that of the IAHR. Eric Sharpe correctly notes, for example, that even though the philological, historical, archaeological, and other techniques in the study of religion were to be conscientiously and accurately cultivated, they were not to be regarded as the ultimate ends towards which that study is addressed. The large human questions about the meaning and purpose of life, that is, were also, and more importantly, the quarry of the student of religion and E.R. Goodenough's presidential address at the inaugural meeting of the ASSR in 1959, according to Sharpe, made that quite clear. Furthermore, writes Sharpe, '... it was equally clear that these were the tones which scholarship would have to be prepared to hear from the USA for some time to come ... [for] from all sides the American position seemed to be crystallising into one in which the old historicism was being weighed in the balance and found seriously wanting, because it was incapable of answering even to its own deepest questions' (Sharpe, 1986, p. 274-5). In contrast to this Sharpe claims that the IAHR, especially in the reaffirmation of its understanding of the character of the academic study of religion at its Marburg conference in 1960, rejected such an approach to the field. '[It] is clear,' he writes, 'that Marburg was in many ways to be a watershed for the simple reason that there methodological discussion established itself for the first time as an integral part of IAHR procedure' (Sharpe, 1986, p. 277).
8. I provide further details on this matter in my essay 'Against Science in the Academic Study of Religion. On the Emergence and Development of the AAR' (forthcoming). It should be noted that in the early years of the existence of the AAR, connections with the IAHR were considered by some as a mark of having achieved the sought for academic status. Michaelson in the mid-sixties wrote: 'During this past summer [1965] the Claremont Colleges hosted the first meeting in the Western Hemisphere of the International Association for the History of Religions, a meeting which brought together scholars in religion from all over the world. This event symbolized the fact that the history of religions is well on its way toward becoming an established discipline in America' (1967, p. 13). This, obviously, was a minority opinion, and an attitude which, if it really did exist amongst others in the AAR, was very short-lived indeed. Eric Sharpe's view of the matter is significantly different. He claims that 'in a great deal of what was said and done at Claremont it was clear that the ideal of disinterested, objective scholarship *for its own sake*, while not abandoned, had been relegated to a position of only relative importance' (Sharpe, 1986, p. 284).
9. It is interesting to note that in the United Kingdom religion was considered inadequate to such a task and that an alternative kind of study was approved to function in this fashion. On this score see Fred Inglis (1993).
10. This was so much the case, in fact, that with the increase in interest in the study of religion in the post-war years there simply were no teachers available who had been trained as *Religions Wissenschaftler*. As Kitagawa points out, the majority of the teachers hired in this period came from the ranks of those trained in seminaries and divinity schools. He writes: 'In a real sense, the

chaotic picture of the undergraduate teaching of the history of religions can be traced to the lack of adequate graduate training centres for *Religions Wissenschaft* in North America. Thus when teachers of world religions are needed at many undergraduate colleges, they usually appoint either philosophers of religion, historians, biblical scholars, or theologians who happen to have personal interests and perhaps had taken two or three courses in the history of religions or comparative religion' (1959, p. 11).

11. It should be noted here, for example, that Murray G. Murphy makes no mention whatsoever of scholars in the History of Religions or *Religions Wissenschaft* in his essay 'On the Scientific Study of Religion in the United States' (1989). *Religions Wissenschaft* as it is practised in the United States and represented by the AAR in his opinion, it appears, does not merit consideration as a scientific undertaking. Neither does Murphy mention the work of the members of The Society for the Scientific Study of Religion which is more directly concerned with the social scientific study of religious phenomena. That may not, however, be simply an oversight since in the opinion of some scholars some sociology is more science than other sociology. Roland Robertson, for example, writes: 'More, perhaps, than anywhere else, the sociology of religion in the USA has occupied an interstitial position between sociology as a very secular discipline and religious studies as both a defence of religion and a critique of Western secularity. To that extent, much of American sociology *of* religion might more aptly be described as sociology *and* religion - a seeking of a conciliation between sociological science and privatized religion ...' (1993, p. 10).

12. A review of all the presidential addresses to the Academy since the change of name from NABI only further supports my critique of Hart here. Of the 27 addresses delivered and published since 1964, 11 deal directly with the problem of the relation of 'theology' to 'religious studies', of which 9 argue strenuously for a kind of merger or integration of the two enterprises, if not a subordination of the latter to the former. The other two argue the opposite, or appear to do so: C. Welch's address, to which I have already referred, argues the matter ambiguously while the address by William C. Clebsch, ('Apples, Oranges, and Manna: Comparative Religion Revisited') clearly favours a scientific or academic approach over the religion-theological in the comparative study of religion. Three addresses - the only addresses not on Western/Christian themes - pay attention to the 'theology'/'religious studies' debate indirectly and all, in my opinion, favour some form of 'integration' of the two enterprises. Of the remaining addresses, 10 are directly concerned, in a rather traditional manner, with Christian religious and theological issues and topics, while the other 3 deal with issues of humanism and hermeneutics which are clearly Western and Christian in their inspiration. I intend in the near future to provide a fuller analysis of these addresses in order to assess their significance with respect to the development of the AAR from the substance of the NABI.

13. Even a cursory review of the scholarly contributions of participants in the annual general meetings of the AAR, and in the various regional meetings of the Academy, provides incontrovertible evidence of the primarily religious,

theological, and crypto-theological concerns of its members. A review of other scholarly associations and societies affiliated in some way or other with the AAR, I think, provides even further evidence in this regard.

14. There are a few programmes of study in the US, however, that do stand out in this regard. In reviewing their undergraduate programme the department of Comparative Religion at Western Michigan writes: 'During the early years of the department, its faculty wrestled with this principal issue: How should religion be studied in a public university? In the course of developing its undergraduate programme, the faculty argued that the study of religion in a public university must occupy a valid and intellectually defensible place within the university curriculum. Thus, the model for the study and teaching of religion could not come from the divinity school or theological seminary' (typescript of 'Ph.D. Proposal in Comparative Religion, Western Michigan University, Draft-October 31, 1994, p. 3). Similar comments regarding the undergraduate programme had already been made by T. Lawson in his 'A Rationale for a Department of Religion in a Public University' (1967). The same comments can be made about the undergraduate programme at the University of Vermont - see in that regard the recent interview with the Chair of Vermont's department in *Religious Studies News*, 9/3, September, 1994, p. 6-7.

15. In 'The Academic Study of Buddhism in the United States: A Current Analysis', Charles S. Prebish comes to a similar conclusion in regard to the attitude of scholars in that area of 'religious studies'. An important conclusion forced upon him by the study, he writes, is that a significant number 'of colleagues who have come to the study of Buddhism, and hence to academe, [have done so] as a result of their strong personal commitment to Buddhism as a religious tradition [and that] for many this has created a tension between scholarship and religious commitment, between Buddhology and personal faith' (Prebish, 1994, p. 278). It is important to note here, moreover, that Prebish used Ray Hart's 'Religious and Theological Studies in American Higher Education' as a model for his own research project.

16. Another 'hint' of the future direction of the field, I think, is to be seen in the emergence of the North American Academy of Religion (NAAR) within the last decade. The Association was founded in order to support the historical, structural, and comparative study of religions and to provide a forum of discussion and debate for those scholars whose interests focused on achieving a scientific explanation of religions and religious phenomena. In part the Association emerged also to represent scholars in the United States in the IAHR since no connection exists between the AAR and the IAHR. Although a connection does exist between the American Society for the Study of Religion (ASSR) and the IAHR, its restrictive membership policy leaves most American scholars interested in this connection without possibility of representation.

XXXV

BIBLIOGRAPHY FRITS STAAL

1953

Review: P.V. Pistorius (1952), *Plotinus and Neoplatonism*, in: *Algemeen Nederlands Tijdschrift voor Wijsbegeerte en Psychologie* 46, pp. 48-51

1954

'Mijn hart neemt velerlei gestalten aan: Aantekeningen naar aanleiding van enkele gedichten van J.H. Leopold: I. Manṣūr al-Hallādj', *Amsterdams Tijdschrift voor Letterkunde* 1, pp. 30-8

'Mijn hart neemt velerlei gestalten aan: II. Ibn 'Arabī', *Amsterdams Tijdschrift voor Letterkunde* 1, pp. 73-81

(met Joh. B.W. Polak) 'Mijn hart neemt velerlei gestalten aan: III. Kanttekeningen bij het gedicht Cheops van J.H. Leopold', *Amsterdams Tijdschrift voor Letterkunde* 1, pp. 120-8

1955

'Parmenides and Indian Thought', *The Philosophical Quarterly* 28, pp. 81-106

'Rationality in Indian and Western Thought', *Mysindia* June 19

'Het Indische Filosofencongres te Peradeniya', *Algemeen Nederlands Tijdschrift voor Wijsbegeerte en Psychologie* 47, pp. 201-3

'Over het cyclische en het rechtlijnige tijdsbegrip', *Amsterdams Tijdschrift voor Letterkunde* 2, pp. 170-88. Also printed separately with 'Bibliografische Aantekening', Amsterdam, pp. 1-21

1956

'An Introduction to the Existentialism of Martin Heidegger', *Journal of the Madras University* 28, pp. 9-35

1958

'Notes on Some Brahmin Communities of South India', *Arts and Letters. The Journal of the Royal India, Pakistan and Ceylon Society* 32, pp. 1-7

1959

'Über die Idee der Toleranz im Hinduismus', *Kairos. Zeitschrift für Religionswissenschaft und Theologie* 1, pp. 215-8

Review

J.J. Poortman (1958), *Ochêma: geschiedenis en zin van het hylisch pluralisme. II. Het hylisch pluralisme in het Oosterse denken*, in: *Bulletin of the School of Oriental and African Studies* 22, pp. 167-9

1960

'Correlations between Language and Logic in Indian Thought', *Bulletin of the School of Oriental and African Studies* 23, pp. 109-22

'Formal Structures in Indian Logic', *Synthese* 12, pp. 279-86

'Means of Formalization in Indian and Western Logic', *Proceedings of the XIIth International Congress of Philosophy*. Firenze, 10, pp. 221-7

'The Construction of Formal Definitions of Subject and Predicate', *Transactions of the Philological Society* 89-103

'Plotinus and St. Augustine: A Note on the Phenomenology of Sage and Saint', in: *Proceedings of the Second Seminar and Conference, Union for the Study of the Great Religions*. Madras: Ganesh & Co, pp. 340-55

'Indian Philosophy at the International Congress of Philosophy', *The Philosophical Quarterly* 33, pp. 132-6

Reviews

D.H.H. Ingalls (1951), *Materials for the Study of Navya-Nyāya Logic*, in: *IIJ* 4, pp. 68-73

D.S. Ruegg (1959), *Contributions à l'histoire de la philosophie linguistique indienne*, in: *Philosophy East and West* 10, pp. 55-7

Chandradhar Sharma (1960), *A Critical Survey of Indian Philosophy*, in: *The Aryan Path* 31, pp. 364-5

1961

Nambudiri Veda Recitation. 's-Gravenhage: Mouton, 102 pp. (Disputationes Rheno-Trajectinae. 5)

Advaita and Neoplatonism: A Critical Study in Comparative Philosophy. Madras: University of Madras, xii + 262 p. (Madras University Philosophical Series. 10)

'The Theory of Definition in Indian Logic', *JAOS* 81, pp. 122-6

'Het indische denken en westerse vooroordelen', *Wijsgerig perspectief op maatschappij en wetenschap* 1, pp. 18-32

Reviews

J.E. van Lohuizen de Leeuw (1961), *De proto-historische culturen van Voor-Indië en hun datering*, in: *JAOS* 81, p. 65

L. Dumont and D. Pocock (eds.) (1960), *Contributions to Indian Sociology*, in: *JAOS* 81, pp. 147-9

1962

'Contraposition in Indian Logic', in: *Proceedings of the 1960 International Congress for Logic, Methodology and Philosophy of Science*. Stanford: Stanford University Press, pp. 634-49

'A Method of Linguistic Description: The Order of Consonants according to Pāṇini', *Language* 38, pp. 1-10

'Negation and the Law of Contradiction in Indian Thought', *Bulletin of the School of Oriental and African Studies* 25, pp. 52-71

'Philosophy and Language', *Essays in Philosophy presented to Dr. T.M.P. Mahadevan*. Madras: Ganesh & Co, pp. 10-25

Reviews

R.C. Zaehner (1960), *Hindu and Muslim Mysticism*, in: *JAOS* 82, pp. 96-8

E.R. Sreekrishna Sarma (1960), *Maṇikaṇa. A Navya-Nyāya Manual*, in: *JAOS* 82, pp. 237-41

1963

Euclides en Pāṇini. Twee methodische richtlijnen voor de filosofie. Amsterdam: Polak en Van Gennep

'Sanskrit and Sanskritization', *Journal of Asian Studies* 22, pp. 261-75. Reprinted 1972

Reviews

H. Scharfe (1961), *Die Logik im Mahābhāṣya*, in: *JAOS* 83, pp. 252-6

B. Shefts (1961), *Grammatical Method in Pāṇini*, in: *Language* 39, pp. 483-8

1964

'Reports on Vedic Rituals and Recitations', *Year Book of the American Philosophical Society*. Philadelphia: The American Philosophical Society, pp. 607-11

'Kanttekening', *Algemeen Nederlands Tijdschrift voor Wijsbegeerte en Psychologie* 56, pp. 125-6

Reviews

R. Gnoli (1960), *The Pramāṇavārttikam of Dharmakīrti. The First Chapter with Autocommentary*, in: *JAOS* 84, pp. 91-2

Paul Hacker (1960), *Prahlāda. Werden und Wandlungen einer Idealgestalt*, I-II, in: *JAOS* 84, pp. 464-7

K.C. Pandey (1963), *Abhinavagupta: An Historical and Philosophical Study*, in: *Algemeen Nederlands Tijdschrift voor Wijsbegeerte en Psychologie* 56, pp. 148-50

1965

'Context-Sensitive Rules in Pāṇini', *Foundations of Language* 1, pp. 63-72. Also published in: *Studies in Indian Linguistics. M.B. Emeneau Felicitation Volume*. Annamalainagar and Poona: Centres of Advanced Study in Linguistics, pp. 332-9

'Reification, Quotation and Nominalization', in: *Logic and Philosophy. Essays in Honour of I.M. Bochenski*. Amsterdam: North Holland, pp. 151-87

'E.W. Beth', *Dialectica* 19, pp. 158-79

'In Memoriam E.W. Beth', *Foundations of Language* 1, pp. 81-2

'Generative Syntax and Semantics', *Foundations of Language* 1, pp. 133-54

'Euclid and Pāṇini', *Philosophy East and West* 15, pp. 99-116

'On the Meaning and Validity of Theories of Rebirth', *Indian Philosophical Annual* 1, pp. 125-6

'Inleiding', *Wijsgerig Perspectief op Maatschappij en Wetenschap* 5 ('Boeddhisme'), Amsterdam: Meulenhoff, pp. 113-14

Review

F. Vos en E. Zürcher, *Spel zonder snaren. Enige beschouwingen over Zen*, in: *Wijsgerig Perspectief op Maatschappij en Wetenschap* 5, p. 165

1966

'Analyticity', *Foundations of Language* 2, pp. 67-93

'Room at the Top in Sanskrit', *IIJ* 9, pp. 165-98

'Pāṇini Tested by Fowler's Automaton', *JAOS* 86, pp. 206-9

'Indian Semantics I', *JAOS* 86, pp. 304-11

'My Approach to Indian Philosophy', *Indian Philosophical Journal* 2, pp. 289-302

(With K. Kuypers), 'Loos Alarm', *Universiteit en Hogeschool* 4

Film: 'Vedic Ritual in South India: Domestic Ritual and Śrauta Ritual', *Stichting Film en Wetenschap*, Utrecht

Reviews

Noam Chomsky (1957) *Syntactic Structures*, (1957) 'Logical Structures in Language', Noam Chomsky and George A. Miller (1958) 'Finite State Languages', Noam Chomsky (1959), 'On Certain Formal Properties of Grammars', 'A Note on Phrase Structure Grammars', (1961) 'On the Notion "Rule of Grammar"', *The Journal of Symbolic Logic* 31, pp. 245-51

1967

Word Order in Sanskrit and Universal Grammar. xi + 98 pp. Dordrecht: Reidel. (Foundation of Language Supplementary Series. 5)

'Some Semantic Relations between Sentoids', *Foundations of Language* 3, pp. 78-100

'Zinloze en zinvolle filosofie', *De Gids* 130,1/2, pp. 49-75

'Naschrift', *De Gids* 130,6/7, pp. 64-6

'Indian Logic', *The Encyclopedia of Philosophy*. New York [etc.]: Macmillan, Vol. IV, pp. 520-3, 568-9

Reviews

Noam Chomsky (1965), *Aspects of the Theory of Syntax*, in: *The Journal of Symbolic Logic* 32, pp. 385-7

Hans-Georg Gadamer (1965), *Wahrheit und Methode. Grundzüge einer philosophischen Hermeneutik*, in: *Foundations of Language* 3, pp. 203-5

1968

(Edited with B. van Rootselaar), *Proceedings of the Third International Congress for Logic, Methodology and Philosophy of Science*. xiii + 554 pp. Amsterdam: North-Holland

'The Twelve Ritual Chants of the Nambudiri Agniṣṭoma', in: J.C. Heesterman, G.H. Schokker and V.I. Subramoniam (eds) *Pratidānam: Studies Presented to F.B.J. Kuiper on the Occasion of his Sixtieth Birthday*. The Hague: Mouton, pp. 409-29. (Janua Linguarum. Series Maior. 34)

'And', *Journal of Linguistics* 4, pp. 79-81

'Acts of Communication', *Common Factor (Oxford)* 5, pp. 26-34

'Meaning, Regular and Irregular', *Foundations of Language* 4, pp. 182-4

'Logica', *Wijsgerig Denken*, Amsterdam: Meulenhoff, pp. 102-25. Reprinted frequently

'Het Hindoeïsme', *De verhouding tussen de levende godsdiensten, Nederlands Gesprek Centrum*, Kampen [etc.]: Kok [etc.], pp. 8-12

'Een wijsgerige visie', *De verhouding tussen de levende godsdiensten, Nederlands Gesprek Centrum*, Kampen [etc.]: Kok [etc.], pp. 41-2

'The Intellectual as Prophet: An Interview with Noam Chomsky', *Delta* 11, pp. 5-23

'In gesprek met Chomsky', *De Gids* 131,6/7, pp. 6-18

'De huidige Amerikaanse filosofie', *De Gids* 131,6/7, pp. 81-6

Review

S.S. Barlingay (1965), *A Modern Introduction to Indian Logic*, in: *The Journal of Symbolic Logic* 33, pp. 603-4

Record Album: (with John Levy) *The Four Vedas, Asch Mankind*. New York: Folkways

1969

'Sanskrit Philosophy of Language', *Current Trends in Linguistics* 5, pp. 499-531. Reprinted 1976.

'Recent Insights in the Nature of Language', *Meanjin Quarterly* 119, pp. 487-94

(With Paul Kiparsky) 'Syntactic and Semantic Relations in Pāṇini', *Foundations of Language* 5, pp. 83-117

(ed.) 'Formal Logic and Natural Languages: A Symposium', in: *Foundations of Language* 5, pp. 256-84

Review

K.N. Jayatilleke (1963), *Early Buddhist Theory of Knowledge*, in: *Foundations of Language* 5, pp. 560-2

1970

'Performatives and Token-Reflexives', *LI* 3, pp. 373-81

'De academicus als nowhere-man', *De Gids* 133,7/8, pp. 96-159

'Cambodia: Sanskrit Inscriptions', *The New York Review of Books* 15/1 (July 2), p. 15

Review

George Cardona (1969), *Studies in Indian Grammarians I: The Method of Description Reflected in the Śiva-Sūtras*, in: *Language* 46, pp. 502-7

1971

'What was Left of Pragmatics in Jerusalem', *Language Sciences* 14 (February), pp. 29-32

'Voorwoord', Gilbert Ryle, *De eenheid van lichaam en geest*. Amsterdam: Athenaeum - Polak & Van Gennep, pp. 9-11

Review

B.K. Matilal (1968), *The Navya-Nyāya Doctrine of Negation: The Semantics and Ontology of Negative Statements in Navya-Nyāya Philosophy*, in: *IIJ* 13, pp. 199-205

1972

'Theory and Practice in the University', *Proceedings of the International Seminar on World Philosophy*. Madras: The Centre of Advanced Study in Philosophy

Reprint of 'Sanskrit and Sanskritization' (1963) in: J.A. Harrison (ed.) *South and Southeast Asia: Enduring Scholarship*. Tucson: University of Arizona Press, pp. 213-7

1973

A Reader on the Sanskrit Grammarians. xxiv + 557 pp. Cambridge and London: MIT Press. (Studies in Linguistics. 1)

'Uncontained Rules of Meaning', *Proceedings of the Fourth International Congress for Logic, Methodology and Philosophy of Science*. Amsterdam [etc.]: North-Holland, pp. 845-62

'The Concept of *pakṣa* in Indian logic', *Journal of Indian Philosophy* 2, pp. 156-66

1974

'The Origin and Development of Linguistics in India', in: Del Hymes (ed.) *Studies in the History of Linguistics: Traditions and Paradigms*. Bloomington: Indiana University Press, pp. 63-74

'Het wetenschappelijk onderzoek van de mystiek', *De Gids* 137,1/2, pp. 3-34, 108-32

(With Y. Bar-Hillel and H. Hiz) 'Problems of Linguistic Semantics from the Standpoint of the Philosophy and the Methodology of Language: Contributions and Discussions', in: Carl H. Heidrich (ed.) *Semantics and Communication. Proceedings of the Third Colloquium of the Institute for Communications Research and Phonetics, University of Bonn*. Amsterdam [etc]: North Holland

Japanese translation of 'The Concept of *pakṣa* in Indian logic' (1973), *Journal of Indian and Buddhist Studies* 22, pp. 1092-1082

Review

Arthur C. Danto (1972), *Mysticism and Morality. Oriental Thought and Moral Philosophy*, *The Journal of Philosophy* 71, pp. 174-81

1975

Exploring Mysticism: A Methodological Essay. xix + 230 pp. Berkeley: University of California Press. Also: *Exploring Mysticism*, 224 pp., Harmondsworth: Penguin Books

'The Concept of Metalanguage and its Indian Background', *Journal of Indian Philosophy* 3, pp. 315-54

1976

'Making Sense of the Buddhist Tetralemma', in: *Philosophy East and West: Essays in Honour of Dr. T.M.P. Mahadevan*. Bombay: Blackie and Son, pp. 122-31

Reprint of 'Sanskrit Philosophy of Language' (1969) in: H. Parret (ed.) *History of Linguistic Thought and Contemporary Linguistics*. Berlin: De Gruyter, pp. 102-36

Film (with Robert Gardner): *Altar of Fire*. Berkeley: Media Center, University of California

1977

Introduzione allo studio del misticismo orientale ed occidentale. 231 pp., Roma: Ubaldini (Italian translation of 'Exploring Mysticism', 1975)

Människan och mystiken. 258 pp., Stockholm: Aldus (Swedish translation of 'Exploring Mysticism', 1975)

'Ṛgveda 10.71 on the Origin of Language', *Revelation in Indian Thought: A Festschrift in Honour of Professor T.R.V. Murti*. Emeryville: Dharma Publishing, pp. 3-14

Review

B.K. Matilal (1971), *Epistemology, Logic and Grammar in Indian Philosophical Analysis*, in: *IIJ* 19, pp. 108-14

1978

Het wetenschappelijk onderzoek van de mystiek. 279 pp., Utrecht-Antwerpen: Spectrum (Dutch translation of 'Exploring Mysticism', 1975)

'On and Around Yoga', *Journal of Indian Philosophy* 6, pp. 177-87

'The Ignorant Brahmin of the Agnicayana', *Annals of the Bhandarkar Oriental Research Institute* 50, pp. 337-48

'Het woord als middel', *De Gids* 141, pp. 353-63

Review

A. Bharati (1977), *The Light at the Center: Context and Pretext of Modern Mysticism*, in: *The Journal of Asian Studies* 37, pp. 388-90

1979

'Oriental Ideas on the Origin of Language', *JAOS* 99, pp. 1-14

'Ritual Syntax', in: M. Nagatomi, B.K. Matilal and J.M. Masson (eds) *Sanskrit and Indian Studies: Essays in Honor of Daniel H.H. Ingalls*. Dordrecht: Reidel, pp. 119-42. (Studies in Classical India. 2)

'The Meaninglessness of Ritual', *Numen. International Journal of the History of Religions* 26, pp. 2-22

Comment: 'Altar of Fire', *American Anthropologist* 81, pp. 346-7

'The Concept of Scripture in the Indian Tradition', in: M. Juergensmeyer and N.G. Barrier (eds) *Sikh Studies: Comparative Perspectives on a Changing Tradition*. Berkeley: Graduate Theological Union, pp. 121-6. (Berkeley Religious Studies Series. 1)

'Sanskrtska Filozofija Jezika', *Dijalog* (Sarajevo) 1/2, pp. 83-115 (Serbo-Croatian translation of 'Sanskrit Philosophy of Language')

Reviews

W. Howard (1977), *Sāmavedic Chant*, in: *JAOS* 99, pp. 347-8

K.H. Potter (1977), *Indian Metaphysics and Epistemology: The Tradition of Navya-Nyāya up to Gaṅgeśa*, in: *The Journal of Philosophy* 76, pp. 347-8

1980

'Rites That Make No Sense', in: J.S. Yadava and V. Gautam (eds) *The Communication of Ideas: 10th International Congress of Anthropological and Ethnological Sciences*. New Delhi, 3, pp. 145-54

'Wat het oosten ons kan leren', *De Gids* 143, pp. 89-95

'The Stamps of Dungarpur', *The Collectors Club Philatelist* 59, pp. 41-5, 100-7, 119-20, 281-6, 309

'A Note on the Stamps of Dungarpur', *India's Stamp Journal* 43, pp. 60-2

1981

'Vedic Religion in Kerala', *The Adyar Library Bulletin* 44-5, pp. 74-89

(With L.W. Lancaster) '1973 Survey of Manuscript Collections in Nepal', *1973 National Geographic Society Research Reports* 14, pp. 385-95

1982

The Science of Ritual. viii + 101 pp. Poona: Bhandarkar Oriental Research Institute

'Ritual, Grammar and the Origins of Science in India', *Journal of Indian Philosophy* 10, pp. 11-43

'What is Happening in Classical Indology?', *The Journal of Asian Studies* 41, pp. 269-91

'The Himalayas and the Fall of Religion', in: D.E. Klimburg-Salter (ed.) *The Silk Route and the Diamond Path. Esoteric Buddhist Art on the Trans-Himalayan Trade-Routes*. Berkeley: University of California Press, pp. 38-51

'Indian Grammarians', *Encyclopedic Dictionary of Semiotics*, ed. T.A. Sebeok

'Indian Logicians', *Encyclopedic Dictionary of Semiotics*, ed. T.A. Sebeok

1983

(With C.V. Somayajipad and M. Itti Ravi Nambudiri) *Agni. The Vedic Ritual of the Fire Altar*, I-II. xxxviii + 716, xvii + 832 pp., 130 Color Plates, 160

Illustrations and Diagrams, 6 Maps, 2 Cassette Recordings. Berkeley: Asian Humanities Press

The Stamps of Jammu & Kashmir. xvii + 286 pp., 16 Plates. New York: Collectors Club

'Moon Chants, Space Fillers and Flow of Milk', in: K.S. Ramamurthi (ed.) *Surabhi: Sreekrishna Sarma Felicitation Volume*. Tirupati: Prof. E.R. Sreekrishna Sarma Felicitation Committee, pp. 17-30

'On the Indian Concept of the Body', in: *Proceedings, II. Conferencia Latino-americana de Sanscritistas*. Mexico City: Universidad Nacional Autónoma de México

Letter: 'Self-Realization', *The New York Review of Books* 30/14 (September 23), pp. 65-6

Reprint of 'The Concept of *pakṣa* in Indian Logic' (1973) in: Charlene McDermott (ed.) *Comparative Philosophy: Selected Essays*. Lanham: University Press of America, pp. 180-92

1983-1984

'Indian Concepts of the Body', *Somatics. Magazine Journal of the Bodily Arts and Sciences* 4/3, pp. 31-41

1984

'The Search for Meaning: Mathematics, Music and Ritual', *American Journal of Semiotics* 2/4:1-57, also appeared as 'Ritmo nel rito', *Conoscenza Religiosa* 2, pp. 197-247

'Ritual, Mantras and the Origin of Language', in: S.D. Joshi (ed.) *Amṛtadhārā. Professor R.N. Dandekar Felicitation Volume*. Delhi: Ajanta Publications, pp. 403-25

'Schat onderschat Schat', *De Gids* 147/7, pp. 503-8

'Over het *Sanskriet* als moeder der Germaanse taaltakken', *De Gids* 147/8, pp. 579-80

'Further Notes on Dungarpur', *India Post* 18/4, pp. 146-7

'Abracadabra voor Intellectuelen', *De Brandende Kwestie*. Amsterdam: Raamgracht, cop., pp. 151-77

1985

Shinpishugi no tankyu. 361 + 21 pp. Tokyo: Hosei University Press (Japanese translation of 'Exploring Mysticism', 1975).

'Mantras and Bird Songs', *JAOS* 105, pp. 549-58

'Substitutions de paradigmes et religions d'Asie', *Cahiers d'extrême Asie* 1, pp. 21-57

'Language and Ritual', in: *Mm. Kuppuswami Sastri Birth Centenary Commemoration Volume*. Madras: G.S. Press, Part 2, pp. 51-62

1986

Over zin en onzin in filosofie, religie en wetenschap. 413 pp. Amsterdam: Meulenhoff - Polak en Van Gennep

'The Sound of Religion', *Numen. International Journal of the History of Religions* 33, pp. 33-64, 185-224

The Fidelity of Oral Tradition and the Origins of Science. MNKAW, Afd. Letterkunde, N.R. 49/8, p. 251-88, Amsterdam [etc]: North Holland

'In the Realm of the Buddha', with photographs by Adelaide de Menil, *Natural History* 95/7, July, pp. 34-45

1987

'De toekomst van de wetenschap in Nederland: open brief aan Minister Deetman', *Elseviers Weekblad*, August 29.

'The ineffable *nirguṇa* brahman', in: K. Schomer and W.H. McLeod (eds) *The Saints: Studies in a Devotional Tradition of India*. Berkeley: University of California Press, pp. 41-6

'Professor Schechner's Passion for Goats', *The Journal of Asian Studies* 46, pp. 105-8

Het Vuuraltaar (Beeld en Geluid). Hilversum: NOS (60 min.)

1988

Universals: Essays in Indian Logic and Linguistics. x + 267 pp. Chicago and London: University of Chicago Press

Een wijsgeer in het Oosten. Op reis door Java en Kalimantan. 119 pp. Amsterdam: Meulenhoff (includes 'De godsdiensten van het oosten zijn niet oosters en ook geen godsdiensten'. Published in *NRC-Handelsblad*, June 11)

'Vedic Mantras', in: H.P. Alper (ed.) *Understanding Mantras*. New York: State University of New York Press, pp. 48-95

'Is there Philosophy in Asia?', in: G.J. Larson and E. Deutsch (eds) *Interpreting Across Boundaries. New Essays in Comparative Philosophy*. Princeton: Princeton University Press, pp. 203-29

'Where was Lake Anodat?', *Journal of the Faculty of Arts of the Chulalongkorn University*, Bangkok, 2531, pp. 129-37

'Rituel et géometrie dans l'hindouisme', *Le grand atlas des religions*. Paris: Encyclopaedia Universalis France S.A., p. 328

Three articles on Indonesia in *Elseviers Weekblad* (The Netherlands), July 16, 23 and 30

Preface to the Second Edition of *Exploring Mysticism*. Berkeley: University of California Press

1989

Rules without Meaning. Ritual, Mantras and the Human Sciences. xxi + 490 pp. New York [etc.]: Lang. (Toronto Studies in Religion. 4)

'The Independence of Rationality from Literacy', *European Journal of Sociology* 30, pp. 301-10

'Moeder vindt beter dat ik geen Sanskriet doe', ('Mother prefers that I don't do Sanskrit') *De Gids* 152/10, pp. 771-92; reprinted in: *Erflaters van de twintigste eeuw.* Amsterdam: Querido, 1991, pp. 111-47

'Anthropologists Against Death', *Visual Anthropology*

1990

Jouer avec le feu. Pratique et théorie du rituel védique. 115 pp., Paris: Institut de Civilisation Indienne

'The Lake of the Yakṣa Chief', in: Tadeusz Skorupski (ed.) *Indo-Tibetan Studies. Papers in honour and appreciation of Professor David L. Snellgrove's contribution to Indo-Tibetan Studies.* Tring: Institute of Buddhist Studies, 275-292-1

'Introduction', in: *Sadao Hasegawa: Paintings and Drawings.* London: GMP Publishers, pp. 4-7

1991

(With S.A. Bonebakker, E. Gene Smith and H.J. Verkuyl) *Baby Krishna. Rapport van de Adviescommisie Kleine Letteren.* 96 pp. The Hague (Report on Asian and Minor Languages for the Ministry of Education and Sciences, Government of the Netherlands)

'The Centre of Space: Construction and Discovery', in: Kapila Vatsyayan (ed.) *Concepts of Space. Ancient and Modern.* New Delhi: Indira Gandhi National Centre for the Arts, pp. 83-100

(With Allan G.Grapard, Burton L.Mack and Ivan Strenski) 'Symposium: Ritual As Such. Frits Staal's Rules without Meaning', *Religion* 21, pp. 205-34

'Voor Jacob Vredenbregt', *Liber Amicorum Jacob Vredenbregt*. Leiden: pp. 127-8

'The Stamp of Tonk', *The London Philatelist* 100, pp. 209-21

'Moeder vindt beter dat ik geen Sanskriet doe' Drie orientalisten: Christiaan Snouck Hurgronje (1857-1936), Willem Caland (1859-1932), 'Robert Hans van Gulik (1910-1967)', in: *Erflaters van de twintigste eeuw*. Amsterdam: Querido, pp. 111-147, 312-13

1992

'Agni 1990. With an Appendix by H.F. Arnold', in: A.W. van den Hoek, D.H.A. Kolff and M.S. Oort (eds) *Ritual, State and History in South Asia. Essays in Honour of J.C. Heesterman*. Leiden [etc.]: Brill, pp. 650-76

'Sūtra', in: Kapila Vatsyayan (ed.) *Kalātattvakośa. A Lexicon of Fundamental Concepts of the Indian Arts*. New Delhi: Indira Gandhi National Centre for the Arts, pp. 303-14

'Vedic Māna', in: Kapila Vatsyayan (ed.) *Kalātattvakośa. A Lexicon of Fundamental Concepts of the Indian Arts*. New Delhi; Indira Gandhi National Centre for the Arts, pp. 355-67

'Heeft de Europese filosofie onze wereld nog iets te bieden?' *De Gids* 155/1, pp. 4-21

'Dilemmas of Early Duttia', *The London Philatelist* 101, pp. 275-83

'Kotah. Stamps and Postal History', *Gibbons Stamp Monthly* 23/6, pp. 63-5

1993

Paperback Edition of *Rules without Meaning*

Concepts of Science in Europe and Asia. 32 pp., Leiden: International Institute for Asian Studies

'Indian Bodies', in: T.P. Kasulis, R.T. Ames and W. Dissanayake (eds) *Self as Body in Asian Theory and Practice*. Albany: State University of New York Press, pp. 59-102

'From Meanings to Trees', *Journal of Ritual Studies* 7, pp. 11-32

'Tradition and Modernity', *Journal of Indian Council of Philosophical Research* 11, pp. 83-88

1994

Articles on: Anubandha, Anuvṛtti, Avyayībhāva, Bahuvrīhi, Bhartṛhari, Dvandva, Guṇa and Vṛddhi, Indian Theories of Meaning, Kāraka, Karmadhāraya, Kātyāyana, Lopa, Padapāṭha, Pāṇini, Paribhāṣā, Patañjali, Pratiśākhya, Samāsa, Sandhi, Sūtra, and Tatpuruṣa in: R.E. Asher *The Encyclopedia of Language and Linguistics*. Oxford: Pergamon/Elsevier Science

Reviews

Guy L. Beck (1993), *Sonic Theology: Hinduism and Sacred Sound*, in: *Religion*

Patrick Olivelle (1992), *Saṃnyāsa Upaniṣads. Hindu Scriptures on Ascetism and Renunciation*, in: *Journal of Asian Studies* pp. 1065-6

1995

Mantras between Fire and Water. Reflections on a Balinese Rite. With an appendix by Dick van der Meij. viii + 112 pp. VKNAW, Afd. Letterkunde N.R. 166, Amsterdam: North Holland

'Changing One's Mind', *Journal of Indian Philosophy* 23/1, pp. 53-5

'The Sanskrit of Science', *Journal of Indian Philosophy* 23/1, pp. 73-127

'Concepts of Science in Europe and Asia' (revised and extended version of 1993 version), *Interdisciplinary Science Reviews* 20/1, pp. 7-19 (with Plate)

'Agni', in: Kapila Vatsyayan (ed.) *Kalātattvakośa. A Lexicon of Fundamental Concepts of the Indian Arts*. New Delhi: Indira Gandhi National Centre for the Arts

'Mantras: Sounds Beyond Language', *Religion*

'The Common Market of Eurasia', Bruxelles

'Ten Geleide' ('By Way of Introduction'), Groningen: Historische Uitgeverij. (*Ergo Cogito*. 5)

'Mysterious Absences in Duttia and Dungarpur', *The London Philatelist* 104/1222, pp. 29-30

1996

'Ritual and the Cosmos', in: Kapila Vatsyayan (ed.) *Concepts of time. Ancient and Modern*. New Delhi: Indira Gandhi National Centre for the Arts, pp. 227-47

1997

(With H.L. Beck and K.M. Schipper), *Voorstel voor een Nationaal Centrum voor Islam Studies aan de Rijksuniversiteit Leiden. Rapport van de Adviescommissie*. 60 pp.

'There is No Religion There', in: Jon R. Stone (ed.) *Reflections and Conversations: Essays in the Academic Study of Religion*. London: Macmillan/St. Martin

'Beyond Relativism', in: Biderman, Shlomo and Rotem (eds) *Relativism and Beyond, Philosophy and Religion Comparative Yearbook*, vol. iv. Leiden: Brill

Reviews

Charles Malamoud (1996), *Cooking the World. Ritual and Thought in Ancient India*, in: *The Book Review Literary Trust* (New Delhi)

ABBREVIATIONS

BEFEO	Bulletin de l'Ecole Française d'Extrême-Orient, Paris
BKI	Bijdragen tot de Taal-, Land- en Volkenkunde, Leiden
GOS	Geakwad's Oriental Series, Baroda
HOS	Harvard Oriental Series, Cambridge, Massachusetts
IIJ	Indo-Iranian Journal, The Hague
IJAL	International Journal of American Linguistics, Chicago
JA	Journal Asiatique, Paris
JAOS	Journal of the American Oriental Society, New Haven
LI	Linguistic Inquiry, Cambridge
MNKAW	Medelingen van de Koninklijke Nederlandse Akademie van Wetenschappen, Amsterdam
OJO	Oud Javaansche Oorkonden, Batavia/'s-Gravenhage
VKI	Verhandelingen van het Koninklijk Instituut voor Taal-, Land- en Volkenkunde, Den Haag/Leiden
VKNAW	Verhandelingen van de Koninklijke Nederlandse Akademie van Wetenschappen, Afdeling Letterkunde, Nieuwe Reeks, Amsterdam
WZKS(O)	Wiener Zeitschrift für die Kunde Süd(-und Ost)asiens, Wien
ZDMG	Zeitschrift der Deutschen Morgenlandischen Gesellschaft, Wiesbaden